1984

HANDBOOK OF APPLICABLE MATHEMATICS

Volume IV: Analysis

HANDBOOK OF APPLICABLE MATHEMATICS

Chief Editor: Walter Ledermann

Editorial Board: Robert F. Churchhouse
Harvey Cohn
Peter Hilton
Emlyn Lloyd
Steven Vajda

Assistant Editor: Carol Jenkins (Mrs van der Ploeq)

Volume I: ALGEBRA
Edited by Walter Ledermann, *University of Sussex*
and Steven Vajda, *University of Sussex*

Volume II: PROBABILITY
Emlyn Lloyd, *University of Lancaster*

Volume III: NUMERICAL METHODS
Edited by Robert F. Churchhouse, *University College Cardiff*

Volume IV: ANALYSIS
Edited by Walter Ledermann, *University of Sussex*
and Steven Vajda, *University of Sussex*

Volume V: GEOMETRY AND COMBINATORICS
Edited by Walter Ledermann, *University of Sussex*
and Steven Vajda, *University of Sussex*

Volume VI: STATISTICS
Edited by Emlyn Lloyd, *University of Lancaster*

HANDBOOK OF

APPLICABLE MATHEMATICS

Chief Editor: Walter Ledermann

Volume IV: Analysis

Edited by

Walter Ledermann

and

Steven Vajda

University of Sussex

A Wiley–Interscience Publication

JOHN WILEY & SONS

Chichester – New York – Brisbane – Toronto – Singapore

Library of Congress Cataloging in Publication Data: (Revised)

Main entry under title:
Handbook of applicable mathematics.
  Vol.   edited by Walter Ledermann and
Steven Vajda.
  'A Wiley–Interscience publication.'
  Includes bibliographies.
  Contents: v. 1. Algebra. — [etc.] —
v. 4. Analysis.
  1. Mathematics—1961–   . I. Ledermann,
Walter, 1911–
QA36.H36   510   79-42724
ISBN 0 471 27704 5 (v. 1)   AACR2

British Library Cataloguing in Publication Data:

Handbook of applicable mathematics.
  Vol. 4: Analysis
  1. Mathematics
  I. Ledermann, Walter   II. Vajda, Steven
  510   QA36
  ISBN 0 471 10141 9

Typeset by J. W. Arrowsmith Ltd, Bristol BS3 2NT, England
and printed in the United States of America by
Vail-Ballou Press, Inc., Binghamton, N.Y.

Contributing Authors

A. M. Arthurs, University of York, Heslington, York, U.K.

J. A. Bather, University of Sussex, Brighton, U.K.

P. J. Bushell, University of Sussex, Brighton, U.K.

D. E. Edmunds, University of Sussex, Brighton, U.K.

W. Ledermann, University of Sussex, Brighton, U.K.

I. Marek, Univerzita Karlova, Prague, Czechoslovakia

I. N. Sneddon, University of Glasgow, U.K.

I. Stewart, University of Warwick, Coventry, U.K.

D. O. Tall, University of Warwick, Coventry, U.K.

P. L. Walker, University of Lancaster, U.K.

G. R. Walsh, University of York, Heslington, York, U.K.

K. J. Whiteman, University of Sussex, Brighton, U.K.

P. Whittle, University of Cambridge, U.K.

D. J. Wright, University of Sussex, Brighton, U.K.

Contents

Introduction
to the
Handbook of Applicable Mathematics

Today, more than ever before, mathematics enters the lives of every one of us. Whereas, thirty years ago, it was supposed that mathematics was only needed by somebody planning to work in one of the 'hard' sciences (physics, chemistry), or to become an engineer, a professional statistician, an actuary or an accountant, it is recognized today that there are very few professions in which an understanding of mathematics is irrelevant. In the biological sciences, in the social sciences (especially economics, town planning, psychology), in medicine, mathematical methods of some sophistication are increasingly being used and practitioners in these fields are handicapped if their mathematical background does not include the requisite ideas and skills.

Yet it is a fact that there are many working in these professions who do find themselves at a disadvantage in trying to understand technical articles employing mathematical formulations, and who cannot perhaps fulfil their own potential as professionals, and advance in their professions at the rate that their talent would merit, for want of this basic understanding. Such people are rarely in a position to resume their formal education, and the study of some of the available textbooks may, at best, serve to give them some acquaintance with mathematical techniques, of a more or less formal nature, appropriate to current technology. Among such people, academic workers in disciplines which are coming increasingly to depend on mathematics constitute a very significant and important group.

Some years ago, the Editors of the present Handbook, all of them actively concerned with the teaching of mathematics with a view to its usefulness for today's and tomorrow's citizens, got together to discuss the problems faced by mature people already embarked on careers in professions which were taking on an increasingly mathematical aspect. To be sure, the discussion ranged more widely than that—the problem of 'mathematics avoidance' or 'mathematics anxiety', as it is often called today, is one of the most serious problems of modern civilization and affects, in principle, the entire community—but it was decided to concentrate on the problem as it affected professional effectiveness. There emerged from those discussions a novel format for presenting mathematics to this very specific audience. The intervening years have been spent in putting this novel conception into practice, and the result is the Handbook of Applicable Mathematics.

THE PLAN OF THE HANDBOOK

The 'Handbook' consists of two sets of books. On the one hand, there are (or will be!) a number of *guide books*, written by experts in various fields in which mathematics is used (e.g. medicine, sociology, management, economics). These guide books are by no means comprehensive treatises; each is intended to treat a small number of particular topics within the field, employing, where appropriate, mathematical formulations and mathematical reasoning. In fact, a typical guide book will consist of a discussion of a particular problem, or related set of problems, and will show how the use of mathematical models serves to solve the problem. Wherever any mathematics is used in a guide book, it is cross-referenced to an article (or articles) in the *core volumes*.

There are 6 core volumes devoted respectively to Algebra, Probability, Numerical Methods, Analysis, Geometry and Combinatorics, and Statistics. These volumes are texts of mathematics—but they are no ordinary mathematical texts. They have been designed specifically for the needs of the professional adult (though we believe they should be suitable for any intelligent adult!) and they stand or fall by their success in explaining the nature and importance of key mathematical ideas to those who need to grasp and to use those ideas. Either through their reading of a guide book or through their own work or outside reading, professional adults will find themselves needing to understand a particular mathematical idea (e.g. linear programming, statistical robustness, vector product, probability density, round-off error); and they will then be able to turn to the appropriate article in the core volume in question and *find out just what they want to know*—this, at any rate, is our hope and our intention.

How then do the content and style of the core volumes differ from a standard mathematical text? First, the articles are designed to be read by somebody who has been referred to a particular mathematical topic and would prefer not to have to do a great deal of preparatory reading; thus each article is, to the greatest extent possible, self-contained (though, of course, there is considerable cross-referencing within the set of core volumes). Second, the articles are designed to be read by somebody who wants to get hold of the mathematical ideas and who does not want to be submerged in difficult details of mathematical proof. Each article is followed by a bibliography indicating where the unusually assiduous reader can acquire that sort of 'study in depth'. Third, the topics in the core volumes have been chosen for their relevance to a number of different fields of application, so that the treatment of those topics is not biased in favour of a particular application. Our thought is that the reader—unlike the typical college student—will already be motivated, through some particular problem or the study of some particular new technique, to acquire the necessary mathematical knowledge. Fourth, this is a handbook, not an encyclopedia—if we do not think that a particular aspect of a mathematical topic is likely to be useful or interesting to the kind of reader we have in mind, we have omitted it. We have not set out to include everything known on a particular topic, and we are *not* catering for the professional mathematician! The Handbook has been written as a contribution to the practice of mathematics, not to the theory.

The reader will readily appreciate that such a novel departure from standard textbook writing—this is neither 'pure' mathematics nor 'applied' mathematics as traditionally interpreted—was not easily achieved. Even after the basic concept of the Handbook had been formulated by the Editors, and the complicated system of cross-referencing had been developed, there was a very serious problem of finding authors who would write the sort of material we wanted. This is by no means the way in which mathematicians and experts in mathematical applications are used to writing. Thus we do not apologize for the fact that the Handbook has lain so long in the womb; we were trying to do something new and we had to try, to the best of our ability, to get it right. We are sure we have not been uniformly successful; but we can at least comfort ourselves that the result would have been much worse, and far less suitable for those whose needs we are trying to meet, had we been more hasty and less conscientious.

It is, however, not only our task which has not been easy. Mathematics itself is not easy! The reader is not to suppose that, even with his or her strong motivation and the best endeavours of the editors and authors, the mathematical material contained in the core volumes can be grasped without considerable effort. Were mathematics an elementary affair, it would not provide the key to so many problems of science, technology and human affairs. It is universal, in the sense that significant mathematical ideas and mathematical results are relevant to very different 'concrete' applications—a single algorithm serves to enable the travelling salesman to design his itinerary, and the refrigerator manufacturing company to plan a sequence of modifications of a given model; and could conceivably enable an intelligence unit to improve its techniques for decoding the secret messages of a foreign power. Given this universality, mathematics cannot be trivial! And, if it is not trivial, then some parts of mathematics are bound to be substantially more difficult than others.

This difference in level of difficulty has been faced squarely in the Handbook. The reader should not be surprised that certain articles require a great deal of effort for their comprehension and may well involve much study of related material provided in other referenced articles in the core volumes—while other articles can be digested almost effortlessly. In any case, different readers will approach the Handbook from different levels of mathematical competence and we have been very much concerned to cater for all levels.

THE REFERENCING AND CROSS-REFERENCING SYSTEM

To use the Handbook effectively, the reader will need a clear understanding of our numbering and referencing system, so we will explain it here. Important items in the core volumes or the guidebooks—such as definitions of mathematical terms or statements of key results—are assigned sets of numbers according to the following scheme. There are six categories of such mathematical items, namely:

   (i) Definitions
  (ii) Theorems, Propositions, Lemmas and Corollaries
 (iii) Equations and other Displayed Formulae
 (iv) Examples
  (v) Figures
 (vi) Tables

Items in any one of these six categories carry a triple designation a.b.c. of arabic numerals, where 'a' gives the *chapter* number, 'b' the *section* number, and 'c' the number of the individual *item*. Thus items belonging to a given category, for example, definitions are numbered in sequence within a section, but the numbering is independent as between categories. For example, in Section 5 of Chapter 3 (of a given volume), we may find a displayed formula labelled (5.3.7) and also Lemma 5.3.7. followed by Theorem 5.3.8. Even where sections are further divided into *subsections*, our numbering system is as described above, and takes no account of the particular subsection in which the item occurs.

As we have already indicated, a crucial feature of the Handbook is the comprehensive cross-referencing system which enables the reader of any part of any core volume or guide book to find his or her way quickly and easily to the place or places where a particular idea is introduced or discussed in detail. If, for example, reading the core volume on Statistics, the reader finds that the notion of a *matrix* is playing a vital role, and if the reader wishes to refresh his or her understanding of this concept, then it is important that an immediate reference be available to the place in the core volume on Algebra where the notion is first introduced and its basic properties and uses discussed.

Such ready access is achieved by the adoption of the following system. There are six core volumes, enumerated by the Roman numerals as follows:

    I Algebra
   II Probability
  III Numerical Methods
  IV Analysis
   V Geometry and Combinatorics
  VI Statistics

A reference to an item will appear in square brackets and will *typically* consist of a pair of entries [see A, B] where A is the volume number and B is the triple designating the item in that volume to which reference is being made. Thus '[see II, (3.4.5)]' refers to equation (3.4.5) of Volume II (Probability). There are, however, two exceptions to this rule. The first is simply a matter of economy!—if the reference is to an item in the same volume, the volume number designation (A, above) is suppressed; thus '[see Theorem 2.4.6]', appearing in Volume III, refers to Theorem 2.4.6. of Volume III.

The second exception is more fundamental and, we contend, wholly natural. It may be that we feel the need to refer to a substantial discussion rather than to a single mathematical item (this could well have been the case in the reference to 'matrix', given as an example above). If we judge that such a comprehensive reference is appropriate, then the second entry B of the reference may carry only two numerals—or even, in an extreme case, only one. Thus the reference '[see I, 2.3]' refers to Section 3 of Chapter 2 of Volume I and recommends the reader to study that entire section to get a complete picture of the idea being presented.

Bibliographies are to be found at the end of each chapter of the core volumes and at the end of each guide book. References to these bibliographies appear in the text as '(Smith (1979))'.

It should perhaps be explained that, while the referencing *within* a chapter of a core volume or *within* a guide book is substantially the responsibility of the author of that part of the text, the cross-referencing has been the responsibility of the editors as a whole. Indeed, it is fair to say that it has been one of their heaviest and most exacting responsibilities. Any defects in putting the referencing principles into practice must be borne by the editors. The successes of the system must be attributed to the excellent and wholehearted work of our invaluable colleague, Carol Jenkins (Mrs van der Ploeq).

CHAPTER 1

Sequences and Series

1.1. SEQUENCES

The word 'sequence' is used in a technical sense in mathematics to describe an unending list of elements

$$s_1, s_2, s_3, \ldots, s_n, \ldots$$

one for each positive integer n. Except in section 1.13, the elements s_n will be real numbers [see I, § 2.6.1], so in modern set-theoretic notation [see I, §§ 1.1 and 1.4] if $\mathbb{R}$ denotes the set of real numbers and $\mathbb{N}$ the positive integers, we may say:

DEFINITION 1.1.1.   A *sequence* of real numbers is a function $s : \mathbb{N} \to \mathbb{R}$.

If we denote $s(1), s(2), s(3), \ldots, s(n), \ldots$ in $\mathbb{R}$ successively by $s_1, s_2, s_3, \ldots, s_n, \ldots$ then from the formal definition we recover the unending list of real numbers. The element $s(n) = s_n$ is called the nth *term* of the sequence. We shall not use the formal definition very often. Sequences were studied before set theoretic notation became standard and several alternative notations are in current usage. The simplest of these is to denote the sequence itself by the list of elements:

$$s_1, s_2, s_3, \ldots, s_n, \ldots.$$

A shorthand notation for this is (s_n). This is particularly useful when s_n is given by a specific formula, say $s_n = 2^n$, when the sequence

$$2, 4, 8, \ldots, 2^n, \ldots$$

may be denoted by (2^n).

The usefulness of the formal definition is that it emphasizes the generality of the notion of a sequence. The terms $s_1, s_2, \ldots, s_n, \ldots$ can be any real numbers, provided that they are clearly specified and they need not be given by a simple formula.

1

EXAMPLE 1.1.1.   Let $s_n = n$th digit of the decimal expansion of π [see § 2.12]:

$$s_1 = 3, s_2 = 1, s_3 = 4, s_4 = 1, s_5 = 5, \ldots$$

Here a general formula for s_n is not known, but for any given value of n, s_n may be calculated by a known computer programme.

EXAMPLE 1.1.2.   Let $t_n = (-1)^n/n^2$. Here the sequence is given by a specific formula.

EXAMPLE 1.1.3.   Put $\tau_n = s_n + t_n$ where s_n, t_n are as in Examples 1.1.1. and 1.1.2, thus

$$\tau_1 = 3 - 1 = 2, \tau_2 = 1 + \tfrac{1}{4} = 1\tfrac{1}{4}, \tau_3 = 4 - \tfrac{1}{9} = 3\tfrac{8}{9}, \tau_4 = 1 + \tfrac{1}{16} = 1\tfrac{1}{16}, \ldots$$

In this case τ_n can be computed from the two preceding examples.

EXAMPLE 1.1.4.   Let $r_n = $ '$\sqrt{2}$ to n decimal places', where for definiteness we do not round up the last place, so

$$r_1 = 1\cdot4, r_2 = 1\cdot41, r_3 = 1\cdot414, r_4 = 1\cdot4142, \ldots$$

1.2.   CONVERGENT SEQUENCES

Suppose we mark the elements of a sequence on a number line. It may happen that, to the limits of accuracy of the drawing, from some element s_N on we find

$$s_N, s_{N+1}, \ldots, s_{N+r}, \ldots$$

are indistinguishable from a specific number s:

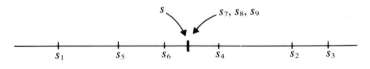

Figure 1.2.1

Let $e_n = s_n - s$; then we can represent this numerically by specifying a permissible error $e > 0$ and say that s_n is indistinguishable from s if the size of e_n is less than e for $n \geq N$. It does not matter whether e_n is positive or negative (corresponding to s_n being greater or less than s). What matters is that the *size* of e_n is less than e. We deal with this by using the *modulus* $|x|$ of a real number x; this is the positive value out of x, $-x$; for instance $|2| = 2$, $|-1| = 1$. We define $|0| = 0$. [See also I, (2.6.5)]. Clearly

$$|xy| = |x||y| \tag{1.2.1}$$

and (by considering the various possible signs of x, y and $x+y$) we may establish the *triangle inequality*:

$$|x+y| \leq |x| + |y|. \tag{1.2.2}$$

Returning to the sequence $s_n = s + e_n$, we can state that s_n is within an error e of s by writing

$$|e_n| < e.$$

If no matter how small e is taken to be we can find N such that

$$n \geq N \quad \text{implies} \quad |e_n| < e,$$

when we say s_n *converges to* s. Of course the smaller the value of e, the larger we may have to take N.

EXAMPLE 1.2.1.   The sequence $((-1)^n/n^2)$ converges to 0, because

$$|(-1)^n/n^2| = 1/n^2$$

and we can make $1/n^2 < e$ provided that

$$n^2 > 1/e.$$

Clearly we can find N such that $N^2 > 1/e$, then if $n \geq N$ we have that

$$|(-1)^n/n^2| = 1/n^2 \leq 1/N^2 < e.$$

Traditionally the Greek letter ε is used for the error and the formal definition of the limit of a sequence is as follows.

DEFINITION 1.2.1.   The sequence (s_n) *converges to the limit* s, if given any $\varepsilon > 0$ there exists a corresponding N such that

$$n \geq N \quad \text{implies} \quad |s_n - s| < \varepsilon.$$

The limit s is variously noted as

$$\lim s_n, \quad \lim_n s_n \quad \text{or} \quad \lim_{n \to \infty} s_n.$$

We also say s_n *tends to* s *as* n *tends to infinity* or in symbols

$$s_n \to s \quad \text{as} \quad n \to \infty.$$

The use of 'infinity' in this phraseology causes a certain amount of trouble to beginners. Perhaps because the word 'infinity' is a noun, the beginner is apt to think that it is the name of a specific number—a number 'larger than any natural number'. Such an interpretation is not intended here; we simply consider the behaviour of s_n as n becomes large. In practice 'large' means bigger than some positive integer N.

EXAMPLE 1.2.2.   Let $s_n = x^n$ where $-1 < x < 1$, then $x^n \to 0$ as $n \to \infty$. We must show that given $\varepsilon > 0$ we can find N such that

$$n \geq N \quad \text{implies} \quad |x^n - 0| < \varepsilon.$$

This is clear when $x = 0$. If $x \neq 0$, then $|x| < 1$ implies $1/|x| > 1$ and we must find N such that $n \geq N$ implies $1/|x|^n > 1/\varepsilon$.

Let $1/|x| = 1 + a$ where $a > 0$.

Then $(1 + a)^2 = 1 + 2a + a^2 > 1 + 2a$ and by induction on n [see I, § 2.1],

$$(1 + a)^n \geq 1 + na$$

for every positive integer n.

Now we make $1 + Na > 1/\varepsilon$ by choosing $N > (\varepsilon^{-1} - 1)/a$. Then if $n \geq N$ we have that

$$1/|x|^n = (1 + a)^n \geq 1 + na \geq 1 + Na > 1/\varepsilon, \quad \text{so} \quad |x|^n < \varepsilon, \quad \text{as required.}$$

EXAMPLE 1.2.3.   If $s_n = $ '$\sqrt{2}$ to n decimal places' as in Example 1.1.4, then

$$s_1 = 1{\cdot}4, \qquad s_2 = 1{\cdot}41, \ldots$$

and clearly

$$s_1 < \sqrt{2} < 1{\cdot}5 = s_1 + 1/10$$
$$s_2 < \sqrt{2} < 1{\cdot}42 = s_2 + 1/10^2$$
$$\vdots$$
$$s_n < \sqrt{2} < s_n + 1/10^n$$
$$\vdots$$

Given $\varepsilon > 0$, choose N such that $1/10^N < \varepsilon$, then

$$n \geq N \quad \text{implies} \quad s_n < \sqrt{2} < s_n + \varepsilon$$

so

$$|s_n - \sqrt{2}| < \varepsilon$$

and $s_n \to \sqrt{2}$ as $n \to \infty$.

In general a decimal expansion [see I, § 2.5]

$$\alpha = a_0 . a_1 a_2 \ldots a_n \ldots ,$$

where a_0 is an integer and $a_1, a_2, \ldots, a_n, \ldots$ are digits ($0 \leq a_i \leq 9$), is defined to be

$$\alpha = \lim_{n \to \infty} \alpha_n,$$

where $\alpha_n = a_0 . a_1 a_2 \ldots a_n$ is the expansion of α to the first n decimal places. In particular, we mention a fact which surprises many beginners:

EXAMPLE 1.2.4.   We have that $0 . 999 \ldots 9 \ldots = 1$, or in words 'point nine recurring' is equal to unity.

Here if $\alpha = 0.999\ldots 9 \ldots$, then

$$\alpha_n = \underbrace{0.99\ldots 9}_{n \text{ places}} = 1 - (1/10)^n,$$

and $\lim_{n \to \infty} \alpha_n = 1$.

This can be verified by referring back to the definition in terms of ε and N, but this is inefficient. A better method is to manufacture new limits from known ones using the very important:

THEOREM 1.2.1   (*Algebra of Limits*).   *If* $\lim s_n = s$, $\lim t_n = t$, *then*

   (i) $\lim (s_n + t_n) = s + t$
   (ii) $\lim (s_n - t_n) = s - t$
   (iii) $\lim s_n t_n = st$
   (iv) $\lim (s_n/t_n) = s/t$ (*all* $t_n \neq 0$, $t \neq 0$).

These results are intuitively obvious and a mathematical proof can be worked out starting with the limits in the first line and using the Definition 1.2.1 to deduce (i)–(iv). For non-mathematicians the effort isn't worth the candle because the technicalities obscure the essential simplicity of the result. Instead we shall demonstrate the power of the Algebra of Limits in action. For instance in Example 1.2.4. we can write

$$\lim \alpha_n = \lim (1 - 1/10^n) = 1 - 0 \text{ (using (ii))}$$

$$= 1.$$

Instead of going back to first principles we can lean on the Algebra of Limits which is itself based on first principles.

EXAMPLE 1.2.5.

$$\lim_{n \to \infty} \frac{4n^2 + n}{2n^2 + 5} = \lim_{n \to \infty} \frac{4 + (1/n)}{2 + (5/n^2)} = \frac{4 + 0}{2 + 0} = 2.$$

EXAMPLE 1.2.6.   Let

$$s_n = 1 + x + x^2 + \ldots + x^{n-1}, \quad \text{where } |x| < 1,$$

[see also Example 1.7.1(vi)]. Then

$$s_n - x s_n = 1 - x^n,$$

whence

$$s_n = \frac{1}{1 - x} - \frac{x^n}{1 - x}.$$

Now $x^n \to 0$ by Example 1.2.2, so by the Algebra of Limits,

$$\lim s_n = \frac{1}{1-x} - \frac{0}{1-x} = \frac{1}{1-x}.$$

It is often possible to gain insight into the nature of a sequence by examining suitable selections of its terms.

DEFINITION 1.2.2.   Let (s_n) be a given sequence. If

$$n_1 < n_2 < \ldots < n_k < \ldots$$

is an increasing infinite sequence of positive integers we say that

$$s_{n_1}, s_{n_2}, \ldots, s_{n_k}, \ldots$$

is a *subsequence* of (s_n); thus the general (kth) term of the subsequence is s_{n_k}. For example,

$$s_2, s_4, s_6, \ldots, s_{2n}, \ldots$$

and

$$s_2, s_3, s_5, s_7, s_{11}, \ldots$$

are subsequences of (s_n), corresponding to the selection of even suffices and prime suffices respectively.

The main result about subsequences is

THEOREM 1.2.2.   *Suppose that $s_n \to l$. Then every subsequence of (s_n) converges to l; that is, all subsequences of a convergent sequence converge, the limit being the same as that of the complete sequence.*

It is very easy to establish the following.

PROPOSITION 1.2.3.   *A convergent sequence is necessarily bounded, that is if $s_n \to l$ (l finite), there exist two numbers A and B such that the inequalities*

$$A \le s_n \le B \tag{1.2.3}$$

hold for all n.

Indeed in the Definition 1.2.1. we may take for ε any positive number we please, say $\varepsilon = 1$; it then follows that

$$|s_n - l| < 1 \quad \text{if } n \ge N,$$

that is, with at most a finite number ($N - 1$) of exceptions all terms of the sequence lie between $l - 1$ and $l + 1$ [see Figure 1.2.2]. If no term of the sequence is greater than $l + 1$ we may put $B = l + 1$, then there will be a greatest term among them, say s_u and we may then put $B = s_u$ (s_2 in Figure 1.2.2). Similarly, we put $A = s_v$ if s_v is the least term which is less than $l - 1$; if there are no such terms we put $A = l - 1$.

Figure 1.2.2

It is important to realize that the converse of Proposition 1.2.3 is false. For example, the oscillating sequence [see § 1.3] $s_n = (-1)^n$, that is

$$1, -1, 1, -1, 1, -1 \ldots \qquad (1.2.4)$$

is bounded but does not converge to a limit. One might be tempted to say that the sequence (1.2.4) has 'two limits', namely -1 and 1; but such a phraseology is quite unacceptable as the Definition 1.2.1 clearly implies that, if a limit exists, then it is unique. However, the notion of a subsequence enables us to describe the situation in a rigorous and meaningful manner.

First we quote, without proof, a celebrated result which is explained more fully in the next chapter [Theorem 2.8.1].

THEOREM 1.2.4 (Bolzano–Weierstrass). *Every (infinite) bounded sequence possesses convergent subsequences.*

For example, the sequence (1.2.4) has the convergent subsequences

$$s_1, s_3, s_5, \ldots \quad \text{and} \quad s_2, s_4, s_6, \ldots$$

having the limits 1 and -1 respectively. It can be shown that, for a given bounded sequence, the set of those real numbers which occur as the limits of convergent sequences, possesses a unique greatest and a unique least number. We are therefore entitled to make the following

DEFINITION 1.2.3. Let $A \le s_n \le B$ $(n = 1, 2, \ldots)$. Then there exists a unique number β such that β is the limit of a subsequence of (s_n); but if $\beta' > \beta$, there is no subsequence which has β' as its limit. We say that β is the *upper limit* or *limit superior* of (s_n) and we write

$$\beta = \overline{\lim} \, s_n \quad \text{or} \quad \beta = \lim \sup s_n,$$

as $n \to \infty$.

Similarly, there exists a unique number α with the property that it is the limit of a subsequence, but no number less than α is the limit of a subsequence. We call α the *lower limit* or *limit inferior* of (s_n) and we write

$$\alpha = \underline{\lim} \, s_n = \lim \inf s_n,$$

as $n \to \infty$.

EXAMPLE 1.2.7. Let $s_n = (1/n) + (-1)^n$, that is

$$s_1 = 0, \, s_2 = \frac{3}{2}, \, s_3 = -\frac{2}{3}, \, s_4 = \frac{5}{4}, \, s_5 = -\frac{4}{5}, \ldots \ldots$$

Then

$$\overline{\lim} \, s_n = 1, \qquad \underline{\lim} \, s_n = -1,$$

subsequences with those limits being

$$s_2, s_4, s_6, \ldots \to 1$$

and

$$s_1, s_3, s_5, \ldots \to -1.$$

This example also illustrates the fact that the upper and lower limits need not coincide with the least upper bound (l.u.b.) and greatest lower bound (g.l.b.) respectively [see I, § 2.6.3]; we have that

$$\text{l.u.b.} \, (s_n) = \tfrac{3}{2}, \qquad \text{g.l.b.} = -1.$$

Incidentally, we note that $\max (s_n) = \tfrac{3}{2}$ while $\min (s_n)$ does not exist.

We conclude this section by listing a few standard formulae, some of which will occur elsewhere in this book; it is understood that, in all cases, n tends to ∞.

EXAMPLE 1.2.8.

(i) $\dfrac{1}{n^\lambda} \to 0 \qquad (\lambda > 0)$

(ii) $n^k x^n \to 0 \qquad (|x| < 1, \, k \text{ fixed})$

(iii) $\left(1 + \dfrac{x}{n}\right)^n \to e^x \qquad$ [see (2.11.4) and Example 4.9.6(i)]

(iv) $\sqrt[n]{a} \to 1 \qquad (a > 0)$

(v) $\sqrt[n]{n} \to 1$

(vi) $n(\sqrt[n]{a} - 1) \to \log a \qquad (a > 0)$

(vii) $\dfrac{2 \cdot 2 \cdot 4 \cdot 4 \ldots 2n \cdot 2n}{1 \cdot 3 \cdot 3 \cdot 5 \ldots (2n-1) \cdot (2n+1)} \to \dfrac{\pi}{2};$

or, equivalently,

(vii)' $\dfrac{(n!)^2 2^{2n}}{(2n)! \sqrt{n}} \to \sqrt{\pi}$

$\left.\begin{array}{c} \\ \\ \\ \\ \\ \\ \end{array}\right\}$ *Wallis' product* [see (4.4.1)]

(viii) $\left(1 + \dfrac{1}{2} + \dfrac{1}{3} + \ldots + \dfrac{1}{n}\right) - \log n \to \gamma,$

where

$$\gamma = 0 \cdot 5772157 \ldots$$

is known as *Euler's* or *Mascheroni's constant.*

1.3. DIVERGENT SEQUENCES

If a sequence (s_n) does not converge to some limit s, then it is said to be *divergent.* The sequence of Example 1.1.1, $s_n = $ 'nth digit of the decimal expansion of π' is divergent. For were it convergent, then it would easily follow that from some point on $s_N, s_{n+1}, \ldots, s_{N+r}, \ldots$ would all have to be the same digit say $s_N = d$, which implies the decimal expansion of π ends up with the digit d recurring. Now

$$0 . ddd \ldots = \frac{d}{9}$$

and this, with a little arithmetic, implies π is a rational number, which it isn't. In this case the value of s_n oscillates amongst the ten possible values $0, 1, 2, \ldots, 9$.

EXAMPLE 1.3.1.   The sequence (n^2) diverges. To prove this, remember a convergent sequence (s_n) must tend to some *real number s.* Were (n^2) to converge to s, then, given $\varepsilon > 0$ we should be able to find N such that $n \geq N$ implies $|n^2 - s| < \varepsilon$; in particular $n \geq N$ implies $n^2 < s + \varepsilon$. This is clearly impossible.

A sequence can diverge in many ways: it can oscillate, it can increase '*beyond all bounds*', or it can mix up these behaviours. For instance, the sequence

$$1, -1, 1, 2, -1, 1, 3, -1, 1, \ldots, n, -1, 1, \ldots$$

has terms which oscillate between 1 and -1, while others increase beyond all bounds; in more rigorous language, this means that, given any positive number M (however great), there exists at least one term s_m such that $s_m > M$.

Of all the possible types of divergence we distinguish just one in which the terms increase beyond all bounds.

DEFINITION 1.3.1.   A sequence (s_n) is said to *diverge* to $+\infty$ if given any real number K, there exists N such that

$$n \geq N \quad \text{implies} \quad s_n > K.$$

EXAMPLE 1.3.2.   The sequence (n^2) diverges to $+\infty$.

The reader should note once again that the symbol $+\infty$ is not meant to denote the name of a specific number. It is a historical term which has remained in the language, and the phrase 'divergent to $+\infty$' simply means that the terms of the sequence grow in size without any bound as explained in the definition. We insert the plus sign in '$+\infty$' because there is an obvious corresponding notion of divergence to $-\infty$ where the terms of the sequence increase in absolute value without any limit and are negative. (For instance $(-n^2)$ diverges to $-\infty$.) This and other forms of divergence will not play any significant role in the rest of our theory. In the case of a sequence (s_n) which diverges to $+\infty$ we write

$$\lim_{n \to \infty} s_n = +\infty$$

or

$$s_n \to +\infty \quad \text{as} \quad n \to \infty.$$

1.4. INCREASING SEQUENCES

We may not be so fortunate as always to be able to compute the explicit limit of a convergent sequence. However, the question of whether a sequence converges or not may often be decided without actually calculating the limit. A case of central importance will be described in this section.

DEFINITION 1.4.1.   A sequence (s_n) is said to be *increasing* if

$$s_1 \leq s_2 \leq \ldots \leq s_n \leq s_{n+1} \leq \ldots.$$

We remark that some texts use the phrase 'monotonic increasing', since they also use the term 'monotonic decreasing' for a sequence

$$s_1 \geq s_2 \geq \ldots \geq s_n \geq s_{n+1} \geq \ldots$$

and then have the word 'monotonic' available to cover both cases [cf. Definition 2.7.1]. We shall use the simpler description 'increasing' on the understanding that it is interpreted in the weak sense, that is, $s_n \leq s_{n+1}$, where equality $s_n = s_{n+1}$ may occur.

The importance of increasing sequences is that, as regards convergence, only two types of behaviour are possible:

THEOREM 1.4.1   (*Increasing Sequence Theorem*).   *If (s_n) is an increasing sequence, then either* (i) *(s_n) is bounded above by some real number K and*

$$\lim s_n = s,$$

where $s \leq K$, or (ii)

$$(s_n) \text{ diverges to } +\infty.$$

EXAMPLE 1.4.1.   Let $s_n = 1 + 1/2^p + \ldots + 1/n^p$, where $p > 1$. Though the limit of the sequence (s_n) is not apparent, we can obtain information about it by grouping terms and considering

$$s_1 = 1$$

$$s_3 = 1 + \left(\frac{1}{2^p} + \frac{1}{3^p}\right) < 1 + \frac{2}{2^p}$$

$$s_7 = 1 + \left(\frac{1}{2^p} + \frac{1}{3^p}\right) + \left(\frac{1}{4^p} + \ldots + \frac{1}{7^p}\right) < 1 + \frac{2}{2^p} + \frac{4}{4^p} \ .$$

When $k = 2^m - 1$,

$$s_k < 1 + \frac{2}{2^p} + \frac{4}{4^p} + \ldots + \frac{2^{m-1}}{2^{p(m-1)}} \ .$$

Hence

$$s_k < 1 + \frac{1}{2^{p-1}} + \left(\frac{1}{2^{p-1}}\right)^2 + \ldots + \left(\frac{1}{2^{p-1}}\right)^{m-1} .$$

Since $0 < (\frac{1}{2})^{p-1} < 1$, we can apply Example 1.2.6 and find that

$$s_k < \frac{1}{1 - (\frac{1}{2})^{p-1}} - \frac{(\frac{1}{2})^m}{1 - (\frac{1}{2})^{p-1}} < \frac{1}{1 - (\frac{1}{2})^{p-1}}.$$

For any n, choose $k = 2^m - 1$ such that $k > n$; then $s_n < s_k$, so for all n,

$$s_n < \frac{1}{1 - (\frac{1}{2})^{p-1}}.$$

Thus $s_n \to s$ where

$$s \leq \frac{1}{1 - (\frac{1}{2})^{p-1}}.$$

In surprising contradistinction to this example, if $p = 1$, the argument given does not work and we find:

EXAMPLE 1.4.2.   If $s_n = 1 + \frac{1}{2} + \frac{1}{3} + \ldots + \frac{1}{n}$, then

$$s_n \to +\infty \quad \text{as} \quad n \to \infty.$$

To prove this, group the terms as follows:

$$s_1 = 1$$

$$s_2 = 1 + \frac{1}{2}$$

$$s_4 = 1 + \frac{1}{2} + \left(\frac{1}{3} + \frac{1}{4}\right)$$

$$s_8 = 1 + \frac{1}{2} + \left(\frac{1}{3} + \frac{1}{4}\right) + \left(\frac{1}{5} + \ldots + \frac{1}{8}\right)$$

$$\vdots$$

$$s_{2^m} = 1 + \frac{1}{2} + \left(\frac{1}{3} + \frac{1}{4}\right) + \ldots + \left(\frac{1}{2^{m-1}+1} + \ldots + \frac{1}{2^m}\right).$$

Then

$$s_1 = 1$$

$$s_2 = 1 + \tfrac{1}{2}$$

$$s_4 > 1 + \tfrac{1}{2} + \tfrac{2}{4}$$

$$s_8 > 1 + \tfrac{1}{2} + \tfrac{2}{4} + \tfrac{4}{8}$$

$$\vdots$$

$$s_{2^m} > 1 + \tfrac{1}{2}m$$

so s_{2^m} increases without limit and since (s_n) is increasing, it follows that $s_n \to +\infty$ as $n \to \infty$.

1.5. DECREASING SEQUENCES

Analogously, we say a sequence (s_n) is *decreasing* if

$$s_1 \geq s_2 \geq \ldots \geq s_n \geq s_{n+1} \geq \ldots .$$

When a decreasing sequence is bounded below by L we may deduce that $s_n \to s$ where $s \geq L$. This follows from the theory of increasing sequences [Theorem 1.4.1] by multiplying by -1. For if (s_n) is decreasing, then $(-s_n)$ is increasing and $s_n \geq L$ gives $-s_n \leq -L$, so

$$-s_n \to \lambda,$$

where $\lambda \leq -L$. Multiplying by -1 again, we find that

$$s_n \to -\lambda \quad \text{where} \quad -\lambda \geq -L.$$

1.6. CAUCHY SEQUENCES

For a sequence which is neither increasing nor decreasing we still have a general principle to appeal to. We may not be able to compute the limit s of a sequence but if $s_n \to s$, then the terms of (s_n) must get close to each other as n increases. We define a *Cauchy sequence* to be a sequence (s_n) with the following property:

given $\varepsilon > 0$ there exists N such that

$$m, n \geq N \text{ implies } |s_m - s_n| < \varepsilon.$$

THEOREM 1.6.1 (*The Cauchy Sequence Theorem*). *A sequence of real numbers (s_n) converges if and only if it is a Cauchy sequence.*

This theorem is a much more subtle result than the earlier 'Increasing Sequence Theorem' because it allows the terms to dot all over the place, up and down in value, as long as they eventually settle 'near' a specific place.

What it tells us is that, if the terms get close to each other in the Cauchy sense, then they must tend to a limit. For this reason the Cauchy Sequence Theorem is also sometimes referred to as '*the general principle of convergence*'. It will prove to be a most useful weapon in our armoury as we proceed with the general theory.

1.7. SERIES

We have already seen examples of sequences such as

$$s_n = 1 + \frac{1}{2^p} + \ldots + \frac{1}{n^p},$$

where each s_n is obtained by successively adding an extra number at each stage; in the present case

$$s_n = s_{n-1} + \frac{1}{n^p}.$$

This is the basic idea of a series:

Given a sequence (a_n) of terms, the *sequence of partial sums* (s_n) is defined by

$$s_1 = a_1$$

$$s_2 = a_1 + a_2$$

$$\vdots$$

$$s_n = a_1 + a_2 + \ldots + a_n,$$

or, recursively,

$$s_n = s_{n-1} + a_n \qquad (n = 2, 3, \ldots).$$

DEFINITION 1.7.1. A *series* consists of a sequence (a_n) of terms from which a sequence (s_n) of *partial sums* is derived, where

$$s_n = a_1 + a_2 + \ldots + a_n.$$

The series is said to *converge to the limit s* if the sequence (s_n) of partial sums converges to s. The number s is called the *sum* of the series. If the series does not converge it is said to *diverge*. In particular if (s_n) diverges to $+\infty$ we shall say the series diverges to $+\infty$.

Thus the concept of convergence for series is reduced to the concept of convergence for sequences, namely their partial sums. Each partial sum involves a finite, but arbitrary, number of terms; unfortunately, it is only in rare cases that we are able to express finite sums by a simple formula. Some instances where this is possible, are listed in the example below.

We shall frequently use the '*sigma notation*'

$$s_n = \sum_{r=1}^{n} a_r = a_1 + a_2 + \ldots + a_n, \qquad (1.7.1)$$

where r is a *dummy suffix* that corresponds to a 'typical' term. The letter r may be replaced by any other convenient letter (but, in the present context, not by n since a_n denotes the final term).

EXAMPLE 1.7.1.

(i)
$$\sum_{r=1}^{n} r = \tfrac{1}{2} n (n+1) \tag{1.7.2}$$

(ii)
$$\sum_{r=1}^{n} (a + (r-1)b) = \frac{n}{2}(2a + (n-1)b) \tag{1.7.3}$$

This series is known as an *arithmetic progression*; it is characterized by the properties that adjacent terms have a constant difference, b in our case.

(iii)
$$\sum_{r=1}^{n} r^2 = \tfrac{1}{6} n (n+1)(2n+1), \tag{1.7.4}$$

(iv)
$$\sum_{r=1}^{n} r^3 = (\tfrac{1}{2} n (n+1))^2 \tag{1.7.5}$$

(v)
$$\sum_{r=1}^{n} \frac{1}{r(r+1)} = \frac{n}{n+1} \tag{1.7.6}$$

(vi)
$$\sum_{r=1}^{n} x^{r-1} = \frac{x^n - 1}{x - 1}, \quad \text{if } x \neq 1. \tag{1.7.7}$$

This series is called a *geometric progression*, consecutive terms having a constant ratio x [see Example 1.2.6].

(vii)
$$\sum_{r=1}^{n} rx^{r-1} = \frac{nx^{n+1} - (n+1)x^n + 1}{(x-1)^2}, \tag{1.7.8}$$

provided that $x \neq 1$.

NOTATION.   The series with terms (a_n) is often denoted by

$$a_1 + \ldots + a_n + \ldots,$$

$$\sum_{n=1}^{\infty} a_n \quad \text{or} \quad \sum_{n} a_n \quad \text{or just } \sum a_n.$$

Historically this notation also doubles for the limit of the series, so we also write

$$s = a_1 + a_2 + \ldots + a_n + \ldots$$

$$s = \sum_{n=1}^{\infty} a_n, \qquad s = \sum_{n} a_n, \qquad s = \sum a_n.$$

This inconsistency of notation is often a cause of confusion for beginners. Another source of difficulty is the use of the word 'term' of a sequence or series. In a sequence (s_n) the nth term is s_n, and convergence of the sequence is convergence of the terms (s_n). For a sum $\sum a_n$, however, the nth term is a_n and convergence of the series $\sum a_n$ means convergence of the sequence of partial sums (s_n) where $s_n = a_1 + \ldots + a_n$. These notational inconsistencies have a long historical precedent and would be difficult to eradicate from the scientific community as a whole, so the reader should learn to cope with them.

Sometimes it is convenient to begin the series with a_0 so that

$$s_0 = a_0$$
$$s_1 = a_0 + a_1$$
$$\vdots$$
$$s_n = a_0 + a_1 + \ldots + a_n$$
$$\vdots$$

If there is any possibility of ambiguity, such a series (and its sum!) will be denoted by

$$\sum_{n=0}^{\infty} a_n \quad \text{or} \quad a_0 + a_1 + \ldots + a_n + \ldots .$$

We note that if a series $\sum a_n$ converges, then $a_n \to 0$. To see this, let

$$s_n = a_1 + \ldots + a_n$$
$$t_n = s_{n-1};$$

then

$$s_n - t_n = a_n.$$

If $s_n \to s$, then also $t_n \to s$ [see Theorem 1.2.2] and by the algebra of limits [see Theorem 1.2.1]

$$a_n \to s - t = 0.$$

We may have $a_n \to 0$ when $\sum a_n$ is divergent (as in the case $a_n = 1/n$, in Example 1.4.2). This means that proving $a_n \to 0$ does not advance our knowledge as to whether $\sum a_n$ converges. However, we can state

PROPOSITION 1.7.1. *If a_n does not tend to zero, then the series $\sum a_n$ diverges.*

EXAMPLE 1.7.2. The series $\sum x^{n-1}$ diverges for $|x| \geq 1$ because x^{n-1} does not tend to zero.

EXAMPLE 1.7.3. If $|x| < 1$, we have that $\sum x^{n-1} = 1/(1-x)$. Since $s_n = 1 + x + \ldots + x^{n-1}$, this is just Example 1.2.6 in another guise. This is the infinite *geometric series*.

Such a computation is usually a luxury however, and numerous tests have been devised to decide whether a series converges without finding its sum. Some of these tests will be discussed in the next two sections.

1.8. SERIES OF NON-NEGATIVE TERMS

If $\sum a_n$ has non-negative terms a_n, then

$$s_n = s_{n-1} + a_n \geq s_{n-1}$$

and the sequence of partial sums (s_n) is increasing. We may therefore use the results stated earlier for increasing sequences:

THEOREM 1.8.1 (*Basic test for series of non-negative terms*).   *The series* $\sum a_n$ $(a_n \geq 0)$ *is convergent if and only if the sequence of partial sums* (s_n) *is bounded above; otherwise* $\sum a_n$ *diverges to* $+\infty$.

This test follows immediately from the Increasing Sequence Theorem 1.4.1.
Often we can deduce results about an unknown series by comparison with a known one.

THEOREM 1.8.2.   (*Comparison test (basic version)*).   *Suppose* $\sum a_n$ *and* $\sum b_n$ *are series of positive terms such that* $a_n \leq b_n$ *for all* n; *then*
   (i) $\sum b_n$ *converges implies* $\sum a_n$ *converges*
   (ii) $\sum a_n$ *diverges implies* $\sum b_n$ *diverges.*

The proof is simple:
   Let

$$s_n = a_1 + \ldots + a_n, \qquad t_n = b_1 + \ldots + b_n$$

then if $\sum b_n$ converges to limit t, the partial sums (t_n) increase to t, so $t_n \leq t$ and

$$s_n = a_1 + \ldots + a_n \leq b_1 + \ldots + b_n = t_n \leq t.$$

Hence by the Increasing Sequence Theorem $s_n \to s$ where $s \leq t$. The proof of the second part is that $\sum a_n$ increases without limit, and since $b_n \geq a_n$ for all n, so must $\sum b_n$ increase without limit.

EXAMPLE 1.8.1.   The series $\sum 1/n^p$ converges for $p > 1$ and diverges for $p \leq 1$. The convergence of $\sum 1/n^p$ for $p > 1$ follows from Example 1.4.1 and the divergence of the series $\sum 1/n$ follows from Example 1.4.2. If $p < 1$ then $n^p < n$; so $1/n^p > 1/n$, and the basic version of the comparison test shows $\sum 1/n^p$ diverges $(p < 1)$. The case in which $p = 1$ is particularly important: the series

$$\sum \frac{1}{n} = 1 + \tfrac{1}{2} + \tfrac{1}{3} + \ldots \tag{1.8.1}$$

is called the *harmonic series*; we have seen that the harmonic series diverges [see Example 1.4.2].
   This basic version can be improved in two ways. First we note that if $\sum b_n$ is a series of positive terms converging to t and $k > 0$, then $\sum kb_n$ is a series

of positive terms which converges to kt. Hence if

$$a_n \le k b_n \qquad (k > 0)$$

then $\sum b_n$ converges implies $\sum a_n$ converges. Secondly the comparative size of a finite number of terms at the beginning of the series is immaterial, we only actually require that there exists a positive integer N such that

$$a_n \le k b_n \quad \text{for } n \ge N.$$

This gives:

THEOREM 1.8.3   (*Comparison test, Mark II*).   *Let $\sum a_n$ and $\sum b_n$ be series in which $a_n \ge 0$ and $b_n > 0$. Suppose that $\sum b_n$ converges and that there exists a positive constant K and an integer N such that*

$$\frac{a_n}{b_n} \le K, \quad \text{if} \quad n \ge N.$$

Then $\sum a_n$ converges.
  Similarly, if $\sum a_n$ diverges and

$$\frac{a_n}{b_n} \le K, \quad \text{if} \quad n \ge N,$$

then $\sum b_n$ diverges.

The actual value of K in this test is immaterial. For instance in comparing

$$\sum \frac{3n^2 + 5n}{4n^4 + 17n + 1} \quad \text{with} \quad \sum \frac{1}{n^2},$$

the actual computation of the number K involved is a waste of time.
  In this type of comparison the following version of the test is most useful:

THEOREM 1.8.4.   (*Comparison test (limit version)*).   *Let $\sum a_n$ and $\sum b_n$ be series in which $a_n \ge 0$ and $b_n > 0$. Suppose that*

$$\lim_{n \to \infty} \left(\frac{a_n}{b_n} \right) = k \ne 0, \infty.$$

Then $\sum a_n$ and $\sum b_n$ either both converge or both diverge.

This test is just a specific application of the previous comparison test; for if

$$\frac{a_n}{b_n} \to k$$

where $k > 0$, then from some number N onwards we must have that

$$\frac{a_n}{b_n} \le 2k$$

and by the Comparison Test, Mark II, if $\sum b_n$ converges, so does $\sum a_n$ and if $\sum a_n$ diverges, so does $\sum b_n$. Because $k \neq 0$, we also have that

$$\frac{b_n}{a_n} \to l$$

where $l = 1/k$; so the roles of a_n, b_n can be reversed giving the complete statement of the limit version of the comparison test.

It is a very useful version, as the following example demonstrates:

EXAMPLE 1.8.2.   The series

$$\sum \frac{3n^2 + 5n}{4n^4 + 17n + 1}$$

is shown to converge by comparing it with the series $\sum (1/n^2)$, which is known to converge [see Example 1.8.1]. For

$$\left(\frac{3n^2 + 5n}{4n^4 + 17n + 1} \right) \Big/ \left(\frac{1}{n^2} \right) = \frac{3n^4 + 5n^3}{4n^4 + 17n + 1}$$

$$= \frac{3 + 5(1/n)}{4 + 17(1/n^3) + (1/n^4)}$$

$$\to \frac{3 + 0}{4 + 0 + 0} = \frac{3}{4},$$

by the Algebra of Limits [see Theorem 1.2.1].

A test which is often helpful is given by

THEOREM 1.8.5   (*The ratio test*).   *Let* $\sum a_n$ *be a series of positive terms*:
(i)   *if* $a_{n+1}/a_n \to l$, *where* $0 \le l < 1$, *then* $\sum a_n$ *converges*;
(ii) *if* $a_{n+1}/a_n \to l > 1$ *or if* $a_{n+1}/a_n \to +\infty$, *then* $\sum a_n$ *diverges*;
(iii) *if* $a_{n+1}/a_n \to 1$, *then the series may converge or diverge*.

Again this is really another version of the comparison test, but its advantage is that we compare successive terms of the *same* sequence rather than having to think up a second sequence whose convergence is known to compare with the given sequence. To see how it comes from the comparison test, we make a subtle comparison with a geometric series [see Example 1.7.3]. In part (i) for example we note that $a_{n+1}/a_n \to l < 1$ means not only that eventually the value a_{n+1}/a_n must be less than 1, but that if we take some k satisfying $l < k < 1$, then eventually we have

$$a_{n+1}/a_n < k \qquad (n \ge N)$$

so

$$a_{n+1} < k a_n \qquad (n \ge N).$$

Hence

$$a_{N+1} < ka_N,$$

$$a_{N+2} < ka_{N+1} < k^2 a_N$$

$$\vdots$$

$$a_{N+r} < k^r a_N.$$

When $k < 1$, we know $\sum k^n$ converges [see Example 1.2.6], and when $n \geq N$, the terms of a_n are term by term smaller than a multiple of the terms of $\sum k^n$, so the Comparison Test, Mark II shows $\sum a_n$ converges.

The reader can easily see in part (ii) that if $a_{n+1}/a_n \to k > 1$, then eventually we must have $a_{n+1}/a_n > 1$, so $a_{n+1} > a_n$ and the terms are increasing in size. This tells us immediately that $\sum a_n$ cannot tend to a limit.

The fact that $a_{n+1}/a_n \to 1$ yields no decision, as stated in (iii), can be demonstrated by considering that if $a_n = 1/n$, then $\sum a_n$ diverges [see Example 1.4.2], but if $a_n = 1/n^2$, then $\sum a_n$ converges [see Example 1.4.1]. In both these cases, $a_{n+1}/a_n \to 1$, so we cannot expect the ratio test to be decisive one way or the other when the limit of the ratio of successive terms is 1.

EXAMPLE 1.8.3.   For any positive real number x, the series

$$1 + \frac{x}{1!} + \frac{x^2}{2!} + \ldots + \frac{x^n}{n!} + \ldots$$

converges by the ratio test, because

$$\frac{a_{n+1}}{a_n} = \frac{x^{n+1}}{(n+1)!} \bigg/ \frac{x^n}{n!} = \frac{x}{n+1} \to 0 \qquad (<1).$$

EXAMPLE 1.8.4.   The series $\sum_0^\infty \{(n+1)/(n+2)\}x^n$ converges for $0 \leq x < 1$ but diverges for $x \geq 1$. Indeed, when $x = 0$ we clearly have convergence because all terms, except the first, are zero. Next when $x > 0$ we have that

$$\frac{a_{n+1}}{a_n} = \frac{n+2}{n+3} x^{n+1} \bigg/ \frac{n+1}{n+2} x^n = \frac{(n+2)^2}{(n+3)(n+1)} x = \frac{\left(1 + \frac{2}{n}\right)^2}{\left(1 + \frac{3}{n}\right)\left(1 + \frac{1}{n}\right)} x.$$

Hence, by the Algebra of Limits,

$$\lim_{n \to \infty} \frac{a_{n+1}}{a_n} = x.$$

Now Theorem 1.8.5 tells us that the series converges if $x < 1$ but that it diverges if $x > 1$. No decision can be reached by that theorem if $x = 1$; however the series then reduces to

$$\Sigma \frac{n+1}{n+2},$$

and is seen to diverge because the terms do not tend to zero.

If often happens that the nth term of the series can be expressed in the form

$$a_n = f(n), \tag{1.8.2}$$

where $f(x)$ is a simple function; we are then concerned with the series

$$\sum_{n=1}^{\infty} f(n).$$

For example,

(i) $$f(x) = \frac{1}{x}, \qquad \Sigma f(n) = \Sigma \frac{1}{n},$$

(ii) $$f(x) = \frac{\cos(\pi x)}{1+x^2}, \qquad \Sigma f(n) = \Sigma \frac{(-1)^n}{1+n^2}$$

[see § 2.12]. This point of view is especially fruitful if $f(x)$ is positive and decreasing. In that case the convergence of the series is closely related to the convergence (that is existence) of the infinite (improper) integral

$$\int_1^{\infty} f(x)\, dx$$

[see § 4.6]. The result is stated in the following:

THEOREM 1.8.6   (*Maclaurin's Integral Test*).   *Let $f(x)$ be a positive and steadily decreasing function. Then the series*

$$\Sigma f(n)$$

converges if and only if the integral

$$\int_1^{\infty} f(x)\, dx$$

is finite.

Remark.   This is one of the very few tests which give a condition that is both necessary and sufficient for convergence.

EXAMPLE 1.8.5.   (i) The series $\Sigma\, 1/n$ diverges [see Example 1.4.2] because the integral

$$\int_1^{\infty} \frac{dx}{x} \qquad \vdots$$

fails to exist [see Example 4.6.1]; indeed

$$\int_1^N \frac{dx}{x} = \log N \to \infty,$$

as $N \to \infty$.

(ii) When $p > 1$, the series $\sum (1/n)^p$ converges [see Example 1.8.1] because

$$\int_1^N \frac{dx}{x^p} = \frac{1}{1-p} \left(\frac{1}{N^{p-1}} - 1 \right) \to \frac{1}{p-1},$$

as $N \to \infty$.

The relationship between summation and integration is of fundamental importance in many branches of mathematics. Theorem 1.8.6 is a particularly simple and striking instance. We sketch the proof based on the following geometrical argument: let A_n be the area beneath the curve

$$y = f(x),$$

bounded by the x-axis and the vertical lines $x = 1$ and $x = n$. Then [see § 4.1]

$$A_n = \int_1^n f(x)\, dx.$$

In Figure 1.8.1 we show the portion of the curve where x varies between 1 and n, together with a collection of rectangles; a typical rectangle, which is

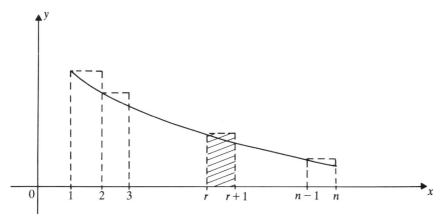

Figure 1.8.1

shaded in the figure, has the interval $[r, r+1]$ as its base, its height being given by

$$a_r = f(r);$$

hence the area of this rectangle [see V, § 1.1.6] is equal to

$$a_r(r+1-r) = a_r.$$

It is evident that the total area of all the rectangles exceeds the area under
the curve. Hence we have the inequality

$$a_1 + a_2 + \ldots + a_{n-1} \geq \int_1^n f(x)\, dx,$$

or more briefly [see (1.7.1)],

$$s_{n-1} \geq \int_1^n f(x)\, dx. \tag{1.8.3}$$

Similarly, we consider a set of rectangles inscribed in the area A_n, as shown
in Figure 1.8.2.

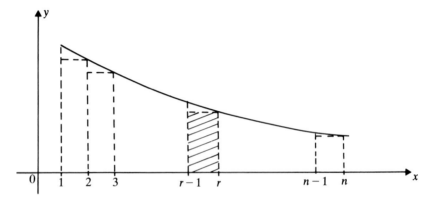

Figure 1.8.2

The rectangles now have areas $a_2, a_3, \ldots, a_n$ respectively, and we deduce
that

$$\int_1^n f(x)\, dx \geq a_2 + a_3 + \ldots + a_n,$$

that is,

$$s_n \leq a_1 + \int_1^n f(x)\, dx. \tag{1.8.4}$$

The Theorem 1.8.6 follows easily from the above inequalities: for if

$$\int_1^\infty f(x)\, dx = A, \tag{1.8.5}$$

say, is finite then

$$\int_1^n f(x)\, dx \leq A$$

for all values of n and (1.8.4) implies that s_n is bounded; hence the series $\sum a_n$ converges because its terms are positive. Conversely, the boundedness of s_n implies that the integral is finite.

More general results can be obtained by considering the area under the curve between $x = N$ and $x = M$ where $M > N$. When the area is known to be finite we take M to be ∞. Thus it can be proved that

$$\int_N^\infty f(x)\, dx < \sum_N^\infty f(n) < \int_{N-1}^\infty f(x)\, dx, \qquad (1.8.6)$$

provided that $f(x)$ is a positive decreasing function such that

$$\sum f(n)$$

converges.

EXAMPLE 1.8.6.   Put $f(x) = 1/x^3$ in (1.8.6), and let $N > 1$. Then we have that

$$\frac{1}{2}\frac{1}{N^2} < \sum_N^\infty \frac{1}{n^3} < \frac{1}{2}\frac{1}{(N-1)^2}.$$

Finally, we mention a test that is sometimes useful when other tests are cumbersome to apply; also it plays an important role in the theory of power series [see Definition 1.10.1]. In its most general form the test involves the notion of upper limit [Definition 1.2.3]; but we shall also give a simplified version which suffices for most cases.

PROPOSITION 1.8.7   (*The nth root test*).   *Let $a_n \geq 0$ and put*

$$\lambda = \overline{\lim}\, (a_n)^{1/n}.$$

Then
   (i) *if $\lambda < 1$, the series $\sum a_n$ converges*
   (ii) *if $\lambda > 1$, the series $\sum a_n$ diverges*   (1.8.7)
   (iii) *if $\lambda = 1$, no conclusion can be drawn.*
   If we make the further assumption that the sequence $(a_n)^{1/n}$ converges, then the statement (1.8.7) holds where

$$\lambda = \lim\, (a_n)^{1/n}.$$

EXAMPLE 1.8.7.   The series

$$\sum_{n=1}^\infty \left(1 - \frac{1}{n}\right)^{n^2}$$

converges because

$$\sqrt[n]{\left(1 - \frac{1}{n}\right)^{n^2}} = \left(1 - \frac{1}{n}\right)^n \to \frac{1}{e} < 1$$

by Example 1.2.8 (iii).

1.9.   SERIES OF POSITIVE AND NEGATIVE TERMS

If $\sum a_n$ has some positive and some negative terms, the tests of the previous section are not applicable as they stand. We may sometimes resolve the question of convergence of $\sum a_n$ by considering $\sum |a_n|$ which has non-negative terms.

DEFINITION 1.9.1.   The series $\sum a_n$ is said to be *absolutely convergent* if $\sum |a_n|$ converges. This definition is useful because we have:

THEOREM 1.9.1   (*The Principle of Absolute Convergence*).   *If $\sum a_n$ is absolutely convergent, then it is convergent.*

Proof.   Let $s_n = a_1 + \ldots + a_n$, $\sigma_n = |a_1| + \ldots + |a_n|$. For $m \geq n$ we have that

$$|s_m - s_n| = |a_{n+1} + \ldots + a_m|$$
$$\leq |a_{n+1}| + \ldots + |a_m|$$
$$= |\sigma_m - \sigma_n|.$$

But (σ_n) is known to be convergent, hence is a Cauchy sequence, and so we now know (s_n) is a Cauchy sequence. Hence (s_n) is convergent by the Cauchy Sequence Theorem 1.6.1.

EXAMPLE 1.9.1. $\sum (-1)^n / n^2$ is absolutely convergent since $\sum 1/n^2$ is convergent [see Example 1.4.1]. Hence $\sum (-1)^n / n^2$ is convergent.

There is one useful test for positive and negative terms which does not depend on absolute convergence:

THEOREM 1.9.2   (*The alternating series test*; *Leibniz's test*).   *Suppose $\sum (-1)^{n-1} a_n$ is a series of terms with alternate signs where $a_n \geq 0$ such that*
   (i) $a_1 \geq a_2 \geq \ldots \geq a_n \geq \ldots$
   (ii) $\lim_{n \to \infty} a_n = 0$,
then $\sum (-1)^{n-1} a_n$ is convergent.

The reason why this test works is not hard to see. If we look at the successive partial sums

$$s_1 = a_1$$
$$s_2 = a_1 - a_2,$$
$$s_3 = a_1 - a_2 + a_3,$$
$$s_4 = a_1 - a_2 + a_3 - a_4,$$
$$\vdots$$

we find that we start off with s_1, then take off something smaller to get s_2, then add something smaller still to get s_3 and so on. The partial sums

$s_1, s_2, s_3, \ldots$ are alternately bobbing up and down, but each time the oscillation up and down decreases by (i) and tends to zero by (ii). We therefore find that (s_n) is a Cauchy sequence and so tends to a limit; in other words the alternating series converges.

EXAMPLE 1.9.2.   The series $\sum (-1)^{n+1}/n$ converges by the alternating series test. This is an example of a sequence which is convergent but not absolutely convergent (since $\sum 1/n$ diverges [see Example 1.4.2]). Such a sequence is called *conditionally convergent*.

The various tests for convergence that we have given will be suitable for a wide range of series. There are stronger tests which will deal with other cases [see Proposition 1.13.1] but a little *ad hoc* common sense can often deal with intricate examples. For further reading consult M. Spivak (1967) and J. A. Green (1958).

1.10. POWER SERIES

A series of the type $\sum c_n x^n$ is called a *power series* in x. This proves to be an important concept; for many useful functions can be defined by power series, for example:

$$\exp x = 1 + \frac{x}{1!} + \frac{x}{2!} + \ldots + \frac{x^n}{n!} + \ldots$$

$$\log_e (1+x) = x - \frac{x^2}{2} + \frac{x^3}{3} - \ldots + (-1)^{n-1} \frac{x^n}{n} + \ldots$$

[see § 2.11]. Of course, these expressions are only valid when the given series converge. All power series converge when $x = 0$.

EXAMPLE 1.10.1.   The series $\sum_{n=0}^{\infty} x^n/n!$ is absolutely convergent for all x [see Example 1.8.3]. (Note we put $0! = 1$ to get the first term.) When $x \neq 0$, we can apply the ratio test [Theorem 1.8.5]; thus

$$\left| \frac{x^{n+1}}{(n+1)!} \right| \bigg/ \left| \frac{x^n}{n!} \right| = |x|/n \to 0;$$

so $\sum x^n/n!$ is always absolutely convergent.

EXAMPLE 1.10.2.   The series $\sum_{n=1}^{\infty} (-1)^{n-1} x^n/n$ is absolutely convergent when $|x| < 1$ and divergent when $|x| > 1$, because if $x \neq 0$, we have that

$$\left| \frac{x^{n+1}}{n+1} \right| \bigg/ \left| \frac{x^n}{n} \right| = \left| \frac{nx}{n+1} \right| = \left| \frac{x}{1+(1/n)} \right| \to |x| \quad \text{as } n \to \infty.$$

The ratio test shows absolute convergence for $|x| < 1$ and divergence for $|x| > 1$. When $x = 1$ the series becomes

$$1 - \tfrac{1}{2} + \tfrac{1}{3} - \tfrac{1}{4} + \ldots,$$

which is convergent [see Example 1.9.2], and when $x = -1$ it is

$$-1 - \tfrac{1}{2} - \tfrac{1}{3} - \tfrac{1}{4} - \ldots,$$

which is divergent (Example 1.4.2 with a minus sign).

EXAMPLE 1.10.3.   The series $\sum n!x^n$ converges only when $x = 0$. For if $x \neq 0$, then

$$\frac{|(n+1)!x^{n+1}|}{|n!x^n|} = (n+1)|x| \to +\infty;$$

so $\sum n!x^n$ is divergent for all x. (In fact when $(n+1)|x| > 1$, i.e. $n > 1/|x| - 1$, then the terms of the series increase in size, so convergence is clearly impossible.) These examples represent various extremes in a spectrum of behaviour for power series. The general situation is described in the following

THEOREM 1.10.1.   *Suppose that the power series $\sum c_n x^n$ converges for at least one non-zero value of x. Then the series either converges absolutely for all x, or else there exists a positive number R such that*
   (i) *$\sum c_n x^n$ converges absolutely for $|x| < R$*
   (ii) *$\sum c_n x^n$ diverges for $|x| > R$*
   (iii) *$\sum c_n x^n$ may converge or diverge for $|x| = R$.*
   (iv) *$1/R = \limsup |c_n|^{1/n}$.*

The number R is called the *radius of convergence* of $\sum c_n x^n$.
   The two possible behaviours for $|x| = R$ in (iii) have already been exhibited in Example 1.10.2. On the other hand the series $\sum x^n/n^2$ has radius of convergence 1 and *converges* for both $x = +1$ and $x = -1$. The series $\sum x^n$ also has radius of convergence 1 and *diverges* for both $x = +1$ and $x = -1$.
   Extending the definition of R we put $R = 0$ when the series converges only for $x = 0$, and we put $R = \infty$ when the series converges for all values of x. We note that item (iv) in Theorem 1.10.1 is an immediate consequence of the nth root test [see Proposition 1.8.7]. For if we put

$$\rho = \limsup |c_n|^{1/n}$$

and apply the test we find that the power series converges when $\rho|x| < 1$ and that it diverges when $\rho|x| > 1$. Hence $R = 1/\rho$.
   The most useful fact about the theorem on convergent power series is that for every x satisfying $|x| < R$ we can define a real number $f(x)$ which is the sum of the series $\sum c_n x^n$. This determines a function f whose domain of definition comprises all x satisfying $|x| < R$.
   As already mentioned in Examples 1.10.1 and 1.10.2 we have the *exponential function*

$$\exp x = 1 + \frac{x}{1!} + \frac{x^2}{2!} + \ldots + \frac{x^n}{n!} + \ldots \quad \text{(for all real } x\text{)}, \qquad (1.10.1)$$

and the function

$$l(x) = x - \frac{x^2}{2} + \frac{x^3}{3} - \ldots + (-1)^{n-1}\frac{x^n}{n} + \ldots \quad (|x| < 1) \quad (1.10.2)$$

where $l(x)$ may be shown to be $\log_e (1+x)$ (see Example 3.6.3).

It should be noted that in determining the radius of convergence we can often use the ratio test as in Examples 1.10.1, 1.10.2 and 1.10.3. If $\sum c_n x^n$ is a power series where $c_n \neq 0$ for all n, then the ratio of the modulus of successive terms (for $x \neq 0$) is

$$\left|\frac{c_{n+1}x^{n+1}}{c_n x^n}\right| = \left|\frac{c_{n+1}}{c_n}\right||x|.$$

By the ratio test, if $\lim_{n\to\infty} |c_{n+1}/c_n|$ exists and is λ then we have that $\sum c_n x^n$ converges absolutely if

$$\lim_{n\to\infty} \left|\frac{c_{n+1}}{c_n}\right||x| < 1;$$

so

$$\lambda |x| < 1$$

or

$$|x| < 1/\lambda.$$

Similarly $\sum c_n x^n$ diverges for $|x| > 1/\lambda$.

Thus the radius of convergence is $1/\lambda$ where

$$1/\lambda = 1 \bigg/ \left(\lim_{n\to\infty} \left|\frac{c_{n+1}}{c_n}\right|\right)$$

$$= \lim_{n\to\infty} \left|\frac{c_n}{c_{n+1}}\right|.$$

This gives us the following

THEOREM 1.10.2 (*Criterion for the radius of convergence*). *If $\sum c_n x^n$ is a power series and*

$$\left|\frac{c_n}{c_{n+1}}\right| \to R \quad \text{as } n \to \infty,$$

then R is the radius of convergence.

As is usual with tests for convergence, this is not universally applicable. For instance the power series

$$\tfrac{1}{2} + x + \tfrac{1}{2}x^2 + x^3 + \tfrac{1}{2}x^4 + \ldots + x^{2m-1} + \tfrac{1}{2}x^{2m} + \ldots$$

has $c_{2m+1} = 1$, $c_{2m} = \frac{1}{2}$ and

$$\left| \frac{c_n}{c_{n+1}} \right| = \begin{cases} \frac{1}{2} & \text{for } n \text{ even.} \\ 2 & \text{for } n \text{ odd.} \end{cases}$$

Thus the criterion is not applicable because $|c_n/c_{n+1}|$ does not tend to a limit. A little common sense in this case shows the series converges for $|x| < 1$ by comparing $\sum |c_n x^n|$ with the convergent series $\sum |x|^n$ and for $|x| \geq 1$ the terms do not tend to zero, so the series diverges. Hence the radius of convergence of this series is 1.

We conclude this section with a brief discussion of the general binomial theorem. For this purpose we define the *binomial coefficients*

$$\binom{\alpha}{n} = \frac{\alpha(\alpha-1)\dots(\alpha-n+1)}{1\cdot 2\dots n}, \tag{1.10.3}$$

where α is an arbitrary real number and n is a positive integer. It is convenient to extend this definition by putting

$$\binom{\alpha}{0} = 1, \quad \text{for all } \alpha, \tag{1.10.4}$$

and

$$\binom{\alpha}{-m} = 0 \tag{1.10.5}$$

$(m = 1, 2, \dots)$. The following properties of the binomial coefficients are readily verified

$$\binom{\alpha-1}{n-1} + \binom{\alpha-1}{n} = \binom{\alpha}{n} \tag{1.10.6}$$

[cf. I, (3.10.5)] and

$$n\binom{\alpha}{n} = \alpha\binom{\alpha-1}{n-1}. \tag{1.10.7}$$

We now consider the series

$$b_\alpha(x) = 1 + \binom{\alpha}{1}x + \dots + \binom{\alpha}{n}x^n + \dots. \tag{1.10.8}$$

If α is a positive integer or zero, say $\alpha = m$, then

$$\binom{m}{n} = 0 \quad \text{for } n > m,$$

and the series terminates; in that case

$$b_m(x) = 1 + \binom{m}{1}x + \dots + \binom{m}{m}x^m,$$

and
$$b_m(x) = (1+x)^m$$

by the elementary binomial theorem [see I, (3.10.1)]. In all other cases (1.10.3) is never zero and we can apply Theorem 1.10.2 in order to determine the radius of convergence of (1.10.8). Thus

$$\binom{\alpha}{n} \Big/ \binom{\alpha}{n+1} = \frac{n+1}{\alpha - n} = \frac{1 + (1/n)}{(\alpha/n) - 1}$$

and

$$\lim \left| \binom{\alpha}{n} \Big/ \binom{\alpha}{n+1} \right| = \left| \frac{1+0}{0-1} \right| = 1.$$

Hence the series (1.10.8), if infinite, has radius of convergence equal to unity and therefore defines a function as x ranges over the open interval $-1 < x < 1$. As a matter of fact, it can be shown that

$$(1+x)^\alpha = 1 + \binom{\alpha}{1} x + \binom{\alpha}{2} x^2 + \ldots + \binom{\alpha}{n} x^n + \ldots \qquad (1.10.9)$$

with the proviso that $|x| < 1$ unless α is a non-negative integer. The formula (1.10.9) is known as the *general binomial theorem*. The proof that indeed

$$b_\alpha(x) = (1+x)^\alpha \qquad (1.10.10)$$

can be based on Taylor's Theorem [see Theorem 3.6.2(i)]. But it is instructive to adopt a more direct procedure, which illustrates some of the more intricate manipulations with power series. First we mention that, within its range of convergence, a power series may be differentiated 'term by term', that is, if

$$f(x) = a_0 + a_1 x + a_2 x^2 + \ldots + a_n x^n + \ldots \quad (|x| < R),$$

then

$$f'(x) = a_1 + 2a_2 x + \ldots + n a_n x^{n-1} + \ldots \quad (|x| < R)$$

[see Theorem 3.2.2]. Hence we may deduce from (1.10.8) that

$$b'_\alpha(x) = \binom{\alpha}{1} + 2\binom{\alpha}{2} x + 3\binom{\alpha}{3} x^2 + \ldots + n\binom{\alpha}{n} x^{n-1} + \ldots,$$

which by virtue of (1.10.7) can be written as

$$b'_\alpha(x) = \alpha \left\{ 1 + \binom{\alpha-1}{1} x + \binom{\alpha-1}{2} x^2 + \ldots + \binom{\alpha-1}{n-1} x^{n-1} + \ldots \right\}.$$

On multiplying this equation by $(1+x)$ and collecting equal powers of x we obtain that

$$(1+x) b'_\alpha(x) = \alpha \left\{ 1 + \left[\binom{\alpha-1}{0} + \binom{\alpha-1}{1} \right] x + \left[\binom{\alpha-1}{1} + \binom{\alpha-1}{2} \right] x^2 + \ldots \right\},$$

$$= \alpha b_\alpha(x),$$

by (1.10.6) and (1.10.8). Hence the function defined by the series must satisfy the differential equation

$$(1+x)b'_\alpha(x) = \alpha b_\alpha(x),$$

which is easily solved by standard methods [see § 7.2]; thus

$$\frac{b'_\alpha(x)}{b_\alpha(x)} = \frac{\alpha}{1+x},$$

$$\frac{d}{dx}\log b_\alpha(x) = \frac{\alpha}{1+x},$$

$$\log b_\alpha(x) = \alpha \log(1+x) + c,$$

since, by inspection, $b_\alpha(0) = 1$, we infer that $c = 0$. Hence

$$b_\alpha(x) = (1+x)^\alpha,$$

as claimed.

EXAMPLE 1.10.4.   (i) When $\alpha = -1$ we find that (1.10.3) reduces to

$$\binom{-1}{n} = \frac{(-1)(-2)\ldots(-n)}{1\,.\,2\ldots n} = (-1)^n.$$

Hence

$$(1+x)^{-1} = 1 - x + x^2 - \ldots + (-1)^n x^n + \ldots, \qquad (1.10.10)$$

which is the geometric series [see Example 1.7.3], with x replaced by $-x$.
   (ii) The formula

$$2(1-t)^{-3} = \sum_{n=0}^{\infty} (n+1)(n+2)t^n \qquad (1.10.11)$$

valid for $-1 < t < 1$, is an instance of the binomial theorem when $x = -t$ and $\alpha = -3$; for

$$\binom{-3}{n} = \frac{(-3)(-4)\ldots(-n-2)}{1\,.\,2\ldots n} = (-1)^n(n+1)(n+2)/2.$$

1.11.   DOUBLE SERIES

Suppose we have a real number $a_{m,n}$ for every pair of positive integers m, n; then we can think of the numbers as being laid out in a rectangular array:

$$
\begin{array}{llll}
a_{1,1} & a_{1,2} \ldots a_{1,n} \ldots \\
a_{2,1} & a_{2,2} \ldots a_{2,n} \ldots \\
\vdots \\
a_{m,1} & a_{m,2} \ldots a_{m,n} \ldots \\
\vdots & \vdots & \vdots
\end{array}
$$

We may attempt to add together all these terms, and this may be done in several different ways. For instance if the sum of each row

$$a_{m,1} + a_{m,2} + \ldots + a_{m,n} + \ldots$$

converges to a sum R_m, say, we could then consider the sum of these rows

$$R_1 + R_2 + \ldots + R_m + \ldots,$$

and if this converges to some value R, we may say that R is the *row sum of the array*. Likewise we could consider the sums of the columns (if convergent):

$$C_n = a_{1,n} + a_{2,n} + \ldots + a_{m,n} + \ldots$$

and then attempt to compute the *column sum of the array*:

$$C = C_1 + C_2 + \ldots + C_n + \ldots.$$

These sums may not converge, or it may happen that one exists but not the other. For instance, in the array

$$\begin{array}{ccccccc}
1 & 1 & 1 & 1 & \ldots & 1 & \ldots \\
-1 & -1 & -1 & -1 & \ldots & -1 & \ldots \\
0 & 0 & 0 & 0 & \ldots & 0 & \ldots \\
0 & 0 & 0 & 0 & \ldots & 0 & \ldots \\
\ldots & \ldots & \ldots & \ldots & \ldots & \ldots & \ldots \\
0 & 0 & 0 & 0 & \ldots & 0 & \ldots
\end{array}$$

the sum of each column is zero, giving a column sum of zero, but the sums of the first two rows do not converge, so the row sum does not exist.

If all the terms of an array have the same sign, however, then this type of difficulty does not arise.

THEOREM 1.11.1.   *If $a_{m,n} \geq 0$ for all m, n, then the row and column sums either both diverge or both converge to the same limit.*

EXAMPLE 1.11.1.   Let $a_{m,n} = a^m b^n$ where $0 \leq a < 1$, $0 \leq b < 1$ then the array is

$$\begin{array}{ccccc}
ab & ab^2 & \ldots & ab^n & \ldots \\
a^2 b & a^2 b^2 & \ldots & a^2 b^n & \ldots \\
\ldots & \ldots & \ldots & \ldots & \ldots \\
a^m b & a^m b^2 & \ldots & a^m b^n & \ldots \\
\ldots & \ldots & \ldots & \ldots & \ldots
\end{array}$$

The sum of the mth row is

$$a^m b + a^m b^2 + \ldots + a^m b^n + \ldots = a^m b(1 + b + \ldots + b^{n-1} + \ldots)$$

$$= a^m b/(1-b)$$

(using the formula for the sum of a geometric progression [see Example 1.2.6]) and the sum by rows is

$$ab/(1-b)+a^2b/(1-b)+\ldots+a^mb/(1-b)+\ldots=\frac{ab}{(1-a)(1-b)}.$$

A similar calculation demonstrates that this equals the sum by columns.

   If the array has different signs for the entries $a_{m,n}$, then, as we have seen, the row and column sums need not give the same result. However, we may consider the related array with entries $|a_{m,n}|$ which, of course, will have the same row and column sums if either converge. We find the following important result:

THEOREM 1.11.2.   *If the row or column sum for the array* $|a_{m,n}|$ *converges, then both the row and column sum of the array* $a_{m,n}$ *converge and have the same value.*

EXAMPLE 1.11.2.   If $a_{m,n}=a^mb^n$ where $|a|<1$, $|b|<1$, then the row and column sums of $|a_{m,n}|$ both converge by the previous example. In this case we may compute the row and column sums of the array a^mb^n to get

$$ab/\{(1-a)(1-b)\}$$

for the sum of the array.

   There are other ways of laying out the terms of the array for computing the sum. For instance we might first add up all the terms in the square array with r rows and r columns to get

$$s_r = \sum_{m=1}^{r} \sum_{n=1}^{r} a_{m,n},$$

and then compute the limit

$$\lim_{r\to\infty} s_r$$

if it exists.

   We could also compute the sum 'diagonally', first taking

$$d_1 = a_{1,1}$$

then

$$d_2 = a_{1,2} + a_{2,1}$$

then

$$d_3 = a_{3,1} + a_{2,2} + a_{1,3}$$

and so on, (these are all diagonals in the array); finally we calculate

$$d_1 + d_2 + d_3 + \ldots.$$

The variations on the manner of organizing the summation process are endless. Fortunately we get a theorem which extends the last one:

THEOREM 1.11.3.   *If the array of moduli* $|a_{m,n}|$ *has a sum by any one method, then the array of* $a_{m,n}$ *has a sum by every method and the sums of the array of* $a_{m,n}$ *are all the same.*

When all the sums are the same, as in the last theorem, we call this value the *sum of the array* and denote it by

$$\sum_{m,n} a_{m,n}.$$

We reserve the notations

$$\sum_n \sum_m a_{m,n} \quad \text{or} \quad \sum_n \left(\sum_m a_{m,n} \right)$$

for the row sum, and

$$\sum_m \sum_n a_{m,n} \quad \text{or} \quad \sum_m \left(\sum_n a_{m,n} \right)$$

for the column sum.

We thus know that if any sum of the array of moduli $|a_{m,n}|$ exists then, all sums of the array of $a_{m,n}$ exist and, in particular,

$$\sum_m \sum_n a_{m,n} = \sum_n \sum_m a_{m,n} = \sum_{m,n} a_{m,n}.$$

An important application of the foregoing results concerns the multiplication of infinite series. Let

$$A = a_0 + a_1 + a_2 + \dots \tag{1.11.1}$$

and

$$B = b_0 + b_1 + b_2 + \dots \tag{1.11.2}$$

be convergent series. We wish to find their product in a manner similar to the multiplication of finitely many terms. The procedure involves forming all possible products $a_i b_j$ $(i, j = 1, 2, \dots)$; these can conveniently be arranged in a doubly infinite array

$$
\begin{array}{ccccc}
a_0 b_0 & a_0 b_1 & a_0 b_2 & \dots & \dots \\
a_1 b_0 & a_1 b_1 & a_1 b_2 & \dots & \dots \\
a_2 b_0 & a_2 b_1 & a_2 b_2 & \dots & \dots \\
\dots & \dots & \dots & \dots & \dots
\end{array}
$$

The product is obtained by adding all the terms of the array. The following result is a straightforward consequence of Theorem 1.11.3.

THEOREM 1.11.4 (*Multiplication Theorem*). *If the series* (1.11.1) *and* (1.11.2) *converge absolutely, then*

$$AB = \sum_{i,j=0}^{\infty} a_i b_j, \tag{1.11.3}$$

where the double sum may be evaluated by any method; in particular, we note summation by squares

$$AB = \lim \left(\sum_{i=0}^{n} a_i \right) \left(\sum_{j=0}^{n} b_j \right), \quad \text{as } n \to \infty,$$

and the summation by diagonals

$$AB = a_0 b_0 + (a_0 b_1 + a_1 b_0) + (a_0 b_2 + a_1 b_1 + a_2 b_0) + \dots$$

$$+ (a_0 b_n + a_1 b_{n-1} + a_2 b_{n-2} + \dots + a_n b_0) + \dots$$

where the terms in each bracket are characterized by the fact that the sum of the indices is constant.

EXAMPLE 1.11.3.   The Example 1.11.2 can be interpreted as an instance of the multiplication theorem, the two series being

$$\frac{a}{1-a} = a + a^2 + a^3 + \dots \quad (|a| < 1)$$

$$\frac{b}{1-b} = b + b^2 + b^3 + \dots \quad (|b| < 1).$$

When we wish to multiply two absolutely convergent power series, say

$$f(x) = a_0 + a_1 x + a_2 x^2 + a_3 x^3 + \dots$$

and

$$g(x) = b_0 + b_1 x + b_2 x^2 + b_3 x^3 + \dots,$$

it is usually most convenient to employ the diagonal method, as this will automatically collect terms in equal powers of x, thus

$$f(x)g(x) = a_0 b_0 + (a_0 b_1 + a_1 b_0)x + (a_0 b_2 + a_1 b_1 + a_2 b_0)x^2$$

$$+ (a_0 b_n + a_1 b_{n-1} + a_2 b_{n-2} + \dots + a_n b_0)x^n + \dots. \tag{1.11.4}$$

EXAMPLE 1.11.4.   Let

$$f(x) = g(x) = (1-x)^{-1} = \sum_{n=0}^{\infty} x^n \quad (|x| < 1).$$

Then $a_i = b_i = 1$ ($i = 0, 1, 2, \dots$). Since the coefficient of x^n in (1.11.4) is the sum of $n + 1$ terms we find that

$$\{f(x)\}^2 = (1-x)^{-2} = \sum_{n=0}^{\infty} (n+1)x^n. \tag{1.11.5}$$

This result may also be obtained from (1.7.8) by letting n tend to infinity, because $x^n \to 0$, as $n \to \infty$.

EXAMPLE 1.11.5.   The series

$$\exp x = 1 + \frac{x}{1!} + \frac{x^2}{2!} + \ldots + \frac{x^n}{n!} + \ldots$$

and

$$\exp y = 1 + \frac{y}{1!} + \frac{y^2}{2!} + \ldots \frac{y^n}{n!} + \ldots$$

converge absolutely for all values of x and y. Forming their product by the diagonal method we obtain that

$$(\exp x)(\exp y) = 1 + \left(\frac{x}{1!} + \frac{y}{1!}\right) + \left(\frac{x^2}{2!} + \frac{x}{1!}\frac{y}{1!} + \frac{y^2}{2!}\right) + \ldots$$

$$+ \left(\frac{x^n}{n!} + \frac{x^{n-1}}{(n-1)!}\frac{y}{1!} + \frac{x^{n-2}}{(n-2)!}\frac{y^2}{2!} + \ldots + \frac{y^n}{n!}\right) + \ldots.$$

The group of terms that are of total degree n in x and y can be written as

$$\frac{1}{n!}\left\{x^n + \binom{n}{1}x^{n-1}y + \binom{n}{2}x^{n-2}y^2 + \ldots + \binom{n}{n-1}xy^{n-1} + y^n\right\} \quad (1.11.6)$$

where

$$\binom{n}{r} = {}_nC_r = \frac{n!}{(n-r)!\,r!} = \frac{n(n-1)\ldots(n-r+1)}{r!}$$

is the binomial coefficient defined in I, (3.8.1). Hence by virtue of the Binomial Theorem we can express (1.11.6) as $(x+y)^n/n!$, and we have proved that $\exp x$ satisfies the fundamental functional equation

$$(\exp x)(\exp y) = \exp(x+y). \quad (1.11.7)$$

1.12.   UNIFORM CONVERGENCE

At the beginning of this chapter the concept of convergence was introduced in relation to sequences of real numbers. Thus we say [Definition 1.2.1] that the sequence

$$s_1, s_2, \ldots, s_n, \ldots \quad (1.12.1)$$

tends to s, if corresponding to every positive number ε (however small), there exists a positive integer $N\ (= N(\varepsilon))$ such that

$$|s_n - s| < \varepsilon \quad \text{if} \quad n \geq N(\varepsilon). \quad (1.12.2)$$

If no such number s exists, then the sequence (s_n) is said to diverge. Evidently, a given sequence of real numbers either converges or diverges.

However, the situation is different when the terms of the sequence are functions of one or several variables [see I, § 1.4]. For the sake of simplicity we shall confine ourselves to a single real variable x which ranges over a finite interval $[a, b]$. Thus we consider a sequence of functions

$$s_1(x), s_2(x), \ldots, s_n(x), \ldots \tag{1.12.3}$$

where $a \leq x \leq b$, and we suppose that, for every fixed value x_0 in that interval, the numerical sequence

$$s_1(x_0), s_2(x_0), \ldots, s_n(x_0), \ldots \tag{1.12.4}$$

converges in the above sense; more precisely, we assume that (1.12.2) holds with the important modification that the integer N will now, in general, depend on both x_0 and ε. Our hypothesis may be described as *point-wise convergence* in the interval $[a, b]$; accordingly, by writing x instead of x_0 we can define the *limit function* $s(x)$ of the sequence $(s_n(x))$ by

$$s(x) = \lim_{n \to \infty} s_n(x) \qquad (a \leq x \leq b). \tag{1.12.5}$$

It is essential to specify the interval with respect to which convergence is discussed; it is usual to assume that this interval is closed and finite, although intervals of the form $[a, \infty)$ sometimes occur [see I, § 2.6.3]. Incidentally, it is not asserted that convergence does not take place for values outside the interval.

EXAMPLE 1.12.1.   Let $s_n(x) = x^n$ $(n = 1, 2, \ldots)$ and suppose that $0 \leq x \leq 1$. It was shown in Example 1.2.2 that

$$s_n(x) \to 0, \quad \text{if } 0 \leq x < 1$$

and, trivially,

$$s_n(x) \to 1, \quad \text{if } x = 1.$$

Hence the limit function is given by

$$s(x) = \begin{cases} 0 & \text{if } 0 \leq x < 1 \\ 1 & \text{if } x = 1 \end{cases}, \tag{1.12.6}$$

and so

$$|s_n(x) - s(x)| = \begin{cases} x^n & \text{if } 0 \leq x < 1 \\ 0 & \text{if } x = 1. \end{cases}$$

In order to ensure that

$$|s_n(x) - s(x)| < \varepsilon \quad \text{if } n \geq N, \tag{1.12.7}$$

we have to choose

$$N > \frac{\log \varepsilon}{\log x} \quad \text{if } 0 < x < 1, \tag{1.12.8}$$

while any value of N will do if $x = 0$ or $x = 1$. As we anticipated, the right-hand side of (1.12.8) depends on x and ε.

There are, however, circumstances in which an integer N can be found which is independent of x and so renders (1.12.7) 'uniform' for all x in $[a, b]$. For instance, if in the preceding example we had taken the interval to be $[0, \frac{1}{2}]$, then

$$|s_n(x) - s(x)| = x^n \le (\tfrac{1}{2})^n$$

and (1.12.7) holds for all x in $[0, \frac{1}{2}]$ provided that

$$N > \frac{\log \varepsilon}{\log (\tfrac{1}{2})},$$

which does not involve x. Thus we are led to the following

DEFINITION 1.12.1.   The sequence $(s_n(x))$ is said to *converge uniformly* to $s(x)$ in the interval $[a, b]$ if, corresponding to every positive ε, there exists an integer $N \ (= N(\varepsilon))$, which does not depend on x, such that

$$|s_n(x) - s(x)| < \varepsilon$$

provided that $n \ge N$.

This definition is rather awkward to apply and it is sometimes more convenient to proceed as follows: suppose the sequence $s_n(x)$ has the limit function $s(x)$ in the interval $[a, b]$ and define

$$\mu_n = \text{l.u.b.} \ |s_n(x) - s(x)| \tag{1.12.9}$$

[see I, § 2.6.3]. When the function $|s_n(x) - s(x)|$ possesses a maximum in $[a, b]$, then

$$\mu_n = \max |s_n(x) - s(x)|. \tag{1.12.10}$$

We then have the following criterion for uniform convergence.

PROPOSITION 1.12.1.   *The sequence $s_n(x)$ tends to $s(x)$ uniformly in $[a, b]$ if and only if*

$$\mu_n \to 0 \quad as \quad n \to \infty, \tag{1.12.11}$$

where μ_n is defined in (1.12.9).

In order to illustrate the significance of (1.12.9) we refer again to Example 1.12.1. Let

$$r_n(x) = s_n(x) - s(x)$$

be the nth remainder function in the interval $[0, 1]$. By (1.12.6)

$$r_n(x) = \begin{cases} x^n & \text{if } 0 \le x < 1 \\ 0 & \text{if } x = 1 \end{cases}$$

[see Figure 1.12.1].

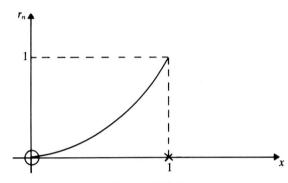

Figure 1.12.1

Evidently, the function $r_n(x)$ has no maximum in $[0, 1]$, but its least upper bound is equal to unity, that is

$$\mu_n = \text{l.u.b. } r_n(x) = 1 \quad (x \in [0, 1]).$$

If we had confined ourselves to the interval $[0, \frac{1}{2}]$ the limit function $s(x)$ would have been zero throughout and so

$$r_n(x) = s_n(x) = x^n.$$

In these circumstances $r_n(x)$ attains its maximum when $x = \frac{1}{2}$, and

$$\mu_n = \max r_n(x) = (\tfrac{1}{2})^n \quad (x \in [0, \tfrac{1}{2}]),$$

[see Figure 1.12.2]

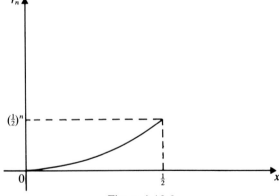

Figure 1.12.2

EXAMPLE 1.12.2.   Consider the sequence

$$s_n(x) = (1-x)x^n \qquad (n = 1, 2, \ldots) \qquad (1.12.12)$$

in the interval $[0, 1]$; see Figure 1.12.3. Clearly $s_n(x) \to 0$ for each x in $[0, 1]$, because $x^n \to 0$ when $0 \le x < 1$ and $1 - x = 0$ when $x = 1$. Thus $s(x) = 0$ and

$$|s_n(x) - s(x)| = s_n(x).$$

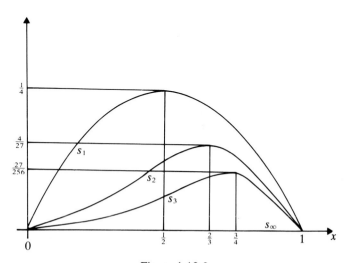

Figure 1.12.3

The maximum of $s_n(x)$ can be found by means of the differential calculus [see § 3.5]: thus

$$s_n'(x) = nx^{n-1} - (n+1)x^n.$$

Note that $s_n'(1) = -1$ for all n.

When $n > 1$, the equation $s_n'(x) = 0$ has the solutions $x = 0$ and $x = n/(n+1)$. The latter corresponds to the maximum of s_n. Hence

$$\mu_n = s_n\!\left(\frac{n}{n+1}\right) = \frac{1}{n+1}\left(\frac{n}{n+1}\right)^n.$$

By Example 1.2.8(iii),

$$\left(1+\frac{1}{n}\right)^n = \left(\frac{n+1}{n}\right)^n \to e, \quad \text{as } n \to \infty,$$

so

$$\frac{1}{n+1}\left(\frac{n}{n+1}\right)^n \to 0\,\frac{1}{e} = 0, \quad \text{as } n \to \infty$$

and we deduce from Proposition 1.12.1 that

$$(1-x)x^n \to 0$$

uniformly in $[0, 1]$.

EXAMPLE 1.12.3. Let

$$t_n(x) = n(1-x)x^n \quad \text{in } [0, 1].$$

As before $t_n(x) \to 0$ for all x in $[0, 1]$, because $nx^n \to 0$ when $0 < x < 1$ [see Example 1.2.8(ii)]; thus the limit function is again equal to zero throughout the interval. However, this time we have that

$$\max |t_n(x)| = \frac{n}{n+1}\left(\frac{n}{n+1}\right)^n \to \frac{1}{e},$$

as $n \to \infty$. It follows that $t_n(x)$ does not tend uniformly to its limit function in $[0, 1]$. Note that $t'_n(1) = -n$.

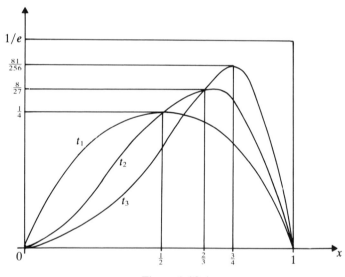

Figure 1.12.4

The concept of uniform convergence is important because it enables us to express some general facts about the nature of the limit function. In particular, we mention without proof the following two main results.

THEOREM 1.12.2. *If each $s_n(x)$ $(n = 1, 2, \ldots)$ is continuous in $[a, b]$ and if $s_n(x) \to s(x)$ uniformly in $[a, b]$, then $s(x)$ is continuous in $[a, b]$.*

THEOREM 1.12.3.   *If each $s_n(x)$ $(n = 1, 2, \ldots)$ can be integrated over $[a, b]$
[see § 4.1] say*

$$j_n = \int_a^b s_n(x)\, dx$$

*and if $s_n(x)$ tends uniformly to $s(x)$ in $[a, b]$, then $s(x)$ can be integrated over
$[a, b]$ and*

$$\int_a^b s(x)\, dx = \lim_{n \to \infty} j_n. \qquad (1.12.13)$$

The result (1.12.13) can be expressed more succinctly as

$$\lim \int_a^b s_n\, dx = \int_a^b (\lim s_n)\, dx. \qquad (1.12.14)$$

In order to illustrate Theorem 1.12.2 we look again at Example 1.12.1.
Evidently each $s_n(x)$ is continuous in $[0, 1]$ but $s(x)$ is not; hence uniform
convergence is ruled out.

Although, under the conditions of Theorem 1.12.2, uniform convergence
ensures the continuity of the limit function, the converse does not hold: the
limit function in Example 1.12.3 is identically zero and therefore continuous;
but, as we saw, the convergence was not uniform.

EXAMPLE 1.12.4.   The sequence

$$u_n(x) = n^2(1 - x)x^n \qquad (n = 1, 2, \ldots)$$

has the limit function $u(x) = 0$ in $[0, 1]$. Hence

$$\int_0^1 u(x)\, dx = 0.$$

On the other hand we find that

$$\int_0^1 u_n(x)\, dx = n^2 \left(\frac{1}{n+1} - \frac{1}{n+2} \right) = \frac{n^2}{(n+1)(n+2)} \to 1, \quad \text{as } n \to \infty.$$

Thus (1.12.14) does not hold in this case. This confirms that the convergence
$u_n(x) \to u(x)$ is not uniform in $[0, 1]$, which could also have been demonstrated
with the aid of Proposition 1.12.1.

The notion of uniform convergence can be transferred to an infinite series

$$\sum_{r=1}^{\infty} u_r(x) \qquad (1.12.15)$$

whose terms are functions of x defined in a given interval $[a, b]$: we form the
sequence of partial sums

$$s_n(x) = \sum_{r=1}^{n} u_r(x) \qquad (n = 1, 2, \ldots),$$

each being regarded as a function in $[a, b]$. If

$$s_n(x) \to s(x), \qquad x \in [a, b], \tag{1.12.16}$$

we say that $s(x)$ is the *sum function* of (1.12.15) in $[a, b]$, and we write

$$s(x) = \sum_{r=1}^{\infty} u_r(x). \tag{1.12.17}$$

If, moreover, the convergence (1.12.16) is uniform in $[a, b]$ we say that the series (1.12.15) *converges uniformly* to $s(x)$ in the interval $[a, b]$.

EXAMPLE 1.12.5.   The series

$$\sum_{r=1}^{\infty} (1-x)^2 x^{r-1}, \tag{1.12.18}$$

defined in $[0, 1]$, has partial sums

$$t_n(x) = \sum_{r=1}^{n} (1-x)^2 x^{r-1} = (1-x)(1-x^n).$$

[see Example 1.7.1(vi)]. Hence the limit function is given by

$$t(x) = \begin{cases} 1-x & \text{if } 0 \le x < 1 \\ 0 & \text{if } x = 1. \end{cases}$$

Thus the formula

$$t(x) = 1 - x$$

holds throughout the interval $[0, 1]$. In order to study the uniformity of convergence, we examine the difference

$$t_n(x) - t(x) = -x^n(1-x),$$

which apart from the sign agrees with the sequence $s_n(x)$ studied in Example 1.12.2. We have seen that $s_n(x)$ tends uniformly to zero, that is $t_n(x)$ tends uniformly to $t(x)$. Thus the series (1.12.18) tends uniformly to $(1-x)$ in $[0, 1]$.

The general results expressed in Theorems 1.12.2 and 1.12.3 can be translated into the language of series as follows:

THEOREM 1.12.4.   *Suppose that the functions* $u_r(x)$ $(r = 1, 2, \ldots)$ *are continuous in* $[a, b]$ *and that the series*

$$\sum_{r=1}^{\infty} u_r(x)$$

converges uniformly to $s(x)$ *in* $[a, b]$. *Then*

(i) $s(x)$ *is continuous in* $[a, b]$,

(ii) $\qquad\qquad \displaystyle\sum_{r=1}^{\infty} \int_a^b u_r(x)\, dx = \int_a^b s(x)\, dx. \tag{1.12.19}$

There is a simple test that allows us to establish uniform convergence in some cases.

PROPOSITION 1.12.5 (*Weierstrass' M-test*). *The series*

$$\sum_{r=1}^{\infty} u_r(x) \qquad\qquad (1.12.20)$$

converges uniformly in $[a, b]$ *if there exists a sequence of positive constants*

$$M_1, M_2, \ldots, M_r, \ldots$$

such that

(i) $|u_r(x)| \le M_r \qquad (r = 1, 2, \ldots)$

for all x *in* $[a, b]$ *and*

(ii) $\sum M_r \quad converges.$ $(1.12.21)$

The test furnishes sufficient conditions for uniform convergence; so uniform convergence may occur although those conditions are not satisfied. However, if the test can be applied, it is noteworthy that we can prove uniform convergence without a knowledge of the sum function. We say the series (1.12.20) is *majorized* by the series (1.12.21).

EXAMPLE 1.12.6.   The series

$$s(x) = \sum_{r=1}^{\infty} (-1)^{r-1} \frac{\cos rx}{r^2}$$

converges uniformly in every interval $[a, b]$, because $|(-1)^{r-1} \cos rx/r^2| \le 1/r^2$ and $\sum 1/r^2$ converges [see Example 1.4.1]. Hence $s(x)$ is a continuous function in $[a, b]$. But this argument does not help us to identify $s(x)$. As a matter of fact, when $[a, b] = [-\pi, \pi]$, it can be shown that

$$s(x) = (\pi^2 - 3x^2)/12$$

[see (20.5.20)].

We conclude this section with an important result about power series.

PROPOSITION 1.12.6.   *Let*

$$\sum_{n=0}^{\infty} c_n x^n \qquad\qquad (1.12.22)$$

be a power series whose radius of convergence is equal to R *[see Theorem 1.10.1] and let* R_1 *be any number satisfying*

$$0 < R_1 < R.$$

Then the series (1.12.22) *converges uniformly in the interval* $[-R_1, R_1]$. On applying Theorem 1.12.4 we can state the following

COROLLARY 1.12.7.   *We have that*

$$\int_0^x \left(\sum_0^\infty c_n t^n \right) dt = \sum_0^\infty \frac{c_n}{n+1} x^{n+1},$$

provided that $|x| < R$.

1.13.   COMPLEX SEQUENCES AND SERIES

In the original definition, the terms of a sequence or series were real numbers, and the question of convergence was discussed in the context of real numbers.

It is, however, quite easy to extend those concepts and results to sequences and series of complex numbers [see I, § 2.7]. We present here the most important definitions and facts [see also § 9.1].

DEFINITION 1.13.1.   The sequence of complex numbers

$$\sigma_1, \sigma_2, \ldots, \sigma_n, \ldots$$

is said to *converge to* σ, if corresponding to every positive real number ε, there exists an integer $N (= N(\varepsilon))$ such that

$$n \geq N \quad \text{implies that} \quad |\sigma_n - \sigma| < \varepsilon \qquad (1.13.1)$$

[cf. Definition 1.2.1]. Of course, the modulus sign in (1.13.1) refers to complex numbers, that is if $\zeta = x + iy$, where x and y are real, then

$$|\zeta| = (x^2 + y^2)^{1/2}$$

[see I, § 2.7.2].

The question of convergence can be decided by examining the real and imaginary parts of the terms:

PROPOSITION 1.13.1.   *Let*

$$\sigma_n = s_n + it_n \qquad (n = 1, 2, \ldots),$$

where s_n *and* t_n *are real. Then the sequence* (σ_n) *converges if and only if the real sequences* (s_n) *and* (t_n) *converge. More precisely, if* $s_n \to s$ *and* $t_n \to t$, *then* $\sigma_n \to s + it$.

EXAMPLE 1.13.1.   Let

$$\sigma_n = \frac{1}{n} (1 + i\sqrt{n^2 - 2}).$$

We can write

$$\sigma_n = \frac{1}{n} + i\sqrt{1 - (2/n^2)}.$$

Since $1/n \to 0$ and $\sqrt{1 - (2/n^2)} \to 1$, we have that $\sigma_n \to i$.

The notion of a Cauchy sequence can be transferred to complex numbers without a change, and we have the analogue of Theorem 1.6.1, thus

THEOREM 1.13.2 (*General Principle of Convergence*). *The sequence σ_n of complex numbers converges, if and only if, for every positive number ε, there exists a positive integer N such that*

$$|\sigma_n - \sigma_m| < \varepsilon$$

provided that $n > N$ and $m > N$.

In a similar manner, we can discuss infinite series of complex terms. If

$$\alpha_1 + \alpha_2 + \alpha_2 + \ldots + \alpha_n + \ldots \tag{1.13.2}$$

is such a series we form the sequence of *partial sums*

$$\sigma_n = \sum_{r=1}^{n} \alpha_r \qquad (n = 1, 2, \ldots) \tag{1.13.3}$$

and, in analogy to Definition 1.7.1 we have

DEFINITION 1.13.2. The series (1.13.2) *converges* if and only if the sequence of partial sums (1.13.3) converges. If $\sigma_n \to \sigma$ we put

$$\sum \alpha_n = \sigma$$

and we call σ the *sum* of (1.13.2).

Alternatively, we may proceed by splitting the terms into real and imaginary parts.

PROPOSITION 1.13.3. *Let $\alpha_n = a_n + ib_n$ $(n = 1, 2, \ldots)$. Then the series*

$$\sum \alpha_n \tag{1.13.4}$$

converges if and only if the real series

$$\sum a_n \quad and \quad \sum b_n \tag{1.13.5}$$

converge; if the sums of these series are a and b respectively, then (1.13.4) converges to $a + ib$.

EXAMPLE 1.13.2. The series

$$\sum \frac{(-1)^{n-1} n + i}{n^2} = \sum \frac{(-1)^{n-1}}{n} + i \sum \frac{1}{n^2}$$

converges, because both series on the right-hand side converge (see Examples 1.8.1 and 1.9.2). Of course, any result that rests on the positiveness or negativeness of real numbers cannot be applied to complex series; for instance the ratio test (Theorem 1.8.5) is not available in the present context. However, the *Principle of Absolute Convergence* (Theorem 1.9.1) still holds; thus we have

THEOREM 1.13.4.   *If $\sum |\alpha_n|$ converges, then $\sum \alpha_n$ converges.*

EXAMPLE 1.13.3.   Consider the series $\sum \alpha_n$, where

$$\alpha_n = \exp\left(-\tfrac{1}{2}n + i\sqrt{n}\right).$$

We recall that, for any real numbers x and y,

$$|\exp(x+iy)| = \exp x$$

[see (9.5.18)]. Hence

$$|\alpha_n| = \exp\left(-\tfrac{1}{2}n\right) = (e^{-1/2})^n.$$

The convergence of $\sum |\alpha_n|$ follows from Example 1.7.3, in which we may put $x = e^{-1/2}$, because $0 < e^{-1/2} < 1$. Hence $\sum \alpha_n$ converges.

We conclude this chapter with a brief discussion of a more subtle technique, which is based on a simple algebraic formula due to N. H. Abel (1826):

PROPOSITION 1.13.5   (*Abel's Summation by Parts*).   *Let $a_1, a_2, \ldots$ and $b_1, b_2, \ldots$ be two sequences of real or complex numbers and put*

$$A_n = a_1 + a_2 + \ldots + a_n \tag{1.13.6}$$

$(n = 1, 2, \ldots)$. *Then, if $n > 1$,*

$$\sum_{r=1}^{n} a_r b_r = A_n b_n + \sum_{r=1}^{n-1} A_r(b_r - b_{r+1}). \tag{1.13.7}$$

The proof of (1.13.7) follows easily from the observation that $a_r = A_r - A_{r-1}$ when $r > 1$ and $a_r = A_r$ when $r = 1$. Hence

$$\sum_{r=1}^{n} a_r b_r = A_1 b_1 + (A_2 - A_1) b_2 + (A_3 - A_2) b_3 + \ldots + (A_n - A_{n-1}) b_n$$

$$= A_1(b_1 - b_2) + A_2(b_2 - b_3) + \ldots + A_{n-1}(b_{n-1} - b_n) + A_n b_n,$$

which is (1.13.7).

Remark.   The formula (1.13.7) is analogous to integration by parts (see § 4.3), namely

$$\int a(x)b(x)\,dx = A(x)b(x) - \int A(x)b'(x)\,dx,$$

where A_n corresponds to $A(x) = \int a(x)\,dx$ and $b'(x)$ corresponds to $b_{n+1} - b_n$.

Several useful tests of convergence are readily derived from (1.13.7). We mention only one such result.

PROPOSITION 1.13.6 (*Dirichlet's Criterion of Convergence*). *The series* $\sum a_r b_r$ *converges if*
   (i) *there exists a number M such that* $|a_1 + a_2 + \ldots + a_n| \leq M$ *for all* n;
   (ii) $b_1 \geq b_2 \geq b_3 \geq \ldots$;
   (iii) $b_n \to 0$ *as* $n \to \infty$.

EXAMPLE 1.13.4. Let θ be a real number such that $\theta \neq 2\pi k$ for any integer k. Then the series

$$\sum \frac{e^{i\theta r}}{r} \tag{1.13.8}$$

converges; this follows from Proposition 1.13.6 with $a_r = e^{i\theta r}$ and $b_r = 1/r$. Indeed it is obvious that the conditions (ii) and (iii) are satisfied. In order to examine (i) we note that

$$A_n = a_1 + a_2 + \ldots + a_n = \sum_{r=1}^{n} e^{i\theta r} = e^{i\theta} \sum_{r=0}^{n-1} e^{i\theta r}$$

$$= e^{i\theta} \frac{e^{i\theta n} - 1}{e^{i\theta} - 1} \quad \text{by (1.7.7)}$$

$$= \frac{e^{i\theta} e^{in\theta/2}}{e^{i\theta/2}} \frac{e^{in\theta/2} - e^{-in\theta/2}}{e^{i\theta/2} - e^{-i\theta/2}}$$

$$= e^{i(n+1)\theta/2} \frac{\sin \frac{1}{2}n\theta}{\sin \frac{1}{2}\theta} \quad \text{by (9.5.11).}$$

Since

$$\left| e^{i(n+1)\theta/2} \right| = 1 \quad [\text{see (9.5.17)}]$$

and

$$\left| \sin \tfrac{1}{2}n\theta \right| \leq 1, \quad [\text{see § 2.12}]$$

we have that

$$A_n \leq \frac{1}{\sin \frac{1}{2}\theta},$$

the right-hand side being finite because $\theta \neq 2\pi k$. Thus the condition (i) is satisfied.

W.L. & D.O.T.

REFERENCES

Green, J. A. (1958). *Sequences and Series*, Routledge and Kegan Paul.
Spivak, M. (1967). *Calculus* (Chapters 21–23), W. A. Benjamin.

CHAPTER 2

Functions of One (Real) Variable

The full power of mathematics as a tool for understanding Nature received its greatest impetus from the development of the infinitesimal calculus by Isaac Newton and Gottfried Wilhelm Leibniz in the 1660s and 1670s. Whilst earlier mathematicians had grasped some of the ideas involved, these two were the first to provide a genuine understanding of the basic techniques. However, their work lacked proper rigour: in particular notions such as 'infinitesimal' appeared self-contradictory, and hence ideas depending on them, such as 'limit'—and thus the whole of calculus—became suspect. This was pointed out at some length by Bishop Berkeley in 1734. The mathematicians of the time paid him scant attention, preferring to follow up the remarkable and important consequences of the mathematics. From this period stem many of the basic ideas and results of classical science, and in this respect the lack of attention to rigour was justified, at the time.

As science progressed, however, the lack of rigour began to be felt in more serious ways, culminating in a crisis over the validity of Fourier's application of trigonometric series to heat conduction. Different mathematicians claimed contradictory theorems, with apparently equally valid proofs.

The resolution of the crisis, begun by Bernard Bolzano and Augustin-Louis Cauchy in the 1820s was essentially completed by Karl Weierstrass in the 1840s. It required very carefully formulated definitions of fundamental concepts such as 'limit' or 'continuity'. In this chapter we shall outline the theoretical framework provided by this approach, and use power series to define a number of standard functions, such as exponential and trigonometric functions. More 'practical' computational methods and problem-solving techniques are treated in Volume III.

The material breaks into three main topics: continuity and limits, differentiability, power series.

2.1. THE DEFINITION OF CONTINUITY

Throughout this chapter we shall be concerned with real-valued functions whose domain D is a subset of the real line [see I, §§ 1.4 and 2.6]; that is,

49

functions

$$f : D \to \mathbb{R}$$

where $D \subseteq \mathbb{R}$. For example [see I, § 14.1], f might be a *polynomial function*

$$f(x) = a_n x^n + a_{n-1} x^{n-1} + \ldots + a_0 \qquad (a_n \neq 0) \tag{2.1.1}$$

with coefficients $a_r \in \mathbb{R}$, in which we take $D = \mathbb{R}$. If $n = 1$, that is

$$f(x) = a_1 x + a_0$$

we say f is *linear*; if $n = 2$, then

$$f(x) = a_2 x^2 + a_1 x + a_0$$

is *quadratic*; if $n = 3, 4, 5, \ldots$, then f is said to be *cubic, quartic, quintic*, etc.

Usually D will be an interval, or a union of several intervals [see I, § 2.6.3]: these possibilities are forced upon us by the need to study classical functions such as $1/x$ or $\log x$.

We shall define an especially useful class of such functions, the *continuous* functions. Loosely speaking the graph of a continuous function must have 'no breaks'; that is, functions like $f(x) = x^2$ are continuous (Figure 2.1.1), as are all polynomial functions; but functions like $f(x) = [x]$ (the greatest integer less than or equal to x) are not because they jump (Figure 2.1.2). The precise definition turns out to be fairly subtle: in some delicate cases it leads to counter-intuitive results, but this is not a serious problem as we shall see.

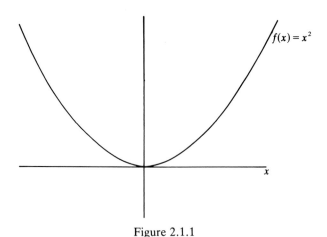

Figure 2.1.1

First we shall define continuity *at a point*.

DEFINITION 2.1.1.   Let $f : D \to \mathbb{R}$ and let $x_0 \in D$. Then f is *continuous at* x_0 if for all $\varepsilon > 0$ there exists $\delta > 0$ (which may depend on ε) such that, whenever $x \in D$ and $|x - x_0| < \delta$, it follows that $|f(x) - f(x_0)| < \varepsilon$.

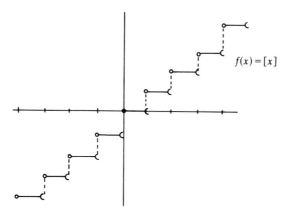

$$f(x) = [x]$$

Figure 2.1.2

The motivation for this definition is very similar to that for convergence of a sequence to a limit [see § 1.2]. In essence, it says that arbitrarily small changes in the *value* of $f(x)$ may be secured by permitting only sufficiently small changes in x. More loosely still, 'small changes in x lead to small changes in $f(x)$'. However, in the precise definition note that we must *first* decide how small (ε) we want changes in $f(x)$ to be, and then say how small (δ) we can permit changes in x. The definition is illustrated geometrically in Figure 2.1.3.

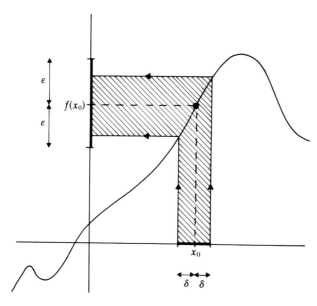

Figure 2.1.3

The advantage of this definition is that it is precise and relatively easy to check mechanically. Any disadvantage stems from its convoluted logical structure.

For example, let us prove that

$$f(x) = x^2 - 3x - 2$$

is continuous at $x = 5$, where $f(5) = 8$.

Clearly the main task will be to estimate the size of $|f(x) - f(5)|$, given that $|x - 5| < \delta$. The easiest way to do this (and it works in most cases rather well) is to change the variable to $u = x - 5$. Then $|x - 5| < \delta$ means 'small values of u', that is, $|u| < \delta$; and this is an easier condition to work with. Then $x = u + 5$, so we have

$$f(x) = (u + 5)^2 - 3(u + 5) - 2$$
$$= u^2 + 7u + 8.$$

Now

$$|f(x) - f(5)| = |u^2 + 7u + 8 - 8|$$
$$= |u^2 + 7u|.$$

We must find δ, in terms of ε, so that $|u| < \delta$ implies $|f(x) - f(5)| < \varepsilon$. Now, if $|u| < \delta$, we have

$$|f(x) - f(5)| = |u^2 + 7u|$$
$$\le |u|^2 + 7|u|$$
$$\le \delta^2 + 7\delta.$$

We are finished if we can make $\delta^2 + 7\delta \le \varepsilon$. To do this, we 'complete the square' [cf. I (9.1.10)] noting that the condition is equivalent to

$$\delta^2 + 7\delta + 49/4 \le \varepsilon + 49/4$$

or

$$(\delta + 7/2)^2 \le \varepsilon + 49/4.$$

Since $\varepsilon > 0$ we can ensure this by making

$$\delta + 7/2 \le \sqrt{\varepsilon + 49/4},$$

that is,

$$\delta \le \sqrt{\varepsilon + 49/4} - 7/2.$$

Any such δ will do: to be explicit we may as well take

$$\delta = \sqrt{\varepsilon + 49/4} - 7/2.$$

Note however, that it is not in general necessary to find the *best* possible δ, or even to give a nice formula for it: all we need is a proof that it exists for any $\varepsilon > 0$.

Continuity *at a point* is of interest mostly as a stepping stone to continuity throughout a domain.

DEFINITION 2.1.2. If $f : D \to \mathbb{R}$ then we say f is *continuous* (*throughout D*) if, for each $x_0 \in D$, it is continuous at x_0.

Functions may be continuous at some points but not at others: for instance $f(x) = [x]$ is continuous at all non-integer values of x, but discontinuous at integers. In contrast $f(x) = x^2$ is continuous over the whole real line. To verify continuity we perform computations similar to that above, but for arbitrary x_0.

2.2. COMBINING CONTINUOUS FUNCTIONS

Not only is continuity necessary for certain very basic theorems to be true, but it (fortunately) holds good for most of the interesting functions of classical mathematics.

It is trivial to check that *constant functions*

$$f(x) = \text{const.}$$

and the *identity function*

$$f(x) = x$$

are continuous throughout $\mathbb{R}$. (For the first, any δ will do; for the second take $\delta = \varepsilon$.) From these we can build up to polynomials or rational functions [see I, § 14.9], making use of the following basic result.

Let $f : D \to \mathbb{R}$, $g : D \to \mathbb{R}$ be two functions with the same domain. Suppose that f and g are continuous at $x_0 \in D$ (respectively continuous throughout D). Then each of the following functions is continuous at x_0 (respectively throughout D):

$f + g$

$f - g$

fg

f/g (except for x_0 such that $g(x_0) = 0$).

Using this result inductively it is easy to prove that any *polynomial function*

$$p(x) = a_n x^n + a_{n-1} x^{n-1} + \ldots + a_0$$

is continuous throughout $\mathbb{R}$; and that any *rational function* $p(x)/q(x)$, with p and q polynomials, is continuous for all $x \in \mathbb{R}$ except when $q(x_0) = 0$ (when the function is not defined in any case).

To apply the above results to functions f and g with distinct domains D and E, it is of course sufficient to restrict both to the common domain $D \cap E$: naturally the results apply only to points lying in $D \cap E$ [see I, § 1.2.1].

We show later that many other standard functions are continuous on their domains of definition.

2.3.   LIMITS OF FUNCTIONS

By analogy with sequences [see § 1.2], we say that $f:D \to \mathbb{R}$ *tends to the limit l as x tends to* x_0 (not necessarily in D) if, for all $\varepsilon > 0$, we can find $\delta > 0$ such that whenever $0 < |x - x_0| < \delta$ and $x \in D$ we have $|f(x) - l| < \varepsilon$.

Note here that we specifically *exclude* $x = x_0$ from consideration. The reason is that we want, say, a function like

$$f(x) = \begin{cases} 0 & x \neq 0 \\ 1 & x = 0 \end{cases}$$

to tend to the limit 0 as $x \to 0$ (Figure 2.3.1): the limit should be able to detect the anomalous value at $x = 0$.

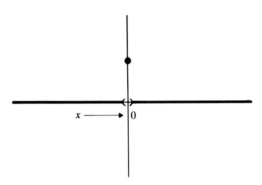

Figure 2.3.1

We write

$$\lim_{x \to x_0} f(x) = l.$$

There is a close connection between limits and continuity, and this permits many results concerning one to be transferred to the other at no extra cost.

DEFINITION 2.3.1.   Let $f:D \to \mathbb{R}$. Then f is *continuous throughout D* if and only if $\lim_{x \to x_0} f(x) = f(x_0)$ for all $x_0 \in D$; and f is *continuous at* $x_0 \in D$ if such an equality holds for a particular x_0.

For example, our results on the continuity of sums, products, etc. of continuous functions imply that if $f:D \to \mathbb{R}$ and $g:D \to \mathbb{R}$, then

$$\lim_{x \to x_0} (f+g)(x) = \lim_{x \to x_0} f(x) + \lim_{x \to x_0} g(x)$$

(in cases where the limits exist), with similar formulae for differences, products, reciprocals, and quotients—where in the last two cases division by zero is excluded as usual.

There are useful conventions concerning the symbol '∞' in this connection. First, if $f(x)$ becomes arbitrarily large and positive as $x \to x_0$ we say that

$$\lim_{x \to x_0} f(x) = \infty;$$

with similar use of $-\infty$ when $f(x)$ becomes large and negative.

Next, if $|f(x) - l|$ becomes arbitrarily small as x becomes arbitrarily large, we say

$$\lim_{x \to \infty} f(x) = l.$$

Again there is a similar use of $-\infty$.

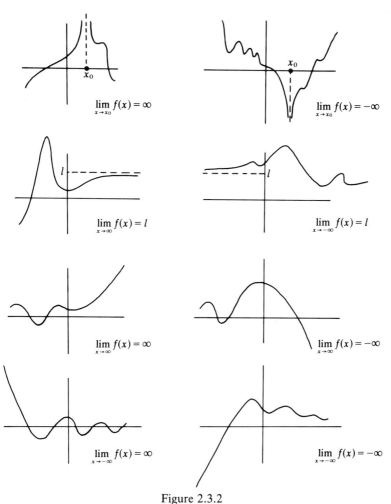

Figure 2.3.2

Finally we can combine the two conventions in the obvious way, leading to expressions

$$\lim_{x \to \pm\infty} f(x) = \pm\infty.$$

Figure 2.3.2 illustrates these conventions graphically.

There are also *left-hand* limits and *right-hand* limits, where the definition is modified to consider only those x satisfying $x < x_0$, or (respectively) $x > x_0$. The notation commonly used is

$$\lim_{x \to x_0-} f(x)$$

$$\lim_{x \to x_0+} f(x).$$

Figure 2.3.3 illustrates various possibilities.

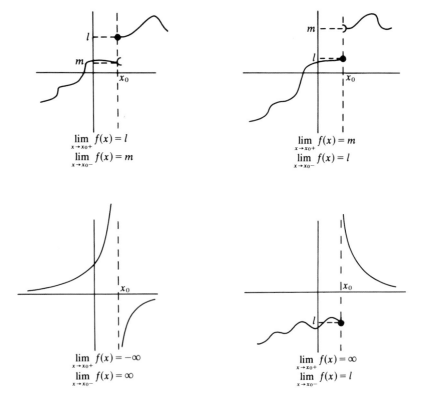

$$\lim_{x \to x_0+} f(x) = l$$
$$\lim_{x \to x_0-} f(x) = m$$

$$\lim_{x \to x_0+} f(x) = m$$
$$\lim_{x \to x_0-} f(x) = l$$

$$\lim_{x \to x_0+} f(x) = -\infty$$
$$\lim_{x \to x_0-} f(x) = \infty$$

$$\lim_{x \to x_0+} f(x) = \infty$$
$$\lim_{x \to x_0-} f(x) = l$$

Figure 2.3.3

There is some useful standard notation concerning the '*order of magnitude*' or '*rate of growth*' of a function near 0 (or ∞).

DEFINITION 2.3.2.   Let f, $g : \mathbb{R} \to \mathbb{R}$ be two functions. Then we write

$$f(x) = O(g(x))$$

(as $x \to \infty$) if $|f(x)/g(x)| \le K$ for some constant K, when x is sufficiently large (i.e. $x \ge x_0$ for some x_0). We write

$$f(x) = o(g(x))$$

(as $x \to \infty$) if

$$\lim_{x \to \infty} f(x)/g(x) = 0$$

and we write

$$f(x) \sim g(x)$$

(as $x \to \infty$), read as '$f(x)$ is *asymptotically equal* to $g(x)$', if

$$\lim_{x \to \infty} f(x)/g(x) = 1.$$

EXAMPLES 2.3.1

(i) $x + \sin x = O(x) \quad$ as $x \to \infty$

(ii) $e^{-x} = o(x^n) \quad$ for any natural number n, as $x \to \infty$

(iii) $x^2 + \dfrac{\sin x}{x+1} \sim x^2 \quad$ as $x \to \infty$.

DEFINITION 2.3.3.   We write

$$f(x) = O(g(x))$$

(as $x \to 0$) if $|f(x)/g(x)| \le K$ for sufficiently small x, that is, $|x| \le x_0$ for some $x_0 > 0$;

$$f(x) = o(g(x))$$

(as $x \to 0$) if

$$\lim_{x \to 0} f(x)/g(x) = 0;$$

and

$$f(x) \sim g(x)$$

(as $x \to 0$) if

$$\lim_{x \to 0} f(x)/g(x) = 1.$$

EXAMPLES 2.3.2

(i) $$e^{-1/x^2} = O(x) \quad \text{as } x \to 0$$

(ii) $$x^2 = o(x) \quad \text{as } x \to 0$$

(iii) $$\sin x \sim x \quad \text{as } x \to 0.$$

These notations are used with variables other than x, of course: in particular when h is a small constant the notation

$$f(h) = O(h^n) \quad (\text{as } h \to 0)$$

means that there is a constant K such that

$$f(h) \leq Kh^n$$

for small enough h; and

$$f(h) = o(h^n) \quad (\text{as } h \to 0)$$

$$\lim_{x \to x_0+} f(x) \neq \lim_{x \to x_0-} f(x)$$

An important type of *discontinuous* function is as follows:

DEFINITION 2.3.4.   A *step function* $f(x)$ is defined by

$$f(x) = \begin{cases} a & \text{if } x \leq c \\ b & \text{if } x > c \end{cases}$$

where a, b, c are constants. It has the graph shown in Figure 2.3.4.

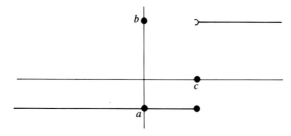

Figure 2.3.4

There is a single *jump discontinuity* at $x = c$:

DEFINITION 2.3.5.   A *discontinuity* of a function $f : D \to \mathbb{R}$ is a point $x_0 \in D$ at which f is not continuous. If

$$\lim_{x \to x_0+} f(x) \neq \lim_{x \to x_0-} f(x)$$

but both limits exist, we say x_0 is a *jump* discontinuity.

If $f : D \to \mathbb{R}$ has only a finite number of discontinuities it is said to be *piecewise continuous*.

2.4. COMPOSITION OF CONTINUOUS FUNCTIONS

Continuity is 'preserved' by composition of functions [see I, (1.4.3)]. More precisely:

PROPOSITION 2.4.1.   *Let $f : D \to \mathbb{R}$ and $g : E \to \mathbb{R}$, and suppose that $x_0 \in E$ is such that $g(x_0) \in D$. If g is continuous at x_0 and f is continuous at $g(x_0)$, then $f \circ g$ is continuous at x_0. In particular if f and g are continuous throughout their domains, then $f \circ g$ is continuous throughout the domain on which it may be defined.*

Thus, for example, since the functions

$$f(x) = x^3 + 2$$

and

$$g(x) = |x|$$

are continuous, so is

$$f \circ g(x) = |x|^3 + 2.$$

This property of continuity is extremely useful when proving that complicated expressions define continuous functions.

2.5. THE SQUEEZE RULE

It should not be imagined that the only possible discontinuities in a function are jumps. For example, the function

$$f(x) = \begin{cases} \sin(1/x) & x \neq 0 \\ 0 & x = 0 \end{cases}$$

has a non-jump discontinuity at zero, because of its wilder and wilder oscillations (Figure 2.5.1).

More delicate is the question of the function (Figure 2.5.2)

$$f(x) = \begin{cases} x . \sin(1/x) & x \neq 0 \\ 0 & x = 0 \end{cases}$$

which is in fact continuous. This follows from the:

PROPOSITION 2.5.1 (*Squeeze rule*).   *If there exist functions h and k such that $h(x) \leq f(x) \leq k(x)$ in some interval containing x_0 as an interior point, and $h(x_0) = f(x_0) = k(x_0)$; and if h and k are continuous at x_0, then so is f.*

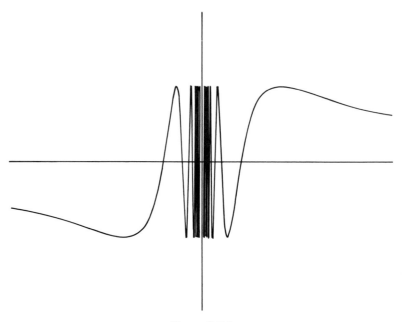

Figure 2.5.1

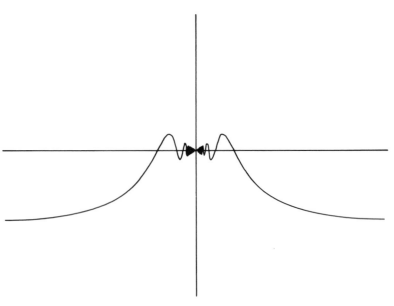

Figure 2.5.2

To apply this to f as above put $h(x) = -|x|$, $k(x) = |x|$ [see I (2.6.5)]. It follows that f is continuous at 0. Elsewhere, continuity follows from section 2.4 once we know sin x is continuous; this we postpone to section 2.12.

2.6.   THE INTERMEDIATE VALUE THEOREM

The following result is used both theoretically, and to show that equations have solutions within certain ranges of values. It corresponds to the intuitive idea that a continuous function has its graph in a single piece (at least if its domain is an interval).

THEOREM 2.6.1   (*Intermediate Value Theorem*).   Let $f:[a, b] \to \mathbb{R}$ be continuous. If ξ lies between $f(a)$ and $f(b)$ then there exists at least one point c with $a \le c \le b$ such that $f(c) = \xi$.

Figure 2.6.1 illustrates this: it is intuitively clear, but in the rigorous context requires a careful proof. Roughly this runs as follows. Assume (without loss of generality) that $f(a) \le f(b)$. Let

$$S = \{x \in [a, b] \,|\, f(u) < \xi \text{ for all } u < x\}.$$

Intuitively S contains all points to the left of the first place where the graph of f crosses the line $y = \xi$. Hence it is not surprising that

$$c = \sup (S)$$

is the desired point [see I, § 2.6.3]. (Figure 2.6.1 shows that this c need not be the only solution of $f(x) = \xi$.)

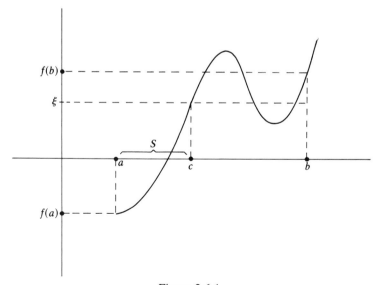

Figure 2.6.1

EXAMPLE 2.6.1. Let

$$f(x) = x^5 - 6x + 3.$$

Then

$$f(-2) = -17$$

$$f(0) = 3$$

so, since $-17 \le 0 \le 3$, we know that $f(x) = 0$ for some c, with $-2 \le c \le 0$. This locates at least one zero of the given quintic polynomial.

For a time, mathematicians contemplated using this property to define continuity, until it was noticed that a function like that in Figure 2.5.1 satisfies the intermediate value property, but clearly should not be considered continuous.

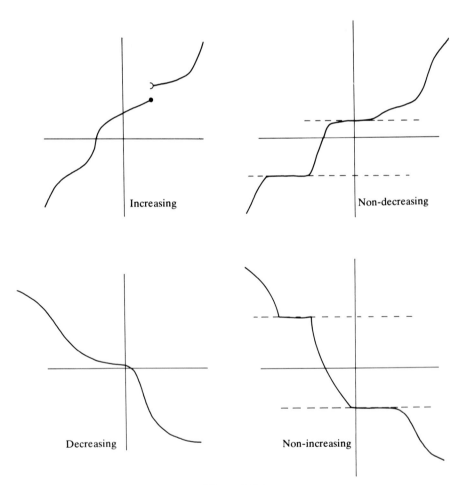

Figure 2.7.1

2.7. MONOTONIC FUNCTIONS AND INVERSES

DEFINITION 2.7.1.   A function $f:D\to\mathbb{R}$ is *monotonic increasing* if whenever $x, y \in D$ and $x < y$ we have

$$f(x) < f(y).$$

If the two inequalities are changed to '$\le$' we obtain the definition of a monotonic *non-decreasing* function; similarly '$>$' and '$\ge$' yield the concepts of a monotonic *decreasing* function and a monotonic *non-increasing* function. Figure 2.7.1 illustrates the graphs of these four types of function.

From I, § 1.4.2 we know that a function has an inverse if and only if it is a bijection. Suppose that $f:[a, b]\to\mathbb{R}$ is a continuous function defined on an interval, and that f is monotonic increasing. Then the image of f is the interval $[f(a), f(b)]\subseteq\mathbb{R}$, by a simple argument using the Intermediate Value Theorem; and f is injective [see I, § 1.4.2] by the monotone property. Hence, if we think of f as a function

$$f:[a, b]\to[f(a), f(b)]$$

it has an *inverse function*

$$f^{-1}:[f(a), f(b)]\to[a, b].$$

The same of course goes for a monotonic *decreasing* function. Figure 2.7.2 illustrates how a continuous monotonic increasing function sets up such a

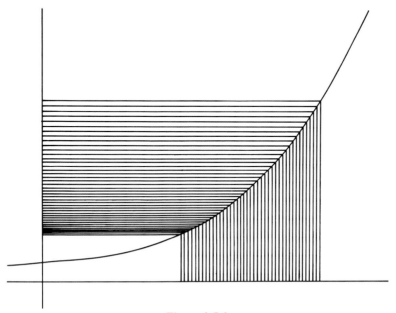

Figure 2.7.2

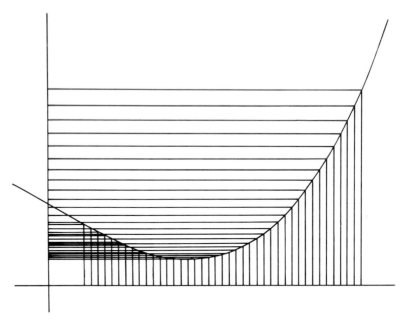

Figure 2.7.3

bijection; Figure 2.7.3 shows that we cannot drop the monotone condition and still have an inverse.

Further, if f is a continuous monotonic increasing (or decreasing) function defined on an interval, and f^{-1} as defined above is its inverse, then f^{-1} is also continuous (and monotonic of the same type). That this is reasonable is shown by Figure 2.7.2: a rigorous proof is harder.

EXAMPLE 2.7.1.   The function

$$f(x) = x^3$$

is monotonic increasing over the whole real line. For

$$x^3 - y^3 = (x - y)(x^2 + xy + y^2)$$

and the last factor may be written

$$x^2 + xy + y^2 = \tfrac{1}{2}(x^2 + y^2 + (x + y)^2)$$

hence is positive. So $x^3 - y^3$ and $x - y$ have the same sign, and are zero only when $x = y$. Hence f is monotonic increasing; it is also continuous. By the above result, it has a continuous inverse function, the *cube root* function, denoted

$$f^{-1}(x) = \sqrt[3]{x} = x^{1/3}.$$

(see Figure 2.7.4).

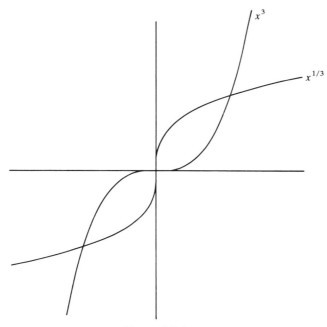

Figure 2.7.4

Similarly the function $f(x) = x^n$, for odd integer n, has a continuous inverse $f^{-1}(x) = \sqrt[n]{x} = x^{1/n}$ defined on the whole real line [see I, § 3.3]. When n is even, however, $f(x) = x^n$ is no longer monotonic increasing over the whole real line: it *decreases* for negative x [see I, § 3.1]. If we restrict it to the set $\mathbb{R}^+ = \{x \in \mathbb{R}: x \geq 0\}$ of positive reals, it becomes monotonic increasing; hence there is a continuous inverse function $f^{-1}(x) = \sqrt[n]{x} = x^{1/n}$ defined only on $\mathbb{R}^+$.

Finally let $r = p/q$ be a rational number in lowest terms, so that $p, q \in \mathbb{Z}$ and $q \neq 0$ [see I, § 2.4.1]. If $r > 0$ we may define

$$x^r = (x^{1/q})^p$$

provided $x^{1/q}$ is defined: that is, for all x if q is odd, and for positive x if q is even. If $r < 0$ we put

$$x^r = 1/x^{-r}$$

noting that x^{-r} is then definable as above. It is easy, though tedious in detail, to check that the usual laws for powers [cf. I, (3.1.1) and (3.1.2)]

$$x^{r+s} = x^r x^s \tag{2.7.1}$$

$$x^{rs} = (x^r)^s \tag{2.7.2}$$

hold, for those values of x for which the functions make sense. In particular, they hold for all positive x and for any rational numbers r, s.

To extend the definition of x^r to all *real* numbers r (for positive x only) it now suffices to take a sequence (r_n) of rationals tending to r as limit, and set

$$x^r = \lim_{n \to \infty} x^{r_n}.$$

Suitable continuity arguments prove this to be independent of the choice of approximating sequence, and that (2.7.1) and (2.7.2) remain valid for all real r, s. An alternative approach using the exponential function will be discussed in section 2.11: it is considerably more elegant.

Functions which can be expressed as the difference between two monotonic functions are important in the theory of integration. To characterize these, we introduce the following notions.

DEFINITION 2.7.2.   Let $[a, b]$ be an interval in $\mathbb{R}$. Define a *partition* of $[a, b]$ to be a set of points $P = \{x_0, \ldots, x_n\}$ such that

$$a = x_0 < x_1 < \ldots < x_n = b.$$

DEFINITION 2.7.3.   Let

$$f : [a, b] \to \mathbb{R}.$$

We say f is of *bounded variation* on $[a, b]$ if there is a constant M such that, for all partitions P, we have

$$\sum_{k=1}^{n} |f(x_k) - f(x_{k-1})| \le M.$$

Functions of bounded variation constitute quite a wide class. For example, it is obvious that a monotonic function on $[a, b]$ is of bounded variation (take $M = f(b) - f(a)$). The following striking result can be proved:

THEOREM 2.7.1.   *A function $f : [a, b] \to \mathbb{R}$ is of bounded variation if and only if there are two monotonic increasing functions $g, h : [a, b] \to \mathbb{R}$ such that*

$$f(x) = g(x) - h(x)$$

for all $x \in [a, b]$.

2.8.   COMPACTNESS PROPERTIES

A number of powerful theoretical results may be gathered under the above heading. In the context of the real line they assert that *closed* intervals $[a, b]$ are unusually well-behaved in connection with continuity and limits [see I, § 2.6.3]. For example, we have the

THEOREM 2.8.1   (*Bolzano–Weierstrass Theorem*).   *Let (a_n) be a sequence of points in a closed interval $[a, b]$. Then there exists a convergent subsequence (a_{n_i}) whose limit is in $[a, b]$.*

The proof of this is so nice that we sketch it. Divide the unit interval in half (Figure 2.8.1(a)). At least one half contains infinitely many points from the sequence: pick a_{n_0} to be one of them. Divide that half into halves: one of these contains infinitely many terms; let a_{n_1} be one of them (Figure 2.8.1(b)).

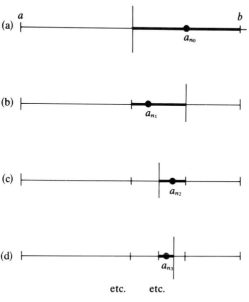

Figure 2.8.1

Repeat the subdivision on that quarter, and proceed inductively (Figure 2.8.1(c)) to construct a_{n_i} for all i. It is intuitively clear that the nest of smaller and smaller intervals 'converges' to a point, and that the sequence (a_{n_i}) converges to the same point.

To state the 'generic' result of this type we introduce the following concept.

DEFINITION 2.8.1.   Let S be a subset of the real line, and let $(T_\alpha)_{\alpha \in A}$ be an indexed family of subsets of S. Then we say that $(T_\alpha)_{\alpha \in A}$ *covers* S if

$$S \subseteq \bigcup_{\alpha \in A} T_\alpha.$$

THEOREM 2.8.2   (*Heine–Borel Theorem*).   *Let* $[a, b]$ *be a closed interval, and let* $(T_\alpha)_{\alpha \in A}$ *be a family of open intervals that covers S. Then there exist finitely many intervals* $T_{\alpha_1}, \ldots, T_{\alpha_n}$ *such that the subfamily* $(T_{\alpha_i})_{1 \le i \le n}$ *covers S.*

Briefly: 'Every open cover of a closed interval has a finite subcover'.

The far-reaching implications of this theorem are not immediately obvious. That the theorem is very delicate is clear if we note that it fails for an open

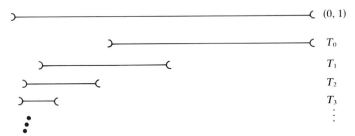

Figure 2.8.2

interval (a, b). For example the sets

$$T_n = (3^{-n-1}, 2^{-n})$$

for $n \in \mathbb{N}$ form a cover of $(0, 1)$, as shown in Figure 2.8.2, but no finite subcover exists.

Here is one consequence of the Heine–Borel theorem.

DEFINITION 2.8.2.   We say that a function $f : D \to \mathbb{R}$ is *bounded* if there exists $M \in \mathbb{R}$ such that $|f(x)| < M$ for all $x \in D$. In this case the numbers

$$\sup_{x \in D} f(x), \qquad \inf_{x \in D} f(x)$$

exist [see I, § 2.6.3]. If we can find $x_1, x_2 \in D$ such that

$$f(x_1) = \sup_{x \in D} f(x)$$

$$f(x_2) = \inf_{x \in D} f(x)$$

then we say that *f attains its bounds.*

We then have a very useful result:

THEOREM 2.8.2.   *Every continuous function $f : D \to \mathbb{R}$, with D a closed interval, is bounded and attains its bounds.*

Figure 2.8.3 illustrates this. It is not true for an open interval D: for example the function

$$f(x) = 1/x$$

is continuous but unbounded on $(0, 1)$; and

$$f(x) = x$$

is bounded on $(0, 1)$ but attains neither upper nor lower bound within that interval.

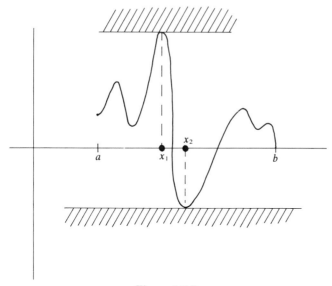

Figure 2.8.3

The above results all generalize to functions of several variables, and lead to the concept of a *compact* topological space [see V, § 5.2].

2.9. DIFFERENTIABLE FUNCTIONS

Differentiability, to be defined in a moment, is a more stringent condition than mere continuity. It has a geometric interpretation: the existence of a tangent to the graph. In physical applications it permits the study of the 'instantaneous rate of change' of a quantity. Mathematically it is best defined as a limiting process, inspired by these interpretations, which then affords a precise formulation of the geometric and physical concepts themselves.

Consider the problem of drawing the tangent to the graph of the function

$$f(x) = x^2$$

at the point where $x = 1$. Obviously the key to the problem is to find the *slope* of the tangent. Now it is easy to find the slope of a *chord*, such as that drawn in Figure 2.9.1 between the points $(1, 1)$ and $(2, 4)$ on the graph: since the end points are distinct it is the expression

$$\frac{4-1}{2-1}$$

which has the value 3 [see V, § 1.2.1]. In general, the slope of the chord from the point $(1, 1)$ to the point $(1+h, (1+h)^2)$ is given by the expression

$$\frac{(1+h)^2 - 1}{1+h-1}$$

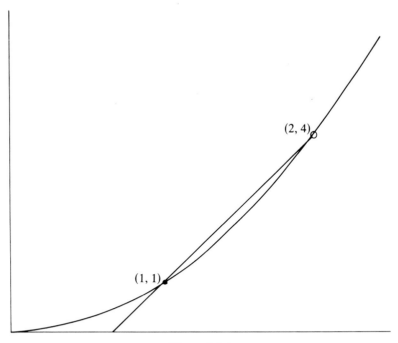

Figure 2.9.1

which simplifies to

$$h + 2$$

when $h \neq 0$. (When $h = 0$ the chord is no longer uniquely defined and the expression for the slope no longer valid.)

Figure 2.9.2 is a computer-drawn diagram showing how the chord behaves for a sequence of decreasing values of h (actually $h = 2^{-n/2}$ for $n = 0, 1, 2, \ldots, 10$). It appears from this figure that the chords approach more and more closely to the tangent line as h approaches zero. For example, Figure 2.9.3 shows the tenth chord in the sequence: to the eye it is indistinguishable from the tangent.

By our above calculation, the slope $h + 2$ of the chord clearly tends to 2 as h tends to zero. Thus we expect the slope of the tangent to be 2.

In the spirit of Greek geometry (but the letter of Descartes) we can even verify this guess. The line through $(1, 1)$ of slope 2 has equation

$$y - 1 = 2(x - 1)$$

[see V, § 1.2.1]. Obviously this cuts the parabola $y = x^2$ at $(1, 1)$; but by solving the two simultaneous equations $y - 1 = 2(x - 1)$, $y = x^2$ we are led to the quadratic

$$x^2 - 1 = 2(x - 1),$$

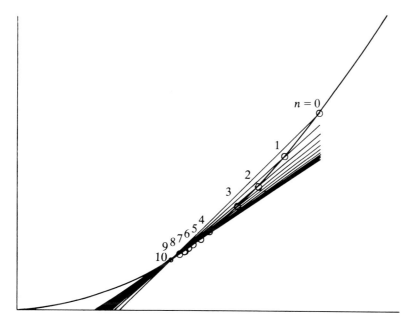

Figure 2.9.2

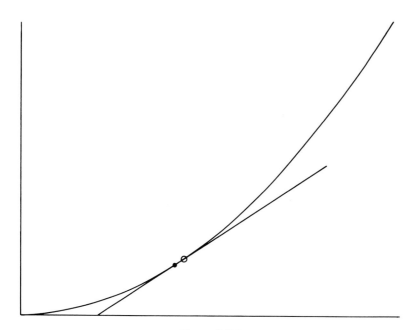

Figure 2.9.3

that is,

$$(x-1)^2 = 0$$

with the *single* root $x = 1$ (of multiplicity 2). Hence $(1, 1)$ is the only point of intersection. A straight line that cuts a parabola in exactly one point is, of course, a tangent in the original sense of Greek geometry [see V, § 1.3.3].

Other examples of the same kind show that it is reasonable to proceed as follows:

DEFINITION 2.9.1.   Let $f: D \to \mathbb{R}$ be a function, and let $x \in D$. We say that f is *differentiable* at x if

$$\lim_{h \to 0} \frac{f(x+h) - f(x)}{h}$$

exists. The value of this limit is defined to be the *derivative* of f at x, denoted variously by

$$f'(x), \qquad Df(x), \qquad Df|_x,$$

and the classical Leibniz notation df/dx. If f is differentiable for all x in D we say simply that f is *differentiable*.

The *tangent* to the graph of f at a point x_0 is then defined to be the line with equation

$$y - f(x_0) = f'(x_0) \cdot (x - x_0). \tag{2.9.1}$$

That is, the line of slope $f'(x_0)$ through the point $(x_0, f(x_0))$.

Finally, as regards physical interpretation, we note that $(f(x+h) - f(x))/h$ is the 'rate of change of f over the interval from x to $x + h$', so its limit for small h is a sensible interpretation of the otherwise meaningless phrase 'the instantaneous rate of change of f at x'.

Note that, since the definition of $\lim_{h \to 0}$ does not involve what happens *at* $h = 0$, the hoary problem that the derivative, in the limit, involves division by zero, is neatly circumvented.

For example, the derivative of $f(x) = x^2$ is given by the expression

$$\lim_{h \to 0} ((x+h)^2 - x^2)/h = \lim_{h \to 0} (2hx + h^2)/h = \lim_{h \to 0} 2x + h = 2x$$

since the case $h = 0$ is not involved in the evaluation of the limit.

Thus

$$f'(x) = 2x$$

is the derivative.

Notice that, if f is differentiable, then $f': D \to \mathbb{R}$ is again a function with domain D.

Detailed discussion of properties of derivatives, and their values for standard functions f, will be given in Chapter 3. Here we shall discuss only the general idea of differentiability.

Every differentiable function is necessarily continuous, because the existence of the above limit implies that

$$\lim_{h \to 0} f(x+h) = f(x).$$

The converse is by no means true. For example, the function $f(x) = |x|$ [see I, (2.6.5)] is not differentiable at $x = 0$: it has a sharp 'corner' at which no unique tangent may be drawn, and the required limit can easily be seen not to exist. For a long time, however, mathematicians inclined to the belief that a continuous function must be differentiable at 'most' points in its domain. This is quite wrong. For an example, define

$$\{x\} = \text{the distance from } x \text{ to the nearest integer}$$

or equivalently

$$\{x\} = \min (x - [x], [x+1] - x).$$

The function

$$f(x) = \sum_{n=1}^{\infty} \{10^n x\}/10^n$$

(defined by an absolutely convergent series [see Definition 1.9.1]) is continuous on the whole real line; but it is differentiable *nowhere*. (For a proof see Spivak (1967) p. 422.) Figure 2.9.4 shows some successive approximations to $f(x)$ obtained by truncating the series expansion: notice the increasing irregularity and the sharp corners, which render its non-differentiability fairly plausible.

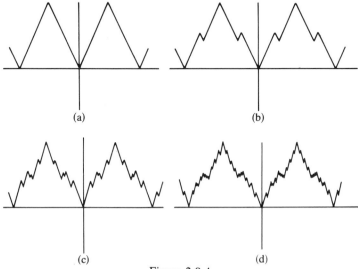

(a)   (b)

(c)   (d)

Figure 2.9.4

2.10. SMOOTH AND ANALYTIC FUNCTIONS

If $f: D \to \mathbb{R}$ is a differentiable function, with derivative $f': D \to \mathbb{R}$, it may or may not be the case that f' is itself differentiable. If it is, then its derivative

$$f'': D \to \mathbb{R}$$

is called the *second derivative* of f.

For example if $f(x) = x^2$ then $f'(x) = 2x$; and this is a differentiable function with derivative $f''(x) = 2$. On the other hand if

$$f(x) = \begin{cases} x^2/2 & x \geq 0 \\ -x^2/2 & x < 0 \end{cases}$$

as in Figure 2.10.1, then $f'(x) = |x|$ which is not differentiable at 0 [see § 2.9]; so $f''(x)$ is not defined at $x = 0$.

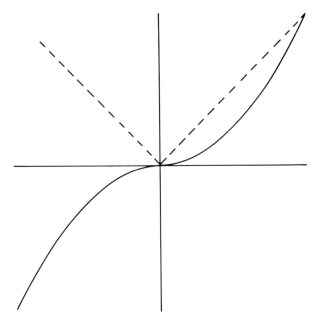

Figure 2.10.1

Similarly f'' may or may not be differentiable ... and in general we define the *n*th *derivative* $f^{(n)}(x)$ inductively:

$$f^{(1)} = f'$$
$$f^{(n+1)} = (f^{(n)})'$$

provided that $f^{(n)}$ is differentiable. Alternative notation for $f^{(n)}(x)$, often used, is

$$D^n f(x), \qquad D^n f|_x, \qquad \frac{d^n f}{dx^n}.$$

DEFINITION 2.10.1.   If $f^{(n)}(x)$ exists (respectively, exists and is continuous) for all $x \in D$, we say that f is *n times differentiable* (respectively *n times continuously differentiable*, or of *class C^n*) in D.

DEFINITION 2.10.2.   If $f^{(n)}(x)$ exists for all n, we say that f is *infinitely differentiable*, *smooth*, or of *class C^∞*. (In the latter terminology, continuous functions are of class C^0.)

DEFINITION 2.10.3.   A function defined by a (convergent) power series

$$f(x) = \sum_{n=0}^{\infty} a_n x^n$$

is said to be *analytic* (within the interval of convergence of the series [see § 1.10]) or of *class C^ω*. Since power series may be differentiated term by term within the interval of convergence [see Theorem 3.2.2], it follows that analytic functions are always smooth.

   If f is analytic, we may differentiate the above series n times and put $x = 0$. The result is that

$$a_n = \frac{1}{n!} f^{(n)}(0).$$

Hence the series expression for $f(x)$ is unique, if it exists, and must take the form

$$\sum_{n=0}^{\infty} \frac{1}{n!} f^{(n)}(0) x^n. \qquad (2.10.1)$$

This is called the *Maclaurin series* for, or *Maclaurin expansion* of, f. Conditions for its convergence are discussed in § 3.6. It is defined as a formal power series for all smooth f, but converges to f only when f is analytic.

EXAMPLE 2.10.1.   The function

$$f(x) = \begin{cases} e^{-1/x^2} & x \neq 0 \\ 0 & x = 0 \end{cases}$$

[see § 2.11] is smooth but not analytic. It is clearly smooth everywhere except perhaps the origin; but detailed calculations (see Spivak (1967) p. 293) show that $f^{(n)}(0) = 0$ for all n, so it is smooth everywhere. It is extremely flat near the origin (Figure 2.10.2). Now its Maclaurin series is

$$0 + 0x + 0x^2 + \ldots$$

which converges—but to 0, not to f. Thus f is smooth but not analytic.

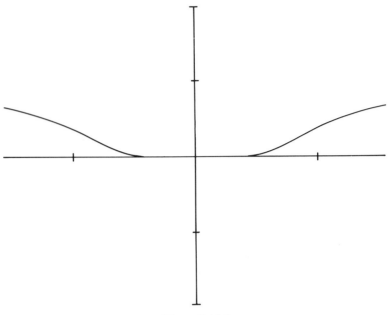

Figure 2.10.2

Most of the standard functions of analysis are analytic: indeed in subsequent sections we shall define many of them by power series. For example [see § 2.12 and V, § 1.2.3]

$$\sin x = \sum_{n=0}^{\infty} (-1)^n \frac{x^{2n+1}}{(2n+1)!}.$$

In applications attention often focuses on the power series *truncated* at some fixed point N, as an approximation to $f(x)$. Thus Figure 2.10.3 plots a number of truncations

$$\sum_{n=0}^{N} (-1)^n \frac{x^{2n+1}}{(2n+1)!}$$

of the series for $\sin x$. Some intuition about the manner in which such truncations approximate to the function can be derived from this figure: in particular note that even a very high degree polynomial, obtained by truncating at large N, is a close fit only over a relatively short interval; but as N grows, so does the length of this interval.

Recent developments (see Volume V, Chapter 8) have shown that for many purposes a function need not be analytic in order that truncations of its Taylor series provide good approximations to it; while even for analytic functions it is possible for *all* truncations to be bad approximations in a precise (and important) sense. This may alter the traditional emphasis on analytic functions as experience in applying the new methods grows.

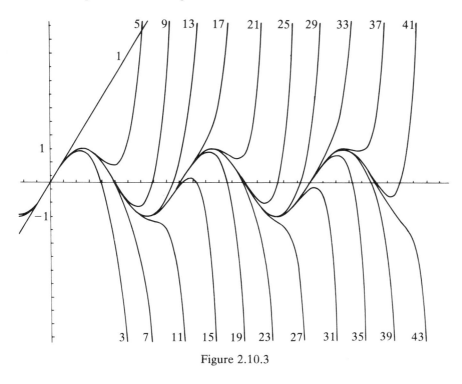

Figure 2.10.3

2.11.   EXPONENTIAL AND LOGARITHMIC FUNCTIONS

We define the exponential function by the power series

$$\exp x = \sum_{n=0}^{\infty} \frac{x^n}{n!}. \tag{2.11.1}$$

By the ratio test this is absolutely convergent for all $x \in \mathbb{R}$ [see Example 1.10.1]. Hence it defines an analytic function

$$\exp : \mathbb{R} \to \mathbb{R}.$$

Since absolutely convergent series may be multiplied in an analogous manner to polynomials [see Theorem 1.11.4] it can be shown by direct computation [see Example 1.11.5] that

$$\exp x \exp y = \exp (x + y) \tag{2.11.2}$$

for all $x, y \in \mathbb{R}$. Then

$$\exp x \exp (-x) = \exp 0 = 1$$

and it follows that $\exp x$ is never zero. Since $\exp 0$ is positive and $\exp$ is continuous, the Intermediate Value Theorem [see Theorem 2.6.1] implies that $\exp x$ is positive for all x.

We have

$$\exp(-x) = 1/\exp x.$$

From the series, $\exp x > 1$ for $x > 0$. Thus, if $x > y$, it follows that

$$\exp x/\exp y = \exp x \exp(-y) = \exp(x - y) > 1$$

so exp is monotonic increasing [see Definition 2.7.1].

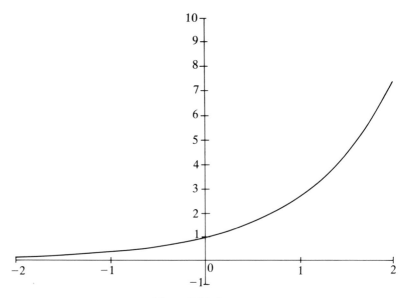

Figure 2.11.1: exp x.

Figure 2.11.1 shows the graph of exp x for $-2 \le x \le 2$. Note that the increase is steeper, the larger x becomes: a tendency even more obvious on the graph for $-10 \le x \le 10$ (Figure 2.11.2). To see why, we differentiate the series term by term: the result is

$$\frac{d}{dx} \exp(x) = \exp(x)$$

so that not only is exp monotonic increasing, but so is its derivative: the slope of the graph increases as x becomes larger.

We define the real number e by

$$e = \exp 1$$

$$= \sum_{n=0}^{\infty} 1/n! \tag{2.11.3}$$

$$\sim 2 \cdot 718281828 \ldots.$$

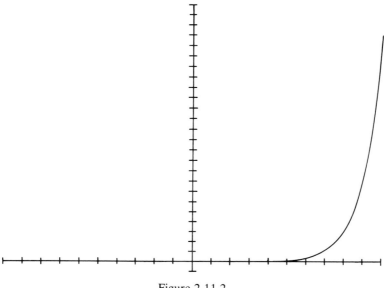

Figure 2.11.2

The decimal digits here do *not* recur: in fact it is easy to prove (Spivak (1967), p. 353) that e is irrational [see I, § 2.6.1]; and (much harder) *transcendental*, that is, not the root of a polynomial with rational coefficients. (See Spivak (1967), p. 362, Stewart (1973), p. 72.)

Now

$$\exp 2 = \exp (1+1) = \exp 1 \, \exp 1 = e^2,$$

and inductively,

$$\exp n = e^n$$

for all integers $n \geq 0$. Further, $\exp(-n) = 1/\exp n = 1/e^n = e^{-n}$, so the result holds also for negative integers. If $p/q \in \mathbb{Q}$ where $p, q \in \mathbb{Z}$ and $q \neq 0$, then

$$(\exp (p/q))^q = \exp p = e^p$$

so

$$\exp (p/q) = e^{p/q}.$$

A continuity argument now shows that for all real numbers r we have

$$\exp r = e^r$$

using the definition of the rth power given in section 2.7. The notation e^x may therefore be used in place of $\exp x$ if desired.

There is an interesting limit associated with the exponential function [see § 1.2]. For any $x \in \mathbb{R}$ we have

$$\lim_{n \to \infty} \left(1 + \frac{x}{n}\right)^n = e^x \tag{2.11.4}$$

and hence, putting $x = 1$,

$$\lim_{n \to \infty} \left(1 + \frac{1}{n}\right)^n = e. \tag{2.11.5}$$

A proof of these results is given in Spivak (1967), [see also Example 4.9.6(i)]. We can show they are plausible by the following non-rigorous argument (which can be rendered rigorous by introducing sufficient technique). Suppose that

$$f(x) = \lim_{n \to \infty} \left(1 + \frac{x}{n}\right)^n$$

exists and is differentiable, and suppose that the right-hand limit is well enough behaved to permit us to differentiate the expression inside the limit to obtain

$$f'(x) = \lim_{n \to \infty} n\left(1 + \frac{x}{n}\right)^{n-1} \cdot \frac{1}{n}$$

$$= \lim_{n \to \infty} \left(1 + \frac{x}{n}\right)^{n-1} = \lim_{n \to \infty} \left\{\left(1 + \frac{x}{n}\right)^n \Big/ \left(1 + \frac{x}{n}\right)\right\}$$

$$= f(x)/1 = f(x).$$

The only solutions of this differential equation are

$$f(x) = k\, e^x$$

for a constant k [see § 7.2]. Putting $x = 0$ we get

$$k = f(0) = \lim_{n \to \infty} \left(1 + \frac{0}{n}\right)^n = \lim_{n \to \infty} 1^n = 1$$

so

$$f(x) = e^x.$$

The above expression may be interpreted in terms of compound interest applied at a rate of $100x\%$ per annum (so that $x = 1$ represents 100%) at intervals of $1/n$ years. As n becomes very large, that is, as the interest is calculated at finer time intervals, the growth of the sum invested becomes exponential. [See I, Chapter 15].

Since $e > 1$ it follows that $\lim_{n \to \infty} e^n = \infty$, hence $\lim_{x \to \infty} \exp x = \infty$. Therefore

$$\lim_{x \to -\infty} \exp(x) = 1/\lim_{x \to -\infty} \exp(-x) = 0.$$

Thus the image of $\mathbb{R}$ under exp is $\{x \in \mathbb{R}: x > 0\}$. Since exp is continuous and monotonic increasing, it has a continuous monotonic increasing inverse [see § 2.7] called the *logarithmic function*

$$\log : \{x \in \mathbb{R} \mid x > 0\} \to \mathbb{R}. \tag{2.11.6}$$

Being the inverse of exp, it satisfies the conditions

$$\exp (\log x) = x \qquad (x > 0)$$
$$\log (\exp x) = x \qquad (x \in \mathbb{R}).$$

The graph of $\log x$ is shown in Figure 2.11.3.

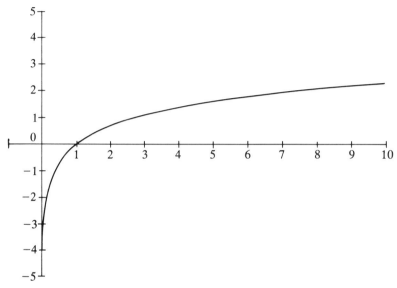

Figure 2.11.3: $\log x$.

In (2.11.2) put $x = \log u$, $y = \log v$. Then

$$\exp (\log u) \exp (\log v) = \exp (\log u + \log v)$$

or

$$uv = \exp (\log u + \log v).$$

Taking the logarithm throughout,

$$\log uv = \log u + \log v \qquad (u, v > 0). \tag{2.11.7}$$

Historically it was this fundamental property that led to the introduction of logarithms (in a modified form) by John Napier in 1614: it permits the solution of multiplication problems using only addition.

We can now find a more elegant approach to the rth power function x^r for $r \in \mathbb{R}$, $x > 0$. We have

$$x^r = (\exp (\log x))^r$$
$$= (e^{\log x})^r$$
$$= e^{r . \log x}$$
$$= \exp (r . \log x).$$

The latter expression defines x^r purely in terms of exp and log, and may be used as an alternative to the procedure sketched in section 2.7.

Taking logarithms in the above we obtain

$$\log(x^r) = r \cdot \log x.$$

The value $\log x$ is often called the *natural logarithm* or logarithm to *base e*. Sometimes the function 'log' is denoted 'ln'. In general if $b \in \mathbb{R}$ and $b > 0$ we define the logarithm to base b of x to be

$$\log_b x = \log x / \log b, \tag{2.11.8}$$

[see also I, § 3.6]. It has the property

$$b^{\log_b x} = x.$$

For arithmetic computation, logarithms to base 10 are often the most convenient, since

$$\log_{10}(10^k \cdot x) = k + \log_{10} x$$

and therefore logarithms need be tabulated only for $1 \le x < 10$. In most scientific and mathematical theorizing, however, it is the natural logarithm that we use.

The type of argument that leads to the definition of x^r can be used in other situations. The following result is cited in II, p. 162:

PROPOSITION 2.11.1.   *Suppose that the function $p(x)$ is continuous for $x > 0$ and satisfies the functional equation*

$$p(nx) = np(x) \tag{2.11.9}$$

for all $x > 0$ and all positive integers n. Then

$$p(x) = \lambda x,$$

where $\lambda = p(1)$.

Proof.   On putting $x = 1$ in (2.11.9) we find that

$$p(n) = np(1) = \lambda n.$$

Next replace n by m and put $x = 1/m$; then

$$p\left(m \frac{1}{m}\right) = \lambda = mp\left(\frac{1}{m}\right)$$

that is

$$p\left(\frac{1}{m}\right) = \lambda \frac{1}{m}.$$

Finally,

$$p\left(\frac{n}{m}\right) = p\left(n \frac{1}{m}\right) = np\left(\frac{1}{m}\right) = \lambda \frac{n}{m}.$$

Thus the result holds for all rational numbers n/m and hence, by continuity, for all positive real numbers x.

It is useful to record some facts about the behaviour of $\exp x$ and $\log x$ when x is positive and either very large or very small. Evidently [see Figures 2.11.1 and 2.11.3]

$$\exp x \to \infty, \quad \text{as } x \to \infty \tag{2.11.10}$$

and

$$\log x \to \infty, \quad \text{as } x \to \infty. \tag{2.11.11}$$

We shall now make these statements more precise: let k be an arbitrary positive integer. When $x > 0$ all terms in the expansion (2.11.1) are positive and we deduce that

$$\exp x > \frac{x^{k+1}}{(k+1)!} \tag{2.11.12}$$

because we have dropped all but one term. We can rewrite (2.11.12) as

$$0 < \frac{x^k}{\exp x} < \frac{(k+1)!}{x}.$$

Since the expression on the right tends to zero as x tends to infinity we obtain that

$$\lim_{x \to \infty} \frac{x^k}{\exp x} = 0. \tag{2.11.13}$$

More generally, if $P(x)$ is any polynomial function [see (2.1.1)] we have that

$$\lim_{x \to \infty} \frac{P(x)}{\exp x} = 0 \tag{2.11.14}$$

which may be paraphrased by the statement that '*exp x tends to infinity faster than any polynomial*'.

Next we shall transform (2.11.13) so that it gives some information about the logarithmic function: put $k = 1$ and $x = \log y^{\alpha}$, where α is an arbitrary positive real number. Then $x \to \infty$ implies that $y \to \infty$, and conversely. Hence (2.11.13) becomes $\alpha \log y / y^{\alpha} \to 0$. The factor α is irrelevant and may be omitted, thus

$$\lim_{y \to \infty} \frac{\log y}{y^{\alpha}} = 0 \quad (\alpha > 0) \tag{2.11.15}$$

which shows that *log y tends to infinity more slowly than any power of y with positive, not necessarily integral exponent*. Clearly [see Figure 2.11.3]

$$\lim_{z \to 0} \log z = -\infty \quad (z > 0). \tag{2.11.16}$$

But

$$\lim_{z \to 0} z^{\alpha} \log z = 0 \qquad (\alpha > 0, z > 0), \qquad (2.11.17)$$

as can be shown by making the substitution $y^{\alpha} = 1/z$ in (2.11.15).

2.12. TRIGONOMETRIC FUNCTIONS

We shall also define the *sine* and *cosine* functions by series:

$$\sin x = \sum_{n=0}^{\infty} (-1)^n \frac{x^{2n+1}}{(2n+1)!} = x - \frac{x^3}{6} + \frac{x^5}{120} - \dots \qquad (2.12.1)$$

and

$$\cos x = \sum_{n=0}^{\infty} (-1)^n \frac{x^{2n}}{2n!} = 1 - \frac{x^2}{2} + \frac{x^4}{24} - \dots \qquad (2.12.2)$$

[see also V, (1.2.9)]. Again both series are absolutely convergent [see Definition 1.9.1] for all $x \in \mathbb{R}$ by the ratio test [see Theorem 1.8.5], so that

$$\sin : \mathbb{R} \to \mathbb{R}$$

$$\cos : \mathbb{R} \to \mathbb{R}$$

are analytic functions. Term-by-term differentiation yields

$$\frac{d}{dx} \sin x = \cos x$$

$$\frac{d}{dx} \cos x = -\sin x.$$

We have

$$\sin 0 = 0$$

$$\cos 0 = 1.$$

By direct computation with the series [see also V, (1.2.17)] it may be shown that

$$\sin (x + y) = \sin x \, \cos y + \cos x \sin y \qquad (2.12.3)$$

$$\cos (x + y) = \cos x \, \cos y - \sin x \sin y. \qquad (2.12.4)$$

From the series,

$$\sin (-x) = -\sin x$$

$$\cos (-x) = \cos x.$$

That is, sin is an *odd* function and cos an *even* function [see V, § 3.6.3]. Hence

$$1 = \cos 0 = \cos (x - x)$$
$$= \cos x \cos (-x) - \sin x \sin (-x)$$
$$= \cos x \cos x + \sin x \sin x$$

which gives the important identity

$$\sin^2 x + \cos^2 x = 1. \tag{2.12.5}$$

(Here we follow tradition and write $\sin^2 x$ for $(\sin x)^2$, etc.) This implies in particular that sin and cos are *bounded*: indeed

$$|\sin x| \le 1 \qquad (x \in \mathbb{R})$$
$$|\cos x| \le 1 \qquad (x \in \mathbb{R}).$$

Now $\cos 0 = 1$, and it can be shown that $\cos 2$, say, is negative. The Intermediate Value Theorem [see Theorem 2.6.1] implies that $\cos x = 0$ for some x between 0 and 2. We define the real number π to be the smallest positive real number such that

$$\cos (\pi/2) = 0.$$

It can be computed that

$$\pi \sim 3 \cdot 14159 \ldots.$$

We also have

$$\sin (\pi/2) = 1.$$

Then the addition formulae (2.12.3), (2.12.4) show that

$$\sin \left(x + \frac{\pi}{2} \right) = \sin x \cos \left(\frac{\pi}{2} \right) + \cos x \sin \left(\frac{\pi}{2} \right)$$
$$= \cos x$$

and similarly

$$\cos \left(x + \frac{\pi}{2} \right) = -\sin x.$$

Using these twice, we have

$$\sin (x + \pi) = -\sin x$$
$$\cos (x + \pi) = -\cos x,$$

and twice again yields

$$\sin (x + 2\pi) = \sin x$$
$$\cos (x + 2\pi) = \cos x.$$

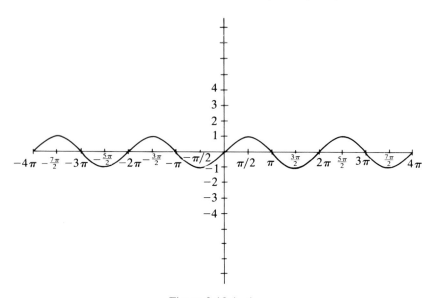

Figure 2.12.1: sin x.

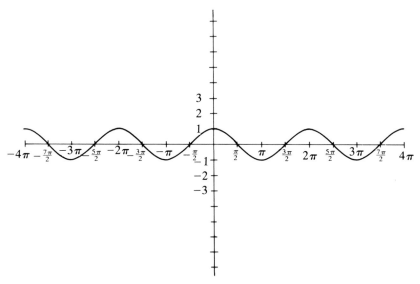

Figure 2.12.2: cos x.

Thus cos and sin are periodic functions of period 2π [see § 20.1]. This periodicity manifests itself in their graphs (Figures 2.12.1 and 2.12.2).

Four other standard functions, the *tangent, secant, cosecant,* and *cotangent,* are defined as follows:

$$\tan x = \sin x / \cos x \qquad (2.12.6)$$

$$\sec x = 1 / \cos x \qquad (2.12.7)$$

$$\operatorname{cosec} x = 1/\sin x \qquad\qquad (2.12.8)$$

$$\cot x = 1/\tan x = \cos x/\sin x. \qquad\qquad (2.12.9)$$

Inasmuch as the zeros of sin are at $x = n\pi$ $(n \in \mathbb{Z})$, and those of cos are at $x = (2n+1)\pi/2$ $(n \in \mathbb{Z})$, it follows that tan and sec tend to infinity at $x = (2n+1)\pi/2$, and cosec and cot tend to infinity at $x = n\pi$. The graphs of the four functions (Figures 2.12.3, 2.12.4, 2.12.5 and 2.12.6) illustrate

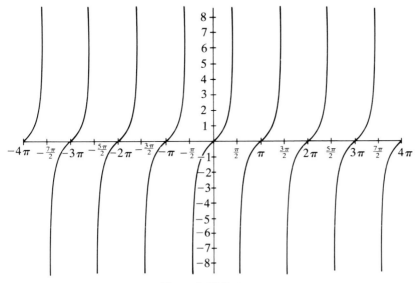

Figure 2.12.3: tan x.

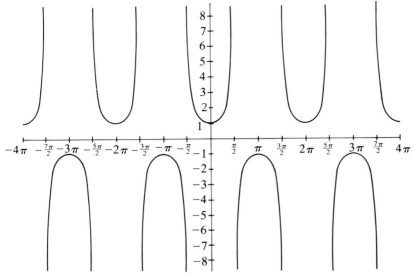

Figure 2.12.4: sec x.

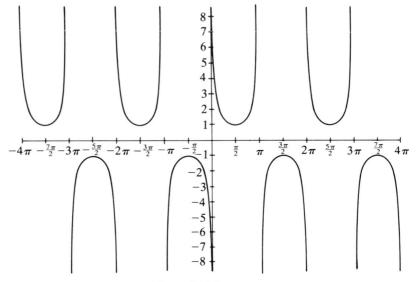

Figure 2.12.5: cosec x.

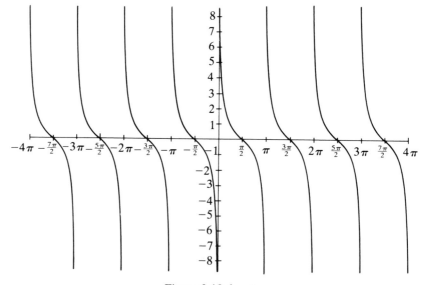

Figure 2.12.6: cot x.

this behaviour. Note that sec and cosec have period 2π, whereas tan and cot have period π.

We can verify that the above power series agree with the customary geometric definitions in trigonometry [see V, § 1.2.3] but we have to use integral calculus (Chapter 4, to which we refer for detailed explanations) as follows: By (2.12.5) we see that for all θ the point

$$P_\theta = (\cos \theta, \sin \theta)$$

lies on the unit circle $x^2 + y^2 = 1$ in the plane [see V, (1.2.8)]. When $\theta = 0$ we
have $P_0 = (1, 0)$; as θ increases P_θ rotates anticlockwise around the circle.

By integration, using some of the above properties of sin and cos, we can
compute the arc length along the circle from P_0 to P_θ [see § 6.1]. It turns out
to be θ. Thus θ is the *radian* measure [see V, § 1.1.9] of the angle P_0OP_θ
(Figure 2.12.7). It follows (by similar triangles [see V, § 1.1.7]) that in a
right-angled triangle ABC (Figure 2.12.8),

$$\sin \theta = BC/AC$$

$$\cos \theta = AB/AC$$

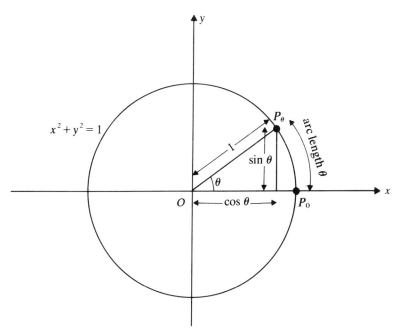

Figure 2.12.7

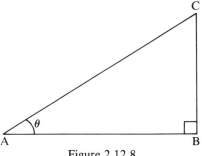

Figure 2.12.8

and thus

$$\tan \theta = BC/AB$$
$$\sec \theta = AC/AB$$
$$\operatorname{cosec} \theta = AC/BC$$
$$\cot \theta = AB/BC.$$

Numerous identities involving trigonometric functions may be deduced from the properties stated above [see V, § 1.2.3]. We content ourselves with listing some representative ones:

$$\sin 2x = 2 \sin x \cos x$$
$$\cos 2x = \cos^2 x - \sin^2 x = 2 \cos^2 x - 1 = 1 - 2 \sin^2 x$$
$$\sin 3x = 3 \sin x - 4 \sin^3 x$$
$$\cos 3x = 4 \cos^3 x - 3 \cos x$$
$$\tan (x + y) = \frac{\tan x + \tan y}{1 - \tan x \tan y}$$
$$\sin mx \sin nx = \tfrac{1}{2}[\cos (m - n)x - \cos (m + n)x]$$
$$\sin mx \cos nx = \tfrac{1}{2}[\sin (m + n)x + \sin (m - n)x]$$
$$\cos mx \cos nx = \tfrac{1}{2}[\cos (m + n)x + \cos (m - n)x].$$

The following formulae are often useful in evaluating integrals [see Theorem 4.3.2 and Example 4.3.11]. In them, $t = \tan \tfrac{1}{2}x$.

$$\cos x = (1 - t^2)/(1 + t^2) \tag{2.12.10}$$
$$\sin x = 2t/(1 + t^2) \tag{2.12.11}$$
$$\tan x = 2t/(1 - t^2). \tag{2.12.12}$$

Finally it follows immediately from the power series representation (2.12.1) of $\sin x$ that

$$\lim_{x \to 0} \frac{\sin x}{x} = 1. \tag{2.12.13}$$

2.13. HYPERBOLIC FUNCTIONS

The hyperbolic functions are in many ways analogous to the trigonometric functions. The *hyperbolic sine* and *hyperbolic cosine* are defined by

$$\sinh x = \sum_{n=0}^{\infty} \frac{x^{2n+1}}{(2n + 1)!} = \tfrac{1}{2}(e^x - e^{-x}) \tag{2.13.1}$$

$$\cosh x = \sum_{n=0}^{\infty} \frac{x^{2n}}{(2n)!} = \tfrac{1}{2}(e^x + e^{-x}). \tag{2.13.2}$$

Further we have the *hyperbolic tangent, secant, cosecant,* and *cotangent*:

$$\tanh x = \sinh x / \cosh x$$

$$\operatorname{sech} x = 1/\cosh x$$

$$\operatorname{cosech} x = 1/\sinh x$$

$$\coth x = \cosh x / \sinh x = 1/\tanh x.$$

Their graphs are shown in Figures 2.13.1–2.13.6.

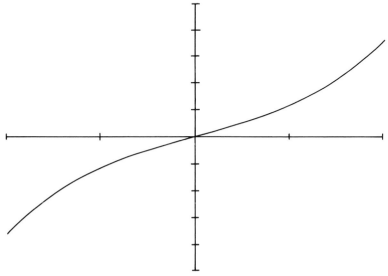

Figure 2.13.1: sinh *x*.

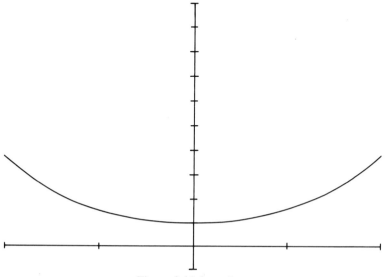

Figure 2.13.2: cosh *x*.

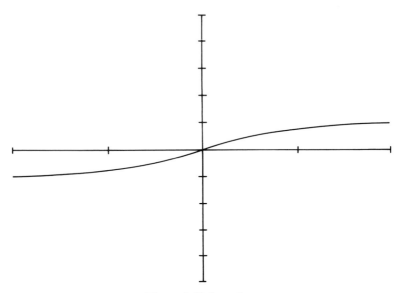

Figure 2.13.3: tanh *x*.

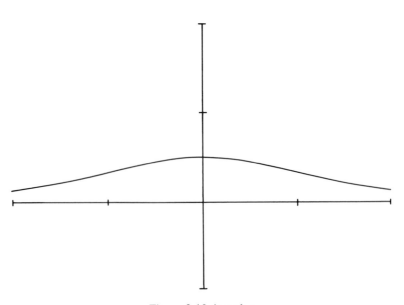

Figure 2.13.4: sech *x*.

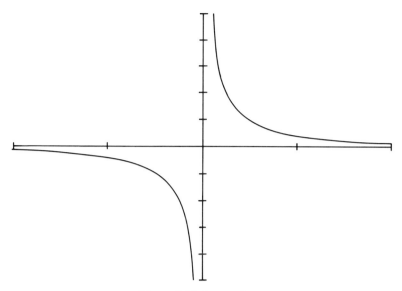

Figure 2.13.5: cosech *x*.

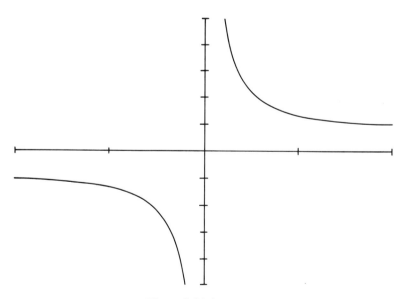

Figure 2.13.6: coth *x*.

Some of the more important identities involving them are:

$$\sinh (x + y) = \sinh x \cosh y + \cosh x \sinh y$$

$$\cosh (x + y) = \cosh x \cosh y + \sinh x \sinh y$$

$$\frac{d}{dx} \sinh x = \cosh x$$

$$\frac{d}{dx} \cosh x = \sinh x$$

$$\cosh^2 x - \sinh^2 x = 1.$$

There are many others, analogous to those listed above for trigonometric functions, but we shall not state them here.

2.14. INVERSE TRIGONOMETRIC AND HYPERBOLIC FUNCTIONS

None of the functions sin, cos, tan is a bijection but each becomes one if its domain and codomain are suitably chosen [see I, Definition 1.4.1 and § 1.4.2]. The usual choice is:

Function	Domain	Codomain
sin	$[-\pi, \pi]$	$[-1, 1]$
cos	$[0, 2\pi]$	$[-1, 1]$
tan	$\left(-\dfrac{\pi}{2}, \dfrac{\pi}{2}\right)$	$\mathbb{R}$

The corresponding inverse functions are denoted

$$\sin^{-1} : [-1, 1] \to [-\pi, \pi]$$

$$\cos^{-1} : [-1, 1] \to [0, 2\pi]$$

$$\tan^{-1} : \mathbb{R} \to \left(-\frac{\pi}{2}, \frac{\pi}{2}\right)$$

or by arc sin, arc cos, arc tan respectively. Their graphs are shown in Figures 2.14.1–2.14.3.

Inverses for sec, cosec, and cot are similarly defined, but are seldom used. In the same way for hyperbolic functions we restrict as follows:

Function	Domain	Codomain
sinh	$\mathbb{R}$	$\mathbb{R}$
cosh	$\mathbb{R}^+$	$[1, \infty]$
tanh	$\mathbb{R}$	$(-1, 1)$

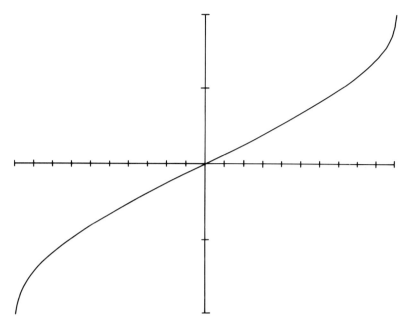

Figure 2.14.1: arc sin x.

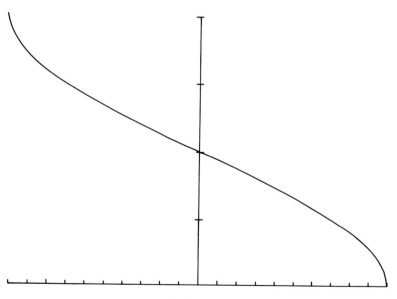

Figure 2.14.2: arc cos x.

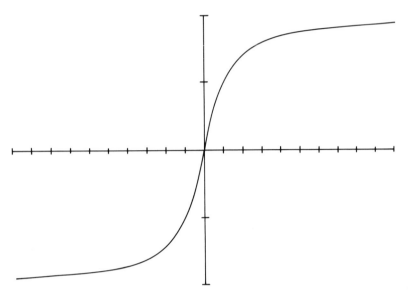

Figure 2.14.3: arc tan x.

and obtain inverses

$$\sinh^{-1}:\mathbb{R}\to\mathbb{R}$$

$$\cosh^{-1}:[1,\infty)\to\mathbb{R}^{+}$$

$$\tanh^{-1}:(-1,1)\to\mathbb{R}.$$

Sometimes these are denoted arg sinh, arg cosh, arg tanh. Inverses for sech, cosech, coth also exist on suitable intervals. The graphs for these inverse functions may be obtained by interchanging the x- and y-axes on Figures 2.14.1–2.14.3. The results are shown in Figures 2.14.4–2.14.6.

The following formulae are worth noting:

$$\sinh^{-1}x=\log(x+\sqrt{x^2+1}) \qquad (x\in\mathbb{R}) \qquad (2.14.1)$$

$$\cosh^{-1}x=\log(x+\sqrt{x^2+1}) \qquad (x\geq1) \qquad (2.14.2)$$

$$\tanh^{-1}x=\tfrac{1}{2}\log((1+x)/(1-x)) \qquad (x\in(-1,1)). \qquad (2.14.3)$$

2.15. ASYMPTOTIC SERIES

A non-convergent series [see § 1.7] may nonetheless prove useful in calculating the values of a function, or discussing certain of its properties. In fact there are a number of methods of assigning a sum to an infinite series which differ from 'convergence to a limit'. An important class of series for which a 'sum' fails to exist for any x, but which give good approximations for large x despite this, was discussed by Poincaré.

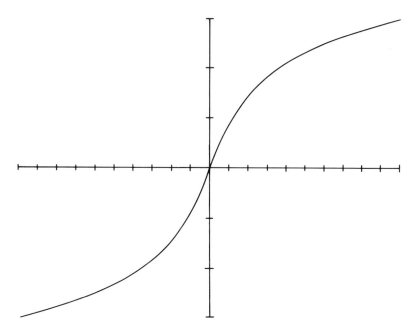

Figure 2.14.4: arg sinh x.

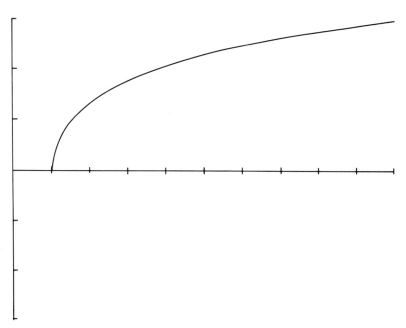

Figure 2.14.5: arg cosh x.

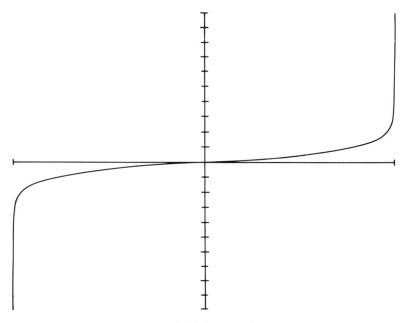

Figure 2.14.6: arg tanh x.

DEFINITION 2.15.1.   A series

$$a_0 + a_1 x^{-1} + a_2 x^{-2} + \ldots$$

is said to be an *asymptotic expansion* of a function $f(x)$ if for any fixed n the expression

$$x^n (f(x) - (a_0 + a_1 x^{-1} + \ldots + a_n x^{-n}))$$

tends to zero as x tends to infinity.
    It follows that for any $\varepsilon > 0$ we can make

$$|f(x) - (a_0 + a_1 x^{-1} + \ldots + a_n x^{-n})| < \varepsilon x^{-n}$$

by taking x sufficiently large; and the right hand side may be made as small as we wish by suitably choosing first ε, and then x.
    The reason why an asymptotic expansion need not converge is that the size of x required may grow with n, so that *no* choice for the size of x produces a small error for all n.

EXAMPLE 2.15.1.   By repeated integration by parts [see § 4.3], it may be shown that if

$$f(x) = \int_x^\infty \frac{e^{x-t}}{t}\, dt,$$

then

$$f(x) = \frac{1}{x} - \frac{1}{x^2} + \frac{2!}{x^3} - \dots + \frac{(-1)^{n-1}(n-1)!}{x^n} + R_n(x)$$

where the 'remainder' term is

$$R_n(x) = (-1)^n n! \int_x^\infty \frac{e^{x-t}}{t^{n+1}} \, dt.$$

Then the series (which diverges for all $x \neq 0$, and is not defined for $x = 0$)

$$\sum_{n=0}^\infty (-1)^{n+1} n! x^{-(n+1)}$$

is an asymptotic expansion of $f(x)$, because it is easy to see that for any fixed n, we have $x^n R_n(x) \to 0$ as $x \to \infty$.

Such expansions may be added or multiplied term-by-term to give asymptotic expansions of the sum or product of the corresponding functions; they may also be integrated term-by-term. However, differentiation term-by-term is not in general valid: a standard example starts with an asymptotic expansion of $e^{-x} \sin(e^x)$.

Different functions can have the same asymptotic expansion: this happens provided that their difference grows more slowly than x^n as $x \to \infty$, for each particular n. That is, f and g have the same asymptotic expansion if

$$\lim_{x \to \infty} x^n(f(x) - g(x)) = 0$$

for all n. This follows at once from the definition. For instance, $f(x)$ and $f(x) + e^{-x}$ have the same asymptotic expansion. On the other hand, two different asymptotic expansions cannot represent the same function.

Asymptotic expansions are important in applications to celestial mechanics, optics, elasticity and numerous other subjects [see Erdélyi (1956) and III, § 7.7.3].

I. S.

REFERENCES

Erdélyi, A. (1956). *Asymptotic Expansions*, Dover, New York.
Spivak, M. (1967). *Calculus*, Benjamin, New York.
Stewart, I. N. (1973). *Galois Theory*, Chapman and Hall, London.

CHAPTER 3

Differential Calculus

3.1. DIFFERENTIABILITY

3.1.1. Introduction

In this introductory paragraph we shall recapitulate briefly the ideas concerning the definition of differentiation which are dealt with at length in section 2.9.

Suppose then that we are given the graph of a function such as that illustrated in Figure 3.1.1. Choose two points A and B on the graph with co-ordinates

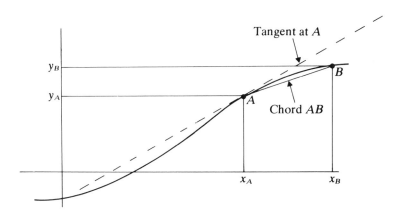

Figure 3.1.1

(x_A, y_A), (x_B, y_B) respectively [see V, § 1.2.1]. The slope of the chord AB is given by the quotient $m = (y_B - y_A)/(x_B - x_A)$. If the graph were a straight line then the chord AB would be a part of it and m would be a constant equal to the slope of the line. In general however, m will vary, depending on the points chosen. We now regard one of our points (say A) as fixed, and the other as variable. We suppose that if B is chosen near to A, then the gradient of AB approaches a limiting value, and that the approximation may be made arbitrarily close by choosing B near enough to A. Stated precisely (as we do below) this involves the notion of a limit, which is dealt with in section 2.3; however in geometrical terms the limiting value is simply the gradient of the tangent at A, provided that the graph at A is sufficiently smooth to enable a tangent to be drawn (Figure 3.1.1 again—see also the examples in section 2.9).

We make these ideas precise, as follows. We drop the suffix A, and write simply (x, y) for (x_A, y_A), where of course $y = f(x)$. We denote the difference $x_B - x_A$ by h: h will be positive or negative according to whether B is to the right or left of A. We now have $x_B = x + h$, and so $y_B = f(x_B) = f(x + h)$. The gradient of the chord is given by

$$m = \frac{y_B - y_A}{x_B - x_A} = \frac{f(x + h) - f(x)}{h}.$$

Moreover the process of taking B close to A simply corresponds to making h small. This leads us to our first definition [cf. Definition 2.9.1]:

DEFINITION 3.1.1
   (i) Suppose that f is defined on the interval (a, b) of $\mathbb{R}$, and that $a < x < b$. We say that f is *differentiable at* x if there is a real number (which we denote temporarily by c) such that

$$\frac{f(x + h) - f(x)}{h} \to c \quad \text{as } h \to 0:$$

equivalently we could say that the limit of $(1/h)(f(x + h) - f(x))$ exists, as $h \to 0$, as described in section 2.3.
   (ii) Let f be as in (i). We say that f is *differentiable on* (a, b) if f is differentiable at each point x of (a, b) as in (i).
   In case (i) we denote the value of the limit c by $f'(x)$ and call it the derivative of f at x. In case (ii) $f'(x)$ is defined for each x in (a, b), so that we have defined a new function f' on (a, b), called the *derived function* or *derivative* of f.

Note.   The following alternative notation (which goes back to Leibniz) is commonly used. The number $x_B - x_A$, which we have called h is denoted δx (the increment in x) and correspondingly $y_B - y_A = f(x + h) - f(x)$ is denoted δy (the increment in y). The slope of the chord AB is then the quotient $\delta y / \delta x$, and its limiting value $c = f'(x)$ is written as dy/dx. Despite some theoretical drawbacks (the same notation dy/dx is used both for the function f' and the real number $f'(x)$; and dy/dx is certainly not a quotient of two real numbers 'dy' and 'dx'), this notation has computational advantages which will be exploited in the next chapter on integration.

To finish this section we refer to the discussion which concludes section 2.9, and in particular to the relation between differentiability and continuity [see § 2.1] which may be summarized as follows.

THEOREM 3.1.1. *A real valued function must be continuous at any point at which it is differentiable. There exist functions which are continuous on an interval, but not differentiable at any point of it.*

These nowhere differentiable functions are not, as is sometimes stated, purely artificial mathematical objects, but have significance in such applications as Brownian motion (in physics, the study of motion of gas particles under continual collisions) and the probabilistic study of random motion.

3.1.2. Calculations from First Principles

Despite the warnings at the end of the preceding subsection, most functions encountered in everyday mathematics are differentiable. In this subsection we give examples of how Definition 3.1.1 applies to some simple functions. This process of applying Definition 3.1.1 directly is called *differentiation from first principles*.

EXAMPLE 3.1.1. Differentiate $f(x) = x^3$, for all $x \in \mathbb{R}$, from first principles.

Suppose a point x is chosen, and h is any (non-zero) real number. Then $f(x+h) = (x+h)^3$, and so

$$\frac{1}{h}(f(x+h) - f(x)) = \frac{1}{h}((x+h)^3 - x^3) = \frac{1}{h}(x^3 + 3x^2h + 3xh^2 + h^3 - x^3),$$

where we have used the Binomial theorem [see § 1.10] to expand $(x+h)^3$. This now simplifies to give

$$\frac{1}{h}(3x^2h + 3xh^2 + h^3) = 3x^2 + 3xh + h^2.$$

If we now take the limit as $h \to 0$, the terms $3xh$ and h^2 tend to zero, and the limit is $3x^2$. We have shown that

$$f'(x) = 3x^2, \quad \text{for all } x \in \mathbb{R}.$$

A similar calculation using the Binomial theorem shows that for any whole number n (positive or negative),

$$\text{if } f(x) = x^n, \quad \text{then } f'(x) = nx^{n-1}. \tag{3.1.1}$$

EXAMPLE 3.1.2. Differentiate $f(x) = \sqrt{x}$, for $x > 0$, from first principles.

If x and h are chosen so that x and $x+h$ are positive, and $h \neq 0$, then

$$\frac{f(x+h) - f(x)}{h} = \frac{\sqrt{x+h} - \sqrt{x}}{h}.$$

We recall the factorization $a^2 - b^2 = (a - b)(a + b)$ for the difference of two squares, and write $a^2 = y$, $b^2 = x$ so that $y - x = (\sqrt{y} - \sqrt{x})(\sqrt{y} + \sqrt{x})$. Using this we have

$$\frac{\sqrt{x+h} - \sqrt{x}}{h} = \frac{(x+h) - x}{h(\sqrt{x+h} + \sqrt{x})} = \frac{h}{h(\sqrt{x+h} + \sqrt{x})} = \frac{1}{\sqrt{x+h} + \sqrt{x}}.$$

As $h \to 0$, this last expression $\to 1/(\sqrt{x} + \sqrt{x}) = 1/2\sqrt{x}$. We have shown that if $f(x) = \sqrt{x}$, then $f'(x) = 1/2\sqrt{x}$ for all $x > 0$.

If we write $f(x) = x^{1/2}$ and $f'(x) = \frac{1}{2}x^{-1/2}$, we see that this conforms to the rule (3.1.1) without, however, being a special case of it. A more complicated argument would show that this result extends to more general values of n, but this is best postponed until the next section, Example 3.2.4.

EXAMPLE 3.1.3

$$f(x) = \sin x, \qquad x \in \mathbb{R}; \qquad f'(x) = \cos x.$$

$$g(x) = \cos x, \qquad x \in R; \qquad g'(x) = -\sin x.$$

These derivatives are deduced from the corresponding power series expansions in section 2.12.

3.2. RULES FOR DIFFERENTIATION

The method of differentiation from first principles, described in section 3.1, becomes highly cumbersome and inefficient when applied to any function beyond the very simplest. To avoid this we notice that most complicated functions are built up from simpler ones by the basic operations of addition and subtraction, multiplication and division, and composition. These operations are dealt with at some length in Chapter 2: here we shall simply summarize that discussion, and illustrate it by means of examples.

Suppose then that we are given two functions f and g which are defined on the whole of $\mathbb{R}$, or some subinterval of it. For example we might take

$$f(x) = 3x, \qquad g(x) = \cos x,$$

both defined on the whole of $\mathbb{R}$. The *sum* (*difference*) of f and g is the function whose value at a point x is the sum (difference) of the values of f and g there:

$$(f \pm g)(x) = f(x) \pm g(x). \tag{3.2.1}$$

In our example

$$(f + g)(x) = 3x + \cos x.$$

The *product* and *quotient* are defined similarly:

$$(f \cdot g)(x) = f(x) \cdot g(x), \tag{3.2.2}$$

$$(f/g)(x) = f(x)/g(x), \tag{3.2.3}$$

where in defining f/g, the points where $g = 0$ must be excluded.

In our example, $(f \cdot g)(x) = 3x \cos x$, and

$$\left(\frac{f}{g}\right)(x) = \frac{3x}{\cos x}, \quad \cos x \neq 0 \quad \left(\text{i.e. } x \neq \pm\frac{\pi}{2}, \pm\frac{3\pi}{2}, \pm\frac{5\pi}{2}, \ldots\right).$$

The *composition* $f \circ g$ is the result of performing first g, then f on x:

$$(f \circ g)(x) = f(g(x)), \tag{3.2.4}$$

that is we start with the real number x, perform g on it to get another real number y say, and then operate with f on y. Notice that if f is not defined for all real values, but only on some subinterval I say, then the values assumed by g must be contained in I in order for $f \circ g$ to be meaningfully defined. The possibly surprising fact that '$f \circ g$' means 'do g, then f' (rather than 'do f, then g') is a logical consequence of writing our functions (f) on the left of the variables (x). In our example,

$$(f \circ g)(x) = f(g(x)) = f(\cos x) = 3 \cos x,$$

while

$$(g \circ f)(x) = g(f(x)) = g(3x) = \cos (3x).$$

Substituting for instance $x = \pi/3$, when $(f \circ g)(x) = 3 \cos (\pi/3) = 3(\frac{1}{2}) = \frac{3}{2}$, while $g \circ f(x) = \cos (3 \cdot \pi/3) = \cos \pi = -1$, should convince us that $f \circ g$ and $g \circ f$ are quite different things. Since composition results from performing two operations successively, it is commonly referred to as being the '*function of a function*'.

The rules for differentiating these combinations of f and g can now be stated very simply:

THEOREM 3.2.1

(i) $\qquad\qquad\qquad (f \pm g)'(x) = f'(x) \pm g'(x) \qquad\qquad\qquad$ (3.2.5)

(ii) *the product rule*:

$$(f \cdot g)'(x) = f'(x) \cdot g(x) + f(x) \cdot g'(x) \tag{3.2.6}$$

(iii) *the quotient rule*:

$$\left(\frac{f}{g}\right)'(x) = \frac{f'(x) \cdot g(x) - f(x) \cdot g'(x)}{(g(x))^2} \tag{3.2.7}$$

(iv) *the function of a function rule or chain rule*:

$$(f \circ g)'(x) = f'(g(x)) \cdot g'(x). \tag{3.2.8}$$

In our example when $f(x) = 3x$, $g(x) = \cos x$, so that $f'(x) = 3$, $g'(x) = -\sin x$,

we have

$$(f+g)'(x) = 3 + (-\sin x) = 3 - \sin x,$$

$$(f \cdot g)'(x) = 3 \cdot \cos x + 3x(-\sin x) = 3 \cos x - 3x \sin x,$$

$$\left(\frac{f}{g}\right)'(x) = \frac{3 \cdot \cos x - 3x(-\sin x)}{(\cos x)^2} = \frac{3 \cos x + 3x \sin x}{(\cos x)^2},$$

$$(f \circ g)'(x) = f'(g(x)) \cdot g'(x) = 3 \cdot g'(x) = -3 \sin x,$$

$$(g \circ f)'(x) = g'(f(x)) \cdot f'(x) = -\sin (f(x)) \cdot f'(x) = -\sin (3x) \cdot 3 = -3 \sin (3x).$$

We shall illustrate the use of these rules (particularly the last of them) further, when we have enlarged our list of standard derivatives. It is worth noticing here however that the slightly illogical notations $(3x/\cos x)'$, or even $d/dx(3x/\cos x)$, in place of $(f/g)'(x)$, are in fact quite useful in practical calculations.

EXAMPLE 3.2.1.   Let $p(x) = a_0 + a_1x + a_2x^2 + \ldots + a_nx^n$, be any polynomial—for instance $p(x) = 7x^6 - \frac{1}{2}x^2 + x - \sqrt{2}$. The rule (3.1.1) together with the sum rule (3.2.5) shows that

$$p'(x) = a_1 + 2a_2x + \ldots + na_nx^{n-1}$$

(the a_0 term, being constant, simply vanishes when differentiated). In the example, $p'(x) = 42x^5 - x + 1$.

EXAMPLE 3.2.2.   Let $h(x) = \tan x = (\sin x/\cos x)$ $(\cos x \neq 0)$. Then from the quotient rule (3.2.7), it follows that

$$h'(x) = \frac{(\cos x)(\cos x) - (\sin x)(-\sin x)}{(\cos x)^2} = \frac{\cos^2 x + \sin^2 x}{\cos^2 x} = \frac{1}{\cos^2 x} = \sec^2 x.$$

Similar calculations for

$$\cot x = \frac{\cos x}{\sin x}, \qquad \sec x = \frac{1}{\cos x} \quad \text{and} \quad \operatorname{cosec} x = \frac{1}{\sin x}$$

give the corresponding entries in the table below.

EXAMPLE 3.2.3.   Let $f(x) = \log x$, $x > 0$. Recall from section 2.11 that log is defined as the inverse of the exponential function:

$$\exp (f(x)) = \exp (\log x) = x, \quad \text{for all } x > 0.$$

Recall also that exp is its own derivative: $\exp' = \exp$. Consequently if we differentiate this relation we obtain from the function of a function rule (3.2.8) that $\exp' (f(x)) \cdot f'(x) = 1$, or $\exp (\log x) \cdot f'(x) = 1$, or $x \cdot f'(x) = 1$. We have shown that if $f(x) = \log x$, then $f'(x) = 1/x$, for all $x > 0$.

The technique used in Example 3.2.3 to differentiate an inverse function is a special case of (3.2.8) which is worth stating separately. Suppose then that, as in section 2.7, f is a monotone function on the subinterval (a, b) of $\mathbb{R}$ and that g is its inverse, $g = f^{-1}$:

$$g(f(x)) = x, \quad \text{for all } x \text{ in } (a, b),$$

or equivalently

$$y = f(x) \quad \text{if and only if } x = g(y).$$

Then on differentiating we get the *inverse function rule*

$$g'(f(x)) \cdot f'(x) = 1, \quad \text{or} \quad g'(y) = \frac{1}{f'(x)}. \tag{3.2.9}$$

It is an easy exercise in the use of this rule to prove the formulae for the derivatives of the inverse sine and tangent functions, given in the table below. The formulae for hyperbolic functions are proved analogously.

EXAMPLE 3.2.4

  (i) The rth power was defined in section 2.7 by

$$x^r = \exp(r \log x),$$

for any real value of r, and $x > 0$. Hence if $f(x) = x^r = \exp(r \log x)$, then by (3.2.8),

$$f'(x) = \exp'(r \log x) \cdot (r \log x)'$$

$$= \exp(r \log x) \cdot \frac{r}{x}$$

$$= x^r \cdot rx^{-1} = rx^{r-1}.$$

This gives the extension of the rule (3.1.1) to all real values of the index.

  (ii) Let a be a positive real number, and

$$f(x) = a^x, \qquad x \in \mathbb{R}.$$

Then

$$f(x) = \exp(x \log a),$$

so by (3.2.8)

$$f'(x) = \exp'(x \log a) \cdot (x \log a)'$$

$$= \exp(x \log a) \cdot \log a$$

$$= a^x \log a.$$

We summarize our results so far in the following table:

Function: $f(x)$	Derived function: $f'(x)$	Function: $f(x)$	Derived function: $f'(x)$
$x^n, n = 0, 1, 2, \ldots$ and $x \in \mathbb{R}$	nx^{n-1}	e^x	e^x
$x^{-m}, m = 1, 2, \ldots, x \neq 0$	$-mx^{-m-1}$	$a^x (a > 0)$	$a^x \log a$
$x^\alpha, \alpha \in \mathbb{R}, x > 0$	$\alpha x^{\alpha-1}$	$\log x, x > 0$	$\dfrac{1}{x}$
$\sin x$	$\cos x$	$\sinh x$	$\cosh x$
$\cos x$	$-\sin x$	$\cosh x$	$\sinh x$
$\tan x$	$\sec^2 x$	$\tanh x$	$\text{sech}^2 x$
$\cot x$	$-\text{cosec}^2 x$	$\coth x$	$-\text{cosech}^2 x$
$\sec x$	$\sec x \tan x$	$\text{sech } x$	$-\text{sech } x \tanh x$
$\text{cosec } x$	$-\text{cosec } x \cot x$	$\text{cosech } x$	$-\text{cosech } x \coth x$
$\sin^{-1} x$	$\dfrac{1}{\sqrt{(1-x^2)}}$	$\sinh^{-1} x$	$\dfrac{1}{\sqrt{(1+x^2)}}$
$\tan^{-1} x$	$\dfrac{1}{1+x^2}$	$\tanh^{-1} x$	$\dfrac{1}{1-x^2}$

Table 3.2.1: Standard derivatives.

Detailed proofs of the validity of the rules (3.2.5)–(3.2.9) for differentiation can be found in any of the texts referred to at the end of this chapter. However because of its great importance, it is worthwhile giving at least an informal account of the reasons why (3.2.8) is valid.

Suppose that f is differentiable at the point a, that $f(a) = b$, and that g is differentiable at b and $g(b) = c$. Denote $f'(a)$ by A, and $g'(b)$ by B. Then

$$\frac{f(x) - f(a)}{x - a} \to A \quad \text{as } x \to a$$

(this is Definition 3.1.1 with a in place of x and $x - a$ in place of h): in other words, if we write ε (the standard notation for a small quantity) for the difference $\{[f(x) - f(a)]/(x - a)\} - A$, then we have on rearranging,

$$f(x) - f(a) = A(x - a) + \varepsilon(x - a),$$

where $\varepsilon \to 0$ as $x \to a$.

This equation shows that for a differentiable function f, $f(x)$ is given approximately by the linear function

$$f(a) + A(x - a), \qquad (3.2.10)$$

when x is near a. Conversely it is easy to see that if

$$f(x) = f(a) + A(x - a) + \varepsilon(x - a)$$

with $\varepsilon \to 0$ as $x \to a$, then f must be differentiable at a. Higher orders of approximation will be considered in section 3.6.

Similar considerations applied to g show that

$$g(y) - g(b) = B(y-b) + \eta(y-b),$$

where $\eta \to 0$ as $y \to b$. Then writing $y = f(x)$, $b = f(a)$, and combining the equations we obtain

$$g(f(x)) - g(f(a)) = B(f(x) - f(a)) + \eta(f(x) - f(a))$$
$$= BA(x-a) + (B\varepsilon + A\eta + \varepsilon\eta)(x-a).$$

Since, as is intuitively reasonable, and not hard to prove, $B\varepsilon + A\eta + \varepsilon\eta \to 0$ as $x \to a$, we deduce that $g \circ f$ is differentiable at a, and its derivative is

$$BA = g'(b) . f'(a) = g'(f(a)) . f'(a),$$

which is (3.2.8) with f and g interchanged.

EXAMPLE 3.2.5.   Let $f(x) = \sin x$, $g(x) = \log x$, $x > 0$. Then

$$(f+g)(x) = \sin x + \log x, \qquad (f+g)'(x) = \cos x + \frac{1}{x},$$

$$(f.g)(x) = \sin x \log x, \qquad (f.g)'(x) = \cos x \log x + \frac{\sin x}{x},$$

$$(f/g)(x) = \frac{\sin x}{\log x}, \qquad (f/g)'(x) = \frac{\cos x . \log x - (1/x)\sin x}{(\log x)^2},$$

$$(f \circ g)(x) = \sin(\log x),$$

$$(f \circ g)'(x) = f'(g(x)) . g'(x) = f'(\log x) . \frac{1}{x} = \cos(\log x) . \frac{1}{x},$$

$$(g \circ f)(x) = \log(\sin x)$$

(defined when $\sin x > 0$, that is, for $2n\pi < x < (2n+1)\pi$ [see § 2.12]);

$$(g \circ f)'(x) = g'(f(x)) . f'(x) = g'(\sin x) . \cos x = \frac{1}{\sin x} . \cos x = \cot x.$$

Note.   The Leibniz notation for the function of a function rule is as follows: Let $y = f(x)$, so that $dy/dx = f'(x)$, and $z = g(y)$, so that $dz/dy = g'(y) = g'(f(x))$. Then $z = g(f(x)) = (g \circ f)(x)$, and (3.2.8) tells us that

$$\frac{dz}{dx} = (g \circ f)'(x) = g'(f(x)) . f'(x) = \frac{dz}{dy} . \frac{dy}{dx}.$$

Needless to say, the *proof* requires an argument such as we sketched earlier, and is not to be achieved by a surreptitious cancelling of the 'dy's'! In this notation the above calculation for the derivative of $\log \sin x$ would appear as follows: Let $y = \sin x$, and $z = \log y = \log \sin x$. Then

$$\frac{dy}{dx} = \cos x, \quad \text{and} \quad \frac{dz}{dy} = \frac{1}{y} = \frac{1}{\sin x},$$

so that

$$\frac{dz}{dx} = \frac{dz}{dy} \cdot \frac{dy}{dx} = \frac{1}{\sin x} \cdot \cos x = \cot x$$

as before. This method is safe, but slower than the version given in Example 3.2.5, since it requires the intermediate stages to be written down explicitly.

We finish this section with a result on the differentiation of power series:

THEOREM 3.2.2.   *Let the power series $\sum_{n=0}^{\infty} a_n(x-a)^n$ have interval of convergence $(a-R, a+R)$, where $R > 0$—i.e. the series is convergent for $|x-a| < R$, and divergent for $|x-a| > R$ [see § 1.10]. Let $f(x)$ be the sum of the power series on $(a-R, a+R)$. Then for all x in $(a-R, a+R)$, f is differentiable and its derivative is given by*

$$f'(x) = \sum_{n=1}^{\infty} na_n(x-a)^{n-1}.$$

Notice that the hypotheses include the extreme case when the interval of convergence is the whole of the real line. Notice also that the series for f' is exactly the result of differentiating the series for f term by term, as we did for polynomials in Example 3.2.1.

COROLLARY 3.2.3.   *Let $f(x)$ be defined as the sum of the power series $\sum_{n=0}^{\infty} a_n(x-a)^n$, as in Theorem 3.2.2. Then f has derivatives of all orders [see § 2.10] at all points of $(a-R, a+R)$, given by*

$$f^{(k)}(x) = \sum_{n=k}^{\infty} n(n-1)\ldots(n-k+1)a_n(x-a)^{n-k}.$$

In particular $f^{(k)}(a) = k!\,a_k$, and

$$f(x) = \sum_{n=0}^{\infty} \frac{f^{(n)}(a)}{n!}(x-a)^n.$$

The corollary is proved by applying Theorem 3.2.2 repeatedly to f, then f', then f'', etc.: this yields the formula for $f^{(k)}(x)$. When $x = a$ is substituted, all terms except the first, where $n = k$, vanish, giving $f^{(k)}(a) = k!\,a_k$, and this gives the series for f, on substituting these values for a_k.

It must be emphasized that this corollary is proved only for functions which are given *a priori* as power series. For other functions a more circumspect approach is required: this will be dealt with in section 3.6.

3.3. THE SIGNIFICANCE OF THE DERIVATIVE

In this section and the next two we describe the information concerning a function which can be obtained from a knowledge of its derivative. Our first result shows what we can deduce from knowing the *sign of f' at a point*.

LEMMA 3.3.1. *Let f be differentiable at a, and $f'(a) > 0$. Then for some $h > 0$, $f(x) > f(a)$ if $x \in (a, a + h)$, while $f(x) < f(a)$ if $x \in (a - h, a)$.*

The proof is easy, since if $f'(a)$ is positive, then $[f(x) - f(a)]/(x - a)$ tends to a positive limit as $x \to a$, and so if $|x - a|$ is small, $f(x) - f(a)$ and $x - a$ have the same sign. It should not be supposed however that f is necessarily monotone [see Definition 2.7.1] on some interval $(a - h, a + h)$. The analogous result with reversed signs throughout is proved in the same way.

Our next result asserts the geometrically evident fact that a differentiable function which assumes equal values at the end points of an interval must have a zero derivative at some intermediate point: Figure 3.3.1 illustrates the point.

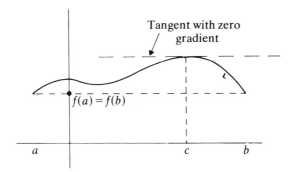

Figure 3.3.1

THEOREM 3.3.2 (*Rolle's Theorem*). *Let f be real-valued, differentiable on the open interval (a, b), and continuous at a and b. Suppose also $f(a) = f(b)$. Then for some point c with $a < c < b$, $f'(c) = 0$.*

(Recall that the notation (a, b) denotes the open interval from a to b, that is, those real numbers which lie between a and b, *excluding* the end points [see I, § 2.6.3].)

If the values of f at a and b are not equal we may apply Theorem 3.3.2 to $g(x) = f(x) - kx$, where $k = [f(b) - f(a)]/(b - a)$ is chosen to make $g(a) = g(b)$. This gives us

THEOREM 3.3.3. (*The Mean Value Theorem*). *Let f be real-valued, differentiable on the interval (a, b), and continuous at a and b. Then there is* *a point c in* (a, b) *for which*

$$f(b) - f(a) = (b - a)f'(c).$$

The fact that the hypotheses require only continuity at a and b rather than differentiability is usually of little practical significance. The result has a simple geometric interpretation: it asserts the existence of a point c where the tangent, whose slope is $f'(c)$, is parallel to the chord joining the end points, whose slope is $[f(b) - f(a)]/(b - a)$; Figure 3.3.2 illustrates this. Looking ahead to Chapter 9, it is worth mentioning that the result is false for complex valued functions.

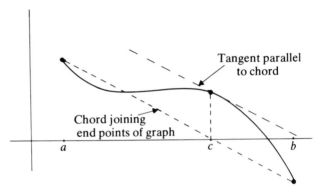

Figure 3.3.2

COROLLARY 3.3.4. *Let f be differentiable on* (a, b), *and* $f'(x) > 0$ *for all* x *in* (a, b). *Then f is monotone increasing on* (a, b). *Correspondingly if* $f'(x) < 0$ *on* (a, b) *then f is monotone decreasing on* (a, b).

The proof is immediate from the mean value theorem, for if $f'(x) > 0$ on (a, b), then for any $x_1 < x_2$ in (a, b),

$$f(x_2) - f(x_1) = (x_2 - x_1)f'(c)$$

for some c in (x_1, x_2), and both $x_2 - x_1$ and $f'(c)$ are positive.

The converse results, namely that if f is monotone increasing (or decreasing) on (a, b) then $f'(x) \geq 0$ (respectively $f'(x) \leq 0$) on (a, b) are easy to prove from the definition 3.1.1(i), but are less useful. Notice that the inequality is weakened from $>$ to $\geq$: the example of $f(x) = x^3$ on $(-1, 1)$ shows that this is inevitable.

EXAMPLE 3.3.1. Let $f(x) = x^3 e^{-x}$, for all $x \in \mathbb{R}$. Then by the product rule (3.2.6) it follows that

$$f'(x) = 3x^2 e^{-x} - x^3 e^{-x} = x^2(3 - x) e^{-x}.$$

Now e^{-x} is always positive, and so $f'(x)$ has the same sign as $x^2(3-x)$, namely positive if $0<x<3$, or $x<0$, negative if $x>3$. Corollary 3.3.4 now shows that f is increasing on $(0, 3)$ and $(-\infty, 0)$, and decreasing on $(3, \infty)$. The function itself is positive for $x>0$, negative for $x<0$.

This example as well as Theorem 3.3.3 and Corollary 3.3.4 themselves suggest that the points where $f'(x)=0$ should be particularly important in discussing the behaviour of f. Such points are called *stationary points* and will be treated in more detail in section 3.5.

3.4. HIGHER DERIVATIVES

We now consider the possibility of repeatedly applying the operation of differentiation to a function. We suppose that, as in Definition 3.1.1(ii), f is differentiable on (a, b), so that $f'(x)$ exists at all points of (a, b). Then f' is again a function which one may (or may not!) be able to differentiate again. The example [see § 2.10]

$$f(x) = \begin{cases} \frac{1}{2}x^2, & x \geq 0 \\ -\frac{1}{2}x^2, & x < 0 \end{cases}$$

shows that f' may exist and be continuous but not differentiable, while the example

$$f(x) = \begin{cases} x^2 \sin(1/x), & x \neq 0 \\ 0, & x = 0 \end{cases}$$

shows that f' may exist at all points without even being continuous: For $f'(0) = 0$ from first principles while by (3.2.6)

$$f'(x) = 2x \sin\left(\frac{1}{x}\right) - \cos\left(\frac{1}{x}\right)$$

which does not approach any limit as $x \to 0$, since $2x \sin(1/x) \to 0$ as $x \to 0$ while $\cos(1/x)$ oscillates between ± 1. We shall suppose however that the functions we shall consider from now on are sufficiently well-behaved for the required derivatives to exist.

DEFINITION 3.4.1. The derivative of f', that is, $(f')'$ or just f'' is called the *second derivative* of f: similarly we may define derivatives of higher orders by $f''^{\cdots'}$ (n dashes) $= (f''^{\cdots'}$ ($n-1$ dashes))$'$. This notation is obviously impossibly cumbersome: instead we write $f^{(n)}$ for $f''^{\cdots'}$ (n dashes) and call it the nth *derivative* of f. It should not be confused with f^n which conventionally denotes the nth power of f. If $f^{(n)}$ exists on (a, b) we say f is n *times differentiable on* (a, b). If $f^{(n)}$ exists and is continuous on (a, b) we say f is n *times continuously differentiable on* (a, b).

The Leibniz notation for f'' is d^2y/dx^2 and for $f^{(n)}$ is d^ny/dx^n. We leave the geometrical significance of higher derivatives (particularly f'') to the next section, and concentrate for the moment on methods for calculating them.

EXAMPLE 3.4.1.   Let $f(x) = 2x^3 - 4x + 1$, $x \in \mathbb{R}$. Then we have successively,

$$f'(x) = 6x^2 - 4, \qquad f''(x) = 12x, \qquad f^{(3)}(x) = 12,$$
$$f^{(4)}(x) = 0 = f^{(n)}(x), \quad \text{all } n \geq 4.$$

In general it is plain that if f is a polynomial of degree k, then f has derivatives of all orders, and in particular, $f^{(n)}(x) = 0$ when $n > k$.

EXAMPLE 3.4.2.   Let $f(x) = x^2 e^x$, $x \in \mathbb{R}$. Then by (3.2.6)

$$f'(x) = x^2 e^x + 2x e^x = (x^2 + 2x) e^x.$$

Similarly

$$f''(x) = (x^2 + 2x) e^x + (2x + 2) e^x = (x^2 + 4x + 2) e^x.$$

It would be possible to prove a formula for $f^{(n)}$ inductively in this example, but this is best left until after Theorem 3.4.1.

The formulae for nth derivatives which correspond to the formulae in Theorem 3.2.1 are important. The rule for sums and differences is immediate:

$$(f \pm g)^{(n)} = f^{(n)} \pm g^{(n)}. \tag{3.4.1}$$

There are no analogues of (3.2.7) and (3.2.8), but for products we have the following famous result:

THEOREM 3.4.1   (*Leibniz' formula*).   *Let f and g be n times differentiable on (a, b). Then $f \cdot g$ is n times differentiable, and*

$$(f \cdot g)^{(n)} = \sum_{r=0}^{n} \binom{n}{r} f^{(r)} g^{(n-r)}. \tag{3.4.2}$$

Here

$$\binom{n}{r} = {}^nC_r = \frac{n!}{r!(n-r)!}$$

is the rth coefficient in the binomial expansion of $(a+b)^n$ [see I, (3.10.1)]. In particular

$$\binom{n}{0} = \binom{n}{n} = 1,$$

while $f^{(0)}$, $g^{(0)}$ are simply f and g. The proof is by induction on n [see I, § 2.1], and exactly parallels the proof of the binomial expansion of $(a+b)^n$.

EXAMPLE 3.4.2   (*continued*).   Find the nth derivative of $f(x) = x^2 e^x$.

Write $g(x) = x^2$, so that $g'(x) = 2x$, $g''(x) = 2$, and all subsequent derivatives are zero: also put $h(x) = e^x$, so that all derivatives $h^{(r)}(x)$ are also $= e^x$. Then by (3.4.2),

$$(g \cdot h)^{(n)} = \sum_{r=0}^{n} \binom{n}{r} g^{(r)} h^{(n-r)}$$

$$= \binom{n}{0} g^{(0)} h^{(n)} + \binom{n}{1} g^{(1)} h^{(n-1)} + \binom{n}{2} g^{(2)} h^{(n-2)};$$

all subsequent terms involve $g^{(r)}$, $r \geq 3$, and are hence $= 0$. Hence

$$f^{(n)}(x) = (g \cdot h)^{(n)}(x) = 1 \cdot x^2 \cdot e^x + n \cdot 2x \cdot e^x + \frac{n(n-1)}{2} \cdot 2 \cdot e^x$$

$$= (x^2 + 2nx + n(n-1)) e^x.$$

This agrees with our earlier result when $n = 2$.

The following examples are very useful in applications of Leibniz formula.

EXAMPLE 3.4.3.   Let $f(x) = x^\alpha$, $\alpha \in \mathbb{R}$, $x > 0$. Then

$$f^{(n)}(x) = \alpha(\alpha - 1)(\alpha - 2) \ldots (\alpha - n + 1) x^{\alpha - n}.$$

The result is immediate by induction (differentiate both sides once more). Notice that if α is a positive integer, $\alpha = k$ say, $k = 1, 2, 3, \ldots$, then $f^{(n)}(x) = 0$ when $n > k$: for other real values of α, all derivatives are non-zero.

EXAMPLE 3.4.4.   Let $f(x) = \sin x$. Then $f'(x) = \cos x$, $f''(x) = -\sin x$, $f'''(x) = -\cos x$, $f''''(x) = \sin x$ and the derivatives repeat themselves in a cycle of length 4:

$$f^{(4n)}(x) = \sin x, \qquad f^{(4n+1)}(x) = \cos x, \qquad f^{(4n+2)}(x) = -\sin x,$$

$$f^{(4n+3)} = -\cos x.$$

This may be neatly summarized in the single formula

$$f^{(r)}(x) = \sin\left(x + r\frac{\pi}{2}\right),$$

as is easily verified. Similar results are valid for cos, but for no other trigonometric function.

EXAMPLE 3.4.5.   Let $g(x) = \log(f(x))$, defined for those values of x for which $f(x) > 0$. We have $g(x) = h(f(x))$ where $h(y) = \log y$ and so by (3.2.8)

$$g'(x) = h'(f(x)) \cdot f'(x) = \frac{f'(x)}{f(x)}.$$

The quotient rule (3.2.7) now gives

$$g''(x) = \frac{f''(x)f(x) - f'(x) \cdot f'(x)}{f(x)^2} = \frac{f''(x)}{f(x)} - \left(\frac{f'(x)}{f(x)}\right)^2.$$

We now come to some results on the evaluation of so-called 'indeterminate forms' for which our knowledge of higher derivatives will be very useful. We first need a more delicate version of the Mean Value Theorem (Theorem 3.3.3):

THEOREM 3.4.2.   *Let f, g be real-valued functions which are differentiable on (a, b) and continuous at a and b. Suppose also that $g'(x) \neq 0$ at all points of (a, b). Then for some point c in (a, b),*

$$\frac{f'(c)}{g'(c)} = \frac{f(b) - f(a)}{g(b) - g(a)}.$$

The proof is by applying the Mean Value Theorem to the function

$$F(x) = (f(x) - f(a))(g(b) - g(a)) - (g(x) - g(a))(f(b) - f(a)).$$

Conversely Theorem 3.3.3 can be deduced from Theorem 3.4.2 by putting $g(x) = x$. Notice that the condition $g'(x) \neq 0$ on (a, b) implies that $g(b) \neq g(a)$ (Theorem 3.3.3 again). It is important, in view of the applications that we shall make, to notice that the hypotheses do not exclude the possibility that g might be differentiable at a or b, or that $g'(a)$, or $g'(b)$ (or both) might then equal zero.

COROLLARY 3.4.3.   *Let x_0 be a point of (a, b), let f and g be differentiable on (a, b) except possibly at x_0 and let f and g be continuous at x_0. Suppose that for some $\delta > 0$, $g'(x) \neq 0$ when $0 < |x - x_0| < \delta$, and that the limit $\lim_{x \to x_0} f'(x)/g'(x)]$ exists and has the value l. Then the limit $\lim_{x \to x_0} [f(x) - f(x_0)]/[g(x) - g(x_0)]$ exists, and has the same value, l.*

Notice that as x tends to x_0, both $f(x) - f(x_0)$ and $g(x) - g(x_0)$ tend to zero, and that the quotient assumes the indeterminate form '0/0'. The result states that the value on the limit (in which the value $x = x_0$ is excluded) has a definite value: it is the same as the limit of $f'(x)/g'(x)$, provided that this latter limit exists. It is important to get these limits the right way round, since there are functions for which $\lim_{x \to x_0} (f(x) - f(x_0)/g(x) - g(x_0))$ exists, while $\lim_{x \to x_0} f'(x)/g'(x)$ does not: $f(x) = x^2 \sin(1/x)$, $g(x) = x$, $x_0 = 0$ gives an example.

The proof of Corollary 3.4.3 follows from Theorem 3.4.2 applied on the interval (x_0, x) if $x > x_0$ or (x, x_0) if $x < x_0$.

COROLLARY 3.4.4

(i) *Let f, g have continuous first derivatives at all points of* (a, b), $x_0 \in (a, b)$, *and suppose* $g'(x_0) \neq 0$. *Then* $\lim_{x \to x_0} [f(x) - f(x_0)]/[g(x) - g(x_0)]$ *exists and equals* $f'(x_0)/g'(x_0)$.

(ii) *Let* f, g *be n-times continuously differentiable on* (a, b), *and suppose* $f^{(r)}(x_0) = g^{(r)}(x_0) = 0$, $1 \leq r \leq n - 1$, $g^{(n)}(x_0) \neq 0$. *Then the limit* $\lim_{x \to x_0} [(f(x) - f(x_0)]/[g(x) - g(x_0)]$ *exists and equals* $f^{(n)}(x_0)/g^{(n)}(x_0)$.

The proof of (i) is immediate from Corollary 3.4.3 while (ii) is Corollary 3.4.3 applied successively to $f^{(n-1)}$, $g^{(n-1)}$; $f^{(n-2)}$, $f^{(n-2)}$; ...; f', g' and f, g.

The results stated in Corollaries 3.4.3 and 3.4.4 are often used in applications when $f(x_0) = g(x_0) = 0$, when they assert the existence of limits of the form $\lim_{x \to x_0} f(x)/g(x)$, where both f, g vanish at x_0.

Both Corollary 3.4.3 and Corollary 3.4.4 are referred to loosely as *de L'Hôpital's rule*—it is without doubt the most misunderstood, misquoted, and misapplied result in the whole of the calculus. The most common opportunity for a misuse of de L'Hôpital's rule is when the limit of a quotient of functions has to be found in circumstances in which the conditions for the rule are not satisfied; when the limit is of the form $\lim_{x \to x_0} f(x)/g(x)$, where $g(x_0) \neq 0$, for example! We finish this section with two examples of its (correct!) use.

EXAMPLE 3.4.6

(i) Find

$$\lim_{x \to 1} \frac{\cos ((\pi/2)x)}{x - 1}.$$

Here $f(x) = \cos ((\pi/2)x)$ and $g(x) = x - 1$ have continuous derivatives on $\mathbb{R}$, $x_0 = 1$, $f(x_0) = 0$, $g(x_0) = 0$, and $g'(x_0) \neq 0$. Hence Corollary 3.4.4(i) applies, and the limit is

$$\frac{f'(1)}{g'(1)} = \frac{-\pi/2 \sin ((\pi/2)x)}{1} \bigg|_{x=1} = -\frac{\pi}{2} \sin \left(\frac{\pi}{2}\right) = -\frac{\pi}{2}.$$

(ii) Find

$$\lim_{x \to 0} \frac{x - \sin x}{x^3}.$$

Here both $f(x) = x - \sin x$ and $g(x) = x^3$ have continuous derivatives of all orders on $\mathbb{R}$, while $f(0) = f'(0) = f''(0) = g(0) = g'(0) = g''(0) = 0$, $f^{(3)}(x) = \cos x$, $f^{(3)}(0) = 1$, $g^{(3)}(x) = 6$. Hence by Corollary 3.4.4(ii) the required limit is

$$\frac{f^{(3)}(0)}{g^{(3)}(0)} = \frac{1}{6}.$$

3.5. STATIONARY POINTS

We now apply our knowledge of differentiation to the investigation of the geometrical properties of the graph of a function. The first result of this kind was Corollary 3.3.4 which said that f was monotone increasing or decreasing on an interval, according as $f'(x)$ was positive or negative there. This suggests that the points where $f'(x) = 0$ should have a special significance, and motivates

DEFINITION 3.5.1.   Let f be differentiable on an interval $[a, b]$. Any point x of $[a, b]$ at which $f'(x) = 0$ is called a *stationary point* (*turning point* or *critical point*) of f.

The most common reason for considering stationary points lies in the search for maximum and minimum points on the graph of f. We define these next, and then consider the relationships between the two ideas.

DEFINITION 3.5.2.   Let f be continuous on an interval $[a, b]$. Let x_0 be a point of $[a, b]$ such that for some positive number h, $f(x) < f(x_0)$ for all x in $[a, b]$ with $|x - x_0| < h$, except $x = x_0$ itself. Then we say f has a *local maximum* at x_0. If instead $f(x) > f(x_0)$ for the same range of values of x, we say f has a *local minimum* at x_0.

Notice at once that the definition of a local maximum does not require the differentiability which is needed in Definition 3.5.1, it simply requires that the value of f at x_0 is greater than those in its immediate vicinity. (The definition does not require continuity either, though we shall not pursue this possibility). It is usual to drop the description 'local' and speak simply of a maximum or minimum of f, but the distinction between a local and an overall maximum or minimum (see Definition 3.5.3 below) is important, and is the reason for giving the extra precision to our definition.

The relationship between the two concepts is summarized in

THEOREM 3.5.1
   (i) *Let f have a local maximum (or minimum) at x_0, $a < x_0 < b$, and suppose that f is differentiable at x_0. Then $f'(x_0) = 0$ (x_0 is a stationary point of f).*
   (ii) *Let f be differentiable on $[a, b]$ and let x_0 be a stationary point of f, at which the sign of f' changes from positive (on the left of x_0) to negative (on the right of x_0). Then f has a local maximum at x_0. If the sign of f' changes from negative to positive, f has a local minimum at x_0.*

The proof of part (i) is immediate from Lemma 3.3.1, for $f'(x_0)$ is assumed to exist, and both $f'(x_0) > 0$ and $f'(x_0) < 0$ are ruled out by the assumption that a maximum or minimum exists at x_0. The proof of (ii) is an easy consequence of Theorem 3.3.4 (the Mean Value Theorem), since for instance if x is to the left of x_0, then $[f(x)-f(x_0)]/[x-x_0]=f'(x_1)$ for some x_1 between x and x_0. But both $f'(x_1)$ and $x-x_0$ are negative, so $f(x)-f(x_0)$ is positive, that is $f(x)>f(x_0)$, and a similar argument works to the right of x_0.

It is important to realize that not all stationary points are maxima or minima—for instance if f is a constant function [see § 2.2] then $f'(x)=0$ at all points, but f has neither maxima nor minima in the sense of our definition. And the well-known example of $f(x)=x^3$ on $\mathbb{R}$, shows that a stationary point can occur at a point (here at $x=0$) where f is monotone increasing [see Definition 2.7.1].

It is equally important (perhaps even more so) to realize that a function can have a local maximum at points other than stationary points. For instance the function $f(x)=x$ on $[0,1]$ has a minimum at 0 and a maximum at 1, despite the fact that $f'(x)=1$ throughout: notice that Theorem 3.5.1(i) does not apply at the end points of intervals. In addition the function $f(x)=|x|$ has a minimum at $x=0$ where the function is not differentiable [see § 2.9] and is thus not stationary. Hence in searching for maxima and minima we have to consider (i) end points of the interval of definition, (ii) points of non-differentiability and (iii) stationary points (zeros of f'). If the function is well behaved then (ii) can be disregarded; (i) generally cannot, however.

Before going further we illustrate these ideas with a simple example.

EXAMPLE 3.5.1. Let $f(x)=\frac{1}{3}x^3-x^2+1$, for all real x. f is differentiable on $\mathbb{R}$, and $f'(x)=x^2-2x=x(x-2)$. Hence $x=0$ and $x=2$ are stationary points, and $f'(x)>0$ if $x<0$ or if $x>2$, while $f'(x)<0$ if $0<x<2$. Hence Theorem 3.5.1(ii) tells us that f has a local maximum at $x=0$, and a local minimum at $x=2$. Here $f(0)=1$, and $f(2)=-\frac{1}{3}$.

If we are interested in the single largest (or smallest) value which our function assumes on an interval, we may use the following definition:

DEFINITION 3.5.3. Let f be continuous on an interval $[a,b]$. Then x_0 is an *overall (global) maximum for f on $[a,b]$* if $f(x) \leq f(x_0)$ for all x in $[a,b]$ and an *overall (global) minimum* if $f(x) \geq f(x_0)$ for all x in $[a,b]$.

Theorem 2.8.2 shows that if f is continuous on the closed bounded interval $[a,b]$, then overall maxima and minima do indeed exist—we repeat that these are not necessarily stationary points. Our Example 3.5.1 shows that a continuous function on $\mathbb{R}$ may have local maxima and minima, but no overall ones.

We now investigate what information may be gained from a knowledge of second, and higher derivatives.

The geometrically intuitive notion of being *concave upwards (convex downwards)* means that for each pair of points x_0, x_1 with $a<x_0<x_1<b$, the chord joining the point $(x_0, f(x_0))$ to $(x_1, f(x_1))$ lies above the graph of f on the interval (x_0, x_1)—see Figure 3.5.1. If the chord lies below the curve we say the graph is *convex upwards (concave downwards)*. In Figure 3.5.1. the

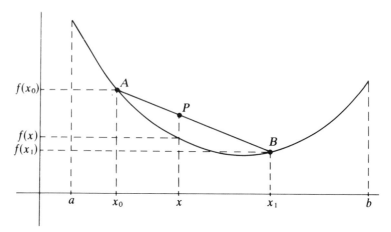

Figure 3.5.1

y-coordinate of the point P on the chord AB is [see V, (1.2.4)]

$$y = \frac{x_1 - x}{x_1 - x_0}f(x_0) + \frac{x - x_0}{x_1 - x_0}f(x_1),$$

so that f is concave upwards if and only if, whenever $x_0 < x < x_1$, we have

$$f(x) < \frac{x_1 - x}{x_1 - x_0}f(x_0) + \frac{x - x_0}{x_1 - x_0}f(x_1). \qquad (3.5.1)$$

Similarly f is convex upwards if and only if (3.5.1) holds with the inequality reversed.

A sufficient condition for f to be concave upwards is contained in

LEMMA 3.5.2.   *Let f be twice differentiable on (a, b), and $f''(x) > 0$ for all x in (a, b). Then the graph of f is concave upwards. Similarly if $f''(x) < 0$, the graph is concave downwards.*

COROLLARY 3.5.3.   *Let x_0 be a stationary point of the twice continuously differentiable function f. Then if $f''(x_0) > 0$, f has a local minimum at x_0: if $f''(x_0) < 0$, f has a local maximum at x_0.*

This follows since if $f''(x_0) > 0$, then $f''(x) > 0$ in the vicinity of x_0, so that Corollary 3.3.4 shows that f' is increasing, and so changes from negative to positive as x increases through x_0. Theorem 3.5.1(ii) now shows that f has a minimum at x_0.

The examples of $f(x) = x^4$, $f(x) = -x^4$, $f(x) = x^3$ which have respectively a minimum, a maximum, and neither, at $x = 0$ should convince the reader that no information can be deduced if $f''(x) = 0$ at a stationary point. For this, knowledge of higher derivatives is required, or (better) Theorem 3.5.1(ii) should be applied directly.

Corollary 3.5.3 is in fact the most widely used method for finding maxima and minima. In Example 3.5.1 for instance, $f''(x) = 2x - 2$, $f''(0) = -2 < 0$, $f''(2) = 2 > 0$. Hence f has a maximum at 0 and a minimum at 2. Its advantage is that it requires only a value of f'' at the point under consideration, rather than the sign of f' on either side. Its disadvantages are twofold: firstly it requires two differentiations which may, particularly for rational functions [see (3.2.7)] involve a lot of manipulation and give scope for error, and secondly as was pointed out above, it fails to provide any information if $f''(x) = 0$.

DEFINITION 3.5.4. A point at which $f''(x)$ changes sign is called a *point of inflexion*. It follows from Lemma 3.5.2 that at a point of inflexion, the graph of f changes from concave upwards to concave downwards, or conversely.

Points of inflexion may be stationary (as $f(x) = x^3$ at $x = 0$), or not (as $f(x) = \sin x$, also at $x = 0$). At a stationary point of inflexion $f'(x)$ has the same sign on either side: the point $x = 0$ in Example 3.3.1 illustrates this. If a potentially unlimited number of derivatives are available, the following result may be used, but we repeat: Theorem 3.5.1(ii) is generally better.

THEOREM 3.5.4. *Let f be n times continuously differentiable on (a, b). If n is even and*

$$f'(x_0) = f''(x_0) = \ldots = f^{(n-1)}(x_0) = 0, \qquad f^{(n)}(x_0) \neq 0$$

then f has a maximum or a minimum at x_0 according as $f^{(n)}(x_0)$ is negative or positive. If n is odd and

$$f''(x_0) = \ldots = f^{(n-1)}(x_0) = 0, \qquad f^{(n)}(x_0) \neq 0$$

then there is a point of inflexion at x_0, which will be stationary if also $f'(x_0) = 0$.

Theorem 3.5.4 follows from Taylors' theorem [Theorem 3.6.2(i)] in the same way that Corollary 3.5.3 is proved via Theorem 3.5.1(ii) from the Mean Value Theorem [Theorem 3.3.3].

3.6. TAYLOR AND MACLAURIN EXPANSIONS

Our definition of the derivative in sub-section 3.1.1 was designed so that the linear function l, given by

$$l(x) = f(x_0) + (x - x_0)f'(x_0),$$

whose graph is the tangent to the graph of f at x_0, gives the closest approximation to f, among all linear functions. A linear function is of course just a polynomial of the first degree [see § 2.1]: in this section we consider the problem of approximating the graph of f by polynomials of higher degree. We shall be concerned with finding the coefficients of the polynomial in such a way that the approximation by a polynomial P_n of degree n is of 'higher than nth order'—that is

$$\frac{f(x) - P_n(x)}{(x-a)^n} \to 0 \quad \text{as } x \to a.$$

Suppose then that we are given a function f which is n times differentiable on (a, b), that x_0 is in (a, b), and that we have to find the coefficients $a_0, a_1, \ldots, a_n$ in the polynomial

$$P_n(x) = a_0 + a_1(x - x_0) + a_2(x - x_0)^2 + \ldots + a_n(x - x_0)^n,$$

so that P_n gives the best approximation to f at x_0. To begin with, we require that $f(x_0) = P_n(x_0) = a_0$, on putting x_0 for x. Then

$$\frac{f(x) - P_n(x)}{x - x_0} = \frac{1}{x - x_0}(f(x) - f(x_0)) - a_1 - a_2(x - x_0) - \ldots - a_n(x - x_0)^{n-1}$$

which tends to $f'(x_0) - a_1$ as $x \to x_0$. Hence we take $a_1 = f'(x_0)$. So far we have simply recovered the linear approximation $f(x_0) + (x - x_0)f'(x_0)$ mentioned above. In the same way we have

$$\frac{f(x) - P_n(x)}{(x - x_0)^2} = \frac{f(x) - f(x_0) - (x - x_0)f'(x_0)}{(x - x_0)^2} - a_2 - a_3(x - x_0) - \ldots - a_n(x - x_0)^{n-2}.$$

An immediate application of de L'Hôpital's rule [Corollary 3.4.4(ii)] shows that the first term on the right approaches $\frac{1}{2}f''(x_0)$, and hence if we take this value for a_2, then $(f(x) - P_n(x))/(x - x_0)^2 \to 0$ as $x \to x_0$. We may continue thus to find the value $(1/2\cdot3)f'''(x_0)$ for a_3, and generally $(1/r!)f^{(r)}(x_0)$ for a_r, $0 \le r \le n$.

This brings us to the following definition, and provides the proof of the theorem which comes after it.

DEFINITION 3.6.1.   Let f be n times differentiable on (a, b), and let x_0 be a point of (a, b). The polynomial

$$P_n(x) = f(x_0) + (x - x_0)f'(x_0) + \frac{(x - x_0)^2}{2!}f''(x_0) + \ldots + \frac{(x - x_0)^n}{n!}f^{(n)}(x_0)$$

is called the nth *Taylor polynomial for f at x_0*.

If it is required to draw attention to x_0 and f explicitly, the more elaborate notation $P_{n,x_0,f}(x)$ may be used.

THEOREM 3.6.1.   *Let f, P_n be as in Definition 3.6.1. Then*

$$\frac{f(x) - P_n(x)}{(x - x_0)^n} \to 0 \quad \text{as } x \to x_0.$$

Theorem 3.6.1 simply asserts that the approximation of $f(x)$ by P_n is of higher order than the nth, that is that the difference $f(x) - P_n(x)$ tends to zero faster than does $(x - x_0)^n$. It is worth pointing out that if f is itself a polynomial of degree at most n, then the expression is exact: $f(x) = P_n(x)$.

EXAMPLE 3.6.1.   Find the Taylor polynomials for cos at $x_0 = 0$, with degrees 2 and 4, and the corresponding approximate values of cos $10°$.

Here $f(x) = \cos x$, $f'(x) = -\sin x$, $f''(x) = -\cos x$, $f'''(x) = \sin x$, $f^{(4)}(x) = \cos x$. Hence $P_2(x) = 1 - \frac{1}{2}x^2$, $P_4(x) = 1 - \frac{1}{2}x^2 + \frac{1}{24}x^4$, are the Taylor polynomials about $x = 0$. The value of x (in radians [see V, § 1.1.9]) is $\pi/18 = 0\cdot1745329$. This gives $P_2(\pi/18) = 0\cdot9847691$, $P_4(\pi/18) = 0\cdot9848078$, the latter being correct to seven decimal places.

If we want to be able to predict the accuracy of the approximation we need a formula for the difference $f(x) - P_n(x)$. We do this in two ways. The first is by an extension of Theorem 3.3.3 and is sometimes called the nth *Mean Value Theorem* and the second involves repeated integration—we refer to the next chapter for the necessary properties.

THEOREM 3.6.2   (*Taylor's theorem*)
   (i) *Let f be an $n + 1$ times differentiable function on (a, b), let x_0 be a point of (a, b) and P_n be the nth Taylor polynomial for f at x_0. Then*

$$f(x) - P_n(x) = \frac{(x - x_0)^{n+1}}{(n+1)!} f^{(n+1)}(x_1),$$

for some point x_1, between x_0 and x.
   (ii) *Let f be $n + 1$ times continuously differentiable on (a, b), and let x_0, and P_n be as in (i). Then*

$$f(x) - P_n(x) = \frac{1}{n!} \int_{x_0}^{x} f^{(n+1)}(t)(x - t)^n \, dt.$$

Notice that (i), like Theorem 3.3.3 from which it is deduced, involves the value of $f^{(n+1)}$ at an unknown point, while (ii) gives the difference explicitly as an integral. In fact both are usually used to give an estimate (rather than an exact value) for the difference, based on an estimate of the size of $f^{(n+1)}$. The proofs of both are complicated rather than deep: we refer the reader to Spivak (1967) for the details.

EXAMPLE   3.6.1   (*continued*).   Here $f(x) = \cos x$, $f^{(5)}(x) = -\sin x$,   so $|f^{(5)}(x)| = |\sin x| \le |x|$. Then

$$|f(x) - P_4(x)| \le \frac{|x^5| \cdot |x|}{5!} = \left(\frac{\pi}{18}\right)^6 \cdot \frac{1}{120} = 2\cdot36 \times 10^{-7},$$

which is an over-estimate of the actual error. A better estimate can be obtained by noticing that the next term in the Taylor expansion is zero, so that $P_4 = P_5$.

Hence the error is bounded by $|x^6||\cos x|/6! < 4 \times 10^{-8}$, which is very close to the actual error.

We now consider under what conditions the Taylor polynomial $P_n(x)$ tends to $f(x)$ as $n \to \infty$. Firstly it is necessary that derivatives of all orders should at least exist at the point x_0 about which the expansion is to be constructed. We may then form the Taylor series, as follows.

DEFINITION 3.6.2.   Let f have derivatives of all orders at x_0. Then the series

$$\sum_{n=0}^{\infty} \frac{(x - x_0)^n}{n!} f^{(n)}(x_0)$$

is called the *Taylor series for f at* x_0. The special case when $x_0 = 0$ is called the *Maclaurin series for f*. [See also § 2.10.]

We emphasize that at this stage we are simply considering a formal aggregate of terms: we have as yet no information on the convergence of the series [see § 1.7] or what its sum might be. The function $f(x) = e^{-1/x^2} (x \neq 0)$, $f(0) = 0$ of Example 2.10.1 has the property that all its derivatives are zero at $x = 0$, and hence the Taylor series about $x = 0$ vanishes identically, so that though convergent, its sum is not equal to $f(x)$. The function

$$f(x) = \int_0^{\infty} \frac{e^{-t}}{1 + x^2 t} \, dt$$

is defined and infinitely differentiable for all real x [see Theorem 4.7.3] and its Taylor series about $x = 0$ is $\sum_{n=0}^{\infty} (-1)^n n! x^{2n}$ which is divergent for all $x \neq 0$ (the formula for the derivatives at $x = 0$ requires the Euler integral for $n!$ [see § 10.2.1]).

A necessary and sufficient condition for the convergence of the Taylor series to the correct value, is that the expression obtained in Theorem 3.6.2 for $f(x) - P_n(x)$ should tend to zero as $n \to \infty$. A sufficient condition for this is given in the next theorem.

THEOREM 3.6.3.   *Let f have derivatives of all orders at all points of* (a, b), *and* x_0 *be a point of* (a, b). *Suppose that there exist positive constants* A, c *such that for all x in* (a, b) *and* $n = 0, 1, 2, \ldots$,

$$|f^{(n)}(x)| \leq A n! c^n.$$

Then the Taylor series for f about x_0 *is convergent with sum* $f(x)$ *on the interval* $(x_0 - 1/c, x_0 + 1/c) \cap (a, b)$.

Notice that the condition on $f^{(n)}$ is certainly satisfied if for instance there exists a constant M such that $|f^{(n)}(x)| \leq M$, all n. The proof of Theorem 3.6.3

follows at once from Theorem 3.6.2, for the error $|f(x) - P_n(x)|$ is given by

$$\frac{|x - x_0|^n}{n!}|f^{(n+1)}(x_1)| \le \frac{|x - x_0|^n}{n!}A(n+1)!c^{n+1}$$

$$= A(n+1)c(|x - x_0|c)^n,$$

and it is easy to show that this tends to zero if $c|x - x_0| < 1$, or $|x - x_0| < 1/c$.

A condition of quite a different kind is given in the following result, for whose proof (which is very delicate) the reader is referred to Apostol (1957) Theorem 13–31.

THEOREM 3.6.4. *Let f have derivatives of all orders at all points of* (a, b), *and suppose for some N,* $f^{(n)}(x) \ge 0$ *for all* $n \ge N$ *and x in* (a, b). *Then the Taylor series for f about* x_0 *converges with sum* $f(x)$ *on any interval* $(x_0 - h, x_0 + h)$ *contained in* (a, b).

We finish this section with an example of the use of these results to establish convergence of a Taylor series to the proper sum [see (1.10.9)].

EXAMPLE 3.6.2.   The binomial function $f(x) = (1 + x)^\alpha$, $\alpha \in \mathbb{R}$, $x > -1$ has derivatives of all orders given by

$$f^{(n)}(x) = \alpha(\alpha - 1) \ldots (\alpha - n + 1)(1 + x)^{\alpha - n}.$$

[see (1.10.9) and cf. Example 3.4.3]. The Taylor series about $x = 0$ (i.e. the Maclaurin series) is then given by

$$\sum_{n=0}^{\infty} \frac{\alpha(\alpha - 1) \ldots (\alpha - n + 1)}{n!} x^n.$$

If we use the expression for $f(x) - P_n(x)$ given by Theorem 3.6.2(i) we get

$$\frac{x^{n+1}}{(n+1)!}f^{(n+1)}(x_1) = \frac{x^{n+1}}{(n+1)!}\alpha(\alpha - 1) \ldots (\alpha - n)(1 + x_1)^{\alpha - n}$$

where x_1 lies between 0 and x. If x is positive, then so is x_1, and so $(1 + x_1)^{\alpha - n} < 1$ as soon as $n > \alpha$. But for $x < 1$, $s_n = (x^{n+1}/(n+1)!)\alpha(\alpha - 1) \ldots (\alpha - n) \to 0$ as $n \to \infty$, since the ratio $s_{n+1}/s_n \to x < 1$ as $n \to \infty$. Hence we have shown that the Taylor series is convergent to $(1 + x)^\alpha$ for $0 \le x < 1$.

The same argument can be extended to the interval $(-\frac{1}{2}, 0)$, but for the full result, namely that the series is convergent to $(1 + x)^\alpha$ on $(-1, 1)$, one must use a more delicate argument either via the integral formula [Theorem 3.6.2(ii)] or by Theorem 3.6.4 as in Apostol (loc. cit.).

Many other expansions are obtained from the binomial expansion by integration, as in the following examples. In each case we use the fact that a convergent power series may be integrated term by term [see Corollary 1.12.7].

EXAMPLE 3.6.3

(i) $$\log(1+t) = \sum_{n=1}^{\infty} \frac{(-1)^{n-1}}{n} t^n, \quad \text{if } |t| < 1.$$

For $\log(1+t) = \int_1^{1+t} du/u = \int_0^t du/(1+u)$, so by the Binomial Theorem [see § 1.10]

$$\log(1+t) = \int_0^t (1 - u + u^2 - u^3 + \ldots + (-1)^n u^n + \ldots) \, du$$

$$= t - \tfrac{1}{2}t^2 + \tfrac{1}{3}t^3 - \tfrac{1}{4}t^4 + \ldots + \frac{(-1)^n}{n+1} t^{n+1} \ldots.$$

(ii) $$\sin^{-1} x = \sum_{n=0}^{\infty} \frac{1 \cdot 3 \cdot \ldots \cdot (2n-1)}{2^n n! (2n+1)} x^{2n+1} \quad \text{if } |x| < 1.$$

For, from Table 4.2.1,

$$\sin^{-1} x = \int_0^x \frac{dt}{\sqrt{1-t^2}} = \int_0^x (1-t)^{-1/2} \, dt$$

$$= \int_0^x \left(1 + \tfrac{1}{2}t^2 + (-\tfrac{1}{2})(-\tfrac{3}{2})\frac{1}{2!} t^4 + \ldots \frac{1 \cdot 3 \cdot \ldots \cdot 2n-1}{2 \cdot 2 \cdot \ldots \cdot 2} \frac{1}{n!} t^{2n} + \ldots \right) dt$$

$$= x + \frac{1}{2 \cdot 3} x^3 + \frac{1 \cdot 3}{2^2} \frac{1}{5 \cdot 2!} x^5$$

$$+ \ldots \frac{1 \cdot 3 \cdot \ldots \cdot (2n-1)}{2^n} \frac{1}{(2n+1)n!} x^{2n+1} + \ldots.$$

(iii) $$\tan^{-1} x = \sum_{n=0}^{\infty} \frac{(-1)^n}{2n+1} x^{2n+1}, \quad \text{if } |x| < 1.$$

For, from Table 4.2.1,

$$\tan^{-1} x = \int_0^x \frac{1}{1+t^2} \, dt = \int_0^x (1 - t^2 + t^4 - \ldots + (-1)^n t^{2n} + \ldots) \, dt$$

$$= x - \tfrac{1}{3}x^3 + \tfrac{1}{5}x^5 - \ldots \frac{+(-1)^n}{2n+1} x^{2n+1} + \ldots.$$

P.L.W.

REFERENCES

Apostol, T. M. (1957). *Mathematical Analysis*, Addison Wesley, Reading, Mass.
Burkill, J. C. (1962). *A First Course in Mathematical Analysis*, C.U.P., Cambridge.
Hardy, G. H. (1962). *A Course of Pure Mathematics*, 10th Ed., C.U.P., Cambridge.
Maddox, I. J. (1977). *Introductory Mathematical Analysis*, Adam Hilger, Bristol.
Scott, D. B. and Tims, S. R. (1966). *Mathematical Analysis*, C.U.P., Cambridge.
Spivak, M. (1967). *Calculus*, Benjamin, New York.

CHAPTER 4

Integral Calculus

4.1. INTEGRABILITY

In this section we shall discuss those geometrical ideas on which the notion of integration rests, and use them to motivate a formal definition of the integral for a suitable class of functions.

Suppose that we are given a function which is defined on an interval $[a, b]$ of the real line, and that we wish to find the area which is enclosed between the graph, and the x-axis, as indicated in Figure 4.1.1.

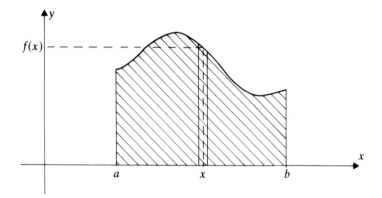

Figure 4.1.1

The method begins by dividing the area into a large number of strips parallel to the y-axis. We then take an approximate value for the area of each strip by multiplying its width by its length, the length being $f(x)$ for any x in the strip. For the class of functions we shall be considering, it is immaterial how the point x is chosen: we shall later (Definition 4.1.2) denote a typical interval by $[x_{i-1}, x_i]$, and the value of x by u_i, so that the area of the rectangle will be $f(u_i)(x_i - x_{i-1})$. Finally we add up the areas of all the strips, and consider whether this sum approaches a limit as the number of strips increases, and the width of all the strips approaches zero [see § 1.2]. If the limit exists, we say that f is *integrable* on $[a, b]$, and that the value of the limit is the *integral* of f between a and b.

It is important to realize that not all functions are integrable, though the class of integrable functions is a wide one which includes all continuous functions, and all monotone functions [see Definitions 2.1.2 and 2.7.1].

The reader who is interested mainly in the techniques of integration might perhaps proceed by now reading through Theorem 4.1.1 below, and then going straight to sections 4.2 and 4.3.

We begin to make these ideas precise by defining the way in which the interval $[a, b]$ is to be divided up.

DEFINITION 4.1.1.   A *dissection*, or *partition*, of an interval $[a, b]$ is a set

$$P = \{x_0, x_1, x_2, \ldots, x_{i-1}, x_i, \ldots, x_n\}$$

where $a = x_0 < x_1 < \ldots < x_{i-1} < x_i < \ldots < x_n = b$. The maximum length of any subinterval is called the *mesh* or *norm* of the dissection, and denoted by $\|P\|$:

$$\|P\| = \max_{1 \le i \le n} (x_i - x_{i-1}).$$

EXAMPLE 4.1.1

$$P_1 = \{0, \tfrac{1}{3}, \tfrac{2}{3}, 1\},$$
$$P_2 = \{0, \tfrac{9}{10}, 1\},$$

and

$$P_3 = \{0, 0 \cdot 01, 0 \cdot 02, 0 \cdot 03, \ldots, 0 \cdot 99, 1\},$$

are all dissections of $[0, 1]$. We have

$$\|P_1\| = \tfrac{1}{3}, \qquad \|P_2\| = \tfrac{9}{10}, \qquad \|P_3\| = 0 \cdot 01.$$

We now describe how to form the approximating sums.

DEFINITION 4.1.2.   Let P be a dissection of $[a, b]$, and for each i, $1 \le i \le n$, let u_i denote any point of $[x_{i-1}, x_i]$. The sum

$$S_P(f) = \sum_{i=1}^{n} f(u_i)(x_i - x_{i-1})$$

is called a *Riemann sum* associated with P.

Notice that $S_P(f)$ is simply the sum of the area of rectangular strips whose bases are the intervals $[x_{i-1}, x_i]$ and whose length is $f(u_i)$. Strictly speaking the sum depends on the choice of u_i as well as the dissection P itself, but we shall not reflect this explicitly in the notation.

EXAMPLE 4.1.2.   Let $f(x) = x$ and $P = \{0, \frac{1}{10}, \frac{2}{10}, \frac{3}{10}, \ldots, \frac{9}{10}, 1\}$ so that $x_i = i/10$. Suppose also that we take $u_i = x_i$. Then

$$S_P(f) = \sum_{i=1}^{10} f\left(\frac{i}{10}\right)\left(\frac{i}{10} - \frac{i-1}{10}\right) = \frac{1}{10} \sum_{i=1}^{10} \frac{i}{10} = \frac{1}{10^2}(1 + 2 + 3 + \ldots + 10)$$

$$= \tfrac{11}{20} = 0 \cdot 55.$$

When the function f is clear from the context, we shall generally write S_P for $S_P(f)$.

It is often convenient, though not essential, to form the dissection as in Example 4.1.2 by choosing a number n ($n = 10$ in the example) and taking P to be the division of $[a, b]$ into n equal parts. In this case it is easy to show that $x_i = a + i(b - a)/n$, $i = 0, 1, 2, \ldots, n$.

We now describe the limiting process by which the integral is obtained.

DEFINITION 4.1.3.   Let f be a real valued function defined on $[a, b]$. Suppose that for any choice of u_i the sums S_P approach a limit L as the mesh $\|P\|$ of P tends to zero. Then we say that f is *integrable* over $[a, b]$, and we say that the limit L is the *integral* of f over $[a, b]$, or in symbols,

$$L = \int_a^b f(x)\, dx.$$

The function f which is to be integrated is called the *integrand*.

Notice that the x is a 'dummy variable'—it may be replaced by any other letter (except a, b, f or d) without altering the meaning: thus

$$\int_a^b f(x)\, dx = \int_a^b f(t)\, dt = \int_a^b f(\theta)\, d\theta, \quad \text{etc.}$$

Sometimes one writes more simply $\int_a^b f$ for $\int_a^b f(x)\, dx$. Notice also that in a region where the function takes negative values, $f(u_i)$ will be negative, and hence that area below the x-axis makes a negative contribution to the total. Finally we emphasize that the definition applies only to functions which are defined on an interval of finite length $[a, b]$, and which are *bounded* on that interval—that is real numbers m, M exist for which $m \le f(x) \le M$ for all x in $[a, b]$. Section 4.6 shows how these conditions may sometimes be relaxed. The above definition is essentially due to Riemann: if we wish to contrast this process with others, such as those in sections 4.8 and 4.9, we shall say that f is *Riemann-integrable*.

The formal definition of the limiting process can be made more explicit by saying that S_P tends to L as $\|P\|$ tends to zero if [cf. Definition 1.2.1]: *for each $\varepsilon > 0$, there is some $\delta > 0$, such that $|S_P - L| < \varepsilon$ for every P with $\|P\| < \delta$, and every possible choice of points (u_i) relative to P.*

The fact which this brings to light, namely that in order to show that a particular function is integrable, one must consider not only every dissection with small enough mesh, but also every possible choice of points (u_i) relative to it, means that verification of the integrability is cumbersome in even the easiest of cases, and we shall only give a single example of it.

EXAMPLE 4.1.3.   Let $f(x) = x$, $a = 0$, $b = 1$. Let P be any dissection of $[0, 1]$ and x_i, u_i have their meanings as in Definition 4.1.2. Then

$$S_P = \sum_{i=1}^{n} f(u_i)(x_i - x_{i-1}) = \sum_{i=1}^{n} u_i(x_i - x_{i-1}),$$

and if we denote by S_1 and S_2 the special sums

$$\sum_{i=1}^{n} x_{i-1}(x_i - x_{i-1}) \quad \text{and} \quad \sum_{i=1}^{n} x_i(x_i - x_{i-1}),$$

we obviously have $S_1 \le S_P \le S_2$. Consequently, in order to show that f is integrable, and that $\int_0^1 f(x)\,dx = \frac{1}{2}$, it will be enough to show that both S_1 and S_2 tend to $\frac{1}{2}$ as the mesh $\|P\|$ tends to zero.

Now on the one hand we have

$$S_1 + S_2 = \sum_{i=1}^{n} (x_i + x_{i-1})(x_i - x_{i-1}) = \sum_{i=1}^{n} (x_i^2 - x_{i-1}^2)$$

$$= (x_1^2 - x_0^2) + (x_2^2 - x_1^2) + \ldots + (x_n^2 - x_{n-1}^2)$$

$$= x_n^2 - x_0^2 = 1^2 - 0^2 = 1.$$

On the other,

$$S_2 - S_1 = \sum_{i=1}^{n} (x_i - x_{i-1})(x_i - x_{i-1}) \le \|P\| \sum_{i=1}^{n} (x_i - x_{i-1}),$$

since each $(x_i - x_{i-1}) \le \|P\|$. But $\sum_{i=1}^{n} (x_i - x_{i-1}) = 1$ (the total length of all subintervals), so that

$$S_2 - S_1 \le \|P\|.$$

It follows that as $\|P\|$ tends to zero, both S_1 and S_2 tend to $\frac{1}{2}$, as required.

The following theorem describes quite wide classes of functions which may be integrated. The proof may be found in any of the references quoted at the end of the chapter.

THEOREM 4.1.1

(i) *Any function f which is real-valued and continuous on an interval* $[a, b]$ *of finite length* [*see* § 2.1] *is integrable over* $[a, b]$. *More generally if there exist numbers M, m for which* $m \le f(x) \le M$ *for all x in* $[a, b]$ (*we say f is* bounded *on* $[a, b]$) *and if f is continuous except at a finite number of points* [*see Definition 2.3.5*], *then f is integrable over* $[a, b]$.

(ii) *Any function which is real-valued and monotone increasing or decreasing on* $[a, b]$ *is integrable over* $[a, b]$.

To finish this section we state without proof some formal properties of the integration process.

THEOREM 4.1.2

(i) *If f is integrable over* $[a, b]$, *and k is any real number, then kf is integrable, and*

$$\int_a^b kf(x)\, dx = k \int_a^b f(x)\, dx.$$

(ii) *If f, g are integrable over* $[a, b]$, *then so is* $f + g$, *and*

$$\int_a^b (f(x) + g(x))\, dx = \int_a^b f(x)\, dx + \int_a^b g(x)\, dx.$$

(iii) *Suppose* $a < c < b$; *then f is integrable over* $[a, b]$ *if and only if it is integrable over both* $[a, c]$ *and* $[c, b]$, *and we have that*

$$\int_a^c f(x)\, dx + \int_c^b f(x)\, dx = \int_a^b f(x)\, dx.$$

(iv) *If f is integrable over* $[a, b]$ *and* $m \le f(x) \le M$ *for all x in* $[a, b]$, *then*

$$m(b - a) \le \int_a^b f(x)\, dx \le M(b - a).$$

(v) *If* $a > b$ *and if f is integrable over* $[b, a]$ *it is conventional (and consistent with Theorem 4.2.2 below) to define*

$$\int_a^b f(x)\, dx = - \int_b^a f(x)\, dx.$$

Parts (i) and (ii) may be combined in the statement that the integral is a linear function of f [see I, § 5.11]. In all except trivial cases, the integral of a product is not equal to the product of the integrals. Rules for integrating products and composite functions will be found in section 4.3. We end this section with the discussion of a special situation which arises in some contexts.

Suppose that $h(x)$ is a function which is zero everywhere in $[a, b]$ except at $x = c$. Then

$$\int_a^b h(x)\, dx = \int_{c-\delta}^{c+\delta} h(x)\, dx.$$

As shown in Figure 4.1.2, the integral on the right-hand side may be made as small as we like, tending to zero as $\delta \to 0$.

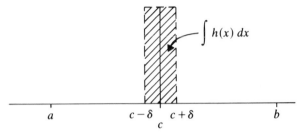

Figure 4.1.2

Thus $\int_a^b h(x)\,dx = 0$. Clearly the same is true of any function which is zero except for a *finite* number of values of x. By Theorem 4.1.2(ii) we have

$$\int_a^b f(x)\,dx = \int_a^b g(x)\,dx$$

if and only if

$$\int_a^b \{f(x) - g(x)\}\,dx = 0$$

and so we have

THEOREM 4.1.3. *If f and g are integrable over $[a, b]$ and if $f(x) = g(x)$ except for a finite number of values of x in $[a, b]$ then*

$$\int_a^b f(x)\,dx = \int_a^b g(x)\,dx.$$

4.2. THE FUNDAMENTAL THEOREM OF CALCULUS

In section 4.1 we introduced the integration process constructively, by a limiting process motivated by the notion of an area. As such it is a new analytical process which is entirely independent of the theory of differentiation in Chapter 3. However the reader who has already had some contact with the calculus will be aware that differentiation and integration are closely connected—indeed in a sense which will shortly be made precise, they are inverse processes, the one having an effect which reverses the other. This section is devoted to explaining this relationship.

Suppose then that we are given a continuous function f on an interval $[a, b]$. Theorems 4.1.1 and 4.1.2(iii) tell us that f is integrable over every subinterval of $[a, b]$, so that if we take a general point x in $[a, b]$, then $\int_a^x f$ (or $\int_a^x f(t)\,dt$, using a new dummy variable to avoid confusion with the upper limit x) is defined and may be used to give a new function $A(x)$ (the area from a to x):

$$A(x) = \int_a^x f(t)\,dt.$$

Figure 4.2.1 illustrates this.

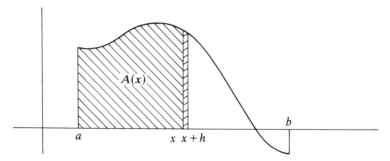

Figure 4.2.1

Suppose now that we take a real number h: Theorem 4.1.2(iii) shows that if $h > 0$,

$$A(x + h) = \int_a^{x+h} f = \int_a^x f + \int_x^{x+h} f = A(x) + \int_x^{x+h} f,$$

and a similar relation holds when $h < 0$. If h becomes small, the integral $\int_x^{x+h} f$ becomes closer to the area of the rectangular strip of width h and height $f(x)$: this may be made precise by using the continuity of f at x. This leads to the approximate relation

$$A(x + h) = A(x) + hf(x),$$

or on re-arranging,

$$\frac{A(x + h) - A(x)}{h}$$

is approximately equal to $f(x)$. Furthermore, and this is the essential point of the argument, the approximation improves as h decreases, or in the limit

$$\lim_{h \to 0} \frac{A(x + h) - A(x)}{h} = f(x).$$

But the limit on the left is simply the limit which in sections 3.1 and 2.9 was used to define the derivative of A. We have arrived at the following fundamental result, in which we replace the function A by the more conventional F.

THEOREM 4.2.1.   *Let f be continuous on* $[a, b]$. *For any x in* $[a, b]$, *let*

$$F(x) = \int_a^x f = \int_a^x f(t) \, dt.$$

Then $F'(x) = f(x)$ *for all x in* $[a, b]$.

Theorem 4.2.1 shows that if one takes a continuous function, integrates it and then differentiates the result, one recovers the original function. However in practice we are more often concerned with integrating derivatives than with differentiating integrals: Theorem 4.2.2 covers this case.

THEOREM 4.2.2. *Let f be continuous on $[a, b]$, and suppose that G is differentiable on $[a, b]$ with $G'(x) = f(x)$ for all x in $[a, b]$. Then*

$$\int_a^b f(x)\, dx = G(b) - G(a).$$

To prove Theorem 4.2.2 we simply let $F(x) = \int_a^x f(t)\, dt$ as in Theorem 4.2.1. We then have $F'(x) = f(x)$ by 4.2.1, while $G'(x) = f(x)$ by hypothesis. Hence $F'(x) - G'(x) = 0$, or in other words, the function $F - G$ has zero gradient throughout $[a, b]$. It follows at once from the mean value theorem [see Theorem 3.3.3] that $F(x) - G(x)$ must be constant on $[a, b]$:

$$F(x) - G(x) = k \quad \text{say.}$$

We find the value of k by putting $x = a$, whence $k = -G(a)$, since $F(a) = 0$. Hence $F(x) = G(x) - G(a)$, and so in particular

$$\int_a^b f(t)\, dt = F(b) = G(b) - G(a) \quad \text{as required.}$$

Both of Theorems 4.2.1 and 4.2.2 are referred to loosely as the *fundamental theorem of calculus*. Theorem 4.2.2 gives the usual rule for integrating a function f over $[a, b]$: first find a function (namely G) whose derivative is f—the value of the integral is then found by taking $G(b) - G(a)$.

   Such a function G is known as an *indefinite integral* of f and one writes $G(x) = \int f(x)\, dx$ without limits—it is determined only to within an additive constant. Notice that the x in '$\int f(x)\, dx$' is not a dummy variable, since the notation denotes a function $G(x)$, while $\int_a^b f(x)\, dx$ (a '*definite integral*') denotes a real number. When the lower limit of integration may be disregarded we use a notation like $\int^2 f(x)\, dx$ or $\int^y f(x)\, dx$.

EXAMPLE 4.2.1 $\int_1^4 x^2\, dx$.   Here $f(x) = x^2$, so a suitable choice for G is $G(x) = \frac{1}{3}x^3$, because $G'(x) = x^2 = f(x)$. Hence the value of the integral is

$$G(4) - G(1) = \frac{1}{3}(4^3 - 1^3) = \frac{1}{3}(63) = 21.$$

EXAMPLE 4.2.2 $\int_0^\pi \sin x\, dx$.   Here $f(x) = \sin x$, $G(x) = -\cos x$,   (since $\cos' = -\sin$), so the integral is equal to

$$G(\pi) - G(0) = -\cos \pi + \cos 0 = 1 + 1 = 2.$$

   It is convenient to use the notation $[G(x)]_a^b$, or $G(x)|_a^b$, to denote $G(b) - G(a)$, so that the above calculation would be written

$$\int_0^\pi \sin x\, dx = [-\cos x]_0^\pi = -\cos \pi + \cos 0 = 2.$$

Since each example of a function and its derivative gives, when reversed, a function and its indefinite integral, we may regard Table 3.2.1 as giving equally well a table of indefinite integrals. We list some examples in table 4.2.1; some of which are alternatives to those listed in Table 3.2.1.

Very extensive tables of definite integrals may be found for instance in Gradshteyn and Ryzhik (1965).

EXAMPLE 4.2.3

$$\int \frac{x^2 + 2x + 3}{x^2 + 5x + 6} \, dx.$$

We write the integrand in partial fractions [see I, § 14.10] as

$$1 - \frac{3x + 3}{x^2 + 5x + 6} = 1 - 3 \frac{x + 1}{(x + 2)(x + 3)} = 1 - 3\left(\frac{-1}{x + 2} + \frac{2}{x + 3}\right)$$

$$= 1 + \frac{3}{x + 2} - \frac{6}{x + 3}.$$

Hence

$$\int \frac{x^2 + 2x + 3}{x^2 + 5x + 6} \, dx = \int \left(1 + \frac{3}{x + 2} - \frac{6}{x + 3}\right) dx$$

$$= x + 3 \log (x + 2) - 6 \log (x + 3) + c.$$

Here the final '+c' is used to indicate that the indefinite integral is determined only to within an additive constant.

EXAMPLE 4.2.4

$$\int \frac{x^2 + 2x + 3}{x^2 + x + 1} \, dx.$$

Here

$$\frac{x^2 + 2x + 3}{x^2 + x + 1} = 1 + \frac{x + 2}{x^2 + x + 1} = 1 + \frac{1}{2} \frac{2x + 1}{x^2 + x + 1} + \frac{3}{2} \frac{1}{x^2 + x + 1},$$

where the middle term is chosen so that its numerator is the derivative of its denominator, and the whole fraction is then the derivative of $\log (x^2 + x + 1)$, as may be verified using the chain rule (3.2.8). In the last term we put $x^2 + x + 1 = (x + \frac{1}{2})^2 + \frac{3}{4}$, so that we can use the formula for the derivative of $\tan^{-1} x/a$, quoted in table (4.2.1) with $a^2 = \frac{3}{4}$. This gives

$$\int \frac{x^2 + 2x + 3}{x^2 + x + 1} \, dx = \int \left(1 + \frac{1}{2} \frac{2x + 1}{x^2 + x + 1} + \frac{3}{2} \frac{1}{(x + \frac{1}{2})^2 + \frac{3}{4}}\right) dx$$

$$= x + \frac{1}{2} \log (x^2 + x + 1) + \frac{3}{2} \frac{1}{\sqrt{\frac{3}{4}}} \tan^{-1} \left(\frac{x + \frac{1}{2}}{\sqrt{\frac{3}{4}}}\right) + c$$

$$= x + \frac{1}{2} \log (x^2 + x + 1) + \sqrt{3} \tan^{-1} \left(\frac{2x + 1}{\sqrt{3}}\right) + c.$$

$h(x)$, or $f'(x)$	$\int h(x)\, dx$, or $f(x)$	$h(x)$, or $f'(x)$	$\int h(x)\, dx$, or $f(x)$
x^m $(m = 0, 1, 2, \ldots)$	$\left.\begin{array}{c} \\ \\ \end{array}\right\} \dfrac{1}{m+1} x^{m+1}$	e^{ax} $(a \neq 0)$	$\dfrac{1}{a}\, e^{ax}$
x^m $(x \neq 0; m = -2, -3, \ldots)$		a^x $(a > 0, a \neq 1)$	$\dfrac{1}{\log a}\, a^x$
x^m $(x > 0; m$ non-integral$)$		$\log x$	$x \log x - x$
x^{-1}, $(x > 0)$	$\log x$	$(a^2 - x^2)^{-1/2}$, $(0 \leq \lvert x \rvert < a)$	$\sin^{-1}\left(\dfrac{x}{a}\right)$
$\sin x$	$-\cos x$	$(x^2 \pm a^2)^{-1/2}$	$\log\left(x + (x^2 \pm a^2)^{1/2}\right)$
$\cos x$	$\sin x$	$(a^2 + x^2)^{-1}$ $(a > 0)$	$\dfrac{1}{a}\tan^{-1}\left(\dfrac{x}{a}\right)$
$\tan x$, $(\cos x > 0)$	$-\log(\cos x)$	$(x^2 - a^2)^{-1}$, $(0 < a < x)$	$\dfrac{1}{2a}\log\left(\dfrac{x - a}{x + a}\right)$
$\cot x$, $(\sin x > 0)$	$\log(\sin x)$		
$\sec x$, $(\sec x + \tan x > 0)$	$\log(\sec x + \tan x)$		
$\cosec x$, $(\tan \frac{1}{2}x > 0)$	$\log(\tan \frac{1}{2}x)$		

Table 4.2.1

4.3. INTEGRATION BY PARTS AND SUBSTITUTION

We now consider the process of finding an indefinite integral for a given function in a more systematic way. Unfortunately there is no general procedure akin to that used to find derivatives in Chapter 3. Instead our principal weapons are two methods, namely integration by parts (Theorem 4.3.1 below) and by substitution, which are obtained from the product and chain rules for derivatives. These may be used separately, or combined together, as in the examples in the remainder of this section. It should be realized at once however that the search may not be successful: there are elementary functions (e^{x^2} is a popular example) which though continuous, and hence formally integrable in the sense of section 4.1, do not have indefinite integrals which are themselves elementary functions. This apparent weakness in fact leads us to define new special functions, as Chapter 10 shows.

We begin by considering the product rule, contained in Theorem 3.2.1. This reads $(f.g)' = f.'g + f.g'$, or in the notation of indefinite integrals,

$$\int (f'(x)g(x) + f(x)g'(x))\, dx = f(x)g(x) + c.$$

This may be re-written in either of the forms

$$\int f(x)g'(x)\, dx = f(x)g(x) - \int f'(x)g(x)\, dx$$

(with an indefinite integral on both sides of the equation, there is no longer any need for the '+c'), or

$$\int_a^b f(x)g'(x)\, dx = [f(x)g(x)]_a^b - \int_a^b f'(x)g(x)\, dx.$$

This gives us our first rule, for 'integration by parts':

THEOREM 4.3.1. *Let f, g have continuous derivatives on $[a, b]$. Then*

$$\int_a^b f(x)g'(x)\, dx = f(b)g(b) - f(a)g(a) - \int_a^b f'(x)g(x)\, dx. \qquad (4.3.1)$$

Successful use of the rule requires a judicious choice of functions f, g as in the following examples.

EXAMPLE 4.3.1 $\int_0^\pi x \cos x\, dx$. Here we choose $f(x) = x$, $g'(x) = \cos x$, so that $f'(x) = 1$, $g(x) = \sin x$. We obtain

$$\int_0^\pi x \cos x\, dx = [x \sin x]_0^\pi - \int_0^\pi 1 . \sin x\, dx$$

$$= 0 - 0 - \int_0^\pi \sin x\, dx = [\cos x]_0^\pi = \cos \pi - \cos 0 = -2.$$

The reader should check for himself what happens if we take $f(x) = \cos x$, $g'(x) = x$ in this example.

EXAMPLE 4.3.2 $\int x^\alpha \log x \, dx$.   First suppose that α may be any real number (positive or negative) except -1. We are looking for a function whose derivative is $x^\alpha \log x$ (for positive values of x since $\log x$ is defined only when $x > 0$). We take $f(x) = \log x$, $g'(x) = x^\alpha$, so that

$$f'(x) = \frac{1}{x}, \qquad g(x) = \frac{1}{\alpha+1} x^{\alpha+1}.$$

We obtain

$$\int x^\alpha \log x \, dx = \frac{1}{\alpha+1} x^{\alpha+1} \log x - \frac{1}{\alpha+1} \int x^{\alpha+1} \cdot \frac{1}{x} \, dx$$

$$= \frac{1}{\alpha+1} x^{\alpha+1} \log x - \frac{1}{\alpha+1} \int x^\alpha \, dx$$

$$= \frac{1}{\alpha+1} x^{\alpha+1} \log x - \frac{1}{(\alpha+1)^2} x^{\alpha+1} + c.$$

In particular, when $\alpha = 0$, we get the standard formula

$$\int \log x \, dx = x \log x - x + c. \tag{4.3.2}$$

In the excluded case, when $\alpha = -1$, we take

$$f(x) = \log x, \qquad g'(x) = \frac{1}{x}, \qquad g(x) = \log x.$$

This gives

$$\int \frac{1}{x} \log x \, dx = (\log x)^2 - \int \frac{1}{x} \log x \, dx,$$

and on combining the integrals we obtain that

$$\int \frac{1}{x} \log x \, dx = \tfrac{1}{2}(\log x)^2 + c. \tag{4.3.3}$$

EXAMPLE 4.3.3.   Let

$$I_n = \int_0^1 x^\beta (\log x)^n \, dx \tag{4.3.4}$$

where β is a real number such that $\beta > -1$ and where n is a non-negative integer. When $n = 0$ we have that

$$I_0 = \int_0^1 x^\beta \, dx = 1/(\beta+1).$$

Assuming that $n > 0$ we can establish a relation between I_n and I_{n-1} as follows: apply integration by parts to (4.3.4) by choosing

$$f(x) = (\log x)^n, \qquad g'(x) = x^\beta, \qquad g(x) = x^{\beta+1}/(\beta+1).$$

Then

$$I_n = \left[\frac{1}{\beta+1} x^{\beta+1} (\log x)^n \right]_0^1 - \frac{n}{\beta+1} \int_0^1 x^{\beta+1} (\log x)^{n-1} \frac{1}{x} \, dx,$$

$$I_n = -\frac{n}{\beta+1} \int_0^1 x^\beta (\log x)^{n-1} \, dx,$$

because $\log 1 = 0$ and

$$\lim_{x \to 0} x^{\beta+1} (\log x)^n = 0,$$

[see (2.11.17)]. Thus

$$I_n = -\frac{n}{\beta+1} I_{n-1}.$$

If $n \geq 2$, the procedure can be repeated so that

$$I_n = -\frac{n}{\beta+1} I_{n-1} = \frac{n(n-1)}{(\beta+1)^2} I_{n-2},$$

and so on until we reach the equation

$$I_n = (-1)^n \frac{n!}{(\beta+1)^n} I_0.$$

Hence finally

$$I_n = (-1)^n \frac{n!}{(\beta+1)^{n+1}}.$$

EXAMPLE 4.3.4 $\int_0^1 \tan^{-1} x \, dx$. Here there is no apparent way of writing the integrand as a product—we use the standard device of taking $f(x) = \tan^{-1} x$, $g'(x) = 1$, so that $f'(x) = 1/(1+x^2)$, $g(x) = x$. This gives

$$\int_0^1 \tan^{-1} x \, dx = [x \tan^{-1} x]_0^1 - \int_0^1 \frac{x}{1+x^2} \, dx.$$

We see by inspection that $x/(1+x^2)$ is the derivative of $\frac{1}{2} \log (1+x^2)$ so that we get

$$1 \cdot \frac{\pi}{4} - 0 - [\tfrac{1}{2} \log (1+x^2)]_0^1 = \frac{\pi}{4} - \tfrac{1}{2} \log 2$$

for the value of the integral.

EXAMPLE 4.3.5

$$\int e^{ax} \cos bx \, dx = (a^2 + b^2)^{-1} e^{ax} (a \cos bx + b \sin bx) + c$$

$$\int e^{ax} \sin bx \, dx = (a^2 + b^2)^{-1} e^{ax} (a \sin bx - b \cos bx) + c.$$

We consider the first integral, and integrate twice by parts to get

$$\int e^{ax} \cos bx \, dx = \frac{1}{b} e^{ax} \sin bx - \frac{a}{b} \int e^{ax} \sin bx \, dx$$

$$= \frac{1}{b} e^{ax} \sin bx + \frac{a}{b^2} e^{ax} \cos bx - \frac{a^2}{b^2} \int e^{ax} \cos bx \, dx,$$

from which we obtain

$$\left(1 + \frac{a^2}{b^2}\right) \int e^{ax} \cos bx \, dx = \frac{1}{b^2} e^{ax} (b \sin bx + a \cos bx) + c,$$

or

$$\int e^{ax} \cos bx \, dx = (a^2 + b^2)^{-1} e^{ax} (a \cos bx + b \sin bx) + c.$$

The other result is proved similarly. A quick alternative method for those familiar with complex numbers is to write $I = \int e^{ax} \cos bx \, dx$, $J = \int e^{ax} \sin bx \, dx$, so that [see (9.5.8)]

$$I + iJ = \int e^{ax} e^{ibx} \, dx = \int e^{(a+ib)x} \, dx = (a+ib)^{-1} e^{(a+ib)x} + c$$

$$= (a^2 + b^2)^{-1}(a - ib) e^{ax} (\cos bx + i \sin bx) + c,$$

and the result follows on multiplying out and equating real and imaginary parts [see I, § 2.7.1].

The second rule is obtained from the chain rule for differentiation of a composite function [see (3.2.8)]. This states that if f, g are differentiable functions for which $f \circ g$ is defined [see I, § 1.4.2], then $(f \circ g)' = (f' \circ g) \cdot g'$, or in the notation of indefinite integrals,

$$\int f'(g(x))g'(x) \, dx = f(g(x)) + c.$$

For definite integrals we obtain

$$\int_a^b f'(g(x))g'(x) \, dx = [f(g(x))]_a^b = f(g(b)) - f(g(a)),$$

which may also be written as

$$[f(y)]_{g(a)}^{g(b)} = \int_{g(a)}^{g(b)} f'(y)\, dy.$$

If we replace f' by another function h, we obtain the following result, known as the rule for *integration by substitution* or the *change of variable* rule.

THEOREM 4.3.2.   *Let g have a continuous derivative on $[a, b]$, and let h be a continuous function for which $h(g(x))$ is defined. Then*

$$\int_a^b h(g(x))g'(x)\, dx = \int_{g(a)}^{g(b)} h(y)\, dy.$$

In order to apply the rule successfully we must first find a suitable function g. Informally we may think of this as putting $y = g(x)$. We then differentiate to get g'—again informally this appears as putting $dy = g'(x)\, dx$—see the note on Leibniz notation below. The final step is to make the appropriate change in the limits.

EXAMPLE 4.3.6.   In Example 4.3.2 we came across the indefinite integral $\int (1/x) \log x\, dx$. This may also be dealt with by substitution, as follows. Consider $\int_1^3 (1/x) \log x\, dx$,

Let $g(x) = \log x$, so that $g'(x) = 1/x$. Then if $h(y) = y$, $h(\log x) = \log x$, and so

$$\int_1^3 \frac{1}{x} \log x\, dx = \int_1^3 h(g(x))g'(x)\, dx = \int_{g(1)}^{g(3)} h(y)\, dy = \int_0^{\log 3} y\, dy$$

$$= [\tfrac{1}{2}y^2]_0^{\log 3} = \tfrac{1}{2}(\log 3)^2.$$

Note.   One may also use the Leibniz notation [see § 3.1] as follows: Let $y = \log x$, so that $dy/dx = 1/x$. Then $\int (1/x) \log x\, dx$ becomes

$$\int y\, dy = \tfrac{1}{2}y^2 + c = \tfrac{1}{2}(\log x)^2 + c,$$

or as a definite integral,

$$\int_{x=1}^{x=3} \frac{1}{x} \log x\, dx = [\tfrac{1}{2}y^2]_{\log 1}^{\log 3} = \tfrac{1}{2}(\log 3)^2.$$

EXAMPLE 4.3.7 $\int_0^\pi \cos x/(1 + \sin^2 x)\, dx$.   Let $g(x) = \sin x$, $g'(x) = \cos x$, $h(y) = 1/(1 + y^2)$. Then $g(0) = 0$, $g(\pi/2) = 1$, and $h(g(x)) = 1/(1 + \sin^2 x)$. It follows that

$$\int_0^{\pi/2} \frac{\cos x\, dx}{1 + \sin^2 x} = \int_0^1 \frac{dy}{1 + y^2} = [\tan^{-1} y]_0^1 = \frac{\pi}{4}.$$

EXAMPLE 4.3.8 $\int_0^1 x^2/\sqrt{1+x}\,dx$. Let $g(x) = 1+x$, $g'(x) = 1$, so that $g(0) = 1$, $g(1) = 2$. Then $x^2 = (g(x)-1)^2$, so we write the integral as

$$\int_0^1 \frac{(g(x)-1)^2}{\sqrt{g(x)}} g'(x)\,dx = \int_1^2 \frac{(y-1)^2}{\sqrt{y}}\,dy,$$

on putting $y = g(x)$. This latter integral may be evaluated by expanding $(y-1)^2 = y^2 - 2y + 1$, so that we get

$$\int_1^2 (y^{3/2} - 2y^{1/2} + y^{-1/2})\,dy = [\tfrac{2}{5}y^{5/2} - \tfrac{4}{3}y^{3/2} + 2y^{1/2}]_1^2$$

$$= \sqrt{2}(\tfrac{2}{5}.2^2 - \tfrac{4}{3}.2 + 2) - (\tfrac{2}{5} - \tfrac{4}{3} + 2)$$

$$= \frac{14\sqrt{2}}{15} - \frac{16}{15},$$

on inserting the value of the limits.

EXAMPLE 4.3.9 $\int_1^3 x\,e^{-x^2}\,dx$. Put $y = g(x) = x^2$, so that $g'(x) = 2x$. We get

$$\tfrac{1}{2}\int_1^3 e^{-x^2}.2x\,dx = \tfrac{1}{2}\int_1^9 e^{-y}\,dy = \tfrac{1}{2}(e^{-1} - e^{-9}).$$

We may sometimes use the rule 'in reverse', that is, proceeding from $\int_{g(a)}^{g(b)} h(g)\,dy$ to $\int_a^b h(g(x)).g'(x)\,dx$, as in the next example.

EXAMPLE 4.3.10 $\int_0^{1/2} dt/(1-t^2)^{3/2}$. Here $h(t) = 1/(1-t^2)^{3/2}$, and we take $t = g(x) = \sin x$, to exploit the fact that $1 - \sin^2 x = \cos^2 x$ [see (2.12.5)]. The limits 0, $\tfrac{1}{2}$ are $\sin 0$, and $\sin \pi/6$ respectively so that with $a = 0$, $b = \pi/6$, we get

$$\int_0^{1/2} \frac{dt}{(1-t^2)^{3/2}} = \int_0^{\pi/6} \frac{\cos x\,dx}{(1-\sin^2 x)^{3/2}} = \int_0^{\pi/6} \frac{\cos x\,dx}{\cos^3 x} = \int_0^{\pi/6} \sec^2 x\,dx$$

$$= [\tan x]_0^{\pi/6} = \frac{1}{\sqrt{3}}.$$

Notice that the requirement in Theorem 4.3.2 that $h(g(x))$ should be defined for x in $[a, b]$ prevents us from taking, for instance, $a = -\pi$, $b = \pi/6$ here, despite the fact that $\sin(-\pi) = \sin 0 = 0$. For when x is in $[-\pi, \pi/6]$, $\sin x$ takes all values in $[-1, \tfrac{1}{2}]$ (in particular $\sin(-\pi/2) = -1$, so that $h(g(x)) = 1/(1-\sin^2 x)^{3/2}$ is not defined when $x = -\pi/2$.

EXAMPLE 4.3.11. The substitution $t = g(x) = \tan \tfrac{1}{2}x$ is often used to reduce the integration of trigonometric functions to the integration of rational functions [see I, § 14.9]. When $\tan \frac{1}{2}x = t$, we find that

$$\sin x = \frac{2 \sin \frac{1}{2}x \cos \frac{1}{2}x}{\sin^2 \frac{1}{2}x + \cos^2 \frac{1}{2}x} = \frac{2 \tan \frac{1}{2}x}{1 + \tan^2 \frac{1}{2}x} = \frac{2t}{1 + t^2},$$

$$\cos x = \frac{\cos^2 \frac{1}{2}x - \sin^2 \frac{1}{2}x}{\cos^2 \frac{1}{2}x + \sin^2 \frac{1}{2}x} = \frac{1 - t^2}{1 + t^2},$$

and

$$\frac{dt}{dx} = g'(x) = \frac{1}{2} \sec^2 \frac{1}{2}x = \frac{1}{2}(1 + \tan^2 \frac{1}{2}x) = \frac{1}{2}(1 + t^2).$$

For instance we obtain

$$\int_0^{1/2\pi} \frac{dx}{2 + \cos x} = \int_0^1 \frac{1}{2 + (1 - t^2)/(1 + t^2)} \cdot \frac{2dt}{1 + t^2},$$

since $\tan (\frac{1}{2} \cdot \pi/2) = \tan \pi/4 = 1$. This latter integral is equal to

$$\int_0^1 \frac{2dt}{2(1 + t^2) + (1 - t^2)} = \int_0^1 \frac{2dt}{3 + t^2} = \left[\frac{2}{\sqrt{3}} \tan^{-1} \frac{t}{\sqrt{3}} \right]_0^1$$

$$= \frac{2}{\sqrt{3}} \tan^{-1} \left(\frac{1}{\sqrt{3}} \right) = \frac{2}{\sqrt{3}} \cdot \frac{\pi}{6} = \frac{\pi}{3\sqrt{3}}.$$

EXAMPLE 4.3.12.   If m and n are two positive integers then

(i) $\qquad \displaystyle\int_0^1 \sin 2\pi nx \sin 2\pi mx \, dx = \begin{cases} 0 & \text{if } m \neq n \\ \frac{1}{2} & \text{if } m = n \end{cases}$

(ii) $\qquad \displaystyle\int_0^1 \cos 2\pi nx \cos 2\pi mx \, dx = \begin{cases} 0 & \text{if } m \neq n \\ \frac{1}{2} & \text{if } m = n \end{cases}$

(iii) $\qquad \displaystyle\int_0^1 \sin 2\pi nx \cos 2\pi mx \, dx = 0.$

Proof.   (i) By virtue of the trigonometric formula [see V (1.2.19)]

$$\sin \alpha \sin \beta = \frac{1}{2}\{\cos (\alpha - \beta) - \cos (\alpha + \beta)\}$$

we can write the integral as

$$\frac{1}{2} \int_0^1 \cos 2\pi (m - n)x \, dx - \frac{1}{2} \int_0^1 \cos 2\pi (m + n)x \, dx. \qquad (4.3.2)$$

But

$$\int_0^1 \cos 2\pi kx \, dx = \frac{1}{2\pi k} [\sin 2\pi kx]_0^1 = 0$$

for any non-zero integer k, so both terms in (4.3.2) are zero unless $m = n$, in which case the value of the integral is equal to

$$\tfrac{1}{2} \int_0^1 \cos 0 \; dx = \tfrac{1}{2}.$$

(ii) Since [see V (1.2.19)]

$$\cos \alpha \cos \beta = \tfrac{1}{2}\{\cos (\alpha + \beta) + \cos (\alpha - \beta)\}$$

the integral may be written as

$$\tfrac{1}{2} \int_0^1 \cos 2\pi(m + \eta)x \; dx + \tfrac{1}{2} \int_0^1 \cos 2\pi(m - n)x \; dx.$$

Again this is clearly zero unless $m = n$, in which case it takes the value $\tfrac{1}{2}$.
(iii) Since

$$\sin \alpha \cos \beta = \tfrac{1}{2}\{\sin (\alpha + \beta) + \sin (\alpha - \beta)\}$$

[cf. V (1.2.17)] the integral takes the form

$$\tfrac{1}{2} \int_0^1 \sin 2\pi(m + n)x \; dx + \tfrac{1}{2} \int_0^1 \sin 2\pi(n - m)x \; dx. \qquad (4.3.3)$$

But for any non-zero integer k we have

$$\int_0^1 \sin 2\pi k x = -\frac{1}{2\pi k}[\cos 2\pi k x]_0^1 = 0$$

and so both terms in (4.3.3) are zero provided that $m \neq n$. When $m = n$ it takes the value

$$\tfrac{1}{2} \int_0^1 \sin 0 \; dx = 0.$$

The substitution $y = 2\pi x$ in (i)–(iii) yields the formulae (iv)–(vi) respectively:
For positive integers m *and* n we have

(iv)
$$\int_0^{2\pi} \sin ny \sin my \; dy = \begin{cases} 0 & \text{if } m \neq n \\ \pi & \text{if } m = n \end{cases}$$

(v)
$$\int_0^{2\pi} \cos ny \cos my \; dy = \begin{cases} 0 & \text{if } m \neq n \\ \pi & \text{if } m = n \end{cases}$$

(vi)
$$\int_0^{2\pi} \sin ny \cos my \; dy = 0.$$

Finally we mention three standard types of integral which occur frequently enough to be worth special mention:

EXAMPLE 4.3.13

$$\int \frac{f'(x)\, dx}{f(x)} = \log\left(f(x)\right) + c, \quad \text{when } f(x) > 0$$

[compare the entries for tan and cot in Table 4.2.1].

EXAMPLE 4.3.14. $\int R(x)\, dx$, where $R(x) = P(x)/Q(x)$ is a quotient of the polynomials [see I, § 14.9]. In this case R is decomposed into partial fractions [see I, § 14.10], that is, a sum of terms of the form $A/(ax+b)$ or $(Ax+B)/ax^2 + bx + c)$, plus a polynomial if the degree of P is at least as great as that of Q. These are then integrated as in Example 4.2.3.

EXAMPLE 4.3.15. $\int R(\sin x, \cos x)\, dx$, where R is a rational function [see I, § 14.9] in sin and cos. The substitution $t = \tan \frac{1}{2}x$ [cf. Example 4.3.11] reduces this to a rational function of t, and hence to Example 4.3.14 above.

4.4. REDUCTION FORMULAE

Suppose that we have to find the value of an integral which depends on an integer parameter, as in the examples $\int_0^{\pi/2} \sin^n x\, dx$, or $\int_0^{\pi/4} \tan^n x\, dx$, $n = 0, 1, 2, \ldots$. It is often possible to express the value of the integral, I_n say, in terms of I_{n-1}, or I_{n-2}, or a linear combination of both; hence knowing the initial values I_0, I_1 we can find a formula for I_n. One such case was already encountered in Example 4.3.3. The following examples illustrate some of the other possibilities.

EXAMPLE 4.4.1

$$\int_0^{\pi/2} \sin^n x\, dx = \int_0^{\pi/2} \cos^n x\, dx = \begin{cases} \dfrac{(n-1)(n-3)\ldots 5 . 3 . 1}{n(n-2)\ldots 6 . 4 . 2}\dfrac{\pi}{2} & \text{if } n \text{ is even,} \\[4mm] \dfrac{(n-1)(n-3)\ldots 4 . 2}{n(n-2)\ldots 5 . 3} & \text{if } n \text{ is odd.} \end{cases}$$

Proof. Let $I_n = \int_0^{\pi/2} \sin^n x\, dx$. If we put $y = (\pi/2) - x$, and use Theorem 4.3.2, we obtain

$$I_n = \int_0^{\pi/2} (\cos y)^n\, dy = \int_0^{\pi/2} (\cos x)^n\, dx.$$

If $n = 0$, $I_0 = \int_0^{\pi/2} dx = \pi/2$, while if $n = 1$

$$I_1 = \int_0^{\pi/2} \sin x\, dx = [-\cos x]_0^{\pi/2} = 0 - (-1) = 1.$$

Suppose then that $n \geq 2$. We have $I_n = \int_0^{\pi/2} \sin^{n-1} x \cdot \sin x \, dx$, and we integrate by parts, using $f(x) = \sin^{n-1} x$, $g'(x) = \sin x$, so that $g(x) = -\cos x$, and $f'(x) = (n-1)\sin^{n-2} x \cdot \cos x$. We get

$$I_n = \int_0^{\pi/2} f(x)g'(x) \, dx = [f(x) \cdot g(x)]_0^{\pi/2} - \int_0^{\pi/2} f'(x) \cdot g(x) \, dx$$

$$= [\sin^{n-1} x(-\cos x)]_0^{\pi/2} - \int_0^{\pi/2} (n-1)\sin^{n-2} x \cos x(-\cos x) \, dx.$$

Since $n \geq 2$, $n - 1 \geq 1$ and the integrated term vanishes at $x = 0$ and $x = \pi/2$. We are left with

$$I_n = (n-1)\int_0^{\pi/2} \sin^{n-2} x \cos^2 x \, dx$$

$$= (n-1)\int_0^{\pi/2} \sin^{n-2} x(1 - \sin^2 x) \, dx = (n-1)\int_0^{\pi/2} (\sin^{n-2} x - \sin^n x) \, dx$$

$$= (n-1)(I_{n-2} - I_n).$$

This gives $(1 + (n-1))I_n = (n-1)I_{n-2}$, or

$$I_n = \frac{n-1}{n} I_{n-2}$$

which is the desired recurrence relation.

We may apply it repeatedly to obtain

$$I_n = \frac{(n-1)}{n} I_{n-2} = \frac{(n-1)(n-3)}{n(n-2)} I_{n-4}$$

$$= \ldots = \frac{(n-1)(n-3)\ldots(n-2r+1)}{n(n-2)\ldots(n-2r+2)} I_{n-2r}.$$

If n is even we take $2r = n$, and obtain

$$I_n = \frac{(n-1)(n-3)\ldots 3 \cdot 1}{n(n-2)\ldots 4 \cdot 2} I_0 = \frac{(n-1)(n-3)\ldots 3 \cdot 1}{n(n-2)\ldots 4 \cdot 2} \cdot \frac{\pi}{2}$$

as required. If n is odd we take $2r = n - 1$, and obtain

$$I_n = \frac{(n-1)(n-3)\ldots 4 \cdot 2}{n(n-2)\ldots 5 \cdot 3} I_1 = \frac{(n-1)(n-3)\ldots 4 \cdot 2}{n(n-2)\ldots 5 \cdot 3}.$$

Thus for instance

$$I_7 = \frac{6 \cdot 4 \cdot 2}{7 \cdot 5 \cdot 3} = \frac{16}{35}, \quad \text{and} \quad I_8 = \frac{7 \cdot 5 \cdot 3 \cdot 1}{8 \cdot 6 \cdot 4 \cdot 2} \cdot \frac{\pi}{2} = \frac{35\pi}{256}.$$

EXAMPLE 4.4.2. $\int_0^{\pi/2} \sin^m x \cos^n x \, dx = I_{m,n}$ satisfies the reduction formula

$$I_{m,n} = \frac{m-1}{m+n} I_{m-2,n}$$

if $m \ge 2$. For

$$I_{m,n} = \int_0^{\pi/2} \sin^{m-1} x (\cos^n x \sin x) \, dx$$

$$= \left[\sin^{m-1} x \left(\frac{-1}{n+1} \cos^{n+1} x \right) \right]_0^{\pi/2}$$

$$+ \frac{1}{n+1} \int_0^{\pi/2} (m-1) \sin^{m-2} x \cos x (\cos x)^{n+1} \, dx,$$

on integrating by parts,

$$= \frac{m-1}{n+1} \int_0^{\pi/2} \sin^{m-2} x \cos^n x (1 - \sin^2 x) \, dx = \frac{m-1}{n+1} (I_{m-2,n} - I_{m,n}).$$

This may be rearranged to give

$$\left(1 + \frac{m-1}{n+1} \right) I_{m,n} = \frac{m-1}{n+1} I_{m-2,n} \quad \text{or} \quad I_{m,n} = \frac{m-1}{m+n} I_{m-2,n}$$

as required. By means of this relation one of the indices (say m) may be reduced either to 1 or to zero. In the first case we have an integral of the form

$$\int_0^{\pi/2} \sin x \cos^n x \, dx = \left[-\frac{1}{n+1} \cos^{n+1} x \right]_0^{\pi/2} = \frac{1}{n+1};$$

in the second we have $\int_0^{\pi/2} \cos^n x \, dx$ whose value is known from Example 4.4.1. It is possible to write down a general formula for $I_{m,n}$ but the different cases make it rather cumbersome.

EXAMPLE 4.4.3. $I_n = \int_0^{\pi/4} \tan^n x \, dx$. Here $I_0 = \int_0^{\pi/4} dx = \pi/4$, while

$$I_1 = \int_0^{\pi/4} \tan x \, dx = [-\log (\cos x)]_0^{\pi/4}$$

$$= -\log \cos \frac{\pi}{4} + \log \cos 0$$

$$= -\log \frac{1}{\sqrt{2}} + \log 1 = \tfrac{1}{2} \log 2.$$

If $n \geq 2$, we may write

$$I_n = \int_0^{\pi/4} \tan^{n-2} x (\tan^2 x) \, dx$$

$$= \int_0^{\pi/4} \tan^{n-2} x (\sec^2 x - 1) \, dx$$

$$= \left[\frac{1}{n-1} \tan^{n-1} x \right]_0^{\pi/4} - \int_0^{\pi/4} \tan^{n-2} x \, dx$$

$$= \frac{1}{n-1} - I_{n-2}.$$

If we apply this repeatedly we obtain

$$I_n = \frac{1}{n-1} - I_{n-2} = \frac{1}{n-1} - \left(\frac{1}{n-3} - I_{n-4} \right)$$

$$= \frac{1}{n-1} - \frac{1}{n-3} + \dots + \frac{(-1)^{r-1}}{n-2r+1} + (-1)^r I_{n-2r}$$

If n is even this results in

$$I_n = \frac{1}{n-1} - \frac{1}{n-3} + \dots + \frac{(-1)^{(n/2)-1}}{1} + (-1)^{(n/2)} I_0,$$

so that I_n gives the difference between $I_0 = \pi/4$ and the partial sums of the series

$$1 - \tfrac{1}{3} + \tfrac{1}{5} - \dots + \frac{(-1)^k}{2k+1}.$$

Similarly if n is odd, I_n is the difference between $I_1 = \tfrac{1}{2}\log 2$ and the partial sums of the series

$$\tfrac{1}{2} - \tfrac{1}{4} + \tfrac{1}{6} - \dots + \frac{(-1)^{k-1}}{2k}$$

[see also Example 4.7.1].

Example 4.4.1 may be used to derive *Wallis' formula*

$$\frac{\pi}{2} = \frac{2^2}{1 \cdot 3} \cdot \frac{4^2}{3 \cdot 5} \cdot \frac{6^2}{5 \cdot 7} \cdot \frac{8^2}{7 \cdot 9} \dots \tag{4.4.1}$$

For, putting $s_n = \int_0^{\pi/2} \sin^n dx$, Example 4.4.1 gives

$$\frac{s_{2m}}{s_{2m-1}} = \frac{(2m-1)^2 (2m-3)^2 \dots 5^2 \cdot 3^2}{2m(2m-2)^2 (2m-4)^2 \dots 4^2 \cdot 2^2} \frac{\pi}{2}$$

and so

$$\lim_{m \to \infty} \frac{2ms_{2m}}{(2m-1)s_{2m-1}} = \lim_{m \to \infty} \frac{(2m-1)(2m-3)^2(2m-5)^2 \ldots 5^2 . 3^2}{(2m-2)^2(2m-4)^2 \ldots 4^2 . 2^2} \frac{\pi}{2}. \quad (4.4.2)$$

Since $\sin^n x$ is non-negative and decreasing with n in $[0, \pi/2]$ we have

$$1 \geq \frac{s_{2m}}{s_{2m-1}} \geq \frac{s_{2m}}{s_{2m-2}} = \frac{2m}{2m-1}$$

and so

$$\frac{2m}{2m-1} \geq \frac{2ms_{2m}}{(2m-1)s_{2m-1}} \geq 1.$$

Thus the limit on the left-hand side of (4.4.2) is 1 and

$$\frac{\pi}{2} = \lim_{m \to \infty} \left\{ \frac{2^2}{1.3} \cdot \frac{4^2}{3.5} \cdot \frac{6^2}{5.7} \cdots \frac{(2m-2)^2}{(2m-3)(2m-1)} \right\}$$

which gives (4.4.1).

4.5. MEAN VALUE THEOREMS

Very often it is not possible to evaluate an integral explicitly, and in this case one must either evaluate it numerically as is done in Chapter 7, Volume III, or one can make an estimate concerning its size, as we show in the following results.

THEOREM 4.5.1 (*The First Mean Value Theorem of integral calculus*)

(i) *Let f, g be integrable on* $[a, b]$, $m \leq f(x) \leq M$, *and* $g(x) \geq 0$ *for all x in* $[a, b]$. *Then*

$$m \int_a^b g(x)\, dx \leq \int_a^b f(x)g(x)\, dx \leq M \int_a^b g(x)\, dx.$$

(ii) *If f is continuous and* $g(x) \geq 0$ *on* $[a, b]$ *then for some point* x_1 *in* $[a, b]$, *we have*

$$\int_a^b f(x)g(x)\, dx = f(x_1) \int_a^b g(x)\, dx.$$

Proof. (i) is immediate if we integrate the inequalities

$$mg(x) \leq f(x)g(x) \leq Mg(x)$$

which are valid at all points of $[a, b]$ by the given restrictions on f, g. For (ii), we choose for m and M the least and greatest values which f attains in $[a, b]$, and notice that (i) says that

$$\frac{\int_a^b f(x)g(x)\, dx}{\int_a^b g(x)\, dx}$$

lies between these values and is hence equal to $f(x_1)$ for some x_1 in $[a, b]$.

THEOREM 4.5.2   (*The Second Mean Value Theorem*).   *Let f be monotone (increasing or decreasing) and g be integrable on $[a, b]$. Then for some x_1 in $[a, b]$*

$$\int_a^b f(x)g(x)\, dx = f(a) \int_a^{x_1} g(t)\, dt - f(b) \int_{x_1}^b g(t)\, dt.$$

For a proof see, for example, Bartle (1976).

The second mean value theorem is most often used in one or other of the two following forms:

COROLLARY 4.5.3

(i) *Let f be positive and decreasing, and g be integrable on $[a, b]$. Then for some x_1 in $[a, b]$,*

$$\int_a^b f(x)g(x)\, dx = f(a) \int_a^{x_1} g(t)\, dt.$$

(ii) *Let f be positive and increasing and g be integrable on $[a, b]$. Then for some x_1 in $[a, b]$,*

$$\int_a^b f(x)g(x)\, dx = f(b) \int_{x_1}^b g(t)\, dt.$$

For the proof of (i), for example, we replace f by f_1 where $f_1(x) = f(x)$ if $a \le x < b$, while $f_1(b) = 0$. This has the effect of removing the third term in Theorem 4.5.2 while leaving the others unaltered and preserving the decreasing character of f.

EXAMPLE 4.5.1.   Let $I = \int_a^b \sin x / x^\alpha\, dx$, where $0 < a < b$, and $\alpha > 0$. Then

$$|I| \le \frac{2}{a^\alpha}$$

For $f(x) = 1/x^\alpha$ is positive and decreasing on $[a, b]$ and $g(x) = \sin x$ is integrable there. Hence by Corollary 4.5.3(i),

$$I = f(a) \int_a^{x_1} \sin t\, dt = \frac{1}{a^\alpha} \int_a^{x_1} \sin t\, dt,$$

and the result follows since $\left| \int_a^{x_1} \sin t\, dt \right| = |\cos a - \cos x_1| \le 2.$

4.6. IMPROPER INTEGRALS

So far we have been considering integrals of the form $\int_a^b f(x)\,dx$, where $[a, b]$ is a finite interval on the real line, and f is a bounded function on $[a, b]$ for which the limiting process outlined in Definition 4.1.3 is successful. We shall now see in what circumstances these restrictions may be removed.

Consider for example the integral $\int_1^A 1/x^2\,dx$ where A is a real number > 1. This has the value $[-1/x]_1^A = 1 - 1/A$: as A becomes large this value approaches 1. More generally, if we consider $\int_1^A 1/x^\alpha\,dx$, where α is any real number > 1, this has the value

$$\left[\frac{1}{1-\alpha} x^{1-\alpha}\right]_1^A = \frac{1}{\alpha - 1}(1 - A^{1-\alpha}),$$

and since $1 - \alpha$ is negative, this approaches $1/(\alpha - 1)$ as A becomes large. This motivates the following definition.

DEFINITION 4.6.1(i). Let a be a real number and f a function which is integrable over each interval $[a, A]$, and for which $\int_a^A f(x)\,dx$ approaches a finite limit L as A increases indefinitely [see § 2.3]. Then we say that $\int_a^\infty f(x)\,dx$ exists and is equal to L, and that f is (*improperly Riemann-*) *integrable* over $[a, \infty)$.

There are similar definitions for $\int_{-\infty}^a f(x)\,dx$ and $\int_{-\infty}^\infty f(x)\,dx$—see Definition 4.6.1(ii) and (iii) below. Notice that $\int_a^\infty f(x)\,dx$ is defined as the result of a double limiting process: firstly for any $A > a$ we have the limits of sums which define $\int_a^A f$ as in section 4.1, and secondly we take a limit as A increases indefinitely: we do *not* introduce a 'dissection of $[a, \infty)$' whatever that might be!

Our discussion preceding Definition 4.6.1 also established the first part of the following result:

EXAMPLE 4.6.1. If $\alpha > 1$, the integral $\int_1^\infty x^{-\alpha}\,dx$ exists, and has the value $(\alpha - 1)^{-1}$.

If $\alpha \le 1$, the integral $\int_1^\infty x^{-\alpha}\,dx$ does not exist. For if $\alpha = 1$, $\int_1^A x^{-1}\,dx = \log A$ which increases indefinitely with A [see § 2.11], while if $\alpha < 1$, $x^{-\alpha} \ge x^{-1}$ on $[1, A]$, so that the integral $\int_1^A x^{-\alpha}\,dx$ cannot exist for these values either [see the Comparison Test, Theorem 4.6.2 below].

DEFINITION 4.6.1(ii). Let a be a real number, and f a function which is integrable over each interval $[B, a]$ with $B < a$, and for which $\int_B^a f(x)\,dx$ approaches a finite limit M as B decreases indefinitely. Then we say that the integral $\int_{-\infty}^a f(x)\,dx$ exists and has the value M, and that f is (*improperly Riemann-*) *integrable* over $(-\infty, a]$.

DEFINITION 4.6.1(iii).   Let f be a function which is integrable over every interval $[a, b]$ of the real line and for which $\int_a^b f(x)\, dx$ approaches a limit N as both b increases indefinitely and a decreases indefinitely. Then we say that the integral $\int_{-\infty}^{\infty} f(x)\, dx$ exists and has the value N, and that f is (*improperly Riemann-*) *integrable* over $(-\infty, \infty)$. Equivalently we could require that for any real number α, both $\int_{\alpha}^{\infty} f(x)\, dx$ and $\int_{-\infty}^{\alpha} f(x)\, dx$ should exist, and then $\int_{-\infty}^{\infty} f(x)\, dx = \int_{-\infty}^{\alpha} f(x)\, dx + \int_{\alpha}^{\infty} f(x)\, dx$.

EXAMPLE 4.6.2.   Let k be a positive real number. Then $\int_a^{\infty} e^{-kx}\, dx$ exists for any real number a, and has the value $(1/k)\, e^{-ka}$. For

$$\int_a^b e^{-kx}\, dx = -\frac{1}{k} [e^{-kx}]_a^b = \frac{1}{k} [e^{-ka} - e^{-kb}],$$

and this approaches $1/k\, e^{-ka}$ as b increases indefinitely.

EXAMPLE 4.6.3.   Let k be a positive real number, and a be any real number. Then

$$\int_0^{\infty} e^{-kx} \cos ax\, dx = \frac{k}{a^2 + k^2},$$

$$\int_0^{\infty} e^{-kx} \sin ax\, dx = \frac{a}{k^2 + a^2}.$$

From Example 4.3.5 we have

$$\int_0^B e^{-kx} \cos ax\, dx = \frac{1}{a^2 + k^2} [e^{-kx}(-k \cos ax + a \sin ax)]_0^B$$

$$= \frac{1}{a^2 + k^2} [k + e^{-Bx}(-k \cos aB + a \sin aB)],$$

which approaches $k/(a^2 + k^2)$ when B increases indefinitely, since $e^{-Bx} \to 0$ and $\cos aB$, $\sin aB$ remain bounded. The other result is proved in the same way.

Most of our results on convergence of integrals may be deduced from the following theorem.

THEOREM 4.6.1   (*The General Principle of Convergence*).   *A necessary and sufficient condition for the existence of $\int_a^{\infty} f(x)\, dx$ is that $\int_A^B f(x)\, dx$ may be made as small as required, by making A and B large enough.*

(In the formal language of limits introduced in § 2.3, this says that for any positive quantity, denoted ε, we may find some real number $x_1 > a$ for which $|\int_A^B f(x)\, dx| < \varepsilon$ whenever $x_1 < A < B$.)

We shall not prove this result, but we shall use it to deduce the following useful corollary.

COROLLARY 4.6.2 (*The Comparison Test*). *Let f, g be functions which are integrable over* $[a, b]$ *for any b greater than a fixed number a. Suppose also that for all* $x \geq a$, $|f(x)| \leq g(x)$ (*in particular g is positive*) *and that the integral* $\int_a^\infty g(x)\, dx$ *exists. Then the integral* $\int_a^\infty f(x)\, dx$ *exists, and*

$$\left| \int_a^\infty f(x)\, dx \right| \leq \int_a^\infty g(x)\, dx.$$

This result follows at once from Theorem 4.6.1 since for any $a < A < B$, we have

$$\left| \int_A^B f(x)\, dx \right| \leq \int_A^B g(x)\, dx$$

so that the left-hand side may be made small by making $\int_A^B g(x)\, dx$ small.

COROLLARY 4.6.3. *Let f be bounded on* $[a, \infty)$ (*that is, for some positive constant M,* $|f(x)| \leq M$ *for all* $x \geq a$). *Then for any real a, and* $k > 0$, $\int_a^\infty e^{-kx} f(x)\, dx$ *exists while, if* $a > 0$ *and* $\alpha > 1$, *then* $\int_a^\infty f(x)/x^\alpha\, dx$ *exists.*

These follow by using 4.6.2 with $g(x) = M e^{-kx}$ or $Mx^{-\alpha}$, and the results of Examples 4.6.2 and 4.6.1.

A more delicate example is provided by the following result.

EXAMPLE 4.6.4. $\int_1^\infty \sin x / x^\alpha\, dx$ exists for any $\alpha > 0$.

(Notice that Corollary 4.6.3 gives this only for $\alpha > 1$). To see this, we take any real numbers A and B with $1 < A < B$. From Example 4.5.1 we see that

$$\left| \int_A^B \frac{\sin x}{x^\alpha} \right| \leq \frac{2}{A^\alpha},$$

which may be made as small as required by making A (and thus B also) sufficiently large. The result then follows from Theorem 4.6.1. The same result is of course valid for $\int_1^\infty \cos x / x^\alpha\, dx$.

We now consider how we may deal with functions which become unbounded at one or more points of their domain of definition. The situation is in many ways analogous to that which leads up to Definition 4.6.1 so we shall proceed at once to the definition.

DEFINITION 4.6.2. Let f be defined on an interval $[a, b]$, and for each positive real number h, suppose that f is integrable over $[a + h, b]$. Then if $\int_{a+h}^b f(x)\, dx$ approaches a limit, L say, as $h \to 0$ we say that f is (*improperly Riemann-*) *integrable* over $[a, b]$, and that

$$L = \int_a^b f(x)\, dx.$$

(Notice that if f is already integrable over $[a, b]$, this is compatible with the former Definition 4.1.3 of $\int_a^b f(x)\, dx$). An obvious modification of this procedure gives the definition when f is integrable only on $[a, b-h]$, and $\int_a^{b-h} f(x)\, dx$ approaches a limit as $h \to 0$.

EXAMPLE 4.6.5. $\int_0^1 x^{-\alpha}\, dx$ exists if and only if $\alpha < 1$: in this case its value is $1/(1-\alpha)$.

For if $0 < h < 1$, $\int_h^1 x^{-\alpha}\, dx = 1/(1-\alpha)(1-h^{1-\alpha})$. This has a limit when $h \to 0$ provided $1-\alpha$ is positive, or $\alpha < 1$, and in this case the limit is $1/(1-\alpha)$.

If we combine this with Example 4.6.1 we see that $\int_0^\infty x^{-\alpha}\, dx$ does not exist for any value of α.

EXAMPLE 4.6.6 $\int_0^1 \log x\, dx = -1$. In this example $\int_h^1 \log x\, dx = [x \log x - x]_h^1 = -h \log h - 1 + h$. But when $h \to 0$, $h \log h$ also $\to 0$, so the result follows [see (2.11.17)].

EXAMPLE 4.6.7. $\int_0^{\frac{1}{2}\pi} \log (\sin \theta)\, d\theta = -\frac{1}{2}\pi \log 2$. Let $I = \int_0^{\frac{1}{2}\pi} \log (\sin \theta)\, d\theta$: the existence of the integral follows since when θ is small, $\sin \theta / \theta$ approaches 1 [see (2.12.13)] and the integral of $\log \sin \theta$ may be compared with the integral of $\log \theta$ in Example 4.6.6. We take it for granted that the usual rules for integration by parts and substitution may be applied to improper integrals, so that we can put $\pi/2 - \theta$ for θ and obtain

$$I = \int_0^{\frac{1}{2}\pi} \log (\cos \theta)\, d\theta.$$

But $\sin \theta = \sin (\pi - \theta)$, so that we also have

$$I = \frac{1}{2} \int_0^\pi \log (\sin \theta)\, d\theta$$

$$= \int_0^{\frac{1}{2}\pi} \log (\sin 2\phi)\, d\phi, \quad \text{on putting } \theta = 2\phi,$$

$$= \int_0^{\frac{1}{2}\pi} \log (2 \sin \phi \cos \phi)\, d\phi$$

$$= \int_0^{\frac{1}{2}\pi} (\log 2 + \log \sin \phi + \log \cos \phi)\, d\phi$$

$$= \log 2 \int_0^{\frac{1}{2}\pi} d\phi + I + I = \log 2 \cdot (\tfrac{1}{2}\pi) + 2I.$$

Hence $I = -\frac{1}{2}\pi \log 2$ as required.

DEFINITION 4.6.3.   The *Cauchy principal value* of $\int_{-\infty}^{\infty} f(x)\, dx$ is defined by

$$\mathcal{P} \int_{-\infty}^{\infty} f(x)\, dx = \lim_{k \to \infty} \int_{-k}^{k} f(x)\, dx.$$

If $\int_{-\infty}^{\infty} f(x)\, dx$ exists then so does $\mathcal{P} \int_{-\infty}^{\infty} f(x)\, dx$ and the two integrals are equal. However, the existence of $\mathcal{P} \int_{-\infty}^{\infty} f(x)\, dx$ does not imply the existence of $\int_{-\infty}^{\infty} f(x)\, dx$.

EXAMPLE 4.6.8. $\int_{-\infty}^{\infty} x\, dx$ does not exist but

$$\mathcal{P} \int_{-\infty}^{\infty} x\, dx = \lim_{k \to \infty} \int_{-k}^{k} x\, dx = \lim_{k \to \infty} 0 = 0.$$

4.7.   INTEGRALS DEPENDING ON A PARAMETER. REPEATED INTEGRALS

It often happens that integrals occur in which the function to be integrated contains, in addition to the variable of integration, one or more further variables as parameters. For instance in Example 4.6.3,

$$\int_{0}^{\infty} e^{-kx} \cos ax\, dx = \frac{k}{a^2 + k^2}, \qquad k > 0,$$

the variable of integration is x, while a, k are real parameters.

It is then of interest to know what properties of the integrand (e.g. differentiability) carry over to the resulting function. In this section we shall quote several such results and show how they may be applied. The first and most fundamental of these concerns the behaviour of the integral when one of the parameters tends to a limit. This can be stated in several forms: we give first the one in which the parameter is an integer which increases indefinitely, and then another version in which the parameter is a real variable which approaches a finite limit.

THEOREM 4.7.1.

(i) *Let* $(f_n(x))$ *be a sequence of functions, each of which is defined for values of* x *which lie in an interval* I *(which may be finite or infinite), and for which* $f_n(x) \to f(x)$ *as* $n \to \infty$, *for each* $x \in I$. *Suppose also that there is a positive function* g, *defined on* I, *for which the integral* $\int_I g(x)\, dx$ *exists, and for which*

$$|f_n(x)| \le g(x) \quad \text{for all } n = 1, 2, 3, \ldots, \text{ and } x \text{ in } I.$$

Then provided the integral $\int_I f(x)\, dx$ *exists, we have*

$$\int_I f_n(x)\, dx \to \int_I f(x)\, dx \quad \text{as } n \to \infty.$$

(ii) *Let* $f(x, y)$ *be a function of two real variables [see § 5.1] which is defined for all values of* x *in an interval* I *(as in* (i)), *and all values of* y *in some interval of the form* $(y_0 - h, y_0 + h)$. *Suppose that* $f(x, y) \to \phi(x)$ *as* $y \to y_0$, *for each* x *in* I, *and that for some positive function* $g(x)$ *defined on* I, $\int_I g(x) \, dx$ *exists, and* $|f(x, y)| \le g(x)$ *for all* x *in* I *and* y *in* $(y_0 - h, y_0 + h)$. *Then provided* $\int_I \phi(x) \, dx$ *exists, we have*

$$\int_I f(x, y) \, dx \to \int_I \phi(x) \, dx \quad as \ y \to y_0.$$

The proof of this result is a consequence of the deeper theory of integration which is outlined in § 4.9, and we shall not attempt it here. In that theory the existence of $\int_I f(x) \, dx$ or $\int_I \phi(x) \, dx$ is guaranteed by the other hypotheses. Notice however that without the existence of the function g, which is said to *dominate* the sequence (f_n), the result may well become false, as Example 4.7.2 below illustrates in various ways. There is a special case which is worth stating separately which is obtained by taking $I = [a, b]$, and $g(x) = M$ (a positive constant) which is of course an integrable function over I.

COROLLARY 4.7.2.   *Let* $(f_n(x))$ *be a sequence of functions defined on* $[a, b]$ *for which* $|f_n(x)| \le M$ *for all* x *in* $[a, b]$ *and* $n = 1, 2, 3, \ldots$. *Suppose that for some integrable function* f, $f_n(x) \to f(x)$ *as* $n \to \infty$ *for each* x *in* $[a, b]$. *Then*

$$\int_a^b f_n(x) \, dx \to \int_a^b f(x) \, dx.$$

Corollary 4.7.2 is often referred to as the *bounded convergence theorem*.

EXAMPLE 4.7.1. $I_n = \int_0^{\pi/4} \tan^n x \, dx \to 0$ as $n \to \infty$.

For we may take $M = 1$ in Corollary 4.7.2, and $f_n(x) = \tan^n x$, which obviously satisfies $|f_n(x)| \le 1$ for all n, and $0 \le x \le \pi/4$ [see § 2.12]. Also $\tan \pi/4 = 1$, while if $0 \le x < \pi/4$, $0 \le \tan x < 1$, so that $f_n(x) \to f(x)$ as $n \to \infty$, where $f(x) = 0$ on $[0, \pi/4)$ while $f(\pi/4) = 1$. This function is integrable over $[0, \pi/4]$ with $\int_0^{\pi/4} f(x) \, dx = 0$ (to see this, partition $[0, 1]$ into N equal parts: any Riemann sum for f must then be $\le 1/N$ and the result follows by making N large); see also Theorem 4.1.3. Hence by Corollary 4.7.2.

$$\int_0^{\pi/4} \tan^n x \, dx \to \int_0^{\pi/4} f(x) \, dx = 0 \quad \text{as } n \to \infty.$$

Referring to Example 4.4.3, we see that we have proved that the series

$$\sum_{k=0}^{\infty} \frac{(-1)^k}{2k+1} \quad \text{and} \quad \sum_{k=0}^{\infty} \frac{(-1)^k}{k+1}$$

are convergent, with sums $\pi/4$ and $\log 2$ respectively.

The following example shows that the assertion of Theorem 4.7.1 need not hold if the conditions are violated.

EXAMPLE 4.7.2

(i) Let $f_n(x)$ be defined for positive x by

$$f_n(x) = \begin{cases} \dfrac{1}{n} & \text{if } 0 \le x \le n^2, \\ 0 & \text{if } x > n^2. \end{cases}$$

Then for each $x \ge 0$, $f_n(x) \to f(x) = 0$ as $n \to \infty$. However

$$\int_0^\infty f_n(x)\, dx = \frac{1}{n} \cdot n^2 = n \to \infty \quad \text{as } n \to \infty.$$

(ii) We recall example 1.12.4: let $f_n(x)$ be defined on $[0, 1]$ by

$$f_n(x) = n^2 x^n (1 - x).$$

Then for each x in $[0, 1)$, $n^2 x^n \to 0$ as $n \to \infty$ [see Example 1.2.8(ii)] while $f_n(1) = 0$: consequently $f_n(x) \to f(x) = 0$ for all x on $[0, 1]$. However

$$\int_0^1 f_n(x)\, dx = n^2 \int_0^1 (x^n - x^{n+1})\, dx$$

$$= n^2 \left(\frac{1}{n+1} - \frac{1}{n+2} \right) = \frac{n^2}{(n+1)(n+2)} \to 1 \quad \text{as } n \to \infty.$$

We now examine the circumstances in which an integral may be differentiated with respect to one of the parameters.

THEOREM 4.7.3.   *Let $f(x, y)$ be a function defined for x in an interval I as in Theorem 4.7.1(i), and y in some interval (c, d). Suppose that for each such value of x and y, f has a partial derivative with respect to y (which we shall denote by $D_2 f(x, y)$, or $\partial f / \partial y$), that $F(y) = \int_I f(x, y)\, dx$ exists for each y, and that for some positive function g defined on I we have that $|D_2 f(x, y)| \le g(x)$ and $\int_I g(x)\, dx$ exists. Then F is differentiable on (c, d) and*

$$F'(y) = \int_I D_2 f(x, y)\, dx.$$

The notation for partial derivatives is discussed fully in section 5.2. Notice (i) that Theorem 4.7.3 says that subject to the given hypotheses we may *differentiate under the integral sign*, and (ii) if I is a finite interval, then the continuity of $D_2 f(x, y)$ is a sufficient condition for the result to hold. The proof of Theorem 4.7.3 follows from Theorem 4.7.1(ii) if we consider

$$\frac{F(y+h) - F(y)}{h} = \int_I \frac{f(x, y+h) - f(x, y)}{h}\, dx.$$

As $h \to 0$,

$$\frac{f(x, y+h) - f(x, y)}{h} \to D_2 f(x, y)$$

while the mean value theorem [see Theorem 3.3.3] shows that this quotient is equal to $D_2 f(x, y + h')$ for some h' with $0 < h'/h < 1$, and this is dominated by $g(x)$.

EXAMPLE 4.7.3.   For $m \geq 0$,

$$\int_0^\infty \frac{\cos mx}{1 + x^2} \, dx = \frac{\pi}{2} e^{-m}.$$

Let $f(m)$ denote $\int_0^\infty \cos mx/(1 + x^2) \, dx$: the integral always exists, since $|\cos mx| \leq 1$ and $\int_0^\infty dx/(1 + x^2)$ exists, and equals $\pi/2$. We begin by integrating by parts to obtain an integral to which Theorem 4.7.3 may be applied. (We cannot immediately differentiate to obtain

$$f'(m) = \int_0^\infty \frac{-x \sin mx}{1 + x^2} \, dx,$$

since $x/(1 + x^2)$ which would have to be $g(x)$, is not integrable over $[0, \infty)$).
    We have, if $m > 0$

$$\int_0^A \frac{\cos mx}{1 + x^2} \, dx = \left[\frac{1}{m} \frac{\sin mx}{1 + x^2} \right]_0^A + \frac{1}{m} \int_0^A \frac{2x \sin mx}{(1 + x^2)^2} \, dx,$$

and so, on letting $A \to \infty$,

$$f(m) = \frac{2}{m} \int_0^\infty \frac{x \sin mx}{(1 + x^2)^2} \, dx.$$

We may now apply Theorem 4.7.3 to differentiate with respect to m, and obtain

$$(mf(m))' = 2 \int_0^\infty \frac{x^2 \cos mx}{(1 + x^2)^2} \, dx$$

(here $g(x) = x^2/(1 + x^2)^2$ which *is* integrable). We may write this as

$$mf'(m) + f(m) = 2 \int_0^\infty \frac{1}{1 + x^2} \left(1 - \frac{1}{1 + x^2} \right) \cos mx \, dx$$

$$= 2f(m) - 2 \int_0^\infty \frac{\cos mx}{(1 + x^2)^2} \, dx,$$

or

$$mf'(m) = f(m) - 2 \int_0^\infty \frac{\cos mx}{(1 + x^2)^2} \, dx.$$

We now differentiate again to obtain

$$mf''(m) + f'(m) = f'(m) + 2 \int_0^\infty \frac{x \sin mx}{(1 + x^2)^2} \, dx,$$

or

$$mf''(m) = mf(m).$$

Hence, if $m > 0$, (or by a similar argument, if $m < 0$) we have $f''(m) = f(m)$.

It is shown in Example 7.4.1 that the only solutions of this differential equation are of the form $f(m) = Ae^m + Be^{-m}$. In our case, if $m > 0$, the integral is bounded by

$$f(0) = \int_0^\infty \frac{dx}{1 + x^2} = \frac{\pi}{2},$$

so that A must be zero, and $f(m) = Be^{-m}$. But $f(0) = \pi/2$, so $B = \pi/2$, and thus $f(m) = (\pi/2) e^{-m}$, if $m > 0$, as required. Notice that since $f(m) = f(-m)$, $f(m) = (\pi/2) e^{-m}$ if $m < 0$.

It sometimes happens that the parameters occur in the limits of the integral, as well as in the integrand. The next theorem covers this case.

THEOREM 4.7.4. *Let I be an interval as in Theorem 4.7.1(i), and $f(x, y)$ be a function which is defined for x in I and $c < y < d$. Suppose that* (i) $\int_I f(x, y)\, dx$ *exists for each y,* (ii) *the partial derivative $D_2(x, y) = (\partial f/\partial y\ (x, y))$ exists for each value of x and y, and satisfies $|D_2 f(x, y)| \le g(x)$, where $\int_I g(x)\, dx$ exists, and* (iii) *that ϕ, ψ are differentiable functions of y on (c, d) which take their values in I. Define*

$$F(y) = \int_{\phi(y)}^{\psi(y)} f(x, y)\, dx \quad for\ c < y < d.$$

Then F is differentiable and

$$F'(y) = f(\psi(y), y)\psi'(y) - f(\phi(y), y)\phi'(y) + \int_{\phi(y)}^{\psi(y)} D_2 f(x, y)\, dx.$$

For a proof, see Apostol (1957).

EXAMPLE 4.7.4

$$F(y) = \int_0^{y^2} \frac{1}{1 + x^4 y}\, dx, \qquad y > 0.$$

Here $I = [0, \infty)$, $\phi(y) = 0$, $\psi(y) = y^2$, $f(x, y) = 1/(1 + x^4 y)$, $(c, d) = (0, \infty)$, $D_2 f(x, y) = -x^4/(1 + x^4 y)^2$, and we may take $g(x) = x^4/(1 + x^4 \delta)^2$ if $y \ge \delta > 0$. Then

$$F'(y) = \frac{1}{1 + (y^2)^4 y}\, 2y + \int_0^{y^2} \frac{-x^4}{(1 + x^4 y)^2}\, dx.$$

Our final theorem of this section concerns repeated integrals: for a more complete treatment, see Chapter 6.

THEOREM 4.7.5.   *Let $f(x, y)$ be a continuous function for $a \leq x \leq b, c \leq y \leq d$ [see Definition 5.1.1]. Then $F_1(x) = \int_c^d f(x, y)\, dy$ is continuous for $a \leq x \leq b$, and $F_2(y) = \int_a^b f(x, y)\, dx$ is continuous for $c \leq y \leq d$, and $\int_a^b F_1(x)\, dx = \int_c^d F_2(y)\, dy$.*

This result says that for continuous functions the result of integrating with respect to both variables is independent of the order in which the integrations are performed [cf. Theorem 6.2.3]. The result may be proved from the definition of the integral as in Bartle (1976) or deduced from Theorem 4.7.3.

4.8.   RIEMANN–STIELTJES INTEGRALS

In the last two sections of this chapter we shall consider two topics which are of a more advanced nature. The first of these is the Riemann–Stieltjes integral in which we generalize the notion of area, which we used to motivate Definitions 4.1.2 and 4.1.3. We retain the notion of a dissection P [see Definition 4.1.1], but the sum associated with P will now be of the form

$$S_P = \sum_{i=1}^n f(u_i)(g(x_i) - g(x_{i-1})),$$

where g is another bounded function defined on $[a, b]$, and as before u_i is some point of $[x_{i-1}, x_i]$. In the important special case when g is monotonically increasing [see Definition 2.7.1], the quantity $g(x_i) - g(x_{i-1})$, which replaces the length $(x_i - x_{i-1})$ of the interval $[x_{i-1}, x_i]$, is always positive.

DEFINITION 4.8.1.   Let f, g be bounded functions on $[a, b]$, and let P be a dissection of $[a, b]$. Then

$$S_P(f, g) = \sum_{i=1}^n f(u_i)(g(x_i) - g(x_{i-1}))$$

is called *a Riemann–Stieltjes sum* associated with P.

Notice that it is not necessary for f or g to be continuous at this stage. The Riemann sums in Definition 4.1.2 obviously correspond to the special case when $g(x) = x$. Intuitively, we are reweighting the interval $[a, b]$, so that instead of associating a subinterval with its length, we associate it with the increase in g over the subinterval.

It would now be possible to define an integration process by requiring that the sums S_P should approach a limit as the mesh $\|P\|$ approaches zero as was done in Definition 4.1.3. However, this leads to a slightly restricted process in which some natural examples of functions are not integrable. Instead we adopt a more subtle definition.

DEFINITION 4.8.2.   Let P, P' be dissections of $[a, b]$. We say P' is a *refinement* of P if all points of P are also in P'. In this case we shall write $P \subset P'$. For instance in Example 4.1.1, P_3 is a refinement of P_2, but not of P_1.

DEFINITION 4.8.3.   Let f, g be bounded functions defined on $[a, b]$. We say that f is *integrable* with respect to g over $[a, b]$ if and only if there is a number L such that for each $\varepsilon > 0$, there is a partition P (depending on ε), such that for every partition $P' \subset P$, we have $|S_{P'}(f, g) - L| < \varepsilon$.

In this case, f is called the *integrand*, g the *integrator*, and the number L is called the *Riemann–Stieltjes integral* (of f with respect to g): we write

$$L = \int_a^b f \, dg = \int_a^b f(x) \, dg(x).$$

It is easy to show that when $g(x) = x$, this definition is in fact equivalent to Definition 4.1.3. The advantage in using Definition 4.8.3 instead of the more immediately obvious '$\lim_{\|P\| \to 0} S_P(f, g)$' is that certain fixed points (namely the points of P) are present in all the dissection P' which are required to give $S_{P'}$ close to L.

EXAMPLE 4.8.1
   (i) We define $H(x)$ for all real numbers x by

$$H(x) = \begin{cases} 1 & \text{if } x \geq 0 \\ 0 & \text{if } x < 0. \end{cases}$$

Suppose f is a bounded function defined on $[-1, 1]$. Then f is integrable with respect to H over $[-1, 1]$ if and only if f is continuous on the left at 0 (that is, the limit of $f(x)$, as x approaches 0 through negative values, exists, and is equal to $f(0)$ [see § 2.3]), and in this case

$$\int_{-1}^1 f(x) \, dH(x) = f(0).$$

This result is an easy consequence of the definition of continuity.
   (ii) It follows from (i) that since H is not continuous on the left at 0, H is not integrable with respect to itself: $\int_1^1 H(x) \, dH(x)$ does not exist.
   (iii) Let $K(x) = \sum_{i=1}^n \alpha_i H_{c_i}(x)$: here α_i ($i = 1, 2, \ldots, n$) and c_i are any real numbers, and

$$H_c(x) = \begin{cases} 1 & \text{if } x \geq c \\ 0 & \text{if } x < c. \end{cases}$$

Notice that K has a jump of amount α_i at c_i.
   Then any continuous function f on $\mathbb{R}$ (more exactly, any f which is continuous on the left at each c_i) is integrable with respect to K, and

$$\int_a^b f(x) \, dK(x) = \sum_{i=1}^n \alpha_i f(c_i), \quad \text{for any } a, b \text{ with } a < c_i \leq b, \text{ all } i. \quad (4.8.1)$$

This example shows the value of the Riemann–Stieltjes integral for unifying both finite sums and integrals into a single conceptual framework.

In addition to the formal properties of the integral, analogous to Theorem 4.1.2(i)–(iv), we note that the integral is linear in g: if f is integrable with respect to g_1 and g_2, then it is integrable with respect to $\beta_1 g_1 + \beta_2 g_2$ ($\beta_1, \beta_2 \in \mathbb{R}$), and

$$\int_a^b f d(\beta_1 g_1 + \beta_2 g_2) = \beta_1 \int_a^b f \, dg_1 + \beta_2 \int_a^b f \, dg_2.$$

We also have the following striking analogue of the formula for integration by parts:

THEOREM 4.8.1.   *Let f, g be bounded functions on $[a, b]$. Then f is integrable with respect to g if and only if g is integrable with respect to f. In this case*

$$\int_a^b f \, dg + \int_a^b g \, df = f(b)g(b) - f(a)g(a).$$

For this proof, the reader is referred to Bartle 1976, Theorem 29.7. It should be noted that this result is sensitive to variations in the definition of the integration process. In particular it fails for the '$\lim_{\|P\| \to 0} S_P(f, g)$' definition mentioned earlier.

In many cases, a Riemann–Stieltjes integral may be reduced to a Riemann integral because of the following result: in fact most cases of practical importance may be dealt with by combining it with the reduction to a finite sum, as in Example 4.8.1(iii).

THEOREM 4.8.2.   *Let g be differentiable on $[a, b]$, and let f and g' be Riemann integrable on $[a, b]$. Then f is integrable with respect to g, and*

$$\int_a^b f(x) \, dg(x) = \int_a^b f(x)g'(x) \, dx.$$

For a proof, see Bartle (1976), Theorem 30.13.

In the important special case when g is monotone increasing [see Definition 2.7.1] on $[a, b]$, all continuous functions [see Definition 2.1.2] are integrable with respect to g, and the analogues of Theorems 4.1.1(i) and 4.1.2 are valid: we shall not restate them in detail, except to note that Theorem 4.1.2(iv) takes the form

LEMMA 4.8.3.   *Let f be bounded, say $|f(x)| \le M$, for x in $[a, b]$, and integrable with respect to the increasing function g. Then*

$$\left| \int_a^b f \, dg \right| \le M(g(b) - g(a)).$$

EXAMPLE 4.8.2 $\int_0^1 x \, d(x^4)$. $g(x) = x^4$ is differentiable on $[a, b]$, and $g'(x) = 4x^3$ is integrable, so by Theorem 4.8.2,

$$\int_0^1 x \, d(x^4) = \int_0^1 x \cdot 4x^3 \, dx = \int_0^1 4x^4 \, dx = \tfrac{4}{5}.$$

EXAMPLE 4.8.3 $\int_0^3 x \, d([x])$. Here $g(x) = [x]$ denotes the greatest integer $\leq x$. When restricted to the interval $[0, 3]$, it is continuous (on the right) at 0, and has jumps of amount 1 at $x = 1, 2, 3$. By comparison with Example 4.8.1(iii), we see that the integral is equal to $1 + 2 + 3 = 6$.

EXAMPLE 4.8.4 $\int_{-\pi/2}^{\pi/2} \sin x \, d(|x|)$. Here $f(x) = \sin x$ is continuous, and $g(x) = |x|$ may be written as the difference of two increasing functions, so that the integral certainly exists. We write this integral as $\int_{-\pi/2}^0 + \int_0^{\pi/2}$, and apply Theorem 4.8.2 on each, to obtain $-\int_{-\pi/2}^0 \sin x \, dx + \int_0^{\pi/2} \sin x \, dx = 2$.

For an example of an increasing function g which does not have an integrable derivative as in Theorem 4.8.2, and is not of the form $\sum \alpha_i H_{c_i}$ as in Example 4.8.1(iii), the reader may consult the next section.

4.9. THE LEBESGUE INTEGRAL

In this final section we shall describe an extension of the theory of integration to a much larger class of functions than we have previously considered. One of the principal weaknesses of the Riemann theory which we have so far considered is that it is not closed under the operation of taking limits: there exist bounded sequences (f_n) of integrable functions which have the property that for some function f, $f_n(x) \to f(x)$ as $n \to \infty$ for all $x \in [a, b]$, but f is not integrable [cf. Theorem 4.7.1]. The class of functions which we shall construct in this section is closed under the formation of limits and is the completion of the class of Riemann integrable functions, in the same way that the real numbers complete the rational numbers [see I, § 2.6.2].

One of the most important features of the theory is the existence of certain small sets such that the behaviour of a function on the small set does not affect the value of its integral: we begin by looking at these (Definitions 4.9.1 and 4.9.2).

DEFINITION 4.9.1. Let E be a set of real numbers. We say that E is *countable* (the terms *denumerable* or *enumerable* are also commonly used [see I, § 1.6]) if E may be put in a $1-1$ correspondence with a subset of the natural numbers $\{1, 2, 3, 4, \ldots\}$.

Notice in particular that any finite set is countable according to this definition. We shall make use of the correspondence to write $E = \{e_1, e_2, \ldots\}$.

EXAMPLE 4.9.1. The set of all integers $\{0, \pm 1, \pm 2, \pm 3, \ldots\}$ is countable.

For make 0 correspond to 1 (we write $0 \rightarrow 1$), and then $1 \rightarrow 2$, $2 \rightarrow 4$, $3 \rightarrow 6$ (and generally $n \rightarrow 2n$ for all positive n), while $-1 \rightarrow 3$, $-2 \rightarrow 5$, $-3 \rightarrow 7$, (and generally $-n \rightarrow 2n + 1$): this gives the required correspondence.

LEMMA 4.9.1

(i) *For each natural number n, suppose a countable set E_n is given. Then the set $\bigcup_1^\infty E_n$ (which contains all the elements of all the sets E_n) is countable (briefly: the union of a countable family of countable sets is countable).*

(ii) *The set of rational numbers is countable.*

(iii) *The set of all real numbers is not countable.*

Proof.   See Weir (1973) section 2.1; also I, § 1.6.

DEFINITION 4.9.2.   Let N be a set of real numbers. We shall say that N is *null* if for each $\varepsilon > 0$, there is a countable family I_n of open intervals whose union contains N, and whose total length is $\leq \varepsilon$. (Notice that the I_n are not required to be disjoint).

Recall that the length of an interval (a, b) is $b - a$ [see I, § 2.6.3]. Hence this definition says that for any positive number ε, we can find a countable family $I_n = (a_n, b_n)$ say, of open intervals such that

(i) each x in N is in some I_n (we write $N \subset \bigcup_1^\infty I_n$), and

(ii) $\sum_{n=1}^\infty (b_n - a_n) \leq \varepsilon$.

Also it is not hard to see that one can drop the requirement that the intervals should be open without altering the class of null sets.

EXAMPLE 4.9.2.   Any countable set is null.

For if $E = \{e_1, e_2, \ldots, e_n \ldots\}$ is countable, and $\varepsilon > 0$ is given, then take

$$I_n = \left(e_n - \frac{\varepsilon}{2^{n+1}}, \ e_n + \frac{\varepsilon}{2^{n+1}} \right).$$

Obviously $E \subset \bigcup_1^\infty I_n$, while the total length of the intervals is

$$\sum_{n=1}^\infty 2 \cdot \frac{\varepsilon}{2^{n+1}} = \varepsilon \left(\tfrac{1}{2} + \tfrac{1}{4} + \tfrac{1}{8} + \ldots \right) = \varepsilon,$$

as required.

EXAMPLE 4.9.3   (*The Cantor Set*).   Not all null sets are countable.

Consider the closed interval $[0, 1]$, and remove from it the open middle third $(\tfrac{1}{3}, \tfrac{2}{3})$ to leave two closed intervals $[0, \tfrac{1}{3}]$ and $[\tfrac{2}{3}, 1]$ each of length $\tfrac{1}{3}$. Next remove the middle third from each of these intervals, which will leave four intervals, namely $[0, \tfrac{1}{9}]$, $[\tfrac{2}{9}, \tfrac{1}{3}]$, $[\tfrac{2}{3}, \tfrac{7}{9}]$ and $[\tfrac{8}{9}, 1]$, each of length $\tfrac{1}{9}$. We continue this process, which results at the n-th stage in our obtaining a set S_n which comprises 2^n intervals each of length $(\frac{1}{3})^n$. The Cantor set E is the set of points common to all the sets S_n. As well as the obvious elements $0, \frac{1}{9}, \frac{2}{9}$, which are end-points of intervals in some S_n, E also contains many other rational numbers (for instance $\frac{1}{4}$). One can prove that E is null (since the total length of the intervals in S_n is $(\frac{2}{3})^n$ which $\to 0$ as $n \to \infty$) but uncountable: see for instance Weir (1973), pp. 20–21 for details. If we define a function f by requiring $f(x) = \frac{1}{2}$ on $(\frac{1}{3}, \frac{2}{3})$; $f(x) = \frac{1}{4}$ on $(\frac{1}{9}, \frac{2}{9})$ and $\frac{3}{4}$ on $(\frac{7}{9}, \frac{8}{9})$; and continuing thus to get for instance $f(x) = 2^{-n}$ on $(3^{-n}, 2.3^{-n})$, we obtain a function on the complement of E [see I, § 1.2.4], which may be extended to the whole of $[0, 1]$ by the requirement of continuity. The resulting function (the *Cantor function*) is continuous, constant on each of the intervals complementary to E, and maps E onto the whole interval $[0, 1]$.

LEMMA 4.9.2.   *The union of a countable collection of null sets is null.*

Proof.   Let $E_1, E_2, \dots, E_n \dots$ be the given null sets. Given $\varepsilon > 0$, use Definition 4.9.2 to find for each E_n a set $I_{n1}, I_{n2}, \dots, I_{nk}, \dots$ of intervals which cover E_n, with total length $< \varepsilon/2^n$. Then the collection of all such intervals (I_{nk}) forms a countable family [see Lemma 4.9.1(i)], covers the union of all the E_n, and has total length $< \varepsilon/2 + \varepsilon/4 + \dots + \varepsilon/2^n + \dots = \varepsilon$.

DEFINITION 4.9.3.   A property which holds except on a null set will be said to hold '*almost everywhere*' or for '*almost all* values of x'. The exceptional set in which the property fails to hold may of course vary from context to context, or from property to property. The phrase 'almost everywhere' is often abbreviated to 'a.e.' or 'p.p.' (presque partout).

We now define our integration process by means of approximations which are quite analogous to that used in section 4.1—we shall point out the parallels (and differences) as we go along. We begin with the simplest class of functions to be integrated.

DEFINITION 4.9.4.   Let $I = [a, b]$ be a bounded closed interval and $P = \{x_0, x_1, \dots, x_n\}$ a partition of it [see Definition 4.1.1]. A function f which is constant on each subinterval (x_{i-1}, x_i) is called a *step function* on $[a, b]$.

We may write a step function neatly using the notion of a *characteristic function*: for any set S, the characteristic function of S (written χ_s) is defined by

$$\chi_s(x) = \begin{cases} 1 & \text{if } x \in S, \\ 0 & \text{if } x \notin S. \end{cases} \tag{4.9.1}$$

Then if $f(x) = c_i$ on the interval $E_i = (x_{i-1}, x_i)$, we may write

$$f = \sum_{i=1}^{n} c_i \chi_{E_i}.$$

Figure 4.9.1 illustrates this.

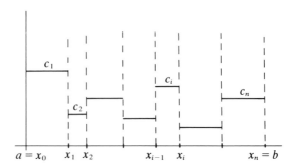

Figure 4.9.1

Notice that the values of f at the points x_i (including the end points a and b) play no part in this definition—f may have any real value, or even be undefined, at these points.

A function which is zero outside a bounded interval I, and equal to a step function on I will be called a step function on the real line.

LEMMA 4.9.3.   *If f, g are step functions, so are $f + g$ and cf for any real number c.*

Proof.   If f is constant on the subintervals determined by the dissection P, and g on those determined by the dissection P', then $f + g$ is constant on the subintervals determined by $P \cup P'$, and is hence a step function. The statement concerning cf is immediate.

DEFINITION   4.9.5.   Let f be a step function, $f = \sum c_i \chi_{E_i}$ as in Definition 4.9.4. We define the *integral* of f, written $\int_I f$ or $\int_I f(x)\,dx$, to be the real number

$$\sum_{i=1}^{n} c_i(x_i - x_{i-1}).$$

It is obvious from Figure 4.9.1 that this definition embodies the natural idea of the integral as the 'area under the graph'. (Notice that if in Definition 4.1.2, we define a step function s to equal $f(u_i)$ on (x_{i-1}, x_i), then the Riemann sum $S_p(f)$ is simply $\int_I s(x)\,dx$.)

A rather subtle point arises here, namely that it is possible to write a step function as $\sum c_i \chi_{E_i}$ in more than one way. This may be got round either by requiring that each $c_i \neq c_{i+1}$, when the function has a discontinuity at x_i [see Definition 2.3.5] and the representation as $\sum c_i \chi_{E_i}$ becomes unique, or by showing that the value of the integral is in any case independent of the representation chosen. We omit the simple arguments necessary to justify these assertions, and also the direct proofs of the following facts.

LEMMA 4.9.4. *Let f, g be step functions on I, and c be any real number. Then*

(i)
$$\int_I (f+g) = \int_I f + \int_I g,$$

(ii)
$$\int_I (cf) = c \int_I f,$$

and

(iii)
$$\text{if } f \le g, \text{ then } \quad \int_I f \le \int_I g.$$

We now extend our integration process by considering increasing sequences of step functions for which the sequence of integrals remains bounded. The first result states that such sequences of functions must in fact converge [see § 1.2] except on a null set.

THEOREM 4.9.5. *Let (S_n) be a sequence of step functions (on an interval, or the whole line) for which*
(i) *$S_n(x) \le S_{n+1}(x)$ for $n = 1, 2, \ldots$ and all x for which both S_n and S_{n+1} are defined, and*
(ii) *there exists a number K such that $\int S_n \le K$ for all n.*
Then there exists a null set N and a real valued function f defined for all $x \notin N$, with $S_n(x) \to f(x)$ for all $x \notin N$. (Notice that f will not usually be a step function).

Proof. See Weir (1973) section 3.2.

DEFINITION 4.9.6
(i) A function defined for almost all x (on the interval I, or on the line) is said to be in the *class M* if there exists a sequence of step functions (S_n) with $\int S_n$ bounded, which is increasing and converges almost everywhere to f. In this case we define $\int_I f$ (or $\int_I f(x)\,dx$) as the value of $\lim_{n\to\infty} \int_I S_n$.
(ii) A function f is said to be in the class L, if there exist functions g, h in M, with $f = g - h$. In this case we define $\int_I f = \int_I g - \int_I h$. Functions of the *class L* are said to be *Lebesgue integrable* over I or the line respectively.

In this definition, as in Definition 4.9.5, there is a problem of consistency: in (i) we have to show that the value of this integral is independent of the choice of sequence (S_n) [see Weir (1973) pp. 34–35] and in (ii) we have to show that it is independent of the choice of g and h. The fact that $\int (f+g) = \int f + \int g$ for functions in M follows by a limiting process from Lemma 4.9.4(i), and the consistency problem for (ii) now follows at once, for if $f = g - h = g_1 - h_1$, then $g + h_1 = g_1 + h$, so $\int g + \int h_1 = \int g_1 + \int h$ and so $\int g - \int h = \int g_1 - \int h_1$ as required.

The reader may be puzzled by the appearance of the two classes M and L in Definition 4.9.6. The reason that M alone will not do is that there exist functions f and g in M with $0 \le f \le g$ but $g - f \notin M$, as the next example shows.

EXAMPLE 4.9.4.   Let f be the function on $[0, 1]$, which is 1 if x is rational, 0 if it is irrational. f vanishes except for a null set, so by taking all $S_n = 0$ we see that $f \in M$, and $\int f = 0$. Let $g = 1 - f$. If g were in M we would have $\int f + \int g = 1$, and so $\int g = 1$. But any positive step functions which is $\leq g$ must be zero except possibly on a finite set, and so must have $\int g = 0$, contradicting the definition of $\int g$ for functions in M.

The Definition 4.9.6 gives us a new class of functions to be integrated. The new process satisfies the same formal properties as previously—in fact the whole of Theorem 4.1.2 holds verbatim for the Lebesgue integral, and we shall not restate it here. It is important to realize that in fact the new process is an extension of the old, as the following theorem which we shall state without proof demonstrates.

THEOREM 4.9.6.   *Let N be a null subset of the bounded interval I, let f be a real-valued function which is bounded on I [see § 2.8] and continuous on $I \backslash N$. Then f is both Riemann and Lebesgue integrable over I, and the two integrals have the same value.*

Since the condition of Theorem 4.9.6 is also necessary for f to be Riemann integrable (though a proof of this fact would take us too far afield), we see that the Lebesgue integral is defined whenever the Riemann integral is, and gives the same value to the integral. The function f in Example 4.9.4 is not Riemann integrable, since by choosing the points u_i in Definition 4.1.2 to be either all rational, or all irrational, we may make the corresponding Riemann sum equal to 1 or 0 respectively, and hence the sum cannot approach a limit. However f is Lebesgue integrable, as pointed out in Example 4.9.4, so the new integral is a genuine extension of the old. How much of an extension will become apparent in the convergence Theorems 4.9.9, 10 and 11 below.

The Lebesgue integral is still intimately bound up with differentiation, though the results here are much deeper and we state them without proof.

THEOREM 4.9.7.   *Let f be Lebesgue integrable over $[a, b]$, and let $F(x) = \int_a^x f$. (We write $\int_a^x f$ in place of the correct, but clumsy, $\int_{[a,x]} f$). Then*

   (i) *F is continuous on $[a, b]$,*
   (ii) *for almost all x in $[a, b]$, F is differentiable, and $F'(x) = f(x)$.*

THEOREM 4.9.8.   *Let D be a countable subset of $[a, b]$, and let F be continuous on $[a, b]$, differentiable at all points of $[a, b] \backslash D$. Let F' be (Lebesgue) integrable over $[a, b]$. Then*

$$\int_a^b F' = F(b) - F(a).$$

If we refer to the Cantor function in Example 4.9.3 we see that it is constant on each interval complementary to the Cantor set E, and hence $\phi' = 0$ except on E. But E is null, and so $\phi' = 0$ for almost all x in $[0, 1]$. Hence we have $\phi' \in L$ and $\int_0^1 \phi' = 0 \neq \phi(1) - \phi(0)$. This example shows that the exceptional set in Theorem 4.9.8 cannot be enlarged to be a null set.

We now come to the main theorems concerning the Lebesgue integral, which show that it has all the properties concerning limiting behaviour which one could reasonably expect. The proofs of the main Theorems 4.9.9, 10 and 11 are again beyond our scope: the reader is referred to Weir (1973) pp. 94–97 for details.

THEOREM 4.9.9 (*Monotone Convergence Theorem*, or *Beppo Levi's Theorem*). *Let f_n be a sequence of positive functions in L, which is increasing almost everywhere (i.e. for a fixed null set N, $f_n(x) \leq f_{n+1}(x)$ for all n, and all x not in N), and whose integrals form a bounded sequence. Then for some function f in L, we have $f_n(x) \to f(x)$ almost everywhere, and*

$$\int f_n \to \int f \quad \text{as } f \to \infty.$$

In order to understand some of the consequences of this result, we recall the definition of the limit inferior of a bounded sequence of real numbers.

DEFINITION 4.9.7.   Let $(x_n) = (x_1, x_2, x_3, \ldots)$ be a bounded sequence of real numbers, that is there exists a real number K such that $-K < x_n < K$ for all n. The inferior limit of (x_n) has already been defined in Definition 1.2.3. An alternative definition is as follows: The *inferior limit* of the sequence (x_n) is the unique number l_1 which has the property that if $\alpha < l_1$, then $x_n < \alpha$ for only finitely many values of n, while if $\alpha > l_1$, then $x_n < \alpha$ for infinitely many values of n (we shall not prove the existence of such a number).

We shall rewrite $l_1 = \liminf x_n$. If (x_n) is a convergent sequence then $\liminf x_n = \lim x_n$.

THEOREM 4.9.10   (*Fatou's Lemma*). *Let f_n be a sequence of positive integrable functions whose integrals $(\int f_n)$ form a bounded sequence. For each x, let $g(x) = \liminf f_n(x)$. Then*

$$\int g \leq \liminf \left(\int f_n \right).$$

(Notice that neither the sequence of functions nor their integrals need be convergent).

EXAMPLE 4.9.5

(i) The Examples 4.7.2(i) and (ii) give examples of functions f_n which converge to zero at each point, while $\int f_n \to \infty$ or to $+1$ respectively: hence strict inequality may occur in Fatou's lemma.

(ii) Let the sequence (f_n) be defined on $[0, 1]$ by

$$f_n(x) = \begin{cases} 1 & \text{on } [0, \frac{1}{3}] \\ 0 & \text{on } (\frac{1}{3}, 1] \end{cases} \quad \text{if } n \text{ is even,}$$

while

$$f_n(x) = \begin{cases} 0 & \text{on } [0, \frac{1}{3}] \\ 1 & \text{on } (\frac{1}{3}, 1] \end{cases} \quad \text{if } n \text{ is odd.}$$

Then $(f_n(x))$ does not converge for any x, but $g(x) = \liminf f_n(x) = 0$ for all x in $[0, 1]$, and so $\int g = 0$. Also $\int f_n$ is alternately $\frac{1}{3}$ and $\frac{2}{3}$, so the sequence of integrals does not converge, but $\liminf (\int f_n) = \frac{1}{3} > \int g$ in accordance with the theorem.

THEOREM 4.9.11 (*Lebesgue's Dominated Convergence Theorem*). *Let* (f_n) *be a sequence of functions in* L *having the properties*
   (i) *for some function* f, $f_n(x) \to f(x)$ *almost everywhere, and*
   (ii) *for some positive function* g *in* L, $|f_n(x)| \le g(x)$ *almost everywhere.*
   Then $f \in L$ *and* $\int |f_n - f| \to 0$ *as* $n \to \infty$, *and in particular*

$$\int f_n \to \int f \quad \text{as } n \to \infty.$$

(Lemma 4.9.2 shows that it is immaterial whether the 'almost everywhere' in (ii) is construed as meaning 'for each n', or 'for all n simultaneously').

The function g is said to *dominate* the sequence f_n, which accounts for the name of the theorem. The Examples 4.7.2(i) and (ii) again show that, in the absence of any dominating function, the result may fail.

Before giving the main applications of the convergence theorems, we shall digress to outline briefly the concept of measurable functions, and measure, which in many treatments precede the construction of the integral. The requirement that a function should be in L (i.e. integrable) imposes two restrictions on it, a regularity condition, and a size condition. The first of these says that the graph shall not be too irregular, and the second says simply that the values of the functions must not be too large (in absolute value) too often, i.e. that $\int |f|$ should be finite. If we require only the first of these conditions, we are left with the concept of a measurable function.

DEFINITION 4.9.8. Let f be a real-valued function defined almost everywhere (on a bounded interval, or on the real line). Then if for every pair of positive real numbers K_1 and K_2, the function g defined by

$$g(x) = \begin{cases} 0 & \text{if } |x| > K_1 \text{ or if } |f(x)| > K_2, \\ f(x) & \text{otherwise,} \end{cases}$$

is integrable, we say f is *measurable*. (Obviously the restriction to values of x with $|x| \le K_1$ is superfluous if f is already defined on a bounded interval).

The function g is obtained from f by 'truncation', as the figure illustrates:

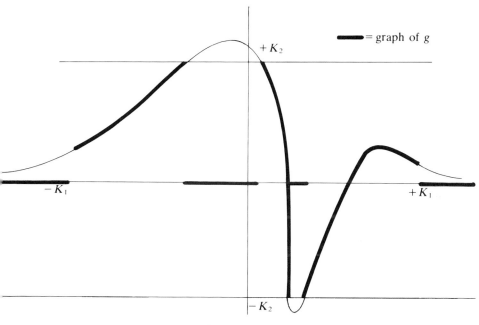

Figure 4.9.2

All sums, products and limits of measurable functions are again measurable, though we shall not stop to prove this. Instead we define measurable sets as follows.

DEFINITION 4.9.9.   Let S be a set of real numbers, and χ_s its characteristic function [see (4.9.1)]. Then we say S is a *measurable set* if and only if χ_s is a measurable function. In the case when χ_s is integrable we define the (Lebesgue) *measure* of S (which we shall write as $m(S)$ or $|S|$) to be the value of $\int \chi_s$. If χ_s is measurable but not integrable we give $m(S)$ the conventional value $+\infty$.

Lebesgue measure has the properties listed in the following theorem.

THEOREM 4.9.12
  (i) $m(S) \geq 0$ *for all measurable sets* S,
  (ii) *if* $S = \bigcup_1^\infty S_n$ *is   a   union   of   disjoint   measurable   sets,   then* $m(S) = \sum_1^\infty m(S_n)$, *(in view of the conventional value $+\infty$ which may occur, we say that a sum is infinite if it involves one or more infinite terms)*,
  (iii) *if I is a bounded interval of length l, then $m(I) = l$ (compare the notation in Lemma 4.9.4)*,
  (iv) *any bounded measurable set has finite measure,*

(v) *if S is measurable, and S_x denotes S translated by the real number x* $(S_x = \{y\colon y = x + s \text{ for some element } s \text{ of } S\})$, *then* $m(S) = m(S_x)$.

(vi) $m(S) = 0$ *if and only if S is null.*

Note. Properties (i) and (ii) define the abstract concept of a measure (not necessarily Lebesgue measure m). (i) (ii) and (iv) define a regular measure on the real line. Properties (iii) and (v) are characteristic of Lebesgue measure.

It is a striking fact that it is quite difficult to construct sets which are not Lebesgue measurable—any of the references to this section will give the details.

We shall not go further in this direction except to say that the whole theory of Riemann–Stieltjes integrals treated in section 4.8 can be extended to a Lebesgue theory of integration with respect to general measures. The reader is referred to Rudin (1966) for details.

We finish this section by making precise our remarks at the beginning of this section that the Lebesgue integrable functions form a completion of the Riemann integrable functions, or the step functions in the same way that the real numbers complete the rational numbers.

DEFINITION 4.9.10.   Let p be a real number ≥ 1. We say that a measurable function f is in the *class* L^p (or L_p) if $|f|^p$ (the pth power of $|f|$) is integrable. The class L of integrable functions is thus identical with L^1 [see Definition 4.9.6(ii)].

We define the *distance between two functions f and g in* L^p to be

$$\|f - g\|_p = \left(\int |f - g|^p \right)^{1/p}.$$

$$(4.9.2)$$

We say that a sequence of functions (f_n) in L^p is a *Cauchy-sequence* in L^p if $\|f_n - f_m\|_p \to 0$ as $m, n \to \infty$ (that is, for each $\varepsilon > 0$, there is some n_0 for which $\|f_n - f_m\|_p < \varepsilon$ when $m, n \geq n_0$)—compare the definition of a Cauchy-sequence of real numbers in section 1.6, where the definition is with respect to the absolute value $|x_n - x_m|$.

Then we have

THEOREM 4.9.13

(i) *The set of step functions (and hence the set of Riemann-integrable functions) is dense in* L^p, *that is, for each* $f \in L^p$ *and each* $\varepsilon > 0$, *there is a step function s with* $\|f - s\|_p < \varepsilon$.

(ii) *The space* L^p *is complete, that is, given any Cauchy sequence* (f_n) *in* L^p, *there is a function f in* L^p *with* $\|f - f_n\|_p \to 0$ [compare Theorem 1.6.1].

The proof of (i) follows from the construction of the space L via step functions in Definition 4.9.6, and that of (ii) from the dominated convergence theorem.

Finally we give the following examples of how the theory applies to particular sequences of functions.

EXAMPLE 4.9.6
   (i) If α is a positive real number

$$\lim_{n \to \infty} \int_0^n \left(1 - \frac{x}{n}\right)^n x^{\alpha - 1} \, dx = \int_0^\infty e^{-x} x^{\alpha - 1} \, dx.$$

   (ii) if p is a real number > -1,

$$\int_0^1 \frac{x^p}{1-x} \log \left(\frac{1}{x}\right) dx = \frac{1}{(p+1)^2} + \frac{1}{(p+2)^2} + \ldots = \sum_{k=1}^\infty (p+k)^{-2}.$$

To see (i) it is sufficient to know that if $n > x$, then $(1 - x/n)^n$ is increasing as n increases, and has limit e^{-x}. We shall assume the value of the limit is known [cf. (2.11.4)]: to see that $(1 - x/n)^n$ increases with n, write

$$\phi(t) = \left(1 - \frac{x}{t}\right)^t \quad (t > x > 0).$$

Then

$$\phi'(t) = \left(1 - \frac{x}{t}\right)^t \log \left(1 - \frac{x}{t}\right) + t\left(1 - \frac{x}{t}\right)^{t-1} \left(\frac{x}{t^2}\right)$$

and this has the same sign as $(1 - x/t) \log (1 - x/t) + x/t = \theta \log \theta + 1 - \theta$ say, where $\theta = 1 - x/t$. Thus we require $\theta \log \theta + 1 - \theta > 0$, or equivalently on rearranging, $\log 1/\theta = -\log \theta < 1/\theta - 1$, which follows from the fact that $\theta < 1$, so $1/\theta > 1$, and if $y > 1$, $\log y = \int_1^y du/u < \int_1^y du = y - 1$. We now put

$$f_n(x) = \begin{cases} \left(1 - \frac{x}{n}\right)^n x^{\alpha-1} & \text{if } 0 \leq x \leq n, \\ 0 & \text{if } x > n. \end{cases}$$

We have shown that $f_n(x)$ increases to $f(x) = e^{-x} x^{\alpha-1}$ as $n \to \infty$, and the result now follows from Theorem 4.9.9. An alternative proof using the fact that $(1 - x/n)^n \leq e^{-x}$ if $0 \leq x \leq n$, and the finiteness of the integral $\int_0^\infty e^{-x} x^{\alpha-1} \, dx$ [see § 10.2.1], may be given using Theorem 4.9.11.

To see (ii) we first notice that $s_n(x) = 1 + x + \ldots + x^n$ is an increasing sequence of positive functions on $[0, 1]$, whose limit is $1/(1-x)$ on $[0, 1)$ [see Example 1.7.3]. Also as $x \to 1$, $[1/(1-x)] \log (1/x) \to 1$ (using for instance de L'Hôpital's rule—Corollary 3.4.4), while the integral is convergent near $x = 0$ since $p > -1$ [compare Example 4.6.5].

An application of (4.9.9) now shows that

$$\int_0^1 s_n(x) x^p \log \left(\frac{1}{x}\right) dx \to \int_0^1 \frac{x^p}{1-x} \log \left(\frac{1}{x}\right) dx.$$

But the integral on the left is

$$\int_0^1 \left(\sum_{k=0}^n x^k \right) x^p \log\left(\frac{1}{x}\right) dx = \sum_{k=0}^n \left(\int_0^1 x^{k+p} \log\left(\frac{1}{x}\right) dx \right)$$

$$= \sum_{k=0}^n \left[\frac{1}{k+p+1} x^{k+p+1} \log\frac{1}{x} \right]_0^1$$

$$+ \int_0^1 \frac{x^{k+p+1}}{k+p+1} \frac{1}{x} dx,$$

and the first term vanishes since $k+p+1>0$, and for any $\delta>0$, $x^\delta \log x \to 0$ as $x \to 0$ [*see* (2.11.7)]. The remaining term gives

$$\sum_{k=0}^n \int_0^1 \frac{x^{k+p}}{(k+p+1)} dx = \sum_{k=0}^n \frac{1}{(k+p+1)^2}.$$

When $n \to \infty$ this gives

$$\sum_{k=0}^\infty \frac{1}{(k+p+1)^2} = \sum_{k=1}^\infty \frac{1}{(k+p)^2},$$

as required.

P. L. W.

REFERENCES

Bartle, R. G. (1976). *The Elements of Real Analysis* (2nd Edn), Wiley, New York.
Gradshteyn, I. S. and Ryzhik I. M. (1965). *Tables of Integrals, Series and Products,* Academic Press, New York.
Maddox, I. J. (1977). *Introductory Mathematical Analysis*, A. Hilger, London.
Rudin, W. (1966). *Principles of Mathematical Analysis*, McGraw-Hill, New York.
Rudin, W. (1966). *Real and Complex Analysis*, McGraw-Hill, New York.
Spivak, M. (1967). *Calculus*, Benjamin, New York.
Weir, A. (1973). *Lebesgue integration and measure*, Cambridge.

CHAPTER 5

Functions of Several
(Real) Variables

5.0. INTRODUCTION

When we attempt to extend the methods and concepts of analysis from functions of one real variable to several, a number of new phenomena occur which require more careful consideration; and in this respect the theory of functions of several variables is more complicated and more delicate to handle. Nevertheless a highly satisfactory theory exists. We shall approach it here, first in the archetypal case of two variables, and then in the general n-variable setting.

The main topics covered include continuity and differentiability; partial derivatives; the derivative as a linear transformation; changes of variable and Jacobians; maxima, minima, and other critical points; the Hessian matrix; constrained extrema, and Lagrange's method of undetermined multipliers. We stress the geometric meaning of the manipulations involved as well as their analytic form.

5.1. TWO VARIABLES

By a real-valued function of two variables we shall mean a function

$$f : D \to \mathbb{R}$$

where D is a subset of the set $\mathbb{R}^2$ of all ordered pairs (x, y), $x, y \in \mathbb{R}$ [see I, §§ 1.2.6, 1.4.1]. That is, f assigns to those ordered pairs (x, y) that belong to the domain D a value

$$f(x, y) \in \mathbb{R}.$$

For example, on $D = \mathbb{R}^2$ we might define [see §§ 2.11, 2.12]

$$f(x, y) = x^2 + y - \tfrac{1}{2} \exp(x - y) + \sin(x)\sin(x + y).$$

Or, on

$$D = \{(x, y) \mid x \neq 0, \, y \neq 0\}$$

175

we might define

$$f(x, y) = 1/xy.$$

The *graph* of such a function is drawn relative to axes (x, y, z) in $\mathbb{R}^3$, as the subset

$$\{(x, y, f(x, y)) | (x, y) \in D\}.$$

Think of the (x, y)-plane lying horizontally, with z measured vertically; above each $(x, y) \in D$ draw a point at height $f(x, y)$. The set of all such points is the graph of f. For example the function with domain the unit disc

$$\{(x, y) | x^2 + y^2 \leq 1\}$$

and values

$$f(x, y) = 2 - x^2 - y^2$$

has the graph shown in Figure 5.1.1.

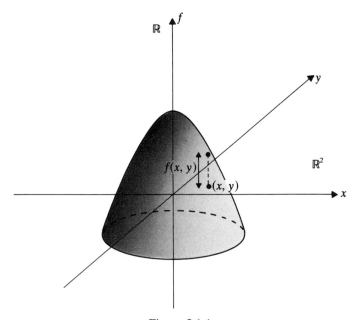

Figure 5.1.1

Continuity of two-variable functions is defined in a manner strictly analogous to that for single-variable functions [see § 2.1]; and it carries the same connotation: small changes in x and y produce small changes in $f(x, y)$. For precision we introduce the *norm*

$$\|(x, y)\| = \sqrt{x^2 + y^2} \tag{5.1.1}$$

which measures the distance from (x, y) to the origin [see I (9.1.2)]. Then we have

DEFINITION 5.1.1.   The function $f: D \to \mathbb{R}$ is *continuous at* $(x_0, y_0) \in D$ if, for all $\varepsilon > 0$, there exists $\delta > 0$ such that whenever $\|(x - x_0, y - y_0)\| < \delta$ it follows that $|f(x, y) - f(x_0, y_0)| < \varepsilon$.

By analogy with section 2.3 we say that $f: D \to \mathbb{R}$ *tends to the limit* l *as* (x, y) *tends to* (x_0, y_0) if, for all $\varepsilon > 0$ there exists $\delta > 0$ such that whenever $\|(x - x_0), (y - y_0)\| < \delta$ it follows that $|f(x, y) - l| < \varepsilon$. If a limit l exists we write

$$\lim_{(x,y)\to(x_0,y_0)} f(x, y) = l.$$

A direct analogue of Definition 2.3.1 furnishes the close connection between limits and continuity in the two-variable case:

DEFINITION 5.1.2.   Let $f: D \to \mathbb{R}$. Then f is *continuous* at a point $(x_0, y_0) \in D$ if and only if

$$\lim_{(x,y)\to(x_0,y_0)} f(x, y) = f(x_0, y_0);$$

and f is *continuous throughout* D if and only if it is continuous at every point of D.

It is clear that the rest of section 2.3 carries over to the two-variable case by direct analogy. In particular we employ $f(x, y) = O(g(x, y))$; $f(x, y) = o(g(x, y))$ and $f(x, y) \sim g(x, y)$ conserving the '*order of magnitude*' or '*rate of growth*' of a function of two variables near $(0, 0)$ or ∞.

Ideas such as *left-hand* or *right-hand* limits do not generalize directly to functions of two variables, because there is no sensible notion of 'left' and 'right' in the plane. However, all is not lost. Let D be a domain in $\mathbb{R}^2$ divided by a curve K into two parts D^+ and D^-, as shown, and let (x_0, y_0) lie on K.

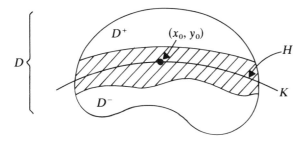

We can define

$$\lim_{(x,y)\to(x_0,y_0)+} f(x, y)$$

as above, provided we modify the definition to include only points $(x, y) \in D^+$;

and similarly for

$$\lim_{(x,y)\to(x_0,y_0)-} f(x, y),$$

using D^-. (The choice of which is $+$ and which $-$ is arbitrary.)

Definition 2.3.5 then has the following two-variable analogue:

DEFINITION 5.1.3. A *discontinuity* of a function $f:D\to\mathbb{R}$ is a point $(x_0, y_0) \in D$ at which f is not continuous. If there is a curve K through (x_0, y_0) dividing D into D^+ and D^- as above, and each of

$$\lim_{(x,y)\to(x_0,y_0)+} f(x, y), \qquad \lim_{(x,y)\to(x_0,y_0)-} f(x, y)$$

exist, but f is not continuous at (x_0, y_0), then (x_0, y_0) is a *jump discontinuity* of f.

If there is a region H containing K such that f is continuous on $D^+ \cap H$ and on $D^- \cap H$, and has jump discontinuities at each $(x_0, y_0) \in K$, we say that f has a *jump discontinuity across K*.

If D can be divided by curves into finitely many subdomains $D_1, \ldots, D_k$, such that f is continuous throughout each D_k, then we say that f is *piecewise continuous* on D. (The values on the dividing curves seldom matter; if they do, we may also require f to have jump discontinuities across these curves.)

Important two-variable functions are *polynomial* functions such as

$$5x^2y - 3x^2 + 2y - 4x + 3$$

which are combinations of terms $x^i y^j$ with real coefficients. The *degree* of a term is $i + j$; and the *degree* of the whole expression is the maximum of the degrees of its terms (in the above example, 3). An expression of degree 2 is said to be *quadratic*; one of degree 3 is *cubic*, and so on. If every term has the same degree the expression is *homogeneous*; for example

$$x^2y - 3y^3 + 2x^3$$

is a homogeneous cubic, whereas the previous expression is a non-homogeneous cubic.

5.2. PARTIAL DERIVATIVES

DEFINITION 5.2.1. Let $F:D\to\mathbb{R}$ be a real-valued function of two variables x and y in the above sense. We define two *partial derivatives* of f with respect to x, y respectively:

$$\frac{\partial f}{\partial x} = \lim_{h\to 0} \frac{f(x+h, y) - f(x, y)}{h}$$

$$\frac{\partial f}{\partial y} = \lim_{h\to 0} \frac{f(x, y+h) - f(x, y)}{h}$$

provided that the limits exist.

Comparing with the definition of an ordinary derivative [see Definition 2.9.1] we see that $\partial f/\partial x$ may be evaluated by 'treating y as a constant and differentiating with respect to x only'; and similarly we may evaluate $\partial f/\partial y$ by treating x as a constant and differentiating with respect to y.

For example, if

$$f(x, y) = x^3 y^2 - 7y^4 + y \cdot \sin x$$

then

$$\frac{\partial f}{\partial x} = 3x^2 y^2 + y \cdot \cos x$$

$$\frac{\partial f}{\partial y} = 2x^3 y - 28y^3 + \sin x.$$

A variety of notations are used for partial derivatives: for instance, instead of $\partial f/\partial x$ and $\partial f/\partial y$ one may encounter $D_x f$, $D_y f$; $D_1 f$, $D_2 f$; f_1, f_2 (where the subscripts 1 and 2 refer to the first and second variables x and y). However in any particular field or text the notation tends to be standardized.

The ordinary derivative has an important geometric interpretation, as the slope of the graph [see § 3.1.1]. There is a similar interpretation of partial derivatives. Consider the function whose graph is Figure 5.1.1. If we differentiate partially with respect to x we first 'treat y as constant'. That is, we *fix* the value of y to be that at the point (x, y) of interest. Fixing y amounts geometrically to slicing Figure 5.1.1 parallel to the x-axis as shown in Figure 5.2.1.

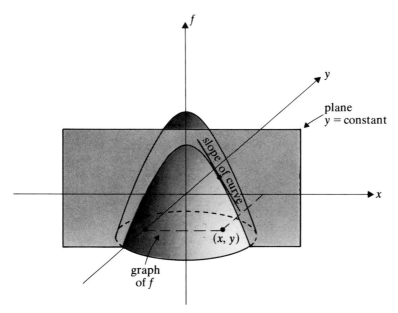

Figure 5.2.1

The plane (y = constant) cuts the graph of f in a curve; this curve is the graph of f varying against x only, with y fixed. Hence the partial derivative with respect to x is the *slope* of this curve at the point (x, y), as shown in Figure 5.2.1.

In other words, $\partial f/\partial x$ is the slope of the graph of f along a direction parallel to the x-axis. Similarly $\partial f/\partial y$ is the slope along a direction parallel to the y-axis.

Note that these two slopes can be quite different.

Note also that their values are strongly tied to the choice of axes. This will be studied in proper detail later, when we consider the effect of changes of variable: these are both more subtle and more commonly useful for partial derivatives than for ordinary ones.

5.3. FUNCTIONS OF *n* VARIABLES

The extension of these ideas to n variables is immediate. A (*real-valued*) *function of n variables* is a function

$$f : D \to \mathbb{R}$$

where $D \subseteq \mathbb{R}^n$, the space of ordered n-tuples $(x_1, \ldots, x_n)$. Important examples are polynomial functions, defined by analogy with the two-variable case [see § 5.1] and (real-valued) functions of two-vector variables, in which case $D \subseteq \mathbb{R}^m \times \mathbb{R}^n$. More generally, if k is a positive integer, then an $\mathbb{R}^k$-*valued function of n variables* is a function.

$$\mathbf{f} : D \to \mathbb{R}^k$$

where $D \subseteq \mathbb{R}^n$. Choosing coordinates $(x_1, \ldots, x_n) \in \mathbb{R}^n$ the former assigns to vectors $(x_1, \ldots, x_n) \in D$ a real number

$$f(x_1, \ldots, x_n)$$

while the latter assigns a vector

$$(f_1(x_1, \ldots, x_n), \ldots, f_k(x_1, \ldots, x_n))$$

and may, if so desired, be thought of as composed of k real-valued functions

$$f_i : D \to \mathbb{R} \qquad (i = 1, \ldots, k).$$

Such *co-ordinate functions* f_i are given a reduced emphasis in a coordinate-free treatment, but are necessary for computation. If we choose in $\mathbb{R}^k$ a basis $\mathbf{e}_1, \ldots, \mathbf{e}_k$ which differs from the standard basis $(1, 0, \ldots, 0)$, $(0, 1, \ldots, 0)$, $\ldots$, $(0, \ldots, 0, 1)$ [see I, § 5.4], there is a corresponding notion of coordinate functions. Namely, write $\mathbf{f}(x_1, \ldots, x_n)$ in the form $f_1(x_1, \ldots, x_n)\mathbf{e}_1 + \ldots + f_k(x_1, \ldots, x_n)\mathbf{e}_k$. Then the f_j ($j = 1, \ldots, k$) are the *coordinate functions of* $\mathbf{f}$ *with respect to* $\mathbf{e}_1, \ldots, \mathbf{e}_k$.

To define continuity we set up a *norm*

$$\|(x_1, \ldots, x_n)\| = \sqrt{x_1^2 + \ldots + x_n^2}$$

on $\mathbb{R}^n$. Again this may be interpreted as '*distance from the origin*'.

DEFINITION 5.3.1. A function $\mathbf{f}:D\to\mathbb{R}^k$ with $D\subseteq\mathbb{R}^n$ is *continuous* at a point $\mathbf{x}_0=(x_{01},\ldots x_{0n})\in D$ if, for all $\varepsilon>0$, there exists $\delta>0$ such that whenever $\mathbf{x}=(x_1,\ldots,x_n)\in D$ and $\|\mathbf{x}-\mathbf{x}_0\|<\delta$, it follows that $\|\mathbf{f}(\mathbf{x})-\mathbf{f}(\mathbf{x}_0)\|<\varepsilon$.

That is, 'small' changes in $\mathbf{x}$ lead to 'small' changes in $\mathbf{f}(\mathbf{x})$, with the same sense as in earlier interpretations of continuity [see §§ 2.1, 5.1].

If $\mathbf{f}:D\to\mathbb{R}^k$ and $f_1,\ldots,f_k$ are the coordinate functions, it follows that $\mathbf{f}$ is continuous at $\mathbf{x}_0$ if and only if each f_i is continuous at x_{0i}.

As usual, $\mathbf{f}$ is *continuous* on D if it is continuous at $\mathbf{x}_0$ for all $\mathbf{x}_0\in D$.

DEFINITION 5.3.2. For a real-valued function $f:D\to\mathbb{R}$, with $D\subseteq\mathbb{R}^n$, we define *n partial derivatives* of f by

$$\frac{\partial f}{\partial x_i}=\lim_{h\to 0}\frac{f(x_1,x_2,\ldots,x_i+h,\ldots x_n)-f(x_1,\ldots,x_n)}{h}$$

provided that the limit exists.

Again it may be computed by treating all variables except x_i as constants, and differentiating with respect to x_i in the usual way. Thus

$$\frac{\partial}{\partial z}(x^3yz-3z^2+x\cdot\sin y)=x^3y-6z,$$

where we use (x,y,z) instead of (x_1,x_2,x_3) as is customary. Other notation for $\partial f/\partial x_i$ often used includes D_if,f_i (not to be confused with coordinate functions), and $f_{,i}$.

DEFINITION 5.3.3. The operator $\partial/\partial x_i$ (equivalently D_i) is called the *i*th *differential operator*.

Because partial derivatives may be evaluated by keeping all variables but one constant, and then differentiating, many basic properties of the derivative carry over: in particular the rules for differentiating sums, products, and quotients apply to partial derivatives [see Theorem 3.2.1]. Specifically, let

$$f:D\to\mathbb{R}$$

$$g:D\to\mathbb{R}$$

be functions of n variables, and define $f+g$, fg, and f/g by

$$(f+g)(x_1,\ldots,x_n)=f(x_1,\ldots,x_n)+g(x_1,\ldots,x_n)$$

$$(fg)(x_1,\ldots,x_n)=f(x_1,\ldots,x_n)g(x_1,\ldots,x_n)$$

$$(f/g)(x_1,\ldots,x_n)=f(x_1,\ldots,x_n)/g(x_1,\ldots,x_n)$$

provided, in the latter, that $g(x_1, \ldots, x_n) \neq 0$. Then for all $i = 1, \ldots, n$,

$$\frac{\partial}{\partial x_i}(f+g) = \frac{\partial f}{\partial x_i} + \frac{\partial g}{\partial x_i} \qquad (5.3.1)$$

$$\frac{\partial}{\partial x_i}(fg) = \left(\frac{\partial f}{\partial x_i}\right)g + f\left(\frac{\partial g}{\partial x_i}\right) \qquad (5.3.2)$$

$$\frac{\partial}{\partial x_i}(f/g) = \frac{\left(\dfrac{\partial f}{\partial x_i}\right)g - f\left(\dfrac{\partial g}{\partial x_i}\right)}{g^2}. \qquad (5.3.3)$$

The analogue of the chain rule (3.2.8), however, is less simple: see section 5.4.

5.4.   THE CHAIN RULE

We now consider the analogue, for partial derivatives, of the chain rule for differentiating the composite of two functions [see I, § 1.4.2], viz.

$$\frac{df}{dx} = \frac{df}{du} \cdot \frac{du}{dx}$$

where $f = f(u)$ and $u = u(x)$. The result (proved, for example, in Protter and Morrey (1964)) is as follows.

THEOREM 5.4.1   (*Chain Rule*).   *Suppose that f is a function of n variables $x_1, \ldots, x_n$, and each x_i is a function of m variables $y_1, \ldots, y_m$. Suppose that f is continuous and that all partial derivatives $\partial f/\partial x_i$ are continuous, and assume that all first partial derivatives $\partial x_i/\partial y_j$ exist $(i = 1, \ldots, n; j = 1, \ldots, m)$. Then*

$$\frac{\partial f}{\partial y_j} = \frac{\partial f}{\partial x_1}\frac{\partial x_1}{\partial y_j} + \frac{\partial f}{\partial x_2}\frac{\partial x_2}{\partial y_j} + \ldots + \frac{\partial f}{\partial x_n}\frac{\partial x_n}{\partial y_j} \qquad (j = 1, \ldots, m).$$

For instance, if $f = f(x, y, z)$ and $x = x(r, s)$, $y = y(r, s)$, $z = z(r, s)$; and the continuity and differentiability hypotheses hold, then

$$\frac{\partial f}{\partial s} = \frac{\partial f}{\partial x}\frac{\partial x}{\partial s} + \frac{\partial f}{\partial y}\frac{\partial y}{\partial s} + \frac{\partial f}{\partial z}\frac{\partial z}{\partial s},$$

and   (5.4.1)

$$\frac{\partial f}{\partial r} = \frac{\partial f}{\partial x}\frac{\partial x}{\partial r} + \frac{\partial f}{\partial y}\frac{\partial y}{\partial r} + \frac{\partial f}{\partial z}\frac{\partial z}{\partial r}.$$

EXAMPLE 5.4.1.   Let

$$f(x, y) = x^2 + y^2$$

where

$$x = r . \cos \theta$$

$$y = r . \sin \theta$$

(that is, changing from Cartesian to polar coordinates [see V, § 1.2.5]). Then

$$\frac{\partial f}{\partial r} = \frac{\partial f}{\partial x} \frac{\partial x}{\partial r} + \frac{\partial f}{\partial y} \frac{\partial y}{\partial r}$$

$$= 2x . \cos \theta + 2y . \sin \theta$$

$$= 2r \cos^2 \theta + 2r \sin^2 \theta$$

$$= 2r,$$

$$\frac{\partial f}{\partial \theta} = \frac{\partial f}{\partial x} \frac{\partial x}{\partial \theta} + \frac{\partial f}{\partial y} \frac{\partial y}{\partial \theta}$$

$$= 2x . (-r \sin \theta) + 2y (r \cos \theta)$$

$$= 0.$$

As a check, we have

$$f(x, y) = x^2 + y^2$$

$$= r^2 \cos^2 \theta + r^2 \sin^2 \theta$$

$$= r^2.$$

So

$$\frac{\partial f}{\partial r} = 2r, \qquad \frac{\partial f}{\partial \theta} = 0$$

as before.

As with the ordinary chain rule, this result has many uses. For example, consider a circular cylinder whose height h and radius r are both varying with time t [see V, § 2.1.1]. Then the volume

$$V = \pi r^2 h$$

also varies with time t. By the chain rule (noting that since V, r and h are functions of the single variable t we have the notational simplification dr/dt instead of $\partial r/\partial t$, etc.) it follows that

$$\frac{dV}{dt} = \frac{\partial V}{\partial r} \frac{dr}{dt} + \frac{\partial V}{\partial h} \frac{dh}{dt}$$

$$= 2\pi rh \frac{dr}{dt} + \pi r^2 \frac{dh}{dt}$$

which allows us to compute dV/dt given the values of r, h, dr/dt, and dh/dt. Many similar problems may be tackled in the same way.

5.5. HIGHER PARTIAL DERIVATIVES

If $f: D \to \mathbb{R}$ has continuous first partial derivatives $\partial f / \partial x_i$, we can consider the effect of partial differentiation of these functions: the second partial derivatives

$$\frac{\partial^2 f}{\partial x_i \, \partial x_j} = \frac{\partial}{\partial x_i} \left(\frac{\partial f}{\partial x_j} \right) \tag{5.5.1}$$

(note that instead of $\partial^2 f / \partial x_i \partial x_i$ we write $\partial^2 f / \partial x_i^2$.) These of course exist if and only if the relevant limits do.

EXAMPLE 5.5.1.   If

$$f(x, y) = x^2 y^3 + y^2 x$$

then

$$\frac{\partial f}{\partial x} = 2xy^3 + y^2$$

$$\frac{\partial f}{\partial y} = 3x^2 y^2 + 2yx$$

$$\frac{\partial^2 f}{\partial x^2} = 2y^3$$

$$\frac{\partial^2 f}{\partial x \, \partial y} = 6xy^2 + 2y$$

$$\frac{\partial^2 f}{\partial y \, \partial x} = 6xy^2 + 2y$$

$$\frac{\partial^2 f}{\partial y^2} = 6x^2 y + 2x.$$

Note that $\partial^2 f / \partial x \partial y$ and $\partial^2 f / \partial y \partial x$ are equal in this case, although their definition as

$$\frac{\partial}{\partial x} \left(\frac{\partial f}{\partial y} \right) \quad \text{and} \quad \frac{\partial}{\partial y} \left(\frac{\partial f}{\partial x} \right)$$

gives no especial reason to expect this. However, it is no accident. It can be proved (Protter & Morrey [1964] p. 691) that if f, $\partial f / \partial x$, $\partial f / \partial y$, $\partial^2 f / \partial x \partial y$ and $\partial^2 f / \partial y \, \partial x$ are all continuous at a point (x_0, y_0), then

$$\frac{\partial^2 f}{\partial x \partial y} (x_0, y_0) = \frac{\partial^2 f}{\partial y \, \partial x} (x_0, y_0). \tag{5.5.2}$$

Without such continuity restrictions this equality need not hold.

Other notation for higher partial derivatives, commonly used, includes in place of $\partial^2 f / \partial x_i \, \partial x_j$

$$D_{ji}f \qquad f_{ji} \qquad f_{,j,i}$$

or similar expressions. Note that the order in the subscripts is the reverse of that in the first expression cited: in all cases the notation means 'first differentiate with respect to x_j, then with respect to x_i'.

Derivatives of higher orders may be defined similarly: for example

$$\frac{\partial^3 f}{\partial x \, \partial y \, \partial z} = \frac{\partial}{\partial x}\left(\frac{\partial^2 f}{\partial y \, \partial z}\right).$$

By repeated use of the above result it follows that (provided that all partial derivatives up to the order concerned exist and are continuous) the order in which the various differentiations are carried out is irrelevant. For example

$$\frac{\partial^4 f}{\partial y \, \partial z^2 \, \partial x} = \frac{\partial^4 f}{\partial z \, \partial x \, \partial z \, \partial y}$$

provided that all fourth-order partial derivatives exist and are continuous.

Higher partial derivatives of composite functions may be evaluated by repeated use of the chain rule (which can become very laborious!).

EXAMPLE 5.5.2. Suppose that

$$u = u(x, y, z)$$

is a function of three variables, but that the third variable is a function of the first two:

$$z = z(x, y).$$

Compute $\partial^2 u / \partial x^2$ (thinking of u as the function $u(x, y, z(x, y))$ of x and y only) in terms of partial derivatives of u (as a function of x, y, and z) and z (as a function of x and y).

A 'traditional' approach to this looks very confusing, because (for example) the symbol 'u' is really being used for two different functions—one of three variables, and one of two. It[*] runs as follows: by the chain rule applied to u,

$$\frac{\partial u}{\partial x} = \frac{\partial u}{\partial x}\frac{\partial x}{\partial x} + \frac{\partial u}{\partial y}\frac{\partial y}{\partial x} + \frac{\partial u}{\partial z}\frac{\partial z}{\partial x}$$

$$= \frac{\partial u}{\partial x} \cdot 1 + \frac{\partial u}{\partial y} \cdot 0 + \frac{\partial u}{\partial z}\frac{\partial z}{\partial x}$$

$$= \frac{\partial u}{\partial x} + \frac{\partial u}{\partial z}\frac{\partial z}{\partial x}.$$

---

[*] Taken from a standard text with minor changes only.

Now we apply it again (noting that $\partial u/\partial x$ and $\partial u/\partial z$ are functions of x, y, z) obtaining

$$\frac{\partial^2 u}{\partial x^2} = \frac{\partial}{\partial x}\left(\frac{\partial u}{\partial x}\right)$$

$$= \frac{\partial}{\partial x}\left(\frac{\partial u}{\partial x} + \frac{\partial u}{\partial z} \cdot \frac{\partial z}{\partial x}\right)$$

$$= \frac{\partial^2 u}{\partial x^2}\frac{\partial x}{\partial x} + \frac{\partial^2 u}{\partial y \partial x}\frac{\partial y}{\partial x} + \frac{\partial^2 u}{\partial z \partial x}\frac{\partial z}{\partial x}$$

$$+ \frac{\partial z}{\partial x}\left(\frac{\partial^2 u}{\partial x \partial z}\frac{\partial x}{\partial x} + \frac{\partial^2 u}{\partial y \partial z}\frac{\partial y}{\partial x} + \frac{\partial^2 u}{\partial z^2}\frac{\partial z}{\partial x}\right) + \frac{\partial u}{\partial z}\frac{\partial^2 z}{\partial x^2}$$

$$= \frac{\partial^2 u}{\partial x^2} + 2\frac{\partial^2 u}{\partial z \partial x}\frac{\partial z}{\partial x} + \frac{\partial^2 u}{\partial z^2}\left(\frac{\partial z}{\partial x}\right)^2 + \frac{\partial u}{\partial z}\frac{\partial^2 z}{\partial x^2}.$$

Note that on the left-hand side we are thinking of u as the function $u(x, y, z(x, y))$ of two variables; but on the right-hand side as $u(x, y, z)$, a function of three variables. Hence expressions such as $\partial u/\partial x$ take on a different meaning according to which side of the equation they inhabit.

For example, suppose that $u(x, y, z) = x + y + z$, and $z(x, y) = x^2 + y^2$. Then, on the left-hand side, we have $u = u(x, y, z(x, y)) = x + y + x^2 + y^2$; so $\partial u/\partial x = 1 + 2x$. But on the right, $u = u(x, y, z) = x + y + z$, so $\partial u/\partial x = 1$.

In traditional terms the problem is traced to the fact that x, y, and z are not *independent* variables: z is related to x and y. While true, this poses problems of its own: the symbol $\partial u/\partial x$ *requires* independent variables for its definition, so the right-hand side, traditionally, makes no sense. Or so it might seem. In fact, *conceptually*, everything makes perfectly good sense: it is the notation that causes trouble, especially the habit of using the same symbols for variables appearing in different functions.

A more civilized approach runs as follows. Introduce three new variables p, q, r. We have a function

$$u = u(p, q, r)$$

of these. Change variable according to

$$\left.\begin{array}{l} p = x \\ q = y \\ r = z(x, y) \end{array}\right\} \tag{5.5.3}$$

where z is a function of two variables x and y. *Define the function v by**

$$v(x, y) = u(x, y, z(x, y)).$$

---

* This is the crucial step!

Now our problem is to compute $\partial^2 v/\partial x^2$ (note the change from u to v here!) in terms of derivatives of u and z. This is done as follows:

$$v(x, y) = u(p, q, r)$$

so

$$\frac{\partial v}{\partial x} = \frac{\partial u}{\partial p}\frac{\partial p}{\partial x} + \frac{\partial u}{\partial q}\frac{\partial q}{\partial x} + \frac{\partial u}{\partial r}\frac{\partial r}{\partial x}$$

$$= \frac{\partial u}{\partial p} \cdot 1 + \frac{\partial u}{\partial q} \cdot 0 + \frac{\partial u}{\partial r}\frac{\partial z}{\partial x}$$

using (5.5.3).

Now

$$\frac{\partial^2 v}{\partial x^2} = \frac{\partial}{\partial x}\left(\frac{\partial v}{\partial x}\right)$$

$$= \frac{\partial}{\partial x}\left(\frac{\partial u}{\partial p}\right) + \frac{\partial}{\partial x}\left(\frac{\partial u}{\partial r} \cdot \frac{\partial z}{\partial x}\right).$$

Here, $\partial u/\partial p$ and $\partial u/\partial r$ are functions of p, q, r, whereas $\partial z/\partial x$ is a function of x and y. So we proceed, using the formula for differentiating a product, and then the chain rule again:

$$\frac{\partial^2 v}{\partial x^2} = \frac{\partial}{\partial x}\left(\frac{\partial u}{\partial p}\right) + \left(\frac{\partial}{\partial x}\left(\frac{\partial u}{\partial r}\right)\right) \cdot \frac{\partial z}{\partial x} + \frac{\partial u}{\partial r} \cdot \frac{\partial^2 z}{\partial x^2}$$

$$= \frac{\partial}{\partial p}\left(\frac{\partial u}{\partial p}\right)\frac{\partial p}{\partial x} + \frac{\partial}{\partial q}\left(\frac{\partial u}{\partial p}\right)\frac{\partial q}{\partial x} + \frac{\partial}{\partial r}\left(\frac{\partial u}{\partial p}\right)\frac{\partial r}{\partial x}$$

$$+ \left(\frac{\partial}{\partial p}\left(\frac{\partial u}{\partial r}\right)\frac{\partial p}{\partial x} + \frac{\partial}{\partial q}\left(\frac{\partial u}{\partial r}\right)\frac{\partial q}{\partial x} + \frac{\partial}{\partial r}\left(\frac{\partial u}{\partial r}\right)\frac{\partial r}{\partial x}\right) \cdot \frac{\partial z}{\partial x} + \frac{\partial u}{\partial r} \cdot \frac{\partial^2 z}{\partial x^2}.$$

Using (5.5.3) we have

$$\frac{\partial p}{\partial x} = 1$$

$$\frac{\partial q}{\partial x} = 0$$

$$\frac{\partial r}{\partial x} = \frac{\partial z}{\partial x}$$

and hence the expression becomes

$$\frac{\partial^2 v}{\partial x^2} = \frac{\partial^2 u}{\partial p^2} + \frac{\partial^2 u}{\partial r \partial p} \cdot \frac{\partial z}{\partial x} + \left(\frac{\partial^2 u}{\partial p \partial r} + \frac{\partial^2 u}{\partial r^2} \cdot \frac{\partial z}{\partial x}\right) \cdot \frac{\partial z}{\partial x} + \frac{\partial u}{\partial r} \cdot \frac{\partial^2 z}{\partial x^2}$$

$$= \frac{\partial^2 u}{\partial p^2} + \left(\frac{\partial^2 u}{\partial r \partial p} + \frac{\partial^2 u}{\partial p \partial r}\right)\frac{\partial z}{\partial x} + \frac{\partial^2 u}{\partial r^2}\left(\frac{\partial z}{\partial x}\right)^2 + \frac{\partial u}{\partial r}\frac{\partial^2 z}{\partial x^2}. \qquad (5.5.4)$$

If u has continuous second partials, we can use the fact that $\partial^2 u/\partial r \partial p = \partial^2 u/\partial p \partial r$ to rewrite this as

$$\frac{\partial^2 u}{\partial p^2} + 2\,\frac{\partial^2 u}{\partial p\,\partial r}\cdot\frac{\partial z}{\partial x} + \frac{\partial^2 u}{\partial r^2}\left(\frac{\partial z}{\partial x}\right)^2 + \frac{\partial u}{\partial r}\,\frac{\partial^2 z}{\partial x^2}.$$

The advantage of the second approach, above, is that it is computationally routine and notationally unambiguous. There is nothing wrong with the more traditional approach, but it requires more careful thought as to what the symbolism *means*. To someone well versed in these calculations, such thought is automatic; but to the student* the whole process may appear baffling.

EXAMPLE 5.5.3.   As a check, we evaluate the above expression for the example $u(x, y, z) = x + y + z$, $z(x, y) = x^2 + y^2$ introduced above. In terms of p, q, r we have

$$u(p, q, r) = p + q + r$$

$$p = x$$

$$q = y$$

$$r = x^2 + y^2 = z.$$

Then

$$\frac{\partial u}{\partial p} = 1 = \frac{\partial u}{\partial r}$$

so that

$$\frac{\partial^2 u}{\partial p^2} = \frac{\partial^2 u}{\partial p\,\partial r} = \frac{\partial^2 u}{\partial r^2} = 0$$

and

$$\frac{\partial^2 z}{\partial x^2} = 2.$$

Thus the expression (5.5.4) yields

$$\frac{\partial^2 v}{\partial x^2} = 2.$$

Now

$$v(x, y) = x + y + x^2 + y^2$$

$$\frac{\partial v}{\partial x} = 1 + 2x$$

$$\frac{\partial^2 v}{\partial x^2} = 2$$

as claimed.

---

* And not only the student! Many standard texts on thermodynamics, for example, become seriously confused over exactly this kind of issue—notably as regards 'Legendre transformations'.

5.6. CRITICAL POINTS

DEFINITION 5.6.1. A *critical, stationary,* or *level* point of a function

$$f : D \to \mathbb{R}$$

of n variables is a point $\mathbf{x} = (x_1, \ldots, x_n)$ at which all first partial derivatives vanish:

$$\partial f / \partial x_i = 0 \qquad (i = 1, \ldots, n).$$

Geometrically, it is a point at which the graph of f is horizontal (that is, all tangent lines are horizontal) as in Figure 5.6.1.

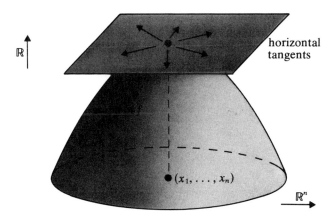

Figure 5.6.1

DEFINITION 5.6.2. The value $f(x_1, \ldots, x_n)$ at such a critical point is called a *critical value* of f (and must not be confused with the critical point $(x_1, \ldots, x_n)$ at which f is evaluated).

EXAMPLE 5.6.1. To find the critical points of

$$f(x, y) = x^3 - 3xy^2 - 3x^2 + 12xy + 3y^2 - 9x - 12y + 15$$

we must solve the equations

$$\frac{\partial f}{\partial x} = 0, \qquad \frac{\partial f}{\partial y} = 0.$$

These yield

$$3x^2 - 3y^2 - 6x + 12y - 9 = 0$$

$$-6xy + 12x + 6y - 12 = 0$$

or

$$x^2 - y^2 - 2x + 4y - 3 = 0$$

$$xy - 2x - y + 2 = 0.$$

We may write these as

$$(x-1)^2 - (y-2)^2 = 0$$

$$(x-1)(y-2) = 0$$

for which the only solutions are

$$x = 1, \qquad y = 2.$$

Hence f has a unique critical point $(1, 2)$.

The critical *value* at f is the number $f(1, 2) = 4$.

We illustrate the graph of f near $(1, 2)$ in Figure 5.6.2. Despite its complicated shape, all tangents at $(1, 2)$ are horizontal.

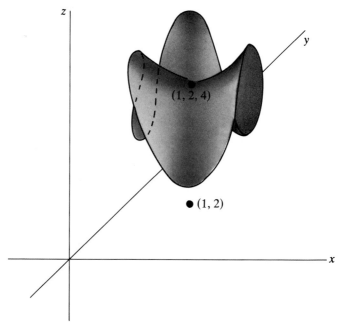

Figure 5.6.2

One application of these ideas is to finding maxima and minima (although many more sophisticated applications are possible—see catastrophe theory, Volume V, Chapter 8).

DEFINITION 5.6.3.   A *local maximum* of f is a point $\mathbf{x} \in D$ such that there exists $\delta > 0$ for which

$$f(\mathbf{x}) \geq f(\mathbf{y}) \tag{5.6.1}$$

whenever $\|\mathbf{x} - \mathbf{y}\| < \delta$. That is, $f(\mathbf{x})$ is greater than or equal to any *nearby* value of f. There is a similar definition of a *local minimum*, with (5.6.1) changed to

$$f(\mathbf{x}) \leq f(\mathbf{y})$$

whenever $\|\mathbf{x} - \mathbf{y}\| < \delta$. A *global maximum* $\mathbf{x}$ of f satisfies

$$f(\mathbf{x}) \geq f(\mathbf{y})$$

for *all* $\mathbf{y} \in D$; a *global minimum* satisfies

$$f(\mathbf{x}) \leq f(\mathbf{y})$$

for all $\mathbf{y} \in D$.

Figure 5.6.3 illustrates various possibilities.

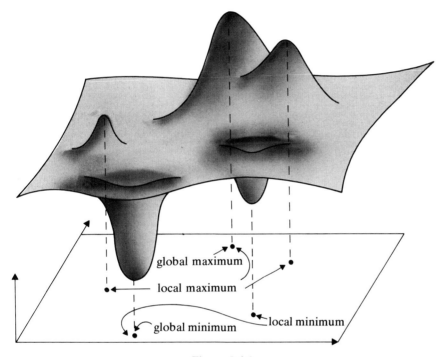

global maximum

local maximum

global minimum

local minimum

Figure 5.6.3

The existence of maxima and minima is often ensured by the following result:

THEOREM 5.6.1.   *Let D be a closed bounded subset of $\mathbb{R}^n$, and let $f : D \rightarrow \mathbb{R}$ be continuous. Then f has at least one global minimum and at least one global maximum in D.*

'*Closed*' here means that limits of sequences in D [see § 1.2] must also lie in D—equivalently, that D includes its boundary. '*Bounded*' means that there exists $M \in \mathbb{R}$ with $\|\mathbf{x}\| \leq M$ for all $\mathbf{x} \in D$. The theorem is proved by using a version of the Heine–Borel theorem [see Theorem 2.8.2] for functions of several variables, and will be discussed further in Chapter 11 (metric spaces). It is an intuitively appealing result, which turns out to be hard to prove rigorously.

The connection with critical points is the following result.

THEOREM 5.6.2.   *Suppose that $f : D \to \mathbb{R}$ is continuous and that its first partial derivatives exist throughout D. Then local maxima and minima of f occur either on the boundary of D, or at critical points of f.*

Thus, to find local maxima and minima, it suffices to look on the boundary of *D*, and at critical points. *Usually* this leads to only finitely many local maxima and minima; hence we may decide on the global maxima and minima by comparing the values of the function at the points in question. (A method often applicable on the boundary is given in section 5.15.)

EXAMPLE 5.6.2.   Find the point on the plane whose equation is

$$4x - 5y + z = 2$$

that is nearest to the origin.

The square of the distance *d* from the origin [see V (2.1.7)] is given by

$$d^2 = x^2 + y^2 + z^2$$
$$= x^2 + y^2 + (2 - 4x + 5y)^2$$
$$= 17x^2 + 26y^2 - 40xy - 16x + 20y + 4 = f(x, y),$$

say. Minimizing *d* is the same as minimizing d^2. So first we calculate the critical points of *f*. We have

$$0 = \partial f / \partial x = 34x - 40y - 16$$
$$0 = \partial f / \partial y = -40x + 52y + 20.$$

Solving in the usual way [see I, § 5.8] we find that

$$x = 4/21, \qquad y = -5/21,$$

from which

$$z = 1/21.$$

On geometrical grounds it is immediately obvious that this point indeed gives a *global* minimum. In a less obvious case to decide this might take much more work. Note that here *x* and *y* may take arbitrary values (the domain of *f* is $\mathbb{R}^2$) so there is no boundary to look at.

Note that in general a critical point may be neither a local maximum nor a local minimum. For example the function

$$f(x, y) = x^2 - y^2$$

has a critical point at the origin: this is a local minimum 'in the *x*-direction' but a local *maximum* 'in the *y*-direction', as shown in Figure 5.6.4. This kind of critical point is called a *saddle*. Figure 5.6.2 shows an even more complicated type of critical point, known as a *monkey-saddle* (a joke that goes back to the days of Hilbert: the monkey has to have a place to put his tail).

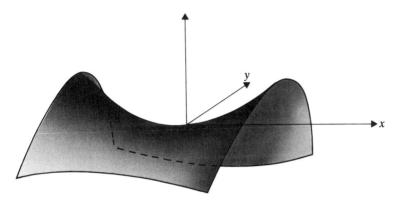

Figure 5.6.4

The nature and classification of types of critical point is highly complex (at least in more than one variable) and only recently has substantial progress been made: [see Volume V, Chapter 8]. However, there is one fairly simple test that *usually* works (for a single function in the absence of special features such as symmetry) and whose failure indicates the need for more sophisticated analysis. To this we turn our attention.

5.7. THE HESSIAN

Here we shall give the test for local maxima, minima, or saddles for functions of two variables only. A detailed explanation of the reasoning behind it is postponed until we have discussed Taylor series expansions [see § 5.8], as is the formulation for functions of more than two variables [see the *Morse Lemma*, V, Lemma 8.2.2].

THEOREM 5.7.1.   *Let* $f : D \to \mathbb{R}$ *be continuous, where* $D \subseteq \mathbb{R}^2$, *and suppose that all third order partial derivatives of f exist and are continuous near the point* (a, b). *Suppose further that* (a, b) *is a critical point of f. Define*

$$p = \partial^2 f / \partial^2 x^2$$

$$q = \partial^2 f / \partial^2 y^2$$

$$r = \partial^2 f / \partial y \, \partial x$$

evaluated at the point (a, b). *Then*:

(1) (a, b) *is a local minimum of f if*

$$pq - r^2 > 0 \quad and \quad p > 0.$$

(2) (a, b) *is a local maximum of f if*

$$pq - r^2 > 0 \quad and \quad p < 0.$$

(3) (a, b) *is a saddlepoint if*

$$pq - r^2 < 0.$$

If $pq - r^2 = 0$ *the test gives no definite decision.*

A proof is given in Protter & Morrey [1964] p. 704: we sketch the general reasoning in section 5.8.

The quantity $pq - r^2$ that features so prominently here is the determinant of the *Hessian matrix*

$$H = \begin{pmatrix} p & r \\ r & q \end{pmatrix} \tag{5.7.1}$$

[see I, § 6.9] and is called the *Hessian* (or *Hessian determinant*).

Let us try out this test on the critical points analyzed above.

EXAMPLE 5.7.1

$$f(x, y) = x^3 - 3xy^2 - 3x^2 + 12xy + 3y^2 - 9x - 12y + 15.$$

We have

$$\partial^2 f/\partial x^2 = 6x - 6$$

$$\partial^2 f/\partial y^2 = -6x + 6$$

$$\partial^2 f/\partial y \partial x = -6y + 12.$$

At the critical point $(a, b) = (1, 2)$ we therefore have

$$p = 0, \qquad q = 0, \qquad r = 0,$$

so $pq - r^2 = 0$ and the test gives no information. (Nor should it, since we have a monkey-saddle; what the test tells us to do is to look more closely.)

EXAMPLE 5.7.2

$$f(x, y) = 17x^2 + 26y^2 - 40xy - 16x + 20y + 4.$$

We have

$$\partial^2 f/\partial x^2 = 34$$

$$\partial^2 f/\partial y^2 = 52$$

$$\partial^2 f/\partial y \partial x = -40.$$

At the critical point $(a, b) = (0, 0)$ we have

$$pq - r^2 = 34 \cdot 52 - (-40)^2$$
$$= 1768 - 1600$$
$$= 168$$

which is positive; and p is positive; so by Theorem 5.7.1 we are in case (1)—a local minimum.

This confirms the conclusion that we arrived at by geometric intuition before.

EXAMPLE 5.7.3

$$f(x, y) = x^2 - y^2.$$

Here we have

$$\partial^2 f / \partial x^2 = 2$$
$$\partial^2 f / \partial y^2 = -2$$
$$\partial^2 f / \partial y \partial x = 0.$$

At the critical point $(a, b) = (0, 0)$ we have

$$p = 2, \qquad q = -2, \qquad r = 0$$

so that

$$pq - r^2 = -4$$

which is negative. By Theorem 5.7.1 we are in case (3)—a saddlepoint.

5.8. TAYLOR EXPANSIONS

Recall that a smooth function of one variable has an associated *Taylor series* expansion [see §§ 2.10 and 3.6]. There is an analogous expansion for functions of several variables, to which we now turn our attention. For simplicity we begin with the two-variable case.

THEOREM 5.8.1 (*Taylor's Theorem*). *Let* $f : D \to \mathbb{R}$ *be a function of two variables, all of whose partial derivatives of order less than or equal to* $p + 1$ *exist and are continuous near* $(a, b) \in D$. *Then for* (x, y) *near* (a, b),

$$f(x, y) = f(a, b) + \sum_{1 \le r+s \le p} \frac{\partial^{r+s} f(a, b)}{\partial x^r \partial y^s} \frac{(x - a)^r}{r!} \frac{(y - b)^s}{s!} + R_p$$

where

$$R_p = \sum_{r+s=p+1} \frac{\partial^{r+s} f(\xi, \eta)}{\partial x^r \partial y^s} \frac{(x - a)^r}{r!} \frac{(y - b)^s}{s!} \tag{5.8.1}$$

and (ξ, η) *is some point on the straight line segment between* (a, b) *and* (x, y), *as in Figure* 5.8.1.

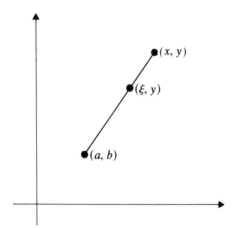

Figure 5.8.1

Here 'near' means 'for all (x, y) such that $\|(a, b) - (x, y)\| < \delta$ for a given $\delta > 0$'. The theorem does not tell us exactly where (ξ, η) lies, but gives enough information for us to estimate the size of the *remainder term* R_p in many cases. The summation sign

$$\sum_{1 \le r+s \le p}$$

requires us to add all possible terms for choices of positive integers r, s (including zero) satisfying the condition $1 \le r + s \le p$.

Noting that [see I (3.8.1)]

$$\frac{1}{r!\,s!} = \frac{1}{(r+s)!}\frac{(r+s)!}{r!\,s!} = \frac{1}{(r+s)!}\binom{r+s}{r}$$

it is easy to write down the first few terms of this expansion:

$$f(x, y) = f(a, b) + \frac{\partial f}{\partial x}(a, b)(x - a) + \frac{\partial f}{\partial y}(a, b)(y - b)$$

$$+ \frac{1}{2}\left[\frac{\partial^2 f}{\partial x^2}(a, b)(x - a)^2 + 2\frac{\partial^2 f}{\partial x\,\partial y}(a, b)(x - a)(y - b)\right.$$

$$\left. + \frac{\partial^2 f}{\partial y^2}(a, b)(y - b)^2\right]$$

$$+ \frac{1}{6}\left[\frac{\partial^3 f}{\partial x^3}(a, b)(x - a)^3 + \frac{3\partial^3 f}{\partial x^2\,\partial y}(a, b)(x - a)^2(y - b)\right.$$

$$\left. + \frac{3\partial^3 f}{\partial x\,\partial y^2}(a, b)(x - a)(y - b)^2 + \frac{\partial^3 f}{\partial y^3}(a, b)(y - b)^3\right] + \ldots.$$

EXAMPLE 5.8.1.   Let $(a, b) = (0, 0)$ and let

$$f(x, y) = \sin x \cos y.$$

Find the Taylor expansion of f up to terms in x and y of order 3.
  We compute, successively, the partial derivatives:

$$\frac{\partial f}{\partial x} = \cos x \cos y$$

$$\frac{\partial f}{\partial y} = -\sin x \sin y$$

$$\frac{\partial^2 f}{\partial x^2} = -\sin x \cos y$$

$$\frac{\partial^2 f}{\partial x \, \partial y} = -\cos x \sin y$$

$$\frac{\partial^2 f}{\partial y^2} = -\sin x \cos y$$

$$\frac{\partial^3 f}{\partial x^3} = -\cos x \cos y$$

$$\frac{\partial^3 f}{\partial x^2 \, \partial y} = \sin x \sin y$$

$$\frac{\partial^3 f}{\partial x \, \partial y^2} = -\cos x \cos y$$

$$\frac{\partial^3 f}{\partial y^3} = \sin x \sin y.$$

Evaluated at $x = 0$, $y = 0$ these become, in the order listed,

$$1, 0, 0, 0, 0, -1, 0, -1, 0.$$

Hence the expansion to order 3 is:

$$f(x, y) = 0 + 1 \cdot x + 0 \cdot y + \tfrac{1}{2}[0x^2 + 2(0)xy + 0y^2]$$
$$+ \tfrac{1}{6}[-1x^3 + 3(0)x^2 y + 3(-1)xy^2 + 0y^3] + \ldots$$
$$= x - \frac{x^3}{6} - \frac{xy^2}{2}.$$

As a check, we may use the series expansion [see (2.12.1) and (2.12.2)]

$$\sin x = x - \frac{x^3}{6} + \ldots$$

$$\cos y = 1 - \frac{y^2}{2} + \ldots$$

to obtain

$$f(x, y) = \left(x - \frac{x^3}{6} + \ldots\right)\left(1 - \frac{y^2}{2} + \ldots\right)$$

$$= x - \frac{x^3}{6} - \frac{xy^2}{2}$$

to terms of order 3.

DEFINITION 5.8.1. If $R_p \to 0$ as $p \to \infty$ we obtain a convergent (double) power series for $f(x, y)$ [see § 1.11], its *Taylor series*. In this case f is said to be *analytic* near (a, b).

If f is merely *smooth* (or of class C^∞), that is, has continuous partial derivatives of all orders [see Definition 5.9.2] then the Taylor series exists as a *formal* power series but (as in the one-variable case) need not converge; or, if it converges, may not converge to f. This has led to an emphasis on analytic functions in the literature, where Taylor series expansions up to some chosen order are customarily used as approximations to f in computations. However, it is neither necessary nor sufficient that a function be analytic for this technique to work: see V, Chapter 8 or Poston and Stewart [1978]. For functions of several variables the Taylor expansion is a delicate tool as well as a powerful one.

For functions of three variables the Taylor expansion takes the form

$$f(x, y, z) = f(a, b, c) + \sum_{1 \le r+s+t \le p} \frac{\partial^{r+s+t} f(a, b, c)}{\partial x^r \, \partial y^s \, \partial z^t} \frac{(x-a)^r}{r!} \frac{(y-b)^s}{s!} \frac{(z-c)^t}{t!} + R_p$$

$$(5.8.2)$$

where

$$R_p = \sum_{r+s+t=p+1} \frac{\partial^{p+1} f(\xi, \eta, \zeta)}{\partial x^r \, \partial y^s \, \partial z^t} \frac{(x-a)^r}{r!} \frac{(y-b)^s}{s!} \frac{(z-c)^t}{t!} \qquad (5.8.3)$$

and (ξ, η, ζ) is on the line joining (a, b, c) to (x, y, z). For more details, and proofs, see Protter and Morrey [1964] p. 697. The reader will easily write down the corresponding expression for four, five, ... n variables.

The Hessian test for maxima, minima, and saddles [see Theorem 5.7.1] may now be rendered plausible. Let $f: D \to \mathbb{R}$ be a function of two variables having a critical point at (a, b), and as before define

$$p = \frac{\partial^2 f}{\partial x^2}(a, b)$$

$$r = \frac{\partial^2 f}{\partial x \, \partial y}(a, b)$$

$$q = \frac{\partial^2 f}{\partial y^2}(a, b).$$

Consider the Taylor expansion to order 2: this is

$$f(x, y) = f(a, b) + 0 \cdot (x - a) + 0 \cdot (y - b)$$
$$+ \tfrac{1}{2}[p(x - a)^2 + 2r(x - a)(y - b) + q(y - b)^2] + \ldots.$$

The constant term $f(a, b)$ affects only the critical *value*, and does not affect the shape of the graph of f near (a, b). There is no linear term because we are at a critical point. The Hessian test says that the quadratic term determines the nature of the critical point, provided that it is *nondegenerate*. More precisely, recall [see I, § 9.1] that the nature of a quadratic form is determined by whether the corresponding quadratic equation has real or complex roots, as follows: If the form is

$$aX^2 + bXY + cY^2$$

then the polynomial equation $au^2 + bu + c = 0$ has two distinct real roots if $b^2 - 4ac > 0$, complex roots if $b^2 - 4ac < 0$, and equal real roots if $b^2 - 4ac = 0$ [see I, Proposition 14.5.4]. The quantity $b^2 - 4ac$ is called the *discriminant* of the form. In our case, the relevant form is

$$\tfrac{1}{2}pX^2 + rXY + \tfrac{1}{2}qY^2$$

(putting $X = x - a$, $Y = y - b$), with discriminant

$$r^2 - pq,$$

the negative of the Hessian. So the sign of the Hessian determines the nature of the form. Negative Hessian corresponds to distinct real roots, which gives a saddle; positive Hessian corresponds to complex roots, and we have a maximum or minimum according as the form is negative definite or positive definite (hence the relevance also of the sign of p). Zero Hessian corresponds to a *degenerate form*, that is a form with two equal roots: the degeneracy leaves room for higher order terms to affect the nature of the critical point, and the test is not sensitive enough to decide.

For functions of n variables the corresponding analysis is most succinctly stated as the *Morse Lemma* [see V, Lemma 8.2.2].

5.9.   THE DERIVATIVE AS A LINEAR TRANSFORMATION

A more sophisticated approach to differentiating functions of several variables unites the various partial derivatives into a single mathematical object, independent of the choice of coordinates in $\mathbb{R}^n$, which provides some useful insight and generalizes effectively to function spaces. We shall consider this now, though not in full abstraction, since it permits a more elegant formulation of some important theorems, and (expressed in matrix language) renders previous results, such as the chain rule, more intuitive.

The derivative Df (or df/dx or f') of a function $f : \mathbb{R} \to \mathbb{R}$ has been defined by

$$Df(x) = \lim_{h \to 0} \frac{f(x + h) - f(x)}{h} \qquad (5.9.1)$$

provided that the limit exists [see Definition 2.9.1]. Geometrically the value of Df at x expresses the slope of the tangent line to the graph of f at the point $(x, f(x))$ as in Figure 5.9.1 [see § 3.1.1].

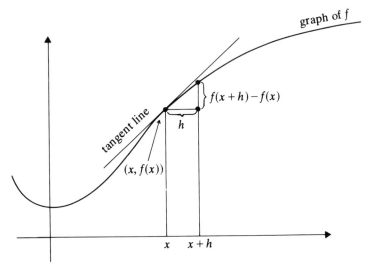

Figure 5.9.1

For convenience let us transfer the origin of coordinates to this point $(x, f(x))$. The tangent line then becomes the graph of a *linear* function $\lambda : \mathbb{R} \to \mathbb{R}$, defined by

$$\lambda(x) = kx$$

where k is the slope of the tangent (so $k = Df(x)$).

In general any linear map $\lambda : \mathbb{R} \to \mathbb{R}$ may be used in this way to define a straight line through $(x, f(x))$, but only one choice, λ_x, of λ yields the *tangent* line. Intuitively, the tangent line is the *best linear approximation* to the graph of f at the given point. How good an approximation it is may be measured by comparing the graphs concerned near x. That is, we must compare

$$f(x + h)$$

and

$$f(x) + \lambda_x(h)$$

for small values of h. The requirement turns out to be that the difference between these,

$$|f(x + h) - f(x) - \lambda_x(h)|,$$

should tend to 0 faster than h does, in the sense that

$$\frac{|f(x+h)-f(x)-\lambda_x(h)|}{|h|} \to 0 \qquad (5.9.2)$$

as $h \to 0$.

(As motivation think of the Taylor expansion of f about x, thus

$$f(x+h) = f(x) + f'(x)h + \tfrac{1}{2}f''(x)h^2 + \dots .$$

Defining $\lambda_x(h) = f'(x)h$, as we should, given the above interpretation of λ_x, we find that

$$\frac{|f(x+h)-f(x)-\lambda_x(h)|}{|h|} = |\tfrac{1}{2}f''(x)h + \dots|$$

which tends to zero with h. This argument provides *motivation* only: it is rigorous only when f is analytic, and we shall wish to apply the ideas we are developing in a much less restrictive context. Nevertheless, it makes the requirement (5.9.2) a sensible one.)

The main object of this otherwise apparently senseless exercise is that (5.9.2) generalizes at once to functions $\mathbf{f}: \mathbb{R}^n \to \mathbb{R}^m$, whereas (5.9.1) does not:

DEFINITION 5.9.1.   We say that a function

$$\mathbf{f}: \mathbb{R}^n \to \mathbb{R}^m$$

is *differentiable* at $\mathbf{x} \in \mathbb{R}^n$ if there exists a linear map [see I, § 5.11]

$$\lambda_{\mathbf{x}}: \mathbb{R}^n \to \mathbb{R}^m$$

such that

$$\lim_{\mathbf{h}\to 0} \frac{\|\mathbf{f}(\mathbf{x}+\mathbf{h})-\mathbf{f}(\mathbf{x})-\lambda_{\mathbf{x}}(\mathbf{h})\|}{\|\mathbf{h}\|} = 0.$$

Note that $\mathbf{h}$ and $\mathbf{x}$ here are vectors in $\mathbb{R}^n$, but we divide out only by the norm $\|\mathbf{h}\|$, which is a real *number* [see I (9.1.2)], so the division makes sense. To generalize (5.9.1) requires division by $\mathbf{h}$, a *vector*, which does *not* make sense: that is why we had to manipulate (5.9.1) into the form (5.9.2).

It is easy to see that $\lambda_{\mathbf{x}}$ is unique if it exists: roughly, the difference between two distinct $\lambda_{\mathbf{x}}$ would be a nonzero linear function of $\mathbf{h}$, hence would tend to zero at the same speed as $\mathbf{h}$, and not faster as the formula requires. See Poston and Stewart (1978) p. 38, or Spivak (1965).

The linear map $\lambda_{\mathbf{x}}$ is defined to be the *value* of the derivative of $\mathbf{f}$ at $\mathbf{x}$, and we use the notation

$$D\mathbf{f}(\mathbf{x}) = \lambda_{\mathbf{x}}. \qquad (5.9.3)$$

This makes the *derivative* a fairly complex object: it is a function $D\mathbf{f}$ with domain $\mathbb{R}^n$, whose codomain is the set of all linear maps $\mathbb{R}^n \to \mathbb{R}^m$. Why?

Because for each $\mathbf{x}$ in $\mathbb{R}^n$ (the domain) the value of $D\mathbf{f}$ at $\mathbf{x}$ is $D\mathbf{f}(\mathbf{x}) = \lambda_{\mathbf{x}}$, which is a linear map from $\mathbb{R}^n \to \mathbb{R}^m$. In other words, the linear map *varies* as $\mathbf{x}$ does. This is unsurprising, for in the case $f : \mathbb{R} \to \mathbb{R}$ all it reflects is that the *slope* of the graph at $(x, f(x))$ varies as x does.

There is a similar geometric interpretation of the derivative in several variables: it is the *best linear approximation* to the graph of the function near the point $\mathbf{x}$ in question. Figure 5.9.2 illustrates this for a function $f : \mathbb{R}^2 \to \mathbb{R}$, where the tangent plane to the graph of f is also the graph of this linear approximation to f.

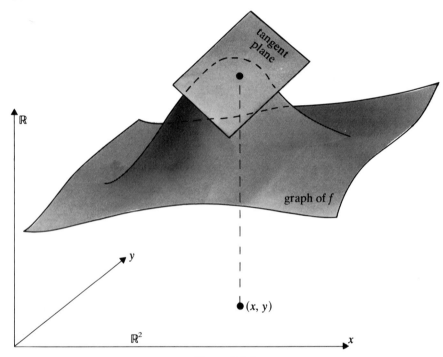

Figure 5.9.2

The derivative, so defined, obeys a number of standard laws, all relatively easy to prove. For details, see Spivak (1965).

(1) If $\mathbf{f} : \mathbb{R}^n \to \mathbb{R}^m$ is constant, then

$$D\mathbf{f}(\mathbf{x}) = \mathbf{0}. \qquad (5.9.4)$$

(2) If $\mathbf{f} : \mathbb{R}^n \to \mathbb{R}^m$ is linear, then

$$D\mathbf{f}(\mathbf{x}) = \mathbf{f} \quad \text{for all } \mathbf{x}. \qquad (5.9.5)$$

(Note. This does *not* mean that $D\mathbf{f} = \mathbf{f}$. The value of the derivative $D\mathbf{f}$ at $\mathbf{x}$ is the linear map $\mathbf{f}$, *not* its value $\mathbf{f}(\mathbf{x})$ at $\mathbf{x}$. Since $\mathbf{f}$ is already linear, it is clearly the best linear approximation to itself near any point $\mathbf{x}$.)

(3) If $\mathbf{f}: \mathbb{R}^n \to \mathbb{R}^m$ has coordinate functions f_i given by $\mathbf{f}(\mathbf{x}) = (f_1(\mathbf{x}), \ldots, f_m(\mathbf{x}))$ ($\mathbf{x} \in \mathbb{R}^n$), then f is differentiable if and only if each f_i is, and then

$$D\mathbf{f}(\mathbf{x}) = (Df_1(\mathbf{x}), \ldots, Df_m(\mathbf{x})). \tag{5.9.6}$$

(4) If $\mathbf{f}, \mathbf{g}: \mathbb{R}^n \to \mathbb{R}^m$ and $k \in \mathbb{R}$, then

$$D(\mathbf{f} + \mathbf{g})(\mathbf{x}) = D\mathbf{f}(\mathbf{x}) + D\mathbf{g}(\mathbf{x}) \tag{5.9.7}$$

$$D(k\mathbf{f})(\mathbf{x}) = kD\mathbf{f}(\mathbf{x}). \tag{5.9.8}$$

(5) If $\mathbf{f}: \mathbb{R}^n \to \mathbb{R}^m$ is differentiable at $\mathbf{x}$, and $\mathbf{g}: \mathbb{R}^m \to \mathbb{R}^p$ is differentiable at $\mathbf{f}(\mathbf{x})$, then $\mathbf{g} \circ \mathbf{f}: \mathbb{R}^n \to \mathbb{R}^p$ (defined by $\mathbf{g} \circ \mathbf{f}(\mathbf{x}) = \mathbf{g}(\mathbf{f}(\mathbf{x}))$) is differentiable at $\mathbf{x}$, and

$$D(\mathbf{g} \circ \mathbf{f})(\mathbf{x}) = D\mathbf{g}(\mathbf{f}(\mathbf{x})) \circ D\mathbf{f}(\mathbf{x}). \tag{5.9.9}$$

Result (5.9.9) is the chain rule [see Theorem 5.4.1] in disguise—as we see in the next section. It says that to obtain the best linear approximation to a composite of two functions, all we need do is compose the best linear approximations to the functions involved (taking these approximations in the neighbourhood of correctly chosen points). This is eminently reasonable, since composition behaves well as regards approximations. This is just one of several theorems concerning the derivative, which may all be summed up as saying that not only is the derivative the best linear approximation available, but that the approximation is excellent for many purposes.

For functions $\mathbb{R}^n \to \mathbb{R}$ there are rules for differentiating products and quotients, strictly analogous to those for functions $\mathbb{R} \to \mathbb{R}$ [see Theorem 3.2.1]. We omit their precise statements.

Higher derivatives may be defined as follows. The space of all linear maps $\mathbb{R}^n$ to $\mathbb{R}^m$ may be identified with $\mathbb{R}^{mn}$ (using matrices as in the next section), so if $\mathbf{f}: \mathbb{R}^n \to \mathbb{R}^m$ then $D\mathbf{f}: \mathbb{R}^{mn} \to \mathbb{R}^m$. The derivative $D(D\mathbf{f})$ is the *second derivative* $D^2\mathbf{f}$ of $\mathbf{f}$ (if it exists); and

$$D^2\mathbf{f}(\mathbf{x}): \mathbb{R}^{m^2 n} \to \mathbb{R}^m. \tag{5.9.10}$$

Third and higher derivatives are then defined recursively by

$$D^{n+1}\mathbf{f}(\mathbf{x}) = D(D^n\mathbf{f}) \tag{5.9.11}$$

provided that $D^n\mathbf{f}$ is differentiable.

DEFINITION 5.9.2.   If $D^r\mathbf{f}$ exists we say f is r *times differentiable*. If further it is continuous then $\mathbf{f}$ is r *times continuously differentiable*, or of *class C^r*. If $\mathbf{f}$ is of class C^r for all r it is *smooth* or of *class C^∞*.

5.10.   THE DERIVATIVE AS A MATRIX

The above definition of the derivative as a linear map is useful for theoretical purposes since it is independent of coordinate choices. For computational purposes, however, it is usual to represent a linear map by a matrix [see I (5.11.11)]. It is here that the partial derivatives come in.

Take a differentiable function

$$\mathbf{f} : \mathbb{R}^n \to \mathbb{R}^m$$

whose coordinate functions are $f_1, \ldots, f_m$. (Note that these are defined relative to the standard basis for $\mathbb{R}^m$ [see I, Example 5.4.2].) The derivative $D\mathbf{f}$ at a point $\mathbf{x} \in \mathbb{R}^n$ is a linear map $\mathbb{R}^n \to \mathbb{R}^m$, and is therefore represented by an $m \times n$ matrix relative to the standard bases for $\mathbb{R}^n$ and $\mathbb{R}^m$. This matrix, it may be proved, has for its (i, j)th entry the first partial derivative

$$\partial f_i / \partial x_j.$$

Hence the matrix of $D\mathbf{f}(\mathbf{x})$ is given by

$$(D\mathbf{f}(\mathbf{x})) = \left(\frac{\partial f_i}{\partial x_j} \right) \qquad (i \leq m, j \leq n). \tag{5.10.1}$$

The proof is not hard: it may be read off from the Taylor expansion (whose linear part, already expressed in terms of partial derivatives, defines the derivative of $\mathbf{f}$) or proved directly (Spivak (1965), Edwards (1973) p. 72). Geometrically it says that the tangent (hyper)plane to the graph of $\mathbf{f}$ must contain the tangent lines in the directions parallel to the axes, as illustrated in Figure 5.10.1.

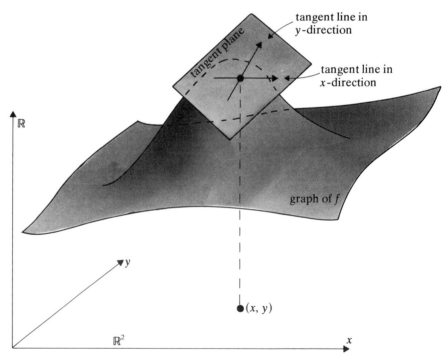

Figure 5.10.1

EXAMPLE 5.10.1.   Let $\mathbf{f}:\mathbb{R}^2 \to \mathbb{R}^3$ be defined by

$$\mathbf{f}(x, y) = (x^2 + y^2, \exp x, y).$$

Find the matrix of the derivative $D\mathbf{f}(x, y)$.

We have coordinate functions

$$f_1(x, y) = x^2 + y^2$$
$$f_2(x, y) = \exp x$$
$$f_3(x, y) = y.$$

So

$$\frac{\partial f_1}{\partial x} = 2x, \qquad \frac{\partial f_1}{\partial y} = 2y$$

$$\frac{\partial f_2}{\partial x} = \exp x, \qquad \frac{\partial f_2}{\partial y} = 0$$

$$\frac{\partial f_3}{\partial x} = 0, \qquad \frac{\partial f_3}{\partial y} = 1$$

and therefore

$$(D\mathbf{f}(x, y)) = \begin{pmatrix} 2x & 2y \\ \exp x & 0 \\ 0 & 1 \end{pmatrix}$$

which is a 3×2 matrix.

EXAMPLE 5.10.2.   Let $\mathbf{f}:\mathbb{R}^4 \to \mathbb{R}^4$ be defined by

$$\mathbf{f}(x, y, z, t) = (yz + t, xz + t, xy + t, x^2 + y^2 + z^2 - t^2).$$

Find the matrix of the derivative.

Now we have

$$f_1(x, y, z, t) = yz + t$$
$$f_2(x, y, z, t) = xz + t$$
$$f_3(x, y, z, t) = xy + t$$
$$f_4(x, y, z, t) = x^2 + y^2 + z^2 - t^2.$$

There are 16 partial derivatives to evaluate, and the matrix required is

$$(D\mathbf{f}(x, y, z, t)) = \begin{pmatrix} 0 & z & y & 1 \\ z & 0 & x & 1 \\ y & x & 0 & 1 \\ 2x & 2y & 2z & -2t \end{pmatrix}.$$

It should be evident that such matrices are relatively rapid to compute.

An important special case occurs for a function $f: \mathbb{R}^3 \to \mathbb{R}$. Then the derivative Df is a vector (equivalently, a 1×3 matrix)

$$(Df) = \left(\frac{\partial f}{\partial x}, \frac{\partial f}{\partial y}, \frac{\partial f}{\partial z}\right). \tag{5.10.2}$$

This vector is often called the *gradient* of f [see (17.3.14)] and denoted

$$\nabla f \quad \text{or} \quad \text{grad} f.$$

The matrix

$$\left(\frac{\partial f_i}{\partial x_j}\right) \qquad (1 \le i \le m; 1 \le j \le n) \tag{5.10.3}$$

of the derivative $D\mathbf{f}(\mathbf{x})$ is often called the *Jacobian matrix* of $\mathbf{f}$, and (especially in older texts) denoted

$$\frac{\partial(f_1, \ldots, f_m)}{\partial(x_1, \ldots, x_n)}.$$

We return once again to the idea of the best linear approximation to a given function which possesses continuous partial derivatives at least of the first order [see (5.9.2) and (5.9.3)].

Let $f: \mathbb{R} \to \mathbb{R}$ be a function of a single variable x. We consider a fixed value of x, and we are interested in the *difference* or *increment* in f, that is

$$\Delta f = f(x+h) - f(x), \tag{5.10.4}$$

as h varies [see Figure 5.9.1]. The definition (5.10.4) applies whether or not h is 'small'; but in practice we are mainly concerned with the case in which h tends to zero. If, by way of motivation, we assume that f has derivatives of sufficiently high order we have the expansion

$$f = f'(x)h + \tfrac{1}{2}f''(x)h^2 + \ldots. \tag{5.10.5}$$

When h is small and $f'(x) \neq 0$, then the first term on the right-hand side dominates the other terms so that

$$f'(x)h$$

provides a good approximation to Δf. For historical reasons it is customary in the present context to put

$$h = dx \tag{5.10.6}$$

and to define

$$df = f'(x) \, dx \tag{5.10.7}$$

as the *differential of* f *at* x. In former times, before mathematical analysis was put on a rigorous basis, dx used to be called an 'infinitesimal' quantity. We do not employ this notion although satisfactory theories are now available.

From our point of view dx is a variable which tends to zero but otherwise possesses no special features except a somewhat bizarre notation. The relationship between Δf and df is summarized by the formula

$$\Delta f = df + o(dx)$$

[see section 2.3]. There is a deceptively simple way in which the chain rule can be 'derived' from (5.10.7). Suppose that x, and therefore also f, is a function of t. On dividing each side by dt we obtain that

$$\frac{df}{dt} = f'(x)\frac{dx}{dt},$$

which is correct but the outcome of an unscrupulous argument.

Next, let $f: \mathbb{R}^2 \to \mathbb{R}$ be a function $f(x, y)$ of two variables. We keep (x, y) fixed and examine the difference

$$\Delta f = f(x + h, y + k) - f(x, y), \tag{5.10.8}$$

as h and k vary. More particularly, we are interested in the case in which h and k simultaneously tend to zero, a situation which may briefly be described by

$$\rho = (h^2 + k^2)^{1/2} \to 0, \tag{5.10.9}$$

the exponent $\frac{1}{2}$ having been introduced to cause ρ to have the same dimension as h and k. We now have the Taylor expansion

$$\Delta f = \frac{\partial f}{\partial x}h + \frac{\partial f}{\partial y}k + \frac{1}{2}\left(\frac{\partial^2 f}{\partial x^2}h^2 + 2\frac{\partial^2 f}{\partial x\,\partial y}hk + \frac{\partial^2 f}{\partial y^2}k^2\right) + \ldots$$

provided that f possesses derivatives of sufficiently high orders. When h and k are small, the linear terms

$$\frac{\partial f}{\partial x}h + \frac{\partial f}{\partial y}k$$

dominate all other terms. We use the conventional notation

$$h = dx, \qquad k = dy, \tag{5.10.10}$$

and we define

$$df = \frac{\partial f}{\partial x}dx + \frac{\partial f}{\partial y}dy \tag{5.10.11}$$

as the *differential* or, more precisely, as the *total differential* of f at (x, y). Again, df is not a mysterious 'infinitesimally small' quantity, but the best linear approximation to Δf in the variables dx and dy. Indeed,

$$\Delta f = df + o(\rho).$$

As was pointed out in section 5.9, the differential is unique in the sense that if

$$f = A \, dx + B \, dy + o(\rho),$$

then

$$A = \frac{\partial f}{\partial x}, \qquad B = \frac{\partial f}{\partial y}.$$

It is now clear how these ideas are generalized to functions of n variables. Suppose that the function

$$f(\mathbf{x}) = f(x_1, x_2, \ldots, x_n)$$

possesses partial derivatives at least of the first order. Let

$$d\mathbf{x} = (dx_1, dx_2, \ldots, dx_n)$$

be another set of n variables. The total differential of f is defined as

$$df = \frac{\partial f}{\partial x_1} \, dx_1 + \frac{\partial f}{\partial x_2} \, dx_2 + \ldots + \frac{\partial f}{\partial x_n} \, dx_n. \qquad (5.10.12)$$

It is convenient to put

$$\rho = \{(dx_1)^2 + (dx_2)^2 + \ldots + (dx_n)^2\}^{1/2}. \qquad (5.10.13)$$

DEFINITION 5.10.1.   The function $f(\mathbf{x})$ is said to be *differentiable* at $\mathbf{x}$ if

$$\Delta f = df + o(\rho). \qquad (5.10.14)$$

It should be noted that a function may possess partial derivatives and yet fail to be differentiable; for the existence of df does not guarantee the truth of (5.10.14).

The following theorem gives sufficient conditions for a function to be differentiable:

THEOREM 5.10.1.   *If the partial derivatives $\partial f/\partial x_1$, $\partial f/\partial x_2$, ..., $\partial f/\partial x_n$ are continuous at $\mathbf{x}$, then f is differentiable at $\mathbf{x}$, that is (5.10.14) holds.*

The formula (5.10.12) for the total differential is frequently used with the tacit understanding that (5.10.14) is true. One application in the experimental sciences is known as the *propagation of errors*: suppose a quantity

$$V = f(x_1, x_2, \ldots, x_n)$$

is to be determined by measuring the values of $x_1, x_2, \ldots, x_n$, which may be subject to observational errors equal to $dx_1, dx_2, \ldots, dx_n$ respectively. Then the resulting total error in V is given by

$$dV = \frac{\partial f}{\partial x_1} \, dx_1 + \frac{\partial f}{\partial x_2} \, dx_2 + \ldots + \frac{\partial f}{\partial x_n} \, dx_n.$$

The idea of a total differential gives rise to a converse problem which, for the sake of simplicity, we shall formulate only for two variables: let $P(x, y)$ and $Q(x, y)$ be two functions which possess continuous partial derivatives of the first order. The expression

$$P\,dx + Q\,dy \qquad\qquad (5.10.15)$$

is called a *Pfaffian form*. The question may be asked whether there exists a function f such that

$$df = P\,dx + Q\,dy. \qquad\qquad (5.10.16)$$

If this equation holds, then by the uniqueness of the differential we must have that

$$P = \frac{\partial f}{\partial x}, \qquad Q = \frac{\partial f}{\partial y}$$

whence

$$\frac{\partial P}{\partial y} = \frac{\partial Q}{\partial x} \qquad\qquad (5.10.17)$$

because each side is equal to $\partial^2 f / \partial x \, \partial y$. The condition (5.10.17) is also sufficient provided that the domain of (x, y) is a simply-connected region of $\mathbb{R}^2$ [see Definition 9.4.2]. Hence we have [see also Theorem 7.10.1]

THEOREM 5.10.2.   *Let D be a simply-connected region of* $\mathbb{R}^2$. *There exists a function* $f: D \to \mathbb{R}$ *such that*

$$df = P\,dx + Q\,dy$$

if and only if $\partial P/\partial y = \partial Q/\partial x$.

5.11.   THE CHAIN RULE IN MATRIX LANGUAGE

We now relate result (5.9.9), the chain rule, to its usual expression in partial derivatives [see Theorem 5.4.1]. Since composition of linear maps corresponds to multiplication of their matrices (relative to a fixed choice of bases) result (5.9.9) translates into matrix terms as follows:

If $\mathbf{f}: \mathbb{R}^n \to \mathbb{R}^m$ is differentiable at $\mathbf{x}$, and $\mathbf{g}: \mathbb{R}^m \to \mathbb{R}^p$ is differentiable at $\mathbf{f}(\mathbf{x})$, then $\mathbf{g} \circ \mathbf{f}: \mathbb{R}^n \to \mathbb{R}^p$ is differentiable at $\mathbf{x}$, and the matrices giving the derivatives are related by

$$(D(\mathbf{g} \circ \mathbf{f})(\mathbf{x})) = (D\mathbf{g}(\mathbf{f}(\mathbf{x})))(D\mathbf{f}(\mathbf{x})) \qquad\qquad (5.11.1)$$

(where the round brackets indicate matrices, and the product is the usual matrix product [see I, § 6.2]).

We illustrate this on (5.4.1). Here we had $f = f(x, y, z)$ where $x = x(r, s)$, $y = y(r, s)$, $z = z(r, s)$. Partial derivatives of f, thought of as the function

$$f(x(r, s), y(r, s), z(r, s))$$

of (r, s) were expressed in terms of partial derivatives of f with respect to x, y, and of z, and x, y, z with respect to r and s.

In matrix terms we proceed as follows. We have two functions

$$f: \mathbb{R}^3 \to \mathbb{R}$$

$$(x, y, z) \to f(x, y, z)$$

and

$$\phi: \mathbb{R}^2 \to \mathbb{R}^3$$

$$(r, s) \to (x(r, s), y(r, s), z(r, s))$$

and we wish to study the composite

$$\psi = f \circ \phi: \mathbb{R}^2 \to \mathbb{R}$$

$$(r, s) \to f(x(r, s), y(r, s), z(r, s)).$$

The derivatives, as matrices, are given by

$$(Df(x, y, z)) = \left(\frac{\partial f}{\partial x}, \frac{\partial f}{\partial y}, \frac{\partial f}{\partial z} \right)$$

$$(D\phi(r, s)) = \begin{pmatrix} \dfrac{\partial x}{\partial r} & \dfrac{\partial x}{\partial s} \\ \dfrac{\partial y}{\partial r} & \dfrac{\partial y}{\partial s} \\ \dfrac{\partial z}{\partial r} & \dfrac{\partial z}{\partial s} \end{pmatrix}$$

$$(D\psi(r, s)) = \left(\frac{\partial \psi}{\partial r}, \frac{\partial \psi}{\partial s} \right).$$

By the chain rule (in matrix formulation),

$$(D\psi) = (D(f \circ \phi)) = (Df)(D\phi).$$

So

$$\left(\frac{\partial \psi}{\partial r}, \frac{\partial \psi}{\partial s} \right) = \left(\frac{\partial f}{\partial x}, \frac{\partial f}{\partial y}, \frac{\partial f}{\partial z} \right) \begin{pmatrix} \dfrac{\partial x}{\partial r} & \dfrac{\partial x}{\partial s} \\ \dfrac{\partial y}{\partial r} & \dfrac{\partial y}{\partial s} \\ \dfrac{\partial z}{\partial r} & \dfrac{\partial z}{\partial s} \end{pmatrix}$$

$$= \left(\frac{\partial f}{\partial x} \frac{\partial x}{\partial r} + \frac{\partial f}{\partial y} \frac{\partial y}{\partial r} + \frac{\partial f}{\partial z} \frac{\partial z}{\partial r}, \; \frac{\partial f}{\partial x} \frac{\partial x}{\partial s} + \frac{\partial f}{\partial y} \frac{\partial y}{\partial s} + \frac{\partial f}{\partial z} \frac{\partial z}{\partial s} \right).$$

Recalling that $\partial\psi/\partial r$ and $\partial\psi/\partial s$ are what, on the left-hand side of (5.4.1) we earlier denoted $\partial f/\partial r$, $\partial f/\partial s$, we recover formula (5.4.1) precisely.

Similar computations work in general, but we shall not make them explicit.

5.12. THE INVERSE FUNCTION THEOREM

If $\mathbf{f} : \mathbb{R}^n \to \mathbb{R}^m$ is a function, then an *inverse* function for $\mathbf{f}$ is a function $\mathbf{g} : \mathbb{R}^m \to \mathbb{R}^n$ such that

$$\mathbf{g}(\mathbf{f}(\mathbf{x})) = \mathbf{x} \qquad (\mathbf{x} \in \mathbb{R}^n)$$
$$\mathbf{f}(\mathbf{g}(\mathbf{y})) = \mathbf{y} \qquad (\mathbf{y} \in \mathbb{R}^m). \tag{5.12.1}$$

The necessary and sufficient condition for $\mathbf{g}$ to exist is that $\mathbf{f}$ be a bijection [see I, § 1.4.2].

A similar definition holds if $\mathbf{f} : U \to V$, where U and V are subsets of $\mathbb{R}^n$ and $\mathbb{R}^m$: we require $\mathbf{g} : V \to U$ to satisfy the conditions (5.12.1) for all $\mathbf{x} \in U, \mathbf{y} \in V$.

In the sequel we shall require the notion of an open subset of $\mathbb{R}^n$ [see § 11.2].

DEFINITION 5.12.1. We say that a subset U is *open* if for all $\mathbf{x} \in U$ there exists $\delta > 0$ such that, whenever $\|\mathbf{x} - \mathbf{y}\| < \delta$, it follows that $\mathbf{y} \in U$. See Figure 5.12.1.

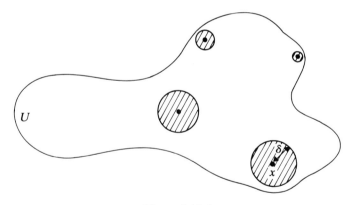

Figure 5.12.1

Intuitively, an open set is one that contains a small ball around each of its points: all points are interior to it. For example in $\mathbb{R}^3$ the set of points

$$U = \{(x, y, z) \mid x^2 + \tfrac{1}{2}y^2 + \tfrac{1}{4}z^2 < 1\}$$

is open: it is an ellipsoid minus its surface [see V, § 2.4.1]. If we change the inequality to '$\leq$', getting an ellipsoid with boundary included, the set is no longer open; small balls around boundary points go outside the set, however small they are (see Figure 5.12.2).

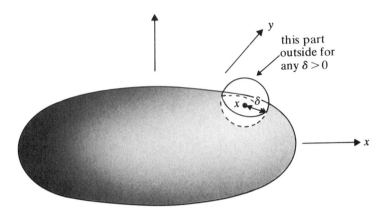

Figure 5.12.2

Our previous theory of functions $\mathbf{f}:\mathbb{R}^n \to \mathbb{R}^m$ can be reworked, without essential changes, to apply to functions $\mathbf{f}:U \to V$ where U and V are open subsets of $\mathbb{R}^n$, $\mathbb{R}^m$ respectively. This is because differentiability is defined *near* a given point. This is useful to do since many important functions are not defined on the whole of $\mathbb{R}^n$; further, a function with domain $\mathbb{R}^n$ may fail to have an inverse, while its restriction to some open subset does have an inverse. (We can see this already in $\mathbb{R}$: the function $f:\mathbb{R}\to\mathbb{R}$ with $f(x)=x^2$ has no inverse, but its restriction to positive reals does have.)

One of the most important theorems concerning the existence of inverses is the following.

THEOREM 5.12.1 (*Inverse Function Theorem*).   *Let* $\mathbf{f}:U \to V$ *where* U *and* V *are open subsets of* $\mathbb{R}^n$, $\mathbb{R}^m$ *respectively. Assume that* $\mathbf{f}$ *is differentiable at* $\mathbf{x} \in U$. *If the linear map* $D\mathbf{f}(\mathbf{x})$ *is non-singular* [*see* I, *Definition* 6.4.2], *then there is an open set* $W \subseteq U$ *such that the restriction* [*see* I (1.4.8)]

$$\mathbf{f}|_W : W \to \mathbf{f}(W)$$

has an inverse $\mathbf{g}$, *and* $\mathbf{g}$ *is differentiable at the point* $\mathbf{f}(\mathbf{x})$. *In fact* $D\mathbf{g}(\mathbf{f}(\mathbf{x})) = (D\mathbf{f}(\mathbf{x}))^{-1}$.

Further, if $\mathbf{f}$ *is r times differentiable at* $\mathbf{x}$, *then* $\mathbf{g}$ *is r times differentiable at* $\mathbf{f}(\mathbf{x})$.

While the statement of this theorem takes a little disentangling, it is not really hard. It says that if the best linear approximation to $\mathbf{f}$ near $\mathbf{x}$ is invertible, then so is $\mathbf{f}$ itself, sufficiently near to $\mathbf{x}$; further the inverse is as smooth as $\mathbf{f}$ is. Once more this tells us that the derivative, as linear approximation, is a good approximation to make: it captures useful information about $\mathbf{f}$ itself.

The function $\mathbf{g}$ whose existence is ensured by the Inverse Function Theorem we shall call a *local* inverse to $\mathbf{f}$, since the conditions (5.12.1) hold only in sufficiently small neighbourhoods of the point concerned.

EXAMPLE 5.12.1. Show that the function

$$\mathbf{f} : \mathbb{R}^2 \to \mathbb{R}^2$$

$$(x, y) \to (x^2 + \sin y, \, xy + 1)$$

has a local inverse near the point $(1, 0)$.

First we compute the derivative as a matrix (5.10.1). We have that

$$(D\mathbf{f}(x, y)) = \begin{pmatrix} 2x & \cos y \\ y & x \end{pmatrix}$$

so that

$$(D\mathbf{f}(1, 0)) = \begin{pmatrix} 2 & 1 \\ 0 & 1 \end{pmatrix}$$

which is nonsingular, since its determinant is 2, which is non-zero. By the Inverse Function Theorem, $\mathbf{f}$ has a local inverse near $(1, 0)$. It would not be easy to give an explicit formula for this local inverse (try it!).

Note that (as in the example) the matrix $(D\mathbf{f}(x, y))$ is non-singular if and only if its determinant is non-zero:

$$\det (D\mathbf{f}(x, y)) \neq 0$$

[see I, Theorem 6.4.2]. In particular this *requires* $m = n$: only for maps $\mathbf{f} : \mathbb{R}^n \to \mathbb{R}^n$ will the Inverse Function Theorem guarantee a local inverse. (In fact, for continuous maps $\mathbf{f} : \mathbb{R}^n \to \mathbb{R}^m$ a local inverse cannot exist unless $m = n$, because the dimension of a space is a topological invariant: this is a very hard theorem to prove, however.)

This determinant is often called the *Jacobian*, or *Jacobian determinant* of $\mathbf{f}$. In the case most often encountered, $m = n = 2$, it takes the form

$$J(f_1, f_2) = \begin{vmatrix} \dfrac{\partial f_1}{\partial x} & \dfrac{\partial f_1}{\partial y} \\ \dfrac{\partial f_2}{\partial x} & \dfrac{\partial f_2}{\partial y} \end{vmatrix} \tag{5.12.2}$$

or, expanded,

$$J(f_1, f_2) = \frac{\partial f_1}{\partial x} \frac{\partial f_2}{\partial y} - \frac{\partial f_1}{\partial y} \frac{\partial f_2}{\partial x}.$$

Hence the Inverse Function Theorem tells us that a function $\mathbf{f} : \mathbb{R}^2 \to \mathbb{R}^2$, defined by $\mathbf{f}(x, y) = (f_1(x, y), f_2(x, y))$, has a local inverse at (x, y) if

$$\frac{\partial f_1}{\partial x} \frac{\partial f_2}{\partial y} \neq \frac{\partial f_1}{\partial y} \frac{\partial f_2}{\partial x} \tag{5.12.3}$$

at (x, y).

It is useful to note some formal properties of the two-rowed Jacobians defined in (5.12.2). If $f, f_1, f_2, f_3, \ldots$ are any functions of x and y whose partial derivatives of the first order exist, then

$$J(f_1, f_2) = -J(f_2, f_1) \tag{5.12.4}$$

$$J(kf_1 f_2) = J(f_1, kf_2) = kJ(f_1, f_2), \tag{5.12.5}$$

where k is a constant. By combining these results we obtain that

$$J(f, kf) = J(kf, f) = 0. \tag{5.12.6}$$

Also,

$$J(f, f_1 + f_2) = J(f, f_1) + J(f, f_2), \tag{5.12.7}$$

[see I, § 6.10]. A less obvious result is the following relationship which involves four arbitrary functions:

$$J(f_1, f_2)J(f_3, f_4) + J(f_2, f_3)J(f_1, f_4) + J(f_3, f_2)J(f_1, f_4) = 0. \tag{5.12.8}$$

The non-vanishing of the determinant is, however, not a *necessary* condition for a local inverse. For example, let $f: \mathbb{R} \to \mathbb{R}$ be defined by $f(x) = x^3$. The derivative at $x = 0$ is zero, nonetheless there is an inverse function $x \to \sqrt[3]{x}$. However, this inverse fails to be differentiable at zero, since its tangent is vertical (Figure 5.12.3). In general if a function with vanishing Jacobian has a local inverse, that inverse cannot be differentiable at the point concerned. (The proof is the chain rule.)

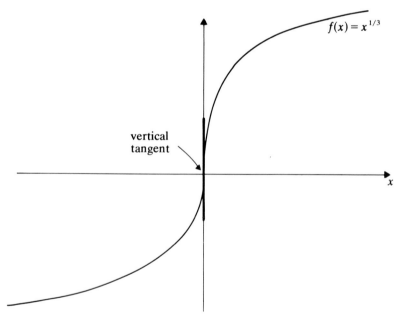

Figure 5.12.3

In modern texts a smooth (infinitely differentiable) function possessing a smooth local inverse is called a *local diffeomorphism*. So the theorem says that a function with non-zero Jacobian is a local diffeomorphism. Such functions are important since they provide smooth *reversible* changes of coordinates. The geometric effect of a local diffeomorphism is to bend coordinates in a smooth way: see Poston and Stewart (1978) p. 48.

5.13. THE IMPLICIT FUNCTION THEOREM

This is a very similar result to the Inverse Function Theorem: indeed each is an easy consequence of the other. It says that if the best linear approximation (derivative) to the set of solutions of an equation

$$\mathbf{f}(\mathbf{x}, \mathbf{y}) = \mathbf{0}$$

is the graph of a function $\mathbf{y} = \mathbf{y}(\mathbf{x})$, then so locally is the set of solutions itself. Here $\mathbf{x} = (x_1, \ldots, x_n)$ and $\mathbf{y} = (y_1, \ldots, y_m)$ may have arbitrarily many components.

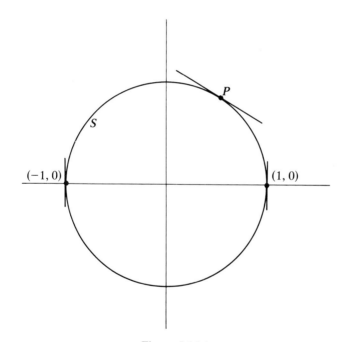

Figure 5.13.1

EXAMPLE 5.13.1. In Figure 5.13.1 we have

$$f(x, y) = x^2 + y^2 - 1$$

so the set of solutions S is a circle [see V (1.2.7)]. The tangent at P is the graph of a function, hence so is S. In fact S is locally the graph of

$$y = +\sqrt{1 - x^2}$$

near P. However, the tangents at $(1, 0)$ and $(-1, 0)$ are vertical, hence not graphs of functions of x; and neither is S the graph of a (single-valued) function near these points.

As with the Inverse Function Theorem, the condition on the derivative is sufficient but not necessary. If

$$f(x, y) = x - y^3$$

then S is the graph of $y = \sqrt[3]{x}$, despite having a vertical tangent (Figure 5.12.3). However, y is not the graph of a *differentiable* function.

If the conditions of the Implicit Function Theorem are satisfied, and if $\mathbf{f}$ is a smooth function, then the function $\mathbf{y}(\mathbf{x})$ that the theorem tells us exists must also be smooth. In formal language:

THEOREM 5.13.1   (*Implicit Function Theorem*).   *If*

$$\mathbf{f} : \mathbb{R}^m \to \mathbb{R}^n$$

is r times differentiable in the neighbourhood of some point $(\mathbf{x}_0, \mathbf{y}_0)$, and if there exists a function $\boldsymbol{\phi} : \mathbb{R}^n \to \mathbb{R}^m$ such that for $(\mathbf{x}, \mathbf{y}) \in \mathbb{R}^m \times \mathbb{R}^n$ we have

$$D\mathbf{f}(\mathbf{x}, \mathbf{y}) = \mathbf{0} \quad \text{if and only if} \quad \mathbf{y} = \boldsymbol{\phi}(\mathbf{x}),$$

then in the neighbourhood of $(\mathbf{x}_0, \mathbf{y}_0)$ there is a function $\boldsymbol{\psi} : \mathbb{R}^n \to \mathbb{R}^m$ such that

$$\mathbf{f}(\mathbf{x}, \mathbf{y}) = \mathbf{0} \quad \text{if and only if} \quad \mathbf{y} = \boldsymbol{\psi}(\mathbf{x}).$$

Further, $\boldsymbol{\psi}$ is r times differentiable.

We say that the function $\boldsymbol{\psi}$ is *defined implicitly* by the equation

$$\mathbf{f}(\mathbf{x}, \mathbf{y}) = \mathbf{0},$$

and the theorem says that locally $\boldsymbol{\psi}$ will be defined provided an analogous function $\boldsymbol{\phi}$ exists when we replace $\mathbf{f}$ by $D\mathbf{f}$, that is, if the *linearized* problem $D\mathbf{f}(\mathbf{x}, \mathbf{y}) = \mathbf{0}$ can be solved by a *function* $\mathbf{f}(\mathbf{x}, \mathbf{y}) = \mathbf{0}$, at least locally. And the linearized problem requires only *linear* algebra to solve it, and for a small number of variables is usually obvious by inspection.

More explicitly, consider the case $m = n = 1$. The solutions to

$$f(x, y) = 0$$

typically form a curve S (Figure 5.13.2). The derivative of f is

$$(Df) = \left(\frac{\partial f}{\partial x}, \frac{\partial f}{\partial y} \right).$$

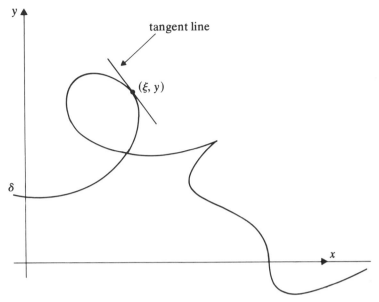

Figure 5.13.2

The tangent line to f at (ξ, η) is the best linear approximation to f there, so its equation is given in matrix form by

$$0 = Df|_{(\xi, \eta)}(x, y) = \left(\frac{\partial f}{\partial x}(\xi, \eta), \frac{\partial f}{\partial y}(\xi, \eta) \right) \binom{x}{y}.$$

Expanding the matrix product, we get the equation of the tangent line in the form

$$x \frac{\partial f}{\partial x}(\xi, \eta) + y \frac{\partial f}{\partial y}(\xi, \eta) = 0.$$

We can solve this for y in terms of x provided that

$$\frac{\partial f}{\partial y} \neq 0$$

and *this* is the condition we need to guarantee that ϕ exists for a function of *two* variables (x, y).

EXAMPLE 5.13.1 (*continued*).   The circle

$$0 = f(x, y) = x^2 + y^2 - 1$$

is locally a graph provided that

$$0 \neq \partial f / \partial y = 2y,$$

that is, $y \neq 0$. And the places where the tangent is vertical are, precisely, those where $y = 0$, namely $(1, 0)$ and $(-1, 0)$.

EXAMPLE 5.13.2.   Take

$$f(x, y) = x^2 - y^2.$$

Then $\partial f/\partial y = 0$ when $y = 0$, that is, only at $(0, 0)$. This time the tangent is not vertical: the curve has a self-intersection with two branches crossing at right angles. (This can happen since $\partial f/\partial x$ is also zero at $(0, 0)$.) Everywhere else, however, the set of zeros of f is locally a graph.

These examples involve functions defining *curves*. Another important case is when $f: \mathbb{R}^3 \to \mathbb{R}$, when the equation

$$f(x, y, z) = 0$$

defines a *surface* in $\mathbb{R}^3$ [see § 6.3]. It is dealt with analogously.

5.14. IMPLICIT DIFFERENTIATION

In applications of calculus the use of the Implicit Function Theorem is usually itself implicit. This is the case, for example, in the technique of 'implicit differentiation', where it is habitually *assumed* that a variable is a function of others, without proof, in cases where the Implicit Function Theorem would *provide* a proof.

For example, suppose that we are looking at the zero-set

$$f(x, y, z) = 0$$

of a smooth function $f: \mathbb{R}^3 \to \mathbb{R}$. *Assuming* that z is a function of x and y, and that this function is differentiable, we may compute its partial derivatives $\partial z/\partial x$ and $\partial z/\partial y$ as follows. To avoid notational difficulties suppose that $z = u(x, y)$: it follows that what we wish to compute is $\partial u/\partial x$ and $\partial u/\partial y$. We know that

$$0 = f(x, y, u(x, y)).$$

Using the chain rule to differentiate with respect to x, we have

$$0 = \frac{\partial f}{\partial x} + \frac{\partial f}{\partial z} \frac{\partial u}{\partial x}$$

so that

$$\frac{\partial u}{\partial x} = -\frac{\partial f}{\partial x} \bigg/ \frac{\partial f}{\partial z}. \tag{5.14.1}$$

Similarly

$$\frac{\partial u}{\partial y} = -\frac{\partial f}{\partial y} \bigg/ \frac{\partial f}{\partial z}. \tag{5.14.2}$$

This is all very well, but it does not establish that a function u exists or is differentiable, and *until this is done the computation makes no sense*. Arguing as for the example above, however, we see that the Implicit Function Theorem applies provided $\partial f/\partial z \neq 0$. Not only does this justify the answer: it tells us to avoid the places where $\partial f/\partial z = 0$, and these are exactly those where the formulae (5.14.1) and (5.14.2) fail to be defined. (If we assume that an implicitly defined function exists and if we compute its derivatives, then we generally obtain formulae that manifestly make no sense under conditions for which the Implicit Function Theorem breaks down—e.g. here the condition $\partial f/\partial z = 0$. This in part explains why the theorem itself is seldom quoted: however the practice is a dangerous one and the student should at least be aware of the possibility of a rigorous check.) See Example 5.15.3, and some more complicated examples of this type of computation which may be found in Protter and Morrey (1964) pp. 686 ff.

5.15. CONSTRAINED EXTREMA AND LAGRANGE MULTIPLIERS

We now consider the problem of finding critical points of a function

$$f: \mathbb{R}^n \to \mathbb{R}$$

subject to *constraints* of the form

$$\phi(\mathbf{x}) = 0 \qquad (\mathbf{x} \in \mathbb{R}^n) \tag{5.15.1}$$

for one or more functions $\phi: \mathbb{R}^n \to \mathbb{R}$.

As an illustration we return to Example 5.6.2 where we found a minimum of

$$f(x, y, z) = x^2 + y^2 + z^2$$

subject to the constraint

$$\phi(x, y, z) = 4x - 5y + z - 2 = 0.$$

In that case we were able to use the constraint to eliminate z as a function of x and y, and then apply our usual techniques to find critical points. In general the equation (5.15.1) may be impossible to solve in closed form for a given variable, and such elimination methods fail. In these circumstances we may use the Implicit Function Theorem to test whether *in principle* a variable is a function of the others; and if it is, follow through the theoretical computations to obtain an explicit answer. The resulting formulae may be summed up as a technique known as *Lagrange's Method of Undetermined Multipliers* [see § 15.1.4]. Once more the use of the Implicit Function Theorem is often implicit too—and here can lead to disaster.

First we shall exemplify the method: then we shall consider when it is valid.

EXAMPLE 5.15.1.   Minimize

$$f(x, y, z) = x^2 + y^2 + z^2$$

subject to

$$\phi(x, y, z) = 4x - 5y + z - 2 = 0.$$

By the Lagrange method we form a new function of *four* variables,

$$F(x, y, z, \lambda) = f(x, y, z) + \lambda\phi(x, y, z)$$
$$= x^2 + y^2 + z^2 + \lambda(4x - 5y + z - 2).$$

We then seek critical points of F, given by

$$0 = \partial F/\partial x = \partial F/\partial y = \partial F/\partial z = \partial F/\partial\lambda.$$

The resulting equations are:

$$2x + 4\lambda = 0$$
$$2y - 5\lambda = 0$$
$$2z + \lambda = 0$$
$$4x - 5y + z - 2 = 0.$$

Note that the last equation is just the constraint $\phi(x, y, z) = 0$ again, so any solution automatically satisfies it. Solving this system of linear equations [see I, § 5.7] we get

$$x = 4/21$$
$$y = -5/21$$
$$z = 1/21$$
$$\lambda = -2/21.$$

Thus we have found the same point as in Example 5.6.2.

In general, to find critical points of f subject to a constraint $\phi = 0$, the method tells us to *form $F = f + \lambda\phi$ and seek critical points of F relative to the original variables plus λ.*

EXAMPLE 5.15.2.   Find the critical points of

$$f(x, y) = x^3 + y^3$$

subject to the constraint

$$\phi(x, y) = x^2 + y^2 - 1 = 0.$$

We form

$$F(x, y) = x^3 + y^3 + \lambda(x^2 + y^2 - 1).$$

Critical points of this are given by

$$0 = \partial F/\partial x = 3x^2 + 2\lambda x$$
$$0 = \partial F/\partial y = 3y^2 + 2\lambda y$$
$$0 = \partial F/\partial \lambda = x^2 + y^2 - 1.$$

There are six solutions:

$x = 0,$	$y = 1,$	$\lambda = -3/2$
$x = 0,$	$y = -1,$	$\lambda = 3/2$
$x = 1,$	$y = 0,$	$\lambda = -3/2$
$x = -1,$	$y = 0,$	$\lambda = 3/2$
$x = 1/\sqrt{2},$	$y = 1/\sqrt{2},$	$\lambda = -3/(2\sqrt{2})$
$x = -1/\sqrt{2},$	$y = -1/\sqrt{2},$	$\lambda = 3/(2\sqrt{2}).$

Figure 5.15.1 illustrates this geometrically, by drawing the graph of f for (x, y)-values on the unit circle (that is, satisfying the constraint). The unit circle is shown in perspective as an ellipse in the horizontal plane, and the value of $f(x, y)$ is measured vertically (zero being the middle of the cylinder in a vertical direction). There are six critical points: three local maxima and three local minima.

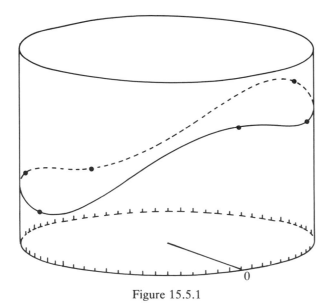

Figure 15.5.1

With several constraints $\phi_1 = \phi_2 = \ldots = \phi_k = 0$, we form

$$F = f + \lambda_1\phi_1 + \ldots + \lambda_k\phi_k \qquad (5.15.2)$$

and seek critical points of F relative to the original variables in addition to $\lambda_1, \ldots, \lambda_k$. The calculations are otherwise identical in form with the examples above.

To see how the Implicit Function Theorem comes in, we shall prove the validity of the Lagrange method in a special case (but a typical one): finding critical points of a three-variable function

$$f(x, y, z)$$

subject to the constraint

$$\phi(x, y, z) = 0. \tag{5.15.3}$$

We assume that (x_0, y_0, z_0) is a critical point.

We wish to solve (5.15.3) for z in terms of (x, y). *By the Implicit Function Theorem this may be done provided that*

$$\frac{\partial \phi}{\partial z}(x_0, y_0, z_0) \neq 0. \tag{5.15.4}$$

If (5.15.4) holds, there exists a function $g(x, y)$ such that $z = g(x, y)$. Define

$$G(x, y) = f(x, y, g(x, y)).$$

This has a critical point at (x_0, y_0). By the chain rule,

$$\frac{\partial G}{\partial x} = \frac{\partial f}{\partial x} + \frac{\partial f}{\partial z}\frac{\partial g}{\partial x} = 0, \qquad \frac{\partial G}{\partial y} = \frac{\partial f}{\partial y} + \frac{\partial f}{\partial z}\frac{\partial g}{\partial y} = 0.$$

But by implicit differentiation of (5.15.3) we also have

$$\frac{\partial z}{\partial x} = \frac{\partial g}{\partial x} = -\frac{\partial \phi}{\partial x}\Big/\frac{\partial \phi}{\partial z}, \qquad \frac{\partial z}{\partial y} = \frac{\partial g}{\partial y} = -\frac{\partial \phi}{\partial y}\Big/\frac{\partial \phi}{\partial z}$$

(note that (5.15.4) allows us to divide out by $\partial \phi/\partial z(x_0, y_0, z_0)$). From these two equations we obtain

$$\frac{\partial f}{\partial x} - \left(\frac{\partial f}{\partial z}\Big/\frac{\partial \phi}{\partial z}\right)\frac{\partial \phi}{\partial x} = 0$$

$$\frac{\partial f}{\partial y} - \left(\frac{\partial f}{\partial z}\Big/\frac{\partial \phi}{\partial z}\right)\frac{\partial \phi}{\partial y} = 0.$$

Obviously the identity

$$\frac{\partial f}{\partial z} - \left(\frac{\partial f}{\partial z}\Big/\frac{\partial \phi}{\partial z}\right)\frac{\partial \phi}{\partial z} = 0$$

also holds.

Set

$$\lambda_0 = -\frac{\partial f}{\partial z}(x_0, y_0, z_0)\Big/\frac{\partial \phi}{\partial z}(x_0, y_0, z_0).$$

Then we have

$$\frac{\partial f}{\partial x}+\lambda_0\frac{\partial \phi}{\partial x}=\frac{\partial f}{\partial y}+\lambda_0\frac{\partial \phi}{\partial y}=\frac{\partial f}{\partial z}+\lambda_0\frac{\partial \phi}{\partial z}=\phi=0.$$

So $(x_0, y_0, z_0, \lambda_0)$ is a critical point of $F = f + \lambda\phi$.
    This completes the proof.

The main implication of this for users of the Lagrange method is that the conditions under which the Implicit Function Theorem holds must be investigated for the problem at hand, and *points where this condition fails* (generally finite in number) *should be studied separately.* Usually this is no problem.
    The following example shows that such checks can be necessary.

EXAMPLE 5.15.3.   Minimize

$$f(x, y) = y$$

subject to the constraint

$$0 = \phi(x, y) = x^2 - y^3.$$

Using Lagrange's multiplier we form

$$F(x, y) = y + \lambda(x^2 - y^3).$$

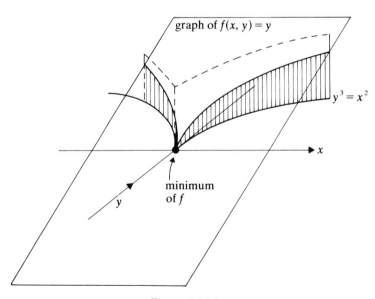

Figure 5.15.2

Critical points of F are given by

$$0 = \partial F / \partial x = 2\lambda x$$
$$0 = \partial F / \partial y = 1 - 3\lambda y^2$$
$$0 = \partial F / \partial \lambda = x^2 - y^3.$$

The first equation tells us that $x = 0$ or $\lambda = 0$. The second equation prohibits $\lambda = 0$, so we must have $x = 0$; but the third equation yields $y = 0$ which contradicts the second. So F has *no* critical points.

Geometrically, however, it is obvious that f has a minimum at $(0, 0)$ when (x, y) lies on the curve $x^2 - y^3 = 0$. See Figure 5.15.2.

What is wrong? Mimicking the above proof, we find that the use of the Implicit Function Theorem here requires $\partial \phi / \partial y$ to be non-zero. So, to find the desired minima, we must look *also* at points where $\partial \phi / \partial y = 0$. In this case the only such point is $(0, 0)$; and this is indeed where the minimum occurs. In general we can also interchange the roles of x and y, and then the only 'bad' points are where $\partial \phi / \partial x = 0$. So in practice the only danger spots are when *both* partial derivatives vanish:

$$\frac{\partial \phi}{\partial x} = 0 = \frac{\partial \phi}{\partial y}.$$

I.S.

REFERENCES

Protter, M. H. and Morrey, C. B. (1964). *College Calculus with Analytic Geometry*, Addison-Wesley, Reading, Mass.

Edwards, C. H. Jr. (1973). *Advanced Calculus of Several Variables*, Academic Press, New York and London.

Poston, T. and Stewart, I. N. (1978). *Catastrophe Theory and Its Applications*, Pitman, London.

Spivak, M. (1965). *Calculus on Manifolds*, Benjamin, New York.

CHAPTER 6

Multiple Integrals

6.1. LINE INTEGRALS

In this section we will define the notion of the integral of a real-valued function along a curve in two or three dimensions—a line integral as it is generally called. This generalizes the ordinary integral on an interval $[a, b]$ of the real line which was studied in Chapter 4, and which now appears as the special case of an integral along the line segment from a to b.

We begin by making precise the intuitive idea of a curve.

DEFINITION 6.1.1. Let I be an interval of the real line and γ a continuous function from I to $\mathbb{R}^3$. Then γ is called a *curve* in $\mathbb{R}^3$ (or sometimes a *space-curve*). The set C of points traced out by γ (i.e. the range of γ) will be called the *path* of γ in $\mathbb{R}^3$. If C lies in some planar subset of $\mathbb{R}^3$, we will call γ a *plane curve*. For t in I we write $\gamma(t) = (\gamma_1(t), \gamma_2(t), \gamma_3(t))$ and call $\gamma_1, \gamma_2, \gamma_3$ the *coordinate* or *component functions* of γ: t itself is called the *parameter*, and I the *parameter-interval* of the curve.

Notice the following points about this definition: (i) the interval I may be open or closed, and bounded or unbounded [see I, § 2.6.3]: however, most of our applications will concern the case when I is a closed bounded interval, (ii) the distinction between γ and C (i.e. between the function and its range) is important in theory, but in applications one speaks loosely of 'the curve C', when no confusion is possible, (iii) when $I = [a, b]$, a closed bounded interval, we say that $\gamma(a)$ is the *initial point* of the curve and $\gamma(b)$ is the *final point*. If $I_1 = [a_1, b_1]$ is another closed bounded interval, and ϕ is a one-to-one continuous mapping of I_1 onto I [see § 2.1] with $\phi(a_1) = a$, $\phi(b_1) = b$ (so that in fact ϕ is an increasing function of t_1, $a_1 \le t_1 \le b_1$ [see § 2.7]), then $\gamma_1 = \gamma \circ \phi$ is another curve, with parameter interval I_1 given by $\gamma_1(t_1) = \gamma(\phi(t_1))$, which has the same path as γ. Such a mapping ϕ is called an *orientation preserving* change of parameter—see the remarks following Definition 6.1.4 for an explanation of why these are important in the theory of line integrals. If ϕ is a one-to-one continuous function on I_1 with $\phi(a_1) = b$, $\phi(b_1) = a$ (so that

ϕ is decreasing on I_1) we say ϕ is *orientation reversing*—Theorem 6.1.1(ii)
gives a simple example of this.

This definition of a curve admits far more functions than we can reasonably
cope with. (There are in fact curves for which C fills out a whole region in
space.) Consequently we shall further restrict γ by the requirement of differentiability.

DEFINITION 6.1.2.   Let γ be a curve, defined on the interval I. We say γ
is *smooth* on I if the three component functions γ_1, γ_2, γ_3 introduced in
Definition 6.1.1 are continuously differentiable on I [see Definition 2.10.1].
We say that γ is an *arc* if γ_1, γ_2, γ_3 are smooth on I except for a finite number
of points at which there are jump discontinuities [see Definition 2.3.5] of the
derivatives (such points appear as corners on the path of γ). When dealing
with smooth curves and arcs, we shall assume that any changes of parameter,
introduced as in remark (ii) above, are continuously differentiable.

EXAMPLE 6.1.1.   Let γ be the plane curve given by

$$\gamma(t) = \begin{cases} \left(\cos \frac{\pi}{2}t, \sin \frac{\pi}{2}t\right), & 0 \le t \le 1 \\ (0, 2-t), & 1 \le t \le 2, \end{cases}$$

as in Figure 6.1.1(i). Then γ is continuous on $[0, 2]$, and is smooth on $[0, 1]$
and $[1, 2]$. Hence γ is an arc.

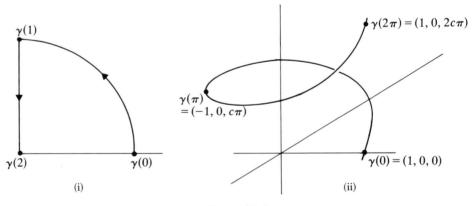

Figure 6.1.1

EXAMPLE 6.1.2.   Let γ be the space curve given by

$$\gamma(t) = (\cos t, \sin t, ct)$$

for a positive constant c, and all real values of t.

Then γ is an infinite ascending spiral, or *helix*, and $\gamma'(t) = (-\sin t, \cos t, c)$
[see § 17.2.5] is continuous for all real t.

Figure 6.1.1(ii) illustrates this curve.

DEFINITION 6.1.3. Let γ be an arc defined on the interval $I = [a, b]$. Then γ_1', γ_2', γ_3' are piecewise continuous on I, and are thus integrable there [see Theorem 4.1.1(i)]. The *arc length*, usually denoted by the letter s, is defined to be

$$s(t) = \int_a^t \sqrt{(\gamma_1'(u)^2 + \gamma_2'(u)^2 + \gamma_3'(u)^2)}\, du.$$

EXAMPLE 6.1.3. The length of the arc in Example 6.1.1 is

$$\int_0^1 \sqrt{\left[\left(\frac{\pi}{2}\right)^2 \left(\sin^2 \frac{\pi}{2} t + \cos^2 \frac{\pi}{2} t\right)\right]}\, dt + \int_1^2 \sqrt{(0^2 + (-1)^2)}\, dt = \frac{\pi}{2} + 1,$$

in accordance with our intuitive ideas of length. Similarly in Example 6.1.2 the length of a single turn of the helix is

$$\int_0^{2\pi} \sqrt{(\sin^2 t + \cos^2 t + c^2)}\, dt = 2\pi\sqrt{(1 + c^2)}.$$

DEFINITION 6.1.4. The *tangent* to a smooth curve at a point t_0 is the vector at the point t_0, in the direction $\gamma'(t_0) = (\gamma_1'(t_0),\ \gamma_2'(t_0),\ \gamma_3'(t_0))$ [see § 17.2.1]. It follows from the above definition of arc length that the derivative of γ with respect to s,

$$\frac{d\gamma}{ds} = \frac{\gamma'(t)}{s'(t)} = (\gamma_1'(t),\ \gamma_2'(t),\ \gamma_3'(t))(\gamma_1'^2 + \gamma_2'^2 + \gamma_3'^2)^{-1/2}$$

is a unit vector in the direction of the tangent [see § 17.2.2]: we shall denote it by $\mathbf{t}$. Since $\mathbf{t}$ is a unit vector, its derivative $d\mathbf{t}/ds$ will be perpendicular to it [see Example 17.2.10]. The length of this derivative is called the *curvature* and denoted by $1/\rho$ (ρ is the *radius of curvature*), so that

$$\frac{d\mathbf{t}}{ds} = \frac{1}{\rho}\mathbf{n} \qquad (6.1.1)$$

where $\mathbf{n}$ is a unit vector (the *principal normal*) which is perpendicular to $\mathbf{t}$. The vector product $\mathbf{b} = \mathbf{t} \wedge \mathbf{n}$ [see § 17.2.3] is a unit vector, perpendicular to both $\mathbf{t}$ and $\mathbf{n}$, called the *binormal*. It is easily shown that $d\mathbf{b}/ds$ is perpendicular to both $\mathbf{b}$ and $\mathbf{t}$, and so must be parallel to $\mathbf{n}$. Hence we may define a quantity called the *torsion* of the curve, and denoted $1/\tau$, by

$$\frac{d\mathbf{b}}{ds} = -\frac{1}{\tau}\mathbf{n}. \qquad (6.1.2)$$

(When γ is a plane curve, $\mathbf{b}$ is a constant vector perpendicular to the plane of the curve, so that $d\mathbf{b}/ds = \mathbf{0}$ and τ is undefined).

Finally one can show that

$$\frac{d\mathbf{n}}{ds} = -\frac{1}{\rho}\mathbf{t} + \frac{1}{\tau}\mathbf{b}, \qquad (6.1.3)$$

completing the set of three formulae for the derivatives of **t**, **n** and **b** with respect to s, known collectively as the *Frenet formulae* [see V, Chapter 12].

EXAMPLE 6.1.4.   In the Example 6.1.2, the curvature is $1/\rho = 1/(1+c^2)$ and the torsion is $1/\tau = c/(1+c^2)$, as is easily verified.
   We now define the integral of a continuous function along an arc.

DEFINITION 6.1.5.   Let γ be an arc, defined on the interval $[a, b]$ and let f be a function which is defined and continuous on some set E which contains the path C of γ. Then the integral of f along γ is defined to be

$$\int_a^b f(\gamma(t))s'(t)\, dt$$

and is written $\int_\gamma f(s)\, ds$ (or sometimes $\int_C f(s)\, ds$ if no confusion is possible).
   It is important to realize that although the value of the integral $\int_\gamma f(s)\, ds$ is defined in terms of a particular parameterization, it is in fact the same for all orientation-preserving changes of parameter [see the remarks following Definitions 6.1.1 and 2]. This fact is an immediate consequence of the change of variables formula for integrals [see Theorem 4.3.2]. If the orientation is reversed, the integral is reversed in sign—Theorem 6.1.2(ii) is an example of this.
   In addition to the usual linearity properties of the integral with respect to f, as in Theorem 4.1.2 parts (i) and (ii), we have the following results which correspond to parts (iii) and (v) of that result.

THEOREM 6.1.1
   (i) *If the arc* γ *is made up of two subarcs* γ_1 *and* γ_2 *(as for instance in Example 6.1.1), then*

$$\int_\gamma f(s)\, ds = \int_{\gamma_1} f(s)\, ds + \int_{\gamma_2} f(s)\, ds.$$

   (ii) *Let the arc* γ *be defined on* $[a, b]$, *and the arc* γ_1 *be defined by* $\gamma_1(t) = \gamma(a+b-t)$: γ_1 *has the same path* C *as* γ, *but traversed in the opposite direction. Then*

$$\int_{\gamma_1} f(s)\, ds = -\int_\gamma f(s)\, ds.$$

EXAMPLE 6.1.5.   Let $f(x, y) = x$, and consider the integral $\int_\gamma f\, ds$ where γ is the arc in Example 6.1.1. For $0 \le t \le 1$, $\gamma(t) = (\cos(\pi/2)t, \sin(\pi/2)t)$, so $x = \cos(\pi/2)t$, and

$$(s'(t))^2 = \left(-\frac{\pi}{2}\sin\frac{\pi}{2}t\right)^2 + \left(\frac{\pi}{2}\cos\frac{\pi}{2}t\right)^2,$$

so $s'(t) = \pi/2$. Similarly if $1 \le t \le 2$, $\gamma(t) = (0, 2-t)$, $\gamma'(t) = (0, -1)$, $s'(t) = 1$, and $x = 0$. Hence from Definition 6.1.5,

$$\int_\gamma f(s)\, ds = \int_0^1 \left(\cos \frac{\pi}{2} t\right)\left(\frac{\pi}{2}\right) dt + \int_1^2 (0)(1)\, dt$$

$$= \left[\sin \frac{\pi}{2} t\right]_0^1 = 1.$$

A kind of line integral which is important, particularly in physical applications, is

$$I = \int_\gamma (f\, dx + g\, dy + h\, dz).$$

Here f, g, h are continuous functions on the path of γ, and the value of the integral is defined to be

$$\int_a^b \{f(\gamma_1(t),\, \gamma_2(t),\, \gamma_3(t))\gamma_1'(t) + g(\gamma_1(t),\, \gamma_2(t),\, \gamma_3(t))\gamma_2'(t)$$

$$+ h(\gamma_1(t),\, \gamma_2(t),\, \gamma_3(t))\gamma_3'(t)\} \, dt,$$

where $\gamma = (\gamma_1, \gamma_2, \gamma_3)$ is the parameterization of the curve, defined on the interval $[a, b]$. We illustrate this type of integral by the following example.

EXAMPLE 6.1.6.   Let γ be the circle $x^2 + y^2 = a^2$, described with the anti-clockwise orientation [see V (1.2.7)]. Evaluate (i) $\int_\gamma x\, dy$, (ii) $\int_\gamma y^2\, dy$.

To evaluate these integrals we put $x = a \cos \theta$, $y = a \sin \theta$, for $0 \le \theta \le 2\pi$ [see V, § 1.3.2].

(i) $\displaystyle \int_\gamma x\, dy = \int_0^{2\pi} a \cos \theta (a \cos \theta\, d\theta) = a^2 \int_0^{2\pi} \tfrac{1}{2}(1 + \cos 2\theta)\, d\theta$

$\displaystyle \qquad = \tfrac{1}{2} a^2 [\theta + \tfrac{1}{2} \sin 2\theta]_0^{2\pi} = \pi a^2.$

(ii) $\displaystyle \int_\gamma y^2\, dy = \int_0^{2\pi} a^2 \sin^2 \theta (a \cos \theta\, d\theta) = a^3 [\tfrac{1}{3}(\sin \theta)^3]_0^{2\pi} = 0.$

Remark.   It is a consequence of Green's theorem—6.2.4 below—that for suitable regions, the area of the region is equal to $\int_\gamma x\, dy$ or $-\int_\gamma y\, dx$ taken around the boundary of the region.

The second integral of Example 6.1.6 is a simple instance of a result which is easily deduced directly from the definition above, namely that for a function $F(x, y)$ having continuous first partial derivatives $\partial F/\partial x$, $\partial F/\partial y$, [see § 5.2] we have

$$\int_\gamma \frac{\partial F}{\partial x}\, dx + \frac{\partial F}{\partial y}\, dy = F(\gamma(b)) - F(\gamma(a)): \qquad (6.1.4)$$

in particular if γ is a closed curve,

$$\int_\gamma \frac{\partial F}{\partial x}\, dx + \frac{\partial F}{\partial y}\, dy = 0. \tag{6.1.5}$$

One final version of the line integral is worth mentioning. This concerns the integration of a vector valued function, $\mathbf{f} = (f_1, f_2, f_3)$, with respect to a vector element of length $\mathbf{ds}$. We write this as $\int_c \mathbf{f}(s) \cdot \mathbf{ds}$, where the dot indicates the scalar product [see (17.2.12)]. This integral may be interpreted in any of the following equivalent (and equal!) ways.

DEFINITION 6.1.6.   Let the arc γ be defined by the co-ordinate functions $(\gamma_1(t), \gamma_2(t), \gamma_3(t))$ defined on $I = [a, b]$, and let $\mathbf{f} = (f_1, f_2, f_3)$ be a continuous vector valued function defined on the path C of γ. Let $\mathbf{t}$ be the unit tangent vector as in Definition 6.1.4. Then we define $\int_\gamma \mathbf{f} \cdot \mathbf{ds}$ by

$$\int_\gamma \mathbf{f} \cdot \mathbf{ds} = \int_\gamma (\mathbf{f} \cdot \mathbf{t})\, ds = \int_a^b (f_1\gamma_1' + f_2\gamma_2' + f_3\gamma_3')\, dt = \int_\gamma (f_1\, dx + f_2\, dy + f_3\, dz).$$

It is important to realize that, as in Definition 6.1.5, the value of the integral is independent of the choice of orientation preserving parameterization.

6.2.   DOUBLE INTEGRALS IN THE PLANE

We have so far considered the problem of integrating a continuous function over a line segment (in Chapter 4) or a smooth image of it, (i.e. a curve), as in the preceding section. We now investigate how to give a meaning to an expression of the form $\int\int_D f(x, y)\, dx\, dy$ where f is a continuous function of two real variables x and y, and D is a suitable set of points in the plane. We first describe what is meant by a 'suitable' set.

DEFINITION 6.2.1.   A set D in the plane is called *normal* (relative to the chosen co-ordinate system) if every line parallel to either the x- or y-axes intersects it in a closed line segment. Figures 6.2.1(i) and (ii) illustrate regions

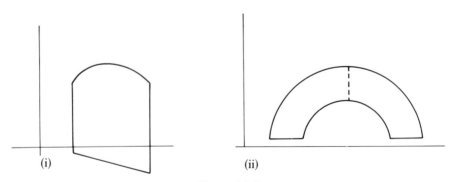

(i)   (ii)

Figure 6.2.1

which are normal and not normal respectively: notice that the dotted line in
(ii) divides the region into two normal sub regions: this will often be the case
in applications (see also the remarks following Theorem 6.2.4).

In the case of a continuous function and a bounded normal region the
integral may be defined by the following procedure, which is analogous to
that used in section 4.1 to define integrals over an interval.

DEFINITION 6.2.2. Let D be a bounded normal region, and let a, b, c, d
be chosen so that $a \leq x \leq b$, $c \leq y \leq d$ for all (x, y) in D. Let
$P_1 = \{x_0, x_1, \ldots, x_m\}$, $P_2 = \{y_0, y_1, \ldots, y_n\}$ be any dissections of $[a, b], [c, d]$ [see
Definition 4.1.1] and let $I_{i,j}$ denote the rectangle with sides $[x_{i-1}, x_i]$ and
$[y_{j-1}, y_j]$, as in Figure 6.2.2.

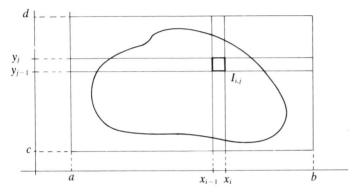

Figure 6.2.2

Let f be a real-valued function defined on D. We define the *approximating
sum*

$$S_{P_1, P_2}(f) = \sum_{i=1}^{m} \sum_{j=1}^{n} f(u_{i,j})(x_i - x_{i-1})(y_j - y_{j-1}),$$

where $u_{i,j}$ is any point of $I_{i,j}$ which is in D. (If $I_{i,j}$ is disjoint from D, set
$f(u_{i,j}) = 0$.) This expansion for $S_{P_1, P_2}(f)$ should be compared with that of $S_P(f)$
in definition 4.1.1; in particular $(x_i - x_{i-1})(y_j - y_{j-1})$ is simply the area of the
rectangle $I_{i,j}$. We say f is *integrable* over D if there is a number L such that
for each $\varepsilon > 0$, there is a $\delta > 0$ for which

$$|S_{P_1, P_2}(f) - L| < \varepsilon$$

whenever both P_1 and P_2 have norm $< \delta$. (Intuitively, $S_{P_1, P_2}(f) \to L$ as the
norms of P_1 and P_2 tend to zero [see § 2.3].)

The number L of Definition 6.2.2 is called the *integral* of f over D and
will be denoted $\iint_D f(x, y) \, dx \, dy$. It should be realized that the double integral
sign ($\iint_D$) and the appearance of the expression '$dx \, dy$' simply emphasizes the

two-dimensional nature of the limiting process which we have used to define the integral: only in later results on repeated integrals [see Theorem 6.2.3] will we consider integration with respect to x and y separately.

DEFINITION 6.2.3.   The *area* of a bounded normal region is defined as $\iint_D dx\, dy$—that is the integral of the constant function 1.

DEFINITION 6.2.4.   If D is not normal, but is made up of a finite disjoint union of normal regions as in Figure 6.2.1(ii), say

$$D = D_1 \cup D_2 \cup \ldots \cup D_n$$

then we may define the integral over D as the sum of the integrals over the regions D_m, $m = 1, 2, \ldots, n$.

The following gives an important class of integrable functions [compare Theorem 4.1.1]:

THEOREM 6.2.1.   *Let f be continuous on a bounded normal region D. Then f is integrable over D.*

The formal properties of this integral are analogous to the corresponding properties of integration of a function of one variable [compare Theorem 4.1.2]:

THEOREM 6.2.2

(i) *If f is integrable over the region D and k is any real number, then kf is integrable over D, and*

$$\iint_D kf(x, y)\, dx\, dy = k \iint_D f(x, y)\, dx\, dy.$$

(ii) *If f, g are integrable over the region D, then so is $f + g$, and*

$$\iint_D (f(x, y) + g(x, y))\, dx\, dy = \iint_D f(x, y)\, dx\, dy + \iint_D g(x, y)\, dx\, dy.$$

(iii) *If D, D' are disjoint regions (each of which comprises a finite number of normal regions as in Definition 6.2.4) and f is integrable over D and D', then f is integrable over $D \cup D'$, and*

$$\iint_{D \cup D'} f(x, y)\, dx\, dy = \iint_D f(x, y)\, dx\, dy + \iint_{D'} f(x, y)\, dx\, dy.$$

(iv) *If f is integrable over D, and $m \leq f(x, y) \leq M$ for all (x, y) in D, then*

$$m \iint_D dx\, dy \leq \iint_D f(x, y)\, dx\, dy \leq M \iint_D dx\, dy.$$

(Notice that these bounds for the integral are of the form $m|D|$, where $|D|$ denotes the area of D as in Definition 6.2.3.)

We now consider the problem of evaluating double integrals, which we do by reducing them to *repeated integrals* with respect to x, y separately [see § 4.7]:

THEOREM 6.2.3.   *Let D be a bounded normal region, and let a, b, c, d be chosen as in Definition 6.2.2. For each x in [a, b], suppose that the line through $(x, 0)$ parallel to the y-axis intersects the region D in the interval $[y_1, y_2]$, where $y_1 = y_1(x)$ and $y_2 = y_2(x)$ depend on x, as in Figure 6.2.3(i). Let x_1, x_2 be defined*

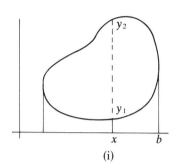
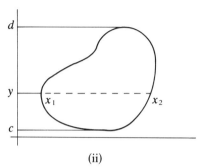

(i)   (ii)

Figure 6.2.3

similarly for $c \le y \le d$, as in Figure 6.2.3(ii). Let f be continuous on D. Then

$$F(x) = \int_{y_1}^{y_2} f(x, y)\, dy \quad \text{and} \quad G(y) = \int_{x_1}^{x_2} f(x, y)\, dx$$

are integrable functions of x in $[a, b]$ and y in $[c, d]$ respectively, and

$$\int_a^b F(x)\, dx = \iint_D f(x, y)\, dx\, dy = \int_c^d G(y)\, dy.$$

Notice that in evaluating $F(x)$ the variable x is regarded as a constant: so too is y in the evaluation of $G(y)$, as in Example 6.2.1 below.

Evidently this theorem reduces the evaluation of each double integral to one or other repeated integrals: which is chosen is simply a matter of convenience, as the following example shows.

EXAMPLE 6.2.1.   Let D be the region contained between the line $y = x$ and the parabola $y = \frac{1}{2}x^2$ as in Figure 6.2.4. Evaluate $\iint_D (x + y)\, dx\, dy$.

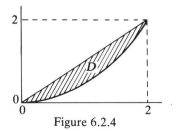

Figure 6.2.4

If we choose to fix x and integrate first with respect to y, then the limits are $y_1 = \frac{1}{2}x^2$ and $y_2 = x$ (using the notation of Theorem 6.2.3). Then

$$F(x) = \int_{\frac{1}{2}x^2}^{x} (x+y)\, dy = [xy + \tfrac{1}{2}y^2]_{\frac{1}{2}x^2}^{x}$$

$$= x(x - \tfrac{1}{2}x^2) + \tfrac{1}{2}(x^2 - \tfrac{1}{4}x^4) = \tfrac{3}{2}x^2 - \tfrac{1}{2}x^3 - \tfrac{1}{8}x^4.$$

Then by Theorem 6.2.3,

$$\iint_D (x+y)\, dx\, dy = \int_0^2 F(x)\, dx = \int_0^2 (\tfrac{3}{2}x^2 - \tfrac{1}{2}x^3 - \tfrac{1}{8}x^4)\, dx$$

$$= \left[\tfrac{1}{2}x^3 - \tfrac{1}{8}x^4 - \frac{1}{8 \cdot 5}x^5\right]_0^2 = 4 - 2 - \tfrac{4}{5} = \tfrac{6}{5}.$$

It is a simple exercise to check that

$$G(y) = \int_y^{\sqrt{2y}} (x+y)\, dx = y + \sqrt{2}\, y^{3/2} - \tfrac{1}{2}y^2$$

and that

$$\int_0^2 G(y)\, dy = \tfrac{6}{5} \quad \text{again.}$$

EXAMPLE 6.2.2.   An interesting special case of Theorem 6.2.3 occurs when the function to be integrated has the form $f(x, y) = f_1(x)f_2(y)$: a product of functions of the variables separately. In this case the integral becomes

$$\iint_D f(x, y)\, dx\, dy = \int_a^b f_1(x) \left\{ \int_{y_1}^{y_2} f_2(y)\, dy \right\} dx,$$

where y_1, y_2 (which will usually depend on x) are as defined in Theorem 6.2.3. If D is a rectangle so that y_1, y_2 are constants, equal to c, d say, then the integral becomes the product of two single integrals

$$\left(\int_a^b f_1(x)\, dx\right)\left(\int_c^d f_2(y)\, dy\right).$$

It is sometimes useful to be able to reduce a double integral of a particular form over a region D to a line integral around the boundary of D. The following result shows how this can be done.

THEOREM 6.2.4   (*Green's Theorem*).   *Let D be a normal region whose boundary is the arc C [see Definitions 6.1.2 and 6.2.1]. Let y_1, y_2 and x_1, x_2 be the functions of x, y respectively, which are defined in Theorem 6.2.3, so that the direction of traverse of C is determined by first taking $y = y_1(x)$ as x increases from a to b, and then $y = y_2(x)$ when x decreases from b back to a. (A corresponding definition can evidently be given in terms of x_1, x_2: in either case this corresponds intuitively to the anti-clockwise direction on C.) Then* if functions P, Q are defined on D, and $P, Q, \partial P/\partial x, \partial Q/\partial y$ are all continuous on D [see § 5.1 and § 5.2] we have

$$\iint_D \left(\frac{\partial P}{\partial x} + \frac{\partial Q}{\partial y} \right) dx\, dy = \int_D (P\, dy - Q\, dx).$$

For a proof we simply have to apply Theorem 6.2.3. For example

$$\iint_D \frac{\partial Q}{\partial y} dx\, dy = \int_a^b F(x)\, dx,$$

where

$$F(x) = \int_{y_1}^{y_2} \frac{\partial Q}{\partial y} dy = Q(x, y_2) - Q(x, y_1).$$

Hence

$$\int_a^b F(x)\, dx = -\int_a^b Q(x, y_1)\, dx + \int_a^b Q(x, y_2)\, dx$$

$$= -\int_c Q(x, y)\, dx$$

with the sign convention on C as explained above. Similarly we can show

$$\iint_D \frac{\partial P}{\partial x} dx\, dy = \int_C P\, dy.$$

Notice that if we take $P(x, y) = x$, or $Q(x, y) = y$, then both $\int_C x\, dy$ and $-\int_C y\, dx$ give the area enclosed by the curve C.

We shall give generalizations of this result in sections 6.3 and 6.4. Notice that it is a two-dimensional analogue of the fundamental theorem of calculus [see Theorem 4.2.2]. It is also worth remarking that if a region D is subdivided into a number of normal subregions, as in Figure 6.2.1(ii), and Theorem 6.2.4 applied to each, then the line integrals along the internal boundaries, which have been introduced to subdivide D, will cancel one another out and make no contribution to the final value of the integral.

To finish this section we shall consider the effect on a double integral of a *change of variables*, thereby giving a two-dimensional analogue of the rule for integration by substitution [see Theorem 4.3.2]. Suppose then that the coordinates x, y are determined as functions of two other variables u and v say: $x = \phi(u, v)$, $y = \psi(u, v)$. We must suppose that the relationship between (u, v) and (x, y) is one-to-one, that is, that each pair (u, v) in a certain region E uniquely determines a pair (x, y) in D, and conversely. Furthermore we shall suppose that the partial derivatives $\partial\phi/\partial u$, $\partial\phi/\partial v$, $\partial\psi/\partial u$, $\partial\psi/\partial v$, are continuous on E and that their determinant

$$J = \frac{\partial\phi}{\partial u} \frac{\partial\psi}{\partial v} - \frac{\partial\phi}{\partial v} \frac{\partial\psi}{\partial u}$$

is never zero on E: we may assume $J>0$ on E since the interchange of u and v changes the sign of J. J is also denoted $\partial(x, y)/\partial(u, v)$, and is called the *Jacobian* of the transformation [see (5.12.2)].

EXAMPLE 6.2.3.   A well known, and important example of such a correspondence is given by polar coordinates, $x = r \cos\theta$, $y = r \sin\theta$, which give a one-to-one correspondence between the region $E = \{(r, \theta): r>0, 0<\theta<2\pi\}$ and $D = \mathbb{R}^2\backslash\{(x, 0) = x \geq 0\}$ [see V, §1.2.5]. In this case $\partial x/\partial r = \cos\theta$, $\partial x/\partial\theta = -r\sin\theta$, $\partial y/\partial r = \sin\theta$, $\partial y/\partial\theta = r\cos\theta$, and

$$J = \partial(x, y)/\partial(r, \theta) = r\cos^2\theta - (-r\sin^2\theta) = r.$$

The effect of such a change of variables on a double integral is as follows.

THEOREM 6.2.5.   *Let* $x = \phi(u, v)$, $y = \psi(u, v)$ *give a one-to-one correspondence between* (u, v) *in a region* E *and* (x, y) *in a region* D, *as above. Let* f *be continuous on* D. *Then*

$$\iint_D f(x, y)\, dx\, dy = \iint_E f(\phi(u, v), \psi(u, v))J\, du\, dv.$$

The proof of this result is beyond our scope: however, as a special case, notice that in taking $f = 1$ we see that the area of D is equal to $\iint_E J\, du\, dv$, that is, that J acts as a kind of local magnification factor for areas, just as the ordinary one-dimensional derivative does for lengths.

EXAMPLE 6.2.4.   Let

$$I = \int_0^1 \int_0^{1-x} \exp\left(\frac{y}{x+y}\right) dy\, dx.$$

We shall sketch the region in the (x, y) plane over which I is taken, and then make the change of variables given by $u = x + y$, $v = y/(x+y)$, which will enable us to evaluate I. (Notice that the integrand of I [see §2.11] cannot be integrated as it stands.)

Firstly, the region of integration is defined by $0 \leq x \leq 1$, and $0 \leq y \leq 1-x$, which determines the triangular region D in Figure 6.2.5(i).

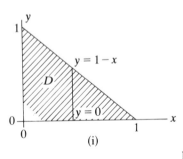

Figure 6.2.5

Our change of variables $u = x + y$, $v = y/(x + y)$ may also be written $y = uv$, $x = u - uv = u(1 - v)$, from which it may be seen that the points of D, which satisfy $0 \leq x$, $0 \leq y$, $0 < x + y \leq 1$, correspond uniquely with the points of E, which satisfy $0 < u \leq 1$, $0 \leq v \leq 1$. (The behaviour of the transformation near $(0, 0)$ in the (x, y) plane is more difficult: if we approach the point $(0, 0)$ along the line $y = mx$ (for a positive constant m), then $u \to 0$, while $v \to mx/(x + mx) = m/(1 + m)$ which may take any fixed value between 0 and 1; conversely all points in E with $u = 0$ map back to the origin in D. This does not affect the validity of the calculation (below) for the value of I, since we notice that the difficult points may be removed entirely by making a further restriction $x + y \geq \delta$ (or $u \geq \delta$ in E)—this is indicated by the dotted line in Figure 6.2.5. Since the integrand, $\exp[y/(x + y)]$, is bounded by $e^1 = e$ [see § 2.11] the integral over the excluded region can be made arbitrarily small by making δ small enough.) The Jacobian is

$$\frac{\partial(x, y)}{\partial(u, v)} = \frac{\partial x}{\partial u}\frac{\partial y}{\partial v} - \frac{\partial x}{\partial v}\frac{\partial y}{\partial u} = (1 - v)u - (-u)v = u.$$

Theorem 6.2.5 now tells us that

$$I = \iint_E e^v u \, du \, dv = \int_0^1 e^v \int_0^1 u \, du \, dv = \tfrac{1}{2}(e - 1).$$

EXAMPLE 6.2.5. We consider

$$I_N = \int_{-N}^N \int_{-N}^N e^{-(x^2 + y^2)} \, dx \, dy,$$

and will use the limiting value of this integral as N becomes large to deduce the value of the important integral

$$\int_{-\infty}^{\infty} e^{-t^2} \, dt = \sqrt{\pi}.$$

First notice that I_N is taken over the square given by $-N \leq x, y \leq N$, so that by Example 6.2.2,

$$I_N = \left(\int_{-N}^N e^{-x^2} \, dx \right)\left(\int_{-N}^N e^{-y^2} \, dy \right) = J_N^2,$$

say, where

$$J_N = \int_{-N}^N e^{-x^2} \, dx.$$

This integral J_N approaches a finite limit L say, as N increases, since if for example $x \geq 1$, $e^{-x^2} \leq e^{-x}$, and the integral $\int_1^{\infty} e^{-x} \, dx$ is convergent [see Example 4.6.2]. Hence also $I_N \to L^2$.

We now introduce $K_N = \iint_{D_N} e^{-(x^2 + y^2)} \, dx \, dy$ where D_N is the circular region given by $x^2 + y^2 \leq 2N^2$.

Then we have

$$I_N \le K_N \le I_{\sqrt{2}N},$$

on comparing the regions over which the integrals are taken (Figure 6.2.6),

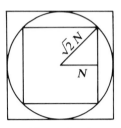

Figure 6.2.6

so that I_N and K_N have a common limit as $N \to \infty$. Now

$$K_N = \iint_{x^2+y^2 \le 2N^2} e^{-(x^2+y^2)} \, dx \, dy = \int_0^{\sqrt{2}N} \int_0^{2\pi} e^{-r^2} \, r \, dr \, d\theta,$$

on changing to polar co-ordinates [see Example 6.2.3]. Hence by Example 6.2.2:

$$K_N = 2\pi \int_0^{\sqrt{2}N} e^{-r^2} \, r \, dr$$

$$= -\pi [e^{-r^2}]_0^{\sqrt{2}N} = \pi(1 - e^{-2N^2}) \to \pi \quad \text{as } N \to \infty.$$

Hence $I_N \to \pi$ also, and thus $J_N \to \sqrt{\pi}$, as required.

6.3. SURFACE INTEGRALS

We now turn to the notion of a surface in three-dimensional space. Just as in section 6.1, a curve was defined as the continuous image of a one-dimensional region (an interval), so here a surface is defined as the continuous image of a two-dimensional region.

DEFINITION 6.3.1.   Let D be a region in the plane, and let $\gamma_1, \gamma_2, \gamma_3$ be continuous real-valued functions defined on D [see § 5.1]. Thus $\gamma = (\gamma_1, \gamma_2, \gamma_3)$ is a continuous function from D to $\mathbb{R}^3$. The set of points S in $\mathbb{R}^3$ described by $\gamma(t, u)$ as (t, u) varies in D is called a surface in $\mathbb{R}^3$. As in Definition 6.1.1 the functions $\gamma_1, \gamma_2, \gamma_3$ will be called the *coordinate*, or *component functions*.

EXAMPLE 6.3.1.   The *hemisphere* in $\mathbb{R}^3$ defined by $x^2 + y^2 + z^2 = r^2$, $z \ge 0$ may be defined as a surface in $\mathbb{R}^3$ by

$$x = \gamma_1(t, u) = t, \qquad y = \gamma_2(t, u) = u, \qquad z = \gamma_3(t, u) = \sqrt{(r^2 - t^2 - u^2)}$$

where (t, u) lies in the circular region $D = \{(t, u): t^2 + u^2 \le r^2\}$.

DEFINITION 6.3.2.   Let S be a surface with coordinate functions γ_1, γ_2, γ_3 as in Definition 6.1.1. We shall say that S is *smooth* if each coordinate function is continuously differentiable on D [see Definition 5.9.2] and that S is *piecewise smooth* if the coordinate functions are continuously differentiable except for a finite number of arcs in D on which the derivatives may have jump discontinuities [see Definition 5.1.3]. All surfaces will be assumed to be piecewise smooth from now on.

Notice that if we fix a value of u, say $u = u_0$ with (t, u_0) in D then as t varies, the point $\gamma(t, u_0)$ describes an arc which lies on the surface S. The tangent to this arc has direction

$$\gamma_t = \frac{\partial \gamma}{\partial t} = \left(\frac{\partial \gamma_1}{\partial t}, \frac{\partial \gamma_2}{\partial t}, \frac{\partial \gamma_3}{\partial t} \right)$$

at a point (t, u_0) [compare Definition 6.1.4]. Similarly if we fix $t = t_0$ then the point $\gamma(t_0, u)$ describes an arc which lies on S, and whose tangent has directions

$$\gamma_u = \frac{\partial \gamma}{\partial u} = \left(\frac{\partial \gamma_1}{\partial u}, \frac{\partial \gamma_2}{\partial u}, \frac{\partial \gamma_3}{\partial u} \right).$$

DEFINITION 6.3.3.   The direction perpendicular to both γ_t and γ_u is called the *normal direction* to the surface: it is defined (to within an ambiguity in sign) by the vector product [see § 17.2.3]

$$\gamma_t \wedge \gamma_u = (J_1, J_2, J_3)$$

say, where

$$J_1 = \frac{\partial \gamma_2}{\partial t} \frac{\partial \gamma_3}{\partial u} - \frac{\partial \gamma_3}{\partial t} \frac{\partial \gamma_2}{\partial u} = \frac{\partial(\gamma_2, \gamma_3)}{\partial(t, u)},$$

and similarly

$$J_2 = \frac{\partial(\gamma_3, \gamma_1)}{\partial(t, u)}, \qquad J_3 = \frac{\partial(\gamma_1, \gamma_2)}{\partial(t, u)}$$

are Jacobians of pairs of γ_1, γ_2, γ_3 with respect to (t, u) [see (5.12.2)].

We say that a point where $\gamma_t \wedge \gamma_u \neq 0$ is a *regular* point of the surface: any other point, where γ_t or γ_u are zero, or parallel, or fail to exist, is called *singular*.

Notice that if t, u are interchanged, the direction of $\gamma_t \wedge \gamma_u$ is reversed. The magnitude of $\gamma_t \wedge \gamma_u$ [see § 17.2.2] will be an important quantity connected with a given surface: we shall denote it by Δ, so that

$$\Delta^2 = J_1^2 + J_2^2 + J_3^2 = |\gamma_t \wedge \gamma_u|^2 = |\gamma_t|^2 |\gamma_u|^2 - (\gamma_t \cdot \gamma_u)^2. \tag{6.3.1}$$

The unit vector in the normal direction is denoted $\mathbf{n}$,

$$\mathbf{n} = \Delta^{-1}(\gamma_t \wedge \gamma_u). \tag{6.3.2}$$

We now consider how to define the area of a curved surface, and begin with an informal discussion to motivate the definition. Consider a point (t, u) in D, and suppose that t, u are increased by small amounts h, k respectively. The corresponding position $\boldsymbol{\gamma}(t, u)$ is increased to $\boldsymbol{\gamma}(t+h, u+k)$ which is approximately

$$\boldsymbol{\gamma}(t, u) + h\boldsymbol{\gamma}_t + k\boldsymbol{\gamma}_u.$$

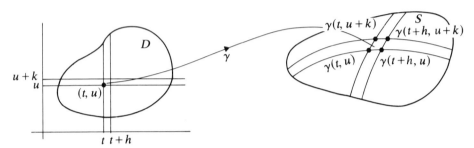

Figure 6.3.1

Thus a small rectangle in D, with area hk is mapped onto a small portion of S whose shape is approximately that of a parallelogram, with sides $h\boldsymbol{\gamma}_t, k\boldsymbol{\gamma}_u$ and whose area is $|h\boldsymbol{\gamma}_t \wedge k\boldsymbol{\gamma}_u| = hk\Delta$. Thus $\Delta = (J_1^2 + J_2^2 + J_3^2)^{1/2}$ acts as a local magnification factor for area, in the same way that $|\boldsymbol{\gamma}'(t)| = (\gamma_1'^2 + \gamma_2'^2 + \gamma_2'^2)^{1/2}$ is a local magnification factor for length in the treatment of curves [see Definition 6.1.3]. This suggests the following definition for surface area:

DEFINITION 6.3.4.   Let S be a surface determined by a function $\boldsymbol{\gamma}$ defined on a region D. Then the *area* of S is defined as the value of the double integral

$$\iint_D \Delta\, dt\, du.$$

(Again the analogy with the formula $\int_I |\boldsymbol{\gamma}'(t)|\, dt$ for arc length is worth noting.)

EXAMPLE 6.3.2.   Let S be the hemisphere defined in Example 6.3.1. Then $\boldsymbol{\gamma}(t, u) = (t, u, (r^2 - t^2 - u^2)^{1/2})$ for $t^2 + u^2 \le r^2$, so $\boldsymbol{\gamma}_t = (1, 0, -t(r^2 - t^2 - y^2)^{-1/2})$, $\boldsymbol{\gamma}_u = (0, 1, -u(r^2 - t^2 - u^2)^{-1/2})$, and

$$\Delta^2 = |\boldsymbol{\gamma}_t|^2 |\boldsymbol{\gamma}_u|^2 - (\boldsymbol{\gamma}_t \cdot \boldsymbol{\gamma}_u)^2 = \left(1 + \frac{t^2}{r^2 - t^2 - u^2}\right)\left(1 + \frac{u^2}{r^2 - t^2 - u^2}\right) - \left(\frac{tu}{r^2 - t^2 - u^2}\right)^2$$

$$= 1 + \frac{t^2 + u^2}{r^2 - t^2 - u^2} = \frac{r^2}{r^2 - t^2 - u^2}.$$

Then the surface area of the hemisphere is by Definition 6.3.4

$$\iint_D r(r^2 - t^2 - u^2)^{-1/2}\, dt\, du$$

This integral is most quickly evaluated by a change to polar coordinates, $t = \rho \cos \theta$, $u = \rho \sin \theta$, $0 \le \rho \le r$, $0 \le \theta \le 2\pi$, and $\partial(t, u)/\partial(\rho, \theta) = \rho$ [see Example 6.2.3]. It becomes

$$\int_0^r \int_0^{2\pi} \frac{r}{(r^2 - \rho^2)^{1/2}} \rho \, d\rho \, d\theta = 2\pi r \int_0^r \frac{\rho \, d\rho}{(r^2 - \rho^2)^{1/2}} = 2\pi r [-(r^2 - \rho^2)^{1/2}]_0^r = 2\pi r^2.$$

The notion of an integral over the surface S is now defined as follows:

DEFINITION 6.3.5.   Let S be a surface determined by a function γ defined on a region D in $\mathbb{R}^2$. Let f be a continuous real-valued function on S. Then the *integral* of f over S, which we shall denote $\iint_S f \, d\sigma$ is defined by

$$\iint_S f \, d\sigma = \iint_D f(\gamma_1(t, u), \gamma_2(t, u), \gamma_3(t, u)) \, \Delta \, dt \, du.$$

It is important to realize that this definition depends intrinsically on f and S, and is independent of the particular parametric function γ chosen to represent S. (We omit the proof of this.) Similarly if $\mathbf{f} = (f_1, f_2, f_3)$ is a continuous vector valued function of t, u in D we may define (as in Definition 6.1.5)

$$\iint_S \mathbf{f} \cdot d\boldsymbol{\sigma} = \iint_S (\mathbf{f} \cdot \mathbf{n}) \, d\sigma,$$

where $\mathbf{n}$ is the unit normal to the surface, as in Definition 6.3.3. The sign ambiguity noted in that definition shows that it is necessary to define a direction to $\mathbf{n}$ at some point of the surface, and extend this continuously to all points of S (we shall assume that S is such that this can be done). In particular if S is a *closed* surface (a piecewise smooth image of a sphere) we shall take $\mathbf{n}$ to be the outward normal direction [see also Theorem 6.4.2].

EXAMPLE 6.3.3.   The *torus* or *anchor-ring* surface, with radii a, r where $0 < r < a$, is given by

$$x = \gamma_1(\theta, \phi) = (a + r \cos \phi) \cos \theta,$$
$$y = \gamma_2(\theta, \phi) = r \sin \phi,$$
$$z = \gamma_3(\theta, \phi) = (a + r \cos \phi) \sin \theta,$$

for $0 \le \theta \le 2\pi$, $0 \le \phi \le 2\pi$.
    Let S be the upper half of T, given by $z \ge 0$, or $0 \le \theta \le \pi$. Find the area of S, and evaluate $\iint_S z \, d\sigma$.
    We have $\gamma = (\gamma_1, \gamma_2, \gamma_3)$ where $\gamma_1, \gamma_2, \gamma_3$ are defined in terms of θ, ϕ, so that

$$\gamma_\theta = \frac{\partial \gamma}{\partial \theta} = (-(a + r \cos \phi) \sin \theta, 0, (a + r \cos \phi) \cos \theta),$$

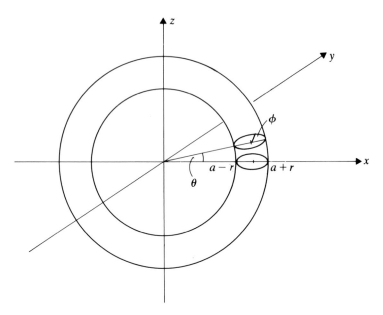

Figure 6.3.2

and

$$\gamma_\phi = \frac{\partial \gamma}{\partial \phi} = (-r \sin \phi \cos \theta, r \cos \phi, -r \sin \phi \sin \theta),$$

and hence $|\gamma_\theta|^2 = (a + r \cos \phi)^2$, $|\gamma_\phi|^2 = r^2$, and $\gamma_\theta \cdot \gamma_r = 0$. It follows that

$$\Delta^2 = |\gamma_\theta|^2 |\gamma_\phi|^2 - (\gamma_\theta \cdot \gamma_\phi)^2 = (a + r \cos \phi)^2 r^2$$

and so $\Delta = r(a + r \cos \phi)$. Hence the area of S is

$$\iint_S d\sigma = \iint_D \Delta \, d\phi \, d\theta = \int_0^\pi \int_0^{2\pi} r(a + r \cos \phi) \, d\phi \, d\theta$$

$$= \pi r \int_0^{2\pi} (a + r \cos \phi) \, d\phi = \pi r \cdot 2\pi a = 2\pi^2 ar.$$

As a consequence we also get the surprising result that the total surface area of T is $4\pi^2 ar = (2\pi a)(2\pi r)$, which is the same as the cylinder of length $2\pi a$ and radius of cross section r.

The integral $\iint_S z \, d\sigma$ is by Definition 6.3.5 equal to

$$\int_0^\pi \int_0^{2\pi} (a + r \cos \phi) \sin \theta \cdot \Delta \, d\phi \, d\theta = r \int_0^\pi \sin \theta \, d\theta \int_0^{2\pi} (a + r \cos \phi)^2 \, d\phi$$

$$= 2r \int_0^{2\pi} (a^2 + 2ar \cos \phi + r^2 \cos^2 \phi) \, d\phi$$

$$= 2r(a^2 2\pi + 2ar \cdot 0 + r^2 \pi) = 2\pi r(2a^2 + r^2).$$

(A consequence of this calculation is that the average value (or *mean*) of z over S, denoted $\bar{z}$, which is defined as $\int \int_S z \, d\sigma / \int \int_S d\sigma$, has been found to be $2\pi r(2a^2 + r^2)/2\pi^2 ar = 2/\pi(a + \frac{1}{2}r^2/a)$.) Our final result is the extension to surface of Green's theorem [Theorem 6.2.4].

THEOREM 6.3.1 (*Stokes's Theorem*). *Let S be a surface whose boundary is a closed arc C* [*see Definition 6.1.2*]. *Suppose also that the normal direction* $\mathbf{n} = (n_1, n_2, n_3)$ *at points of S* [*see Definition 6.3.3*] *is chosen so that the direction of traverse of C corresponds to an anti-clockwise circuit of S, relative to the direction of* $\mathbf{n}$ (compare the statement of Theorem 6.2.4). *Let* $\mathbf{f} = (f_1, f_2, f_3)$ *be a vector-valued function on S, having continuous first partial derivatives on S* [*see Definition 5.9.2*]. *Then*

$$\iint_S \left\{ n_1\left(\frac{\partial f_3}{\partial y} - \frac{\partial f_2}{\partial z}\right) + n_2\left(\frac{\partial f_1}{\partial z} - \frac{\partial f_3}{\partial x}\right) + n_3\left(\frac{\partial f_3}{\partial x} - \frac{\partial f_1}{\partial y}\right) \right\} d\sigma = \int_C \mathbf{f} \cdot \mathbf{ds}.$$

(In terms of the vector differential operator ∇ [see (17.3.14)] this equation can be written more neatly as

$$\iint_S (\nabla \wedge \mathbf{f}) \cdot \mathbf{d\sigma} = \int_C \mathbf{f} \cdot \mathbf{ds}.)$$

The proof of this result comes from writing the surface integral over S as an integral over the parameter region D, and in the same way, the line integral around C as a line integral around the boundary of D. Green's theorem [Theorem 6.2.4] is then applied to obtain equality. For the details of the proof, see Apostol (1957).

EXAMPLE 6.3.4. Let C, S be as in Theorem 6.3.1, and suppose that

$$\mathbf{f} = (f_1, f_2, f_3) = \left(\frac{\partial F}{\partial x}, \frac{\partial F}{\partial y}, \frac{\partial F}{\partial z}\right)$$

for some real-valued function F with continuous derivatives of degree ≤ 2. Then for instance

$$\frac{\partial f_3}{\partial y} = \frac{\partial^2 F}{\partial y \partial z}, \qquad \frac{\partial f_2}{\partial z} = \frac{\partial^2 F}{\partial z \partial y},$$

and the conditions on F imply that these mixed derivatives are equal. Hence each term on the left-hand side in the equation in Stokes's theorem vanishes, and we conclude that $\int_C \mathbf{f} \cdot \mathbf{ds} = 0$.

6.4. VOLUME INTEGRALS

In this section we extend the idea of integration from two to three dimensions. The development is parallel to that of section 6.2 and we refer the reader to that section for motivation.

DEFINITION 6.4.1.   A set V in three-dimensional space is called *normal* (relative to the chosen coordinate system) if every line parallel to the x-, y- or z-axes intersects it in a closed line segment [compare Definition 6.2.1].

DEFINITION 6.4.2.   Let V be a bounded normal region and a, b, c, d, e, f be chosen so that $a \leq x \leq b$, $c \leq y \leq d$, $e \leq z \leq f$ for all (x, y, z) in V. Let P_1, P_2, P_3 be dissections of $[a, b]$ $[c, d]$, $[e, f]$ respectively [see Definition 4.1.1], and let $I_{i,j,k}$ be the rectangular block with sides $[x_{i-1}, x_i]$, $[y_{j-1}, y_j]$, $[z_{k-1}, z_k]$ (where as in Definition 6.2.2, x_i, y_j, z_k denote typical points of P_1, P_2, P_3). Let g be a real-valued function on V. Then the *approximating sum* $S_{P_1,P_2,P_3}(g)$ is defined as

$$\sum_{i=1}^{l} \sum_{j=1}^{m} \sum_{k=1}^{n} g(u_{i,j,k})(x_i - x_{i-1})(y_j - y_{j-1})(z_k - z_{k-1}),$$

when $u_{i,j,k}$ is any point of $I_{i,j,k}$ which is in V (or $g(u_{i,j,k}) = 0$ if no point of $I_{i,j,k}$ is in V).

DEFINITION 6.4.3.   We say f is *integrable* over V if there is a number L such that for each $\varepsilon > 0$, there is a $\delta > 0$ such that $|S_{P_1,P_2,P_3}(g) - L| < \varepsilon$ whenever P_1, P_2, P_3 are dissections with norm $< \delta$. In this case we say that L is the *integral* of f over V, and write

$$L = \iiint_D g(x, y, z) \, dx \, dy \, dz.$$

In particular the *volume* of V is defined as $\iiint_V dx \, dy \, dz$.

As in section 6.2, Theorems 6.2.1 and 6.2.2, a continuous function is integrable over a bounded normal region, and the integral has the familiar properties of linearity, etc. Furthermore we have the following obvious analogue of Theorem 6.2.3:

THEOREM 6.4.1.   *Let V be a bounded normal region and let a, b, c, d, e, f be as in Definition 6.4.2. For each point (x, y) with $a \leq x \leq b$, $c \leq y \leq d$, let $[z_1, z_2]$ be the interval in which the line through (x, y) parallel to the z-axis intersects V. Let g be continuous on V, and define $F(x, y)$ by*

$$F(x, y) = \int_{z_1}^{z_2} g(x, y, z) \, dz.$$

Then

$$\iiint_V g(x, y, z) \, dx \, dy \, dz = \iint_D F(x, y) \, dx \, dy,$$

where D is the set of (x, y) for which some (x, y, z) is in V (D is the projection of V onto the (x, y)-plane).

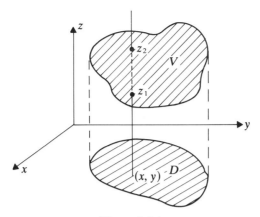

Figure 6.4.1

This result enables us to reduce a triple integral to a double one, and hence, via Theorem 6.2.3 to a product of integrals with respect to a single variable. Similar results may be obtained by interchanging the roles of x, y, z in the statement of the theorem.

EXAMPLE 6.4.1.   Evaluate $\iiint_V xyz \, dx \, dy \, dz$ where V is the region given by $x \geq 0$, $y \geq 0$, $z \geq 0$, $x + y + z \leq 1$.

With the notation of Theorem 6.4.1, D is the region given by

$$x \geq 0, \, y \geq 0, \qquad x + y \leq 1, \quad \text{and} \quad z_1 = 0, \qquad z_2 = 1 - x - y.$$

Hence the required integral is

$$\iint_D xy \left(\int_{z_1}^{z_2} z \, dz \right) dx \, dy = \tfrac{1}{2} \iint_D xy(1 - x - y)^2 \, dx \, dy.$$

We now use Theorem 6.2.3 to write this as

$$\tfrac{1}{2} \int_0^1 x \left\{ \int_0^{1-x} y(1 - x - y)^2 \, dy \right\} dx$$

$$= \tfrac{1}{2} \int_0^1 x \int_0^{1-x} \{(1 - x)(1 - x - y)^2 - (1 - x - y)^3\} \, dy \, dx$$

$$= \tfrac{1}{2} \int_0^1 x[-(1 - x)\tfrac{1}{3}(1 - x - y)^3 + \tfrac{1}{4}(1 - x - y)^4]_0^{1-x} \, dx$$

$$= \tfrac{1}{2} \int_0^1 x[(\tfrac{1}{3} - \tfrac{1}{4})(1 - x)^4] \, dx = \frac{1}{2 \cdot 3 \cdot 4} \int_0^1 (1 - x)x^4 \, dx$$

(putting $(1 - x)$ for x)

$$= \frac{1}{2 \cdot 3 \cdot 4} (\tfrac{1}{5} - \tfrac{1}{6}) = \frac{1}{2 \cdot 3 \cdot 4 \cdot 5 \cdot 6} = \frac{1}{6!}.$$

The three-dimensional result which is parallel to 6.2.4 (Green's Theorem) is known as *Gauss' theorem*, and relates the value of a volume integral to an integral over its boundary.

THEOREM 6.4.2   (*Gauss' Theorem*).   *Let V be a normal region in 3-dimensional space, whose boundary is the closed surface S. Suppose that we take* **n** *as the outward normal on S* [*see Definition* 6.3.5]. *Let* $\mathbf{f} = (f_1, f_2, f_3)$ *be a vector valued function on V with continuous first partial derivatives there. Then*

$$\iiint_V \left(\frac{\partial f_1}{\partial x} + \frac{\partial f_2}{\partial y} + \frac{\partial f_3}{\partial z} \right) dx \, dy \, dz = \iint_S \mathbf{f} \cdot \mathbf{n} \, d\sigma.$$

In terms of the vector differential operator ∇ [*see* (17.3.14)] *this may be written more briefly as*

$$\iiint_V (\nabla \cdot \mathbf{f}) \, dx \, dy \, dz = \iint_S \mathbf{f} \cdot d\boldsymbol{\sigma}.$$

Finally we consider the formula for a change of variables in a triple integral, which is analogous to Theorem 6.2.5.

THEOREM 6.4.3.   *Let* $x = \phi(u, v, w)$, $y = \psi(u, v, w)$, $z = \xi(u, v, w)$ *give a one-to-one correspondence between* (x, y, z) *in a region V and* (u, v, w) *in a region U, and suppose that* ϕ, ψ, ξ *have continuous first partial derivatives for which the Jacobian determinant*

$$J = \begin{vmatrix} \dfrac{\partial \phi}{\partial u} & \dfrac{\partial \phi}{\partial v} & \dfrac{\partial \phi}{\partial w} \\[2mm] \dfrac{\partial \psi}{\partial u} & \dfrac{\partial \psi}{\partial v} & \dfrac{\partial \psi}{\partial w} \\[2mm] \dfrac{\partial \xi}{\partial u} & \dfrac{\partial \xi}{\partial v} & \dfrac{\partial \xi}{\partial w} \end{vmatrix} \quad \text{is positive.}$$

Let f be a continuous function on V. Then

$$\iiint_V f(x, y, z) \, dx \, dy \, dz$$

$$= \iiint_U f(\phi(u, v, w), \psi(u, v, w), \xi(u, v, w)) J \, du \, dv \, dw.$$

For our next example we introduce spherical polar coordinates.

DEFINITION   6.4.4   (Spherical   polar   coordinates   [see   also   V, § 2.1.5]).   For a point $P = (x, y, z)$ in three-dimensional space, let $r \geq 0$ be defined by $r^2 = x^2 + y^2 + z^2$: thus r is simply the Euclidean distance from (x, y, x) to the origin, 0 [see V (2.1.7)]. Now define θ by $z = r \cos \theta$: for $r \neq 0$ this determines θ uniquely as an angle in the range $0 \leq \theta \leq \pi$, while θ is indeterminate at $(0, 0, 0)$. Geometrically, θ is the angle between the vector OP and the (positive) z-axis.

Thirdly the angle ϕ is defined by $x = r \sin \theta \cos \phi$, $y = r \sin \theta \sin \phi$: ϕ is determined to within a multiple of 2π for all points with $\sin \theta \neq 0$, that is, points not on the z-axis. Geometrically, ϕ is the angle through which the xz-plane must be rotated to contain P, as the diagram indicates.

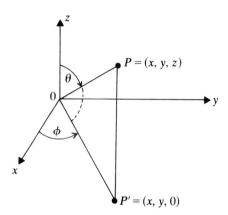

Figure 6.4.2

The Jacobian determinant of this transformation is

$$J = \frac{\partial(x, y, z)}{\partial(r, \theta, \phi)} = \begin{vmatrix} \sin \theta \cos \phi & \sin \theta \sin \phi & \cos \theta \\ r \cos \theta \cos \phi & r \cos \theta \sin \phi & -r \sin \theta \\ -r \sin \theta \sin \phi & r \sin \theta \cos \phi & 0 \end{vmatrix} = r^2 \sin \theta,$$

on expanding the determinant, for instance by its last column [see I § 6.9].

EXAMPLE 6.4.2.   Evaluate $\iiint_V (x^2 + y^2 + z^2)^{1/2} \, dx \, dy \, dz$, where V is the sphere $x^2 + y^2 + z^2 \leq a^2$.

We change to spherical polar coordinates, as above. The function to be integrated is $(x^2 + y^2 + z^2)^{1/2}$ which is simply r in the new coordinates, so that the integral becomes

$$\int_0^a \int_0^\pi \int_0^{2\pi} r \cdot r^2 \sin \theta \cdot d\phi \, d\theta \, dr = \int_0^a r^3 \, dr \int_0^\pi \sin \theta \, d\theta \int_0^{2\pi} d\phi$$

$$= \tfrac{1}{4} a^4 . 2 . 2\pi$$

$$= \pi a^4.$$

6.5.   EXTENSIONS TO HIGHER DIMENSION

Our discussions in the first four sections of this chapter have been limited to one, two and three dimensions, since these are the cases which are of most frequent application and have the most obvious geometrical content. However the formal similarities between the sections may have suggested that a unified treatment of functions of any number of variables is possible, and that there are results of that theory which contain Green's, Stokes' and Gauss' theorems as special cases. For this general theory, the reader is referred to Rudin (1964) or O'Neill (1966).

P. L. W.

REFERENCES

Apostol, T. M. (1957). *Mathematical Analysis*, Addison-Wesley, Reading, Mass.
Ledermann, W. (1966). *Multiple Integrals*, Routledge and Kegan Paul, London.
O'Neill, B. (1966). *Differential Geometry*, Academic Press, New York.
Rudin, W. (1964). *Principles of Mathematical Analysis*, McGraw-Hill, New York.
Rutherford, D. E. (1957). *Vector Methods*, Oliver & Boyd, Edinburgh.

CHAPTER 7

Ordinary Differential Equations

7.1. INTRODUCTION

Suppose that y is a real-valued function of a real variable t. Any equation which implies a relation between y and its derivatives with respect to t is called an ordinary differential equation. The term 'ordinary' indicates that there is only one independent variable and distinguishes such equations from partial differential equations, discussed in Chapter 8.

A simple example is

$$\frac{dy}{dt} = ky. \tag{7.1.1}$$

A solution of this equation is any function y of t for which (7.1.1) is valid. The most general such function is $y(t) = c \exp(kt)$, where c is an arbitrary constant [see § 7.2 and § 2.11].

Equation (7.1.1) models any system in which the rate of change of a certain quantity is proportional to the quantity present at any instant. For example if the rate of growth of the size of a population is proportional to its size, we arrive at (7.1.1) and deduce that the population will grow exponentially. The independent variable would be time in this case and $y(0) = c$ would be the initial size of the population. If $y(0) = y_0$ is a known quantity we would wish to find the solution of (7.1.1) for which

$$y(0) = y_0. \tag{7.1.2}$$

The problem of solving an equation such as (7.1.1) with a condition such as (7.1.2) is called an *initial value problem*. We have seen that the solution of this simple problem is $y(t) = y_0 \exp(kt)$.

Usually a system is described by the measurement of n quantities such as positions, velocities, temperatures, masses, in physical applications or say prices, costs and quantities in economics. The state of the system at time t is represented, therefore, by a vector $\mathbf{x}(t) = (x_1(t), x_2(t), \ldots, x_n(t))$ of n real numbers. A theory of how the system changes with time often leads to a system of ordinary differential equations

$$\frac{dx_i}{dt} = f_i(t, x_1, x_2, \ldots, x_n) \qquad (1 \le i \le n) \tag{7.1.3}$$

with a prescribed initial value

$$\mathbf{x}(t_0) = (c_1, c_2, \ldots, c_n). \tag{7.1.4}$$

The theory of such systems is described in §§ 7.9–7.11.

Equations (7.1.3) are called *first-order systems*, since only first-order derivatives are involved. However, much of classical applied mathematics involves second-order equations. The simplest and most important example is

$$\frac{d^2y}{dx^2} + k^2 y = 0, \tag{7.1.5}$$

where we have taken x to be the independent variable. The general solution of this equation is

$$y(x) = A \cos kx + B \sin kx \tag{7.1.6}$$

where A and B are arbitrary constants [see Example 7.4.1]. This follows from the elementary properties of the trigonometric functions,

$$\frac{d}{dx} \cos kx = -k \sin kx, \qquad \frac{d}{dx} \sin kx = \cos kx$$

[see § 2.12].

Since (7.1.6) involves two arbitrary constants we can solve (7.1.5) subject to the initial value conditions

$$y(x_0) = c_1 \qquad y'(x_0) = c_2. \tag{7.1.7}$$

Some calculation shows that the solution of (7.1.5) and (7.1.7) is

$$y(x) = c_1 \cos k(x - x_0) + (c_2/k) \sin k(x - x_0).$$

The general nth-order ordinary differential equation could be written as

$$F\left(x, y, \frac{dy}{dx}, \ldots, \frac{d^n y}{dx^n}\right) = 0 \tag{7.1.8}$$

where $d^n y/dx^n$ is the highest-order derivative present.

If y and its derivatives occur only to the first degree as algebraic quantities, the equation is said to be *linear*, otherwise it is *nonlinear*. Thus, the general linear nth-order equation has the form

$$a_0(x) \frac{d^n y}{dx^n} + a_1(x) \frac{d^{n-1} y}{dx^{n-1}} + \ldots + a_n(x) y = f(x). \tag{7.1.9}$$

The equations

$$\left(\frac{dy}{dx}\right)^2 + y = 0 \quad \text{and} \quad y\frac{dy}{dx} + x = 0$$

are non-linear because of the appearance of the second-degree terms $(dy/dx)^2$ and $y(dy/dx)$ respectively.

Linear equations are discussed in sections 7.2–7.8. In particular section 7.8 considers the problem of solving a second order linear equation given its initial and final value in an interval $a \le x \le b$. Such *boundary value problems* are of great interest in classical applied mathematics.

Non-linear equations are much less amenable to solution in terms of known elementary functions. The standard tricks which work in some very special cases are given in section 7.10.2. These methods were devised in the eighteenth and nineteenth centuries, using much skill and ingenuity. The entire approach to solving nonlinear problems was changed by the work of Henri Poincaré at the beginning of this century. Instead of trying to find individual exact solutions he concentrated on using the differential equation to reveal the general behaviour of all possible solutions. This qualitative theory yields even more interesting information than some complicated formula for a solution. Particular solutions are best investigated using modern computing techniques [see III, Chapter 8]. A brief introduction to the qualitative theory is given in section 7.11.

7.2.   LINEAR EQUATIONS OF FIRST ORDER

The general linear equation of the first order is

$$a(x)\frac{dy}{dx} + b(x)y = c(x). \tag{7.2.1}$$

Dividing both sides of the equation by $a(x)$, we have

$$\frac{dy}{dx} + p(x)y = f(x), \tag{7.2.2}$$

where $p(x) = \{b(x)/a(x)\}$ and $f(x) = \{c(x)/a(x)\}$.

If $p(x)$ is identically zero, that is, if

$$\frac{dy}{dx} = f(x), \tag{7.2.3}$$

then the solution of the equation is

$$y(x) = \int f(x)\,dx + c,$$

where c is an arbitrary constant of integration [see § 4.2]. It may be possible to evaluate the indefinite integral in terms of elementary functions by the techniques of the theory of integration. Much effort has been spent on such problems in the past and there are extensive tables of integrals which should be consulted if the integration seems to be difficult; see, for example, Table 4.2.1, or Gradshteyn and Ryzhik (1965).

Suppose that we are interested in the solution of equation (7.2.3) for which $y(a) = \alpha$. The constant c is then determined and

$$y(x) = \int_a^x f(s) \, ds + \alpha.$$

The value of $y(x)$ at a given x could now be determined by the techniques of numerical integration described in Volume III, Chapter 8. If the integral occurs frequently in analysis it is possible that its values have been tabulated and indeed that it is regarded by some as an 'elementary function'.

EXAMPLE 7.2.1.   The solution of the initial-value problem

$$\sqrt{x} \frac{dy}{dx} = \exp(-x), \qquad y(1) = 0,$$

is

$$y(x) = \int_1^x (1/\sqrt{s}) \exp(-s) \, ds = \gamma(\tfrac{1}{2}, x) - \gamma(\tfrac{1}{2}, 1),$$

where

$$\gamma(\alpha, x) = \int_0^x s^{\alpha-1} \exp(-s) \, ds,$$

the *incomplete Gamma function*, tabulated by Pearson (1922).

If $p(x)$ is not identically zero in equation (7.2.2), let

$$P(x) = \int p(x) \, dx.$$

It is easy to verify that the solution of the equation is

$$y(x) = \exp\{-P(x)\}\left[\int f(x) \exp\{P(x)\} \, dx + c\right],$$

where c is an arbitrary constant of integration.

EXAMPLE 7.2.2.   If

$$\frac{dy}{dx} + (1/x)y = x^2,$$

then

$$P(x) = \int (1/x) \, dx = \log x,$$

$$\exp\{P(x)\} = \exp\{\log x\} = x$$

and

$$y(x) = (1/x)\{(x^4/4) + c\}.$$

7.3. LINEAR EQUATIONS OF NTH ORDER

7.3.1. General Theory of the Homogeneous Equation

The general homogeneous linear equation of nth order can be written in the standard form

$$\frac{d^n y}{dx^n} + p_1(x) \frac{d^{n-1}y}{dx^{n-1}} + \ldots + p_{n-1}(x) \frac{dy}{dx} + p_n(x)y = 0. \qquad (7.3.1)$$

The equation is homogeneous in the algebraic sense that if $y(x)$ is a solution then so is $\lambda y(x)$, where λ is an arbitrary constant.

We will suppose that x lies in an interval I, $a \leq x \leq b$ [see I, § 2.6.3] and that the coefficient functions $p_1, p_2, \ldots, p_n$ are continuous functions in I [see § 2.1]. Denote by $y^{(k)}(a)$ the kth derivative of $y(x)$ evaluated at $x = a$. Then we have

THEOREM 7.3.1. *There exists a unique solution of equation* (7.3.1) *for which*

$$y(a) = c_1, \ y^{(1)}(a) = c_2, \ldots, \ y^{(n-1)}(a) = c_n, \qquad (7.3.2)$$

where $c_1, c_2, \ldots, c_n$ *are n given constants.*

In particular, if y is the solution of equation (7.3.1) for which y and its first $n - 1$ derivatives are zero at $x = a$, then y is identically zero in I.

It is clear that the sum of any two solutions of the equation is again a solution. This fundamental property of any linear homogeneous equation determines the structure of its set of all possible solutions. To describe this set of solution functions concisely we need the language of linear algebra.

DEFINITION 7.3.1. Suppose that $y_1, y_2, \ldots, y_n$ are functions defined in I and that $\alpha_1, \alpha_2, \ldots, \alpha_n$ are constants. If

$$\alpha_1 y_1(x) + \alpha_2 y_2(x) + \ldots + \alpha_n y_n(x) = 0$$

for all x in I only when $\alpha_1 = \alpha_2 = \ldots = \alpha_n = 0$, then the functions are said to be *linearly independent* in I; otherwise, they are said to be *linearly dependent* in I [cf. I, § 5.3].

EXAMPLE 7.3.1. The functions

$$y_1(x) = 1, \qquad y_2(x) = \tfrac{1}{2}x, \qquad y_3(x) = 1 - x,$$

are linearly dependent in any given interval I, since

$$y_1(x) - 2y_2(x) - y_3(x) = 0$$

for all x in I.

EXAMPLE 7.3.2.   The functions

$$y_1(x) = 1, \qquad y_2(x) = x, \qquad y_3(x) = x^2$$

are linearly independent in any interval I, $a \le x \le b$ $(a \ne b)$, since

$$\alpha_1 + \alpha_2 x + \alpha_3 x^2 = 0$$

for *all* x in I is impossible for constants α_1, α_2, α_3 except when $\alpha_1 = \alpha_2 = \alpha_3 = 0$.

DEFINITION 7.3.2.   A *fundamental set of solutions* of equation (7.3.1) is any set of n linearly independent solutions.

The set of solutions of the equation is described in the following simple but striking theorem.

THEOREM 7.3.2.   *There exists a fundamental set of solutions for equation (7.3.1) and every solution can be expressed as a linear combination of solutions in the fundamental set.*

In practical terms this means that to find all possible solutions of equation (7.3.1) we need to find only n linearly independent solutions $y_1, y_2, \ldots, y_n$, say. The solution

$$y(x) = \alpha_1 y_1(x) + \alpha_2 y_2(x) + \ldots + \alpha_n y_n(x),$$

which involves n arbitrary constants $\alpha_1, \alpha_2, \ldots, \alpha_n$ is called the *general solution*. These constants can be chosen to satisfy various initial or boundary conditions such as (7.3.2).

We will see later that the set of solutions for a non-linear equation can be much more complicated.

It is useful to have a criterion for the linear independence of a set of n solutions of equation (7.3.1).

DEFINITION 7.3.3.   Let $y_1, y_2, \ldots, y_n$ be n solutions of equation (7.3.1). Their *Wronskian determinant*,

$$W(x) \equiv W(y_1(x), y_2(x), \ldots, y_n(x))$$

is defined to be the $n \times n$ determinant whose kth row is

$$(y_1^{(k-1)}(x) \; y_2^{(k-1)}(x) \; \ldots \; y_n^{(k-1)}(x)),$$

with the convention that $y^{(0)}(x) = y(x)$ [see I, § 6.9].

One finds that

$$W(x) = W(x_0) \exp\left\{-\int_{x_0}^{x} p_1(s)\, ds\right\} \tag{7.3.3}$$

where x_0 is any point in I. This is *Liouville's formula*. Moreover we have the important theorem;

THEOREM 7.3.3.   *The n solutions $y_1, y_2, \ldots, y_n$ of equation (7.3.1) are linearly independent in I if and only if their Wronskian determinant is not identically zero in I.*

In view of Liouville's formula, it is only necessary to check that $W(x_0) \neq 0$ for a single point x_0 in I!

EXAMPLE 7.3.3.   It is easily verified that the equation

$$(x-1)y'' - xy' + y = 0$$

has solutions $y_1(x) = x$ and $y_2(x) = \exp x$. Their Wronskian

$$W(y_1(x), y_2(x)) = \det \begin{pmatrix} x & \exp x \\ 1 & \exp x \end{pmatrix} = (x-1) \exp x$$

$$\neq 0, \quad \text{except at } x = 1.$$

From Theorem 7.3.2, the general solution of the equation is

$$y(x) = \alpha_1 x + \alpha_2 \exp x$$

for x in any interval.

Notice that Theorem 7.3.3 applies to the above example only if the interval I under consideration does not contain $x = 1$. For the standard form (7.3.1) of the equation is

$$y'' - \{x/(x-1)\}y' + \{1/(x-1)\}y = 0$$

and neither $p(x) = -\{x/(x-1)\}$ nor $q(x) = \{1/(x-1)\}$ is continuous at $x = 1$.

7.3.2.   General Theory of the Non-homogeneous Equation

Let L be the differential operator defined by

$$Ly = \frac{d^n y}{dx^n} + p_1(x)\frac{d^{n-1}y}{dx^{n-1}} + \ldots + p_{n-1}(x)\frac{dy}{dx} + p_n(x)y. \tag{7.3.4}$$

The general non-homogeneous linear equation of nth order is written briefly as

$$Ly = f, \tag{7.3.5}$$

where f is a function not identically zero in the interval I under consideration. The set of solutions of the equation is easy to describe.

THEOREM 7.3.4.   *Let v be any particular solution of equation (7.3.5) and let u be the general solution of the associated homogeneous equation*

$$Ly = 0. \tag{7.3.6}$$

The general solution of equation (7.3.5) is $y = u + v$.

We call u the *complementary function* and v a *particular integral* of (7.3.5). Let $y_1, y_2, \ldots, y_n$ be a fundamental set of solutions of equation (7.3.6). If $\alpha_1, \alpha_2, \ldots, \alpha_n$ are constants then

$$\alpha_1 y_1(x) + \alpha_2 y_2(x) + \ldots + \alpha_n y_n(x)$$

is a particular solution of equation (7.3.6). A particular integral of (7.3.5) can be found by the *method of variation of constants* or variation of parameters, that is, we let

$$v(x) = \alpha_1(x)y_1(x) + \alpha_2(x)y_2(x) + \ldots + \alpha_n(x)y_n(x)$$

and find functions $\alpha_1(x), \alpha_2(x), \ldots, \alpha_n(x)$ such that

$$Lv = f.$$

To illustrate the method consider equation (7.3.5) when $n = 2$. Let y_1 and y_2 be a fundamental set of solutions for equation (7.3.6). Let

$$v(x) = \alpha_1(x)y_1(x) + \alpha_2(x)y_2(x)$$

and suppose that

$$\alpha_1'(x)y_1(x) + \alpha_2'(x)y_2(x) = 0. \tag{7.3.7}$$

Then $v'(x) = \alpha_1(x)y_1'(x) + \alpha_2(x)y_2'(x)$ and hence

$$Lv = \alpha_1(x)Ly_1 + \alpha_2(x)Ly_2 + \alpha_1'(x)y_1'(x) + \alpha_2'(x)y_2'(x)$$

$$= \alpha_1'(x)y_1'(x) + \alpha_2'(x)y_2'(x)$$

$$= f$$

if we suppose also that

$$\alpha_1'(x)y_1'(x) + \alpha_2'(x)y_2'(x) = f(x). \tag{7.3.8}$$

But we can solve the linear equations (7.3.7) and (7.3.8) for $\alpha_1'(x)$ and $\alpha_2'(x)$ [see I. § 5.9] and we find that

$$\alpha_1'(x) = -f(x)y_2(x)/(y_1(x)y_2'(x) - y_1'(x)y_2(x))$$

and

$$\alpha_2'(x) = f(x)y_1(x)/(y_1(x)y_2'(x) - y_1'(x)y_2(x)).$$

The functions $\alpha_1(x)$ and $\alpha_2(x)$ can be found by integration of these formulae.

EXAMPLE 7.3.4.   Consider the non-homogeneous linear second order equation

$$Ly = \frac{d^2y}{dx^2} - y = x \exp(x).$$

The associated homogeneous equation has the linearly independent solutions $y_1(x) = \exp(x)$ and $y_2(x) = \exp(-x)$, found by the methods described below in section 7.4.1. Therefore, equations (7.3.7) and (7.3.8) become

$$\alpha_1'(x) \exp(x) + \alpha_2'(x) \exp(-x) = 0$$

and

$$\alpha_1'(x) \exp(x) - \alpha_2'(x) \exp(-x) = x \exp(x).$$

Solving these equations

$$\alpha_1'(x) = (\tfrac{1}{2})x \quad \text{and} \quad \alpha_2'(x) = -(\tfrac{1}{2})x \exp(2x).$$

Therefore,

$$\alpha_1(x) = (\tfrac{1}{4})x^2$$

and

$$\alpha_2(x) = \{-(\tfrac{1}{4})x + (\tfrac{1}{8})\} \exp(2x).$$

It follows that the general solution of the inhomogeneous equation is

$$y(x) = \alpha_1 \exp(x) + \alpha_2 \exp(-x) + \{(x^2/4) - (x/4) + (\tfrac{1}{8})\} \exp(x).$$

If we know a fundamental set of solutions for equation (7.3.6) it is possible to write down a formula for a particular integral of equation (7.3.5), although this may not be the simplest method of finding such a particular integral in practice.

THEOREM 7.3.5.   *Let $y_1, y_2, \ldots, y_n$ be a fundamental set of solutions for the equation $Ly = 0$ and let $W(s)$ be their Wronskian determinant. Let $G(x, s)$ be the determinant formed by deleting the last row of $W(s)$ and replacing it by $(y_1(x)\, y_2(x)\, \ldots\, y_n(x))$. Then*

$$v(x) = \int_a^x \{G(x, s)/W(s)\}f(s)\, ds \tag{7.3.9}$$

is the particular integral of the equation $Ly = f$ for which

$$v(a) = 0, \; v^{(1)}(a) = 0, \ldots, v^{(n-1)}(a) = 0.$$

Formula (7.3.9) can be simplified by a sensible choice of the fundamental set $y_1, y_2, \ldots, y_n$. If y_k is the solution of $Ly = 0$ for which $y_k^{(k-1)}(a) = 1$ and $y_k^{(i)}(a) = 0$ for $0 \le i \le n - 1$, $i \ne k - 1$, then $W(a) = 1$, the determinant of the identity matrix [see I, 6.2.8]. From Liouville's formula (7.3.3),

$$W(x) = \exp\left\{-\int_a^x p_1(s)\, ds\right\}.$$

In particular, if $p_1(x) = 0$ for x in I, then $W(x) = 1$ for x in I.
Notice that, when $n = 2$,

$$W(s) = \det\begin{pmatrix} y_1(s) & y_2(s) \\ y_1'(s) & y_2'(s) \end{pmatrix} \quad \text{and} \quad G(x, s) = \det\begin{pmatrix} y_1(s) & y_2(s) \\ y_1(x) & y_2(x) \end{pmatrix}$$

so that the general solution of

$$\frac{d^2 y}{dx^2} + p_1(x)\frac{dy}{dx} + p_2(x)y = f(x)$$

is

$$y(x) = \alpha_1 y_1(x) + \alpha_2 y_2(x) + \int_a^x \{y_1(s)y_2(x) - y_1(x)y_2(s)\}\{f(s)/W(s)\}\, ds, \quad (7.3.10)$$

or

$$y(x) = \left\{\alpha_1 - \int_a^x [y_2(s)f(s)/W(s)]\, ds\right\}y_1(x) + \left\{\alpha_2 + \int_a^x [y_1(s)f(s)/W(s)]\, ds\right\}y_2(x).$$

The last form of the solution can be useful when one is interested in the behaviour of $y(x)$ as $x \to \infty$, say.

EXAMPLE 7.3.5.   Suppose that

$$\frac{d^2 y}{dx^2} + \lambda^2 y = f(x), \qquad y(0) = 1, \qquad y'(0) = 0.$$

If $f(x) \equiv 0$, the solution of this problem is $y(x) = \cos \lambda x$ [see § 7.4.1]. If f is in some sense small for large x, how does this affect the solution? Let us suppose that $\int_0^\infty |f(x)|\, dx$ is finite and for later use let

$$A = \int_0^\infty \sin \lambda s\{f(s)/\lambda\}\, ds$$

and

$$B = \int_0^\infty \cos \lambda s\{f(s)/\lambda\}\, ds.$$

The homogeneous equation associated with our problem has a fundamental set of solutions $y_1(x) = \cos \lambda x$ and $y_2(x) = \{\sin \lambda x/\lambda\}$ with Wronskian $W(x) = 1$.

Therefore, using (7.3.10),

$$y(x) = \cos \lambda x + \int_0^x \{\cos \lambda s \sin \lambda x - \sin \lambda s \cos \lambda x\}\{f(s)/\lambda\}\, ds$$

$$= \left\{1 - \int_0^x \sin \lambda s [f(s)/\lambda]\, ds\right\} \cos \lambda x + \left\{\int_0^x \cos \lambda s [f(s)/\lambda]\, ds\right\} \sin \lambda x$$

$$= (1 - A) \cos \lambda x + B \sin \lambda x + r(x),$$

where

$$r(x) = \left\{\int_x^\infty \sin \lambda s [f(s)/\lambda]\, ds\right\} \cos \lambda x - \left\{\int_x^\infty \cos \lambda s [f(s)/\lambda]\, ds\right\} \sin \lambda x.$$

Now $r(x) \to 0$ as $x \to \infty$ and hence, for large x, the solution behaves very much like the function

$$u(x) = (1 - A) \cos \lambda x + B \sin \lambda x = C \cos (\lambda x + \delta), \quad \text{say.}$$

Thus the presence of a non-zero f produces a change in amplitude and a phase shift in the solution of the same problem with $f(x) \equiv 0$.

7.3.3. Reduction of the Order of a Homogeneous Equation

Let u be a known solution of equation (7.3.1). Substitution of $y = uv$ into (7.3.1) will give an equation of order $n - 1$ for dv/dx, as will be illustrated in the following example.

EXAMPLE 7.3.6. One solution, found by inspection, of the equation

$$x^2 y''' - (3x^2 + 2x)y'' + (3x^2 + 4x + 2)y' - (x^2 + 2x + 2)y = 0$$

is $u(x) = \exp x$.

If $y(x) = v \exp x$ is a solution, where v must be determined, then

$$x^2 v''' - 2xv'' + 2v' = 0$$

an equation of Euler's type discussed in section 7.5 below. One finds the two solutions $v_1(x) = x$ and $v_2(x) = x^2$.

Therefore, the general solution of our equation is

$$y(x) = \{\alpha_1 + \alpha_2 x + \alpha_3 x^2\} \exp x.$$

7.4. LINEAR EQUATIONS WITH CONSTANT COEFFICIENTS

7.4.1. The Homogeneous Equation

We turn to the problem of finding explicit solutions to nth order linear equations, considering first the case when the coefficient functions are constants. The general such homogeneous equation is

$$Ly = \frac{d^n y}{dx^n} + a_1 \frac{d^{n-1} y}{dx^{n-1}} + \ldots + a_{n-1} \frac{dy}{dx} + a_n y = 0. \qquad (7.4.1)$$

DEFINITION 7.4.1.   The *characteristic polynomial* of equation (7.4.1) is

$$p(\lambda) = \lambda^n + a_1 \lambda^{n-1} + \ldots + a_{n-1}\lambda + a_n.$$

It is easy to verify that $y(x) = \exp(\lambda x)$ is a solution of equation (7.4.1) if and only if

$$p(\lambda) = 0. \qquad (7.4.2)$$

Therefore our present problem reduces to finding the n roots of the algebraic equation (7.4.2). The latter problem is discussed in Volume I, Chapter 14, and numerical computation of the roots in Volume III, Chapter 5.

THEOREM 7.4.1.   *If λ is a root of equation (7.4.2) of multiplicity k, then $\exp(\lambda x)$; $x \exp(\lambda x), \ldots, x^{k-1} \exp(\lambda x)$ are k linearly independent solutions of equation (7.4.1).*

It may happen that $p(\lambda)$ has complex roots. If the coefficients $a_1, a_2, \ldots, a_n$ are real numbers and if $\lambda = \mu + i\nu$ is a complex root of multiplicity k then $\bar{\lambda} = \mu - i\nu$ will be a complex root of the same multiplicity [see I, Proposition 14.5.3]. The functions

$$\exp \mu x \cos \nu x, \, x \exp \mu x \cos \nu x, \ldots, x^{k-1} \exp \mu x \cos \nu x$$

and

$$\exp \mu x \sin \nu x, \, x \exp \mu x \sin \nu x, \ldots, x^{k-1} \exp \mu x \sin \nu x$$

provide $2k$ real linearly independent solutions.
    By these means we find n linearly independent solutions of equation (7.4.1) and hence its general solution.
    If the characteristic polynomial is of high degree it may be difficult to find all its roots. However it may be possible to decide whether or not the roots have negative real parts from a test such as the Routh–Hurwitz Criterion. The somewhat complicated details are given in Cronin (1981).

EXAMPLE 7.4.2.   The characteristic polynomial of the equation

$$y''' - 3y' + 2y = 0$$

is

$$p(\lambda) = \lambda^3 - 3\lambda + 2 = (\lambda + 2)(\lambda - 1)^2.$$

Therefore, the general solution of the equation is

$$y(x) = (\alpha_1 + \alpha_2 x) \exp x + \alpha_3 \exp(-2x).$$

An immediate consequence of Theorem 7.4.1 which is of some importance in the theory of control systems is the following *stability* result.

THEOREM 7.4.2.   *Every solution of equation (7.4.1) tends to zero as x tends to infinity if and only if every root of the characteristic polynomial has a negative real part.*

EXAMPLE 7.4.1.   Equation (7.1.5) has characteristic polynomial

$$p(\lambda) = \lambda^2 + k^2 = (\lambda - ik)(\lambda + ik)$$

and so the general solution of (7.1.5) is

$$y(x) = \alpha_1 \exp(ikx) + \alpha_2 \exp(-ikx)$$

$$= A \cos kx + B \sin kx$$

where, by (9.5.8), $A = \alpha_1 + \alpha_2$ and $B = i(\alpha_1 - \alpha_2)$.

7.4.2.   The Non-homogeneous Equation

To find the general solution of the non-homogeneous equation

$$Ly = f \tag{7.4.3}$$

we must find the general solution of the associated homogeneous equation (7.4.1), as just described, and any particular integral of equation (7.4.3). [See Theorem 7.3.4 above.]

The particular integral may sometimes be found by inspection:

EXAMPLE 7.4.3.   The homogeneous equation associated with the equation

$$\frac{d^2y}{dx^2} = x + y \tag{7.4.4}$$

is $d^2y/dx^2 - y = 0$. This has characteristic polynomial

$$p(\lambda) = \lambda^2 - 1 = (\lambda - 1)(\lambda + 1)$$

and so the complementary function of (7.4.4) is

$$\alpha_1 \exp x + \alpha_2 \exp(-x).$$

By inspection, a particular integral of (7.4.4) is $y = -x$, and so the general solution of (7.4.4) is

$$\alpha_1 \exp x + \alpha_2 \exp(-x) - x.$$

The particular integral could be found by using the formula (7.3.9) but the calculations involved are usually very heavy. We describe next a method of finding a particular integral based on the operational calculus.

We define the *differential operator* D^k by

$$D^k y = \frac{d^k y}{dx^k} \qquad (k = 1, 2, 3, \ldots).$$

The sum and products of such operators are defined as follows;

$$\{D^m + D^n\}y = D^m y + D^n y, \qquad \{D^m D^n\}y = D^{m+n}y$$

and

$$\{\alpha D^m\}y = \alpha D^m y \quad \text{for any constant } \alpha, \text{ and } n, m = 1, 2, 3, \ldots.$$

The symbols $D, D^2, \ldots$ can be added and multiplied as if D were an algebraic indeterminate. This is not mysterious but simply a consequence of identities such as

$$D^m D^n = D^{m+n} \quad \text{and} \quad \alpha D^m + \beta D^m = (\alpha + \beta)D^m.$$

Using the characteristic polynomial, equation (7.4.3) can be written as

$$p(D)y = f.$$

The operational calculus provides rules for making sense of the 'solution' of this equation,

$$y = [p(D)]^{-1}f.$$

Some of these rules are listed in Table 7.4.1; further information can be found in Bateman (1966).

$p(D)$	f	$[p(D)]^{-1}f$
D	f	$\int f(x)\, dx$
$D^n \quad (n = 1, 2, 3, \ldots)$	f	$\int_0^x \dfrac{(x-u)^{n-1}}{(n-1)!} f(u)\, du$
$D^n \quad (n = 1, 2, 3, \ldots)$	1	$x^n/n!$
$D^n \quad (n = 1, 2, 3, \ldots)$	$\exp(ax)$	$\exp(ax)/a^n$
$p(D), \quad p(a) \neq 0$	$\exp(ax)$	$\exp(ax)/p(a)$
$p(D) = F(D^2) + G(D)H(D^2),$ where $F(-a^2) \neq 0$ and $H(-a^2) = 0$	$\sin ax$	$\sin ax/F(-a^2)$
	$\cos ax$	$\cos ax/F(-a^2)$
$p(D)$	$\exp(ax)g(x)$	$\exp(ax)[p(a+D)]^{-1}g$
$p(D)$	$\alpha f + \beta g$	$\alpha[p(D)]^{-1}f + \beta[p(D)]^{-1}g$

Table 7.4.1

EXAMPLE 7.4.4
$$(D^2 + 6D + 9)y = 50 \exp(2x).$$

Here $p(D) = (D+3)^2$ and hence
$$y(x) = 50 \exp(2x)/(2+3)^2 = 2 \exp(2x).$$

EXAMPLE 7.4.5
$$(D^2 - 4D + 4)y = 50 \exp(2x).$$

Here $p(D) = (D-2)^2$ and $p(2) = 0$. However, viewing $\exp(2x)$ as $\exp(2x)g(x)$ with $g(x) = 1$, we have
$$y(x) = 50 \exp(2x)[p(2+D)]^{-1} . 1 = 50 \exp(2x)[D^2]^{-1} . 1 = 25x^2 \exp(2x).$$

EXAMPLE 7.4.6
$$(D^3 + 4D^2 + 5D + 2)y = \cos 2x.$$

Here $p(D) = (D+4)(D^2+4) + D - 14$ and we have
$$y(x) = (D+14)[(D+14)(D+4)(D^2+4) + D^2 - 14^2]^{-1} \cos 2x$$
$$= (D+14)\{\cos 2x\}/\{-4 - 14^2\}$$
$$= (1/200)\{2 \sin 2x - 14 \cos 2x\}.$$

Even more remarkable manipulations with the operator D are possible. A mathematical account of this fascinating theory is given in Mikusinski (1959).

EXAMPLE 7.4.7
$$(D^2 - 1)y = x^3.$$

We proceed formally and write
$$y(x) = -(1 - D^2)^{-1}x^3$$
$$= -(1 + D^2 + D^4 + D^3 + \ldots)x^3 \qquad [\text{compare (1.10.10)}]$$
$$= -x^3 - 6x, \quad \text{since } D^m x^3 = 0 \text{ for } m \geq 4.$$

It is easy to verify that this is indeed a particular integral of the equation. Its general solution is therefore, by Example 7.4.1 and Theorem 7.3.4,
$$y(x) = \alpha_1 \exp x + \alpha_2 \exp(-x) - (x^3 + 6x).$$

7.5.   EULER'S EQUATION

There is no general method of finding explicit solutions to linear nth order equations whose coefficient functions are not constants. However, the equation
$$t^n \frac{d^n y}{dt^n} + a_1 t^{n-1} \frac{d^{n-1} y}{dt^{n-1}} + \ldots + a_{n-1} t \frac{dy}{dt} + a_n y = 0,$$

usually called *Euler's equation*, can be transformed into an equation with constant coefficients by the change of independent variable $x = \log t$. Alternatively, one can seek solutions of the form $y(t) = t^{\alpha}$ directly, by substitution into the equation.

EXAMPLE 7.5.1.   Consider the equation

$$t^2 \frac{d^2y}{dt^2} - 2t \frac{dy}{dt} + 2y = 0.$$

If $x = \log t$, then using the chain rule [see (3.2.8)]

$$\frac{d}{dt} = \frac{dx}{dt} \frac{d}{dx} = \frac{1}{t} \frac{d}{dx}$$

and

$$\frac{d^2}{dt^2} = \frac{d}{dt} \left\{ \frac{1}{t} \frac{d}{dx} \right\} = -\frac{1}{t^2} \frac{d}{dx} + \frac{1}{t^2} \frac{d^2}{dx^2}.$$

The equation becomes

$$\frac{d^2y}{dx^2} - 3 \frac{dy}{dx} + 2y = 0,$$

which has the characteristic polynomial

$$p(\lambda) = \lambda^2 - 3\lambda + 2 = (\lambda - 2)(\lambda - 1).$$

Therefore, the general solution of the equation is

$$y(t) = \alpha_1 \exp x + \alpha_2 \exp (2x)$$
$$= \alpha_1 t + \alpha_2 t^2.$$

7.6. LINEAR EQUATIONS OF SECOND ORDER

7.6.1. Introduction

We have seen in section 7.3 that the problem of finding the general solution of the linear second order equation

$$\frac{d^2y}{dx^2} + p(x) \frac{dy}{dx} + q(x)y = r(x) \tag{7.6.1}$$

reduces to finding two linearly independent solutions of the equation

$$Ly = \frac{d^2y}{dx^2} + p(x) \frac{dy}{dx} + q(x)y = 0. \tag{7.6.2}$$

But if we know one solution $u(x)$, say, with $u(x) \neq 0$, and let $y(x) = u(x)v(x)$, we find that

$$u\frac{d^2v}{dx^2}+\left(2\frac{du}{dx}+p(x)u\right)\frac{dv}{dx}=0.$$

Solving this first order equation for dv/dx as described above in section 7.2, and integrating, we find that

$$v(x)=\int (1/u(x))^2 \exp\left\{-\int p(x)\,dx\right\}dx. \qquad (7.6.3)$$

Moreover, the Wronskian determinant (Definition 7.3.3)

$$W(u(x),u(x)v(x))=\exp\left\{-\int p(x)\,dx\right\}\neq 0,$$

so that the solutions u, uv are linearly independent.

Our problem reduces to finding one solution of the homogeneous equation (7.6.2)!

Second-order linear equations have been studied extensively because of their importance in classical applied mathematics. We will describe two general methods of solution applicable to special types of equation. First we list a few of the famous equations whose solutions have been studied by these methods. This may give the reader some feeling for the equations which he may expect to find in the references such as Kamke (1955). In each case a second solution can be found, either from (7.6.3) or as described below.

1. *Legendre's equation.*

$$(1-x^2)y''-2xy'+n(n+1)y=0 \qquad (n=0,1,2,\ldots)$$

$$y(x)=P_n(x),$$

Legendre's polynomial of degree n [see § 10.3.2].

2. *Bessel's equation.*
(a) $$x^2y''+xy'+(x^2-\nu^2)y=0$$

$$y(x)=J_\nu(x),$$

a Bessel function of the first kind [see § 10.4.2].

(b) $$xy''+(2\nu+1)y'+xy=0$$

$$y(x)=x^{-\nu}J_\nu(x).$$

(c)   *Airy's equation*

$$y''+xy=0$$

$$y(x)=\sqrt{x}J_{1/3}(\tfrac{2}{3}x^{3/2}).$$

3. *Weber's equation.*

$$y'' + (\nu + \tfrac{1}{2} - \tfrac{1}{4}x^2)y = 0$$

$$y(x) = D_\nu(x),$$

a parabolic cylinder function [see Whittaker and Watson, 1927].

4. The *confluent hypergeometric equation.*

$$x^2 y'' + \{-\tfrac{1}{4}x^2 + kx + \tfrac{1}{4} - m^2\}y = 0.$$

5. The *hypergeometric equation.*

$$x(1-x)y'' + \{c - (a+b+1)x\}y' - aby = 0.$$

7.6.2.   Solution by Power Series

If

$$f(x) = a_0 + a_1(x - x_0) + a_2(x - x_0)^2 + \dots,$$

where the infinite sum converges for $|x - x_0| < R$ and diverges for $|x - x_0| > R$, we will say that f has a *power series expansion at x_0 with radius of convergence* R. Convergence for all x is equivalent to an infinite radius of convergence. If the series converges only for $x = x_0$, we say that the radius of convergence is zero [see § 1.10 and § 9.5].

The theory of power series is discussed in Chapter 9 of this volume. The natural setting for the theory is within the theory of complex valued functions of a complex variable. However, since most of the differential equations of applied mathematics have real coefficients we restrict ourselves to real equations and to the search for real solutions.

DEFINITION 7.6.1.   The point x_0 is called an *ordinary point* of equation (7.6.2) with radius of convergence R if $p(x)$ and $q(x)$ have power series expansions at x_0 of positive radii of convergence R_1 and R_2 respectively, and R is the smaller of R_1 and R_2.

THEOREM 7.6.1.   *Let x_0 be an ordinary point of equation (7.6.2) with radius of convergence R. Then the unique solution of the initial value problem*

$$y'' + p(x)y' + q(x)y = 0, \qquad y(x_0) = c_1, \qquad y'(x_0) = c_2$$

has a power series expansion at x_0 with radius of convergence R.

To find the power series solution at an ordinary point we let

$$y(x) = a_0 + a_1(x - x_0) + a_2(x - x_0)^2 + \dots,$$

substitute this power series in the equation for y and equate to zero the coefficient of $(x - x_0)^k$ for $k = 0, 1, 2, \dots$. This will yield recurrence relations for the coefficients $a_0, a_1, a_2, \ldots$ [see I, § 14.12]. We can calculate the first few coefficients, beginning with $a_0 = c_1$ and $a_1 = c_2$, and in some cases we may be able to find an expression for a_n for every $n = 0, 1, 2, \ldots$ [see I, § 14.13].

EXAMPLE 7.6.1. The origin is an ordinary point with infinite radius of convergence for the equation

$$y'' - xy = 0.$$

If

$$y(x) = a_0 + a_1 x + a_2 x^2 + \ldots = \sum_0^\infty a_n x^n,$$

we find that

$$\sum_0^\infty (n+2)(n+1)a_{n+2} x^n - \sum_1^\infty a_{n-1} x^n = 0.$$

Therefore, $2 \cdot 1 a_2 = 0$, the term independent of x, and

$$(n+2)(n+1)a_{n+2} - a_{n-1} = 0 \quad \text{for } n = 1, 2, 3, \ldots.$$

The solution with $y(0) = 1$ and $y'(0) = 0$ is found to be

$$y_1(x) = 1 + \frac{x^3}{2 \cdot 3} + \frac{x^6}{2 \cdot 3 \cdot 5 \cdot 6} + \ldots$$

$$= 1 + \sum_1^\infty \frac{1}{2 \cdot 5 \cdot 8 \ldots (3n-1)3^n} \frac{x^{3n}}{n!},$$

and the solution with $y(0) = 0$ and $y'(0) = 1$ is

$$y_2(x) = x + \frac{x^4}{3 \cdot 4} + \frac{x^7}{3 \cdot 4 \cdot 6 \cdot 7} + \ldots$$

$$= x + \sum_{n=1}^\infty \frac{1}{4 \cdot 7 \cdot 10 \ldots (3n+1)3^n} \frac{x^{3n+1}}{n!}.$$

If x is not an ordinary point it is called a *singular point*. The power series method of solution can be modified to include a special class of singular points which occur in important equations such as Bessel's equation and the hypergeometric equation.

DEFINITION 7.6.2. The point x_0 is called a *regular singular point* of equation (7.6.2) if it is not an ordinary point and if $(x - x_0)p(x)$ and $(x - x_0)^2 q(x)$ have power series expansions at x_0 with positive radii of convergence.

At a regular singular point

$$p(x) = \frac{\lambda}{(x - x_0)} + p_0 + p_1(x - x_0) + p_2(x - x_0)^2 + \ldots \tag{7.6.4}$$

and

$$q(x) = \frac{\mu}{(x-x_0)^2} + \frac{\nu}{(x-x_0)} + q_0 + q_1(x-x_0) + \ldots, \qquad (7.6.5)$$

where at least one of the numbers λ, μ, ν is not zero.

DEFINITION 7.6.3.   The equation

$$f(c) \equiv c(c-1) + \lambda c + \mu = 0 \qquad (7.6.6)$$

is called the *indicial equation* at the regular singular point x_0. Let c_1 and c_2 denote the roots of this equation. Their dependence on λ and μ is summarized in Figure 7.6.1.

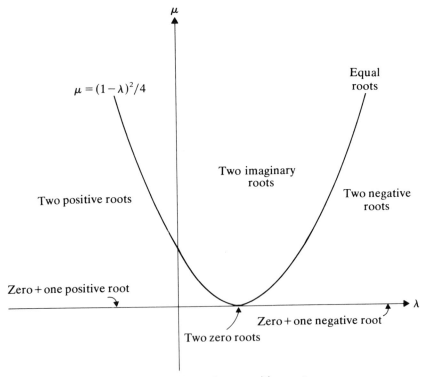

μ

$\mu = (1-\lambda)^2/4$

Equal roots

Two imaginary roots

Two positive roots

Two negative roots

Zero + one positive root

Zero + one negative root

λ

Two zero roots

One negative and one positive root

Figure 7.6.1

$$c_1, c_2 = (1-\lambda)/2 \pm [\{(1-\lambda)/2\}^2 - \mu]^{1/2}.$$

It can be shown that provided $|c_1 - c_2|$ is not a positive integer, there exist functions $a_n(c)$ for $n = 1, 2, 3, \ldots$ such that, if a_0 is not zero and

$$y(x, c) = (x-x_0)^c \{a_0 + a_1(c)(x-x_0) + a_2(c)(x-x_0)^2 + \ldots\} \qquad (7.6.7)$$

then [see (7.3.4)]

$$Ly(x, c) = a_0 f(c)(x - x_0)^{c-2}. \tag{7.6.8}$$

Notice that, if $c_1 = c_2$, then $f(c) = (c - c_1)^2$ and differentiation of equation (7.6.8) with respect to c gives

$$L \, \partial y(x, c)/\partial c = a_0 (x - x_0)^{c-2} \{(c - c_1)^2 \log (x - x_0) + 2(c - c_1)\}. \tag{7.6.9}$$

THEOREM 7.6.2. *If $|c_1 - c_2|$ is not a positive integer, then equation (7.6.2) has two linearly independent solutions of the form $y(x, c_1)$ and $y(x, c_2)$ when $c_1 \neq c_2$, and $y(x, c_1)$ and $\partial y(x, c_1)/\partial c$ when $c_1 = c_2$.*

THEOREM 7.6.3. *Suppose that $c_1 = c_2 + m$, where m is a positive integer.*
    Case A. *The function $y(x, c_2)$ has a_0 and a_m arbitrary constants and is therefore the general solution of equation (7.6.2).*
    Case B. *If $Y(x, c) = (c - c_2) y(x, c)$, then equation (7.6.2) has two linearly independent solutions of the form $Y(x, c_2)$ and $\partial Y(x, c_2)/\partial c$.*

It might appear in Case B that $Y(x, c_2)$ is identically zero, but it will be found that

$$a_n(c) = A_n(c)/(c - c_2)$$

where $A_n(c_2)$ is not zero for every n.
    If the indicial equation has complex roots $\alpha \pm i\beta$, two linearly independent solutions are found by taking the real and imaginary parts of $y(x, \alpha + i\beta)$, remembering that

$$x^{\alpha + i\beta} = x^\alpha \{\cos (\beta \log x) + i \sin (\beta \log x)\}$$

using (9.5.8).
    The procedure for finding the solutions is called the *method of Frobenius*. It can involve heavy calculations and for clarity we will apply the method to a simple equation which illustrates the various possibilities by suitable choices of two parameters.

EXAMPLE 7.6.2.   Consider the equation

$$xy'' + (\alpha - x)y' - \beta y = 0. \tag{7.6.10}$$

Inspection of the standard form

$$y'' + \{(\alpha/x) - 1\}y' - (\beta/x)y = 0 \tag{7.6.11}$$

shows that the origin is a regular singular point with indicial equation

$$c(c - 1 + \alpha) = 0.$$

It is best to substitute

$$y(x) = x^c \{a_0 + a_1 x + a_2 x^2 + \ldots\} = \sum_{n=0}^{\infty} a_n x^{c+n} \qquad (a \neq 0)$$

into equation (7.6.10), rather than into the standard form (7.6.11). We find that

$$x \sum_{n=0}^{\infty} (n+c)(n+c-1)a_n x^{n+c-2} + (\alpha - x) \sum_{n=0}^{\infty} (n+c)a_n x^{n+c-1}$$

$$- \beta \sum_{n=0}^{\infty} a_n x^{n+c} = 0,$$

that is

$$\sum_{n=0}^{\infty} \{(n+c)(n+c-1) + \alpha(n+c)\}a_n x^{n+c-1} - \sum_{n=0}^{\infty} \{(n+c) + \beta\}a_n x^{n+c} = 0.$$

Equating to zero the coefficient of x^{n+c-1} for $n = 0, 1, 2, \ldots$ we obtain

$$\{c(c-1) + \alpha c\}a_0 = 0$$

and

$$(n+c)(n+c+\alpha-1)a_n = (n+c+\beta-1)a_{n-1} \qquad (n = 1, 2, 3, \ldots). \tag{7.6.12}$$

(a) If $\alpha = \frac{1}{2}$ and $\beta = 1$, the indicial equation has the two roots $c = 0$ and $c = \frac{1}{2}$. Further,

$$a_n(c) = a_0\{(c+\tfrac{1}{2})(c+1+\tfrac{1}{2}) \ldots (c+n-1+\tfrac{1}{2})\}$$

for $n = 1, 2, 3, \ldots$.

Taking $a_0 = 1$, $c = 0$ and $c = \frac{1}{2}$ we obtain the linearly independent solutions

$$y_1(x) = 1 + \sum_{n=1}^{\infty} \{2^{2n}n!/(2n+2)!\}x^n$$

and

$$y_2(x) = x^{1/2}\left\{1 + \sum_{n=1}^{\infty} (x^n/n!)\right\} = x^{1/2} \exp x.$$

(b) If $\alpha = 1$ and $\beta = 1$, the indicial equation has the double root $c = 0$ and

$$a_n(c) = a_0\{(c+1)(c+2) \ldots (c+n)\}$$

for $n = 1, 2, 3 \ldots$. Therefore, taking $a_0 = 1$,

$$y(x, c) = x^c\left\{1 + \sum_{n=1}^{\infty} \{x^n[(c+1)(c+2) \ldots (c+n)]\}\right\}.$$

Setting $c = 0$ in $y(x, c)$ and in $\partial y(x, c)/\partial c$ gives the linearly independent solutions

$$y_1(x) = 1 + \sum_{1}^{\infty} \{x^n/n!\} = \exp x$$

and

$$y_2(x) = y_1(x) \log x - \sum_{n=1}^{\infty} \left\{ 1 + \frac{1}{2} + \ldots + \frac{1}{n} \right\} x^n / n!.$$

Usually the simplest way to find $a_n'(c)$ is to differentiate $\log a_n(c)$. In this example

$$a_n'(c) = -\left\{ \frac{1}{c+1} + \frac{1}{c+2} + \ldots + \frac{1}{c+n} \right\} a_n(c).$$

(c) If $\alpha = -1$ and $\beta = -1$, the indicial equation has the roots $c_1 = 2$ and $c_2 = 0$. Equation (7.6.12) becomes

$$(n+c)(n+c-2)a_n = (n+c-2)a_{n-1} \qquad (n = 1, 2, 3, \ldots)$$

and with $c = c_2 = 0$,

$$n(n-2)a_n = (n-2)a_{n-1}, \qquad (n = 1, 2, 3, \ldots).$$

Therefore, $a_1 = a_0$, a_2 is arbitrary and

$$a_n = (a_{n-1}/n) \quad \text{for } n = 3, 4, 5, \ldots.$$

Thus,

$$y(x, c_2) = a_0(1+x) + a_2 \left(x^2 + \frac{x^3}{3} + \frac{x^4}{3 \cdot 4} + \ldots \right)$$

$$= a_0(1+x) + 2a_2\{\exp x - (1+x)\}.$$

The general solution is

$$y(x) = \alpha_1(1+x) + \alpha_2 \exp x.$$

(d) If $\alpha = -1$ and $\beta = 1$, the indicial equation again has the roots $c_1 = 2$ and $c_2 = 0$, but equation (7.6.12) becomes

$$(n+c)(n+c-2)a_n = (n+c)a_{n-1}.$$

When $c = c_2 = 0$ this gives $a_1 = -a_0$ and $2a_1 = 0$, which contradicts our fundamental assumption that $a_0 \neq 0$.

Let

$$A_n(c) = (c - c_2)a_n(c) = c a_n(c)$$

and

$$Y(x, c) = x^c \{c a_0 + A_1(c)x + A_2(c)x^2 + \ldots\}.$$

Then

$$A_1(c) = c a_0 / (c-1), \quad A_2(c) = a_0 / (c-1)$$

and

$$A_n(c) = a_0 / \{(c-1)(c+1)(c+2) \ldots (c+n-2)\}$$

for $n = 3, 4, 5, \ldots.$

Taking $a_0 = -1$ and setting $c = c_2 = 0$ in $Y(x, c)$ and $\partial Y(x, c)/\partial c$ gives the linearly independent solutions

$$y_1(x) = x^2 \left\{ 1 + \sum_1^\infty (x^n/n!) \right\} = x^2 \exp(x)$$

and

$$y_2(x) = y_1(x) \log x - \left\{ 1 - x - x^2 + \frac{1}{2} \frac{x^3}{1!} + \dots \right.$$

$$\left. \dots + \left(\frac{1}{2} + \frac{1}{3} + \dots + \frac{1}{n-2} \right) \frac{x^n}{(n-2)!} + \dots \right\}.$$

7.6.3. Solution by Definite Integrals

The general solution of the homogeneous linear equation

$$x \frac{d^2 y}{dx^2} + (\alpha + \beta + x) \frac{dy}{dx} + \alpha y = 0 \tag{7.6.13}$$

can be written in the form

$$y(x) = \alpha_1 \int_0^1 e^{-xu} u^{\alpha-1} (1-u)^{\beta-1} \, du + \alpha_2 \int_1^\infty e^{-xu} u^{\alpha-1} (1-u)^{\beta-1} \, du,$$

provided that α and β are both positive.

There are several advantages in having solutions in this form. It is often possible to find the power series expansion at any fixed point; this describes the behaviour of the solution near the point of interest. Moreover, the behaviour of the solution as $x \to \infty$, say, can be found, something which a power series solution does not reveal.

For example, for the solution $y_1(x)$ of the equation (7.6.13) above, near the origin we have

$$y_1(x) = \int_0^1 u^{\alpha-1} (1-u)^{\beta-1} \sum_0^\infty \frac{(-xu)^n}{n!} \, du$$

$$= \sum_0^\infty \left\{ \int_0^1 u^{\alpha+n-1} (1-u)^{\beta-1} \, du \right\} \frac{(-x)^n}{n!}$$

$$= \sum_0^\infty \frac{\Gamma(\alpha+n)\Gamma(\beta)}{\Gamma(\alpha+\beta+n)} \frac{(-x)^n}{n!},$$

where the Eulerian integral has been evaluated in terms of Gamma functions [see § 10.2].

Further, making the substitution $t = xu$ in the integrand,

$$y_1(x) = x^{-\alpha} \int_0^x e^{-t} t^{\alpha-1} \{ 1 - (t/x) \}^{\beta-1} \, dt,$$

and it is not hard to show that

$$y_1(x) \sim x^{-\alpha} \Gamma(\alpha) \quad \text{as } x \to \infty,$$

where $\sim$ means that the quotient $\{y_1(x)/x^{-\alpha}\Gamma(\alpha)\}$ tends to unity as x tends to infinity [see § 2.3].

To explain how to find such definite integral solutions we consider the equation

$$My \equiv p_0(x)\frac{d^2y}{dx^2} + p_1(x)\frac{dy}{dx} + p_2(x)y = 0. \tag{7.6.14}$$

The *adjoint equation* is defined to be

$$M^*y \equiv \frac{d^2}{dx^2}\{p_0(x)y\} - \frac{d}{dx}\{p_1(x)y\} + p_2(x)y = 0 \tag{7.6.15}$$

and M^* is called the *adjoint differential operator* to M.

The definition of M^* leads to *Green's identity*

$$\int_a^b [\phi(M\psi) - (M^*\phi)\psi]\,dx = \left[\phi p_0\frac{d\psi}{dx} - \psi\frac{d}{dx}(\phi p_0) + p_1\phi\psi\right]_a^b$$

$$= [P(\phi, \psi)]_a^b \quad \text{say.} \tag{7.6.16}$$

Suppose that $K(x, u)$ satisfies the partial differential equation [see § 8.1]

$$P_0(x)\frac{\partial^2}{\partial x^2}K(x, u) + P_1(x)\frac{\partial}{\partial x}K(x, u) + P_2(x)K(x, u)$$

$$= p_0(u)\frac{\partial^2}{\partial u^2}K(x, u) + p_1(u)\frac{\partial}{\partial u}K(x, u) + p_2(u)K(x, u),$$

which we can abbreviate to

$$L_x K(x, u) = M_u K(x, u),$$

and that

$$y(x) = \int_a^b K(x, u)\phi(u)\,du,$$

where ϕ is some given function. Then

$$L_x y(x) = \int_a^b [L_x K(x, u)]\phi(u)\,du$$

$$= \int_a^b [M_u K(x, u)]\phi(u)\,du$$

$$= \int_a^b K(x, u)[M_u^*\phi(u)]\,du + [P(\phi, K)]_a^b,$$

using Green's identity. Therefore, $L_x y(x) = 0$, provided that

$$\text{(i)} \quad M_u^* \phi(u) = 0 \quad \text{and} \quad \text{(ii)} \quad [P(\phi, K)]_a^b = 0.$$

In practice we have to guess a kernel function $K(x, u)$ which leads to an operator M_u for which we can solve $M_u^* \phi(u) = 0$, and then choose a and b to satisfy condition (ii).

For the equation (7.6.13) above,

$$L_x y = x \frac{d^2 y}{dx^2} + (\alpha + \beta + x) \frac{dy}{dx} + \alpha y.$$

Take $K(x, u) = \exp(xu)$; then

$$L_x[\exp(xu)] = \{xu^2 + (\alpha + \beta + x)u + \alpha\} \exp(xu)$$

$$= \{(u^2 + u)x + (\alpha + \beta)u + \alpha\} \exp(xu)$$

$$= M_u[\exp(xu)],$$

where

$$M_u \phi(u) = (u^2 + u) \frac{d\phi}{du} + \{(\alpha + \beta)u + \alpha\} \phi.$$

Therefore

$$M_u^* \phi(u) = -\frac{d}{du} \{(u^2 + u)\phi\} + \{(\alpha + \beta)u + \alpha\} \phi$$

$$= -(u^2 + u) \frac{d\phi}{du} + \{(\alpha - 1) + (\alpha + \beta - 2)u\} \phi$$

and the solution of $M_u^* \phi = 0$ is

$$\phi(u) = u^{\alpha - 1} (1 + u)^{\beta - 1}.$$

If

$$y(x) = \int_a^b \exp(xu) \phi(u) \, du,$$

then

$$L_x y(x) = [u^\alpha (1 + u)^\beta \exp(xu)]_a^b.$$

A solution is found by taking $b = 0$ and $a = -1$,

$$y(x) = \int_{-1}^0 \exp(xu) u^{\alpha - 1} (1 + u)^{\beta - 1} \, du,$$

and replacing u by $-u$ we obtain the solution $y_1(x)$ above. Similarly, we find the second solution for positive x by taking $b = -1$ and $a = -\infty$.

The method is even more powerful if the theory of contour integration for functions of a complex variable is employed. There is a good introduction to the theory in Burkill (1962) and Murray (1974) and many examples of its use in Whittaker and Watson (1927). Contour integration is discussed in Chapter 9.

7.6.4. Asymptotic Forms of Solutions to Linear Second Order Equations

Consider the linear second order equation

$$\frac{d^2y}{dx^2} + (\lambda + q(x))y = 0. \tag{7.6.17}$$

If $q(x) \to 0$ as $x \to \infty$, we might expect the solutions of the equation to behave for large values of x like solutions of the equation

$$\frac{d^2y}{dx^2} + \lambda y = 0. \tag{7.6.18}$$

However, this happens only if $q(x) \to 0$ sufficiently rapidly, as the following cautionary example shows. Every solution of the equation $y'' + y = 0$ is bounded, since its general solution is $y(x) = \alpha_1 \cos x + \alpha_2 \sin x$, but the equation

$$y'' + \{1 - (8 \sin 2x)/(2x + \sin 2x)\}y = 0$$

has the unbounded solution $y(x) = (2x + \sin 2x) \cos x$.

Before stating some positive results it is useful to introduce some notation. The symbol $o(1)$ is used to denote any function which tends to zero as x approaches some specified limit [see § 2.3]. Thus, $f(x) = g(x) + o(1)$ as $x \to \infty$ means that $f(x) - g(x) \to 0$ as $x \to \infty$. It follows, for example, that $\sin x = x\{1 + o(1)\}$ as $x \to 0$ and $x^3 + x = x^3\{1 + o(1)\}$ as $x \to \infty$.

The following theorem can be extremely useful. It gives approximate solutions to equation (7.6.17) when x is large.

THEOREM 7.6.4

(i) *If $q(x) \to 0$ as $x \to \infty$, λ is not zero and $\int^\infty |q(x)|\, dx$ is finite, the equation has two solutions of the form*

$$y(x) = exp\ (\pm ix\sqrt{\lambda})\{1 + o(1)\} \quad as\ x \to \infty.$$

(ii) *If $q(x) \to 0$ as $x \to \infty$, λ is not zero and $\int^\infty |q'(x)|\, dx$ is finite, the equation has two solutions of the form*

$$y(x) = exp\left(\pm i \int^x [\lambda + q(s)]^{1/2}\, ds\right)\{1 + o(1)\} \quad as\ x \to \infty.$$

If in addition $\int^\infty |q(s)|^2\, ds$ is finite, then

$$y(x) = exp\left(\pm i\left[x\sqrt{\lambda} + (1/2\sqrt{\lambda})\int^x q(s)\, ds\right]\right)\{1 + o(1)\}$$

as $x \to \infty$.

EXAMPLE 7.6.3.   The equation $y'' + \{1 + (1/x^2)\}y = 0$ has two solutions of the form

$$y(x) = \exp\left(\pm i[x - 1/(2x)]\right)\{1 + o(1)\},$$

that is

$$y_1(x) = \cos\{x - 1/(2x)\}\{1 + o(1)\}$$

and

$$y_2(x) = \sin\{x - 1/(2x)\}\{1 + o(1)\}, \quad \text{as } x \to \infty.$$

EXAMPLE 7.6.4.   The equation $y'' + \{-1 + (1/x^2)\}y = 0$ has two solutions of the form

$$y(x) = \exp\left(\pm[x + (1/2x)]\right)\{1 + o(1)\} \quad \text{as } x \to \infty.$$

Finally, we remark that if $\int^\infty x|q(x)|\,dx$ is finite then the general solution of the equation

$$\frac{d^2y}{dx^2} + q(x)y = 0$$

is $y(x) = \{\alpha_1 + \alpha_2 x\}\{1 + o(1)\}$ as $x \to \infty$.

A comprehensive survey of results of this kind is given in Cesari (1963). A good introduction to the theory is contained in Bellman (1953).

7.6.5. Some Useful Transformations

The equation

$$\frac{d^2u}{dt^2} + p'(t)\frac{du}{dt} + q(t)u = 0 \tag{7.6.19}$$

is transformed by the substitution

$$u(t) = v(t)\exp\left(-\tfrac{1}{2}\int^t p(s)\,ds\right)$$

into the simpler form

$$\frac{d^2v}{dt^2} + \{q(t) - \tfrac{1}{2}p'(t) - \tfrac{1}{4}p^2(t)\}v = 0, \tag{7.6.20}$$

to which the results of the last section may be applicable.

EXAMPLE 7.6.5.   Consider *Bessel's equation*

$$\frac{d^2u}{dt^2} + (1/t)\frac{du}{dt} + \{1 - (\nu^2/t^2)\}u = 0.$$

If $u(t) = v(t)t^{-1/2}$, then

$$\frac{d^2v}{dt^2} + (1 - (\nu^2 - \tfrac{1}{4})/t^2)v = 0.$$

Therefore, by results of the last section, Bessel's equation has two solutions of the form

$$u_1(t) = t^{-1/2} \cos\{t + (\nu^2 - \tfrac{1}{4})/(2t)\}\{1 + o(1)\}$$

and

$$u_2(t) = t^{-1/2} \sin\{t + (\nu^2 - \tfrac{1}{4})/(2t)\}\{1 + o(1)\}$$

as $t \to \infty$.

The equation

$$\frac{d^2v}{dt^2} \pm Q^2(t)v = 0 \qquad (7.6.21)$$

is transformed by the substitutions

$$x = \int Q(t)\, dt \quad \text{and} \quad y(x) = v(t)\sqrt{Q(t)}$$

into the form

$$\frac{d^2y}{dt^2} + \{\pm 1 - \tfrac{1}{2}P'(x) - \tfrac{1}{4}P^2(x)\}y = 0 \qquad (7.6.22)$$

where

$$P(x) = [Q'(t)/Q^2(t)].$$

If $Q(t) \to \infty$ as $t \to \infty$, then $x \to \infty$ as $t \to \infty$ and the results of the last section may be applicable to equation (7.6.22).

EXAMPLE 7.6.6.   Consider the equation

$$\frac{d^2v}{dt^2} - t^2v = 0.$$

Here $Q(t) = t$, $x = \tfrac{1}{2}t^2$ and $P(x) = 1/t^2 = 1/(2x)$. The equation for $y(x) = v(t)t^{1/2}$ is therefore

$$\frac{d^2y}{dx^2} + \{-1 + 3/(16x^2)\}y = 0.$$

From Theorem 7.6.4 we conclude that the given equation has two solutions of the form

$$v(t) = t^{-1/2} \exp(\pm\tfrac{1}{2}t^2)\{1 + o(1)\}$$

as $t \to \infty$.

7.7. QUALITATIVE THEORY OF THE SECOND-ORDER LINEAR EQUATION

7.7.1. Introduction

Most differential equations cannot be solved explicitly in terms of elementary functions. The intensive study of a few equations of great importance in the classical theories of heat, light, sound and gravitational and electromagnetic fields led to the methods described in section 7.6. However during the nineteenth century the direction of research work gradually shifted from methods of exact solution to the study of the general nature of a solution, deduced directly from the equation. This qualitative theory attempts to answer questions such as the following: is a solution always bounded, are there functions bounding a solution from above or from below, is a solution monotone [see § 2.7] or does a solution have a finite or infinite number of zeros?

We will consider such questions for the second-order linear equation in the standard form

$$y'' + q(x)y = 0 \qquad (a \leq x < \infty). \tag{7.7.1}$$

Many results are known in two special, but frequently occurring cases, namely when $q(x)$ is either always positive or is always negative. Since

$$yy'' = -q(x)y^2,$$

the sign of $y(x)y''(x)$ is the same as the sign of $-q(x)$. Thus, for example, suppose that q is always positive; if $y(x) > 0$, then $y''(x) < 0$ and the graph of a solution is convex down; if $y(x) < 0$, then $y''(x) > 0$ and the graph is convex up [see Lemma 3.5.2].

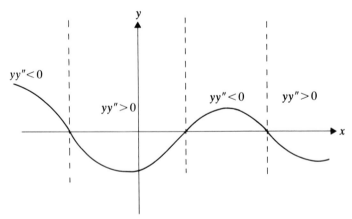

Figure 7.7.1

It is instructive to analyse two simple examples with well-known explicit solutions [see § 7.4.1].

The equation

$$y'' - y = 0 \qquad (0 \le x < \infty) \qquad (7.7.2)$$

has the general solution

$$y(x) = \alpha_1 \exp x + \alpha_2 \exp(-x),$$

a function with at most one zero, which is eventually monotone and which tends to plus or minus infinity with x unless α_1 is zero [see § 2.11].

The equation

$$y'' + y = 0 \qquad (0 \le x < \infty) \qquad (7.7.3)$$

has the general solution

$$y(x) = \alpha \cos(x + \delta),$$

a function with infinitely many zeros, which remains bounded as x tends to infinity [see § 2.12].

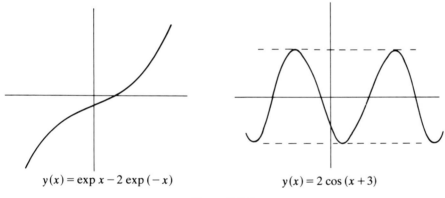

$$y(x) = \exp x - 2 \exp(-x) \qquad\qquad y(x) = 2 \cos(x + 3)$$

Figure 7.7.2

Roughly speaking, if q is always positive we should expect bounded oscillating solutions and if q is always negative we should expect monotone solutions some of which may be unbounded.

In the next two sections we give some precise results of this kind.

7.7.2. Oscillatory Behaviour

The equation

$$y'' + p(x)y' + q(x)y = 0 \qquad (a \le x < \infty) \qquad (7.7.4)$$

has the solution $y(x) \equiv 0$ for $a \le x < \infty$. Throughout this section we exclude this trivial solution from consideration.

Notice that, from the general theory in section 7.3.1, any non-trivial solution which is zero at x_0 must have a non-zero derivative at x_0. Therefore, the graph of a solution which meets the x-axis at some x_0 greater than a, must cross the x-axis at x_0.

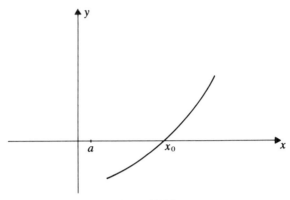

Figure 7.7.3

One of the oldest results on the zeros of equation (7.7.4) is *Sturm's Separation Theorem*;

THEOREM 7.7.1.   *If y_1 and y_2 are linearly independent solutions of equation (7.7.4) then y_2 has exactly one zero between any two successive zeros of y_1.*

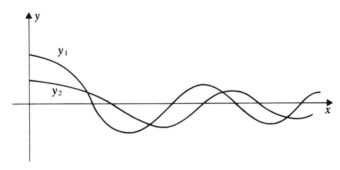

Figure 7.7.4

It follows that if equation (7.7.4) has one solution which has infinitely many zeros, then so does every solution of the equation.

DEFINITION 7.7.1.   The equation (7.7.4) is said to be *oscillatory* if every solution has infinitely many zeros. Otherwise it is said to be *non-oscillatory*.

Again using the Sturm separation theorem, we can deduce that if equation (7.7.4) is non-oscillatory then every solution has at most one zero in $\alpha \le x < \infty$, for some $\alpha \ge a$.

Clearly equation (7.7.2) is non-oscillatory and equation (7.7.3) is oscillatory.

THEOREM 7.7.2.   *If $q(x) \leq 0$ for $x \geq a$, then equation (7.7.1) is non-oscilla-tory and every solution has at most one zero in $a < x < \infty$.*

If $q(x) > 0$ for $x \geq 0$ we expect equation (7.7.1) to be oscillatory. However, this is not always so, as the following example shows.

EXAMPLE 7.7.1.   *The Euler equation*

$$y'' + (k/x^2)y = 0 \qquad (1 \leq x < \infty) \qquad (7.7.5)$$

is oscillatory if $k > \frac{1}{4}$ and non-oscillatory if $k \leq \frac{1}{4}$. This follows from inspection of the general solution found by the methods of section 7.5. If $\beta = \{|k - \frac{1}{4}|\}^{1/2}$, the general solution is

$$\alpha \sqrt{x} \cos (\beta x + \delta), \quad \text{if } k > \tfrac{1}{4},$$

$$\alpha_1 x^{1/2+\beta} + \alpha_2 x^{1/2-\beta}, \quad \text{if } k < \tfrac{1}{4},$$

and

$$\sqrt{x}(\alpha_1 + \alpha_2 \log x), \quad \text{if } k = \tfrac{1}{4}.$$

The example suggests that if the equation is oscillatory then $q(x)$ must not tend to zero too rapidly as x tends to infinity. One such criterion is used in the next theorem.

THEOREM 7.7.3.   *If $q(x) > 0$ for $x \geq a$ and if $\int_a^\infty q(x)\,dx$ is infinite, then equation (7.7.1) is oscillatory.*

EXAMPLE 7.7.2.   Both the equations

$$y'' + xy = 0 \qquad (1 \leq x < \infty)$$

and

$$y'' + (1/x)y = 0 \qquad (1 \leq x < \infty)$$

are oscillatory, since both

$$I_1 = \int_1^x x\,dx = \tfrac{1}{2}(x^2 - 1)$$

and

$$I_2 = \int_1^x (1/x)\,dx = \log x$$

tend to infinity with x.

There remains the case $q(x) > 0$ for $x \geq a$ and $\int_a^\infty q(x)\,dx$ finite. The results of Example 7.7.1 can be extended to a wide class of equations by using *Sturm's Comparison Theorem*;

THEOREM 7.7.4.   *Let y_1 and y_2 be solutions of the equations*

$$y_1'' + q_1(x)y_1 = 0 \qquad (7.7.6)$$

and

$$y_2'' + q_2(x)y_2 = 0, \qquad (7.7.7)$$

respectively. If q_1 and q_2 are continuous functions and if

$$q_2(x) > q_1(x) > 0 \qquad (7.7.8)$$

for $a \leq x \leq b$, then y_2 has at least one zero between any two successive zeros of y_1 in $a \leq x \leq b$.

COROLLARY. *If equation (7.7.6) is oscillatory in $a \leq x < \infty$ and if the inequalities (7.7.8) hold for $a \leq x < \infty$, then equation (7.7.7) is oscillatory.*

Combining the corollary with Example 7.7.1 we can state a useful test for oscillatory behaviour.

THEOREM 7.7.5.   *If*

$$q(x) \geq k/x^2 \qquad (a \leq x < \infty)$$

with $k > \frac{1}{4}$, then equation (7.7.1) is oscillatory.
    If

$$q(x) < 1/(4x^2) \qquad (a \leq x < \infty),$$

then equation (7.7.1) is non-oscillatory.

EXAMPLE 7.7.3.   *If $q(x) = (\gamma/x^{3/2})$, where $\gamma > 0$, then $q(x) \geq (1/x^2)$ for $x \geq (1/\gamma^2)$, and hence from Theorem 7.7.5 the equation*

$$y'' + (\gamma/x^{3/2})y = 0 \qquad (a \leq x < \infty)$$

is oscillatory.

The *eigenvalues* of Sturm–Liouville problems, discussed below in section 7.8.2, are related often to the zeros of solutions of differential equations. By considering special cases of equations (7.7.6) and (7.7.7) in Sturm's Comparison Theorem, we can deduce interesting estimates of the distance between successive zeros of a solution.

THEOREM 7.7.6.   *If $M \geq q(x) > 0$ for $a \leq x < \infty$, then the distance between successive zeros of a solution of equation (7.7.1) is at least $(\pi/\sqrt{M})$, and if $q(x) \geq m > 0$ for $a \leq x < \infty$ then the distance is at most $(\pi/\sqrt{m})$.*

EXAMPLE 7.7.4.   Let y be a solution of the oscillatory equation

$$y'' + \{1 + (k^2 - \tfrac{1}{4})/x^2\}y = 0 \qquad (0 < x < \infty).$$

If $k > \frac{1}{2}$, any interval of length π contains at least one zero of y and, if $0 \le k < \frac{1}{2}$, then any such interval contains at most one zero.

This result has applications to the theory of Bessel functions; see Example 7.6.5 and section 10.4.

7.7.3. Bounds for Solutions

In section 7.6.4 and in Example 7.7.1 we gave examples of oscillatory equations with an unbounded solution. The next few theorems give conditions which ensure that various types of oscillatory equations have only bounded solutions.

THEOREM 7.7.6. *If*

$$0 < m \le q(x) \le M < \infty \qquad (a \le x < \infty)$$

and if q has a continuous second derivative with $\int_a^\infty |q''(x)| \, dx$ finite, then all solutions of equation (7.7.1), together with their first derivatives, are bounded for $a \le x < \infty$.

EXAMPLE 7.7.5. Every solution of the oscillatory equation

$$y'' + \{2 + \cos(\exp[-x])\}y = 0$$

is bounded in $0 \le x < \infty$.

The next theorem applies when $q(x)$ is approximately a positive constant for large values of x.

THEOREM 7.7.7. *Let*

$$q(x) = \lambda + R(x) + S(x),$$

where $\lambda > 0$, the integrals $\int^\infty |R(x)| \, dx$ and $\int^\infty |S'(x)| \, dx$ are both finite and $S(x) \to 0$ as $x \to \infty$.
Then every solution of equation (7.7.1) is bounded.

EXAMPLE 7.7.6. Every solution of the oscillatory equation

$$y'' + \{1 + (m/x) + (k/x^2)\}y = 0$$

is bounded in $1 \le x < \infty$.

Results have been obtained for equations with unbounded functions q. We give one such theorem.

THEOREM 7.7.8. *If $q(x)$ is non-decreasing and if $q(x)$ tends to infinity with x, then every solution of equation (7.7.1) is bounded and at least one solution tends to zero as x tends to infinity.*

EXAMPLE 7.7.7.   Every solution of the oscillatory equation

$$y'' + xy = 0$$

is bounded in $0 \leq x < \infty$.

Many other equations can be dealt with by using the following comparison theorem which says that a small perturbation of an equation does not destroy the boundedness of solutions.

THEOREM 7.7.9.   *If every solution of equation (7.7.1) is bounded and if $\int^{\infty} |r(x)| \, dx$ is finite, then every solution of*

$$y'' + \{q(x) + r(x)\}y = 0 \qquad (a \leq x < \infty)$$

is bounded.

EXAMPLE 7.7.8.   From Theorems 7.7.8 and 7.7.9, every solution of the oscillatory equation

$$y'' + \{x + (\sin x / x^2)\}y = 0$$

is bounded in $1 \leq x < \infty$.

The theorems given so far all lead to bounds of the form

$$|y(x)| \leq K < \infty \qquad (a \leq x < \infty).$$

It is possible sometimes to obtain more precise bounds of the form

$$z_2(x) \leq y(x) \leq z_1(x) \qquad (a \leq x \leq b).$$

This is difficult to achieve if y is oscillating, but comparatively easy for solutions of non-oscillatory equations with q always negative.

THEOREM 7.7.10.   *Consider the initial-value problem*

$$Ly \equiv y'' + p(x)y' + q(x)y = f(x),$$

$$y(a) = c_1 \quad and \quad y'(a) = c_2,$$

where p, q, f are continuous functions in $a \leq x \leq b$, and

$$q(x) \leq 0 \qquad (a \leq x \leq b).$$

Let z_1 and z_2 be functions such that

$$Lz_1 \geq f \quad (a \leq x \leq b), \qquad z_1(a) \geq c_1, \qquad z_2'(a) \geq c_2$$

and

$$Lz_2 \leq f \quad (a \leq x \leq b), \qquad z_2(a) \leq c_1, \qquad z_2'(a) \leq c_2.$$

Then

$$z_2(z) \leq y(z) \leq z_1(x) \qquad (a \leq x \leq b).$$

In practice, one tries to find simple functions such as polynomials for z_1 and z_2.

EXAMPLE 7.7.9.   Suppose that

$$Ly = y'' - xy = 0, \qquad y(0) = 0, \qquad y'(0) = 1 \qquad (0 \le x \le 1).$$

If

$$z(x) = x + \alpha x^3 + (x^4/12),$$

then

$$Lz = 6\alpha x + x^2 - x\{x + \alpha x^3 + (x^4/12)\}$$
$$= x\{6\alpha - x^3[\alpha + (x/12)]\}.$$

Now, if $\alpha = 0$,

$$Lz = -(x^5/12) \le 0 \quad \text{in } 0 \le x \le 1,$$

and if $6\alpha \ge \alpha + (1/12)$, that is $\alpha \ge (1/60)$,

$$Lz \ge x\{\alpha(1 - x^3) + (1 - x^4)/12\} \ge 0$$

in $0 \le x \le 1$.

Using Theorem 7.7.10, we deduce that

$$x + (x^4/12) \le y(x) \le x + (x^3/60) + (x^4/12)$$

for $0 \le x \le 1$. In particular,

$$1 \cdot 08 \le y(1) \le 1 \cdot 10.$$

(In Example 7.6.1, we found that

$$y(x) = x + \frac{x^4}{3 \cdot 4} + \frac{x^7}{3 \cdot 4 \cdot 6 \cdot 7} + \frac{x^{10}}{3 \cdot 4 \cdot 6 \cdot 7 \cdot 9 \cdot 10} + \dots)$$

Our last theorem on this subject asserts that a non-oscillatory equation must have an unbounded solution and a second solution which is small by comparison.

THEOREM 7.7.11.   *Every non-oscillatory equation (7.7.1) has two solutions u and v such that $\int^\infty dx/u^2(x)$ is finite, and $\int^\infty dx/v^2(x)$ is infinite.*

COROLLARY 1.   *If u and v are linearly independent solutions of a non-oscillatory equation (7.7.1), then $\int^\infty dx/(u^2(x) + v^2(x))$ is finite.*

COROLLARY 2.   *If equation (7.7.1) is non-oscillatory then it has a solution u such that the inequality*

$$|u(x)| \le Kx^{1/2} \qquad (x_0 \le x < \infty)$$

is impossible for any finite K and any $x_0 \ge a$.

EXAMPLE 7.7.10. Here are some non-oscillatory equations with their 'large' and 'small' solutions;

(i) $y'' - y = 0$, $u(x) = \exp x$, $v(x) = \exp(-x)$

(ii) $y'' = 0$, $u(x) = x$, $v(x) = 1$

(iii) $y'' + (1/4x^2)y = 0$, $u(x) = x^{1/2}\log x$, $v(x) = x^{1/2}$.

7.8. STURM–LIOUVILLE THEORY

7.8.1. Fourier Series

One of the most ancient problems pursued by mathematicians is a satisfactory mathematical account of the nature of musical sounds. Little progress was made from the time of Pythagoras until the early eighteenth century when the study of a vibrating string, such as a violin string, was taken up by the leading mathematicians of the time, using the comparatively new methods of the calculus. The *wave equation*, introduced by D'Alembert, for the displacement of a string fixed at its ends and with a given initial displacement, was

$$\frac{\partial^2 \phi}{\partial x^2} = \frac{1}{c^2}\frac{\partial^2 \phi}{\partial t^2} \qquad (0 < x < l, 0 < t < \infty),$$

where

$$\phi(0, t) = \phi(l, t) = 0 \qquad (t > 0)$$

and

$$\phi(x, 0) = f(x), \qquad \frac{\partial}{\partial t}\phi(x, 0) = 0 \qquad (0 \le x \le l).$$

Euler sought a solution by the method of *separation of the variables* [see § 8.2]: this consists in seeking solutions of the form

$$\phi(x, t) = u(x)v(t).$$

It then follows that

$$\{u''(x)/u(x)\} = \{v''(t)/c^2 v(t)\} = -\lambda,$$

where λ is any constant independent of both x and t.

We have now two ordinary differential equations,

$$u''(x) + \lambda u(x) = 0 \tag{7.8.1}$$

and

$$v''(t) + \lambda c^2 v(t) = 0, \tag{7.8.2}$$

together with the boundary conditions

$$u(0) = u(l) = 0, \tag{7.8.3}$$

$$v'(0) = 0 \tag{7.8.4}$$

and

$$u(x)v(0) = f(x), \qquad 0 \le x \le l. \tag{7.8.5}$$

The general solution of equation (7.8.1) is

$$u(x) = \alpha \cos (x\sqrt{\lambda}) + \beta \sin (x\sqrt{\lambda})$$

and to satisfy the boundary conditions at the two points $x = 0$ and $x = l$, we must have

$$u(0) = \alpha = 0$$

and

$$u(l) = \alpha \cos (l\sqrt{\lambda}) + \beta \sin (l\sqrt{\lambda}) = 0.$$

Now, if β is zero, the solution u is identically zero, a worthless solution. Therefore, we must take α to be zero, β to be non-zero and choose λ so that

$$\sin (l\sqrt{\lambda}) = 0. \tag{7.8.6}$$

Since $\sin x$ vanishes only when x is an integer multiple of π, we must choose λ to be one of the numbers λ_n where

$$\lambda_n = (n\pi/l)^2 \qquad (n = 1, 2, 3, \ldots).$$

Solving equations (7.8.2) and (7.8.4) we have

$$\phi_n(x, t) = b_n \sin (x\sqrt{\lambda_n}) \cos (ct\sqrt{\lambda_n}),$$

where b_n is an arbitrary constant. This function satisfies equations (7.8.1) to (7.8.4), but not equation (7.8.5) in general. Euler now wrote

$$\phi(x, t) = \sum b_n \sin (x\sqrt{\lambda_n}) \cos (ct\sqrt{\lambda_n})$$

as a solution of the original problem, but without specifying the number of terms in the summation. To solve the vibrating string problem completely we need to choose the constants b_n so that

$$f(x) = \sum_{n=1}^{\infty} b_n \sin (x\sqrt{\lambda_n}). \tag{7.8.6}$$

Expressions such as (7.8.6) and more generally

$$f(x) = \tfrac{1}{2}a_0 + \sum_{n=1}^{\infty} (a_n \cos (n\pi x/l) + b_n \sin (n\pi x/l)), \tag{7.8.7}$$

occurred in various astronomical problems about the same time and later in the work of Fourier on the conduction of heat. Both Euler and Fourier [see

(20.5.36) and (20.5.37)] arrived by various arguments at the formulae

$$a_n = (2/l) \int_0^l f(x) \cos (n\pi x/l) \, dx$$

and   (7.8.8)

$$b_n = (2/l) \int_0^l f(x) \sin (n\pi x/l) \, dx,$$

for $n = 0, 1, 2, 3, \ldots.$

The meaning and validity of (7.8.7) provoked vigorous debate throughout the eighteenth century and attempts to put such *Fourier expansions* above criticism have been one of the most fruitful sources of modern analysis. There are functions, continuous in $0 < x < \pi$, whose Fourier series do not converge at a dense set of points in the interval! Even today, conditions on f, both necessary and sufficient to ensure that the Fourier series for f converges to f at every point in the interval, are not known.

Many other boundary value problems in applied mathematics lead to ordinary differential equations with boundary values assigned at two points. A solution exists only for certain values of a parameter in the equation, and to satisfy all the boundary conditions a function has to be expressed as a sum of such solutions. The permissible values of the parameter $\{\lambda_n\}$ are called *eigenvalues* of the problem, the associated solutions $\{\phi_n\}$ are called *eigenfunctions* or *eigensolutions* and the expansion

$$f(x) = \sum_1^\infty c_n \phi_n(x)$$

is called the *eigenfunction expansion* for f.

In the case of Fourier series the eigenfunctions are trigonometric functions. In other important instances they are Bessel functions, Legendre polynomials or Hermite functions. See §§ 10.3, 10.4 and 10.5.2.

7.8.2.   Eigenfunction Expansions

A general theory of eigenfunction expansions associated with second-order differential equations was initiated by Sturm and Liouville early in the nineteenth century and became a part of the general spectral theory of compact symmetric operators in Hilbert space [see § 11.6.3] developed a century later.

Let

$$Ly = -\frac{d}{dx}\left\{ p(x)\frac{dy}{dx}\right\} + q(x)y$$ (7.8.9)

where $p(x) > 0$ for $a < x < b$.

Consider the two-point boundary value problem

$$Ly = \lambda \rho(x)y,$$ (7.8.10)

$$\left.\begin{array}{l} \alpha_1 y'(a) + \beta_1 y(a) = 0 \\ \alpha_2 y'(b) + \beta_2 y(b) = 0 \end{array}\right\}, \tag{7.8.11}$$

where $\rho(x) > 0$ for $a < x < b$, $\alpha_1^2 + \beta_1^2 \neq 0$ and $\alpha_2^2 + \beta_2^2 \neq 0$.

Suppose that p', q and ρ are all continuous functions in $a \leq x \leq b$. Then it can be shown that:

(i) Equations (7.8.10) and (7.8.11) have a solution only when λ takes certain values $\lambda_1 < \lambda_2 < \lambda_3 \ldots$. These eigenvalues form an infinite sequence of real numbers which tend to infinity.

(ii) The corresponding eigenfunctions $\{y_n\}$ can be chosen so that

$$\int_a^b \rho(x) y_n^2(x)\, dx = 1.$$

They satisfy the fundamental orthogonality relations

$$\int_a^b \rho(x) y_m(x) y_n(x)\, dx = 0 \qquad (m \neq n). \tag{7.8.12}$$

[see § 10.5.0].

(iii) If

$$\int_a^b \rho(x) f^2(x)\, dx < \infty \tag{7.8.13}$$

and

$$c_n = \int_a^b \rho(x) f(x) y_n(x)\, dx,$$

then

$$\sum_1^\infty c_n^2 = \int_a^b \rho(x) f^2(x)\, dx \tag{7.8.14}$$

and

$$\lim_{N \to \infty} \left\{ \int_a^b \rho(x) \left[f(x) - \sum_{n=1}^N c_n y_n(x) \right]^2 dx \right\} = 0. \tag{7.8.15}$$

Equation (7.8.14) is known as *Parseval's equality* [see also § 20.6.3].

(iv) If f'' is continuous in $a \leq x \leq b$ and if f satisfies the boundary conditions (7.8.11), then the eigenfunction expansion

$$f(x) = \sum_1^\infty c_n y_n(x) \tag{7.8.16}$$

is valid, and the series converges uniformly in $a \leq x \leq b$ [see § 1.12].

It is evident that, when (7.8.16) is valid with uniform convergence in the interval, then (7.8.15) is also valid. However, (7.8.15) holds for a much wider class of functions than those mentioned in (iv) above. In this case we say that the eigenfunction expansion for f converges to f in mean square with the weight function ρ.

The theory can be modified to include different kinds of boundary conditions, infinite intervals, and equations whose coefficient functions may be continuous in $a < x < b$ but unbounded as x approaches either a or b.

EXAMPLE 7.8.1. *Chebyshev's equation* can be written as

$$-\frac{d}{dx}\left\{\sqrt{1-x^2}\,\frac{dy}{dx}\right\} = (\lambda/(1-x^2))y \qquad (-1 < x < 1).$$

If we admit only those solutions which remain bounded as x approaches either -1 or 1, then the solution in power series method of section 7.6.2 shows that λ must be n^2 where $n = 0, 1, 2, \ldots$, with corresponding solutions

$$y_n(x) = (2/\pi)^{1/2} \cos{(n \cos^{-1} x)} \qquad (n = 1, 2, 3, \ldots)$$

and

$$y_0(x) = (1/\pi)^{1/2}.$$

Each y_n is a polynomial in x of degree n, the nth Chebyshev polynomial [see § 10.5.3] and

$$\int_{-1}^{1} y_m(x)y_n(x)\{1-x^2\}^{-1/2}\,dx = \begin{cases} 1, & m = n, \\ 0, & m \neq n. \end{cases}$$

If f is such that

$$\int_{-1}^{1} \{f(x)\}^2\{1-x^2\}^{-1/2}\,dx < \infty$$

and

$$c_n = \int_{-1}^{1} f(x)y_n(x)\{1-x^2\}^{-1/2}\,dx,$$

then equations (7.8.14) and (7.8.15) are valid.

7.9. FIRST-ORDER LINEAR SYSTEMS

7.9.1. General Theory

Many problems reduce to the study of a system of first-order linear equations in the form

$$\frac{dy_i}{dt} = a_{i,1}(t)y_1 + a_{i,2}(t)y_2 + \ldots + a_{i,n}(t)y_n \qquad (7.9.1)$$

where $i = 1, 2, 3, \ldots, n$. Here t is the independent variable and $y_1, y_2, \ldots, y_n$

are n functions of t. We assume always that the n^2 functions $a_{ij}(t)$, $1 \le i, j \le n$, are continuous functions of t in some interval of interest, usually either $a \le t \le b$ or $a \le t < \infty$ [see § 2.1].

The equations (7.9.1) can be written concisely as

$$\frac{d}{dt}\mathbf{y}(t) = \mathbf{A}(t)\mathbf{y}(t) \tag{7.9.2}$$

where $\mathbf{y} = (y_1, y_2, \ldots, y_n)$, a column vector, and $\mathbf{A}(t) = (a_{ij}(t))$, an $n \times n$ matrix [see I, § 5.7].

THEOREM 7.9.1. *If $a_{i,j}(t)$, $1 \le i$, $j \le n$ are continuous functions of t for $a \le t \le b$, then there exists a unique solution of equation (7.9.2) which is such that*

$$\mathbf{y}(a) = \mathbf{\alpha}, \tag{7.9.3}$$

where $\mathbf{\alpha}$ is a given constant n-vector.

The general theory of equation (7.9.2) is very similar to that of the nth-order homogeneous linear equation (7.3.1) discussed in section 7.3.1. Indeed the latter equation is a special case of equation (7.9.2). For, if we let

$$y_k(t) = y^{(k-1)}(t), \qquad 1 \le k \le n,$$

then equation (7.3.1) can be written as

$$\frac{d}{dt}\begin{pmatrix} y_1 \\ y_2 \\ \vdots \\ y_{n-1} \\ y_n \end{pmatrix} = \begin{pmatrix} 0 & 1 & 0 & \cdots & 0 & 0 \\ 0 & 0 & 1 & \cdots & 0 & 0 \\ \vdots & \vdots & \vdots & \cdots & \vdots & \vdots \\ & & & \cdots & 0 & 1 \\ -p_n & -p_{n-1} & -p_{n-2} & & -p_2 & -p_1 \end{pmatrix}\begin{pmatrix} y_1 \\ y_2 \\ \vdots \\ y_{n-1} \\ y_n \end{pmatrix}. \tag{7.9.4}$$

Thus, an nth-order homogeneous linear equation for one unknown function is equivalent to a first-order homogeneous linear system for n unknown functions. The number n is called the *dimension* of the system.

DEFINITION 7.9.1. A *fundamental set of solutions* for the n-dimensional system (7.9.2) in $a \le t \le b$ is any set of n solution vectors which are linearly independent in $a \le t \le b$ [see I, Definition 5.3.2]. A *fundamental matrix* for the system is a matrix whose columns form a fundamental set of solutions.

THEOREM 7.9.2. *There exists a fundamental set of solutions for the system (7.9.2) and every solution can be expressed as a linear combination of vectors .in a fundamental set.*

The theorem implies that the solution set of the system forms an n-dimensional subspace of a suitable vector space [see I, § 5.5], and that if

$y_1, y_2, \ldots, y_n$ is a fundamental set, then the general solution of the system of equations is

$$y = c_1 y_1 + c_2 y_2 + \ldots + c_n y_n.$$

If $U(t)$ is a fundamental matrix for the system then the unique solution of the initial value problem (7.9.2) and (7.9.3) is

$$y(t) = U(t)U^{-1}(a)\alpha. \qquad (7.9.5)$$

Notice that $U(t)$ satisfies the matrix differential equation

$$\frac{d}{dt}U(t) = A(t)U(t). \qquad (7.9.6)$$

It can be shown that the determinant of $U(t)$ is

(i) $$\det U(t) = \{\det U(a)\} \exp\left\{\int_a^t \text{tr } A(s)\, ds\right\}, \qquad (7.9.7)$$

where the trace of the matrix [cf. I (6.2.4)]

$$\text{tr } A(s) = a_{11}(s) + a_{22}(s) + \ldots + a_{nn}(s),$$

and that [see I, Definition 6.12.1]

(ii) if the columns of $U(t)$ are solutions of (7.9.2), then $U(t)$ is a fundamental matrix in $a \le t \le b$ if and only if $\det U(t) \ne 0$ for $a \le t \le b$.

Equation (7.9.7) is a generalization of Liouville's formula (7.3.3) for the nth-order homogeneous linear equation.

The theory of the non-homogeneous system

$$\frac{d}{dt}y = A(t)y + b(t), \qquad (7.9.8)$$

where $b(t)$ is a column whose entries are continuous functions, is summarized in the following theorem.

THEOREM 7.9.3

(i) *Let* v *be any particular solution of (7.9.8) and let* u *be the general solution of the associated homogeneous system (7.9.2). The general solution of (7.9.8) is*

$$y = u + v.$$

(ii) *If* $U(t)$ *is a fundamental matrix for (7.9.2) then*

$$v(t) = U(t) \int_a^t U^{-1}(s)b(s)\, ds$$

is the particular solution of (7.9.8) for which $v(a) = 0$.

7.9.2. Linear Systems with Constant Coefficients

There are several ways of treating systems with constant coefficients, that is

$$\frac{d}{dt}\mathbf{y} = \mathbf{A}\mathbf{y} \qquad (0 \leq t < \infty), \qquad (7.9.9)$$

where $\mathbf{A}$ is a matrix of constants.

We can define the exponential function for a matrix by

$$\exp \mathbf{A} = \mathbf{I} + \mathbf{A} + \frac{1}{2!}\mathbf{A}^2 + \dots,$$

[cf. (2.11.1)] and then show that the solution of (7.9.9) and (7.9.3) is

$$\mathbf{y}(t) = \exp\left((t-a)\mathbf{A}\right)\boldsymbol{\alpha}. \qquad (7.9.10)$$

A fundamental matrix, $\mathbf{U}(t)$, for (7.9.9) such that $\mathbf{U}(0) = \mathbf{I}$, the unit matrix, is given by

$$\mathbf{U}(t) = \exp(t\mathbf{A}),$$

and equation (7.9.7) yields the beautiful identity

$$\det(\exp \mathbf{A}) = \exp(\operatorname{tr} \mathbf{A}),$$

valid for any $n \times n$ matrix $\mathbf{A}$.

Of course, equation (7.9.10) does not give a useful formula for a solution since it involves an infinite sum of powers of the matrix $\mathbf{A}$.

The most efficient way to solve the initial value problem for (7.9.9) is probably the *Laplace transform method* [see § 13.4 for details of the Laplace transform]. Let

$$\hat{\mathbf{y}}(s) = \int_0^\infty \exp(-st)\mathbf{y}(t)\, dt,$$

and assume that the integrand vanishes at infinity. Then

$$\int_0^\infty \exp(-st)\frac{d}{dt}\mathbf{y}(s)\, ds = [\exp(-st)\mathbf{y}(s)]_0^\infty + s\hat{\mathbf{y}}(s)$$

$$= s\hat{\mathbf{y}}(s) - \mathbf{y}(0),$$

and

$$\int_0^\infty \exp(-st)\mathbf{A}\mathbf{y}(s)\, ds = \mathbf{A}\hat{\mathbf{y}}(s).$$

Multiplying equation (7.9.9) by $\exp(-st)$ and integrating with respect to t over $0 \leq t < \infty$, we obtain

$$s\hat{\mathbf{y}}(s) - \mathbf{y}(0) = \mathbf{A}\hat{\mathbf{y}}(s)$$

and hence

$$(s\mathbf{I} - \mathbf{A})\hat{\mathbf{y}}(s) = \mathbf{y}(0), \qquad (7.9.11)$$

a system of linear equations for the functions $\hat{y}_1(s), \ldots, \hat{y}_n(s)$. The solution of these equations is equivalent to finding the inverse of the matrix $s\mathbf{I} - \mathbf{A}$ [see I, § 6.4] since

$$\hat{\mathbf{y}}(s) = (s\mathbf{I} - \mathbf{A})^{-1}\mathbf{y}(0). \tag{7.9.12}$$

Taking $\mathbf{y}(0) = \boldsymbol{\alpha}$, we can use tables of inverse Laplace transforms to find from (7.9.12) the solution of (7.9.9), denoted by $\mathbf{y}(t, \boldsymbol{\alpha})$, which is such that $\mathbf{y}(0, \boldsymbol{\alpha}) = \boldsymbol{\alpha}$. The solution of (7.9.9) such that $\mathbf{y}(a) = \boldsymbol{\alpha}$ is given by

$$\mathbf{y}(t) = \mathbf{y}(t - a, \boldsymbol{\alpha}).$$

EXAMPLE 7.9.1.   Taking Laplace transforms of the system

$$\dot{y}_1 = 2y_1 - 3y_2,$$

$$\dot{y}_2 = y_1 - 2y_2,$$

we obtain

$$s\hat{y}_1(s) - y_1(0) = 2\hat{y}_1(s) - 3\hat{y}_2(s),$$

$$s\hat{y}_2(s) - y_2(0) = \hat{y}_1(s) - 2\hat{y}_2(s).$$

Thus

$$(s - 2)\hat{y}_1(s) + 3\hat{y}_2(s) = y_1(0)$$

and

$$-\hat{y}_1(s) + (s + 2)\hat{y}_2(s) = y_2(0).$$

Solving these linear equations [see I, § 5.8 and (6.4.11)] we find that

$$\hat{y}_1(s) = \{(s + 2)y_1(0) - 3y_2(0)\}/(s^2 - 1)$$

and

$$\hat{y}_2(s) = \{y_1(0) + (s - 2)y_2(0)\}/(s^2 - 1).$$

Taking inverse Laplace transforms as described in section 13.4.2 we find that

$$y_1(t) = (3/2)\{y_1(0) - y_2(0)\} \exp t + (1/2)\{-y_1(0) + 3y_2(0)\} \exp(-t)$$

and

$$y_2(t) = (1/2)\{y_1(0) - y_2(0)\} \exp t + (1/2)\{-y_1(0) + 3y_2(0)\} \exp(-t).$$

For systems of low dimension and for certain types of matrix $\mathbf{A}$, equation (7.9.9) can be solved by a direct method. Suppose that

$$\mathbf{y}(t) = \exp(\lambda t)\mathbf{u}$$

is a solution for some constant λ and constant vector $\mathbf{u}$. Then

$$\lambda \exp(\lambda t)\mathbf{u} = \frac{d}{dt}\mathbf{y}(t) = \mathbf{A}\mathbf{y}(t) = \exp(\lambda t)\mathbf{A}\mathbf{u},$$

and dividing by the non-zero factor exp (λt), we have

$$\mathbf{A}\mathbf{u} = \lambda \mathbf{u}.$$

Thus $\mathbf{u}$ is an eigenvector of A with associated eigenvalue λ [see I, § 7.1]. It is shown in Volume I [see Theorem 7.2.3], that the eigenvalues are the roots of the nth degree polynomial $\det(\lambda \mathbf{I} - \mathbf{A})$. The eigenvectors are then found by solving sets of linear equations.

THEOREM 7.9.4.   *If the matrix* $\mathbf{A}$ *has n linearly independent eigenvectors* $\mathbf{u}_1, \mathbf{u}_2, \ldots, \mathbf{u}_n$ *with associated eigenvalues* $\lambda_1, \lambda_2, \ldots, \lambda_n$ *then the general solution of* (7.9.9) *is*

$$\mathbf{y}(t) = \sum_{1}^{n} c_i \exp(\lambda_i t)\mathbf{u}_i. \tag{7.9.13}$$

If some of the eigenvalues are complex numbers, then complex constants c_i can be chosen so that $\mathbf{y}(t)$ is a real valued function containing n real arbitrary constants.

If the n eigenvalues are distinct numbers then the eigenvectors will be linearly independent [see I, Theorem 7.4.2] and Theorem 7.9.4 applies.

EXAMPLE 7.9.2.   In matrix notation, the system of Example 7.9.1 becomes

$$\frac{d}{dt}\begin{pmatrix} y_1 \\ y_2 \end{pmatrix} = \begin{pmatrix} 2 & -3 \\ 1 & -2 \end{pmatrix}\begin{pmatrix} y_1 \\ y_2 \end{pmatrix} = \mathbf{A}\begin{pmatrix} y_1 \\ y_2 \end{pmatrix}.$$

Elementary calculations [see I, § 7.3] show that the eigenvalues of $\mathbf{A}$ are 1 and -1 with associated eigenvectors $(3, 1)'$ and $(1, 1)'$ respectively. Since these vectors are linearly independent the general solution of the system is

$$y(t) = c_1 \exp t\begin{pmatrix} 3 \\ 1 \end{pmatrix} + c_2 \exp(-t)\begin{pmatrix} 1 \\ 1 \end{pmatrix}.$$

Notice that

$$y_1(0) = 3c_1 + c_2$$

and

$$y_2(0) = c_1 + c_2.$$

If we are given $y_1(0)$ and $y_2(0)$ we must solve these linear equations for c_1 and c_2. It is easy to check that

$$c_1 = (1/2)\{y_1(0) - y_2(0)\} \quad \text{and} \quad c_2 = (1/2)\{-y_2(0) + 3y_2(0)\}.$$

The behaviour of solutions of (7.9.9) for large values of t depends on the eigenvalues of $\mathbf{A}$. Suppose that $\mathbf{A}$ has the n linearly independent eigenvectors $\mathbf{u}_1, \mathbf{u}_2, \ldots, \mathbf{u}_n$ with associated eigenvalues $\lambda_1, \lambda_2, \ldots, \lambda_n$ and that the real part of λ_i is negative for $1 \le i \le k$ and positive for $k+1 \le i \le n$. Let M be the subspace spanned by $\mathbf{u}_1, \mathbf{u}_2, \ldots, \mathbf{u}_k$ [see I (5.5.6)] and suppose that $\boldsymbol{\alpha}$ is in M. Then

$$\boldsymbol{\alpha} = c_1\mathbf{u}_1 + c_2\mathbf{u}_2 + \ldots + c_k\mathbf{u}_k,$$

where the $c_1, c_2, \ldots, c_k$ are uniquely determined. The unique solution to the initial value problem (7.9.9) and (7.9.3) is

$$\mathbf{y}(t) = \sum_1^k c_i \exp{(\lambda_i[t - a])}\mathbf{u}_i.$$

Therefore, $\mathbf{y}(t)$ is in M for all values of t and $\mathbf{y}(t)$ tends to zero as t tends to infinity. However, if $\boldsymbol{\alpha}$ is not in M then the solution which passes through $\boldsymbol{\alpha}$ never reaches M and the distance from $\mathbf{y}(t)$ to the origin tends to infinity with t. The subspace M is called the *stable manifold* for the system [see Hirsch and Smale, 1974]. Geometrically speaking, M is a subspace through the origin and any solution curve meeting M must lie in M and must approach the origin as t tends to infinity.

It may happen that $\mathbf{A}$ does not have n linearly independent eigenvectors. Using the Jordan canonical form for $\mathbf{A}$ [see I, § 7.6] and some rather heavy linear algebra, a general form for the solution of (7.9.9) can be found. We omit the details which can be found in Coddington and Levinson (1955) Chapter 3. The main result of the theory is the existence of a stable manifold associated with eigenvalues with negative real part, and the following theorem.

THEOREM 7.9.5.   *Every solution of equation (7.9.9) tends to zero as t tends to infinity if and only if every eigenvalue of* $\mathbf{A}$ *has a negative real part.*

EXAMPLE 7.9.3.   Every solution of the system

$$\frac{d}{dt}\begin{pmatrix} y_1 \\ y_2 \\ y_3 \end{pmatrix} = \begin{pmatrix} -1 & 2 & 0 \\ 0 & -4 & 3 \\ 0 & 0 & -2 \end{pmatrix}\begin{pmatrix} y_1 \\ y_2 \\ y_3 \end{pmatrix}$$

tends to zero as t tends to infinity, since the eigenvalues of the matrix of the system are -1, -4 and -2.

7.9.3.   Asymptotic Form of Solutions

In this section we are interested in finding the approximate form of solutions to the system

$$\frac{d}{dt}\mathbf{y} = (\mathbf{A} + \mathbf{B}(t))\mathbf{y} \qquad (0 \le t < \infty) \tag{7.9.14}$$

for large values of t, where $\mathbf{A}$ is a constant matrix and $\mathbf{B}(t)$ is, in some sense, small.

DEFINITION 7.9.2.   The *norm*, $|\mathbf{C}|$, of an $m \times n$ matrix $\mathbf{C} = (c_{i,j})$ is defined by

$$|\mathbf{C}| = \sum_{i=1}^{m} \sum_{j=1}^{n} |c_{i,j}|.$$

This norm function measures the magnitude of a matrix in a convenient way for our purpose. However, several other definitions of a matrix norm are used and the reader is warned to check its form when consulting the literature [see III, § 6.2].

THEOREM 7.9.6.   *If every solution of the equation*

$$\frac{d}{dt}\mathbf{y} = \mathbf{Ay} \qquad (0 \le t < \infty)$$

is bounded and if

$$\int^{\infty} |\mathbf{B}(t)|\, dt < \infty,$$

then every solution of equation (7.9.14) is bounded [see § 4.2].

The asymptotic forms of solutions to equation (7.9.14) depend very strongly on the nature of the eigenvalues of $\mathbf{A}$. The formulae are comparatively simple when $\mathbf{A}$ has n distinct eigenvalues.

THEOREM 7.9.7.   *Let $\mathbf{A}$ have distinct eigenvalues $\lambda_1, \lambda_2, \ldots, \lambda_n$ with associated eigenvectors $\mathbf{u}_1, \mathbf{u}_2, \ldots, \mathbf{u}_n$. If*

$$\int^{\infty} |\mathbf{B}(t)|\, dt < \infty,$$

then equation (7.9.14) has n linearly independent solutions $\mathbf{y}_1, \mathbf{y}_2, \ldots, \mathbf{y}_n$ such that

$$\mathbf{y}_k(t) = \exp(\lambda_k t)\{\mathbf{u}_k + \mathbf{v}_k(t)\}$$

where $|\mathbf{v}_k(t)| \to 0$ as $t \to \infty$, $1 \le k \le n$.

For more general results covering the case of repeated eigenvalues and with different conditions on $\mathbf{B}(t)$, see Coddington and Levinson (1955), Chapter 3.

EXAMPLE 7.9.4.   Consider the system

$$\frac{d}{dt}\begin{pmatrix} y_1 \\ y_2 \end{pmatrix} = \begin{pmatrix} 2 & -3 \\ 1 & -2 \end{pmatrix}\begin{pmatrix} y_1 \\ y_2 \end{pmatrix} + \begin{pmatrix} 0 & \exp(-t) \\ (1+t^2)^{-1} & 0 \end{pmatrix}\begin{pmatrix} y_1 \\ y_2 \end{pmatrix}.$$

Here,

$$\int_0^\infty |\mathbf{B}(t)|\, dt = \int_0^\infty \{\exp(-t) + (1+t^2)^{-1}\}\, dt < \infty,$$

and using the results of Example 7.9.2 and Theorem 7.9.7 we deduce that the system has solutions $\mathbf{y}_1(t)$ and $\mathbf{y}_2(t)$ such that

$$\mathbf{y}_1(t) = \exp t\left\{\begin{pmatrix} 3 \\ 1 \end{pmatrix} + \mathbf{v}_1(t)\right\}$$

and

$$\mathbf{y}_2(t) = \exp(-t)\left\{\begin{pmatrix} 1 \\ 1 \end{pmatrix} + \mathbf{v}_2(t)\right\},$$

where $\mathbf{v}_1(t)$, $\mathbf{v}_2(t) \to 0$ as $t \to \infty$.

7.10. NON-LINEAR FIRST-ORDER DIFFERENTIAL EQUATIONS

7.10.1. Introduction

The results and methods of the theory of non-linear differential equations are entirely different from those of the linear theory. We will illustrate this claim by considering various types of first-order equations which can be solved explicitly in terms of elementary functions. These equations were the object of much study in the eighteenth century. Ingenious methods were devised to solve very special equations occurring in mechanics and optics but no general theory was established.

Consider the differential equation

$$\frac{d}{dt}y = y^2. \tag{7.10.1}$$

The right-hand side of this equation is a continuous function for $-\infty < t, y < \infty$ [see § 2.1]. However, the equation has the solution

$$y(t, \alpha) = (\alpha - t)^{-1}$$

which does not exist for every finite t-interval, since $y(t, \alpha)$ tends to plus infinity or minus infinity as t approaches α from below and above respectively. This phenomenon, the existence of solutions with such hidden singularities, cannot occur for linear equations. Moreover it is easy to check that

$$y(t) = y(t, \alpha) + y(t, \beta)$$

is not a solution in any t-interval. Therefore the fundamental property of linear homogeneous equations, that the sum of any two solutions is a solution [see § 7.3.1] is not valid for (7.10.1).

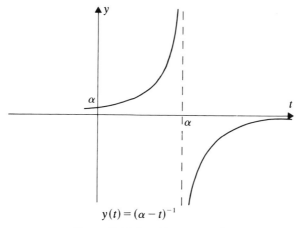

$$y(t) = (\alpha - t)^{-1}$$

Figure 7.10.1: $y(t) = (\alpha - t)^{-1}$.

Now consider the initial value problem

$$\frac{d}{dt}y = (3/2)y^{1/3}, \qquad y(0) = 0. \tag{7.10.2}$$

For a linear equation such a problem would have a unique solution. However, (7.10.2) has many solutions, we could take

$$y(t) \equiv 0 \quad \text{for } -\infty < t < \infty$$

or

$$y(t, c) = \begin{cases} (t-c)^{3/2}, & \text{for } c \le t < \infty, \\ 0, & \text{for } -\infty < t < c, \end{cases}$$

where c is any positive number.

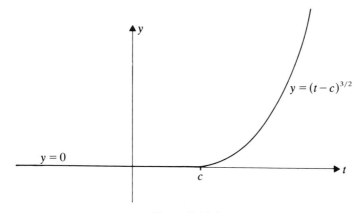

Figure 7.10.2

The uniqueness property for linear initial value problems can break down in a more spectacular way. There is an equation of the form

$$\frac{d}{dt}y = f(t, y)$$

where f is an everywhere continuous function such that at least two solutions have the property

$$y(t_0) = y_0,$$

for any t_0 and y_0.

It is often said of non-linear equations of order n that their *general* solution is a solution containing n arbitrary constants, all other solutions being *singular* solutions. However the concept of a 'general solution' is not easy to define and such classifications have lost favour in this century.

Here is a first-order equation with a solution containing n arbitrary constants! It is easy to check that the equation

$$\left(\frac{dy}{dt}\right)^2 + y^2 = 1$$

has the solutions

$$y_1(t) = \sin(t - \alpha), \qquad y_2(t) = -1, \qquad y_3(t) = 1.$$

These solutions can be pieced together as in Figure 7.10.3 to provide a solution containing the arbitrary constants $\alpha_1, \alpha_2, \ldots, \alpha_n$.

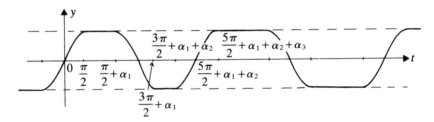

Figure 7.10.3

7.10.2. Special Equations

An equation such as

$$f(x, y) = c \tag{7.10.3}$$

may define y as a function of x. For example the equation

$$x^2 + y^2 = r^2 \tag{7.10.4}$$

implies that near the point $(r/\sqrt{2}, r/\sqrt{2})$

$$y = \{r^2 - x^2\}^{1/2},$$

and near the point $(r/\sqrt{2}, -r/\sqrt{2})$

$$y = -\{r^2 - x^2\}^{1/2}$$

[see Theorem 5.13.1].

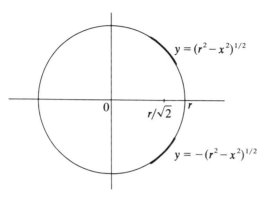

Figure 7.10.4

It may not be possible to solve (7.10.3) for y explicitly, but the existence of such a solution function may be guaranteed by the Implicit Function Theorem (Theorem 5.13.1) for sufficiently smooth functions f. In this case, differentiation with respect to x of (7.10.3) gives the differential equation

$$\frac{\partial f}{\partial x} + \frac{\partial f}{\partial y}\frac{dy}{dx} = 0.$$

For example, from (7.10.4), we have

$$2x + 2y\frac{dy}{dx} = 0.$$

Now consider the problem of finding a solution of the non-linear equation

$$M(x, y) + N(x, y)\frac{dy}{dx} = 0. \qquad (7.10.5)$$

If there exists a function f such that

$$\frac{\partial f}{\partial x} = M, \qquad \frac{\partial f}{\partial y} = N$$

then equation (7.10.5) is said to be *exact*. A solution is defined implicitly by the equation $f(x, y) = c$ [see also Theorem 5.10.2].

$$f(x, y) = c.$$

THEOREM 7.10.1.   *Equation (7.10.5) is exact if and only if*

$$\frac{\partial M}{\partial y} = \frac{\partial N}{\partial x}.$$

EXAMPLE 7.10.1.   The equation

$$(x^2 + y^2 - 1) + 2xy\frac{dy}{dx} = 0$$

is exact and has the solution

$$(x^3/3) + xy^2 - x = c.$$

It may happen that equation (7.10.5) is not exact but when multiplied by $\mu(x, y)$ gives an exact equation

$$\mu M + \mu N \frac{dy}{dx} = 0. \tag{7.10.6}$$

Such an *integrating factor* μ is determined by the partial differential equation

$$\frac{\partial}{\partial y}(\mu M) = \frac{\partial}{\partial x}(\mu N),$$

which will be harder to solve, in general, than equation (7.10.5). However, various special cases can be treated. For example,

$$\mu = \exp\left\{\int \left(\frac{\partial M}{\partial y} - \frac{\partial N}{\partial x}\right)\frac{dx}{N}\right\}$$

is an integrating factor provided that the integrand is a function of x only. Many such results are found in older texts such as Boole (1865) or Bateman (1918).

There are two simple cases of equation (7.10.5) which should be noted:

1. *Separated variables.*
   The equation

$$M(x) + N(y)\frac{dy}{dx} = 0$$

   has the solution

$$\int M(x)\,dx + \int N(y)\,dy = c.$$

2. *Homogeneous equations.*
   If

$$\frac{dy}{dx} = F(y/x),$$

   the substitution $u = (y/x)$ will produce an equation with separated variables and the solution

$$\int \frac{du}{F(u) - u} + \log x = c.$$

EXAMPLE 7.10.2.   From 1 the equation

$$2x + 2y\frac{dy}{dx} = 0$$

has the solution

$$x^2 + y^2 = c.$$

EXAMPLE 7.10.3.   If we put $u = (y/x)$ in the equation

$$\frac{dy}{dx} = (y/x)^2 + (y/x)$$

then $y = xu$ and hence

$$x\frac{du}{dx} + u = u^2 + u.$$

Therefore,

$$(1/x) + (1/u^2)\frac{du}{dx} = 0$$

and

$$\log x - (1/u) = c,$$

which gives

$$y = x\{\log x - c\}^{-1}.$$

Unfortunately these methods will not solve one of the simplest non-linear equations, namely

$$\frac{d}{dx} y = r(x) + q(x)y + p(x)y^2. \tag{7.10.7}$$

A special case of this equation, known as *Riccati's equation*, is

$$\frac{du}{dx} + bu^2 = cx^m. \tag{7.10.8}$$

If we put $y = xu$, we obtain

$$x\frac{dy}{dx} - ay + by^2 = cx^n, \tag{7.10.9}$$

where $a = 1$ and $n = m + 2$.

Equation (7.10.9) has a known explicit solution when $(n \pm 2a)/2n$ is a positive integer or zero, and Riccati's equation when $m = -2$ or when m is $-4k/(2k + 1)$ for any integer k.

For example the solution of

$$u' + u^2 = x^{-(4/3)}$$

is

$$6x^{1/3} + \log\{(3ux^{2/3} - 3 - ux^{1/3})/(3ux^{2/3} + 3 + ux^{1/3})\} = c.$$

Riccati's equation is of special interest, since the change of dependent variable in (7.10.8),

$$w = \exp\left(b \int u \, dt\right)$$

leads to the second-order linear equation

$$\frac{d^2 w}{dx^2} - bcx^m w = 0.$$

7.11. NON-LINEAR FIRST-ORDER SYSTEMS

7.11.1.   General Theory

In section 7.10 we tried to give the flavour of early work in non-linear theory. In this century the search for explicit solutions of special equations has been replaced by the development of a qualitative theory for general systems of equations. This theory has a vast and rapidly growing literature. The problems arising in the theory of ordinary differential equations, placed in a more abstract and general setting, have led to the theory of dynamical systems. New disciplines such as global analysis and differential topology are vital tools in its development. We can give only a brief outline of the beginnings of the subject. However, we hope that it will be clear that the qualitative theory and modern numerical analysis, taken together, give powerful weapons for an attack on any non-linear problem.

Let $\mathbf{x}$ denote a point in $\mathbb{R}^n$, real Euclidean space of n dimensions [see I, Example 5.2.2], let t be a real number and let $\mathbf{y}(t)$ be an $\mathbb{R}^n$-valued function of t. Finally, let $\mathbf{f}(t, \mathbf{y})$ be an $\mathbb{R}^n$-valued function of t and $\mathbf{y}$ [see § 5.3]. We consider the system of differential equations

$$\dot{\mathbf{y}}(t) = \mathbf{f}(t, \mathbf{y}(t)) \tag{7.11.1}$$

where the dot denotes differentiation with respect to t, and we impose the condition

$$\mathbf{y}(t_0) = \boldsymbol{\alpha} \tag{7.11.2}$$

on a solution of the system.

In the absence of any means of solving this problem in terms of elementary functions it is of vital importance to know when a solution can exist and whether or not it is uniquely determined by condition (7.11.2).

DEFINITION 7.11.1.   Let I be an interval, $a \le t \le b$, and let Ω be a subset of $\mathbb{R}^n$. If there exists a positive constant K such that

$$|\mathbf{f}(t, \mathbf{x}) - \mathbf{f}(t, \mathbf{y})| \le K|\mathbf{x} - \mathbf{y}| \qquad (7.11.3)$$

for each t in I and $\mathbf{x}, \mathbf{y}$ in Ω, then $\mathbf{f}$ is said to satisfy a *Lipschitz condition* or to be *Lipschitz continuous* in Ω.

In this definition, $|\mathbf{x}|$ denotes the Euclidean length of the n-vector $\mathbf{x}$, that is,

$$|\mathbf{x}| = \{x_1^2 + x_2^2 + \ldots + x_n^2\}^{1/2}.$$

The inequality (7.11.3) implies that $\mathbf{f}(t, \mathbf{x})$ is a continuous function of $\mathbf{x}$ in Ω, but not all such continuous functions satisfy a Lipschitz condition of the form (7.11.3). The importance of Lipschitz continuous functions stems from the following existence and uniqueness of solutions theorem for our problem (7.11.1) and (7.11.2).

THEOREM 7.11.1.   *Let I be an interval containing t_0 and let Ω be an open set in $\mathbb{R}^n$ containing $\boldsymbol{\alpha}$ [see Definition 5.12.1]. If $\mathbf{f}$ satisfies a Lipschitz condition of the form (7.11.3) then there exists a sub-interval J of I, containing t_0, such that equation (7.11.1) has a unique solution for t in J, satisfying (7.11.2).*

Since J may be an interval very much smaller in length than I, this theorem is called a *local existence theorem* and only guarantees the existence of a unique solution near t_0.

The following theorem provides a useful means of identifying Lipschitz continuous functions.

THEOREM 7.11.2.   *If $\mathbf{f}(\mathbf{y})$ has co-ordinate functions $f_1(\mathbf{y}), f_2(\mathbf{y}), \ldots, f_n(\mathbf{y})$ and if*

$$|\partial f_i(\mathbf{y})/\partial y_j| \le M < \infty, \qquad 1 \le i, j \le n,$$

for all in Ω, then $\mathbf{f}$ is Lipschitz continuous in Ω.

EXAMPLE 7.11.1.   Consider the one-dimensional system

$$\dot{y} = (3/2)y^{1/3}, \qquad y(t_0) = \alpha \quad (\alpha > 0). \qquad (7.11.4)$$

Here $f(t, y) = (3/2)y^{1/3}$ and $\partial f(t, y)/\partial y = \{1/2y^{2/3}\}$. The partial derivative is bounded in any closed set which does not contain the point $y = 0$, and for t in any bounded interval. Provided that $\alpha \ne 0$, the existence of a unique local solution follows from Theorems 7.11.1 and 7.11.2. It is easy to check that this solution is

$$y(t) = (t - t_0 + \alpha^{2/3})^{3/2}, \qquad t_0 - \alpha^{2/3} < t < t_0 + \alpha^{2/3}.$$

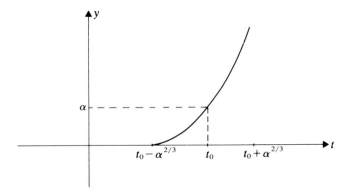

Figure 7.11.1

If $\alpha = 0$, we saw in section 7.10.1 that (7.11.4) has infinitely many solutions. Notice that because of this we cannot extend the interval of definition of the solution when $\alpha \neq 0$ to the left of $t_0 - \alpha^{2/3}$ and still claim that it is unique.

How far can we extend the local solution provided by Theorem 7.11.1? Unfortunately this is a difficult problem with little general theory. The possibility of hidden singularities as described in section 7.10.1 and the above example illustrate the complexity of the situation.

7.11.2. Autonomous Systems

DEFINITION 7.11.2.   The system of differential equations

$$\dot{\mathbf{y}} = \mathbf{f}(\mathbf{y}) \tag{7.11.5}$$

is said to be *autonomous*, since the function $\mathbf{f}$ does not depend explicitly on t.

EXAMPLE 7.11.3.   The first of the following systems is autonomous, the second is not;

(i)
$$\frac{d}{dt}\begin{pmatrix} y_1 \\ y_2 \end{pmatrix} = \begin{pmatrix} y_2^2 \\ y_1 \end{pmatrix},$$

(ii)
$$\frac{d}{dt}\begin{pmatrix} y_1 \\ y_2 \end{pmatrix} = \begin{pmatrix} ty_1 + y_2^2 \\ t^3 + y_1 \end{pmatrix}.$$

A system such as (7.11.1) can be transformed into an equivalent autonomous system at the price of an increase in dimension of the system [see § 7.9.1]. Let

$$\mathbf{z} = \begin{pmatrix} \mathbf{y} \\ t \end{pmatrix} \quad \text{and} \quad \mathbf{F}(\mathbf{z}) = \begin{pmatrix} \mathbf{f}(t, \mathbf{y}) \\ 1 \end{pmatrix};$$

then equation (7.11.1) is clearly equivalent to

$$\dot{\mathbf{z}} = \mathbf{F}(\mathbf{z}). \tag{7.11.6}$$

EXAMPLE 7.11.4.   The equivalent autonomous system to example (ii) above is

$$\frac{d}{dt}\begin{pmatrix} z_1 \\ z_2 \\ z_3 \end{pmatrix} = \begin{pmatrix} z_3 z_1 + z_2^2 \\ z^3 + z_1 \\ 1 \end{pmatrix}.$$

The general initial value problem for an autonomous system reduces to a special case, for if $\mathbf{z}(t)$ is a solution of (7.11.6) for which $\mathbf{z}(0) = \alpha$, then it is easy to verify that $\mathbf{w}(t) = \mathbf{z}(t - t_0)$ is a solution for which $\mathbf{w}(t_0) = \alpha$.

7.11.3.   Qualitative Theory

In view of the results of section 7.11.2 we confine ourselves to the study of the autonomous initial value problem

$$\dot{\mathbf{y}} = \mathbf{f}(\mathbf{y}), \qquad \mathbf{y}(0) = \mathbf{x}. \qquad (7.11.7)$$

Henceforward we assume that $\mathbf{f}$ is continuous in $\mathbb{R}^n$ [see Definition 5.3.1] and that (7.11.7) has a unique solution for each $\mathbf{x}$ in $\mathbb{R}^n$. We denote this solution by $\mathbf{y}(t, \mathbf{x})$ and we suppose that it exists for $-\infty < t < \infty$. It can be shown that $\mathbf{y}(t, \mathbf{x})$ is a continuous function of $(t, \mathbf{x})$ in $\mathbb{R}^{n+1}$.

In general, as t increases, the point $\mathbf{y}(t, \mathbf{x})$ describes a curve in $\mathbb{R}^n$, passing through $\mathbf{x}$ when $t = 0$. The tangent vector to a solution curve in the direction of increasing t is $\mathbf{f}(\mathbf{y})$. The magnitude or length of the vector $\mathbf{f}(\mathbf{y})$ measures the rate of turning of the solution curve at the point $\mathbf{y}$ as t increases.

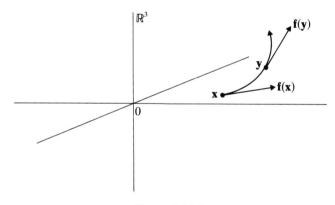

Figure 7.11.2

However, if $\mathbf{f}(\mathbf{x}) = \mathbf{0}$, the unique solution of (7.11.7) is

$$y(t, \mathbf{x}) = \mathbf{x} \qquad (-\infty < t < \infty)$$

and the solution curve degenerates to a single point.

DEFINITION 7.11.3.   If $\mathbf{f}(\mathbf{x}) = \mathbf{0}$, $\mathbf{x}$ is called a *critical point* or a *stationary point* of the system (7.11.7) [see § 5.6].

The collection of all possible solution curves to (7.11.7) is called a *flow* in $\mathbb{R}^n$. The qualitative theory of the system consists of methods of building up a clear picture of the flow. There is an extensive theory in two dimensions which we consider briefly in section 7.11.4. In higher dimensions the solution curves are not confined to a plane and their nature and relative dispositions can be extremely complicated. The theory of such systems is a rapidly growing area of contemporary mathematics.

Many of the terms used in qualitative theory are derived from dynamics and astronomy or from the theory of electric fields and fluid flow. A solution curve is called a *solution path* or *orbit*. The space containing the flow is referred to as *phase space* or *state space* and the collection of solution paths as the *phase portrait* of the system.

If $\mathbf{x}$ is not a stationary point and if $\mathbf{y}(t_0 + \tau_0, \mathbf{x}) = \mathbf{y}(t_0, \mathbf{x})$ for some non-zero number τ_0, it can be shown that there must exist a least positive number τ such that $\mathbf{y}(t + \tau, \mathbf{x}) = \mathbf{y}(t, \mathbf{x})$ for $-\infty < t < \infty$. In this event, $\mathbf{y}(t, \mathbf{x})$ is called a *periodic solution* with *period* τ, and its solution path is a closed path.

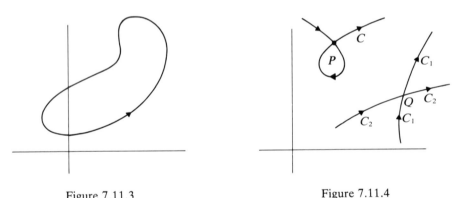

Figure 7.11.3   Figure 7.11.4

Figure 7.11.3 shows a simple closed path in $\mathbb{R}^2$. Notice that curves C, C_1 and C_2 intersecting as shown in Figure 7.11.4 cannot consist of solution paths unless P and Q are stationary points. Otherwise our initial value problem would not have a unique solution at P or at Q.

The behaviour of the flow very near critical points and the relative positions of the critical points and closed paths largely determine the overall picture of the flow in $\mathbb{R}^n$.

7.11.4.   Qualitative Theory of Two-Dimensional Systems

In this section we consider the qualitative theory of (7.11.7) for two-dimensional systems. The phase space in this case is a plane and the traditional

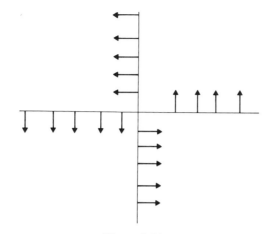

Figure 7.11.5

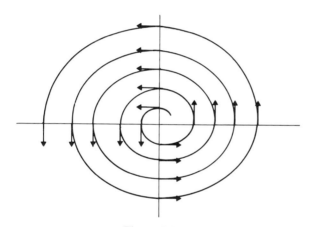

Figure 7.11.6

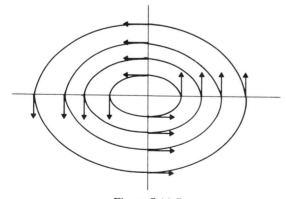

Figure 7.11.7

notation is

$$\dot{x} = P(x, y), \qquad \dot{y} = Q(x, y) \qquad (7.11.8)$$

and

$$x(0) = x_0, \qquad y(0) = y_0. \qquad (7.11.9)$$

We can reach a rough idea of the phase portrait by the *method of isoclines.* This consists of plotting the vector $(P(x, y), Q(x, y))$ at a number of points and guessing the curves which have these vectors as tangents. This has the obvious drawback that inaccuracies in computing and plotting may lead to ambiguity of interpretation. For example the tangent vectors in Figure 7.11.5 look compatible with the flow in either Figure 7.11.6 or Figure 7.11.7. However, the method is useful in practice and usually provides inspiration for analytical techniques.

The critical points of (7.11.8) are the points of intersection of the curves $P(x, y) = 0$ and $Q(x, y) = 0$. By changing the origin of coordinates, if necessary, we can suppose that the origin is a critical point. To examine the flow near $(0, 0)$ we suppose that P and Q have Taylor expansions

$$P(x, y) = ax + by + \alpha x^2 + \beta xy + \gamma y^2 + \cdots$$
$$Q(x, y) = cx + dy + \alpha' x^2 + \beta' xy + \gamma' y^2 + \cdots \qquad (7.11.10)$$

valid near $(0, 0)$ [see Theorem 5.8.1].

Neglecting the second and higher-degree terms we are led to consider the linear approximation to (7.11.8),

$$\dot{x} = ax + by, \qquad \dot{y} = cx + dy \qquad (7.11.11)$$

or in matrix notation [see I, § 5.7],

$$\frac{d}{dt}\begin{pmatrix} x \\ y \end{pmatrix} = \begin{pmatrix} a & b \\ c & d \end{pmatrix}\begin{pmatrix} x \\ y \end{pmatrix} = \mathbf{A}\begin{pmatrix} x \\ y \end{pmatrix}.$$

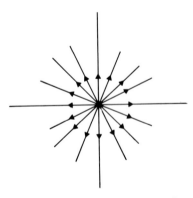

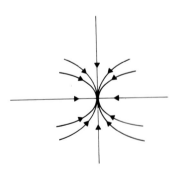

Figure 7.11.8: Node, $\mu = \lambda > 0$.   Figure 7.11.9: Node, $\lambda < \mu < 0$.

The linear system (7.11.11) is solved best by the eigenvalue-eigenvector method described in section 7.9.2. For example, if $ad - bc \neq 0$, the origin is called a *simple critical point* of (7.11.11) and the behaviour of solutions depends on the nature of the eigenvalues λ, μ of **A**. Some possibilities are illustrated in Figures 7.11.8 to 7.11.13.

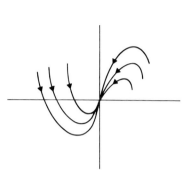

Figure 7.11.10: Node, $\mu = \lambda$.

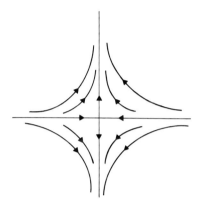

Figure 7.11.11: Saddle point, $\lambda < 0 < \mu$.

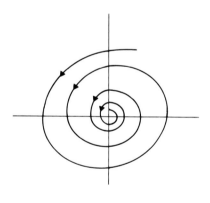

Figure 7.11.12:
Focus, $\lambda = \alpha + i\beta$, $\mu = \alpha - i\beta$ $(\alpha, \beta \neq 0)$.

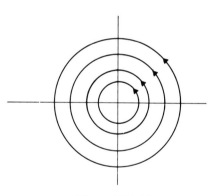

Figure 7.11.13:
Centre, $\lambda = i\beta$, $\mu = -i\beta$ $(\beta \neq 0)$.

With suitable assumptions on P and Q it can be shown that, comparing the systems (7.11.11) and (7.11.8), a centre for (7.11.11) may be a centre or a focus for (7.11.8), a node for (7.11.11) may be a node or a focus for (7.11.8), but a focus or a saddle point for (7.11.11) remain unchanged for (7.11.8). The general effect on the flow of moving from (7.11.11) to (7.11.8) near $(0, 0)$ is illustrated in Figure 7.11.14.

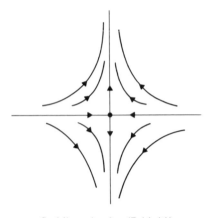

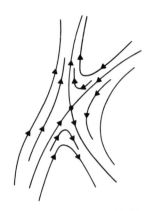

Saddle point for (7.11.11)   Saddle point for (7.11.8)

Figure 7.11.14

EXAMPLE 7.11.5.   The critical points of the system

$$\dot{x} = x - y$$
$$\dot{y} = 4x^2 + 2y^2 - 6$$
(7.11.2)

occur when

$$y - x = 0$$

and

$$4x^2 + 2y^2 - 6 = 0.$$

Putting $y = x$ in the second equation gives

$$6x^2 - 6 = 0$$

and hence

$$x^2 = 1.$$

The critical points are $P = (1, 1)$ and $Q = (-1, 1)$. To examine the flow near P we put

$$x = 1 + u \quad \text{and} \quad y = 1 + v.$$

The system becomes

$$\dot{u} = u - v$$
(7.11.3)

and

$$\dot{v} = 8u + 4v + 4u^2 + 2v^2.$$

The linear approximation to this system is

$$\frac{d}{dt}\begin{pmatrix}u\\v\end{pmatrix}=\begin{pmatrix}1&-1\\8&4\end{pmatrix}\begin{pmatrix}u\\v\end{pmatrix}=\mathbf{A}\begin{pmatrix}u\\v\end{pmatrix}.\qquad(7.11.4)$$

The eigenvalues of $\mathbf{A}$ are found to be $(1/2)\{5\pm i\sqrt{23}\}$ [see I, § 7.3] and hence P is a focus for (7.11.4) (see Figure 7.11.12).

As we remarked above, it can be shown that P is a focus for (7.11.2). Similar analysis shows that Q is a saddle point for (7.11.2).

Finally we note that $x(t)$ is increasing to the right of the line $y=x$, since then $\dot{x}=x-y>0$. Similarly, $x(t)$ is decreasing to the left of the line $y=x$, and $y(t)$ is increasing and decreasing outside and inside the ellipse $4x^2+2y^2-6=0$ respectively.

With a little thought we can deduce that the flow for (7.11.2) has the behaviour shown in Figure 7.11.15.

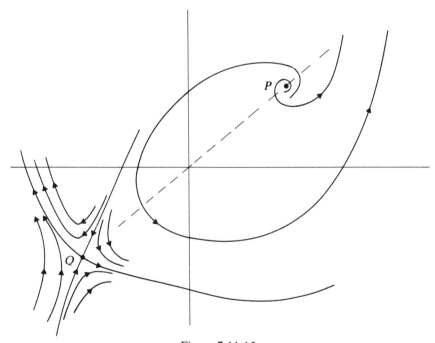

Figure 7.11.15

If $ad-bc=0$, but not every entry in $\mathbf{A}$ is zero, the system (7.11.11) has a line of critical points. For example, if $b\neq0$, every point on $y=-(a/b)x$ is a critical point.

If the expansions (7.11.10) have no linear terms then we must consider a non-linear approximation to (7.11.8).

These analytical methods help to determine the flow near critical points. To complete the phase portrait we must follow solutions paths as t goes to

either plus or minus infinity. In particular, in applications one is interested
often in the existence and position of closed paths. Suppose that **x** is not a
critical point and that **y**$(t, \mathbf{x})$ remains bounded as t tends to infinity. Then
either the solution path is a closed path or it approaches a critical point or a
closed path as t tends to infinity. The possibilities are illustrated in Figure
7.11.16. A closed path which is approached by another path in this way is
called a *limit cycle*.

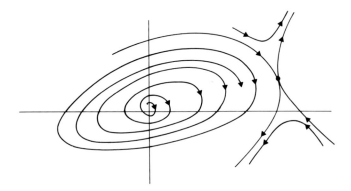

Figure 7.11.16

The existence of such limit cycles is of fundamental importance in many
applications. For example, a limit cycle might represent the stable periodic
fluctuations in the numbers of competing species in a mixed population. For
a wide range of starting values such a mixed population might eventually
settle down to the regular periodic behaviour described by a limit cycle. The
corresponding flow would show paths starting outside the limit cycle spiraling
in towards it and paths starting inside spiraling outwards as in Figure 7.11.17.

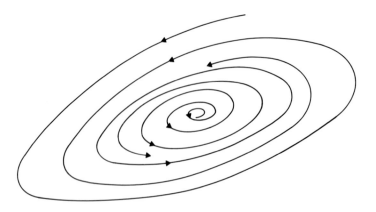

Figure 7.11.17

It can be shown that a closed path must go round at least one critical point and that a system without critical points can have neither closed paths nor any bounded solutions.

Finally we state the celebrated *criterion of Poincaré and Bendixon* for the existence of a limit cycle: if K is a closed and bounded set containing no critical points and if every path crosses the boundary of K from outside K to inside K, then K contains a limit cycle.

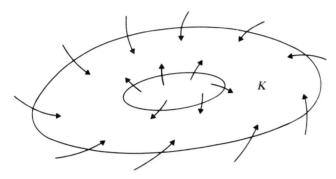

Figure 7.11.18

A typical such K is shown in Figure 7.11.18. The arrows indicate the direction of paths on the boundary of K. The limit cycle in K must go round a critical point of the system which must lie, therefore, inside the smaller ellipse.

The Poincaré–Bendixon criterion, and other useful results, are deduced from the Poincaré–Bendixon Theorem which describes the possible structure of the 'limit sets' of a flow. For precise definitions and many results and applications see Hirsch and Smale (1974). In particular, this reference gives a detailed account of the classical *equations of Volterra and Lotka* describing the growth of two species when one, the predators, use the other, the prey, as their food supply.

The reader will find many worked examples and a fairly comprehensive treatment of two-dimensional systems in the following texts; Sansone and Conti (1964) and Andronov et al. (1973).

7.11.5. Stability Theory

Speaking loosely, we might say that a system is stable if eventually it comes to rest, or stays near its rest point or stays near a periodic motion. The definitions of Lyapunov stability, *Lyapunov asymptotic stability* and orbital stability attempt to make these ideas precise. There is a good survey of Lyapunov stability in Hahn (1967), and recent developments are described in Cronin (1980) and Hirsch and Smale (1974).

The last reference gives also a brief introduction to structural stability. This theory considers the effect on a flow of changing slightly the differential equations governing the system. There is always the possibility of error in choosing the form and parameters of a mathematical model for a physical system. A good model will be structurally stable in the sense that the flow is not dramatically affected by small changes in the model. The study of structural stability uses powerful abstract mathematical theories and is a lively subject of contemporary research work.

P.J.B.

REFERENCES

Andronov, A. A., Leontovich, E. A., Gordon, I. I. and Maier, A. G. (1973). *Qualitative Theory of Second-Order Dynamic Systems*, Wiley, New York.

Bateman, H. (1966). *Differential Equations*, Chelsea, New York (First published 1918).

Bellman, R. (1953). *Stability Theory of Differential Equations*, McGraw-Hill, New York.

Boole, G. (1959). *A Treatise on Differential Equations*, 5th ed., Chelsea, New York.

Burkill, J. C. (1962). *The Theory of Ordinary Differential Equations*, 3rd ed. (University Mathematical Texts), Oliver and Boyd.

Cesari, L. (1963). *Asymptotic Behaviour and Stability Problems in Ordinary Differential Equations*, Springer, Berlin.

Coddington, E. A. and Levinson, N. (1955). *Theory of Ordinary Differential Equations*, McGraw-Hill, New York.

Cronin, J. (1980). *Differential Equations, Introduction and Qualitative Theory*, Dekker.

Gradshteyn and Ryzlik (1965). *Table of Integrals, Series and Products*, Academic Press, New York.

Hahn, W. (1967). *Stability of Motion*, Springer-Verlag.

Hirsch, M. and Smale, S. (1974). *Differential Equations, Dynamical Systems and Linear Algebra*, Academic Press.

Kamke, E. (1955). *Differentialgleichungen; Lösungsmethoden und Lösungen*, 2 vols., Chelsea, New York.

Mikusinski, J. (1959). *Operational Calculus*, Pergamon, London.

Murray, J. D. (1974). *Asymptotic Analysis*, Oxford University Press.

Pearson, K. (1922). *Tables of the Incomplete Γ-Function*, H.M.S.O., London.

Sansone, G. and Conti, R. (1964). *Non-Linear Differential Equations*, Pergamon, Oxford.

Whittaker, E. T. and Watson, G. N. (1927). *A Course of Modern Analysis*, Cambridge University Press.

CHAPTER 8

Partial Differential Equations

8.0. INTRODUCTION

This subject is not only tremendously broad but has undergone a radical transformation over the last few decades. It arose, centuries ago, because the description of those physical processes that lend themselves to mathematical analysis often involves a partial differential equation, such as Laplace's equation, the wave equation and the equation governing the conduction of heat. Many special techniques were devised to deal with various specific problems, and it is probably fair to say that until the present century the body of knowledge concerning partial differential equations resembled not so much a theory as a chaotic assembly of apparently unrelated facts. Today the position is different: the subject is still very much lacking in structure as compared with, say, group theory [see I, Chapter 8], but some order has been brought into the chaos, and broad lines of development are clearly visible. That this is so is largely due to the systematic use of modern pure mathematics, especially functional analysis [see Chapter 19]. These modern techniques have enabled a breakaway to be made from the ancient preoccupation with exact solutions of partial differential equations, a preoccupation made all the more agonizing because only rarely is it possible to obtain a useful explicit formula for a solution of a problem involving a partial differential equation. They have also brought about great progress in the qualitative description of solutions of wide classes of problems. Very often the essential features of a solution which are revealed by this qualitative information are all that is needed in a problem; and should numerical values of a solution be required, knowledge of the general behaviour of the solution may facilitate computation of an approximation to the solution. The progress has been especially remarkable in non-linear problems, which are almost entirely resistant to the more traditional techniques. Naturally, it will always be essential to be familiar with the classical procedures relating to exact solutions, but nowadays this knowledge has to be supplemented by modern methods, which are often abstract in nature.

In a short article such as this it is not possible to be at all comprehensive. We shall attempt to describe some of the more important results and techniques connected with linear problems, and to indicate a few of the promising developments in non-linear theory.

8.1. PRELIMINARIES

A partial differential equation is an equation which involves a function and its derivatives up to a certain order. To be a little more precise, let Ω be an open subset of Euclidean n-space $\mathbb{R}^n$ [see Definition 5.12.1], with points represented by $\mathbf{x} = (x_1, \ldots, x_n)$, and let $\boldsymbol{\alpha} = (\alpha_1, \ldots, \alpha_n)$, where each α_i is a non-negative integer. Such an $\boldsymbol{\alpha}$ is called a *multi-index*, and we write [see I (3.7.1)]

$$|\boldsymbol{\alpha}| = \sum_{i=1}^{n} \alpha_i, \qquad \boldsymbol{\alpha}! = \alpha_1! \alpha_2! \ldots \alpha_n!,$$

$$\mathbf{x}^{\boldsymbol{\alpha}} = x_1^{\alpha_1} x_2^{\alpha_2} \ldots x_n^{\alpha_n}, \qquad |\mathbf{x}| = \left(\sum_{k=1}^{n} x_k^2 \right)^{1/2},$$

$$D_j = \frac{\partial}{\partial x_j}, \qquad D^{\boldsymbol{\alpha}} = \frac{\partial^{|\boldsymbol{\alpha}|}}{\partial x_1^{\alpha_1} \ldots \partial x_n^{\alpha_n}}.$$

When $|\boldsymbol{\alpha}| = 0$, $D^{\boldsymbol{\alpha}} f = f$ for any function f on Ω.

The use of this notation enables us to write in a concise way expressions which would appear overwhelmingly complex when written out in the traditional way.

Given any non-negative integer m, let $N(m)$ be the number of distinct multi-indices $\boldsymbol{\alpha}$ with $|\boldsymbol{\alpha}| \leq m$. Suppose that to each such $\boldsymbol{\alpha}$ there corresponds a complex number $a_{\boldsymbol{\alpha}}$ [see I, § 2.7]. We denote by $(a_{\boldsymbol{\alpha}})_{|\boldsymbol{\alpha}| \leq m}$ the ordered $N(m)$-tuple of complex numbers $a_{\boldsymbol{\alpha}}$ in which the ordering is decided by the rule that $a_{\boldsymbol{\alpha}}$ comes before $a_{\boldsymbol{\beta}}$ if $|\boldsymbol{\alpha}| < |\boldsymbol{\beta}|$ or if $|\boldsymbol{\alpha}| = |\boldsymbol{\beta}|$ and $\alpha_i < \beta_i$, where i is the least integer with $\alpha_i \neq \beta_i$. For example, if $m = 3$ and $n = 2$, then $(a_{\boldsymbol{\alpha}})_{|\boldsymbol{\alpha}| \leq m}$ is the 6-tuple $(a_{(0,0)}, a_{(0,1)}, a_{(1,0)}, a_{(0,2)}, a_{(1,1)}, a_{(2,0)})$.

DEFINITION 8.1.1.   A *partial differential equation in* Ω is an equation of the form

$$F(\mathbf{x}, (D^{\boldsymbol{\alpha}} u(\mathbf{x}))_{|\boldsymbol{\alpha}| \leq m}) = 0, \qquad \mathbf{x} \in \Omega \tag{8.1.1}$$

where F is a given function and u is the unknown function [see § 5.3]; a real- or complex-valued function u on Ω is said to be a *classical solution* or simply a *solution* of (8.1.1) if $D^{\boldsymbol{\alpha}} u(\mathbf{x})$ exists for all $\mathbf{x} \in \Omega$ and for all $\boldsymbol{\alpha}$ with $|\boldsymbol{\alpha}| \leq m$, and (8.1.1) holds.

The equation is said to be *of order* k if k is the largest integer such that, loosely speaking, F genuinely depends on $D^{\boldsymbol{\alpha}} u$ for some $\boldsymbol{\alpha}$, say $\boldsymbol{\alpha}_0$, with $|\boldsymbol{\alpha}_0| = k$. In the case when F is differentiable [see § 5.9], this means that $\partial F(\mathbf{x}, (p_{\boldsymbol{\alpha}})_{|\boldsymbol{\alpha}| \leq m}) / \partial p_{\boldsymbol{\alpha}_0}$ is not identically zero; and in all specific cases it will be clear what is meant. To avoid tiresome discussions we shall henceforth assume that any equation of the form of (8.1.1) has order m.

Some examples may help to clarify the position, and one can do no better than to go back to the three great equations of mathematical physics. These are *Laplace's equation* (or the *harmonic equation*)

$$\Delta_n u \equiv \sum_{i=1}^{n} D_i^2 u = 0, \tag{8.1.2}$$

(which is a special case of *Poisson's equation* $\Delta_n u = f$, f being a given function on Ω), the *heat equation* (or *diffusion equation*)

$$D_1 u - \sum_{i=2}^{n} D_i^2 u = 0, \tag{8.1.3}$$

and the *wave equation*

$$D_1^2 u - \sum_{i=2}^{n} D_i^2 u = 0. \tag{8.1.4}$$

All these equations are of second order, and turn up with great frequency in physical problems. Laplace's equation and Poisson's equation occur in electrostatics, the theory of current flow, and in the steady irrotational two-dimensional flow of an inviscid incompressible fluid; the heat equation governs the distribution of temperature in a heat conducting material; and the wave equation appears in the theory of vibrating strings or membranes.

These famous equations are also *linear*, in the sense that the derivatives of u occur linearly in them. More generally, we shall say that equation (8.1.1) is *linear* if it can be written in the form

$$\sum_{|\alpha| \leq m} a_\alpha D^\alpha u = f \tag{8.1.5}$$

where the coefficients a_α and f are given functions of $\mathbf{x}$. The equation (8.1.5) is of order k if k is the largest integer such that there exists α with $|\alpha| = k$ and a_α not identically zero. As in the case of the more general equation (8.1.1), we shall in future assume that (8.1.5) is of order m. It will also be convenient to refer to the *differential operator* $\sum_{|\alpha| \leq m} a_\alpha D^\alpha$ and to denote it by L or $P(\mathbf{x}, D)$, the latter notation being intended to emphasize the fact that the operator is a polynomial in D [see (2.1.1)]. Equation (8.1.5) will thus often be written as $Lu = f$ or $P(\mathbf{x}, D)u = f$, and if all the a_α are constant we may write the equation as $P(D)u = f$.

By no means all partial differential equations of scientific interest are linear, an outstandingly important class of non-linear equations being formed by the *quasi-linear* equations. These are equations of the form of (8.1.1) in which F depends linearly on the $D^\alpha u$ for $|\alpha| = m$, and can be written as

$$\sum_{|\alpha|=m} a_\alpha(\mathbf{x}, (D^\beta u(\mathbf{x}))_{|\beta| \leq m-1}) D^\alpha u(\mathbf{x}) = b(\mathbf{x}, (D^\beta u(\mathbf{x}))_{|\beta| \leq m-1}). \tag{8.1.6}$$

All equations which arise from variational principles are of this kind, a particularly interesting one being the *minimal surface equation* [see (5.10.2)]

$$\sum_{i=1}^{n} D_i \left(\frac{D_i u}{\sqrt{1 + |\text{grad } u|^2}} \right) = 0.$$

Non-linear equations which are not even quasi-linear seem to turn up principally in questions of differential geometry and related subjects; perhaps the most notable example is the *Monge–Ampère equation* in the plane:

$$(D_1^2 u)(D_2^2 u) = (D_1 D_2 u)^2.$$

Now we turn to the kind of questions which are commonly asked about partial differential equations. Does equation (8.1.1) have a solution in Ω, possibly satisfying some additional conditions, and if so, is the solution unique? Are there only finitely many distinct solutions? Can the solution be made unique by the imposition of further conditions? Can anything be said about the qualitative behaviour of solutions? Does the solution depend continuously upon the data of the problem?

The answers to these questions depend very much on the kind of equation studied, and on any additional conditions which are imposed. To illustrate this it is helpful to study certain specific equations in the plane $\mathbb{R}^2$.

EXAMPLE 8.1.1.   The wave equation

$$D_1^2 u - D_2^2 u = 0$$

has infinitely many solutions of the form

$$u(x_1, x_2) = f(x_1 + x_2) + g(x_1 - x_2),$$

f and g being any suitably smooth functions [see § 2.10]. If we require that u and $D_2 u$ should take on prescribed smooth values when $x_2 = 0$, say $u_0(x_1)$ and $u_1(x_1)$ respectively, then any solution of the form given above must satisfy

$$f(x_1) + g(x_1) = u_0(x_1), \qquad f'(x_1) - g'(x_1) = u_1(x_1),$$

and hence [see § 7.1]

$$f(x_1) = \tfrac{1}{2}\{u_0(x_1) + u_2(x_1)\}, \qquad g(x_1) = \tfrac{1}{2}\{u_0(x_1) - u_2(x_1)\},$$

where

$$u_2(x_1) = \int_0^{x_1} u_1(y) \, dy.$$

A solution of the problem is thus

$$u(x_1, x_2) = \tfrac{1}{2}\{u_0(x_1 + x_2) + u_2(x_1 + x_2)\} + \tfrac{1}{2}\{u_0(x_1 - x_2) - u_2(x_1 - x_2)\},$$

and it can be shown that this solution is unique. It clearly depends continuously on the given functions u_0 and u_1.

EXAMPLE 8.1.2. Consider Laplace's equation in an infinite strip:

$$D_1^2 u + D_2^2 u = 0 \quad \text{for } 0 < x_1 < 1,\ x_2 > 0.$$

Again there are infinitely many solutions—just take $\sin ax_1 \cosh ax_2$ for any real a, for example [see § 2.12 and § 2.13] but if we impose the boundary conditions

$$\left. \begin{array}{l} u(x_1, 0) = \exp\left(-\sqrt{n}\right) \sin\left(4n+1\right)\pi x_1 \\[4pt] D_2 u(x_1, 0) = 0 \end{array} \right\} \quad \text{for } 0 \le x_1 \le 1,$$

$$u(0, x_2) = u(1, x_2) = 0 \quad \text{for } x_2 \ge 0,$$

n being any positive integer, then it can be shown that there is a unique solution, namely,

$$u_n(x_1, x_2) = \exp\left(-\sqrt{n}\right) \sin\left(4n+1\right)\pi x_1 \cosh\left(4n+1\right)\pi x_2.$$

However, by choosing n large enough, the given boundary function $\exp\left(-\sqrt{n}\right) \sin\left(4n+1\right)\pi x_1$ (and any finite number of its derivatives) can be made as small as we please, while

$$u_n(\tfrac{1}{2}, x_2) = \exp\left(-\sqrt{n}\right) \cosh\left(4n+1\right)\pi x_2$$

becomes arbitrarily large for any fixed $x_2 > 0$ if n is chosen sufficiently big. Small changes in boundary data thus give rise to large changes in the solution, and the solution does not depend continuously on the boundary data.

In both of these examples it will be noticed that there are infinitely many solutions of the partial differential equation when no additional conditions are imposed. This is consistent with the belief held for many years that, for linear equations with smooth coefficients at least, there always will be solutions provided that no side conditions have to be fulfilled. The reasonableness of such a belief is suggested by the fact that for linear *ordinary* differential equations with smooth coefficients, solutions really do always exist [see § 7.3]. However, Hans Lewy (1957) destroyed this belief when he gave an example of a linear partial differential equation, with very well-behaved coefficients, but which has no solutions at all in any open set Ω. The equation is (with $n = 3$)

$$D_1 u + iD_2 u - 2i(x_1 + ix_2)D_3 u = f,$$

where f is a certain infinitely differentiable real-valued function [see Definition 2.10.2]. The position is much better for linear equations with constant coefficients, for it can be shown that if $P(D) = \sum_{|\alpha| \le m} a_\alpha D^\alpha$ and f is any given infinitely differentiable function which vanishes outside some compact subset of $\mathbb{R}^n$ [see V, § 5.2], then there is an infinitely differentiable function u such that $P(D)u = f$ throughout $\mathbb{R}^n$. Even for linear equations with variable coefficients substantial progress has been made since Lewy found his surprising example, and the work of Hörmander and others has led to the discovery of classes of such equations for which one can be certain that solutions really exist. For further details of all these questions we refer to Hörmander (1963) and Nirenberg (1973).

We conclude this section with a brief account of the most natural generalization of the initial-value problems for ordinary differential equations [see § 7.1]. This is the *Cauchy problem*, and to express this in the most suggestive form we suppose that one of the co-ordinates, say x_n, is singled out and written as t, the remaining co-ordinates being represented by $\hat{\mathbf{x}} = (x_1, \ldots, x_{n-1})$. We shall also write D_t for $\partial/\partial t$, while $D_{\hat{\mathbf{x}}}^{\hat{\alpha}}$ (with $\hat{\alpha} = (\alpha_1, \ldots, \alpha_{n-1})$) will stand for the derivative of order $|\hat{\alpha}|$ in the variables $x_1, \ldots, x_{n-1}$. The Cauchy problem asks whether there is a solution u of the problem

$$\left. \begin{aligned} D_t^k u(\mathbf{x}) &= G(\hat{\mathbf{x}}, t, (D_{\hat{\mathbf{x}}}^{\hat{\alpha}} D_t^j u(\mathbf{x}))_{|\hat{\alpha}|+j \leq k, j < k}), \\ D_t^j u(\hat{\mathbf{x}}, 0) &= \phi_j(\hat{\mathbf{x}}) \qquad (0 \leq j < k) \end{aligned} \right\} \tag{8.1.7}$$

the 'initial data' $\phi_0, \ldots, \phi_{k-1}$ being given on some subset of the hyperplane $t = 0$. The basic result concerning (8.1.7) is the *Cauchy–Kovalevsky theorem*:

THEOREM 8.1.1.   *If $G, \phi_0, \ldots, \phi_{k-1}$ are analytic near the origin* [see § 9.1] *then* (8.1.7) *has a unique analytic solution defined in some neighbourhood of the origin.*

Problem (8.1.7) is but a special case of the general Cauchy problem, in which data are given not necessarily on a hyperplane but on an *analytic hypersurface S*, by which we simply mean that near any point $\mathbf{x}_0 \in S$ the points of S satisfy an equation of the form

$$x_i = \psi(x_1, \ldots, x_{i-1}, x_{i+1}, \ldots, x_n)$$

for some i, $1 \leq i \leq n$, ψ being an analytic function. For example, a sphere is an analytic hypersurface. The Cauchy problem in its general form then asks for a function u such that

$$\left. \begin{aligned} F(\mathbf{x}, (D^\alpha u(\mathbf{x}))_{|\alpha| \leq k}) &= 0, \\ D_\nu^j u(\mathbf{x}) &= \phi_j(\mathbf{x}) \quad \text{for all } \mathbf{x} \in S, \text{ and all } j, 0 \leq j < k. \end{aligned} \right\} \tag{8.1.8}$$

Here $D_\nu u$ is the *normal derivative* of u on S [see Definition 6.3.3], and the functions $F, \phi_0, \ldots, \phi_{k-1}$ are given and analytic; the functions ϕ_j are said to form the *Cauchy data* of u on S. The equation

$$F(\mathbf{x}, (D^\alpha u(\mathbf{x}))_{|\alpha| \leq k}) = 0$$

reduces to the first equation in (8.1.7) provided that the equation $F = 0$ can be solved for $D_t^k u$ to give $D_t^k u$ as an analytic function G of the remaining variables. This condition of solubility is the so-called *non-characteristic condition*, and we can illustrate its significance by consideration of the case in which F is *affine*, that is, using the notation introduced above for (8.1.7),

$$F(\hat{\mathbf{x}}, t, (D_{\hat{\mathbf{x}}}^{\hat{\alpha}} D_t^j u(\mathbf{x}, t))_{|\hat{\alpha}|+j \le k}) = \sum_{|\hat{\alpha}|+j \le k} a_{\hat{\alpha}j}(\hat{\mathbf{x}}, t) D_{\hat{\mathbf{x}}}^{\hat{\alpha}} D_t^j u - f(\hat{\mathbf{x}}, t)$$

$$= Lu - f,$$

say; while S contains the origin and near the origin coincides with the hyperplane $t = 0$.

DEFINITION 8.1.2. The *principal symbol* of L at $\mathbf{x}$ is defined to be

$$\sigma_{\mathbf{x}}(L, \boldsymbol{\xi}) = \sum_{|\hat{\alpha}|+j = k} a_{\hat{\alpha}j}(\hat{\mathbf{x}}, t)(\hat{\boldsymbol{\xi}})^{\hat{\alpha}} \xi_n^j,$$

where $\boldsymbol{\xi} = (\xi_1, \ldots, \xi_n)$, $\hat{\boldsymbol{\xi}} = (\xi_1, \ldots, \xi_{n-1})$, and each ξ_i $(1 \le i \le n)$ is real. A vector $\boldsymbol{\xi}$ is called *characteristic for L at* $\mathbf{x}$ if $\sigma_{\mathbf{x}}(L, \boldsymbol{\xi}) = 0$, and the hypersurface S above is said to be characteristic for L at $\mathbf{x}$ if the normal $\nu(\mathbf{x})$ to S at $\mathbf{x}$ is a characteristic vector; S is *non-characteristic* if it is not characteristic at any point.

We see that if S is non-characteristic then $a_{0k}(\hat{\mathbf{x}}, 0) \ne 0$ near the origin, and thus by continuity $a_{0k}(\hat{\mathbf{x}}, t) \ne 0$ for all sufficiently small t. Because of this we can solve the equation $F = 0$ for $D_t^k u$, and obtain

$$D_t^k u = a_{0k}^{-1} \left[\sum_{|\hat{\alpha}|+j \le k, j < k} a_{\hat{\alpha}j} D_{\hat{\mathbf{x}}}^{\hat{\alpha}} D_t^j u - f \right].$$

The Cauchy–Kovalevsky theorem may now be applied to this form of the equation.

If the non-characteristic condition is not satisfied, various difficulties may emerge. For example, the line $t = 0$ is characteristic for the equation $D_x D_t u = 0$ in $\mathbb{R}^2$, and if u is a solution of this equation with $u(x, 0) = \phi_0(x)$, $D_t u(x, 0) = \phi_1(x)$, then $D_x \phi_1 = 0$, so that ϕ_1 must be constant. Hence the Cauchy problem is not soluble unless ϕ_1 is constant. But even if ϕ_1 is constant, uniqueness is lost, for any function of the form $\phi_0(x) + f(t)$, where $f'(0) = \phi_1$, is a solution of the problem.

We shall return to the matter of characteristics later, when we discuss hyperbolic equations. For the moment, let us note some important facts about the Cauchy–Kovalevsky theorem:

  (i) It is a *local* existence theorem only; that is, it gives a solution only in some neighbourhood of any point on S. These local solutions can be patched together to give a solution in a neighbourhood of S.
  (ii) It is of limited use, because the solution need not depend continuously on the data.
  (iii) It is false if the hypothesis of analyticity is weakened: this is shown by the Lewy example.

(iv) It leaves open the question as to whether there might be *non-analytic* solutions of the equations. If the equation is linear, a theorem of Holmgren (see, for example, Trèves (1975)) shows that there cannot be any non-analytic solutions.

For a more detailed discussion of all these topics we refer to the books by Hörmander (1963) and Trèves (1975).

8.2.   THE LAPLACE OPERATOR

We now leave the general operators discussed in section 8.1 and consider more specific operators, because it is from a study of these special cases that we obtain a feeling for what might be possible in more general circumstances. In this section we deal with Laplace's operator

$$\Delta_n = \sum_{k=1}^{n} D_k^2,$$

[sometimes denoted ∇^n—see III, § 9.5] and give some of the principal results which relate to solutions of Laplace's equation $\Delta_n u = 0$, together with a few of the many methods of solution of the equation.

DEFINITION 8.2.1.   Let Ω be an open subset of $\mathbb{R}^n$ [see Definition 5.12.1]. A real-valued function u which is twice continuously differentiable in Ω (we shall henceforth abbreviate this by writing $u \in C^k(\Omega)$ whenever u is k times continuously differentiable in Ω [see Definition 5.9.2]) is said to be *harmonic* in Ω if $\Delta_n u = 0$ in Ω [see also § 9.13].

One of the most basic properties of harmonic functions is the *mean-value property*: the value of a harmonic function at a point $\mathbf{x}$ is equal to the mean of its values on any small enough sphere or ball centred at $\mathbf{x}$. More formally, let $B(\mathbf{x}, \delta)$ be the open ball with centre $\mathbf{x}$ and radius δ, that is, $B(\mathbf{x}, \delta) = \{\mathbf{y} \in \mathbb{R}^n : |\mathbf{x} - \mathbf{y}| < \delta\}$ where $|\mathbf{z}| = (\sum_{k=1}^{n} z_k^2)^{1/2}$; let $\overline{B(\mathbf{x}, \delta)}$ be the corresponding closed ball, and denote by $\partial B(\mathbf{x}, \delta)$ the boundary of $B(\mathbf{x}, \delta)$ [see § 11.2]. The result mentioned above is:

THEOREM 8.2.1   (*Mean Value Theorem*).   Let u be harmonic in Ω, let $x \in \Omega$ and let $\delta > 0$ be so small that $\overline{B(\mathbf{x}, \delta)} \subset \Omega$. Then

$$u(\mathbf{x}) = \frac{1}{\omega_n \delta^{n-1}} \int_{\partial B(\mathbf{x},\delta)} u(\mathbf{y}) \, d\sigma(\mathbf{y}) = \frac{n}{\omega_n \delta^n} \int_{B(\mathbf{x},\delta)} u(\mathbf{z}) \, d\mathbf{z},$$

where $\omega_n = 2\pi^{n/2}/\Gamma(n/2)$ is the area of the unit sphere in $\mathbb{R}^n$, Γ is the gamma function [see § 10.2.1], and $d\sigma(\mathbf{y})$ is the surface measure [see Definition 6.3.5].

The proof of this is comparatively simple, and merely involves the use of Green's theorem [see Theorem 6.2.4]. A little more work shows that in fact the mean value property *characterizes* harmonic functions: any function u which is continuous in Ω and has the mean value property at all points of Ω is actually harmonic in Ω, from which it follows fairly quickly that harmonic functions are infinitely differentiable and even analytic. Another direct consequence of the mean value theorem is the celebrated *maximum principle*.

THEOREM 8.2.2 (*Maximum Principle*). *Let u be harmonic in Ω and suppose that there is a point $\mathbf{x}_0 \in \Omega$ such that $u(\mathbf{x}_0) = \sup_{\mathbf{x} \in \Omega} u(\mathbf{x})$ [see I, § 2.6.3]. Then if Ω is connected, u is constant on Ω.*

Of course, the same conclusion holds if u takes on its *minimum* at some point of Ω. The hypothesis that Ω is connected is essential, for if we take Ω to be the union of two disjoint open balls, the function u which is identically zero in one of the balls and equal to 1 throughout the other ball satisfies the hypotheses of the theorem but is not constant throughout Ω.

The maximum principle has many important consequences, some of which we give below; $\bar{\Omega}$ and $\partial\Omega$ will stand for the closure and boundary of Ω respectively [see § 11.2]:

COROLLARY 8.2.3. *Let Ω be bounded and connected [see V, § 5.2], let u be harmonic in Ω and continuous in $\bar{\Omega}$. Then the maximum and minimum values of u in $\bar{\Omega}$ are always assumed on $\partial\Omega$.*

This implies that if such a function is constant on $\partial\Omega$, it is constant throughout Ω.

COROLLARY 8.2.4. *The Dirichlet problem for Poisson's equation $\Delta_n u = f$ in a bounded domain (an open, connected set) Ω has at most one solution.*

By the *Dirichlet problem* we mean the question that asks whether there is a function $u \in C^2(\Omega)$ which is continuous on $\bar{\Omega}$ and such that $\Delta_n u = f$ in Ω, $u = g$ on $\partial\Omega$, where f and g are given continuous functions. The uniqueness of solutions follows immediately; for if u and v are two solutions, then $\Delta_n(u - v) = 0$ in Ω and $u - v = 0$ on $\partial\Omega$, so that by (i), $u = v$ throughout Ω. Note that the boundedness of Ω is crucial here: take Ω to be the infinite strip $\{(x_1, x_2) : x_1 \in \mathbb{R}, 0 < x_2 < \pi\}$ in the plane, and consider the problem $\Delta_n u = 0$ in Ω, $u = 0$ on $\partial\Omega$. This has two distinct solutions $u_1 = 0$, $u_2(x_1, x_2) = \exp x_1 \sin x_2$. However, the Dirichlet problem in an unbounded domain Ω will have at most one solution if suitable restrictions are made on the growth of the solution; this is the content of theorems of *Phragmén–Lindelöf* type, for which we refer to the book by Protter and Weinbeger (1967).

COROLLARY 8.2.5. *Let u be bounded and harmonic in $\mathbb{R}^n$. Then u is constant.*

This is *Liouville's theorem*, which can be improved so that u need be bounded only from above or from below.

To obtain further information it is convenient to study a certain special solution of Laplace's equation. It is easy to verify that the function u defined by

$$u(\mathbf{x}) = \begin{cases} |\mathbf{x}|^{2-n}, & \text{if } n > 2, \\ \log |\mathbf{x}|, & \text{if } n = 2, \end{cases}$$

is harmonic throughout the whole of $\mathbb{R}^n$ except at 0; it is, of course, singular at the origin [see § 9.8]. For technical reasons we introduce a multiple K of this function, namely

$$K(\mathbf{x}) = \begin{cases} -(n-2)^{-1}\omega_n^{-1}|\mathbf{x}|^{2-n}, & \text{if } n > 2, \\ (2\pi)^{-1} \log |\mathbf{x}|, & \text{if } n = 2, \end{cases} \tag{8.2.1}$$

and call it the *fundamental solution* of Laplace's equation. One of the most important properties of K is that $\Delta K = \delta$, the *Dirac 'δ-function'*, by which we mean that

$$\int_{\mathbb{R}^n} K(\mathbf{x})\Delta_n \phi(\mathbf{x}) \, d\mathbf{x} = \phi(0)$$

for all infinitely differentiable functions ϕ which are identically zero outside some compact subset of $\mathbb{R}^n$. (The function ϕ defined by

$$\phi(\mathbf{x}) = \begin{cases} \exp\left(\dfrac{1}{|\mathbf{x}|^2 - 1}\right) & \text{if } |\mathbf{x}| < 1, \\ 0 & \text{if } |\mathbf{x}| \geq 1 \end{cases}$$

is such a function.) Using this we obtain a method by which solutions of Poisson's equation $\Delta_n u = f$ in $\mathbb{R}^n$ may be constructed. All that has to be done is to take u to be the *convolution* $f * K$ of f and K; the convolution is the function defined by

$$(f * K)(\mathbf{x}) = \int_{\mathbb{R}^n} f(\mathbf{y})K(\mathbf{x} - \mathbf{y}) \, d\mathbf{y}, \qquad \mathbf{x} \in \mathbb{R}^n. \tag{8.2.2}$$

It turns out that, under suitable conditions on f, $f * K$ is a twice continuously differentiable function which satisfies Poisson's equation in an open set Ω; it is enough to suppose that $\int_{\mathbb{R}^n} |f(\mathbf{y})| \, d\mathbf{y} < \infty$, that f should be *Hölder-continuous* in Ω (that is, f should satisfy an inequality of the form

$$|f(\mathbf{x}) - f(\mathbf{y})| \leq C|\mathbf{x} - \mathbf{y}|^\lambda \tag{8.2.3}$$

for some λ, $0 < \lambda \leq 1$), and that (if $n = 2$)

$$\int_{|\mathbf{y}| > 1} |f(\mathbf{y})| \log |\mathbf{y}| \, d\mathbf{y} < \infty.$$

Naturally, the more smooth f is made, the smoother $f * K$ becomes, and if instead of the Hölder condition we require that f be infinitely differentiable in Ω, then $f * K$ is infinitely differentiable in Ω.

We thus have a means of constructing a solution of Poisson's equation. However, most commonly what is needed is a solution which satisfies certain boundary conditions, that is, conditions to be fulfilled on $\partial\Omega$. We have already met the Dirichlet problem, in which the values of the solution are prescribed on the boundary of Ω; another frequently occurring problem is the *Neumann problem*, which requires the normal derivative of the solution to take on specified values on $\partial\Omega$. More formally, the Neumann problem asks for a function $u \in C^2(\Omega) \cap C(\bar\Omega)$ such that

$$\Delta_n u = f \quad \text{in } \Omega, \qquad D_\nu u = g \quad \text{on } \partial\Omega$$

where f and g are given continuous functions. Unlike the Dirichlet problem, the Neumann problem does not have a unique solution, for if u is a solution so is $u + c$, where c is any constant. Since by Green's theorem [see Theorem 6.2.4], if $\partial\Omega$ is smooth enough,

$$\int_\Omega f(x)\, dx = \int_\Omega \Delta_n(u)(x)\, dx = \int_{\partial\Omega} \frac{\partial u}{\partial\nu}\, d\sigma = \int_{\partial\Omega} g\, d\sigma$$

we also see that the Neumann problem will have no solution at all unless f and g are *compatible* in the sense that

$$\int_\Omega f(x)\, dx = \int_{\partial\Omega} g\, d\sigma.$$

The Dirichlet and Neumann problems occur very often in practical situations; and so do *intermediate* or *mixed* problems, in which the unknown function u may be specified on part of the boundary $\partial\Omega$, while $\partial u/\partial\nu$ is given on the remainder of $\partial\Omega$; or problems in which a linear combination of u and $\partial u/\partial\nu$ is specified on $\partial\Omega$. For this reason we shall devote a good deal of space to methods of finding a solution of such problems.

First note that the Dirichlet problem

$$\Delta_n u = f \quad \text{in } \Omega, \qquad u = g \quad \text{on } \partial\Omega \tag{8.2.4}$$

can be reduced to the case where either f or g is zero, for if u_1 and u_2 are functions in $C^2(\Omega) \cap C(\bar\Omega)$ such that

$$\Delta_n u_1 = f \quad \text{in } \Omega, \qquad u_1 = 0 \quad \text{on } \partial\Omega, \tag{8.2.5}$$

$$\Delta_n u_2 = 0 \quad \text{in } \Omega, \qquad u_2 = g \quad \text{on } \partial\Omega, \tag{8.2.6}$$

then clearly $u_1 + u_2$ is a solution of (8.2.4). Moreover, these two simpler problems (8.2.5) and (8.2.6) are virtually equivalent, for if we can solve (8.2.4) then provided that f and $\partial\Omega$ are smooth enough we can extend g to a function G which is in $C^2(\bar\Omega)$, and then find the function u_1 which satisfies

$$\Delta_n u_1 = \Delta_n G \quad \text{in } \Omega, \qquad u_1 = 0 \quad \text{on } \partial\Omega.$$

Then $G - u_1$ solves (8.2.6). Conversely, if we can solve (8.2.6) and wish to solve (8.2.5), define a function F by

$$F(x) = \begin{cases} f(x), & x \in \Omega \\ 0, & \text{otherwise,} \end{cases}$$

and let u_2 be the solution of

$$\Delta_n u_2 = 0 \quad \text{in } \Omega, \qquad u_2 = F * K \quad \text{on } \partial\Omega,$$

K being the fundamental solution of Laplace's equation. Then $F * K - u_2$ solves (8.2.5). Similar remarks apply to the Neumann problem.

There still remains the matter of obtaining an explicit solution of any one of the boundary-value problems mentioned. To do this requires a good deal of co-operation from the equations, the domain Ω and the boundary data; and given this, one of the most frequently used methods is that of *separation of variables*, by which is meant that we look for a solution u in the form $u(x_1, \ldots, x_n) = u_1(x_1) \ldots u_n(x_n)$. The procedure is best illustrated by examples.

EXAMPLE 8.2.1. Suppose that we wish to find a solution of Laplace's equation in the unit square Q, the values of the solution being prescribed on the boundary of Q. We thus need a function u which satisfies

$$\left. \begin{array}{l} D_1^2 u + D_2^2 u = 0 \quad \text{when } 0 < x_1 < 1 \text{ and } 0 < x_2 < 1, \\[4pt] u(x_1, 0) = f_1(x_1), \qquad u(x_1, 1) = f_2(x_1) \quad \text{when } 0 \leq x_1 \leq 1, \\[4pt] u(0, x_2) = f_3(x_2), \qquad u(1, x_2) = f_4(x_2) \quad \text{when } 0 \leq x_2 \leq 1, \end{array} \right\} \quad (8.2.7)$$

the functions f_1, f_2, f_3 and f_4 being given, and (for simplicity) supposed zero at the appropriate corners of the square. It is clear that if we can find a solution of the problem

$$\left. \begin{array}{l} D_1^2 u + D_2^2 u = 0 \quad \text{for } 0 < x_1 < 1 \text{ and } 0 < x_2 < 1, \\[4pt] u(x_1, 0) = f_1(x_1) \quad \text{for } 0 \leq x_1 \leq 1, \\[4pt] u(x_1, x_2) = 0 \quad \text{on the other three sides of the square,} \end{array} \right\} \quad (8.2.8)$$

and of the three similar problems in each of which u is required to be equal to f_i on the appropriate side of the square and to be zero on the remaining three sides, then by adding all four solutions we obtain a solution of (8.2.7). (This method of addition of solutions of simpler problems to obtain a solution of a more complicated problem is known as the *principle of superposition*; it clearly is possible only for linear equations.)

To find a solution of (8.2.8) we try the separation of variables method; that is, look for a solution u of the form $u(x_1, x_2) = X_1(x_1)X_2(x_2)$. Then $X_1'' X_2 + X_1 X_2'' = 0$ inside the square, dashes denoting differentiation with respect to the appropriate variable. At points at which u is non-zero we therefore have

$$X_1''/X_1 = -X_2''/X_2,$$

and since the left-hand side is a function of x_1 only while the right-hand side depends on x_2 alone, both must be constant. Thus for some constant λ,

$$X_1'' + \lambda X_1 = 0, \qquad X_2'' - \lambda X_2 = 0.$$

Since the boundary conditions require that $X_1(x_1)$ be zero when $x_1 = 0$ and when $x_1 = 1$, we conclude that $\lambda > 0$ and that $X_1(x_1)$ is a multiple of $\sin(\sqrt{\lambda}x_1)$, where $\lambda = n^2\pi^2$ for some positive integer n [compare Example 7.4.1]. (These values of λ are called the *eigenvalues* of the boundary-value problem for X_1, the corresponding functions $\sin n\pi x$ being *eigenfunctions* of that problem.) As for X_2, we see that since it must satisfy $X_2'' = n^2\pi^2 X_2$, $X_2(1) = 0$, it must be a multiple of $\sinh n\pi(1 - x_2)$. It follows that all the conditions of (8.2.8) save the one concerning $u(x_1, 0)$ are satisfied by the function $\sin n\pi x_1 \sinh n\pi(1 - x_2)$ for any positive integer n, and of course by any finite linear combinations of such functions; it remains to satisfy this last condition. To do this we try to represent u as an infinite series of functions of this kind, and tentatively write

$$u(x_1, x_2) = \sum_{n=1}^{\infty} a_n \sin n\pi x_1 \sinh n\pi(1 - x_2). \qquad (8.2.9)$$

The condition $u(x_1, 0) = f_1(x_1)$ leads to

$$f_1(x_1) = \sum_{n=1}^{\infty} a_n \sin n\pi x_1 \sinh n\pi,$$

so that $a_n \sinh n\pi$ must be the nth Fourier coefficient b_n of f_1 in a Fourier sine series expansion [see (20.5.2)]; that is,

$$b_n = 2\int_0^1 f_1(x) \sin \pi n x \, dx.$$

We therefore have a *formal* solution of the problem; it is merely formal at this stage because it naturally remains to verify that the convergence of the series (8.2.9) is good enough for the function so defined to satisfy the differential equation, and we also should investigate whether or not the function u takes on the boundary values continuously. If f_1 is continuous on $[0, 1]$ it can be shown that (8.2.9) gives a solution of the problem (8.2.8) which is continuous on $\bar{Q}$; it can be represented as

$$u(x_1, x_2) = \int_0^1 K(\xi, x_1, x_2)f_1(\xi) \, d\xi,$$

where

$$K(\xi, x_1, x_2) = 2\sum_{n=1}^{\infty} \frac{\sinh n\pi(1 - x_2)}{\sinh n\pi} \sin n\pi x_1 \sin n\pi\xi,$$

the interchange of summation and integration being justified by uniform convergence [see § 1.12]. The maximum principle may be used to show this is the only solution of (8.2.8) which is continuous on $\bar{Q}$. In fact, the conditions on f_1 can be weakened; for example, if f_1 is bounded and continuous except at a finite number of points, then (8.2.9) gives a function which is harmonic on Q and which, when extended to $x_2 = 0$ by setting $u(x_1, 0) = f_1(x_1)$, is continuous at points of continuity of f_1. Use of the *Phragmén–Lindelöf principle* (see Weinberger (1965), p. 107) gives uniqueness of this solution.

The use of the method of separation of variables is not confined to problems expressed in terms of rectangular Cartesian co-ordinates $x_1, x_2, \ldots x_n$:

EXAMPLE 8.2.2. Suppose we have to solve Laplace's equation in the unit disc in the plane, that is, in $\{(x_1, x_2): x_1^2 + x_2^2 < 1\}$, the values of the solution being prescribed on the boundary of the disc. It is natural to express the equation in terms of plane polar coordinates r and θ [see V, § 1.2.5], and if we do this it takes the form

$$\frac{\partial^2 u}{\partial r^2} + \frac{1}{r}\frac{\partial u}{\partial r} + \frac{1}{r^2}\frac{\partial^2 u}{\partial \theta^2} = 0. \tag{8.2.10}$$

We now look for solutions in the form $u(r, \theta) = R(r)\Theta(\theta)$, and with dashes denoting differentiation with respect to the appropriate variables the equation becomes

$$r(rR'' + R')/R = -\Theta''/\Theta.$$

As before, we conclude that for some constant λ,

$$r(rR'' + R')/R = \lambda = -\Theta''/\Theta.$$

Because we are trying to solve the problem in the disc we are interested in finding solutions which are periodic in θ with period 2π, and we accordingly look for solutions of $\Theta'' + \lambda\Theta = 0$ in the interval $(-\pi, \pi)$ which have this periodicity. This is possible if and only if $\lambda = n^2$ ($n = 0, 1, 2, \ldots$), and corresponding to these eigenvalues are eigenfunctions $\cos n\theta$ and $\sin n\theta$ [see Example 8.2.1]. The equation for R now becomes

$$r^2 R'' + rR' - n^2 R = 0,$$

which has solutions $R(r) = a + b \log r$ if $n = 0$, $R(r) = ar^n + br^{-n}$ otherwise, a and b being arbitrary constants [see § 7.5]. The terms involving $\log r$ or r^{-n} blow up as the origin $r = 0$ is approached, and therefore have to be excluded if we require a solution which is continuous in the disc. (Terms of this sort would, of course, be retained if we had to solve the equation in an annulus around the origin.) We thus have solutions $r^n \sin n\theta$ and $r^n \cos n\theta$ of Laplace's equation in the disc, and to find a solution of the equation which satisfies the condition

$$u(1, \theta) = f(\theta)$$

on the boundary of the disc, f being a given continuous function, we try a series of the form

$$u(r, \theta) = \tfrac{1}{2}a_0 + \sum_{n=1}^{\infty} r^n(a_n \cos n\theta + b_n \sin n\theta).$$

This leads to consideration of the equation [see (20.4.22)]

$$f(\theta) = \tfrac{1}{2}a_0 + \sum_{n=1}^{\infty} (a_n \cos n\theta + b_n \sin n\theta),$$

from which we see that a_n and b_n should be the Fourier coefficients of f:

$$a_n = \frac{1}{\pi} \int_{-\pi}^{\pi} f(\phi) \cos n\phi \, d\phi, \qquad b_n = \frac{1}{\pi} \int_{-\pi}^{\pi} f(\phi) \sin n\phi \, d\phi.$$

A short computation shows that with these values of a_n and b_n, not only do we obtain a solution of the Dirichlet problem for Laplace's equation which is continuous in the closed disc, but also that in the open disc (that is, when $r < 1$) the solution is given by ·

$$u(r, \theta) = \frac{1 - r^2}{2\pi} \int_{-\pi}^{\pi} \frac{f(\phi)}{1 + r^2 - 2r \cos (\theta - \phi)} \, d\phi. \qquad (8.2.11)$$

This is *Poisson's integral formula* for the solution of the Dirichlet problem [see (8.2.17)]. In a disc of radius r_0 exactly the same process gives

$$u(r, \theta) = \frac{r_0^2 - r^2}{2\pi} \int_{-\pi}^{\pi} \frac{f(\phi)}{r_0^2 + r^2 - 2r_0 r \cos (\theta - \phi)} \, d\phi.$$

We have seen in these last two examples, each of which is set in the plane, how the method of separation of variables may lead to Fourier series. In higher dimensions, multiple Fourier series may appear [cf. § 20.7]:

EXAMPLE 8.2.3.   If we try to solve a Dirichlet problem for Laplace's equation in a cube, say

$$D_1^2 u + D_2^2 u + D_3^2 u = 0 \quad \text{for } 0 < x_i < \pi \ (i = 1, 2, 3),$$

$$u = 0 \quad \text{on the faces } x_1 = 0, \, x_1 = \pi, \, x_2 = 0, \, x_2 = \pi, \, x_3 = \pi$$

$$u(x_1, x_2, 0) = f(x_1, x_2),$$

and use the method of separation of variables, then we arrive at a formal solution

$$u(x_1, x_2, x_3) = \sum_{n=1}^{\infty} \sum_{m=1}^{\infty} a_{mn} \frac{\sinh \{(\pi - x_3)\sqrt{m^2 + n^2}\}}{\sinh \pi \sqrt{m^2 + n^2}} \sin mx_1 \sin nx_2,$$

where the a_{mn} are the Fourier coefficients of f:

$$a_{mn} = \frac{4}{\pi^2} \int_0^{\pi} \int_0^{\pi} f(x_1, x_2) \sin mx_1 \sin nx_2 \, dx_1 \, dx_2.$$

Subject to suitable conditions on f it can be verified that this formal solution is a genuine solution of the problem.

Laplace's equation in a cylinder or ball may be solved by the same process:

EXAMPLE 8.2.4.   Suppose that we have the cylindrical case, for which it is obviously a good idea to express the Laplacian in cylindrical polar coordinates (r, θ, z). If the resulting expression is independent of θ it is called an *axisymmetric* form of Laplace's equation [see V, § 2.1.5]. Let us consider the following Dirichlet problem:

$$\left.\begin{array}{l} \dfrac{\partial^2 u}{\partial r^2} + \dfrac{1}{r}\dfrac{\partial u}{\partial r} + \dfrac{1}{r^2}\dfrac{\partial^2 u}{\partial \theta^2} + \dfrac{\partial^2 u}{\partial z^2} = 0 \quad \text{for } 0 < r < r_0,\, 0 < z < \pi, \\[2mm] u(r, \theta, 0) = u(r, \theta, \pi) = 0 \quad \text{for } 0 \le r \le r_0,\, 0 \le \theta \le 2\pi, \\[2mm] u(r_0, \theta, z) = f(\theta, z) \quad \text{for } 0 \le \theta \le 2\pi,\, 0 \le z \le \pi, \end{array}\right\} \quad (8.2.12)$$

f being a given function. If we look for a solution in the form $R(r)\Theta(\theta)Z(z)$ we rapidly find that $\Theta(\theta)$ is of the form $\sin m\theta$ or $\cos m\theta$ $(m = 0, 1, 2, \ldots)$ and that $Z(z)$ takes the form $\sin nz$ $(n = 1, 2, \ldots)$. As for R, it must satisfy the differential equation

$$R'' + \frac{1}{r}R' - \left(\frac{m^2}{r^2} + n^2\right)R = 0,$$

which is related to Bessel's equation of order m [see Example 7.6.5]:

$$\frac{d^2 v}{dt^2} + \frac{1}{i}\frac{dv}{dt} - \left(\frac{m^2}{t^2} - 1\right)v = 0.$$

If we impose the natural requirement that R should remain bounded as $r \to 0$, then we see that

$$R(r) = I_m(nr),$$

where I_m is the Bessel function with imaginary argument,

$$I_m(r) = \sum_{k=0}^{\infty} \frac{1}{k!\,(k+m)!}(r/2)^{m+2k}$$

[see § 10.4.1 and Weinberger (1965)]. As before it now follows that we take as solution of the Dirichlet problem the function u given by

$$u(r, \theta, z) = \frac{1}{2}\sum_{n=1}^{\infty} a_{n0}\frac{I_0(nr)}{I_0(nr_0)}\sin nz$$

$$+ \sum_{n=1}^{\infty}\sum_{m=1}^{\infty}\frac{I_m(nr)}{I_m(nr_0)}(a_{nm}\cos m\theta + b_{nm}\sin m\theta)\sin nz$$

where

$$a_{nm} = \frac{2}{\pi^2} \int_0^\pi \int_0^{2\pi} f(\theta, z) \sin nz \cos m\theta \, d\theta \, dz,$$

$$b_{nm} = \frac{2}{\pi^2} \int_0^\pi \int_0^{2\pi} f(\theta, z) \sin nz \sin m\theta \, d\theta \, dz.$$

If, for example, f is continuous for $0 \le \theta \le 2\pi$, $0 \le z \le \pi$ and $f(\theta, 0) = f(\theta, \pi) = 0$, then it can be verified that the above expression for u really is a solution of the problem and that it is continuous on the closed cylinder.

EXAMPLE 8.2.5.   In spherical polar co-ordinates, Laplace's equation becomes

$$\frac{\partial^2 u}{\partial r^2} + \frac{2}{r} \frac{\partial u}{\partial r} + \frac{1}{r^2 \sin \theta} \frac{\partial}{\partial \theta}\left(\sin \theta \frac{\partial u}{\partial \theta}\right) + \frac{1}{r^2 \sin^2 \theta} \frac{\partial^2 u}{\partial \phi^2} = 0, \qquad (8.2.13)$$

so that if we look for a solution in the form $u(r, \theta, \phi) = R(r)\Theta(\theta)\Phi(\phi)$ we find that $\Phi(\phi) = \sin m\phi$ or $\cos m\phi$ $(m = 0, 1, 2, \ldots)$, and that

$$r^2\left(R'' + \frac{2}{r} R'\right) - \lambda R = 0,$$

$$\frac{d}{d\theta}\left(\sin \theta \frac{d\Theta}{d\theta}\right) - \frac{m^2}{\sin \theta} \Theta + \lambda \Theta \sin \theta = 0,$$

for some constant λ. Put $t = \cos \theta$ and set $\Theta(\theta) = P(\cos \theta)$; the function P then satisfies the differential equation

$$\frac{d}{dt}\left[(1 - t^2) \frac{dP}{dt}\right] - \frac{m^2}{1 - t^2} P + \lambda P = 0 \quad \text{for } |t| < 1, \qquad (8.2.14)$$

and under natural conditions on P at the end-points $t = \pm 1$ (see Weinberger (1965), p. 189), we are led to take λ to be of the form $n(n + 1)$ $(n = 0, 1, 2, \ldots)$. Equation (8.2.14) then has a solution

$$P_n^{(m)}(t) = (1 - t^2)^{m/2} \frac{d^m}{dt^m} P_n(t),$$

where

$$P_n(t) = \frac{1}{2^n n!} \sum_{k=0}^n (-1)^{n-k} \binom{n}{k} \frac{(2k)!}{(2k - n)!} t^{2k-n} \qquad (8.2.15)$$

is the *Legendre polynomial* of degree n [see § 10.3.2]; the functions $P_n^{(m)}$ are called the *associated Legendre functions*. The corresponding function R is given by $R(r) = r^n$ or r^{-n-1}, and thus we have found the harmonic functions

$$\begin{matrix} r^n \\ r^{-n-1} \end{matrix} P_n^{(m)}(\cos \theta) \begin{matrix} \sin \\ \cos \end{matrix} m\phi$$

which may be used in series form as before to obtain a solution of boundary value problems for Laplace's equation in a ball or annular region, and to obtain a Poisson integral representation for the solution of the Dirichlet problem in a ball, just as in the two-dimensional case.

Other well-known special functions may turn up if we apply the method of separation of variables to equations involving the Laplace operator:

EXAMPLE 8.2.6. Consider the *Schrödinger eigenvalue problem* (in spherical polar coordinates r, θ, ϕ)

$$\Delta_n u + \frac{a}{r} u + \lambda u = 0 \quad \text{for } r > 0, \tag{8.2.16}$$

where u is continuous and bounded in $\mathbb{R}^3$, a is a positive constant and λ is the eigenvalue. Then, if we look for a solution of the form $R(r)\Theta(\theta)\Phi(\phi)$ we find that

$$R'' + \frac{2}{r} R' + \left(\lambda + \frac{a}{r} - \frac{n(n+1)}{r^2} \right) R = 0 \quad \text{for } r > 0.$$

The changes of variable $l = a/(2\sqrt{-\lambda})$, $t = 2r\sqrt{-\lambda}$, $R(r) = \tilde{R}(t)$ lead to

$$\tilde{R}'' + \frac{2}{t} \tilde{R}' + \left(-\frac{1}{4} + \frac{l}{t} - \frac{n(n+1)}{t^2} \right) \tilde{R} = 0,$$

and the condition that $\tilde{R}$ should be continuous and bounded turns out to imply that this is possible only if l is an integer, $l > n$, and that in that case,

$$\tilde{R}(t) = t^n e^{-t/2} \frac{d^{2n+1}}{dt^{2n+1}} L_{l+n}(t),$$

where L_m is the *Laguerre polynomial* given by

$$L_m(t) = \sum_{k=0}^{m} (-1)^k \binom{m}{k} m(m-1) \ldots (k+1) t^k,$$

[see § 10.5.1]. The *Hermite polynomials* [*see* § 10.5.2] also appear in connection with the Schrödinger operator (see Courant and Hilbert (1962)).

We have noticed that for the Dirichlet problem for Laplace's equation in a disc an integral representation of the solution was obtained. This leads us to consider the Green's function of such problems.

DEFINITION 8.2.2. The *Green's function* of the Laplace operator in a domain Ω is a function $G(\mathbf{x}, \mathbf{y})$ which, for all $\mathbf{x} \in \Omega$ and all $\mathbf{y} \in \bar{\Omega}$, satisfies the conditions:

(i) for each $\mathbf{x} \in \Omega$, the function (of $\mathbf{y}$) with values $G(\mathbf{x}, \mathbf{y}) - K(\mathbf{x} - \mathbf{y})$ is harmonic on Ω and continuous on $\bar{\Omega}$ (here K stands for the fundamental solution);

(ii) for each $\mathbf{x} \in \Omega$ and each $\mathbf{y} \in \partial\Omega$, $G(\mathbf{x}, \mathbf{y}) = 0$.

If such a function G exists it can be shown to be unique. It turns out that if Ω has a smooth enough boundary then G does exist and has the properties that $G(\mathbf{x}, \mathbf{y}) = G(\mathbf{y}, \mathbf{x})$ for all $\mathbf{x}, \mathbf{y} \in \Omega$, and that for each $\mathbf{x} \in \Omega$, $G(\mathbf{x}, \cdot)$ is infinitely differentiable on $\bar{\Omega} - \{\mathbf{x}\}$. The importance of the Green's function is that if it is known it is possible to construct a solution of the Dirichlet problem. For example, the solution of (8.2.5) is given by

$$(\tilde{f} * K)(\mathbf{x}) + \int_{\Omega} \{G(\mathbf{x}, \mathbf{y}) - K(\mathbf{x}, \mathbf{y})\} f(\mathbf{y}) \, d\mathbf{y},$$

where $\tilde{f}(\mathbf{x}) = f(\mathbf{x})$ if $\mathbf{x} \in \Omega$, $\tilde{f}(\mathbf{x}) = 0$ otherwise; and the solution of (8.2.6) is

$$\int_{\partial\Omega} g(\mathbf{y}) \frac{\partial}{\partial \nu_{\mathbf{y}}} G(\mathbf{x}, \mathbf{y}) \, d\sigma(\mathbf{y}), \tag{8.2.17}$$

this last expression being known as the *Poisson integral formula* for the solution of the Dirichlet problem.

Naturally the main question, from a practical point of view, is whether the Green's function can be found for a given domain Ω, and it turns out that in various standard situations this function can be determined.

EXAMPLE 8.2.7. When Ω is a ball, say $\Omega = B(0, 1) \subset \mathbb{R}^n$, G is given by

$$G(\mathbf{x}, \mathbf{y}) = \begin{cases} \omega_n^{-1}(2-n)^{-1}\left[|\mathbf{x}-\mathbf{y}|^{2-n} - \left| \dfrac{\mathbf{x}}{|\mathbf{x}|} - |\mathbf{x}|\mathbf{y} \right|^{2-n}\right] & \text{if } \mathbf{x} \neq 0, \\ \omega_n^{-1}(2-n)^{-1}[|\mathbf{y}|^{2-n} - 1] & \text{if } \mathbf{x} = 0, \end{cases}$$

when $n > 2$; if $n = 2$ we have

$$G(\mathbf{x}, \mathbf{y}) = \begin{cases} (2\pi)^{-1}\left[\log|\mathbf{x}-\mathbf{y}| - \log\left| \dfrac{\mathbf{x}}{|\mathbf{x}|} - |\mathbf{x}|\mathbf{y} \right| \right] & \text{if } \mathbf{x} \neq 0, \\ (2\pi)^{-1}\log|\mathbf{y}| & \text{if } \mathbf{x} = 0. \end{cases}$$

Use of this last expression in (8.2.17) provides the Poisson integral formula which was given earlier for the solution of the Dirichlet problem in the unit disc [see (8.2.11)].

The Green's function can be found fairly easily for the other special domains, such as boxes, annular rings and ellipsoids; see Courant and Hilbert (1962) I, pp. 377–387 for details. In general, however, the determination of the Green's function for a given domain is a difficult problem.

Going back to methods of solution of boundary-value problems involving the Laplacian, it should be pointed out that for two-dimensional problems a powerful tool is available in the shape of the theory of functions of a complex variable, or more precisely, conformal mappings [see § 9.12], the essential point being that harmonic functions go into harmonic functions under a conformal mapping.

EXAMPLE 8.2.8.   Suppose we wish to solve the problem

$$D_1^2 u + D_2^2 u = 0 \quad \text{in } \Omega,$$

$$u(x_1, x_2) = \phi(x_1, x_2) \quad \text{when } (x_1, x_2) \in \partial\Omega,$$

where Ω is a bounded domain in $\mathbb{R}^2$ and ϕ is a given, well-behaved function. If Ω has a pleasant shape we may be able to apply the method of separation of variables and so obtain a solution; but failing this we may be in some difficulty. Suppose, however, that we introduce the complex variable $z = x_1 + ix_2$, identify $\mathbb{R}^2$ with the complex plane, and assume that there is a conformal one-to-one map f of a domain Ω_1 onto Ω such that $f(\partial\Omega_1) = \partial\Omega$. In $\bar{\Omega}_1$ denote the points by (ξ_1, ξ_2) and introduce the complex variable $\zeta = \xi_1 + i\xi_2$. Then the conformal map is expressed by $z = f(\zeta)$, or $z = x_1(\xi_1, \xi_2) + ix_2(\xi_1, \xi_2)$. If we put $U(\xi_1, \xi_2) = u(x_1(\xi_1, \xi_2), x_2(\xi_1, \xi_2))$, the properties of conformal mappings ensure that

$$\frac{\partial^2 U}{\partial \xi_1^2} + \frac{\partial^2 U}{\partial \xi_2^2} = 0 \quad \text{in } \Omega_1,$$

and that U is prescribed on $\partial\Omega_1$. The point of all this is that if Ω_1 happens to be the kind of domain appropriate for the use of the method of separation of variables, such as a circular disc or the interior of a rectangle, then we can solve the problem in Ω_1, and then use the fact that there is an analytic function g inverse to f, that is, such that $f(g(z)) = z$, to obtain a solution $u(x_1, x_2) = U(\xi_1(x_1, x_2), \xi_2(x_1, x_2))$ of the original problem.

The same approach can be used to handle the Neumann problem, in which the normal derivative of u is specified on $\partial\Omega$. For examples illustrating in detail the use of this method we refer to Weinberger (1965).

Complex variable techniques can also be used to handle two-dimensional problems on *unbounded domains*.

EXAMPLE 8.2.9.   If we consider the Dirichlet problem in the exterior of a circle,

$$D_1^2 u + D_2^2 u = 0 \quad \text{for } x_1^2 + x_2^2 > 1,$$

$$u(x_1, x_2) = x_1 \quad \text{for } x_1^2 + x_2^2 = 1,$$

and require that u should be bounded, then introduction of the complex variable $z = x_1 + ix_2$ and use of the inversion $\zeta = 1/z$ transforms this problem into

$$\frac{\partial^2 U}{\partial \xi_1^2} + \frac{\partial^2 U}{\partial \xi_2^2} = 0 \quad \text{for } \xi_1^2 + \xi_2^2 < 1,$$

$$U(\xi_1, \xi_2) = \xi_1 \quad \text{for } \xi_1^2 + \xi_2^2 = 1,$$

where $U(\xi_1, \xi_2) = u(x_1(\xi_1, \xi_2), x_2(\xi_1, \xi_2))$. This latter problem has the unique solution $U(\xi_1, \xi_2) = \xi_1$, and so $u(x_1, x_2) = x_1/(x_1^2 + x_2^2)$ is a solution of the original problem. It is, in fact, the only *bounded* solution (the function $u(x_1, x_2) = x_1$ satisfies all the requirements of the problem save that of boundedness).

In a similar way, problems in half-planes can be solved by the use of appropriate bilinear maps. See Weinberger (1965) for examples.

For problems in three (or more) dimensions these important techniques are not available, although it is worth mentioning *Kelvin's inversion formula*, which enables problems in exterior domains to be reduced to problems in bounded domains. This rests on the observation that if $U(\xi_1, \xi_2, \xi_3)$ is harmonic then the function

$$\frac{1}{r} U\left(\frac{x_1 - a_1}{r^2}, \frac{x_2 - a_2}{r^2}, \frac{x_3 - a_3}{r^2}\right),$$

where a_1, a_2, a_3 are constants and $r^2 = (x_1 - a_1)^2 + (x_2 - a_2)^2 + (x_3 - a_3)^2$, is harmonic in x_1, x_2, x_3. In other words, the inversion $\xi_i = (x_i - a_i)/r^2$ $(i = 1, 2, 3)$ achieves the reduction mentioned. We refer to Weinberger (1965) for details of the application of this method.

Finally, one of the most powerful methods of solution of boundary value problems of this type involves the use of integral transforms [see Chapter 13].

8.3. THE HEAT OPERATOR

This is the operator

$$\frac{\partial}{\partial t} - \sum_{k=1}^{n} \frac{\partial^2}{\partial x_k^2}$$

which we shall often write as $D_t - \sum_{k=1}^{n} D_k^2$ or $D_t - \Delta_n$, in terms of co-ordinates $t \in \mathbb{R}$ and $\mathbf{x} = (x_1, \ldots, x_n) \in \mathbb{R}^n$. Perhaps the simplest problem naturally associated with this operator is that which seeks to determine a function $u(\mathbf{x}, t)$ such that

$$D_t u - \Delta_n u = 0 \quad \text{for all } t > 0 \text{ and all } \mathbf{x} \in \mathbb{R}^n,$$

$$u(\mathbf{x}, 0) = f(\mathbf{x}) \quad \text{for all } \mathbf{x} \in \mathbb{R}^n,$$

where f is a given function. This corresponds to the *initial-value* or *Cauchy problem* of finding the temperature distribution in a heat conducting solid which occupies all of $\mathbb{R}^n$, given that the temperature at the initial time $t = 0$ at the point $\mathbf{x}$ is $f(\mathbf{x})$. One method of solution is to take the Fourier transform with respect to the $\mathbf{x}$-variable [see § 13.2 and § 13.2.6 in particular]. Thus write

$$\hat{u}(\boldsymbol{\xi}, t) = \int_{\mathbb{R}^n} e^{-2\pi i \mathbf{x} \cdot \boldsymbol{\xi}} u(\mathbf{x}, t)\, d\mathbf{x}, \qquad \hat{f}(\boldsymbol{\xi}) = \int_{\mathbb{R}^n} e^{2\pi i \mathbf{x} \cdot \boldsymbol{\xi}} f(\mathbf{x})\, d\mathbf{x},$$

where $\mathbf{x} \cdot \boldsymbol{\xi} = \sum_{k=1}^{n} x_k \xi_k$. Proceeding formally we find that $\hat{u}$ satisfies the initial-value problem

$$\frac{\partial}{\partial t} \hat{u}(\boldsymbol{\xi}, t) + 4\pi^2 |\boldsymbol{\xi}|^2 \hat{u}(\boldsymbol{\xi}, t) = 0 \quad \text{for } t > 0,\ \boldsymbol{\xi} \in \mathbb{R}^n,$$

$$\hat{u}(\boldsymbol{\xi}, 0) = \hat{f}(\boldsymbol{\xi}) \quad \text{for } \boldsymbol{\xi} \in \mathbb{R}^n,$$

the solution of which is

$$\hat{u}(\boldsymbol{\xi}, t) = \hat{f}(\boldsymbol{\xi})\, e^{-4\pi^2 |\boldsymbol{\xi}|^2 t}.$$

By familiar properties of the Fourier transform we can recover u from this, and it turns out that u is a convolution, namely [see (13.2.32)]

$$u(\mathbf{x}, t) = (f * H_t)(\mathbf{x}), \tag{8.3.1}$$

where

$$H_t(\mathbf{x}) = (4\pi t)^{-n/2}\, e^{-|\mathbf{x}|^2/(4t)} \qquad (t > 0,\ \mathbf{x} \in \mathbb{R}^n).$$

The function $H(x, t) \equiv H_t(x)$ is called the *heat kernel* (or *Gauss kernel*). We thus have a *formal* solution (8.3.1) of our original initial-value problem; it can be shown, under appropriate conditions on f, that this really is a solution of the problem, the most simple method to do this being direct verification that formula (8.3.1) has all the properties needed. In particular, it emerges that if f is merely bounded and continuous, (8.3.1) gives a solution of the problem which is continuous on $\mathbb{R}^n \times [0, \infty)$ and infinitely differentiable on $\mathbb{R}^n \times (0, \infty)$; the initial data are smoothed out by the equation. (In fact, given any $T > 0$ the solution $f * H_t$ makes sense for $t < T$ even if all we require of f is that $|f(\mathbf{x})| \leq \text{const. exp}\{T|\mathbf{x}|^2\}$. Moreover, the solution is unique within the class of solutions such that given any $\varepsilon > 0$, there is a positive constant C such that [see (5.10.2)]

$$|u(\mathbf{x}, t)|,\ |\text{grad}_{\mathbf{x}}\, u(\mathbf{x}, t)| \leq C\, \exp\{\varepsilon |\mathbf{x}|^2\}.$$

If we extend H by defining $H(\mathbf{x}, t) = 0$ whenever $t \leq 0$, the resulting function, still denoted by H for simplicity, plays a rôle analogous to that of the fundamental solution K of the Laplace equation. Thus it turns out that

$$\frac{\partial}{\partial t} H - \Delta_n H = \delta,$$

where δ is the Dirac function, while if f is a function which is integrable over $\mathbb{R}^{n+1}$ and is infinitely differentiable in some open subset Ω of $\mathbb{R}^{n+1}$ (points of

$\mathbb{R}^{n+1}$ being denoted by $(\mathbf{x}, t)$, with $\mathbf{x} \in \mathbb{R}^n$, $t \in \mathbb{R}$), then $u = f * H$ is an infinitely differentiable solution of $\partial u / \partial t - \Delta_n u = f$ in Ω.

Another problem naturally associated with the heat equation is that of heat flow in a bounded domain $\Omega \subset \mathbb{R}^n$ over a finite or infinite time interval. In such a case we are led not to a pure initial-value problem, as in the situation last discussed, but rather to an initial-boundary-value problem, in which the temperature throughout the body is specified at some initial instant $(t = 0)$, and for all subsequent times in the interval boundary conditions are laid down. For example, the temperature on the boundary of the body might be specified, or the normal derivative of the temperature might be required to be zero on the boundary (corresponding to insulation of the boundary). Just as for the Dirichlet problem for Laplace's equation, the uniqueness of solutions of such problems follows from a maximum principle:

THEOREM 8.3.1   (*Maximum Principle*).   *Let Ω be a bounded domain in $\mathbb{R}^n$, let $0 < T < \infty$ and let $u(\mathbf{x}, t)$ be a real-valued continuous function on $\bar{\Omega} \times [0, T]$ which is in $C^{2,1}(\Omega \times (0, T))$ (twice continuously differentiable with respect to the $\mathbf{x}$ variables, once continuously differentiable with respect to t). Then the maximum and minimum of u on $\bar{\Omega} \times [0, T]$ must occur either on $\Omega \times \{0\}$ or on $\partial\Omega \times [0, T]$.*

The set $\Gamma = (\Omega \times \{0\}) \cup (\partial\Omega \times [0, T])$ is called the *parabolic boundary* of the cylinder $Q = \Omega \times (0, T)$; it plays the same part as does the boundary in problems for Laplace's equation. From the maximum principle it follows that there is at most one function $u \in C(\bar{Q}) \cap C^{2,1}(Q)$ such that

$$\frac{\partial u}{\partial t} - \Delta_n u = 0 \quad \text{in } Q, \qquad u = \phi \quad \text{on } \Gamma,$$

ϕ being a given continuous function. It should be noticed that the values of u on the *whole* of the boundary of Q cannot be specified; once they are given on Γ the values on the rest of ∂Q, that is, on the top $\Omega \times \{T\}$ of the cylinder Q, are automatically determined.

The method of separation of variables can be used to handle initial-boundary-value problems of the kind mentioned.

EXAMPLE 8.3.1.   Consider the problem

$$\frac{\partial u}{\partial t} - \Delta_n u = 0 \quad \text{on } \Omega \times (0, \infty),$$

$$u(\mathbf{x}, 0) = f(\mathbf{x}) \quad \text{for all } \mathbf{x} \in \Omega,$$

$$u(\mathbf{x}, t) = 0 \quad \text{for all } \mathbf{x} \in \partial\Omega \text{ and all } t > 0.$$

Here Ω is a bounded domain in $\mathbb{R}^n$ and f is a given continuous function on $\bar{\Omega}$ which is zero on $\partial\Omega$. We look for solutions in the form $u(\mathbf{x}, t) = X(\mathbf{x})T(t)$

and find that $T(t) = C \exp(\lambda t)$ and $\Delta_n X = \lambda X$ in Ω, for some constants λ and C. Moreover, the conditions on $\partial\Omega$ lead us to require that X should vanish on $\partial\Omega$. Further progress by way of an explicit solution is impossible unless Ω is of a special shape. For example, if Ω is the one-dimensional interval $(0, \pi)$, we have the problem

$$\frac{d^2 X}{dx^2} = \lambda X \quad \text{for } 0 < x < \pi, X(0) = X(\pi) = 0,$$

which has eigenvalues n^2 with corresponding eigenfunctions $\sin nx$ ($n = 1, 2, \ldots$). This suggests that we should look for a solution of the original problem of the form

$$u(x, t) = \sum_{n=1}^{\infty} a_n \exp\{-n^2 t\} \sin nx,$$

and that to satisfy the initial condition we should take the a_n to be the Fourier sine coefficients of f:

$$a_n = \frac{2}{\pi} \int_0^{\pi} f(x) \sin nx \, dx.$$

By this means we obtain a formal solution of the problem; under appropriate conditions on f it can be shown that this really is a genuine solution (see, for example, Weinberger (1965)) and that u satisfies the initial condition in the sense that $\lim_{t \to 0} u(x, t) = f(x)$. It is enough, for instance, that f should be continuous on $[0, \pi]$ with $f(0) = f(\pi) = 0$, and that f'^2 should be integrable over $(0, \pi)$. Indeed, under these conditions it turns out that u is infinitely differentiable with respect to both x and t, even though the function f may have no derivatives beyond the first. This shows again that the process of heat conduction described by the heat equation is a smoothing process. Moreover, this initial-boundary-value problem, in common with more general ones for the heat equation, is *well-posed* in the sense that the solution depends continuously on the data. The *backward* problem, however, in which the values of u at some final instant T are specified together with some boundary conditions, is not well-posed because of the smoothing effect of the heat operator; the temperature distribution at $t = T$ cannot have originated from an earlier temperature distribution unless it is analytic.

Another technique useful in the solution of initial-boundary-value problems is that of the *Laplace transform* [see § 13.4 and § 13.4.8 in particular]. Suppose we consider the one-dimensional heat equation

$$\frac{\partial u}{\partial t} - \frac{\partial^2 u}{\partial x^2} = 0 \quad \text{for } x > 0, t > 0,$$

with initial condition $u(x, 0) = f(x)$ for $x > 0$, and boundary condition $\partial u / \partial x(0, t) = u(0, t)$ for $t > 0$; we also ask that u be bounded.

Physically, this problem corresponds to the determination of the temperature distribution in a semi-infinite conducting rod when heat is removed at a rate proportional to the temperature at the end. Introduce the Laplace transform U:

$$U(x, s) = \int_0^\infty \exp(-st)u(x, t)\, dt.$$

Formal manipulations lead us to consider the ordinary differential equation for U,

$$\frac{\partial^2 U}{\partial x^2} - sU(x, s) = -f(x),$$

under the conditions that U should be bounded and that

$$\frac{\partial U}{\partial x}(0, s) = U(0, s).$$

It follows that

$$U(x, s) = \frac{1}{2\sqrt{s}} \int_0^\infty \exp(-(x-y)\sqrt{s})\, f(y)\, dy$$

$$+ \frac{\sqrt{s}-1}{2\sqrt{s}(1+\sqrt{s})} \int_0^\infty \exp(-(x+y)\sqrt{s})f(y)\, dy.$$

Use of the inverse Laplace transform [see § 13.4.2] plus a little work (see Weinberger (1965)) shows that a formal solution of our problem is

$$u(x, t) = \int_0^\infty K(x, y, t)f(y)\, dy,$$

where

$$K(x, y, t) = \frac{1}{2\sqrt{\pi t}}(e^{-(x-y)^2/(4t)} + e^{-(x+y)^2/(4t)}) - e^{x+y+t}\, \mathrm{erfc}\left(\frac{x+y}{2\sqrt{t}} + \sqrt{t}\right),$$

and erfc is the *complementary error function*

$$\mathrm{erfc}(x) = \frac{2}{\sqrt{\pi}} \int_x^\infty e^{-y^2}\, dy.$$

Mild conditions on f, for example that $\int_0^\infty |f(x)|\, dx < \infty$, are enough to make it possible to show that this formal solution is a genuine solution of the problem.

Just as for the Laplacian, a Green's function may be introduced for problems involving the heat operator. We refer to Folland (1976) and Weinberger (1965) for details of this.

8.4.   THE WAVE OPERATOR

In terms of coordinates $t \in \mathbb{R}$ and $\mathbf{x} = (x_1, \ldots, x_n) \in \mathbb{R}^n$ this is the operator

$$\frac{\partial^2}{\partial t^2} - \sum_{k=1}^n \frac{\partial^2}{\partial x_k^2},$$

sometimes written as $\partial^2/\partial t^2 - \Delta_n$ or $D_t^2 - \Delta_n$. The fundamental problem for the wave equation is the Cauchy problem, or initial-value problem, in which data are given on a hypersurface $S \subset \mathbb{R}^{n+1}$. Our earlier discussion in section 8.1 indicates that S should be non-characteristic if strange results are to be avoided. However, if $n \geq 2$ this is not enough on its own. For, with $n \geq 2$ and any positive number k, if we consider the function u defined by

$$u(\mathbf{x}, t) = \exp(-k/2) \sin kx_1 \sinh kx_2,$$

then u satisfies the wave equation and the initial conditions

$$u(x_1, 0, x_3, \ldots, x_n, t) = 0, \qquad D_2 u(x_1, 0, x_3, \ldots, x_n, t) = k \exp(-k/2) \sin kx_1$$

on the non-characteristic hyperplane $x_2 = 0$. As $k \to \infty$ the Cauchy data tend uniformly to zero, as do all their derivatives, but $u(\mathbf{x}, t)$ makes infinite oscillations if $x_2 > 1/2$ and $x_1 \neq 0$. It turns out that to avoid this kind of pathology, and to have a well-posed Cauchy problem on S, it is necessary and sufficient that S should be *space-like*, by which we mean that its normal $\boldsymbol{\nu} = (\nu_1, \ldots, \nu_n, \nu_0) = (\hat{\boldsymbol{\nu}}, \nu_0)$ is such that $|\nu_0| > |\hat{\boldsymbol{\nu}}|$ at every point; in other words, $\boldsymbol{\nu}$ lies inside the *light cone* $\{(\boldsymbol{\xi}, \tau) \in \mathbb{R}^n \times \mathbb{R} : \tau^2 = |\boldsymbol{\xi}|^2\}$.

We deal in more detail with the important special case in which S is the hyperplane $t = 0$. As the wave operator is unchanged if we replace t by $-t$, it is enough to consider solutions in the half-space $t > 0$. The following theorem is of great importance:

THEOREM 8.4.1. *Let u be a twice continuously differentiable solution of the wave equation in the infinite strip $\{(\mathbf{x}, t) \in \mathbb{R}^n \times \mathbb{R} : 0 \leq t \leq T\}$, and suppose that $u = D_t u = 0$ on the ball $B = \{(\mathbf{x}, 0) : |\mathbf{x} - \mathbf{x}_0| \leq t_0\}$ in the hyperplane $t = 0$, where $\mathbf{x}_0 \in \mathbb{R}^n$ and $0 < t_0 < T$. Then u is zero in the conical set*

$$\mathscr{C} = \{(\mathbf{x}, t) : 0 \leq t \leq t_0, |\mathbf{x} - \mathbf{x}_0| \leq t_0 - t\}.$$

This result implies that the value of a solution u of the wave equation at a point $(\mathbf{x}_0, t_0)$ depends only on the Cauchy data of u on the ball B cut out of the hyperplane $t = 0$ by the *light cone* $\mathscr{C}$ with vertex $(\mathbf{x}_0, t_0)$. To put it another way, the Cauchy data on a subset of the hyperplane $t = 0$ influence the values of u only at those points inside the light cones issuing from points

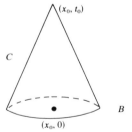

Figure 8.4.1

of the subset. The same holds if the hyperplane $t = 0$ is replaced by any space-like hypersurface.

We have already seen in section 8.1 that the solution of the Cauchy problem

$$D_t^2 u = \Delta_n u \quad \text{for } t > 0, \mathbf{x} \in \mathbb{R}^n,$$

$$u(\mathbf{x}, 0) = f(\mathbf{x}), \qquad D_t u(\mathbf{x}, 0) = g(\mathbf{x}), \qquad \mathbf{x} \in \mathbb{R}^n \left.\right\} \qquad (8.4.1)$$

can be found easily when $n = 1$; it is

$$u(x, t) = \tfrac{1}{2}\{f(x + t) + f(x - t)\} + \int_{x-t}^{x+t} g(s) \, ds,$$

provided that $f \in C^2(\mathbb{R})$ and $g \in C^1(\mathbb{R})$. This solution may be thought of as the superposition of two plane waves travelling along the x-axis with speeds 1 and -1 without changing shape. The solution is unique and evidently depends continuously on the data; we have a well-posed problem. Note that $u(x_0, t_0)$ is affected by a change in f only if $f(x_0 + t_0)$ or $f(x_0 - t_0)$ is changed; in other words and in physical terms, thinking of u as a displacement, the effect of the initial displacement is propagated at speed 1. In contrast to this, $u(x_0, t_0)$ is influenced by a change of g anywhere in the interval $[x_0 - t_0, x_0 + t_0]$; that is, the effect of a change of initial velocity is propagated at all speeds up to 1. Here we see in very concrete terms the results about regions of dependence and influence alluded to above.

DEFINITION 8.4.1. The interval $[x_0 - t_0, x_0 + t_0]$ on the line $t = 0$ is the *domain of dependence* on the initial data for u at the point (x_0, t_0); its end-points are the points at which the *characteristic lines* through (x_0, t_0) (with slopes ± 1) cut the x-axis. The values of f and g at a point y influence u at (x, t) only if $|x - y| \leq t$; the *domain of influence* of the data at a point $(y, 0)$ is then the part of the upper half of the $\mathbf{x}, t$ plane bounded by the characteristic lines through $(\mathbf{y}, 0)$.

To find a solution of (8.4.1) when $n > 1$ is less easy, but is possible using the method of *spherical means*:

DEFINITION 8.4.2. Given any $\mathbf{x} \in \mathbb{R}^n$ and any $r \in \mathbb{R}$ the *spherical mean* $M_\phi(\mathbf{x}; r)$ of a continuous real-valued function ϕ on $\mathbb{R}^n$ is defined to be

$$M_\phi(\mathbf{x}; r) = \frac{1}{\omega_n} \int_{\partial B(0,1)} \phi(\mathbf{x} + r\mathbf{y}) \, d\sigma(\mathbf{y}); \qquad (8.4.2)$$

if ϕ also depends on a parameter t we shall denote its spherical mean by $M_\phi(\mathbf{x}; r, t)$.

Now let u be a solution of (8.4.1) and form its spherical mean $M_u(\mathbf{x}; r, t)$, regarding u as a function of $\mathbf{x}$ which also depends on a parameter t. It turns out that *if n is odd*, M_u satisfies the one-dimensional wave equation with initial conditions similar to those given in (8.4.1). We can therefore find M_u, and then recover u from the result that

$$u(\mathbf{x}, t) = \lim_{r \to 0} M_u(\mathbf{x}; r, t);$$

what emerges is an explicit formula for the solution of (8.4.1), namely,

$$u(\mathbf{x}, t) = \frac{1}{1 \cdot 3 \cdot 5 \ldots (n-2)\omega_n} \left[\frac{\partial}{\partial t} \left(\frac{1}{t} \frac{\partial}{\partial t} \right)^{(n-3)/2} \left\{ t^{n-2} \int_{\partial B(0,1)} f(\mathbf{x} + t\mathbf{y}) \, d\sigma(\mathbf{y}) \right\} \right.$$

$$\left. + \left(\frac{1}{t} \frac{\partial}{\partial t} \right)^{(n-3)/2} \left\{ t^{n-2} \int_{\partial B(0,1)} g(\mathbf{x} + t\mathbf{y}) \, d\sigma(\mathbf{y}) \right\} \right]. \tag{8.4.3}$$

This is the solution of (8.4.1), provided that n is odd, $n \geq 3$, $f \in C^{(n+3)/2}(\mathbb{R}^n)$ and $g \in C^{(n+1)/2}(\mathbb{R}^n)$.

We still have to deal with the case in which n is even, but this can be treated by the *method of descent*, which relies on the observation that if u is a solution of the wave equation in $\mathbb{R}^{n+1} \times \mathbb{R}$ which does not depend upon x_{n+1}, then it is a solution of the wave equation in $\mathbb{R}^n \times \mathbb{R}$ when viewed as a function of $x_1, x_2, \ldots, x_n$ and t. To solve (8.4.1) when n is even, we think of f and g as functions on $\mathbb{R}^{n+1}$ which do not depend upon x_{n+1}, write down the solution (8.4.3) and check that it does not depend upon x_{n+1}. By this means it turns out that if n is even, $f \in C^{(n+4)/2}(\mathbb{R}^n)$ and $g \in C^{(n+2)/2}(\mathbb{R}^n)$, then the solution of (8.4.1) is

$$u(\mathbf{x}, t) = \frac{2}{1 \cdot 3 \cdot 5 \ldots (n-1)\omega_{n+1}} \left[\frac{\partial}{\partial t} \left(\frac{1}{t} \frac{\partial}{\partial t} \right)^{(n-2)/2} \left\{ t^{n-1} \int_{B(0,1)} \frac{f(\mathbf{x} + t\mathbf{y})}{\sqrt{1 - |\mathbf{y}|^2}} \, d\mathbf{y} \right\} \right.$$

$$\left. + \left(\frac{1}{t} \frac{\partial}{\partial t} \right)^{(n-2)/2} \left\{ t^{n-1} \int_{B(0,1)} \frac{g(\mathbf{x} + t\mathbf{y})}{\sqrt{1 - |\mathbf{y}|^2}} \, d\mathbf{y} \right\} \right]. \tag{8.4.4}$$

We now have a complete solution of the Cauchy problem (8.4.1) for all values of n. (For a full derivation of these results see Folland (1976).) In all cases we see that the value of u at a point $(\mathbf{x}_0, t_0)$ depends only on the values of f and g on the ball $|\mathbf{x} - \mathbf{x}_0| \leq t_0$ cut out of the hyperplane $t = 0$ by the light cone with vertex $(\mathbf{x}_0, t_0)$. The formulae also show that when n is odd and ≥ 3, $u(\mathbf{x}_0, t_0)$ depends only on the values of f, g and their first few derivatives on the sphere $|\mathbf{x} - \mathbf{x}_0| = t_0$ rather than on the whole ball $|\mathbf{x} - \mathbf{x}_0| \leq t_0$; when n is even, $u(\mathbf{x}_0, t_0)$ depends on the values of f and g on the whole ball, with greater weight attached to the values near the boundary because of the factor $1/\sqrt{1 - |\mathbf{y}|^2}$. These conclusions embody what is called *Huyghen's principle*, which asserts in particular that a listener at distance d from a musical instrument hears at time t exactly what has been played at time $t - (d/c)$ (c is the velocity of sound) rather than a mixture of all the notes played up to this time; in an even-dimensional world he would not hear sharp, clear notes but a mixture of those previously played.

So far we have considered pure initial-value problems, in which the space variable $\mathbf{x}$ ranges throughout $\mathbb{R}^n$. More often $\mathbf{x}$ is allowed to vary throughout some subset of $\mathbb{R}^n$, and we then have an *initial-boundary-value problem*:

EXAMPLE 8.4.1.   If we are concerned with the movement of a vibrating string tied at its end-points and set into motion in a prescribed way, we end up with a problem of the form

$$D_t^2 u - D_x^2 u = 0 \quad \text{for } 0 < x < l, \, t > 0,$$

$$\left.\begin{array}{l} u(x, 0) = f(x) \quad \text{and} \quad D_t u(x, 0) = g(x) \quad \text{for } 0 \le x \le l, \\ u(0, t) = u(l, t) = 0 \quad \text{for } t \ge 0, \end{array}\right\} \tag{8.4.5}$$

where $f(0) = f(l) = g(0) = g(l) = 0$. Beginning with the standard solution

$$u(x, t) = p(x + t) + q(x - t)$$

for some functions p and q, and applying the initial conditions we rapidly find (for some constant K) that

$$p(x) = \tfrac{1}{2} f(x) + \tfrac{1}{2} \int_0^x g(y) \, dy + K, \qquad q(x) = \tfrac{1}{2} f(x) - \tfrac{1}{2} \int_0^x g(y) \, dy - K$$

for $0 \le x \le l$, giving

$$u(x, t) = \tfrac{1}{2} [f(x + t) + f(x - t)] + \tfrac{1}{2} \int_{x-t}^{x+t} g(y) \, dy \tag{8.4.6}$$

provided that $0 \le x + t \le l$ and $0 \le x - t \le l$; we also need $f \in C^2$, $g \in C^1$. This formula gives us the solution in the triangular region of the x, t plane specified by

$$0 \le t \le x, \, t \le l - x.$$

To obtain the solution for larger times we use the boundary conditions to obtain

$$p(t) + q(-t) = 0, \qquad p(l + t) + q(l - t) = 0 \quad \text{for } t \ge 0.$$

These can be written as

$$q(z) = -p(-z) \quad \text{for } z \le 0, \qquad p(z) = -q(2l - z) \quad \text{for } z \ge l.$$

Since we know $p(z)$ when $0 \le z \le l$ we can now find $q(z)$ when $-l \le z \le 0$:

$$q(z) = -\tfrac{1}{2} f(-z) - \tfrac{1}{2} \int_0^{-z} g(y) \, dy - K.$$

We now have $q(z)$ when $-l \le z \le l$, from which we can find $p(z)$ when $l \le z \le 3l$:

$$p(z) = \begin{cases} -\tfrac{1}{2} f(2l - z) + \tfrac{1}{2} \displaystyle\int_0^{2l-z} g(y) \, dy + K, & l \le z \le 2l, \\[2ex] \tfrac{1}{2} f(z - 2l) + \tfrac{1}{2} \displaystyle\int_0^{z-2l} g(y) \, dy + K, & 2l \le z \le 3l. \end{cases}$$

Repetition of this process will give $p(z)$ for all $z \ge 0$ and $q(z)$ for all $z \le l$, which means that $u(x, t)$ is determined for all x, t such that $0 \leq x \leq l$, $t \geq 0$. This sound complicated, but an examination of what we have done reveals that the solution of (8.4.5) is given by (8.4.6) for all $x \in [0, l]$ and all $t \geq 0$, where the functions $f(x)$ and $g(x)$ are as specified for $0 \leq x \leq l$, and are extended to other values so that they are both odd about 0 and about l; in other words, set

$$f(x) = -f(-x), \qquad g(x) = -g(-x)$$

for x negative, and

$$f(x) = -f(2l - x), \qquad g(x) = -g(2l - x),$$

so that f and g become, when extended from the interval $[0, l]$, odd periodic functions of period $2l$ [see § 20.1]. The function u given by (8.4.6) will be in C^2 and will be a genuine solution of our problem if the *extended* functions f and g are in C^2, though this condition is not necessary. The problem is well-set; the solution depends continuously upon the data.

To clarify what is going on in this problem, and to explain the rôle of the end-points 0 and l it may be helpful to consider the case in which $g = 0$ and $f(x) = 0$ except in a small neighbourhood of some point x_0, $0 < x_0 < l$, where it is positive. For fixed times t such that $t < x_0$, $t < l - x_0$ it follows from (8.4.6) that $u(x, t) = 0$ unless x is near $x_0 - t$ or $x_0 + t$; thus for small values of t the set of points (x, t) at which $u(x, t) > 0$ is concentrated around the two charac-teristics $x \pm t = x_0$ passing through $(x_0, 0)$, so that we may speak of the disturb-ance as being propagated along these characteristics. Now suppose that $x_0 < l/2$. At time $t = x_0$ the left characteristic intersects the boundary $x = 0$.

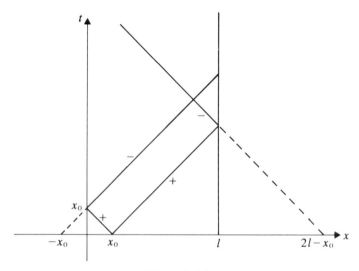

Figure 8.4.2

If we take $x < x_0$, to find $u(x, t)$ when $t > x$ we have to extend f as an odd function of x, and as this extended function is zero for $-l < x < 0$ except near $x = -x_0$, it follows that until t is near to $x_0 + x$, $u(x, t) = \frac{1}{2}[f(x+t) + f(x-t)] = 0$. When t is close enough to $x_0 + x$ the term $\frac{1}{2}f(x-t) = -\frac{1}{2}f(t-x)$ will give a negative value for u, so that we have a negative value propagated along the characteristic $x - ct = -x_0$. As this characteristic meets the boundary $x = 0$ at the same time $t = x_0$ as the left characteristic through $(x_0, 0)$ we say that this left characteristic has been *reflected* on $x = 0$ as a right characteristic. Similarly the right characteristic through $(x_0, 0)$ is reflected from $x = l$ as a left characteristic propagating a negative value, which will ultimately be reflected from $x = 0$ as a right characteristic propagating a positive value, and so on. Thus the influence of a displacement at x_0 is concentrated around the characteristics through $(x_0, 0)$ and their reflections.

A similar kind of discussion carried out for the case in which $f = 0$ and $g(x)$ is zero save in the vicinity of x_0, with $\int_0^l g(x)\, dx = 2$, shows that $u(x, t)$ is constant except in the neighbourhood of the characteristics through $(x_0, 0)$ and their reflections, the constant being 1 or -1 in the parallelograms formed by the characteristics and zero in the triangles.

We therefore see that a change in g affects $u(x, t)$ only at the points inside the parallelograms, while a change in f affects $u(x, t)$ only at the points on the boundaries of these parallelograms. Hence any disturbance at $(x_0, 0)$ will influence $u(x, t)$ only in the parallelograms and on their boundaries; this set is the *domain of influence* of $(x_0, 0)$.

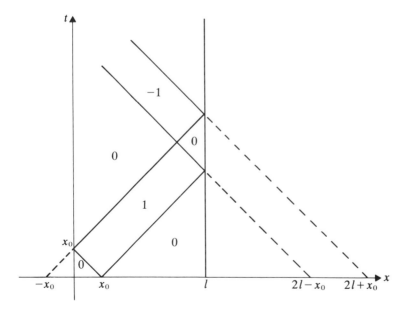

Figure 8.4.3

For a more detailed discussion of these topics we refer to Bers, John and Schechter (1964), Courant and Friedrichs (1948) and Weinberger (1965).

To solve initial-value or initial-boundary-value problems involving the wave operator we may naturally employ some or all of the techniques already mentioned in section 8.3: separation of variables, Fourier series, Fourier transforms and Laplace transforms.

EXAMPLE 8.4.2.   To illustrate the procedure when more than one method has to be used, consider the following problem, expressed in terms of polar coordinates r, θ [see V, § 1.2.5]:

$$\frac{\partial^2 u}{\partial t^2} - \frac{\partial^2 u}{\partial r^2} - \frac{1}{r}\frac{\partial u}{\partial r} - \frac{1}{r^2}\frac{\partial^2 u}{\partial \theta^2} = 0 \quad \text{for } r > 0, 0 < \theta < 2\pi, t > 0,$$

$$u(r, \theta, 0) = 0, \qquad \frac{\partial u}{\partial t}(r, \theta, 0) = f(r, \theta),$$

$$\frac{\partial u}{\partial \theta}(r, 0, t) = \frac{\partial u}{\partial \theta}(r, 2\pi, t) = 0.$$

This is related to the motion of sound waves when a rigid semi-infinite wall is present at $\theta = 0$ (independence of a third coordinate z is assumed). If we form the Laplace transform of u with respect to t [see § 13.4.1]:

$$U(r, \theta, s) = \int_0^\infty e^{-st} u(r, \theta, t) \, dt,$$

we find that U must satisfy

$$\frac{\partial^2 U}{\partial r^2} + \frac{1}{r}\frac{\partial U}{\partial r} + \frac{1}{r^2}\frac{\partial^2 U}{\partial \theta^2} - s^2 U = -f(r, \theta),$$

$$\frac{\partial U}{\partial \theta}(r, 0, s) = \frac{\partial U}{\partial \theta}(r, 2\pi, s) = 0,$$

and instead of boundary conditions with respect to the variable r at 0 and infinity we require that U be bounded. To this new problem we apply the method of separation of variables, which shows that if f is zero there are solutions of the form $J_{n/2}(irs) \cos \frac{1}{2}n\theta$ ($n = 0, 1, 2, \ldots$), $J_{n/2}$ being the Bessel function of order $n/2$ [see § 10.4.2]. It is therefore natural to express U and f as cosine series [see § 20.5.3]:

$$U(r, \theta, s) = \frac{1}{2}u_0(r, s) + \sum_{n=1}^{\infty} u_n(r, s) \cos \frac{1}{2}n\theta,$$

$$f(r, \theta) = \frac{1}{2}f_0(r) + \sum_{n=1}^{\infty} f_n(r) \cos \frac{1}{2}n\theta,$$

where

$$u_n(r, s) = \frac{1}{\pi} \int_0^{2\pi} U(r, \theta, s) \cos \tfrac{1}{2} n\theta \, d\theta,$$

$$f_n(r) = \frac{1}{\pi} \int_0^{2\pi} f(r, \theta) \cos \tfrac{1}{2} n\theta \, d\theta.$$

The coefficients u_n satisfy the ordinary equation

$$\frac{\partial^2}{\partial r^2} u_n + \frac{1}{r} \frac{\partial}{\partial r} u_n - \left(\frac{n^2}{4r^2} + s^2 \right) u_n = -f_n,$$

and by the standard procedures of ordinary differential equations it turns out that the solution of this equation which is regular [see § 9.1] at 0 and infinity, assuming that f_n is regular and goes to zero at infinity, is

$$u_n(r, s) = \int_0^{\infty} G_n(r, \rho, s) f_n(\rho) \, d\rho,$$

where

$$G_n(r, \rho, s) = \begin{cases} I_{n/2}(\rho s) K_{n/2}(rs) & \text{if } \rho \leq r, \\ I_{n/2}(rs) K_{n/2}(\rho s) & \text{if } \rho > r, \end{cases}$$

I and K being the modified Bessel functions of the first and second kinds respectively [see § 10.4.3]. We are now faced with the task of recovering U from its Fourier coefficients; this needs a little work, and we have to refer to Weinberger (1965) for the details. What emerges is that the solution u is given by

$$u(r, \theta, t) = \int_0^{\infty} \int_0^{2\pi} \Gamma(r, \theta, \rho, \phi, t) f(\rho, \phi) \rho \, d\rho \, d\phi,$$

where Γ is a rather complicated but explicitly known function. For an interpretation of this solution in terms of diffraction effects see Weinberger (1965).

Other kinds of initial-boundary-value problems occur in the scattering of waves by an obstacle. In such problems we have to find a solution of the wave equation $D_t^2 u - \Delta u = 0$ in the exterior of some bounded region $\Omega \subset \mathbb{R}^n$, the solution having zero normal derivative on the boundary of Ω, and having initial values corresponding to a plane wave coming in from infinity in the direction of the x_1-axis. If we split the incoming wave into components of the form $\exp i\lambda(x_1 - t)$ we reduce the problem to that of finding a complex-valued function v which satisfies the *reduced wave equation*

$$\Delta_n v + \lambda^2 v = 0 \tag{8.4.7}$$

in the exterior of Ω, has zero normal derivative on $\partial\Omega$, and which satisfies the *Sommerfeld radiation condition*

$$\lim_{r\to\infty} r\left(\frac{\partial}{\partial r} - i\lambda\right) w(r\xi) = 0 \quad \text{uniformly for } |\xi| = 1, \, \xi \in \mathbb{R}^n ; \qquad (8.4.8)$$

here $w(\mathbf{x}) = v(\mathbf{x}) - e^{i\lambda x_1}$. For a full discussion of this important problem see Courant and Hilbert (1962), I.

8.5.   ELLIPTIC, PARABOLIC AND HYPERBOLIC LINEAR EQUATIONS

For simplicity we start with linear *second-order* equations, that is, equations of the form

$$Lu \equiv \sum_{i,j=1}^{n} a_{ij} D_i D_j u + \sum_{i=1}^{n} b_i D_i u + cu = f, \qquad (8.5.1)$$

where the a_{ij}, b_i, c and f are given real-valued functions on an open set $\Omega \subset \mathbb{R}^n$, and $a_{ij} = a_{ji}$ for all i and j.

DEFINITION 8.5.1.   Equation (8.5.1) is said to be *elliptic* at a point $\mathbf{x}_0 \in \Omega$ if there is a positive constant $\mu(\mathbf{x}_0)$ such that for all $\xi = (\xi_i) \in \mathbb{R}^n$,

$$\sum_{i,j=1}^{n} a_{ij}(x_0)\xi_i\xi_j \geq \mu(\mathbf{x}_0)|\xi|^2.$$

If (8.5.1) is elliptic at each point of Ω we say that it is *elliptic in* Ω; and if in addition the a_{ij} are bounded in Ω and the number $\mu(\mathbf{x}_0)$ may be chosen to be independent of the particular point $\mathbf{x}_0$ in Ω, the equation is said to be *uniformly elliptic in* Ω.

The equation (8.5.1) is thus elliptic at $\mathbf{x}_0 \in \Omega$ if the quadratic form $Q(\mathbf{x}_0, \xi) = \sum_{i,j=1}^{n} a_{ij}(\mathbf{x}_0)\xi_i\xi_j$ is positive definite [see I, § 9.2]. If we make a smooth change of variable $\mathbf{x} \mapsto \mathbf{y}$, given by $y_i = g_i(x_1, \ldots, x_n)$ $(i = 1, \ldots, n)$ [see Definition 5.9.2], then a simple computation shows that (8.5.1) is taken into an equation of the form

$$\sum_{i,j=1}^{n} \tilde{a}_{ij} \frac{\partial^2 v}{\partial y_i \partial y_j} + \sum_{i=1}^{n} \tilde{b}_i \frac{\partial v}{\partial y_i} + \tilde{c}v = \tilde{f},$$

where the coefficients $\tilde{a}_{ij}$ are given by

$$\tilde{a}_{ij} = \sum_{k,l=1}^{n} a_{kl} D_k y_i D_l y_j$$

and $v(y) = u(x)$. Thus at the point $\mathbf{x}_0$ the coefficients of the *principal part* of (8.5.1), namely the part involving only the second-order derivatives, are transformed like the coefficients of the quadratic form $Q(\mathbf{x}_0, \xi)$ when the ξ_i are transformed to parameters η_i by the linear map $\xi_i = \sum_{j=1}^{n} D_j y_i(\mathbf{x}_0)\eta_j$ [see I, § 5.11]. It is a familiar fact [see I, § 9.2] that the quadratic form Q can be transformed by a suitable linear transformation into the canonical form

$$\sum_{i=1}^{n} k_i \eta_i^2,$$

where the coefficients k_i take only the values 1, -1 or 0. This leads to our classification procedure. Equation (8.5.1) is then elliptic at $\mathbf{x}_0$ if all the k_i are 1; if all the k_i are -1 then evidently the equation $-Lu = -f$ is elliptic at $\mathbf{x}_0$.

DEFINITION 8.5.2. The equation (8.5.1) is called *properly hyperbolic*, or *normally hyperbolic*, or simply *hyperbolic* at $\mathbf{x}_0$ if none of the k_i is zero and all of them save one are of the same sign. (If more than one k_i is equal to 1 and more than one equal to -1, none being zero, the equation is called *ultrahyperbolic* at $\mathbf{x}_0$.) The remaining case is when at least one of the k_i is zero; the equation is then said to be *parabolic* at $\mathbf{x}_0$. As in the case of elliptic equations (8.5.1) is called *hyperbolic in* Ω or *parabolic in* Ω if it is hyperbolic or parabolic at each point of Ω.

The reason for this terminology is that when $n = 2$ the curve $Q(\mathbf{x}_0, \boldsymbol{\xi}) = 1$ is an ellipse, a parabola or a hyperbola [see V, § 1.3] according as the equation is elliptic, parabolic or hyperbolic at $\mathbf{x}_0$. Notice that only the coefficients of the second-order derivatives affect the classification of the equation.

The classic representatives of these types of equation are Laplace's equation, the wave equation and the heat equation, which are respectively elliptic, hyperbolic and parabolic in any domain Ω. By way of contrast, *Tricomi's equation* in $\mathbb{R}^2$,

$$D_1^2 u + x_2 D_2^2 u = 0, \tag{8.5.2}$$

is elliptic in the upper half-plane $x_2 > 0$, hyperbolic in the lower half-plane and parabolic on the x_1-axis. The natural well-posed problems for our three kinds of equations are those which we have already considered in sections 8.2, 8.3 and 8.4. For elliptic equations we study boundary-value problems in which a linear combination of the unknown function and its normal derivative is specified on the boundary; while for parabolic and hyperbolic equations initial-value and initial-boundary-value problems have to be examined. Of course, when we come to think about *how* we should solve these various kinds of problems we may bring into play the battery of techniques mentioned earlier: separation of variables, Fourier series, and Fourier and Laplace transforms. The difficulty is that unless we are lucky, and receive some co-operation from the coefficients of the equation and the shape of the domain, none of these methods may apply, and we shall be unable to write down an explicit solution of the problem. It is for this reason that more theoretical methods have to be used, beginning with the questions of whether the problem has a solution at all, and whether it is unique. At least the knowledge that a unique solution exists which depends continuously on the data will ensure that a computer-aided numerical solution of the problem may sensibly be attempted, should such detailed information be required. If no solution exists, clearly any numerical work would be a waste of time.

Our discussion from now on will therefore be somewhat theoretical, and we begin with elliptic equations. For these a maximum principle holds, as already explained in section 8.2 for Laplace's equation. The version which we shall give is due to E. Hopf, and is usually referred to in the literature as the *strong* maximum principle:

THEOREM 8.5.1   (*Maximum Principle*).   *Let Ω be a domain in $\mathbb{R}^n$ (not necessarily bounded), let equation (8.5.1) be uniformly elliptic in Ω, let the coefficients a_{ii}, b_i and c be bounded $(i = 1, 2, \ldots, n)$, and let $u \in C^2(\Omega)$ be such that $Lu \geq 0$ (≤ 0) in Ω. Then*:

   (i) *if $c = 0$ and u attains its maximum (minimum) in Ω, u is a constant*;
   (ii) *if $c < 0$ and u attains a non-negative maximum (non-positive minimum) in Ω, u is a constant.*

The condition $c \leq 0$ is essential for the validity of the theorem since the function $u(\mathbf{x}) = \exp -|\mathbf{x}|^2$ satisfies $\Delta_n u + (2n - 4|\mathbf{x}|^2)u = 0$ throughout $\mathbb{R}^n$, yet has an absolute maximum at $\mathbf{0}$.

To examine what happens when u has a maximum or minimum at a boundary point, we need a restriction on Ω.

DEFINITION 8.5.3.   The domain Ω is said to satisfy an *interior ball condition* at $\mathbf{x}_0 \in \partial\Omega$ if there is a ball $B \subset \Omega$ such that $\mathbf{x}_0 \in \partial B$.

THEOREM 8.5.2   (*Boundary Point Theorem*).   *Let L be uniformly elliptic in Ω, let a_{ii}, b_i and c be bounded and let $u \in C^2(\Omega)$ be such that $Lu \geq 0$ in Ω. Suppose that Ω satisfies an interior ball condition at a point $\mathbf{x}_0 \in \partial\Omega$, that u is continuous at $\mathbf{x}_0$, and that $u(\mathbf{x}) < u(\mathbf{x}_0)$ for all $\mathbf{x} \in \Omega$. Then if $c = 0$, the outer normal derivative $\partial u/\partial \nu$ $(\mathbf{x}_0)$ of u at $\mathbf{x}_0$, if it exists, is positive. If $c \leq 0$ the same conclusion holds provided that $u(\mathbf{x}_0) \geq 0$.*

These results (the proofs of which may be found in Gilbarg and Trudinger (1977) or Protter and Weinberger (1967)) enable us to establish the uniqueness and continuous dependence on the data of solutions of the usual boundary-value problems, such as the Dirichlet and the Neumann problems [see (8.2.1) and (8.2.2)].

As for the existence of solutions of boundary-value problems involving L, fundamental work was done in 1934 by Schauder, who established the existence of a solution of the Dirichlet problem in a bounded domain Ω by means of certain *a priori* estimates for solutions. To describe these results we need some notation.

DEFINITION 8.5.4.   Given any λ, $0 < \lambda \leq 1$, and any integer $m \geq 0$, we shall denote by $C^{m+\lambda}(\bar{\Omega})$ the space of all real-valued functions u which are m-times continuously differentiable on $\bar{\Omega}$ and are such that for all indices $\boldsymbol{\alpha}$ with

$|\alpha| = m$, $D^{\alpha}u$ is Hölder-continuous with exponent λ on $\bar{\Omega}$ [see (8.2.3)]; we shall say that the boundary $\partial\Omega$ is *of class* $C^{2+\lambda}$ if given any point $x_0 \in \partial\Omega$, there is a neighbourhood of x_0 in which $\partial\Omega$ may be given by an equation $y_n = \phi(y_1, \ldots, y_{n-1})$, where $(y_1, \ldots, y_n)$ are Cartesian coordinates with y_n along the outward normal to $\partial\Omega$ at x_0, and ϕ is twice continuously differentiable with second derivatives which are Hölder-continuous with exponent λ.

THEOREM 8.5.3. *Let* Ω *be bounded, let the coefficients* a_{ij}, b_i, c *and* f *be in* $C^{\lambda}(\bar{\Omega})$ *for some* λ, $0 < \lambda \leq 1$, *suppose that* L *is uniformly elliptic in* Ω, *let* $\partial\Omega$ *be of class* $C^{2+\lambda}$ *and let* g *be a given function in* $C^{2+\lambda}(\partial\Omega)$. *Then if* $c \leq 0$ *there is precisely one function* $u \in C^{2+\lambda}(\bar{\Omega})$ *such that* $Lu = f$ *in* Ω *and* $u = g$ *on* $\partial\Omega$.

If the condition $c \leq 0$ is dropped it turns out that the problem can be reduced to one to which the Fredholm–Riesz–Schauder theory of compact linear operators in a Banach space can be applied [see § 19.3.6]; conditions on f may be required for a solution to exist. It should also be noted that the existence of solutions can be established even for domains Ω which do not have boundaries of class $C^{2+\lambda}$; however, the solution which exists need not be in $C^{2+\lambda}(\bar{\Omega})$ for in general it will fail to have the right amount of smoothness up to the boundary.

The situation as regards other kinds of boundary-value problems for equation (8.5.1) is also well developed, and like the Dirichlet problem the existence theory for such questions depends on certain *a priori* estimates, this time due to Fiorenza. The basic existence theorem is the following:

THEOREM 8.5.4. *Let* Ω *be bounded with boundary of class* $C^{2+\lambda}$ *for some* λ, $0 < \lambda \leq 1$, *let the coefficients* a_{ij}, b_i, c *and* f *be in* $C^{\lambda}(\bar{\Omega})$, *suppose that* L *is uniformly elliptic in* Ω *and let* B *be the boundary operator defined by* $Bu = \sum_{i=1}^{n} g_i D_i u + gu$, *where* g_i *and* g *belong to* $C^{1+\lambda}(\partial\Omega)$, $g > 0$, *and there is a positive constant* ν_0 *such that* $\sum_{i=1}^{n} g_i(x) \cos(n, x_i) \geq \nu_0$ *on* $\partial\Omega$, *where* n *denotes the outward normal to* $\partial\Omega$ *at* x. *Then given any* $h \in C^{1+\lambda}(\partial\Omega)$, *there is exactly one function* $u \in C^{2+\lambda}(\bar{\Omega})$ *such that* $Lu = f$ *in* Ω, $Bu = h$ *on* $\partial\Omega$.

For proofs of these results we refer to Gilbarg and Trudinger (1977), where various extensions will also be found. Theorems 8.5.3 and 8.5.4 show that for equations with Hölder-continuous coefficients the main boundary-value problems have solutions with well-defined differentiability properties. If the coefficients are not Hölder-continuous it is no longer reasonable to expect that such smooth solutions should exist. It nevertheless turns out that solutions of a sort do exist, but we shall defer consideration of this topic until we come to consider elliptic equations of arbitrary order.

Turning now to parabolic second-order equations we shall consider equations of the form

$$Mu \equiv \sum_{i,j=1}^{n} a_{ij} D_i D_j u + \sum_{i=1}^{n} b_i D_i u + cu - \frac{\partial u}{\partial t} = f,$$

for which a strong maximum principle holds just as for the corresponding elliptic equation.

THEOREM 8.5.5. (*Maximum Principle*).   *Let Q be a domain in $\mathbb{R}^{n+1}$, and let the coefficients a_{ij}, b_i and c be bounded and continuous in $\bar{Q}$ with $c \leq 0$, $a_{ij} = a_{ji}$, and let the matrix (a_{ij}) be uniformly positive definite in $\bar{Q}$. Given any $(\mathbf{x}_0, t_0) \in Q$, denote by $S(\mathbf{x}_0, t_0)$ the set of all points $(\mathbf{x}, t)$ in Q that can be connected to $(\mathbf{x}_0, t_0)$ by a continuous curve in Q along which the t-coordinate is non-decreasing from $(\mathbf{x}, t)$ to $(\mathbf{x}_0, t_0)$. Suppose that $u \in C^{2,1}(Q)$ is such that $Mu \geq 0$ in Q, and that u attains a positive maximum in Q at a point $(\mathbf{x}_0, t_0) \in Q$. Then $u(\mathbf{x}, t) = u(\mathbf{x}_0, t_0)$ for all $(\mathbf{x}, t) \in S(\mathbf{x}_0, t_0)$.*

For a proof of this important result see Friedman (1964). In most applications the domain Q will be of the form $\Omega \times (0, T)$, where Ω is a domain in $\mathbb{R}^n$, and in such cases the set $S(\mathbf{x}_0, t_0)$ will be given by $\{(\mathbf{x}, t): \mathbf{x} \in \Omega, 0 < t \leq t_0\}$. The uniqueness of solutions of initial-boundary-value problems can be established by use of this theorem and its variants, and as for existence, we have the following (see, for example, Friedman (1964) and Ladyzhenskaya *et al.* (1968)):

THEOREM 8.5.6.   *Let $Q = \Omega \times (0, T)$ where Ω is a bounded domain in $\mathbb{R}^n$ and $T < \infty$. Let the boundary of Ω be of class $C^{2+\lambda}$ for some λ, $0 < \lambda \leq 1$, let the coefficients of M be in $C^{\lambda}(\bar{Q})$, let $a_{ij} = a_{ji}$, suppose that the matrix (a_{ij}) is uniformly positive definite in $\bar{Q}$, and let $f \in C^{\lambda}(\bar{Q})$ and $g \in C^{2+\lambda}(\Gamma)$, where Γ is the parabolic boundary of Q. Suppose also that the consistency condition $Mg = f$ holds on $\partial\Omega \times \{0\}$. Then there is a unique function u in $C^{2+\lambda}(\bar{Q})$ such that $Mu = f$ in Q, $u = g$ on Γ.*

This copes with Dirichlet boundary conditions; for other kinds of boundary conditions and for versions of the theorem for domains with less smooth boundaries see Ladyzhenskaya *et al.* (1968).

As for hyperbolic equations, we refer to Courant and Hilbert (1962), for a detailed discussion of the second-order case, and shall deal immediately with equations of arbitrary order, beginning with those having constant coefficients.

DEFINITION 8.5.5.   Let $P(D)$ be the differential operator

$$P(D) = \sum_{|\alpha| \leq m} a_\alpha D^\alpha \qquad (8.5.3)$$

where the a_α are constants. Define P_m to be the polynomial in $\boldsymbol{\xi}$,

$$P_m(\boldsymbol{\xi}) = \sum_{|\alpha| = m} i^{|\alpha|} a_\alpha \boldsymbol{\xi}^\alpha$$

where $\boldsymbol{\xi} = (\xi_1, \ldots, \xi_n)$, $\boldsymbol{\xi}^\alpha = \prod_{j=1}^{n} \xi_j^{\alpha_j}$, and each ξ_j is a complex number. Given any vector $\mathbf{n} \in \mathbb{R}^n$ we say that P is *hyperbolic* with respect to $\mathbf{n}$ if $P_m(\mathbf{n}) \neq 0$ and there is a real number τ_0 such that $P(\xi + i\tau\mathbf{n}) \neq 0$ if $\xi \in \mathbb{R}^n$ and $\tau < \tau_0$.

It turns out that if P is hyperbolic with respect to $\mathbf{n}$ then so is P_m; if P is homogeneous, so that $P = P_m$, then P is hyperbolic with respect to $\mathbf{n}$ if and only if $P(\mathbf{n}) \neq 0$ and the equation $P(\xi + \tau\mathbf{n}) = 0$ has only real roots τ when ξ is real; and from this latter result it follows that if P is homogeneous and hyperbolic with respect to $\mathbf{n}$ then it is proportional to a polynomial with real coefficients. Our definition is in agreement with that given earlier in this section for second-order operators [see Definition 8.5.2].

The significance of the notion of hyperbolicity becomes apparent when one wishes to study the Cauchy problem for the equation

$$P(D)u = f. \tag{8.5.4}$$

Suppose H is a half-space defined by an inequality $\langle \mathbf{x}, \mathbf{n} \rangle \equiv \sum_{j=1}^{n} x_j n_j > 0$ and suppose that the boundary of half-space ∂H, $\langle \mathbf{x}, \mathbf{n} \rangle = 0$, is non-characteristic with respect to $P(D)$. Provided that P is hyperbolic with respect to $\mathbf{n}$ it can be shown that there is exactly one solution of the Cauchy problem for $P(D)u = f$ in which the values of u and its first $(m-1)$ derivatives normal to ∂H are prescribed on the boundary. Of course we have to be rather more precise about f and the Cauchy data, but for this and for details of the proof we refer to Hörmander (1963), where it is also shown how to deal with operators with variable coefficients. For a discussion of the Cauchy problem for hyperbolic *systems* of equations see Bers, John and Schechter (1964); initial-boundary-value problems for hyperbolic equations are handled in Trèves (1975).

Let us now return to elliptic and parabolic equations, this time of arbitrary order.

DEFINITION 8.5.6.   The operator $P(\mathbf{x}, D) = \sum_{|\alpha| \leq m} a_\alpha(\mathbf{x}) D^\alpha$ with complex-valued coefficients a_α is said to be *elliptic* at a point $\mathbf{x}_0$ in a domain $\Omega \subset \mathbb{R}^n$ if for all $\xi \in \mathbb{R}^n$, $\xi \neq 0$, $\sum_{|\alpha| = m} a_\alpha(\mathbf{x}_0) \xi^\alpha \neq 0$.

If P is elliptic, it turns out that if the a_α with $|\alpha| = m$ are real, then m is even; the same conclusion holds for any elliptic P if $n > 2$, and we shall accordingly here discuss only the case with m even, say $m = 2k$.

DEFINITION 8.5.7.   The operator P is called *strongly elliptic* at $\mathbf{x}_0$ if for all $\xi \in \mathbb{R}^n \setminus \{0\}$,

$$(-1)^k \operatorname{Re} \sum_{|\alpha| = 2k} a_\alpha(\mathbf{x}_0) \xi^\alpha > 0. \tag{8.5.5}$$

If P is elliptic (or strongly elliptic) at each point of Ω it is said to be *elliptic* (or *strongly elliptic*) in Ω; if there is a positive constant c such that for all

$\mathbf{x} \in \Omega$ and all $\boldsymbol{\xi} \in \mathbb{R}^n$,

$$(-1)^k \operatorname{Re} \sum_{|\alpha| = 2k} a_\alpha(\mathbf{x}) \boldsymbol{\xi}^\alpha \geq c |\boldsymbol{\xi}|^{2k}, \tag{8.5.6}$$

and if the a_α with $|\alpha| = 2k$ are bounded in Ω, we say that P is *uniformly strongly elliptic* in Ω. The *polyharmonic operator* $(-1)^k \Delta_n^k$ is uniformly strongly elliptic in any domain Ω.

No really effective maximum principle has been found for elliptic operators of order greater than 2. However, the Dirichlet problem and other boundary-value problems for these higher-order equations are in quite good shape. Thus if Ω is a bounded domain in $\mathbb{R}^n$ with sufficiently smooth boundary, and the coefficients a_α are smooth enough, the work of Agmon, Douglis and Nirenberg (1959) enables one to deal with the Dirichlet problem

$$Pu = f \quad \text{in } \Omega,$$

u and its first $(k-1)$ normal derivatives specified on $\partial\Omega$; a Fredholm alternative situation arises in the sense that either there is a solution given any smooth f, or else there is a solution if and only if f satisfies a finite number of orthogonality relations. The same holds if instead of the Dirichlet boundary conditions given above we have the requirements

$$B_j u \equiv \sum_{|\alpha| \leq k_j} b_{j\alpha} D^\alpha u = \phi_j \quad \text{on } \partial\Omega \ (j = 1, \ldots, k)$$

where $k_j < k$ $(j = 1, \ldots, k)$; naturally the B_j cannot be entirely arbitrary but should satisfy a *complementing condition* which links them to P (see Agmon, Douglis and Nirenberg (1959)).

If the coefficients of P are not smooth the methods of Agmon, Douglis and Nirenberg do not work and we have to cast about for another method. One is to hand in the shape of the so-called *Hilbert space approach*, which we shall briefly describe for the simple case of the special Dirichlet problem which asks for a solution of the equation in *divergence form*

$$Lu \equiv \sum_{|\alpha|, |\beta| \leq k} D^\alpha(a_{\alpha\beta} D^\beta u) \tag{8.5.7}$$

in a bounded domain $\Omega \subset \mathbb{R}^n$, subject to the boundary conditions

$$D^\beta u = 0 \quad \text{on } \partial\Omega, \text{ for all } \beta \text{ such that } |\beta| \leq k - 1. \tag{8.5.8}$$

DEFINITION 8.5.8.   For any real number $p > 1$ we introduce the *Sobolev space* $W^{k,p}(\Omega)$ defined by

$$W^{k,p}(\Omega) = \{u \in L^p(\Omega) : D^\alpha u \in L^p(\Omega) \text{ for all } \alpha, |\alpha| \leq k\};$$

here the derivatives $D^\alpha u$ are supposed to be taken in the *sense of distributions*, by which we mean that $D^\alpha u$ is the unique function (apart from sets of measure zero) such that for all $\phi \in C_0^\infty(\Omega)$ (the set of infinitely differentiable functions which vanish outside some compact subset of Ω)

$$\int_\Omega u D^\alpha \phi \, d\mathbf{x} = (-1)^{|\alpha|} \int_\Omega \phi D^\alpha u \, d\mathbf{x}.$$

[See also Example 19.2.4.]

There is no implication that $D^\alpha u$ is the usual or classical αth derivative of u, which may well not exist; if it does, then it coincides with $D^\alpha u$ as defined above. Endowed with the norm [see § 11.6]

$$\|u\|_{k,p} = \left(\sum_{|\alpha| \le k} \|D^\alpha u\|_p^p \right)^{1/p},$$

where $\|v\|_p = (\int_\Omega |v(\mathbf{x})|^p \, d\mathbf{x})^{1/p}$, $W^{k,p}(\Omega)$ becomes a reflexive, separable Banach space [see Definitions 19.2.5, 19.2.6 and 19.2.20] and if $p = 2$ it is even a Hilbert space [see Definition 19.2.10], with inner product (in case all the functions are real, which we shall assume for definiteness)

$$(u, v)_{k,2} = \left(\sum_{|\alpha| \le k} (D^\alpha u, D^\alpha v)_2 \right)^{1/2},$$

where

$$(u, v)_2 = \int_\Omega u(\mathbf{x}) v(\mathbf{x}) \, d\mathbf{x}.$$

By $\mathring{W}^{m,p}(\Omega)$ we shall mean the closure in $W^{m,p}(\Omega)$ of $C_0^\infty(\Omega)$ [see Definition 19.2.5].

The Hilbert space method of handling the Dirichlet problem given above replaces the problem by a weaker problem which gives rise to a functional equation in the Hilbert space $\mathring{W}^{m,2}(\Omega)$; this new problem can easily be shown to have a solution, and then under appropriate smoothness conditions on the coefficients and Ω it may be shown that the solution of the weaker problem is actually a solution of the original problem. In more detail what we do is to introduce the *Dirichlet form*

$$a(u, \phi) = \sum_{|\alpha|,|\beta| \le k} (-1)^{|\alpha|} (a_{\alpha\beta} D^\beta u, D^\alpha \phi)_2 \tag{8.5.9}$$

and to consider the *generalized Dirichlet problem*: does there exist $u \in \mathring{W}^{m,2}(\Omega)$ such that for all *test functions* $\phi \in C_0^\infty(\Omega)$,

$$a(u, \phi) = (f, \phi)_2?$$

This arises from the original problem if we multiply the equation $Lu = f$ by ϕ, integrate over Ω and then integrate by parts [see § 4.3]. The space $\mathring{W}^{m,2}(\Omega)$ is used because the elements of it may be thought of as vanishing on the boundary in some sense. This generalized problem can easily be solved if we assume that the $a_{\alpha\beta}$ are in $L^\infty(\Omega)$ and that $f \in L^2(\Omega)$. For use of the Riesz representation theorem [see Theorem 19.2.2] quickly shows that the new problem is equivalent to the question of proving that an equation of the form

$$Tu = w,$$

in which T is a map from $\mathring{W}^{k,2}(\Omega)$ to itself induced by the Dirichlet form, has a solution no matter what element w of $\mathring{W}^{k,2}(\Omega)$ we take. If we assume that L is uniformly strongly elliptic in Ω and has suitably continuous coefficients $a_{\alpha\beta}$ for $|\alpha| = |\beta| = k$, it turns out that *Gårding's inequality* holds: there exist constants $c > 0$, $c_1 \geq 0$ such that for all $u \in \mathring{W}^{k,2}(\Omega)$,

$$a(u, u) \geq c\|u\|_{k,2}^2 - c_1\|u\|_2^2. \tag{8.5.10}$$

If $c_1 = 0$ it can be shown that T maps $\mathring{W}^{m,2}(\Omega)$ onto itself, and is injective, so that the generalized Dirichlet problem always has a unique solution. If $c_1 > 0$ a Fredholm alternative situation arises (see Friedman (1969)).

Once the existence of a solution of the generalized Dirichlet problem is known, there comes the question of its connection, if any, with the original Dirichlet problem, and for a proof of the fact that it really is a solution of that problem under suitable hypotheses on the coefficients and the boundary we refer to Friedman (1969).

This deals with the Dirichlet problem, at least for zero boundary data—if the boundary data are not smooth one simply makes an appropriate subtraction to reduce to the zero case (see Friedman (1969)). For other types of boundary-value problems the same kind of technique can be used, the general philosophy being that the boundary conditions are catered for by use of an appropriate space V; $\mathring{W}^{m,2}(\Omega) \subset V \subset W^{m,2}(\Omega)$. All this, and much else, is treated in the books Agmon (1965), Friedman (1969), Lions and Magenes (1972) and Nečas (1967).

Finally, we come to parabolic problems. A general definition of parabolicity for equations of arbitrary order is somewhat difficult to give, particularly since numerous different versions appear in the literature (see Eidelman (1969), Friedman (1964) and Ladyzenskaya *et al.* (1968)). Because of this we shall concentrate on operators of the form

$$\mathscr{L}u = \frac{\partial u}{\partial t} + \sum_{|\alpha| \leq 2k} a_\alpha(\mathbf{x}, t)D^\alpha u \tag{8.5.11}$$

with coefficients defined in a cylinder $\bar{Q}$, where $Q = \Omega \times (0, T)$, $T < \infty$, and Ω is a domain in $\mathbb{R}^n$.

DEFINITION 8.5.9.   The operator $\mathscr{L}$ is said to be *parabolic* at $(\mathbf{x}_0, t_0)$ if $\sum_{|\alpha| \leq 2k} a_\alpha(\mathbf{x}, t_0)D^\alpha u$ is strongly elliptic at x_0; $\mathscr{L}$ is parabolic on a set if it is parabolic at each point of the set; and if the coefficients a_α are bounded in $\bar{Q}$ and there is a positive constant c such that

$$(-1)^k \operatorname{Re} \sum_{|\alpha| \leq 2k} a_\alpha(\mathbf{x}, t)\boldsymbol{\xi}^\alpha \geq c|\boldsymbol{\xi}|^{2k} \tag{8.5.12}$$

for all $(\mathbf{x}, t) \in \bar{Q}$ and for all $\boldsymbol{\xi} \in \mathbb{R}^n$, then $\mathscr{L}$ is said to be *uniformly parabolic* in $\bar{Q}$.

If $\Omega = \mathbb{R}^n$ we may study the Cauchy problem for $\mathscr{L}$; if Ω is bounded, it can be proved that under suitable conditions there are solutions of the usual initial-boundary-value problems, and we shall spend a little time to outline what happens in this latter case. The typical kind of problem is

$$\mathscr{L}u = f \quad \text{in } Q,$$

$$\frac{\partial^j u}{\partial \nu^j} = \phi_j \quad \text{on } \{(\mathbf{x}, t): \mathbf{x} \in \partial\Omega, 0 < t \leq T\} \quad \text{for} \quad 0 \leq j \leq m - 1,$$

$$u(\mathbf{x}, 0) = \psi(\mathbf{x}) \quad \text{for all} \quad \mathbf{x} \in \Omega.$$

For simplicity let us suppose that all the ϕ_j and ψ are zero and that f is continuous on $\bar{Q}$. One way of tackling the problem would be to adapt the method of procedure used for the case when $k = 1$ and mentioned earlier in this section. However, this requires a good deal of smoothness of the coefficients, and if this smoothness is not present the Hilbert space approach may be used. What happens is that the problem is replaced by an *evolution equation*

$$\frac{du}{dt} + A(t)u = f(t) \tag{8.5.13}$$

in the Hilbert space $L^2(\Omega)$. Here for each t, $f(t)$ is to be understood as the function $f(\mathbf{x}, t)$ belonging to $L^2(\Omega)$, and $A(t)$ is the operator with domain $W^{2k,2}(\Omega) \cap \mathring{W}^{k,2}(\Omega)$ and given by $A(t)v(\mathbf{x}) = \sum_{|\alpha| \leq 2k} a_\alpha(\mathbf{x}, t) D^\alpha v(\mathbf{x})$; $u(t)$ is to be thought of as a function with values in $L^2(\Omega)$, so that for each t, it is a function $u(\mathbf{x}, t)$ belonging to $L^2(\Omega)$. The original initial-boundary-value problem is replaced by an initial-value problem for the evolution equation, in which one looks for a solution such that $u(0) = 0$; this initial condition is intended to correspond to the initial condition $u(\mathbf{x}, 0) = 0$ for all $\mathbf{x} \in \Omega$, while the condition that $u \in \mathring{W}^{k,2}(\Omega)$ is a weakened version of the boundary conditions involving the normal derivatives of u. Under appropriate conditions on the coefficients a_α and f, and on the boundary $\partial\Omega$, the theory of *semi-groups of operators* (see Friedman (1969)) may be used to establish the existence of a solution of this new problem, which may be shown to be a solution of the original problem. Again, all this may be adapted to cope with different kinds of boundary conditions; for a detailed exposition we refer to Friedman (1969) and Lions and Magenes (1972).

To conclude this section it is appropriate to warn of the dangers which may lie ahead if we have to deal with an equation which cannot be categorized as elliptic, parabolic or hyperbolic. Difficulties arise if, for example, we consider an equation of the form

$$\sum_{i,j=1}^{n} a_{ij} D_i D_j u + \sum_{i=1}^{n} b_i D_i u + cu = f \tag{8.5.14}$$

in some bounded domain $\Omega \subset \mathbb{R}^n$, and instead of requiring it to be elliptic in Ω, we suppose that $\sum_{i,j=1}^{n} a_{ij}(x)\xi_i\xi_j \geq 0$ for all x in Ω and all ξ in $\mathbb{R}^n$. Such equations are said to be *degenerate-elliptic* or *elliptic-parabolic*, and in the absence of further restrictions on the coefficients Dirichlet data cannot be prescribed on all of $\partial\Omega$. A striking illustration of the kind of problem which may arise is provided by the following, in the plane, due to Fichera (1960). Consider the equation

$$x_2^2 D_1^2 u - 2x_1 x_2 D_1 D_2 u + x_1^2 D_2^2 u - 2x_1 D_1 u - 2x_2 D_2 u + cu = 0,$$

where $c < 0$.

Fichera shows that if Ω is any disc centred at the origin then $u = 0$ is the only solution of the equation in Ω, so that specification of any boundary data on $\partial\Omega$ other than 0 is inadmissible. By way of contrast, if Ω is the star-shaped domain indicated in the diagram, specification of arbitrary continuous Dirichlet data on the whole of $\partial\Omega$ is allowed, and moreover existence of some kind of solution corresponding to these data can be proved. Note that such a solution would have to be identically zero in the largest disc contained in the star-like domain.

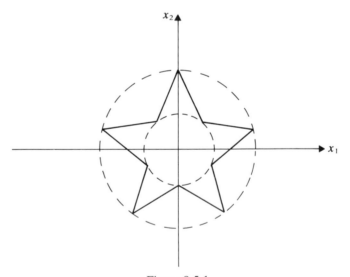

Figure 8.5.1

For some details of what is known about equations for which ellipticity degenerates, either in the domain or on its boundary, we refer to Baouendi and Goulaouic (1969), Kohn and Nirenberg (1965) and Murthy and Stampacchia (1968). For other equations which change type, such as the Tricomi equation, see Bitsadze (1964).

8.6. NON-LINEAR PROBLEMS

8.6.0. Introduction

This is such a vast subject that all that can be attempted in a short article such as this is to indicate a few areas in which progress has been made.

We begin with elliptic equations, in which there is a sharp difference between the methods used to handle second-order equations and those of higher order. In both cases, however, we study only *quasi-linear* equations, which are equations in which the derivatives of highest order occur linearly. There are good reasons for this: the equations which turn up in applications often come from variational arguments and are therefore quasi-linear; and regrettably there has been very little progress to date in the theory of non-linear equations which are not even quasi-linear, the study of the Monge–Ampère equation in the plane

$$(D_1^2 u)(D_2^2 u) = (D_1 D_2 u)^2 \tag{8.6.1}$$

being a notable exception (Miranda (1970)).

Taking the second-order case first, we consider equations of the form

$$\sum_{i,j=1}^{n} a_{ij}(\mathbf{x}, u, D_1 u, \ldots, D_n u) D_i D_j u = b(\mathbf{x}, u, D_1 u, \ldots, D_n u) \tag{8.6.2}$$

in an open subset Ω of $\mathbb{R}^n$ [see § 11.2]. The functions a_{ij} and b are assumed to be continuously differentiable functions of their arguments [see Definition 5.9.2], and the equation is *elliptic* in the sense that $\sum_{i,j=1}^{n} a_{ij}(\mathbf{x}, u, \mathbf{p})\xi_i\xi_j > 0$ for all $\mathbf{x} \in \Omega$, all real u, and all ξ and $\mathbf{p}$ in $\mathbb{R}^n$, $\xi \neq \mathbf{0}$. The classical Dirichlet problem for (8.6.2) requires that we should prove there is a function $u \in C^2(\Omega) \cap C(\bar{\Omega})$ satisfying (8.6.2) in Ω and taking on prescribed continuous boundary values on $\partial\Omega$. For uniformly elliptic equations, that is, equations for which there are positive constants ν, μ such that

$$\nu|\xi|^2 \leq \sum_{i,j=1}^{n} a_{ij}(\mathbf{x}, u, \mathbf{p})\xi_i\xi_j \leq \mu|\xi|^2$$

for all $\mathbf{x} \in \Omega$, all $u \in \mathbb{R}$ and all $\xi, \mathbf{p} \in \mathbb{R}^n$, existence of a solution of the Dirichlet problem in a bounded domain Ω can be proved, the procedure being given in full in the book by Ladyzhenskaya and Ural'tseva (1968). For non-uniformly elliptic equations the position is very much more complex, but thanks to the work of Serrin (1969) in particular, striking progress has been made and existence of solutions has been established in numerous cases by the use of techniques which go back to S. Bernstein in the early years of this century; perhaps the most remarkable result, and one which shows how the equation and the domain have to co-operate in order that there should be a solution of the problem, is that relating to the equation of constant mean curvature:

$$(1 + |\text{grad } u|^2)\Delta_n u - \sum_{i,j=1}^{n} (D_i u)(D_j u)D_{ij}u = n\Lambda(1 + |\text{grad } u|^2)^{3/2}$$

where Λ is constant. It turns out that if $\partial\Omega$ is of class C^2 the Dirichlet problem for this equation is soluble for arbitrarily given continuous boundary data if and only if the mean curvature H of $\partial\Omega$ satisfies $H \geq n/(n-1)\,|\Lambda|$ at each point of $\partial\Omega$. When $n = 2$ and $\Lambda = 0$, so that the equation is the minimal surface equation, this condition is that Ω should be convex [see § 15.2.2]. See also Nitsche (1975), Osserman (1969) and Stampacchia (1966).

In contrast to the Dirichlet problem, there has been relatively little progress in dealing with the Neumann problem for equation (8.6.2).

The higher-order equations we shall consider are of the form

$$\sum_{|\alpha|\leq m} D^\alpha A_\alpha(\mathbf{x}, u, Du, \ldots, D^m u) = f \tag{8.6.3}$$

where $D^r u$ is intended to stand for all derivatives $D^\gamma u$ with $|\gamma| = r$, and the equation is studied in a bounded domain $\Omega \subset \mathbb{R}^n$. It is supposed that the coefficients A_α are continuous and have *polynomial growth*, that is to say, there is a constant c such that for all $\mathbf{x} \in \Omega$ and all $\boldsymbol{\eta} \in \mathbb{R}^{n+1}$,

$$|A_\alpha(\mathbf{x}, \boldsymbol{\eta})| \leq c(1 + |\boldsymbol{\eta}|^{p-1}) \tag{8.6.4}$$

for some p, $1 < p < \infty$. For simplicity let us consider the special Dirichlet problem in which we look for a solution u of (8.6.3) such that $D^\beta u = 0$ on $\partial\Omega$ for all β with $|\beta| \leq m - 1$. The techniques used for the second-order problems do not work here, partly because of the lack of a maximum principle, and in any case the lack of assumed smoothness of the coefficients is such that it is appropriate to think of a solution of a generalized Dirichlet problem; we replace the original Dirichlet problem by the question: does there exist an element $u \in \overset{\circ}{W}{}^{m,p}(\Omega)$ such that

$$a(u, v) = \int_\Omega fv \, d\mathbf{x}$$

for all $v \in C_0^\infty(\Omega)$? Here a is the non-linear Dirichlet form

$$a(u, v) = \sum_{|\alpha|\leq m} (-1)^{|\alpha|} \int_\Omega A_\alpha(\mathbf{x}, u, \ldots, D^m u) D^\alpha v \, d\mathbf{x},$$

and as in the linear case discussed in section 8.5, boundary conditions are built into the space $X \equiv \overset{\circ}{W}{}^{m,p}(\Omega)$. A little work shows that the form a generates a non-linear map T from X to its dual X^* [see § 19.2.2] by the rule $(Tu)(v) = a(u, v)$, and consequently the existence of a solution is assured if it can be shown that $T(X) = X^*$. Now in the linear case, ellipticity gives rise to Gårding's inequality (8.5.10), which under further conditions on the coefficients implies that $a(u, u) \geq \text{const.}\, \|u\|_{m,2}^2$, and by anology with this it is natural to impose on a the conditions

$$\left.\begin{array}{l} a(u, u-v) - a(v, u-v) \geq 0 \quad \text{for all } u, v \in X, \\[2mm] a(u, u)/\|u\|_{m,p} \to \infty \quad \text{as } \|u\|_{m,p} \to \infty. \end{array}\right\} \tag{8.6.5}$$

The first of these, when translated in terms of the operator T, says that

$(Tu - Tv, u - v) \geq 0$ (where $(.,.)$ refers to the pairing between X and X^*), that is, it requires T to be *monotone* [cf. Definition 19.5.7]. The second condition, amounting to $(Tu, u)/\|u\|_{m,p} \to \infty$ as $\|u\|_{m,p} \to \infty$, is that T should be *coercive* [cf. Definition 19.5.8]. We are thus led to the theory of monotone operators, a chapter of non-linear functional analysis which has been developed rapidly by F. E. Browder and others in recent years (see Brézis (1973) and Browder (1966) and (1976)). According to that theory, a monotone, coercive map with mild continuity properties maps X onto X^* [cf. Theorem 19.5.4] so that provided that the structure of the A_α is such that conditions (8.6.5) are fulfilled, the generalized Dirichlet problem will have a solution, and this solution can be linked with the original Dirichlet problem given enough regularity of the boundary and the coefficients.

Extensions of these results are possible: see Browder (1970) for less restrictive hypotheses on the coefficients, Hess (1975) and Simader (1976) for the case in which Ω is unbounded and Gossez (1974) for a relaxation of the polynomial growth condition (8.6.4) on the A_α. See also Fučik (1980).

Second-order quasi-linear parabolic equations of the form

$$\frac{\partial u}{\partial t} - \sum_{i,j=1}^{n} a_{ij}(x, t, u, D_1 u, \ldots, D_n u) D_i D_j u = b(x, t, u, D_1 u, \ldots, D_n u)$$

(8.6.6)

may be handled in the same kind of way as the corresponding elliptic equations; see Edmunds and Peletier (1971), Ladyzhenskaya *et al.* (1968) and Serrin (1971) for further details. The article by Browder (1970) shows how initial-boundary-value problems for higher-order parabolic equations of the form

$$\frac{\partial u}{\partial t} - \sum_{|\alpha| \leq m} D^\alpha A_\alpha(\mathbf{x}, t, u, Du, \ldots, D^m u) = f$$

may be approached via the theory of non-linear semi-groups of operators.

For work on the non-linear wave equation $\partial^2 u/\partial t^2 - \Delta_n u + f(|u|^2) = 0$ see Browder (1962).

We have seen how certain boundary-value problems and initial-boundary-value problems can be tackled. The following boundary-value problem is of a kind which turns up in applications (see Lions (1969)), but which cannot be dealt with by the previous methods:

$$-\Delta_n u + u = f \quad \text{in } \Omega,$$

$$u \geq 0, \quad \frac{\partial u}{\partial \nu} \geq 0 \quad \text{and} \quad u \frac{\partial u}{\partial \nu} = 0 \quad \text{on } \partial\Omega.$$

Here Ω is a bounded domain in $\mathbb{R}^n$ and $\partial u/\partial \nu$ is the normal derivative of u. The unusual feature of this is that at each point of $\partial\Omega$ either u or $\partial u/\partial \nu$ is zero, but the subset of $\partial\Omega$ on which u is zero is one of the unknowns of the problem. Such problems, which may involve elliptic or parabolic operators, come under the heading of *variational inequalities*, for the theory of which we refer to Lions (1969) and Stampacchia (1969). Typically a variational inequality arises when one seeks to minimize a functional not over a linear space, which would give rise to one of the standard problems already discussed, but rather over a convex set [see § 15.2]; these inequalities are of importance in the theories of elasticity and of stopping processes, for example.

Another very important topic is that of *bifurcation theory*, in which one studies an equation of the form $F(\mathbf{x}, \lambda) = 0$, λ being a parameter. The kind of phenomenon which occurs is that as λ varies there is a family of solution $\mathbf{x}(\lambda)$ but that at some critical value of λ the family may disappear or may divide into several branches. This happens when a straight rod lying on a table is compressed by forces acting at its ends. The rod keeps its shape when the forces are small; in other words, the only solution of the equations of elasticity is the trivial one; as the forces increase a critical value is reached beyond which the rod buckles. The same phenomenon occurs when a viscous liquid is placed in the space between two concentric circular cylinders, the outer one of which is then made to rotate. At low angular velocities the liquid moves round in the obvious manner, but as the speed of rotation is increased a critical speed is reached beyond which a new and totally different form of fluid motion is seen. To analyze these questions of non-uniqueness of solutions of partial differential equations some topological apparatus is commonly used, notably the theory of the *topological degree* of a mapping. For this see Krasnosel'skii (1964a), (1964b) and Zeidler (1977), and for various accounts of bifurcation theory and its applications to the theory of partial differential equations see Krasnosel'skii (1964a), (1964b), Nirenberg (1974) and Temam and Foias (1977).

Finally we give a short list of equations having special interest, together with details of where precise information on these equations may be found.

8.6.1. The Navier-Stokes Equations

These describe the motion of a viscous incompressible fluid occupying part of $\mathbb{R}^n$ ($n = 2$ or 3), and are

$$\frac{\partial \mathbf{u}}{\partial t} - \nu \Delta_n \mathbf{u} + \sum_{i=1}^{n} u_i D_i \mathbf{u} = \mathbf{f} - \operatorname{grad} \mathbf{p},$$

$$\operatorname{div} \mathbf{u} = 0,$$

where $\mathbf{u} = (u_1, \ldots, u_n)$ represents the velocity of the fluid, $\mathbf{p}$ is the pressure, $\mathbf{f}$ is an external force and ν is a positive constant connected with the viscosity of the fluid. Initial-value and initial-boundary-value problems arise for this system of equations, the questions of interest including the existence, uniqueness and smoothness of solutions, bifurcation of solutions and the behaviour of solutions as the viscosity parameter ν tends to zero. These matters are comprehensively discussed in Ladyzhenskaya (1963), Temam (1977) and Temam and Foias (1977).

8.6.2. The Korteweg–de Vries Equation

In one space variable x this is the equation

$$\frac{\partial u}{\partial t} = 6u \frac{\partial u}{\partial x} - \frac{\partial^3 u}{\partial x^3},$$

and one studies the Cauchy problem for the equation or looks for periodic or almost periodic solutions. Currently equations of this kind, which may be described as non-linear wave equations, are the objects of intense study, one reason for this being that a connection has been discovered between them and the spectral theory of auxiliary linear operators, and this connection enables one in a sense to integrate the non-linear equations. Thus the Korteweg–de Vries (K–dV) equation has been reduced to the inverse scattering problem for the Schrödinger operator $-d^2/dx^2 + q(x)$, and knowledge of the theory of this problem enables one to solve the Cauchy problem for the K–dV equation. This method enables one to look at the asymptotic behaviour of solutions in time, and to obtain particular solutions which correspond to the interaction of a finite number of *solitons*, which are solitary waves of the form $u(x - ct)$. For a survey, with many references, of this important topic, see Dubrovin, Matveev and Novikov (1976).

8.6.3. The Porous Medium Equation

This is the degenerate parabolic equation (in one space dimension)

$$\frac{\partial u}{\partial t} = \frac{\partial^2}{\partial x^2}(u^m),$$

m (>1) being a constant. It represents the flow of a homogeneous gas through a homogeneous isotropic porous medium, and has attracted much attention recently, particularly with regard to the existence of similarity solutions and solutions which are not smooth. See Atkinson and Peletier (1971) and the references contained in this paper for work on this topic.

8.6.4. The Equations of Elasticity

Both the classical linearized equations and the non-linear versions have been studied in some detail. We refer to Ball (1977), Fichera (1972) and Knops and Payne (1971); in these works extensive bibliographies will be found.

8.6.5. Nonlinear Diffusion Equations

These turn up in such varied fields as population genetics, the theory of combustion and propagation of nerve impulses. (Of course the porous medium equation mentioned above is a particular nonlinear diffusion equation.) For example, the celebrated Hodgkin–Huxley equation

$$\frac{\partial u}{\partial t} = \frac{\partial^2 u}{\partial x^2} + F(u),$$

where F is a nonlinear function, is of this type; it arises in a model of the propagation of a voltage pulse through the nerve axon of a giant squid. All that need be said here is that such equations have been subjected to a good deal of mathematical analysis lately; the paper by Aronson and Weinberger (1975) and the references contained in it will give some idea of the flavour of work in this area.

8.6.6. Pseudo-Differential Operators

This topic has been left until the end because it is of comparatively recent origin and also requires more mathematical sophistication than most of the other subjects mentioned.

An idea of the theory may be gained by beginning with the linear partial differential operator of order m,

$$P(\mathbf{x}, D) = \sum_{|\alpha| \le m} a_\alpha(\mathbf{x}) D^\alpha.$$

We may represent the action of P on a sufficiently well-behaved function u defined on $\mathbb{R}^n$ by use of the Fourier transform [see § 13.2], and find that

$$P(\mathbf{x}, D)u = \int_{\mathbb{R}^n} \int_{\mathbb{R}^n} \exp\left(2\pi i(\mathbf{x}-\mathbf{y}) . \boldsymbol{\xi}\right) P(\mathbf{x}, \boldsymbol{\xi}) u(\mathbf{y}) \, d\mathbf{y} \, d\boldsymbol{\xi}, \qquad (8.6.7)$$

where $P(\mathbf{x}, \boldsymbol{\xi})$ is a polynomial in $\boldsymbol{\xi}$ of order m, the *symbol* of P. Pseudo-differential operators are operators given by the expression (8.6.7), but in which the symbol $P(\mathbf{x}, \boldsymbol{\xi})$ need not be a polynomial in $\boldsymbol{\xi}$, but may be an infinitely differentiable function which behaves like a sum of homogeneous functions in $\boldsymbol{\xi}$ for large $|\boldsymbol{\xi}|$. For accounts of the theory of such operators and their applications see Friedrichs (1970), Hörmander (1971), Nirenberg (1973) and Trèves (1980).

D. E. E.

REFERENCES

Agmon S. (1965). *Lectures on Elliptic Boundary Value Problems*, Van Nostrand, London.

Agmon, S., Douglis, A. and Nirenberg, L. (1959). Estimates Near the Boundary for Solutions of Elliptic Partial Differential Equations Satisfying General Boundary Conditions I, *Comm. Pure Appl. Math.* **12**, 623–727.

Aronson, D. G. and Weinberger, H. F. (1975). *Nonlinear Diffusion in Population Genetics, Combustion and Nerve Propagation, Partial Differential Equations and Related Topics*, Springer Lecture Notes 446.

Atkinson, F. V. and Peletier, L. A. (1971). Similarity of Profiles of Flows Through Porous Media, *Arch. Rat. Mech. Anal.* **42**, 369–379.

Ball, J. M. (1977). Convexity Conditions and Existence Theorems in Nonlinear Elasticity, *Arch. Rat. Mech. Anal.* **63**, 337–403.

Baouendi, M. S. and Goulaouic, C. (1969). Régularité et Théorie Spectrale pour une Classe d'Opérateurs Elliptiques Dégénérés, *Arch. Rat. Mech. Anal.* **34**, 361–379.

Bers, L., John, F. and Schechter, M. (1964). *Partial Differential Equations*, Interscience, New York.

Bitsadze, A. V. (1964). *Equations of Mixed Type*, Pergamon, London.

Brézis, H. (1973). *Opérateurs Maximaux Monotones et Semi-groupes de Contractions dans des Espaces de Hilbert*, North-Holland, Amsterdam.

Browder, F. E. (1962). On Non-Linear Wave Equations, *Math. Zeit.* **80**, 249–264.

Browder, F. E. (1966). *Problèmes non Linéares*, Montreal University Press, Montreal.

Browder, F. E. (1970). Existence Theorems for Nonlinear Partial Differential Equations, *Proc. Symp. Pure Math.* Vol. 16, American Mathematical Society, Providence.

Browder, F. E. (1976). Nonlinear Operators and Nonlinear Equations of Evolution in Banach Spaces, *Proc. Symp. Pure Math.* Vol. 18, Part 2, American Mathematical Society, Providence.

Courant, R. and Hilbert, D. (1953 and 1962). *Methods of Mathematical Physics, I and II*, Interscience, New York.

Courant, R. and Friedrichs, K. O. (1948). *Supersonic Flow and Shock Waves*, Interscience, New York.

Dubrovin, B. A., Matveev, V. B. and Novikov, S. P. (1976). Non-Linear Equations of Korteweg–de Vries type, Finite-Zone Linear Operators and Abelian Varieties, *Russ. Math. Surveys* **31**, 59–146.

Edmunds, D. E. and Peletier, L. A. (1971). Quasilinear Parabolic Equations, *Ann. Sc. Norm. Sup. Pisa* **25**, 397–421.

Eidelman, S. D. (1969). *Parabolic Systems*, North-Holland, Amsterdam.

Fichera, G. (1960). On a Unified Theory of Boundary Value Problems for Elliptic-Parabolic Equations of Second Order, *Boundary Problems in Differential Equations*, University of Wisconsin Press, Madison, pp. 97–120.

Fichera, G. (1972). Existence Theorems in Elasticity, In *Handbuch der Physik*, Vol. VIa/2, Edited by C. Truesdell, Springer-Verlag, Berlin.

Folland, G. B. (1976). *Partial Differential Equations*, Princeton University Press.

Friedman, A. (1964). *Partial Differential Equations of Parabolic Type*, Prentice-Hall, New Jersey.

Friedman, A. (1969). *Partial Differential Equations*, Holt, Rinehart and Winston, New York.

Friedrichs, K. O. (1970). *Pseudo-Differential Operators: An Introduction*, Lecture Notes, Courant Institute of Mathematical Sciences, New York University, New York.

Fučik, S. (1980). *Solvability of nonlinear equations and boundary value problems*, Society of Czechoslovak Mathematicians and Physicists and D. Reidel Publ. Co., Prague and Dordrecht.

Gilbarg, D. and Trudinger, N. S. (1977). *Elliptic Partial Differential Equations of Second Order*, Die Grundlehren der Math. Wissenschaften, Vol. 224, Springer-Verlag, Berlin.

Gossez, J. P. (1974). Nonlinear Elliptic Boundary-Value Problems for Equations with Rapidly (or Slowly) Increasing Coefficients, *Trans Amer. Math. Soc.* **190**, 163–204.

Hess, P. (1975). Problèmes aux Limites Non Linéares dans des Domaines non bornés, *C.R. Acad. Sci. Paris* **281**, 555–557.

Hörmander, L. (1963). *Linear Partial Differential Equations*, Die Grundlehren der Math. Wissenschaften, Vol. 116, Springer-Verlag, Berlin.

Hörmander, L. (1971). On the Existence and the Regularity of Solutions of Linear Pseudo-Differential Equations, *Enseignement Math.* **17**, 99–163.

Knops, R. and Payne, L. E. (1971). *Uniqueness Theorems in Linear Elasticity*, Springer-Verlag, Berlin.

Kohn, J. J. and Nirenberg, L. (1965). Non-Coercive Boundary Value Problems, *Comm. Pure Appl. Math.* **18**, 443–492.

Krasnosel'skii, M. A. (1964a). *Topological Methods in the Theory of Nonlinear Integral Equations*, Macmillan, New York.

Krasnosel'skii, M. A. (1964b). *Positive Solutions of Operator Equations*, Noordhoff, Groningen.

Ladyzhenskaya, O. A. (1963). *The Mathematical Theory of Viscous Incompressible Flow*, Gordon and Breach, New York.

Ladyzhenskaya, O. A. and Ural'tseva, N. N. (1968). *Linear and Quasilinear Elliptic Equations*, Academic Press, New York.

Ladyzhenskaya, O. A., Ural'tseva, N. N. and Solonnikov, V. (1968). Linear and Quasilinear Parabolic Equations, *Amer. Math. Soc. Translations*, Vol. 23.

Lewy, H. (1957). An Example of a Smooth Linear Partial Differential Equation without Solution, *Ann. of Math.* **66**, 155–158.

Lions, J. L. (1969). *Quelques Méthodes de Résolution de Problèmes aux Limites*, Dunod, Paris.

Lions, J. L. and Magenes, E. (1972). *Non-Homogeneous Boundary Value Problems and Applications*, Vol. I, Springer-Verlag, Berlin.

Miranda, C. (1970). *Partial Differential Equations of Elliptic Type*, Springer-Verlag, Berlin.

Murthy, M. K. V. and Stampacchia, G. (1968). Boundary Value Problems for some Degenerate Elliptic Equations, *Ann. Mat. Pura Appl.* **80**, 1–122.

Nečas, J. (1967). *Les Méthodes Directes en Théorie des Équations Elliptiques*, Masson-Academia, Paris–Prague.

Nirenberg, L. (1973). Lectures on Linear Partial Differential Equations, *Regional Conference Series in Mathematics*, Vol. 17, American Mathematical Society, Providence.

Nirenberg, L. (1974). Topics in Nonlinear Functional Analysis, *Courant Institute Lecture Notes*, New York University.

Nitsche, J. C. C. (1975). Vorlesungen über Minimalflächen, *Die Grundlehren der Mathematischen Wissenschaften*, Vol. 199, Springer-Verlag, Berlin.

Osserman, R. (1969). *A Survey of Minimal Surfaces*, Van Nostrand, New York.

Protter, M. H. and Weinberger, H. F. (1967). *Maximum Principles in Differential Equations*, Prentice-Hall, Englewood Cliffs, N.J.

Serrin, J. (1969). The Problem of Dirichlet for Quasilinear Elliptic Differential Equations with Many Independent Variables, *Phil. Trans.* **264**, 413–496.

Serrin, J. (1971). Gradient Estimates for Solutions of Nonlinear Elliptic and Parabolic Equations, *Contributions to Nonlinear Functional Analysis*, Edited by E. H. Zarantonello, Academic Press, New York.

Simader, C. G. (1976). Another Approach to the Dirichlet Problem for Very Strongly Nonlinear Elliptic Equations, Ordinary and Partial Differential Equations, *Springer Lecture Notes in Mathematics*, Vol. 564, 425–437.

Stampacchia, G. (1966). *Equations Elliptiques du Second Ordre à Coefficients Discontinus*, University of Montreal Press.

Stampacchia, G. (1969). Variational Inequalities, Theory and Applications of Monotone Operators, *Proc. NATO Advanced Study Institute*, Ediz. Oderisi, Gubbio.

Temam, R. (1977). *Navier–Stokes Equations, Theory and Numerical Analysis*, North-Holland, Amsterdam.

Temam, R. and Foias, C. (1977). Structure of the Set of Stationary Solutions of the Navier–Stokes Equations, *Comm. Pure Appl. Math.* **30**, 149–164.

Trèves, F. (1975). *Basic Linear Partial Differential Equations*, Academic Press, New York.

Trèves, F. (1980). *Introduction to pseudo-differential and Fourier integral operators. I, II*. Plenum Press, New York and London.

Weinberger, H. F. (1965). *A First Course in Partial Differential Equations*, Blaisdell, New York.

Zeidler, E. (1977). *Vorlesungen über Nichtlineare Funktionanalysis I, II, III*, Teubner, Leipzig.

CHAPTER 9

Functions of a Complex Variable

The theory of functions of a complex variable—that is, complex *analysis*—may be said to have originated in the researches of Augustin-Louis Cauchy (1789–1857); although some of his main results were anticipated in unpublished work of Carl Friedrich Gauss (1777–1855). The theory is in many ways simpler and more complete than that of real functions: for example, every differentiable function is infinitely differentiable and has a power series expansion. In other ways it is more complicated: integration, for example, involves the values of the function along some *path* in the complex plane; further, the value of the integral may depend on the choice of path. The complexities can be handled, and a very beautiful theory results.

Moreover, this theory is widely applicable in the sciences. In some ways this is a surprise: a complex number is a very artificial construct, introduced for mathematical elegance rather than for 'physical' reasons, whereas a 'real' number has direct relevance to physical measurement [see I, §§ 2.6 and 2.7]. But the mathematical elegance of complex analysis lends it considerable mathematical power; power which can be turned, so to speak, to real problems. Very early on it was recognized that a complex number could capture both the amplitude and phase of a sinusoidally varying current of electricity in a single mathematical object; and this observation turned out to be of importance in electrical engineering. Complex analysis also turned up in the two-dimensional Laplace Equation [see § 8.2], of relevance to fluid dynamics and other branches of physics, and played a considerable part in its early development. Later applications abound. And, in addition, it turns out that complex analysis can be used to calculate *real* definite integrals—a manifestly useful exercise. The whole story is a good example of a generalization first carried out for purely mathematical reasons—very sound ones, but having little to do with applications—which repaid the effort put into its development by yielding results of practical significance.

Here we shall develop some of the fundamental concepts and techniques of complex analysis, with especial regard to contour integration and power series methods. The topics covered include the Cauchy–Riemann equations,

Cauchy's Theorem on integration round a closed loop, Taylor and Laurent series, residues, and conformal mappings.

9.1. DIFFERENTIATION OF COMPLEX FUNCTIONS

Continuity and limits of complex functions are defined by direct analogy with the real case [see §§ 2.1–2.3], replacing the absolute value $|x|$ of a real number x by the absolute value $|z| = \sqrt{(x^2 + y^2)}$ of a complex number $z = x + iy$ [see I, § 2.7.1]. For example, a sequence (a_n) of complex numbers tends to a (complex) *limit* l if, for all $\varepsilon > 0$, we can find N such that

$$|a_n - l| < \varepsilon$$

for all $n > N$. Pictorially, this means that the points a_n in the complex plane approach the point l more and more closely as n increases sufficiently (Figure 9.1.1).

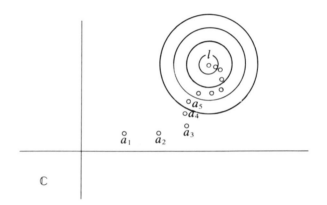

Figure 9.1.1

It is easy to show that the complex sequence (a_n) converges to the limit l if and only if the two *real* sequences (b_n) and (c_n) given by

$$b_n = \mathrm{Re}\,(a_n)$$

$$c_n = \mathrm{Im}\,(a_n)$$

converge respectively to $\mathrm{Re}\,(l)$ and $\mathrm{Im}\,(l)$ [see Proposition 1.13.3]. In other words, we can evaluate the limit of a complex sequence by considering real and imaginary parts separately [see also § 1.13].

EXAMPLE 9.1.1. To work out

$$\lim_{n \to \infty} \left(\frac{5}{n} + \frac{n^2 + 1}{2n^2} i \right)$$

we may argue that

$$\lim_{n\to\infty} \frac{5}{n} = 0, \qquad \lim_{n\to\infty} \frac{n^2+1}{2n^2} = \frac{1}{2},$$

so the desired limit is

$$0 + \tfrac{1}{2}i = \tfrac{1}{2}i.$$

Exactly analogous remarks apply to limits

$$\lim_{z\to a} f(z)$$

of a function $f : \mathbb{C} \to \mathbb{C}$. (Recall from I, § 2.7.1 that $\mathbb{C}$ denotes the set of complex numbers.)

DEFINITION 9.1.1. We say that a complex function $f : \mathbb{C} \to \mathbb{C}$ is *differentiable* at a point $z \in \mathbb{C}$ if the limit

$$\lim_{h\to 0} \frac{f(z+h)-f(z)}{h} \tag{9.1.1}$$

exists. Differentiable functions are also called *regular*. We denote its value, if it does exist, by $f'(z)$ or df/dz; and then f' becomes a complex function, the *derivative* of f, (whose domain is the set of all $z \in \mathbb{C}$ for which f is differentiable at z). In fact, we need not assume that f has domain $\mathbb{C}$: if $f : D \to \mathbb{C}$ where D is a subset of $\mathbb{C}$, similar remarks apply.

It is technically convenient to assume that the domain D is an *open* subset of the complex plane: that is to say that, given any point $z_0 \in D$ there exists a disc $|z - z_0| < r$, with $r > 0$, that also lies in D. In future the word 'domain' will carry the connotation 'open'.

The resulting theory of differentiation closely resembles the real case. The formulae for differentiating sums, products, and quotients look exactly the same; and so does the Chain Rule for composites of functions [see (3.2.8)]. The derivative of a polynomial function is computed exactly as for a real function [see Example 3.2.1]: for example if $f(z) = z^n$ then $f'(z) = nz^{n-1}$. For this reason we shall not elaborate on these results, but instead go on to the more interesting properties of complex differentiation, which do not have real counterparts.

Of these a fundamental one is a relation between the real and imaginary parts of the derivative. Let $z = x + iy \in \mathbb{C}$, and let

$$f(z) = u(z) + iv(z)$$

where $u(z)$ and $v(z)$ are the real and imaginary parts of $f(z)$. For example, if $f(z) = z^3$, then

$$f(z) = (x + iy)^3$$
$$= x^3 - 3xy^2 + i(3x^2y - y^3)$$

so that $u(x, y) = x^3 - 3xy^2$, $v(x, y) = 3x^2y - y^3$.

It is a simple matter to show that f is continuous [see § 2.1] if and only if each of u and v is a continuous function of two real variables [see Definition 5.1.1]. Conditions for differentiability of f are not so straightforward.

THEOREM 9.1.1.   *Let $f:D \to \mathbb{C}$ be a function whose domain D is a subset of $\mathbb{C}$; let $f = u + iv$, where u and v are the real and imaginary parts of f. Suppose that u and v have continuous first order partial derivatives throughout D. Then f is differentiable in D if and only if the equations*

$$\frac{\partial u}{\partial x} = \frac{\partial v}{\partial y}, \qquad \frac{\partial v}{\partial x} = -\frac{\partial u}{\partial y} \qquad\qquad (9.1.2)$$

hold for all $x + iy \in D$. Further,

$$f'(z) = \frac{\partial u}{\partial x} + i\frac{\partial v}{\partial x}.$$

For example, with $f(z) = z^3$ we have [see § 5.2]

$$\frac{\partial u}{\partial x} = 3x^2 - 3y^2 = \frac{\partial v}{\partial y}$$

$$\frac{\partial v}{\partial x} = 6xy = -\frac{\partial u}{\partial y}.$$

Equations (9.1.2) are called the *Cauchy–Riemann equations*. They are proved in any text on complex analysis; for example Titchmarsh (1932), Ahlfors (1966), Tall (1977), Jameson (1970).

There is nothing resembling these equations for real functions. As a result, differentiability takes on a very different flavour for complex functions, and it is customary to use a different word: henceforth we call a differentiable complex function *analytic*, although some authors use the term *holomorphic*. (The word 'analytic' refers to an equivalent, though apparently stronger property: having a power series expansion. See section 9.6.)

EXAMPLE 9.1.2.   Is the function $f:\mathbb{C} \to \mathbb{C}$ defined by

$$f(x + iy) = x^2 + y^2 + 2ixy$$

analytic?

We have, in the above notation,

$$u(x, y) = x^2 + y^2$$

$$v(x, y) = 2xy.$$

These are certainly continuous. However,

$$\frac{\partial v}{\partial x} = 2y \neq -\frac{\partial u}{\partial y} = -2x.$$

So f is not analytic.

EXAMPLE 9.1.2.   Is the function $f : \mathbb{C} \to \mathbb{C}$ defined by

$$f(x + iy) = \sin x \cosh y + i \cos x \sinh y$$

analytic? [see §§2.12, 2.13].
   Now

$$u(x, y) = \sin x \cosh y$$
$$v(x, y) = \cos x \sinh y.$$

These are continuous, and

$$\frac{\partial u}{\partial x} = \cos x \cosh y = \frac{\partial v}{\partial y}$$

$$\frac{\partial v}{\partial x} = -\sin x \sinh y = -\frac{\partial u}{\partial y}.$$

So f is analytic.

9.2.   PATHS IN THE COMPLEX PLANE

   The integral of a real function between upper and lower limits is uniquely defined. But for a complex function, where the upper and lower limits are points in the complex plane, the value of the integral depends in general on the path by which we move from one point to the other. For this reason we first make precise the idea of a path.

DEFINITION 9.2.1.   A *path* in the complex plane is a continuous map

$$\gamma : [a, b] \to \mathbb{C},$$

where $[a, b]$ is a closed interval on the real line,

$$[a, b] = \{x \in \mathbb{R} \mid a \le x \le b\}.$$

Its *initial* point is $\gamma(a)$, its *terminal* point is $\gamma(b)$, and its *track* is the set of points

$$\{\gamma(t) \mid a \le t \le b\}.$$

Less formally we speak of a path γ from z_1 to z_2, these being the initial and terminal points respectively. We say that t *parametrizes* the path.

   For an intuitive picture think of t as representing time. Then a path is a moving point on the plane, and its track is a curve drawn on the plane by the moving point. Figure 9.2.1 illustrates this.

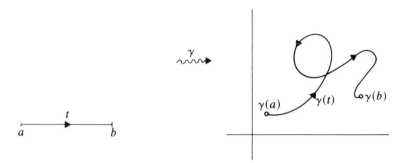

Figure 9.2.1

The distinction between paths and tracks is important. Different paths may have the same track, as shown in Figure 9.2.2; and the *direction* along the track defined by increasing t is especially important since it affects the value of an integral along the path.

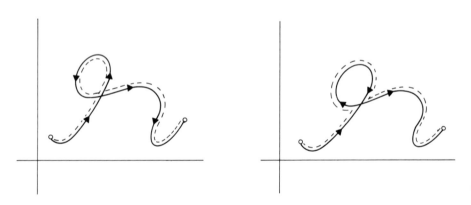

Figure 9.2.2

The word 'continuous' in the definition of path refers to a function $\mathbb{R} \to \mathbb{C}$, not quite our usual situation: it may be defined in the obvious way, or we can insist that if $\gamma(t) = u(t) + iv(t)$ then u and v are continuous functions of t in the usual sense. The latter is very easy to check, since we have great familiarity with real functions.

EXAMPLE 9.2.1. If we define $\gamma(t) = (t^3 - 4) + it^2$ for $0 \le t \le 1$ then γ is a path. The initial point is -4, the terminal point is $-3 + i$, and the track is illustrated in Figure 9.2.3.

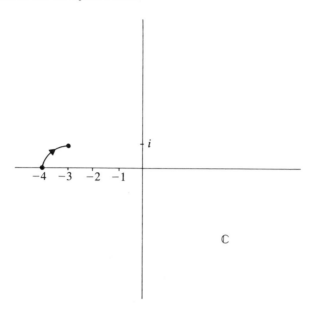

Figure 9.2.3

Although different paths can have the same track, it is sometimes useful to refer to the geometry of the track in cases where we can standardize our choice of path. Except where the contrary is stated, we will take the paths to be given by the following standard choices of γ in the following cases [see V, § 1.3.2 and § 2.2.1]:

(i) The line segment from z_0 to z_1: $\gamma(t) = z_0(1-t) + z_1 t$, $0 \leq t \leq 1$.
(ii) A polygon with vertices $z_0, z_1, \ldots, z_n$:
$\gamma(t) = z_m(m+1-t) + z_{m+1}(t-m)$, $m \leq t \leq m+1$, $0 \leq m \leq n-1$.
(iii) The *unit circle* $\gamma(t) = \cos t + i \sin t$, $0 \leq t \leq 2\pi$.
(iv) The circle centre z_0, radius r: $\gamma(t) = z_0 + r(\cos t + i \sin t)$, $0 \leq t \leq 2\pi$.

If the terminal point of a path γ is the same as the initial point of a path δ, then we can form the *composite* path $\gamma + \delta$ defined by

$$(\gamma + \delta)(t) = \begin{cases} \gamma(t) & a \leq t \leq b \\ \delta(t-b+c) & b \leq t \leq d-c+b \end{cases} \tag{9.2.1}$$

where γ, δ are defined on $[a, b]$ and $[c, d]$ respectively, and $\gamma(b) = \delta(c)$. Intuitively we get $\gamma + \delta$ by sticking the tracks together, and adjusting the clock on δ so that the time intervals adjoin; then we run round both in turn. Given a sequence of paths $\gamma_1, \ldots, \gamma_n$ whose end points fit properly, we can similarly define $\gamma_1 + \ldots + \gamma_n$. See Figure 9.2.4. We can also define the *reverse* path $-\gamma$ by setting $-\gamma(t) = \gamma(a+b-t)$.

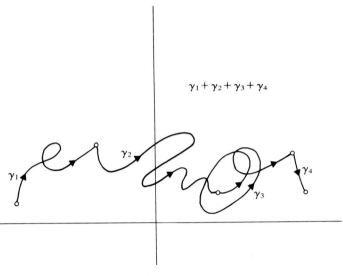

$$\gamma_1 + \gamma_2 + \gamma_3 + \gamma_4$$

Figure 9.2.4

If the track of γ lies inside a subset S of the complex plane, we say that γ is a *path in S*.

9.3.   INTEGRATION

We now define the integral of a complex function $f: D \to \mathbb{C}$, where D is a domain. This is done by analogy with the Riemann integral

$$\int_a^b g(x)\, dx$$

of a real function g [see § 4.1]. Because of the way the interval $[a, b]$ is subdivided in defining the Riemann integral, it turns out to be necessary to consider a path γ in D, and define the complex integral of a function $f(z)$ *along* γ.

Provided γ is sufficiently well behaved (has a continuous derivative [see Definition 2.10.2]) we can give a formula for the integral in terms of real functions only. Although it would be possible to use this formula to *define* the integral, it would obscure the analogy with the real case: however, the formula remains a preferable way to *compute* the integral.

This dependence of integrals upon paths is one of the most striking features of complex analysis, and is where many of its greatest differences from the real case arise.

We begin by recalling the definition of the Riemann integral

$$\int_a^b f(x)\, dx$$

of a real-valued function f [see Definition 4.1.3]. The idea is to subdivide the interval $[a, b]$ using a partition P given by points t_r such that

$$a = t_0 < t_1 < \ldots < t_n = b.$$

We then choose points s_r in the subintervals $[t_{r-1}, t_r]$ and form the *Riemann sum*

$$S(P) = \sum_{r=1}^{n} f(s_r)(t_r - t_{r-1}) \tag{9.3.1}$$

which approximates the area under the graph of f as shown in Figure 9.3.1. We define the *mesh* of P to be the maximum value of the lengths of the subintervals. The integral is then the limit of $S(P)$ as the mesh tends to zero, provided that limit exists: functions for which it exists are said to be *integrable*, and every continuous function has that property.

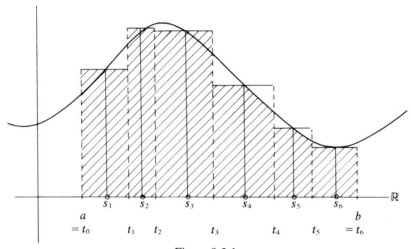

Figure 9.3.1

To find an analogous definition for the integral of a complex function is not hard: the only genuine problem is deciding what a partition P should look like. This problem has an obvious solution if we work in terms of a fixed path γ between $z_1 = \gamma(a)$ and $z_2 = \gamma(b)$. For we can then subdivide the *interval* $[a, b]$ as before, and use these points to subdivide the track of the path.

For precision, we assume that $f : D \to \mathbb{C}$ is continuous, and that $\gamma : [a, b] \to D$ is a path in D for which γ' exists and is continuous (that is, γ is a *smooth* path). In terms of real functions this means that if $\gamma(t) = x(t) + iy(t)$ then x and y are differentiable real functions [see § 3.1].

Subdivide $[a, b]$ by a partition P as before: $a = t_0 < t_1 < \ldots < t_n = b$. Choose points s_r in $[a, b]$ such that $t_{r-1} \le s_r \le t_r$ for $r = 1, \ldots, n$. Form the sum

$$\sum_{r=1}^{n} f(\gamma(s_r))(\gamma(t_r) - \gamma(t_{r-1})). \tag{9.3.2}$$

If we write $\xi_r = \gamma(s_r)$, and $z_r = \gamma(t_r)$, this sum becomes

$$\sum_{r=1}^{n} f(\xi_r)(z_r - z_{r-1})$$

and the analogy with the real case is clearer: the z_r subdivide the track of γ and the ξ_r lie on γ between the z_r as shown in Figure 9.3.2. It is harder to think of the sum as approximating an 'area', however, since the numbers

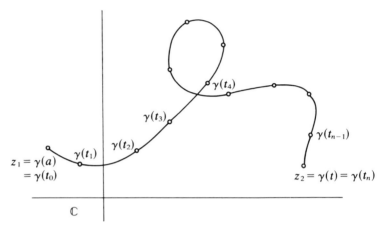

Figure 9.3.2

involved are complex: instead we should think of it as a natural extension of the real case. If the sum (9.3.2) tends to a limit as the mesh of P tends to zero, then we define that limit to be the *integral of f along* γ,

$$\int_{\gamma} f \quad \text{or} \quad \int_{\gamma} f(z)\, dz.$$

In fact we can show that under the given hypotheses on f and γ, the limit does exist; and we can even evaluate it. The detailed calculation may be found in Titchmarsh (1932), p. 71: its result is that the integral of f along γ is given by

$$\int_{a}^{b} f(\gamma(t))\gamma'(t)\, dt$$

which, as an integral of a complex function of a real variable t, involves no choice of path for t: it may be defined directly by writing $f(\gamma(t))\gamma'(t) = U(t) + iV(t)$, for real functions U and V, and the integral then becomes

$$\int_{a}^{b} U(t)\, dt + i \int_{a}^{b} V(t)\, dt,$$

an expression involving only real functions under the integral signs.

EXAMPLE 9.3.1.   Let $\gamma(t) = t^2 + it$, $f(z) = z^2$, $0 \leqslant t \leqslant 1$. Evaluate $\int_\gamma f(z)\, dz$.
This is equal to

$$\int_0^1 f(\gamma(t))\gamma'(t)\, dt = \int_0^1 (t^2 + it)^2(2t + i)\, dt$$

$$= \int_0^1 (2t^5 - 4t^3)\, dt + i \int_0^1 (5t^4 - t^2)\, dt$$

$$= -\tfrac{2}{3} + \tfrac{2}{3}i.$$

EXAMPLE 9.3.2.   Integrate $1/z$ around the unit circle.

Here $\gamma(t) = \cos t + i \sin t$, $0 \leq t \leq 2\pi$. By the formula, we have

$$\int_\gamma \frac{dz}{z} = \int_0^{2\pi} \frac{1}{\cos t + i \sin t} \cdot (-\sin t + i \cos t)\, dt$$

$$= \int_0^{2\pi} i \cdot dt$$

$$= 2\pi i.$$

It is easy to extend the definition of the complex integral to paths that are
only *piecewise* smooth; that is paths

$$\gamma = \gamma_1 + \ldots + \gamma_n \tag{9.3.3}$$

where the γ_r are smooth. All we do is set

$$\int_\gamma f = \int_{\gamma_1} f + \ldots + \int_{\gamma_n} f. \tag{9.3.4}$$

For convenience, such a path γ will be called a *contour*. Accordingly (9.3.4)
is a *contour integral*.

9.4.   CAUCHY'S THEOREM

The theorem we shall now expound is without doubt the central result in
complex analysis: the influence it has on the subsequent development of the
subject is enormous. That its potential is so great may not be immediately
clear, however. It asserts merely that, under certain conditions, the integral
of a complex function around a *closed* contour is zero. However, one consequence is that integrals round certain paths may be replaced by integrals
round quite different paths, and this yields a 'geometric' approach to calculating them which has no counterpart for real functions.

We begin by setting up a few topological notions (without proofs) that are
relevant to the theorem.

DEFINITION 9.4.1.   A contour is *closed* if its initial and terminal points coincide (Figure 9.4.1); such a contour is also called a *loop*. A closed contour is *simple* if it does not cross itself, (Figure 9.4.2).

Figure 9.4.1   Figure 9.4.2

A famous—and difficult—result is the *Jordan curve theorem*:

THEOREM 9.4.1.   *Every simple closed contour possesses an inside and an outside which, together with the contour, yield the whole complex plane.*

The proof of this is more subtle than one might expect, and we shall make no attempt even to indicate it here: the result is however extremely plausible.

Given any closed contour and any point not on it, we can count how many times the contour winds around the point in an anticlockwise direction. We call this the *winding number* of the contour relative to the point. If γ is the contour and z_0 the point, then the winding number is given by the integral

$$W(\gamma, z_0) = \frac{1}{2\pi i} \int_\gamma \frac{1}{z - z_0} \, dz. \tag{9.4.1}$$

(To see this is plausible, move the origin by putting $w = z - z_0$. By Example 9.3.2 above, the integral of dw/w along the unit circle is $2\pi i$. Obviously the integral along a contour that winds n times around the unit circle is $2n\pi i$. This shows that the formula makes sense for paths round a circle: the extension to arbitrary paths is not especially hard.) The winding number is always an *integer* for closed paths γ.

Cauchy's theorem comes in an amazing variety of versions, and we shall here record two of the more general.

THEOREM 9.4.2.   *If f is differentiable in a domain D and γ is a closed contour in D whose winding number relative to any point not in D is zero, then $\int_\gamma f = 0$.*

THEOREM 9.4.3.   *If f is differentiable in a domain D and γ is a simple closed contour in D whose inside lies in D, then $\int_\gamma f = 0$.*

(The point of Theorem 9.4.3 is that D should have no 'holes' lying inside the track of γ. Thus Figure 9.4.3 is ruled out.)

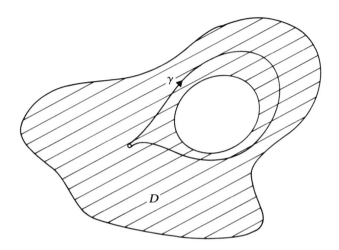

Figure 9.4.3

Cauchy's theorem *fails* for contours which have non-zero winding number about a point outside the domain, or for simple contours whose inside does not lie fully within D. The simplest example is our previous Example 9.3.2: the integral of $1/z$ around the unit circle is *not* zero. Note that the domain of $1/z$ is the set of *non-zero* complex numbers; the contour winds around the point 0, not in the domain, and indeed 0 is inside the contour but not in the domain. Thus Cauchy's theorem, in either form, does not hold for completely general contours.

Fortunately, the greatest interest lies in the situations where the theorem fails, because we can 'control' to some extent the amount of failure. [see § 9.9].

DEFINITION 9.4.2.   Let us say that a domain D is *simply connected* if the inside of every simple closed contour in D is also in D. Thus D has no 'holes' at all.

One consequence of Theorem 9.4.2 is that if γ and δ are paths in a simply connected domain, with the same initial points and the same terminal points, then $\int_\gamma f = \int_\delta f$. To see this, let $\gamma - \delta$ be the closed contour obtained by composing γ with the reverse of δ. By Theorem 9.4.2, $\int_{\gamma - \delta} f = 0$; hence $\int_\gamma f = \int_\delta f$. [See Figure 9.4.4.]

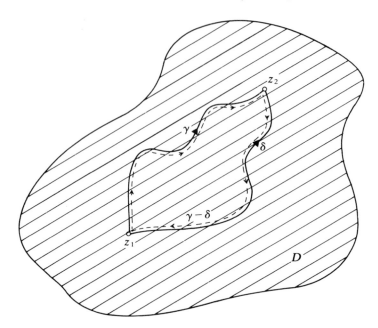

Figure 9.4.4

Simply connected domains are very well behaved. If f is a differentiable function defined on a simply connected domain D, then it may be shown that f has a *primitive* or *antiderivative* F in D. This means that the derivative F' equals f.

To see this, let z_0 be any point in the simply connected domain, and w any other point, and define

$$F(w) = \int_\gamma f(z)\, dz \qquad (9.4.2)$$

where γ is *any* path from z_0 to w inside the domain. By our above remark, (9.4.2) is independent of γ; and it is easy to show that $F' = f$.

It then follows that for any path γ in a simply connected domain, with initial point z_0 and terminal point z_1, we have

$$\int_\gamma f(z)\, dz = F(z_1) - F(z_0), \qquad (9.4.3)$$

a situation highly reminiscent of the real case.

In a non-simply connected domain, however, this result fails, for the same reason that Cauchy's theorem does.

9.5.   POWER SERIES

Weierstrass discovered that complex function theory could be approached using power series

$$\sum_{n=0}^{\infty} a_n z^n$$

to define functions $f(z)$. This idea proved very fruitful. We can define convergence of such series exactly as in the real case [see § 1.10]; and ideas such as absolute convergence [see § 1.9] carry over. The first important result is that power series converge on discs:

THEOREM 9.5.1.   *For any power series $\sum a_n z^n$ there exists a real number $R \geq 0$ (possibly $R = \infty$) such that*:
(i) *If $|z| < R$ then the series converges absolutely*,
(ii) *If $|z| > R$ the series diverges.*

(Exactly what happens for $|z| = R$ depends intricately on the series and on z.)
For example the power series $\sum_{n=0}^{\infty} z^n$ converges for $|z| < 1$, and diverges for $|z| > 1$. When it converges its value is $1/(1-z)$ [see (1.10.9)]. We call R the *radius of convergence*. It can be calculated using the formula

$$\frac{1}{R} = \limsup |a_n|^{1/n}. \tag{9.5.1}$$

The proof of this involves only the comparison test for series, and is quite easy [see Theorem 1.10.1].
The disc

$$\{z \in \mathbb{C}: |z| < R\} \tag{9.5.2}$$

is called the *disc of convergence* of the series.
A power series may be differentiated or integrated term by term within the disc of convergence (but not necessarily on its boundary). So, if

$$f(z) = \sum_{n=0}^{\infty} a_n z^n$$

has radius of convergence R, then for $|z| < R$,

$$f'(z) = \sum_{n=0}^{\infty} n a_n z^{n-1} \tag{9.5.3}$$

and

$$\int_{\gamma} f(z)\, dz = \sum_{n=0}^{\infty} \frac{a_n}{n+1} z^{n+1} \tag{9.5.4}$$

where γ is *any* contour from the origin to the point z, that stays within the disc of convergence. (That the integral is independent of γ follows by Cauchy's theorem.) By induction it follows that a power series may be integrated or differentiated arbitrarily many times within the disc of convergence.

Thus, for example, since we know that

$$1/(1-z) = 1 + z + z^2 + \ldots$$

it follows by taking derivatives that

$$1/(1-z)^2 = 1 + 2z + 3z^2 + \ldots,$$

$$2/(1-z)^3 = 2 + 6z + 12z^2 + \ldots,$$

and so on.

Numerous important real functions, such as the trigonometric and exponential functions, may be defined by *real* power series [see §§ 2.11, 2.12]. By permitting the variable in these series to become complex we can extend the definition to the complex case.

For example, if $z \in \mathbb{C}$ we define

$$\exp z = e^z = \sum_{n=0}^{\infty} \frac{z^n}{n!} \tag{9.5.5}$$

$$\sin z = \sum_{n=0}^{\infty} (-1)^n \frac{z^{2n+1}}{(2n+1)!} \tag{9.5.6}$$

$$\cos z = \sum_{n=0}^{\infty} (-1)^n \frac{z^{2n}}{(2n)!} \tag{9.5.7}$$

Each of these series has radius of convergence $R = \infty$. Thus the exponential, sine, and cosine functions are defined for all complex z. The familiar properties of these functions, such as

$$\exp(z_1 + z_2) = \exp z_1 \exp z_2$$

may be proved by manipulating series as in the real case [see (1.11.7) and (2.11.2)].

By expanding $\exp(iz)$ in a power series (put iz for z in the series for $\exp(z)$) and separating into real and imaginary parts, we obtain the important result

$$\exp(iz) = \cos z + i \sin z. \tag{9.5.8}$$

Changing z to nz and noting that $e^{inz} = (e^{iz})^n$ we obtain *De Moivre's formula* [see also I (2.7.29)]

$$(\cos z + i \sin z)^n = \cos(nz) + i \sin(nz). \tag{9.5.9}$$

Changing iz to $-iz$, we get

$$e^{-iz} = \cos z - i \sin z,$$

and it follows that

$$\cos z = \tfrac{1}{2}(e^{iz} + e^{-iz}) \tag{9.5.10}$$

$$\sin z = \frac{1}{2i}(e^{iz} - e^{-iz}).\qquad(9.5.11)$$

Complex hyperbolic functions are defined as in the real case [see § 2.13]

$$\cosh z = \tfrac{1}{2}(e^z + e^{-z})\qquad(9.5.12)$$

$$\sinh z = \tfrac{1}{2}(e^z - e^{-z}).\qquad(9.5.13)$$

Comparing with the formulae (9.5.10) and (9.5.11) for sine and cosine, we find that

$$\cosh(iz) = \cos z\qquad(9.5.14)$$

$$\sinh(iz) = i . \sin z.\qquad(9.5.15)$$

Thus there is an intimate link between trigonometric and hyperbolic functions for complex z. In particular we may evaluate the real and imaginary parts of the function $\sin z$ as follows:

$$\sin(x + iy) = \sin x \cos(iy) + \cos x \sin(iy)$$

$$= \sin x \cosh y - i \cos x \sinh(-y)$$

$$= \sin x \cosh y + i . \cos x \sinh y.\qquad(9.5.16)$$

There is a similar formula for the cosine.

Suppose that r is any real number ≥ 0, and θ any real number whatsoever. Consider the complex number

$$r e^{i\theta} = r(\cos\theta + i \sin\theta) = r . \cos\theta + ir . \sin\theta.$$

Geometrically, this lies distance r from the origin, and the line joining it to the origin makes an angle θ radians with the real axis. Thus (r, θ) are the *polar* coordinates of the point concerned [see I, § 2.7.3], as in Figure 9.5.1.

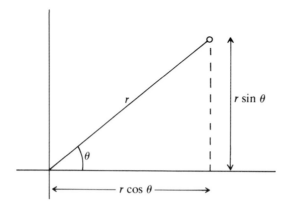

Figure 9.5.1

It follows that every complex number z can be put into the form $r\,e^{i\theta}$. Here $r = |z|$ is unique; whereas θ is unique only up to integer multiples of 2π. If we insist that $0 \le \theta < 2\pi$ then θ may be defined uniquely also. We call θ the *argument* of z, and write

$$\theta = \arg z$$

[see also I, § 2.7.3].

Since $\exp{(iy)} = \cos y + i \sin y$ it follows that

$$|\exp{(iy)}| = \cos^2 y + \sin^2 y = 1. \tag{9.5.17}$$

Therefore

$$|\exp{(x+iy)}| = |\exp x \,.\, \exp{(iy)}| = |\exp x||\exp{(iy)}| = |\exp x|. \tag{9.5.18}$$

We have not yet discussed the complex logarithmic function. This is much more interesting, and we shall briefly explain why. As in the real case, we want the logarithm to be defined by the equation

$$\exp \log{(z)} = z.$$

Writing

$$z = r \exp{(i\theta)}$$

$$= \exp{(\log r + i\theta)}$$

(where $\log r$ is the usual *real* logarithm [see § 2.11]) it follows that we should define

$$\log z = \log r + i\,.\,\arg z + 2n\pi i$$

for arbitrary integer n. (This is because $\theta = \arg z + 2n\pi$, determined only up to integer multiples of 2π.) Thus the logarithmic function becomes *multi-valued*. In the sense of set theory a 'genuine' function must be *single*-valued [see I, Definition 1.4.1]. It is not sufficient to choose one value, say $\log{(r)} + i\,.\,\arg{(z)}$, because this has 'bad' continuity properties near the real axis: instead we must suitably extend our function concept. This can be done, and the best way is to introduce 'Riemann surfaces' [see § 9.17 and Titchmarsh (1932) p. 146] to take care of the problem. At the moment, however, it is preferable for us to think of the logarithm as a function whose definition is slightly ambiguous, and whose precise value on any given occasion must be *chosen* by some stated rule from among the infinitely many possible values (9.5.17). The rule may vary to suit the occasion.

9.6.   TAYLOR SERIES

As in the real case [see § 3.6], a complex function may often be expanded in a Taylor series. The following result, of independent interest, is crucial to a proper investigation.

THEOREM 9.6.1   (*Cauchy's Integral Formula*).   *If* $z_0 \in \mathbb{C}$, *and* f *is differentiable inside a disc centre* z_0, *radius* R; *and if* $\gamma(t) = z_0 + r e^{it}$ $(0 \le t \le 2\pi)$ *is a circular path with* $r < R$, *then for* $|z_0 - w| < r$ *we have*

$$f(w) = \frac{1}{2\pi i} \int_\gamma \frac{f(z)}{z - w} \, dz.$$

Using this result, we may prove

THEOREM 9.6.2.   *Suppose that* f *is differentiable for* $|z - z_0| < R$. *Then*

$$f(z_0 + h) = \sum_{n=0}^{\infty} a_n h^n$$

where the series converges absolutely for $|h| < R$. *Further, taking* γ *as above, we have*

$$a_n = \frac{1}{2\pi i} \int_\gamma \frac{f(z)}{(z - z_0)^{n+1}} \, dz.$$

Essentially this follows by putting $w = z_0 + h$ in the Cauchy integral formula, and expanding $1/(z - w)$ in powers of h: a detailed proof may be found in [Titchmarsh (1932) p. 83].

This theorem implies that any differentiable function has *some* power series expansion in some disc; however the formula for the coefficient a_n is not yet the familiar one for a Taylor series. But once we know that a power series exists, it is trivial to find the coefficients, using the standard term-by-term differentiation: this leads to the alternative formula

$$a_n = \frac{f^{(n)}(z_0)}{n!}.$$

We then have the standard Taylor expansion:

THEOREM 9.6.3   (*Taylor series*).   *If* f *is differentiable in a domain* D, *then* f *is differentiable arbitrarily many times in* D. *If* $z_0 \in D$ *and the disc* $|z - z_0| < r$ *lies inside* D, *then there exists a convergent power series expansion*

$$f(z_0 + h) = \sum_{n=0}^{\infty} \frac{f^{(n)}(z_0)}{n!} h^n$$

for $|h| < r$.

The result here is far stronger than in the real case. A differentiable real function need not be twice differentiable; and even a *smooth* real function may not possess a valid Taylor series (see § 2.10). Yet in the complex case, *differentiability implies that the function is infinitely differentiable and has a valid Taylor expansion.* To emphasize this property, formally much stronger than differentiability, we call such a function *analytic*. The change is of emphasis *only*: the properties 'analytic' and 'differentiable' are logically equivalent, by the theorem above. But this is a highly non-trivial equivalence!

Putting $z_0 + h = z$ in the above, we obtain a formula for $f(z)$ as a power series in $(z - z_0)$. We call this the *Taylor expansion of f about z_0*.

9.7. LAURENT SERIES

The Taylor series expansion is too limited for many applications. A useful generalization was given by Laurent, who considered 'power series' involving negative powers as well as positive. The benefits which accrue are hinted at by the following example. The function $f(z) = \exp(-1/z^2)$ is very badly behaved as regards Taylor series expanson. We have seen (in § 2.10) that, restricted to the real line, its Taylor series is $0 + 0x + 0x^2 + \ldots$ which does not converge to $f(x)$; and on the whole complex plane it is, if such a statement makes sense, even less capable of being represented by a Taylor series! The natural series representation is obtained by starting with the series for exp z and replacing z by $-1/z^2$, which gives

$$f(z) = 1 - z^{-2} + \frac{1}{2!} z^{-4} - \frac{1}{3!} z^{-6} + \ldots$$

and this is a series of the 'negative powers' type. It converges for all z for which $-1/z^2$ is defined, namely $z \neq 0$.

The general series of this type can be written in the form

$$\sum_{n=-\infty}^{\infty} a_n (z - z_0)^n,$$

where this is to be thought of as a compact notation for

$$\left(\sum_{n=0}^{\infty} a_n (z - z_0)^n \right) + \left(\sum_{n=1}^{\infty} a_{-n} (z - z_0)^{-n} \right)$$

and hence converges if and only if the two bracketed series converge. We know that power series converge inside a disc, [see Theorem 9.5.1]. Consequently power series with negative powers alone should converge *outside* a disc (for instance, that for $\exp(-1/z^2)$ converges outside the disc $|z| = 0$) and those with both positive and negative powers should converge in the region *between* two concentric circles. Such a region is called an *annulus*: more precisely, if R_1 and R_2 are real numbers or ∞, with $0 \leq R_1 < R_2 \leq \infty$, and if $z_0 \in \mathbb{C}$, then

$$\{z \in \mathbb{C} : R_1 \leq |z - z_0| \leq R_2\} \qquad (9.7.1)$$

is an annulus.

THEOREM 9.7.1 (*Laurent's theorem*). *If f is differentiable in the annulus* $R_1 \le |z - z_0| \le R_2$ *where* $0 \le R_1 < R_2 \le \infty$ *then*

$$f(z_0 + h) = \sum_{n=0}^{\infty} a_n h^n + \sum_{n=1}^{\infty} b_n h^{-n}$$

where $\sum a_n h^n$ *converges for* $|h| < R_2$, $\sum b_n h^{-n}$ *converges for* $|h| > R_1$, *and in particular both series converge on the interior of the given annulus.*

Further, if $C_r(t) = z_0 + r e^{it}$ *where* $R_1 < r < R_2$, $0 \le t \le 2\pi$, *then*

$$a_n = \frac{1}{2\pi i} \int_{C_r} \frac{f(z)}{(z - z_0)^{n+1}} \, dz,$$

$$b_n = \frac{1}{2\pi i} \int_{C_r} f(z)(z - z_0)^{n-1} \, dz.$$

{*Remark.* In the more compact notation, we set $c_n = a_n$ $(n \ge 0)$ and $c_{-n} = b_n$ $(n \ge 1)$ and then

$$f(z_0 + h) = \sum_{n=-\infty}^{\infty} c_n h^n$$

which converges on the interior of the annulus; and then

$$c_n = \frac{1}{2\pi i} \int_{C_r} \frac{f(z)}{(z - z_0)^{n+1}} \, dz$$

for all $n \in \mathbb{Z}$.}

Note that we can no longer assert that $a_n = f^{(n)}(z_0)/n!$ since $f(z)$ need not be differentiable for $|z - z_0| < R_1$ under our hypotheses.

DEFINITION 9.7.1. The *Laurent series of* $f(z)$ *about* z_0 is the series

$$\sum_{n=-\infty}^{\infty} c_n h^n \tag{9.7.2}$$

where $h = z - z_0$ and c_n is as defined above. We also refer to it as the Laurent *expansion* of $f(z)$.

EXAMPLE 9.7.1. Let $f(z) = \exp z + \exp(1/z)$. We have

$$\exp z = \sum_{n=0}^{\infty} \frac{1}{n!} z^n \quad \text{for all } z$$

$$\exp(1/z) = \sum_{n=0}^{\infty} \frac{1}{n!} z^{-n} \quad \text{for } z \ne 0$$

and so

$$f(z) = \sum_{m=-\infty}^{\infty} c_m z^m$$

where

$$c_m = 1/m! \qquad (m \geq 1)$$

$$c_0 = 2$$

$$c_m = 1/(-m)! \qquad (m \leq -1)$$

and the expansion is valid for $z \neq 0$.

EXAMPLE 9.7.2. Let $f(z) = 1/z + 1/(1-z)$. In a similar way, we have $f(z) = \sum c_m z^m$ where

$$c_m = \begin{cases} 0 & (m < -1) \\ 1 & (m \geq -1) \end{cases}$$

and the series converges absolutely for $0 < |z| < 1$.

9.8. POLES AND ZEROS

DEFINITION 9.8.1. A *zero* of a function f, differentiable in a domain D, is a point z_0 such that $f(z_0) = 0$.

If we expand in a Taylor series about z_0 we get

$$f(z) = \sum_{n=0}^{\infty} a_n (z - z_0)^n$$

for $|z - z_0| < R$. Either all a_i are zero, in which case $f(z) \equiv 0$; or there exists m such that

$$a_0 = a_1 = \ldots = a_{m-1} = 0$$

$$a_m \neq 0.$$

In this case we say that z_0 is a *zero of order m* (z_0 is a *simple zero* if $m = 1$). It follows that

$$f(z) = (z - z_0)^m g(z) \qquad (9.8.1)$$

where $g(z_0) \neq 0$ and g is differentiable for $|z - z_0| < R$. From this we can prove that a zero of finite order is *isolated*; that is, there exists a disc around it containing no other zeros.

Simple zeros have a nice property: they are preserved by small perturbations. For example, let p be a polynomial with a simple zero z_0, and consider a small perturbation $p(z) - \varepsilon q(z)$ where q is a polynomial of small degree. Then this has a zero $z_0(\varepsilon)$ which is close to z_0. In fact, more precisely, we have

$$|z_0(\varepsilon) - z_0 + \varepsilon q(z_0)/p'(z_0)| = O(\varepsilon^2)$$

as $\varepsilon \to 0$ [see § 2.3]. That is, for small ε, $z_0(\varepsilon) = z_0 - \varepsilon q(z_0)/p'(z_0)$ to order 2 in ε.

DEFINITION 9.8.2.   If f is differentiable in a *punctured disc*

$$0 < |z - z_0| < R \qquad (9.8.2)$$

we say that z_0 is an *isolated singularity* of f.

We can use the Laurent expansion to study such singularities. There is a Laurent series

$$f(z) = \sum_{n=0}^{\infty} a_n (z - z_0)^n + \sum_{n=1}^{\infty} b_n (z - z_0)^{-n}$$

valid for $0 < |z - z_0| < R$; and this series can behave in three radically different ways.

   (i) *All $b_n = 0$.* By defining $f(z_0) = a_0$ we obtain a function which is differentiable on the whole disc $|z - z_0| < R$, with Taylor series

$$\sum_{n=0}^{\infty} a_n (z - z_0)^n. \qquad (9.8.3)$$

In this case z_0 is said to be a *removable singularity*. It arises more from our domain of definition of f than from any intrinsic feature of f.

EXAMPLE 9.8.1.   Consider

$$f(z) = \frac{\sin z}{z} \qquad (z \neq 0).$$

Around $z_0 = 0$ we have by (9.5.6),

$$f(z - z_0) = 1 - \frac{z^2}{3!} + \frac{z^4}{5!} - \cdots$$

and so by defining $f(0) = 1$ we get a function differentiable for all $z \in \mathbb{C}$.

   (ii) *Only finitely many b_n non-zero.* Then

$$f(z) = \frac{b_m}{(z - z_0)^m} + \ldots + \frac{b_1}{z - z_0} + \sum_{n=0}^{\infty} a_n (z - z_0)^n \qquad (9.8.4)$$

where $b_m \neq 0$. In this case we say that z_0 is a *pole of order m* of f.

EXAMPLE 9.8.2.   By (9.5.6) we have

$$f(z) = z^{-4} \cdot \sin z \qquad (z \neq 0)$$

$$= \frac{1}{z^3} - \frac{1}{3! \, z} + \sum (-1)^n \frac{z^{2n+1}}{(2n+5)!}$$

has a pole of order 3 at $z_0 = 0$.

   Poles of orders $1, 2, 3, \ldots$ are often called *simple, double, triple, $\ldots$* poles.

   (iii) *Infinitely many b_n non-zero.* Then we say that z_0 is an *isolated essential singularity*.

EXAMPLE 9.8.3

$$f(z) = \sin(1/z) \qquad (z \neq 0)$$

$$= \frac{1}{z} - \frac{1}{3! \, z^3} + \frac{1}{5! \, z^5} - \cdots$$

has an isolated essential singularity at $z_0 = 0$.

The removable singularities are not very interesting; the essential ones too complicated for further discussion. That leaves poles, about which we can say more. The following criterion is often useful:

THEOREM 9.8.1.   *If f is differentiable in $0 < |z - z_0| < R$, then f has a pole of order m at z_0 if and only if*

$$\lim_{z \to z_0} (z - z_0)^m f(z) = l \neq 0.$$

EXAMPLE 9.8.4.   Let

$$f(z) = \frac{5z + 3}{(1 - z)^3 \sin^2 z} \qquad (0 < |z| < 1).$$

Then using the binomial expansion of $(1 - z)^3$ and (9.5.6) we have $\lim_{z \to 0} z^2 f(z) = 3 \neq 0$, so there is a double pole at the origin. Also, $\lim_{z \to 1} (z - 1)^3 f(z) = -8/(\sin^2 1) \neq 0$, so there is a triple pole at $z_0 = 1$.

In this section we have only considered poles and zeros which are *finite* points. For a discussion of the behaviour of an analytic function at ∞, see § 9.16.

9.9.   RESIDUES

The tasks to which complex analysis may be set include the explicit computation of definite integrals and the summation of series. The basic idea is to use Cauchy's theorem to exploit the exceptional nature of the term $b_1/(z - z_0)$ in the Laurent expansion of an analytic function.

DEFINITION 9.9.1.   If f has an isolated singularity at z_0 and Laurent expansion

$$f(z_0 + h) = \sum_{n=0}^{\infty} a_n h^n + \sum_{n=1}^{\infty} b_n h^{-n} \qquad (0 < |h| < R)$$

we define the *residue of f at z_0* to be

$$\mathrm{res}\,(f, z_0) = b_1.$$

Now the integral of z^n round any circle containing the origin is zero except for $n = -1$, when it is $2\pi i$ [cf. Example 9.3.2]. Integrating term by term it follows that

$$\text{res}\,(f, z_0) = \frac{1}{2\pi i} \int_\gamma f(z)\, dz \qquad (9.9.1)$$

where $\gamma(t) = z_0 + r e^{it}$ $(0 \le t \le 2\pi)$ and $0 < r < R$. This shows the relevance of residues to integration. More generally we have:

THEOREM 9.9.1 (*Cauchy's Residue Theorem*).   *Let S be a domain containing a simple closed contour γ and the points inside γ. If f is differentiable in S except for finitely many isolated singularities at $z_1, \ldots, z_n$ inside γ, then*

$$\int_\gamma f(z)\, dz = 2\pi i \sum_{r=1}^{n} \text{res}\,(f, z_r).$$

Again we sketch a proof. A suitable contour winds around γ and around each singularity, as in Figure 9.9.1. Each straight part is described twice in opposite directions, so the contributions cancel. By Cauchy's theorem the integral around this contour is zero. But the integral around each circle is given in terms of the residue at the corresponding singularity. Keeping an eye on the signs, we get the result.

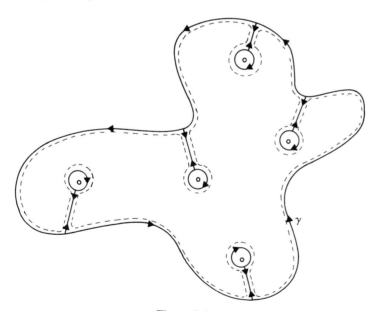

Figure 9.9.1

The residue theorem can be used to calculate integrals (and not just integrals round simple closed contours). For it to be of much use we must find ways of calculating residues. The following two lemmas are very useful in this respect.

LEMMA 9.9.2.   *If z_0 is a simple pole of f then*

$$\text{res}\,(f,\,z_0) = \lim_{z \to z_0}\,(z - z_0)f(z).$$

If $f(z) = p(z)/q(z)$, where $p(z_0) \neq 0$, $q(z_0) = 0$, $q'(z_0) \neq 0$, then

$$\text{res}\,(f,\,z_0) = p(z_0)/q'(z_0).$$

We sketch a proof. We have

$$f(z) = \frac{b_1}{z - z_0} + \sum_{n=0}^{\infty} a_n(z - z_0)^n$$

and so

$$(z - z_0)f(z) = b_1 + \sum_{n=0}^{\infty} a_n(z - z_0)^{n+1}$$

which tends to b_1 as $z \to z_0$.

For the second part, note that

$$\lim_{z \to z_0} \frac{(z - z_0)p(z)}{q(z)} = \lim_{z \to z_0} p(z) \bigg/ \left(\frac{q(z) - q(z_0)}{z - z_0} \right)$$

since $q(z_0) = 0$, and this is equal to $p(z_0)/q'(z_0)$ [see (9.1.1)].

EXAMPLE 9.9.1.   If

$$f(z) = \frac{\cos\,(\pi z)}{1 - z^{976}}$$

then

$$\text{res}\,(f,\,1) = \frac{\cos\,(\pi)}{-976 \cdot 1^{975}} = \frac{1}{976}.$$

LEMMA 9.9.3.   *If z_0 is a pole of f of order m then*

$$\text{res}\,(f,\,z_0) = \lim_{z \to z_0} \left\{ \frac{1}{(m-1)!} \frac{d^{m-1}}{dz^{m-1}}((z - z_0)^m f(z)) \right\}.$$

The proof is similar to that of Lemma 9.9.2.

EXAMPLE 9.9.2.   Consider

$$f(z) = \left(\frac{z+1}{z-1} \right)^3$$

which has a triple pole at $z_0 = 1$. Then

$$(z-1)^3 f(z) = (z+1)^3$$

and so

$$\frac{1}{2!} \frac{d^2}{dz^2}((z-1)^3 f(z)) = \frac{6}{2!}(z+1)$$

which tends to $3 \cdot 2 = 6$ as $z \to 1$. So res $(f, 1) = 6$.

On occasion another technique may be brought into play: working out the appropriate part of the Laurent series. (It is a waste of time to work out the whole thing, because the point about residues is that we don't need the whole thing, but only b_1.) For instance,

$$f(z) = 1/(z^2 \sin z)$$

$$= 1 / \left(z^2 \left(z - \frac{z^3}{6} + \dots \right) \right)$$

$$= \frac{1}{z^3} \left(1 - \frac{z^2}{6} + \dots \right)^{-1}$$

$$= \frac{1}{z^3} \left(1 + \frac{z^2}{6} + \dots \right)$$

$$= \frac{1}{z^3} + \frac{1}{6z} + \dots$$

so that res $(f, 0) = \frac{1}{6}$.

9.10. EVALUATION OF DEFINITE INTEGRALS

We now consider a number of techniques for the calculation of various kinds of definite integral.

9.10.1. Integrals of the form $\int_0^{2\pi} Q(\cos t, \sin t)\, dt$

Let $\gamma(t) = e^{it}$ $(0 \le t \le 2\pi)$, the unit circle. If

$$z = \gamma(t) = e^{it}$$

then by (9.5.10) and (9.5.11) we have

$$\cos t = \frac{1}{2} \left(z + \frac{1}{z} \right)$$

$$\sin t = \frac{1}{2i} \left(z - \frac{1}{z} \right),$$

from which we get

$$\int_0^{2\pi} Q(\cos t, \sin t)\, dt = \int_\gamma Q\left(\frac{1}{2}\left(z+\frac{1}{z}\right), \frac{1}{2i}\left(z-\frac{1}{z}\right)\right)\frac{dz}{iz}$$

$$= 2\pi i \Sigma \qquad \text{(by Theorem 9.9.1),}$$

where Σ is the sum of the residues of

$$\frac{1}{iz} Q\left(\frac{1}{2}\left(z+\frac{1}{z}\right), \frac{1}{2i}\left(z-\frac{1}{z}\right)\right) \qquad\qquad (9.10.1)$$

inside γ.

EXAMPLE 9.10.1.   Consider

$$\int_0^{2\pi} (\cos^3 t + \sin^2 t)\, dt.$$

Then (9.10.1) becomes

$$\frac{1}{iz}\left(\frac{1}{8}\left(z+\frac{1}{z}\right)^3 - \frac{1}{4}\left(z-\frac{1}{z}\right)^2\right) = \frac{1}{8i}z^2 - \frac{1}{4i}z + \frac{3}{i} + \frac{1}{2iz} + \frac{3}{iz^2} - \frac{1}{4iz^3} + \frac{1}{iz^4}$$

which has just one pole inside C, of residue $1/2i$. So the integral is equal to $2\pi i/2i = \pi$.

If Q is at all complicated the computations can become very tedious. Sometimes integrals of this kind can be found from the real and imaginary parts of an integral

$$\int_\gamma g(z)\, dz$$

with a suitable choice of g.

EXAMPLE 9.10.2

$$\int_\gamma \frac{e^z}{z}\, dz = 2\pi i$$

since it follows from (9.5.5) that e^z/z has residue 1 at $z = 0$. Therefore

$$\int_0^{2\pi} \frac{\exp(\cos t + i\sin t)}{e^{it}} i\, e^{it}\, dt = 2\pi i$$

so

$$\int_0^{2\pi} \exp(\cos t + i\sin t)\, dt = 2\pi,$$

so

$$\int_0^{2\pi} e^{\cos t}(\cos(\sin t) + i\sin(\sin t))\, dt = 2\pi$$

and equating real and imaginary parts yields

$$\int_0^{2\pi} e^{\cos t} \cos (\sin t) \, dt = 2\pi$$

$$\int_0^{2\pi} e^{\cos t} \sin (\sin t) \, dt = 0.$$

9.10.2. Integrals of the form $\int_{-\infty}^{\infty} f(x) \, dx$

The real integral

$$\int_{-\infty}^{\infty} f(x) \, dx \qquad\qquad (9.10.2)$$

is defined to be equal to

$$\lim \int_{x_1}^{x_2} f(x) \, dx, \quad \text{as } x_2 \to \infty, \, x_1 \to -\infty \qquad (9.10.3)$$

provided the limit exists [see § 4.6]. The techniques we are about to discuss allow the calculation of

$$\lim_{R \to \infty} \int_{-R}^{R} f(x) \, dx \qquad\qquad (9.10.4)$$

which is known as the *Cauchy principal value* of the integral and will be written

$$\mathscr{P} \int_{-\infty}^{\infty} f(x) \, dx.$$

If (9.10.3) exists then so does (9.10.4) and the two are equal. But the Cauchy principal value may exist when (9.10.3) does not [see Example 4.6.8].

It follows that when we use the technique below, we must take into account the convergence of (9.10.3). This, in part, leads to condition (ii) of the next theorem.

THEOREM 9.10.1. *Suppose*

(i) *f is analytic in a domain containing the upper half-plane, that is where* $\text{Im}(z) \geq 0$, *except for a finite number of poles, none of which lie on the real axis,*

(ii) *if* $S_R(t) = R \exp(it)$ $(0 \leq t \leq \pi)$ *then for sufficiently large R there is a constant A such that*

$$|f(z)| \leq A/R^2$$

when z lies on the track of S_R.

Then

$$\int_{-\infty}^{\infty} f(x) \, dx = 2\pi i \Sigma$$

where Σ *is the sum of the residues at the poles of f in the upper half-plane.*

Proof.   Choose R large enough for (ii) to be satisfied, and so that all the poles lie inside $S_R \cup [-R, R]$.

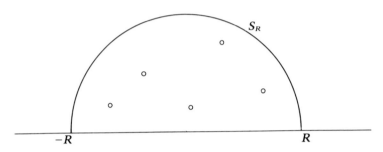

Figure 9.10.1

Then by Cauchy's theorem

$$\int_{-R}^{R} f(x)\, dx + \int_{S_R} f(z)\, dz = 2\pi i \Sigma$$

with Σ as stated. Now let $R \to \infty$. Then

$$\left| \int_{S_R} f(z)\, dz \right| \le \frac{A}{R^2} \pi R = \pi A / R$$

which tends to 0 as $R \to \infty$. Hence

$$\lim_{R \to \infty} \int_{-R}^{R} f(x)\, dx = 2\pi i \Sigma,$$

that is,

$$\mathcal{P} \int_{-\infty}^{\infty} f(x)\, dx = 2\pi i \Sigma.$$

However, (ii) tells us that $|f(x)| \le A/x^2$ for large $|x|$, and so it follows from real analysis that

$$\int_{-\infty}^{\infty} f(x)\, dx$$

exists. It is therefore equal to the Cauchy principal value, and so to $2\pi i \Sigma$ as claimed.

Remark.   Condition (ii) is certainly satisfied if $f(z) = p(z)/q(z)$ where p and q are polynomials such that q has no real zeros and the degree of q is greater than or equal to $2 +$ the degree of p.

EXAMPLE 9.10.3.   Consider

$$\int_{-\infty}^{\infty} \frac{dx}{(x^2+a^2)(x^2+b^2)},$$

where $a>0$, $b>0$, $a \neq b$. By the above remark this satisfies (ii), and (i) is obviously true. Now the only poles of $1/(z^2+a^2)(z^2+b^2)$ in the upper half-plane are simple poles at ia, ib. The residue at ia is, by Lemma 9.9.2,

$$\lim_{z\to ia} \frac{z-ia}{(z^2+a^2)(z^2+b^2)} = \frac{1}{2ia(b^2-a^2)}$$

and similarly that at ib is

$$\frac{1}{2ib(a^2-b^2)}.$$

Hence the value of the integral is

$$2\pi i\left(\frac{1}{2ia(b^2-a^2)} + \frac{1}{2ib(a^2-b^2)}\right) = \frac{\pi}{ab(a+b)}.$$

Remark.   In the proof of Theorem 9.10.1 we did not require $f(z)$ to be real for z on the real axis.

EXAMPLE 9.10.4.   Let

$$f(z) = e^{iz}/(z^2+a^2)(z^2+b^2).$$

We see that on S_R we have $|e^{iz}| = |e^{-y+ix}| = e^{-y} \leq 1$ for $y \geq 0$ [see 9.5.18] so that (ii) holds. As before we have simple poles at ia, ib; but now the residues are

$$e^{-a}/2ia(b^2-a^2) \quad \text{and} \quad e^{-b}/2ib(a^2-b^2).$$

Hence

$$\int_{-\infty}^{\infty} \frac{e^{ix}}{(x^2+a^2)(x^2+b^2)} dx = \pi\left(\frac{e^{-a}}{a(b^2-a^2)} + \frac{e^{-b}}{b(a^2-b^2)}\right)$$

and equating real and imaginary parts we obtain

$$\int_{-\infty}^{\infty} \frac{\cos x}{(x^2+a^2)(x^2+b^2)} dx = \frac{\pi}{b^2-a^2}\left(\frac{e^{-a}}{a} - \frac{e^{-b}}{b}\right)$$

$$\int_{-\infty}^{\infty} \frac{\sin x}{(x^2+a^2)(x^2+b^2)} dx = 0.$$

Of these, the second is obvious since the integrand is an odd function [see § 2.12], but the first result is far from being obvious.

We can try to generalize this method in (at least) two ways: by getting a better estimate for $\int_{S_R} f(z)\,dz$; or by allowing f to have poles on the real axis. The first we deal with in section 9.10.3, the second in section 9.10.4.

9.10.3.   Integrals of the form $\int_{-\infty}^{\infty} f(x)\,e^{ix}\,dx$

THEOREM 9.10.2.   *Suppose*

(i) *f is analytic in a domain containing the upper half-plane, except for a finite number of poles, none on the real axis,*

(ii) *for sufficiently large R there exists a constant A such that*

$$|f(z)| \le A/R \quad \text{for } |z| = R.$$

Then

$$\int_{-\infty}^{\infty} f(x)\,e^{ix}\,dx = 2\pi i \Sigma',$$

where Σ' is the sum of the residues of $f(z)\,e^{iz}$ at poles in the upper half-plane.

Proof.   It is possible to use the same contour as in section 9.10.2 and prove that

$$\lim_{R \to \infty} \int_{S_R} f(z)\,e^{iz}\,dz = 0,$$

but this only calculates the Cauchy principal value; and the convergence problem raised thereby is much harder because all we know is that f behaves like $1/x$ for large x, which on its own does not imply the existence of

$$\int_{-\infty}^{\infty} f(x)\,e^{ix}\,dx.$$

By more delicate arguments this obstacle may be overcome, but it is easier to sidestep the whole question by using a different contour, as shown:

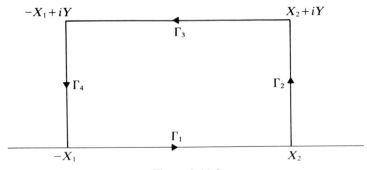

Figure 9.10.2

We prove that as $X_1, X_2, Y \to \infty$ each of

$$\int_{\Gamma_r} f(z) \, e^{iz} \, dz$$

for $r = 2, 3, 4$, tends to zero. It will then follow as before that

$$\lim_{X_1, X_2 \to \infty} \int_{-X_1}^{X_2} f(x) \, e^{ix} \, dx = 2\pi i \Sigma'$$

which is what we want.

Now

$$\left| \int_{\Gamma_2} f(z) \, e^{iz} \, dz \right| = \left| \int_0^Y f(X_2 + it) \, e^{iX_2 - t} i \, dt \right|$$

$$\leq \int_0^Y \frac{A}{X_2} e^{-t} \, dt$$

$$\leq A/X_2$$

for sufficiently large X_2; and similarly

$$\left| \int_{\Gamma_4} f(z) \, e^{iz} \, dz \right| \leq A/X_1$$

for sufficiently large X_1. Also

$$\left| \int_{\Gamma_3} f(z) \, e^{iz} \, dz \right| = \left| -\int_{-X_1}^{X_2} f(t + iY) \, e^{it - Y} \, dt \right|$$

$$\leq \left| \int_{-X_1}^{X_2} \frac{A}{Y} e^{-Y} \, dt \right|$$

$$= A Y^{-1} e^{-Y} (X_1 + X_2).$$

Now for fixed X_1 and X_2 this tends to zero as $Y \to \infty$. Then letting X_1, X_2 tend to ∞ we obtain the desired result.

EXAMPLE 9.10.5.   Take

$$\int_{-\infty}^{\infty} \frac{x^3 e^{ix}}{(x^2 + a^2)(x^2 + b^2)} \, dx \qquad (a, b > 0, \, a \neq b)$$

which does not satisfy the conditions for 9.10.1, but does for 9.10.2. The integrand has simple poles in the upper half-plane at ia and ib. Calculating the residues in the usual way, applying the above result, and equating imaginary parts yields

$$\int_{-\infty}^{\infty} \frac{x^3 \sin x}{(x^2 + a^2)(x^2 + b^2)} \, dx = \frac{\pi}{b^2 - a^2} (b^2 e^{-b} - a^2 e^{-a}).$$

Equating real parts shows that the corresponding cosine integral is zero (but once more this is obvious on other grounds).

9.10.4. Poles on the Real Axis

If $f(z)$ has poles on the real axis we make 'indentations' in the contour; i.e. draw small semicircles as in the diagram. Suppose these semicircles have

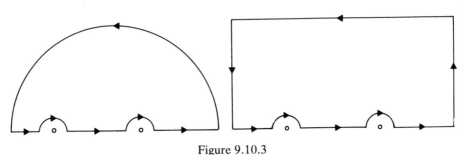

Figure 9.10.3

radii $\varepsilon_1, \varepsilon_2, \ldots .$ Then we proceed as above, but letting $\varepsilon_1, \varepsilon_2, \ldots \to 0$ at the same time as $R, X_1, X_2, Y \to \infty$. There is a problem here similar to that in section 9.10.2: all we calculate is a Cauchy principal value

$$\mathscr{P} \int_a^b f(x)\, dx = \lim_{\varepsilon \to 0} \left(\int_a^{x_0 - \varepsilon} f(x)\, dx + \int_{x_0 + \varepsilon}^b f(x)\, dx \right)$$

for a pole at x_0. This again may exist even though

$$\int_a^b f(x)\, dx = \lim_{\varepsilon_1 \to 0} \int_a^{x_0 - \varepsilon_1} f(x)\, dx + \lim_{\varepsilon_2 \to 0} \int_{x_0 + \varepsilon_2}^b f(x)\, dx$$

does not. Thus

$$\int_{-1}^1 \frac{dx}{x}$$

does not exist, but

$$\mathscr{P} \int_{-1}^1 \frac{dx}{x} = 0.$$

Consequently we have a convergence problem to discuss, once the Cauchy principal value has been obtained. Apart from this, the hypotheses and conclusions of section 9.10.2 and section 9.10.3 remain valid even when poles occur on the real axis, provided that Σ and Σ' are summed only over the non-real poles in the upper half-plane. Rather than state a cumbersome general theorem we content ourselves with a typical example:

EXAMPLE 9.10.6

$$\int_{-\infty}^{\infty} \frac{e^{ix}}{x}\, dx.$$

There is a real pole at 0, so we take as contour

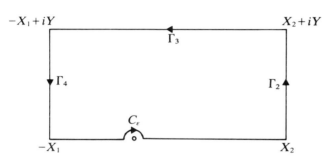

Figure 9.10.4

As before, the integrals along Γ_2, Γ_3, and Γ_4 tend to zero. Since there are no poles inside the contour,

$$\lim_{X_1, X_2 \to 0} \left\{ \int_{-X_1}^{-\varepsilon} \frac{e^{ix}}{x}\, dx + \int_{\varepsilon}^{X_2} \frac{e^{ix}}{x}\, dx + \int_{C_\varepsilon} \frac{e^{iz}}{z}\, dz \right\} = 0.$$

But

$$\frac{e^{iz}}{z} = \frac{1}{z} + \sum_{n=1}^{\infty} \frac{z^{n-1}}{n!}$$

$$= \frac{1}{z} + \phi(z)$$

where ϕ is analytic. Hence $|\phi(z)| \le M$ in a neighbourhood of 0, and so

$$\lim_{\varepsilon \to 0} \int_{C_\varepsilon} \frac{e^{iz}}{z}\, dz = \lim_{\varepsilon \to 0} \int_{C_\varepsilon} \left(\frac{1}{z} + \phi(z) \right) dz$$

$$= \lim_{\varepsilon \to 0} \left\{ -\int_0^\pi \frac{1}{\varepsilon\, e^{it}} i\varepsilon\, e^{it}\, dt \right\} + \lim_{\varepsilon \to 0} M\pi\varepsilon$$

$$= -i\pi.$$

Therefore

$$\mathscr{P} \int_{-\infty}^{\infty} \frac{e^{ix}}{x}\, dx = i\pi.$$

Equating real and imaginary parts,

$$\mathscr{P}\int_{-\infty}^{\infty} \frac{\cos x}{x}\, dx = 0,$$

$$\mathscr{P}\int_{-\infty}^{\infty} \frac{\sin x}{x}\, dx = \pi.$$

The first integral exists only as a principal value, since $\cos x/x$ behaves like $1/x$ for small x [see (2.12.2)]. But for the second we have

$$\mathscr{P}\int_{-\infty}^{\infty} \frac{\sin x}{x}\, dx = \lim_{\varepsilon \to 0}\left(\int_{-\infty}^{-\varepsilon} \frac{\sin x}{x}\, dx + \int_{\varepsilon}^{\infty} \frac{\sin x}{x}\, dx\right)$$

$$= 2\lim_{\varepsilon \to 0}\int_{\varepsilon}^{\infty} \frac{\sin x}{x}\, dx$$

and this limit does exist, and hence by definition equals

$$2\int_{0}^{\infty} \frac{\sin x}{x}\, dx$$

because $\sin x/x \to 1$ as $x \to 0$ [see (2.12.13)]. Hence we can remove the $\mathscr{P}$ from the second expression above. Further, we have proved that

$$\int_{0}^{\infty} \frac{\sin x}{x}\, dx = \frac{\pi}{2}.$$

9.10.5. Integrals of the Form $\int_{-\infty}^{\infty} (e^{ax}/\phi(e^{x}))\, dx$

For integrals of this type we integrate around a contour of the form

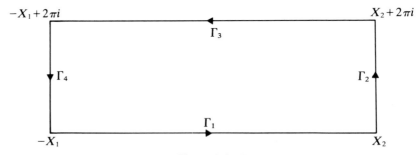

Figure 9.10.5

Now on Γ_3, if we substitute $z = -t + 2\pi i$, then $-X_2 \le t \le X_1$, and so

$$\int_{\Gamma_3} \frac{e^{az}}{\phi(e^{z})}\, dz = \int_{-X_2}^{X_1} \frac{e^{-at+2\pi ia}}{\phi(e^{-t})}(-1)\, dt$$

$$= e^{2\pi ia}\int_{X_1}^{-X_2} \frac{e^{-at}}{\phi(e^{-t})}\, dt$$

which, putting $t = -x$, is equal to

$$-e^{2\pi i a} \int_{-X_1}^{X_2} \frac{e^{ax}}{\phi(e^x)} \, dx.$$

If ϕ is such that $\int_{\Gamma_j} (e^{az}/\phi(e^z)) \, dz \to 0$ as $X_1, X_2 \to \infty$ (for $j = 2, 4$) then we obtain

$$(1 - \exp(2\pi a i)) \int_{-\infty}^{\infty} \frac{\exp(ax)}{\phi(e^x)} \, dx = 2\pi i \Sigma''$$

where Σ'' is the sum of the residues of $\exp(az)/\phi(z)$ at poles between the lines $\mathrm{Im}(z) = 0$, $\mathrm{Im}(z) = 2\pi$.

EXAMPLE 9.10.7. Consider

$$\int_{-\infty}^{\infty} \frac{\exp(ax)}{e^{2x} + 1} \, dx \qquad (0 < a < 1).$$

The relevant singularities are at $i\pi/2$, $3i\pi/2$, with corresponding residues

$$-\frac{1}{2} \exp\left(\frac{i\pi a}{2}\right), \qquad -\frac{1}{2} \exp\left(\frac{3i\pi a}{2}\right).$$

Hence the value of the integral is

$$\frac{2\pi i}{1 - e^{2\pi i a}} \left(-\frac{1}{2} \exp\left(\frac{i\pi a}{2}\right) - \frac{1}{2} \exp\left(\frac{3\pi i a}{2}\right)\right).$$

Putting $k = \exp(i\pi a/2)$ this is easily seen to be equal to

$$\frac{\pi i}{k + (1/k)} = \frac{\pi}{2 \sin(\pi a/2)},$$

[see (9.5.11)].

9.11. SUMMATION OF SERIES

Let f be a function which is analytic at $z = n$ for an integer n. The functions $\cot(\pi z)$ and $\mathrm{cosec}(\pi z)$ [see § 2.12] have simple poles at $z = n$, where $n \in \mathbb{Z}$; and one may check that

$$\mathrm{res}\,(f(z) \cot \pi z, n) = \frac{f(n)}{\pi},$$

$$\mathrm{res}\,(f(z) \, \mathrm{cosec}\, \pi z, n) = \frac{(-1)^n f(n)}{\pi}.$$

This suggests a method for summing certain series, as follows: We let C_N be the square whose vertices are

$$\{(N + \tfrac{1}{2})(\pm 1 \pm i)\}$$

parametrized, as usual, in the anti-clockwise direction as shown.

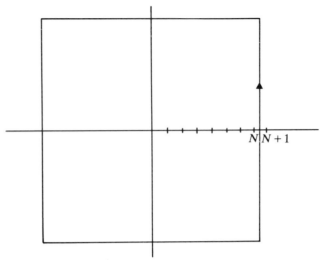

Figure 9.11.1

We claim that cosec πz and cot πz are bounded on C_N, where the bound is *independent of N*. First note that on the two sides parallel to the real axis we have $z = x + iy$ where $|y| \geq \frac{1}{2}$. In this case

$$|\text{cosec } \pi z| = (\tfrac{1}{2}|e^{i\pi z} - e^{-i\pi z}|)^{-1}$$

$$\leq (\tfrac{1}{2}|e^{-\pi y} - e^{\pi y}|)^{-1}$$

$$= (\sinh |\pi y|)^{-1}$$

$$\leq (\sinh \pi/2)^{-1}$$

[see § 2.13] and

$$|\cot \pi z| = \left| \frac{e^{i\pi z} + e^{-i\pi z}}{e^{i\pi z} - e^{-i\pi z}} \right|$$

$$\leq \left| \frac{|e^{i\pi z}| + |e^{-i\pi z}|}{|e^{i\pi z}| - |e^{-i\pi z}|} \right|$$

$$= \left| \frac{e^{-\pi y} + e^{\pi y}}{e^{-\pi y} - e^{\pi y}} \right|$$

$$= \coth |\pi y|$$

$$\leq \coth \pi/2.$$

On the other two sides, we have $z = \pm N + \frac{1}{2} + it$, and so

$$|\text{cosec } \pi z| = |\sin \pi z|^{-1}$$

$$= |\cos i\pi t|^{-1}$$

$$= (\cosh |\pi t|)^{-1}$$

$$\leq 1$$

and

$$|\cot \pi z| = |\tan it|$$

$$= \left| \frac{1 - e^{-2t}}{1 + e^{-2t}} \right|$$

$$\leq 1.$$

Hence there is a constant M such that $|\cot \pi z| \leq M$, $|\operatorname{cosec} \pi z| \leq M$, for z on any C_N.

Suppose now that for large enough $|z|$ we have

$$|f(z)| \leq A/|z|^2.$$

Then we claim that

$$\Sigma^* = 0$$

where Σ^* is the sum of the residues of $f(z) \cot \pi z$.

By Cauchy's theorem we have

$$\int_{C_N} f(z) \cot (\pi z) \, dz = 2\pi i \Sigma_N^*$$

where Σ_N^* is the sum of the residues of $f(z) \cot \pi z$ inside C_N. As $N \to \infty$, $\Sigma_N^* \to \Sigma^*$, so it is sufficient to show that the integral tends to zero. But we can estimate the integral easily:

$$\left| \int_{C_N} f(z) \cot (\pi z) \, dz \right| \leq \frac{A}{N^2} M(8N + 4)$$

for large enough N. As $N \to \infty$, this tends to zero, as claimed.

Now Σ^* often takes the form of an infinite series: the fact that it is zero allows us to sum certain related series. This is best illustrated by an example:

EXAMPLE 9.11.1. Let

$$f(z) = 1/z^2.$$

At an integer $n \neq 0$ the function $z^{-2} \cot \pi z$ has a simple pole with residue $1/(n^2 \pi)$, whereas at the origin it has a triple pole with residue $-\pi/3$. Hence

$$0 = \Sigma^* = -\pi/3 + \sum_{n=-\infty}^{\infty} 1/(n^2 \pi)$$

(where $n \neq 0$ in the infinite sum)

$$= -\pi/3 + \frac{2}{\pi} \sum_{n=1}^{\infty} 1/n^2.$$

Hence

$$\sum_{n=1}^{\infty} 1/n^2 = \pi^2/6, \tag{9.11.1}$$

a theorem originally proved by Euler by a different method [see (20.5.21)].

If we use cosec πz instead of cot πz a similar theorem applies, and allows us to sum series of the form $\sum (-1)^n f(n)$. For instance, using $f(z) = z^{-2}$ and arguing much as above, we can prove that

$$\sum_{n=1}^{\infty} (-1)^{n+1}/n^2 = \pi^2/12 \qquad (9.11.2)$$

[see also (20.5.22)].

9.12. CONFORMAL TRANSFORMATIONS

We can think of a complex function f as mapping one copy of the complex plane on to another one. Curves in the first plane are thus transformed into (possibly) different curves in the second. It transpires that a *differentiable* function f transforms the curves in a way that keeps all angles between curves unchanged provided that the value of f' is non-zero. Such a transformation is said to be *conformal*. There are interesting applications of conformal transformations to fluid dynamics and potential theory, as well as more theoretical uses: we outline these in section 9.13.

Let $f: D \rightarrow \mathbb{C}$, where D is a domain. We use (x, y) as co-ordinates in D, and (u, v) as co-ordinates in the image $\mathbb{C}$; putting $z = x + iy$, $w = u + iv$. Thus f transforms a subset D of the (x, y)-plane into a subset of the (u, v)-plane. A path γ in D, with

$$\gamma(t) = x(t) + iy(t) \qquad (a \le t \le b)$$

transforms to a path

$$f\gamma(t) = u(x(t), y(t)) + iv(x(t), y(t)) \qquad (a \le t \le b) \qquad (9.12.1)$$

in the (u, v)-plane.

A straightforward computation (see [Titchmarsh (1932) p. 189]) shows:

THEOREM 9.12.1. *Let $f: D \rightarrow \mathbb{C}$ be analytic. Then f is conformal at all $z_0 \in D$ such that $f'(z_0) \neq 0$.*

EXAMPLE 9.12.1. Let $f(z) = z^3$. We have

$$u(x, y) + iv(x, y) = (x + iy)^3$$
$$= (x^3 - 3xy^2) + i(3x^2 y - y^3)$$

and so

$$u(x, y) = x^3 - 3xy^2$$
$$v(x, y) = 3x^2 y - y^3.$$

Consider the paths

$$\gamma_1(t) = 1 + it$$
$$\gamma_2(t) = it + 1$$

which are, respectively, the lines $x = 1$, $y = 1$, and hence meet at right angles. The paths $\Gamma_1 = f\gamma_1$ and $\Gamma_2 = f\gamma_2$ are given by

$$\Gamma_1(t) = (1 - 3t^2) + i(3t - t^3)$$
$$\Gamma_2(t) = (t^3 - 3t) + i(3t^2 - 1).$$

If we sketch these curves we get the following picture.

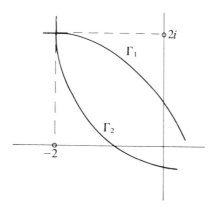

Figure 9.12.1

Now Γ_1 and Γ_2 meet at right angles at $(-2, 2)$.

If $f'(z_0) = 0$, it is not true that f is conformal at z_0. For example if $f(z) = z^2$, then the positive half of the real axis and the 'positive' half (from 0 through i and outwards) of the imaginary axis transform, respectively, into the positive and negative halves of the real axis. Originally they met at an angle of $\pi/2$, but after transformation they meet at π. In fact, if z_0 is a zero of f' of order m, then the angles between paths meeting at z_0 is multiplied by $m + 1$ on transforming by f.

We can find out a little about how f affects lengths. If z_0, $z \in \mathbb{C}$ and f is analytic at z_0 then the ratio of the distance between $f(z)$ and $f(z_0)$, and the distance between z and z_0, is

$$\frac{|f(z) - f(z_0)|}{|z - z_0|} = \left| \frac{f(z) - f(z_0)}{z - z_0} \right|$$

which tends to $|f'(z_0)|$ as $z \to z_0$ [see (9.1.1)]. Hence near to z_0 the distances are multiplied by $|f'(z_0)|$.

EXAMPLE 9.12.2

$$f(z) = 1/z.$$

This time we have

$$u(x, y) = x/(x^2 + y^2)$$

$$v(x, y) = -y/(x^2 + y^2).$$

If c is real and positive the circle

$$\gamma_c(t) = c\,e^{it} \qquad (0 \le t \le 2\pi) \tag{9.12.2}$$

transforms into

$$\Gamma_c(t) = c^{-1} e^{-it} \qquad (0 \le t \le 2\pi). \tag{9.12.3}$$

Hence the system of concentric circles (9.12.2), as c varies, maps into the system of concentric circles (9.12.3). However, points inside the circles of (9.12.2) map to points outside circles of (9.12.3).

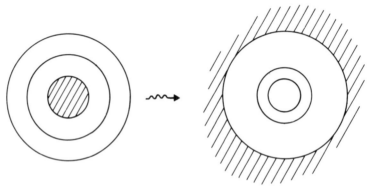

Figure 9.12.2

Further, the lines $x = ky$ ($k \in \mathbb{R}$) through the origin are given by

$$\delta_k(t) = t + kit$$

and transform to

$$\Delta_k(t) = (t + kit)^{-1} = \frac{1}{1 + k^2}\left(\frac{1}{t}\right) - \frac{ik}{1 + k^2}\left(\frac{1}{t}\right)$$

which also represents lines through the origin. Now Γ_c and Δ_k meet at right angles, as do γ_c and δ_k.

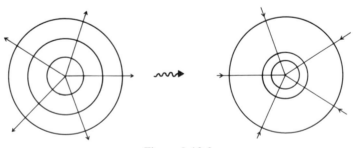

Figure 9.12.3

EXAMPLE 9.12.3

$$f(z) = \sin z.$$

We have

$$u(x, y) = \sin x \cosh y$$

$$v(x, y) = \cos x \sinh y.$$

Corresponding to the lines $x = c$ $(c \in \mathbb{R})$ we obtain the confocal hyperbolae [see V, § 1.3.4]

$$\frac{u^2}{\sin^2 c} - \frac{v^2}{\cos^2 c} = 1$$

and corresponding to the lines $y = d$ $(d \in \mathbb{R})$ the confocal ellipses [see V, § 1.3.2]

$$\frac{u^2}{\cosh^2 d} + \frac{v^2}{\sinh^2 d} = 1.$$

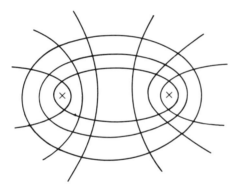

Figure 9.12.4

The two systems of straight lines meet at right angles, hence so do the systems of conics, except where $f'(z) = 0$, at $f(z) = \pm 1$.

9.13. POTENTIAL THEORY

The two-dimensional *Laplace equation*

$$\frac{\partial^2 \phi}{\partial x^2} + \frac{\partial^2 \phi}{\partial y^2} = 0$$

for a function $\phi(x, y)$ [see § 8.1] is important in potential theory, with applications in particular to fluid dynamics. It is closely connected with complex function theory, as the following argument demonstrates. Let

$f: D \to \mathbb{C}$ be analytic, with $z = x + iy$,

$$f(z) = u(x, y) + iv(x, y).$$

Then we have

$$f'(z) = \frac{\partial u}{\partial x} + i\frac{\partial v}{\partial x} = \frac{\partial v}{\partial y} - i\frac{\partial u}{\partial y}$$

as in section 9.1. By the remarks following Theorem 9.6.3 f'' exists throughout D. If we let $f'(z) = U + iV$ then the Cauchy–Riemann equations (9.1.2) show that

$$\frac{\partial U}{\partial x} = \frac{\partial V}{\partial y}, \qquad \frac{\partial V}{\partial x} = -\frac{\partial U}{\partial y}.$$

Now

$$U = \frac{\partial u}{\partial x} = \frac{\partial v}{\partial y}, \qquad V = \frac{\partial v}{\partial x} = -\frac{\partial u}{\partial y},$$

and so we get

$$\frac{\partial^2 u}{\partial x^2} = \frac{\partial}{\partial x}\left(\frac{\partial u}{\partial x}\right) = \frac{\partial U}{\partial x} = \frac{\partial V}{\partial y} = -\frac{\partial}{\partial y}\left(\frac{\partial u}{\partial y}\right) = -\frac{\partial^2 u}{\partial y^2}$$

and therefore $u(x, y)$ satisfies the Laplace equation. Similarly $v(x, y)$ does.

For instance, consider $f(z) = z \exp z$. Then

$$u(x, y) = x \exp x \cos y - y \exp x \sin y$$

$$v(x, y) = y \exp x \cos y + x \exp x \sin y$$

and it may be verified directly that these functions satisfy Laplace's equation.

Solutions of Laplace's equation are called *harmonic* or *potential functions* [see Definition 8.2.1]. Pairs of functions u, v obtained from an analytic function f as above are called *harmonic conjugates*.

Now the lines $u = $ constant, $v = $ constant are orthogonal (mutually perpendicular) in the (u, v)-plane, and so by conformality the lines

$$u(x, y) = \text{constant}$$

$$v(x, y) = \text{constant}$$

in the (x, y)-plane are orthogonal. In potential theory, if u is harmonic, the lines $u(x, y) = $ constant are called *equipotential lines*, and the set of orthogonal curves $v(x, y) = $ constant are called *stream-lines*. In the case of fluid flow described by the Laplace equation the stream-lines represent the paths along which the fluid flows. If we are given u in a domain D and wish to find the stream-lines given by v, we can often use complex integration. We have, for a fixed point $z_0 \in D$, and for any $z_1 \in D$,

$$f(z_1) - f(z_0) = \int_{z_0}^{z_1} f'(z) \, dz$$

$$= \int_{z_0}^{z_1} \left(\frac{\partial u}{\partial x} - i \frac{\partial u}{\partial y} \right) dz.$$

For example, if we take $u(x, y) = x^2 - y^2$, which is harmonic, then picking $z_0 = 0$ for convenience we have

$$f(z_1) = \int_0^{z_1} (2x + 2iy) \, dz$$

$$= \int_0^{z_1} 2z \, dz$$

$$= z_1^2.$$

Hence $f(x + iy) = (x + iy)^2$, so that $v(x, y) = \text{Im}\,(f(x + iy)) = 2xy$. So the stream-lines are given by the equations $2xy = \text{constant}$, or equivalently

$$xy = \text{constant}.$$

Often, as in this case, one can *guess* what $v(x, y)$ ought to be—a process dignified by the name 'inspection'.

Conformal mapping and complex function theory played a part in the design of aircraft in the early days of aviation. In particular the transformation from the z-plane to the w-plane given by

$$\frac{w - 2}{w + 2} = \left(\frac{z - 1}{z + 1} \right)^2$$

maps a circle in the z-plane, passing through -1 and containing $+1$ in its interior, into a 'bent teardrop' shape, resembling the cross-section of an aeroplane's wing, and known as a *Joukowski aerofoil* after the discoverer of the transformation.

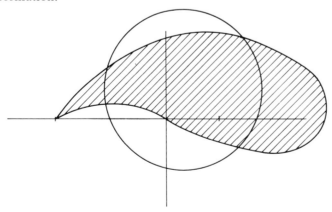

Figure 9.13.1

This is used in the following way. It is quite easy to solve the Laplace equation and find stream-lines for the flow of a fluid around a circular disc. Now apply the Joukowski transformation: the disc maps to an aerofoil, the stream-lines round the disc map to stream-lines around the aerofoil. From this one may calculate properties of the flow, in particular the amount of 'lift' imparted to the aircraft. More subtle transformations give more accurate information.

9.14. THE METHOD OF STEEPEST DESCENT

This technique, also known as the *saddlepoint method*, provides asymptotic approximations to integrals of the form

$$\Phi(s) = \int_A^B \exp\left(sf(z)\right) dz$$

as *s* tends to infinity along the positive real axis. It may be generalized to other types of integral, but this is the simplest case. The method involves deforming the contour of integration from *A* to *B* into one along which certain estimates may be made more easily. We therefore assume that *f* is analytic throughout an open set *D* containing this contour and all necessary deformations of the contour.

We shall sketch the method and state the result without proof: details may be found in Kyrala (1972) p. 162 ff.

The absolute value of the integrand $\exp\left(sf(z)\right)$ for real *s*, which is clearly crucial to any asymptotic estimate, is determined by the real part of $f(z)$. We let

$$u(x, y) = \text{Re } f(z)$$

$$v(x, y) = \text{Im } f(z)$$

so that

$$f(z) = u + iv.$$

We consider critical points of *f*, that is, points *z* at which $f'(z) = 0$. Now, by Theorem 9.1.1,

$$f'(z) = u_x + iv_x$$

where the subscripts denote partial derivatives. It follows that $f'(z) = 0$ if and only if $u_x = v_x = 0$. By the Cauchy–Riemann equations (9.1.2) of Theorem 9.1.1, $v_x = -u_y$, so $f'(z) = 0$ if and only if $u_x = u_y = 0$. Hence $z = x + iy$ is a critical point of *f* if and only if (x, y) is a critical point of $u: R^2 \to R$. [See § 5.6.]

Such a critical point will be non-degenerate [see § 5.8] if and only if the Hessian

$$H(x, y) = \begin{vmatrix} u_{xx} & u_{xy} \\ u_{yx} & u_{yy} \end{vmatrix} \neq 0.$$

By the Cauchy–Riemann equations,

$$H(x, y) = -(u_{xx})^2 - (v_{xx})^2.$$

Also,

$$f''(z) = u_{xx} + iv_{xx}.$$

Hence z is a non-degenerate critical point if and only if $f''(z) \neq 0$. In this case, $H(x, y)$ is negative, so the critical point of u is a Morse saddle (see V, § 8).

From a (non-degenerate) saddle, there must emerge two curves in the surface which descend the valleys on either side of the saddle as fast as possible, as in Figure 9.14.1. Along these curves the absolute value of Re $(f(z))$ decreases as rapidly as possible; hence the integral $\Phi(s)$ may be closely approximated by its behaviour near the saddlepoint, provided the contour of integration follows these curves of steepest descent (hence the name).

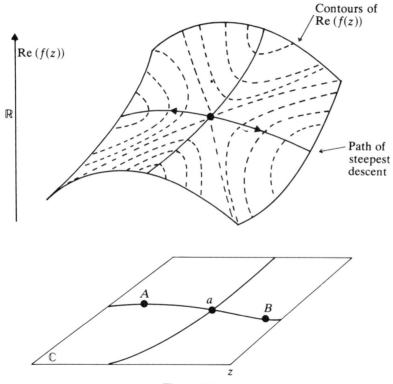

Figure 9.14.1

The upshot of the analysis is as follows. Suppose that $f(z)$ has a saddlepoint at $z = a$, so that $f'(a) = 0$, $f''(a) \neq 0$. (This latter implies non-degeneracy.) Suppose that A and B lie on either side of the saddlepoint, on the paths of steepest descent (with suitable modifications to the result, one of them may be *at* the saddlepoint). Then for large real positive s, the asymptotic expression

$$\Phi(s) \sim i \exp{(sf(a))}(2\pi/sf''(a))^{1/2}$$

holds. (That is, the ratio of the two sides tends to 1 as s tends to infinity.)

To apply this method we need to know the equations of the lines of steepest descent. The contours of the graph of $u(x, y) = \mathrm{Re}\,(f(z))$ are the level curves

$$\mathrm{Re}\,(f(z)) = \text{constant}.$$

Lines of steepest descent will cut these at right angles (this is geometrically obvious: head downhill as fast as possible and you move perpendicular to contours) and hence, by Cauchy–Riemann, are given by the equation

$$\mathrm{Im}\,(f(z)) = \text{constant}.$$

The constant will be equal to $\mathrm{Im}\,(f(a))$, since the lines pass through $z = a$.

EXAMPLE 9.14.1.   Find an asymptotic expression for $\int_{-1}^{1} \exp{(-sz^2)}\, dz$ as s tends to positive infinity.

Here we are in the above situation, with $f(z) = -z^2$. Then $f'(z) = -2z$, $f''(z) = -2$. Critical points occur only at $z = 0$. Now $u(x, y) = -x^2 + y^2$, $v(x, y) = -2xy$. The u-constant and v-constant contours are as shown in Figure 9.14.2.

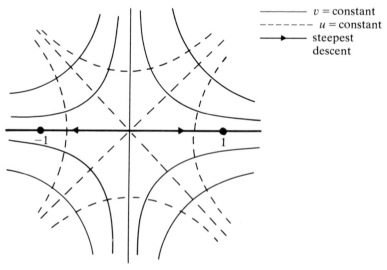

Figure 9.14.2

The lines of steepest descent from the origin are the two halves of the x-axis, and the contour of integration lies along them. Hence, by steepest descent, we obtain the asymptotic expression

$$i \exp (sf(0))(2\pi/sf''(0))^{1/2} = i \cdot e^{s \cdot 0}(2\pi/-2s)^{1/2}$$
$$= (\pi/s)^{1/2}.$$

As a check, recall that the area under a Gaussian curve $y = \exp(-sx^2)$ from $x = -\infty$ to $+\infty$ is given exactly by $(\pi/s)^{1/2}$. As s increases, the 'tails' of the curve contribute less and less to the integral, as shown in Figure 9.14.3.

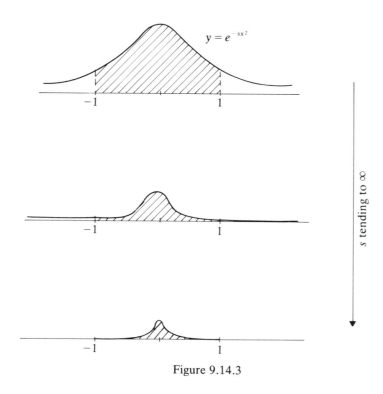

Figure 9.14.3

9.15. THE METHOD OF STATIONARY PHASE

This is closely related to the method of steepest descent, and is used to evaluate asymptotically integrals of the form

$$\Psi(\omega) = \int_a^b \exp (i\omega\phi(t))f(t) \, dt$$

where ϕ, f are real analytic functions. A rigorous justification involves Fourier transforms [see § 13.2], which are beyond the scope of this chapter; but the main step is to argue that for large ω the only significant contributions to the integral occur near stationary points of ϕ. (Physically, $\omega\phi(t)$ represents the phase of a wave; if the frequency is very high, the contributing waves interfere destructively except near stationary points of the phase, where the waves are 'nearly in phase'.)

Suppose that s is a point of stationary phase; that is, $\phi'(s) = 0$. We may Taylor-expand ϕ about s to second order [see § 2.10]:

$$\phi(t) \sim \phi(s) + \tfrac{1}{2}\phi''(s) \cdot (t-s)^2.$$

(This will be a sufficiently good approximation *provided* $\phi''(s) \neq 0$; as in section 9.14, there is a non-degeneracy condition.) Substitute this approximation for $\phi(t)$ into the integral. If ω is large enough so that the main contribution comes from t-values near to s, then the integral is approximately given by

$$\Psi(\omega) \sim \exp\,(i\omega\phi(s))f(s) \int_{-\infty}^{\infty} \exp\,(i\omega\phi''(s)(t-s)^2/2)\,dt.$$

Evaluating the integral on the right (which is of a standard form, called a *Fresnel integral*, and equal to $(1+i)(\pi/\omega\phi''(s))^{1/2}$) we obtain the asymptotic expression

$$\Psi(\omega) \sim \exp\,(i\omega\phi(s))f(s) \cdot (1+i)(\pi/\omega\phi''(s))^{1/2}.$$

If ϕ has several stationary points s, the contributions from these are summed.

EXAMPLE 9.15.1.   Find an asymptotic expression for

$$\int_{-4}^{4} \frac{e^{i\omega(t^3-3t)}}{t^2+1}\,dt.$$

In the method of stationary phase, we have $\phi(t) = t^3 - 3t$, $f(t) = 1/(1+t^2)$. Then $\phi'(t) = 0$ if and only if $3t^2 - 3 = 0$, that is, $t = 1$ or -1. Adding the two contributions (and noting that both points are within the range of integration—in general points outside the range must be ignored) we obtain the approximation

$$\frac{\exp\,(i\omega(-2))}{2}(1+i)\left(\frac{\pi}{6\omega}\right)^{1/2} + \frac{\exp\,(i\omega(-4))}{2}(1+i)\left(\frac{\pi}{-6\omega}\right)^{1/2}$$

$$= \frac{1}{2}\left(\frac{\pi}{6\omega}\right)^{1/2}[(\exp\,(-2i\omega) + \exp\,(-4i\omega)) + i(\exp\,(-2i\omega) - \exp\,(-4i\omega))].$$

Stationary Phase methods may be generalized to cases where ϕ has degenerate critical points, using results from catastrophe theory [see V, Chapter 8, and Poston and Stewart (1978) Chapter 12, § 6].

9.16.   THE EXTENDED COMPLEX PLANE, OR RIEMANN SPHERE

We now consider a way of describing the behaviour of a complex function 'at infinity' by adjoining to the complex plane $\mathbb{C}$ an extra point '∞'. That a single point is required is due to the geometry of the plane. The idea is due to Riemann and has a good geometric realization.

We think of $\mathbb{C}$ as being embedded as the (x, y)-plane in $\mathbb{R}^3$, so that a point $x + iy \in \mathbb{C}$ is identified with $(x, y, 0) \in \mathbb{R}^3$. Let

$$S^2 = \{(\xi, \eta, \zeta) \in \mathbb{R}^3 : \xi^2 + \eta^2 + \zeta^2 = 1\}$$

be the 'unit sphere', Figure 9.16.1. A line joining the 'North pole' $(0, 0, 1)$ to $(x, y, 0)$ cuts S^2 in a unique point (ξ, η, ζ); so we have a one-to-one correspondence between $\mathbb{C}$ and the points of S^2 other than the North pole. As the reader may verify,

$$(\xi, \eta, \zeta) \quad \text{corresponds to} \quad \left(\frac{\xi}{1-\zeta}\right) + i\left(\frac{\eta}{1-\zeta}\right).$$

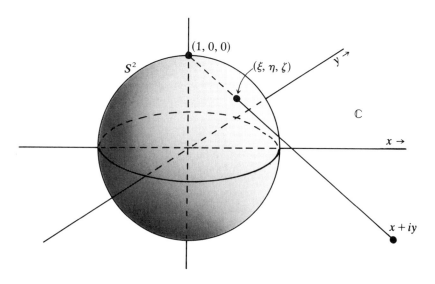

Figure 9.16.1

As (ξ, η, ζ) gets near to $(0, 0, 1)$ it follows (as is obvious geometrically) that $|x + iy|$ becomes very large: thus it is reasonable to introduce the symbol ∞ to correspond to $(0, 0, 1) \in S^2$. Then we have a one-to-one correspondence between S^2 and $\mathbb{C} \cup \{\infty\}$, which latter we call the *extended complex plane*. It may be identified with S^2 which is then known as the *Riemann Sphere*. We can now think of $\mathbb{C} \cup \{\infty\}$ either as a plane plus an extra point, or as a sphere: correspondingly we think of $\mathbb{C}$ as a plane, or as a sphere without a North pole. Both viewpoints are valuable, depending on the problem in hand.

Since $\{x + iy : |x + iy| > R\}$ corresponds to a 'cap' between a line of latitude and the North pole it makes sense to think of $\{z : |z| > R\}$ as a 'neighbourhood of ∞'. Such neighbourhoods get smaller as R gets larger. Doing this leads to a concept of continuity on S^2 agreeing with geometrical intuition.

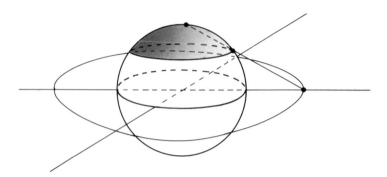

We now consider the *behaviour of an analytic function at* ∞.

DEFINITION 9.16.1.   Suppose that $f(z)$ is analytic in $\{z \in \mathbb{C}: |z| > R\}$. Then we can define

$$g(z) = f(1/z) \qquad (0 < |z| < 1/R).$$

Since $g'(z) = -z^{-2} f'(1/z)$ it follows that $g(z)$ is analytic for $0 < |z| < 1/R$, and therefore has an isolated singularity at 0 [see Definition 9.8.2]. We shall say that *f has a removable singularity, pole of order m*, or *isolated essential singularity at* ∞ if and only if $g(z)$ has the corresponding singularity at 0 [see § 9.8]. (Note that this is a definition, not a theorem.)

EXAMPLES 9.16.1
 (1) $f(z) = 1/z$ ($|z| > 0$). Then $g(z) = z$ ($|z| > 0$) so that f has a removable singularity at ∞.
 (2) $f(z) = z$. Then $g(z) = 1/z$ ($|z| > 0$) so that f has a simple pole at ∞.
 (3) $f(z) = \exp z$. Then $g(z) = \exp(1/z)$ ($|z| > 0$) and f has an isolated essential singularity at ∞.
 (4) $f(z) = 1/\sin z$ ($z \neq n\pi$, $n \in \mathbb{Z}$). Then $g(z) = 1/\sin(1/z)$. We can't say that f has an isolated essential singularity at ∞, because f is not analytic in $\{z: |z| > R\}$ for any $R > 0$. However, f certainly has some sort of singularity at ∞! In general if $f(z)$ has a sequence of isolated singularities $z_1, z_2, \ldots$ such that $z_n \to \infty$ then we say that f has an *essential* (but not isolated) *singularity* at ∞.

If f has a removable singularity at ∞ then g has a removable singularity at 0, so may be considered analytic in $|z| < 1/R$. Thus we may say that f is *analytic at* ∞, and define $f(\infty) = g(0) = \lim_{z \to 0} g(z)$, so that $f(\infty) = \lim_{z \to \infty} f(z)$. For instance if $f(z) = 1/z$ then $g(0) = 0$ so $f(\infty) = 0$.

Further we say that f has a *zero of order m at* ∞ if g has a zero of order m at 0. Thus if $f(z) = 1/z^m$ ($|z| > 0$, $m \in \mathbb{Z}$, $m > 0$) then $g(z) = z^m$ and f has a zero of order m at ∞.

In general any statement about the behaviour of f at ∞ can be translated into one about that of g at 0—and this is how to prove such a statement.

9.17. RIEMANN SURFACES

With the exception of log z, which was briefly discussed at the end of section 9.5, all functions treated in this chapter were assumed to be single-valued.

It is beyond the scope of this book to embark upon a complete theory of multi-valued functions, and we have to be content with a brief informal description.

The simplest case of a multi-valued function is the square root. We write

$$w = z^{1/2} \tag{9.17.1}$$

to denote the set of complex numbers w which, for a given value of z, satisfy

$$w^2 = z. \tag{9.17.2}$$

In other words we regard (9.17.1) and (9.17.2) as expressing equivalent relationships between z and w. Suppose that z_0 is an arbitrary non-zero complex number. We can write z_0 uniquely in the form

$$z_0 = r_0 \exp(i\theta_0) \qquad (0 \le \theta_0 < 2\pi). \tag{9.17.3}$$

Then

$$w_0 = \sqrt{r_0} \exp(i\theta_0/2) \tag{9.17.4}$$

is a solution of (9.17.2), where $\sqrt{r_0}$ denotes the positive value of the square root of r_0. The other solution is

$$w_1 = \sqrt{r_0} \exp(i(\theta_0 + 2\pi)/2), \tag{9.17.5}$$

and we have that

$$w_1 = \exp(i\pi)\sqrt{r_0} \exp(i\theta_0/2).$$

Hence

$$w_1 = -w_0,$$

because $\exp(i\pi) = -1$ [see (9.5.8)]. There are no other solutions of (9.17.2). If we adopt the convention that arg z_0 $(= \theta_0)$ lies between 0 and 2π, as indicated in (9.17.3), then w_0 and w_1 are each completely determined by z_0. The solution w_0 is sometimes called the *principal value* of $z^{1/2}$; it corresponds to the 'positive square root' when z_0 happens to be a positive real number $(\theta_0 = 0)$. Every complex number, except zero, has two distinct square roots. The square root of zero has only one value, namely zero.

The structure of a multi-valued function $f(z)$ can be illuminated by an ingenious geometrical device, known as the *Riemann surface* of $f(z)$, which we shall denote by $\mathcal{R}$. The Riemann surface of $z^{1/2}$ consists of two *sheets* R_0, R_1 which are copies of the z-plane; they are placed upon each other in such a way that zero is a common point but every non-zero number is entered both in R_0 and R_1, in positions that correspond to the same point in the z-plane. A typical point of R_0 is given by

$$R_0: z_0 = r \exp(i\theta) \qquad (0 \le \theta < 2\pi); \tag{9.17.6}$$

it lies 'underneath' its counterpart in R_1 which is represented by

$$R_1: z_1 = r \exp(i\theta) \qquad (2\pi \le \theta < 4\pi). \tag{9.17.7}$$

We note that when $r = 0$, then $z_0 = z_1 = 0$, which confirms that R_0 and R_1 coalesce at the origin. If we define

$$w = r^{1/2} \exp(i\theta) \tag{9.17.8}$$

and let $z \ (= r \exp(i\theta))$ range over the complete Riemann surface, then w will attain each of the values w_0 and w_1 defined in (9.17.4) and (9.17.5) precisely once, that is w is a single-valued function on $\mathscr{R}$. The values which $z^{1/2}$ takes on R_0 and R_1 respectively, are called the *branches* of $z^{1/2}$; the function is analytic at each interior point of the branches R_0 and R_1. But in order to ensure that w is continuous on $\mathscr{R}$, we have to stipulate how R_0 and R_1 are connected apart from their attachment at 0. For this purpose we make a *cut* along the positive real axis [see Figure 9.17.1] and we postulate that on crossing the cut we pass from R_0 to R_1 or from R_1 to R_0, as indicated by the scheme

$$\ldots \to R_0 \to R_1 \to R_0 \to R_1 \to \ldots \tag{9.17.9}$$

If we follow a closed path round 0 the branches are interchanged when we have completed the circuit. For this reason the point 0 is called a *branch point* of $z^{1/2}$.

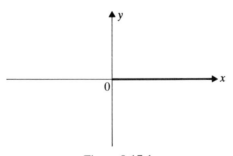

Figure 9.17.1

In general, a multi-valued function will have more than two, possibly infinitely many, branches which are joined at suitably placed cuts, and the following definition applies:

DEFINITION 9.17.1.   A point b is called a *branch point* of a multi-valued function if the branches of the function are permuted when z moves round b in a closed path.

The function

$$\log z = \log |z| + i(\arg z + 2\pi n) \qquad (n \in \mathbb{Z}) \qquad (9.17.10)$$

has infinitely many branches. Its Riemann surface consists of sheets

$$R_0, R_1, R_{-1}, R_2, R_{-2}, \ldots,$$

which can be placed into one-to-one correspondence with the integers $0, \pm 1,$ ± 2. Each sheet contains all complex numbers with the exception of zero. A single cut may be made along the positive real axis in order to connect the sheets. The points of R_n $(n \in \mathbb{Z})$ are given by

$$R_n : \{z, z \neq 0, 2\pi n \le \arg z < 2(n + 1)\}.$$

We pass from R_n to R_{n+1} by crossing the cut in the direction of the positive y-axis. The point 0, though not a member of $\mathcal{R}$ is a branch point of $\log z$.

It should be remarked that there is some arbitrariness in choosing the position of the cut; for example, the negative real axis could have been used to construct an equivalent Riemann surface for $\log z$.

Further details and illuminating examples can be found in Spiegel (1964), Chapter 1.

I.S.

REFERENCES

Ahlfors, L. V. (1966). *Complex Analysis*, McGraw-Hill.

Jameson, G. J. O. (1970). *A First Course on Complex Functions*, Chapman and Hall.

Kyrala, A. (1972). *Applied Functions of a Complex Variable*, Wiley.

Poston, T. and Stewart, I. N. (1978). *Catastrophe Theory and Its Applications*, Pitman.

Spiegel, M. R. (1964). *Theory and Problems of Complex Variables*, Schaum.

Tall, D. O. (1977). *Functions of a Complex Variable*, Routledge and Kegan Paul.

Titchmarsh, E. C. (1932). *The Theory of Functions*, Oxford University Press.

CHAPTER 10

Special Functions

10.1. JACOBIAN ELLIPTIC FUNCTIONS AND ELLIPTIC INTEGRALS

10.1.0. Introduction

The problem of finding the length of arc of an ellipse [see § 6.1] gives rise to an integral which cannot be reduced to integrals of elementary type. However, this integral and a large class of similar integrals, known as elliptic integrals, can be reduced to standard forms which may then be evaluated numerically with the help of tables. Thus the usual techniques of integration are effectively extended to include this large new class of integrals.

Circular functions are regarded as more natural and fundamental than their inverses—the latter arise from elementary integrals [see Table 4.2.1]. In the same way, it is found that elliptic functions are more fundamental than their inverses, the elliptic integrals. Thus it is necessary to have a working knowledge of elliptic functions in order to evaluate elliptic integrals; this is provided in the opening section of the text. The main part of the text deals with the theory and numerical evaluation of elliptic integrals.

Elliptic functions and integrals arise in many branches of pure and applied mathematics, for example, geometry, mechanics, hydrodynamics, electromagnetic theory and elasticity. The present text is confined mainly to real variables; the extension to complex variables is particularly useful for dealing with certain conformal transformations, details of which may be found in Bowman (1953), Copson (1935) and Milne-Thomson (1950). I am grateful to Dover Publications, Inc. for permission to quote from L. M. Milne-Thomson *Jacobian Elliptic Function Tables.*

10.1.1. Jacobian Elliptic Functions

Elliptic functions are most easily understood by considering the analogy with circular functions. For example, sin u can be defined by the integral

$$u = \int_0^x \frac{dt}{(1-t^2)^{1/2}}, \qquad -1 \le x \le 1; \qquad (10.1.1)$$

427

we have [cf. Table 4.2.1]

$$u = \sin^{-1} x, \qquad x = \sin u.$$

It is important to notice that the integral is equal to the *inverse* of the sine function. By analogy with equation (10.1.1), if

$$u = \int_0^x \frac{dt}{[(1-t^2)(1-k^2 t^2)]^{1/2}}, \qquad -1 \le x \le 1, 0 < k < 1, \qquad (10.1.2)$$

we write

$$u = \mathrm{sn}^{-1} x, \qquad x = \mathrm{sn}\, u.$$

The integral (10.1.2) is an *elliptic integral of the first kind with modulus k*, and sn *u* is the corresponding *elliptic function*. (Elliptic functions are pronounced letter by letter; thus *s, n, u*.)

Pursuing the analogy further, a special case of equation (10.1.1) arises when $x = 1$, namely [see § 2.12]

$$\frac{\pi}{2} = \int_0^1 \frac{dt}{(1-t^2)^{1/2}}.$$

Correspondingly, when $x = 1$ in equation (10.1.2), we have

$$K = \int_0^1 \frac{dt}{[(1-t^2)(1-k^2 t^2)]^{1/2}}, \qquad 0 < k < 1, \qquad (10.1.3)$$

which is called the *complete elliptic integral of the first kind with modulus k*.

From equations (10.1.2) and (10.1.3), we can deduce the following properties of sn *u*.

(i) sn *u* is a function of *u* and *k* [see § 5.1]; it is written sn (u, k) when the modulus needs to be emphasized.
(ii) sn $0 = 0$, sn $K = 1$, sn $(-K) = -1$, where *K* is a function of *k*.
(iii) sn *u* is an odd function of *u*, that is, sn $(-u) = -\mathrm{sn}\,(u)$, and increases steadily from -1 to 1 as *u* increases from $-K$ to *K*.

These properties can be summarized by saying that sn *u* is analogous to sin *u*, with *K* playing the role of $\pi/2$ [see § 2.12]. So far, sn *u* is defined only for $-K \le u \le K$; we shall later extend the definition to all values of *u*, real or complex.

Having defined sn *u*, we next define the elliptic functions cn *u* and dn *u* by

$$\mathrm{cn}\, u = (1 - \mathrm{sn}^2 u)^{1/2}, \qquad \mathrm{dn}\, u = (1 - k^2 \mathrm{sn}^2 u)^{1/2}, \qquad (10.1.4)$$

where the positive square roots are taken when $-K \le u \le K$. Thus cn *u* and dn *u* are even functions of *u*, that is, cn $(-u) = \mathrm{cn}\,(u)$ and dn $(-u) = \mathrm{dn}\,(u)$, and

$$\mathrm{cn}\, 0 = 1, \qquad \mathrm{dn}\, 0 = 1,$$

$$\mathrm{cn}\, K = 0, \qquad \mathrm{dn}\, K = (1 - k^2)^{1/2} = k', \qquad (10.1.5)$$

where k' is called the *complementary modulus*.

The functions sn u, cn u and dn u are the three basic *Jacobian elliptic functions*. In the limiting cases $k = 0$ and $k = 1$, we have respectively,

$$k = 0: \quad \text{sn } u = \sin u, \quad \text{cn } u = \cos u, \quad \text{dn } u = 1;$$

$$k = 1: \quad \text{sn } u = \tanh u, \quad \text{cn } u = \text{dn } u = \text{sech } u.$$

In terms of the original analogy, cn u corresponds to $\cos u$, and, although dn u has no direct analogue, it is also similar in some respects to $\cos u$. From equations (10.1.4), we can deduce the important identities

$$\left.\begin{array}{c} \text{sn}^2 u + \text{cn}^2 u = 1, \\[4pt] \text{dn}^2 u + k^2 \text{sn}^2 u = 1, \\[4pt] k^2 \text{cn}^2 u + k'^2 = \text{dn}^2 u, \\[4pt] \text{cn}^2 u + k'^2 \text{sn}^2 u = \text{dn}^2 u. \end{array}\right\} \quad (10.1.6)$$

It is now clear that any one of the elliptic functions sn u, cn u, dn u can be expressed in terms of any other.

Differentiating equation (10.1.2) with respect to x [see Theorem 4.7.3] and then making the substitution $x = \text{sn } u$, we find

$$\frac{d}{du} \text{sn } u = \text{cn } u \text{ dn } u. \quad (10.1.7)$$

Equations (10.1.4) now yield

$$\frac{d}{du} \text{cn } u = -\text{sn } u \text{ dn } u, \quad \frac{d}{du} \text{dn } u = -k^2 \text{sn } u \text{ cn } u. \quad (10.1.8)$$

Analogous to the formulae for $\sin(u+v)$, etc., there are formulae for sn $(u+v)$, etc. which we give without proof. These enable us to extend the definitions of sn u, etc. to all real values of u.

$$\left.\begin{array}{l} \text{sn}(u+v) = \dfrac{\text{sn } u \text{ cn } v \text{ dn } v + \text{sn } v \text{ cn } u \text{ dn } u}{\Delta}, \\[12pt] \text{cn}(u+v) = \dfrac{\text{cn } u \text{ cn } v - \text{sn } u \text{ sn } v \text{ dn } u \text{ dn } v}{\Delta}, \\[12pt] \text{dn}(u+v) = \dfrac{\text{dn } u \text{ dn } v - k^2 \text{sn } u \text{ sn } v \text{ cn } u \text{ cn } v}{\Delta}, \end{array}\right\} \quad (10.1.9)$$

where

$$\Delta = 1 - k^2 \text{sn}^2 u \text{ sn}^2 v.$$

As well as giving the obvious formulae for sn $2u$, cn $2u$ and dn $2u$, equations

(10.1.9) also yield the following useful results:

$$\frac{1-\operatorname{cn} 2u}{1+\operatorname{cn} 2u}=\frac{\operatorname{sn}^2 u \, \operatorname{dn}^2 u}{\operatorname{cn}^2 u}, \qquad \frac{1-\operatorname{dn} 2u}{1+\operatorname{dn} 2u}=\frac{k^2 \operatorname{sn}^2 u \, \operatorname{cn}^2 u}{\operatorname{dn}^2 u},$$

$$\operatorname{sn}^2\left(\frac{u}{2}\right)=\frac{1-\operatorname{cn} u}{1+\operatorname{dn} u}, \qquad \operatorname{cn}^2\left(\frac{u}{2}\right)=\frac{\operatorname{dn} u+\operatorname{cn} u}{1+\operatorname{dn} u},$$

$$\operatorname{dn}^2\left(\frac{u}{2}\right)=\frac{k'^2+\operatorname{dn} u+k^2 \operatorname{cn} u}{1+\operatorname{dn} u}.$$

Let $v = K$ in equations (10.1.9). This gives

$$\operatorname{sn}(u+K)=\frac{\operatorname{cn} u}{\operatorname{dn} u},$$

$$\operatorname{cn}(u+K)=-\frac{k' \operatorname{sn} u}{\operatorname{dn} u},$$

$$\operatorname{dn}(u+K)=\frac{k'}{\operatorname{dn} u}.$$

Hence

$$\operatorname{sn} 2K = 0, \qquad \operatorname{cn} 2K = -1, \qquad \operatorname{dn} 2K = 1. \qquad (10.1.10)$$

Now let $v = 2K$ in equations (10.1.9). Then, using equations (10.1.10), we find

$$\operatorname{sn}(u+2K)=-\operatorname{sn} u,$$

$$\operatorname{cn}(u+2K)=-\operatorname{cn} u,$$

$$\operatorname{dn}(u+2K)=\operatorname{dn} u.$$

Thus sn u and cn u are *periodic* functions of u with period $4K$, and dn u is periodic with period $2K$ [see § 20.1].

When the Jacobian elliptic functions are considered as functions of a complex variable, it turns out that they possess complex periods as well as real periods. This property is characteristic of elliptic functions in general, and is made precise by the following definitions.

DEFINITION 10.1.1.   A period is called *primitive* if no submultiple of it is a period. A function $f(w)$, of a complex variable w, which has two primitive periods whose ratio is not real is called *doubly-periodic*.

We can now give a general definition of an elliptic function:

DEFINITION 10.1.2.   An *elliptic function* is a doubly-periodic function which is analytic [see § 9.1] (except at poles), and which has no other singularities in the finite part of the complex plane [see § 9.8].

Pairs of primitive periods of sn u, cn u and dn u are as follows:

$$\text{sn } u: \quad 4K, 2iK',$$

$$\text{cn } u: \quad 4K, 2K + 2iK',$$

$$\text{dn } u: \quad 2K, 4iK',$$

where K' is the same function of k' as K is of k. These pairs are not unique; for example, dn u also has the primitive periods $2K$, $2K + 4iK'$. Although the graphs of sn u and cn u resemble those of sin u and cos u, respectively [see Figures 2.12.1 and 2.12.2], the above formulae show that it is not true that one is merely a displacement of the other through the quarter-period K. Note also that dn $u > 0$ for all real values of u.

Appendix A summarizes and extends the results we have just obtained for the values of sn u, cn u and dn u when u lies outside the interval $[0, K]$.

We have assumed so far that $0 < k < 1$; but it often happens, in integrals such as (10.1.2), for example, that k lies outside this interval—it may even be imaginary. Appendix B shows how to change the modulus of sn u, cn u and dn u in the following ways:

(i) To its reciprocal
(ii) From imaginary to real
(iii) To a larger value
(iv) To a smaller value

Finally, we consider in more detail the case when $f(w)$ is an elliptic function of the complex variable w. Any elliptic function $f(w)$ with periods Ω, Ω' takes the same value at the points w and $w + m\Omega + m'\Omega'$, where m and m' are any integers. For this reason, it is convenient to consider a parallelogram with vertices at the points w, $w + \Omega$, $w + \Omega'$, $w + \Omega + \Omega'$, called a *period-parallelogram* for $f(w)$. It is sufficient to study the properties of $f(w)$ in one such parallelogram. The whole of the w-plane can be sub-divided, though not uniquely, into period-parallelograms; corresponding points are said to be *congruent*. For example, period-parallelograms for the three basic Jacobian elliptic functions are defined by the following sets of vertices:

$$\text{sn } w: \quad \pm 2K \pm iK',$$

$$\text{cn } w: \quad \pm 2K, 2iK', 4K + 2iK',$$

$$\text{dn } w: \quad \pm K \pm 2iK'.$$

The functions sn w, cn w and dn w each have two simple zeros and two simple poles in a period-parallelogram, as shown in the following table; the respective residues at the poles are also given [see §§ 9.8, 9.9]. They have no other zeros or poles, except at congruent points.

	Zeros	Poles	Residues
sn w	$0, 2K$	$iK', 2K + iK'$	$k^{-1}, -k^{-1}$
cn w	$K, -K$	$iK', 2K + iK'$	$-ik^{-1}, ik^{-1}$
dn w	$K + iK', K - iK'$	$iK', -iK'$	$-i, i$

10.1.2. Elliptic Integrals

The theory of Jacobian elliptic functions finds its major practical application in the evaluation of elliptic integrals, of which equation (10.1.2) is a typical example. An elliptic integral is an integral of the form

$$\int R(x, X^{1/2})\, dx,$$

where $R(x, y)$ is a rational function of x and y (that is, a ratio of two polynomials, each of them in both x and y), and X is a cubic or quartic in x with no repeated factor. If X has a repeated factor then the integral can be evaluated in terms of elementary functions.

There are three standard types of elliptic integral, which may be written as

$$\int \frac{dt}{[(1-t^2)(1-k^2t^2)]^{1/2}}, \tag{10.1.11}$$

$$\int \left(\frac{1-k^2t^2}{1-t^2}\right)^{1/2} dt, \tag{10.1.12}$$

and

$$\int \frac{t^2\, dt}{(1+vt^2)[(1-t^2)(1-k^2t^2)]^{1/2}}, \qquad v \neq 0. \tag{10.1.13}$$

These are called *elliptic integrals of the first, second and third kind*, respectively. It will be shown later that every elliptic integral can be reduced to the sum of a finite number of elementary integrals and integrals of these three types.

It is sometimes convenient to use alternative forms of the standard integrals. For example, writing

$$t = \sin\theta,$$

and taking the limits of integration for θ to be 0 and ϕ, the integrals become, respectively,

$$F(k, \phi) = \int_0^\phi \frac{d\theta}{(1-k^2\sin^2\theta)^{1/2}} = \text{sn}^{-1}(\sin\phi), \tag{10.1.14}$$

$$E(k, \phi) = \int_0^\phi (1 - k^2 \sin^2 \theta)^{1/2} \, d\theta, \tag{10.1.15}$$

$$\Pi(k, \nu, \phi) = \int_0^\phi \frac{\sin^2 \theta \, d\theta}{(1 + \nu \sin^2 \theta)(1 - k^2 \sin^2 \theta)^{1/2}}. \tag{10.1.16}$$

Alternatively, the expressions for the three standard integrals are made more concise by means of the substitution

$$t = \operatorname{sn} v,$$

with $v = u$ when $t = \sin \phi$ in (10.1.11)–(10.1.13); or equivalently

$$\sin \theta = \operatorname{sn} v,$$

with $v = u$ when $\theta = \phi$ in (10.1.14)–(10.1.16). Then

$$F(k, \phi) = u, \tag{10.1.17}$$

$$E(k, \phi) = E(u) = \int_0^u \operatorname{dn}^2 v \, dv, \tag{10.1.18}$$

$$\Pi(k, \nu, \phi) = \Pi(u) = \int_0^u \frac{\operatorname{sn}^2 v \, dv}{1 + \nu \operatorname{sn}^2 v}. \tag{10.1.19}$$

The *complete elliptic* integrals take the following forms in the various notations.

First kind

$$K = F\left(k, \frac{\pi}{2}\right) = \int_0^1 \frac{dt}{[(1 - t^2)(1 - k^2 t^2)]^{1/2}} = \int_0^{\pi/2} \frac{d\theta}{(1 - k^2 \sin^2 \theta)^{1/2}}.$$

Second kind

$$E = E\left(k, \frac{\pi}{2}\right) = E(K) = \int_0^1 \left(\frac{1 - k^2 t^2}{1 - t^2}\right)^{1/2} dt$$

$$= \int_0^{\pi/2} (1 - k^2 \sin^2 \theta)^{1/2} \, d\theta = \int_0^K \operatorname{dn}^2 v \, dv.$$

Third kind

$$\Pi\left(k, \nu, \frac{\pi}{2}\right) = \Pi(K) = \int_0^1 \frac{t^2 \, dt}{(1 + \nu t^2)[(1 - t^2)(1 - k^2 t^2)]^{1/2}}$$

$$= \int_0^{\pi/2} \frac{\sin^2 \theta \, d\theta}{(1 + \nu \sin^2 \theta)(1 - k^2 \sin^2 \theta)^{1/2}}$$

$$= \int_0^K \frac{\operatorname{sn}^2 v \, dv}{1 + \nu \operatorname{sn}^2 v}.$$

There are several useful relations connecting K, E, K' and E', where E' is the same function of k' as E is of k. For example,

$$\text{(i)} \quad \frac{dK}{dk} = \frac{E - k'^2 K}{kk'^2}, \qquad \text{(ii)} \quad \frac{dE}{dk} = -\frac{(K - E)}{k},$$

$$\text{(iii)} \quad \frac{dK'}{dk} = -\frac{(E' - k^2 K')}{kk'^2}, \qquad \text{(iv)} \quad \frac{dE'}{dk} = \frac{k(K' - E')}{k'^2},$$

$$\text{(v)} \quad KE' + K'E - KK' = \frac{\pi}{2} \quad (\textit{Legendre's formula}).$$

Most of the effort in the evaluation of elliptic integrals lies in the algebraic manipulation required to reduce them to the standard forms. Systematic methods of reduction are considered in later sections; meanwhile, we present a few simple examples involving elliptic functions and integrals.

10.1.3. Examples

EXAMPLE 10.1.1   *Simple Pendulum.*   The equation of motion of a simple pendulum of length l is

$$\ddot{\psi} = -\omega^2 \sin \psi,$$

where ψ is the angle that the pendulum makes with the downward vertical, and $\omega^2 = g/l$. Assume that $\psi = 0$ when $t = 0$, and that $\dot{\psi} = 0$ when $\psi = \alpha$ ($< \pi/2$). The usual approximation $\sin \psi \approx \psi$ [see (2.12.13)] leads to simple harmonic motion, but we have *exactly*

$$\dot{\psi}^2 = 2\omega^2 (\cos \psi - \cos \alpha) = 4\omega^2 \left(\sin^2 \frac{\alpha}{2} - \sin^2 \frac{\psi}{2} \right).$$

Therefore

$$\frac{d\psi}{dt} = \pm 2\omega \left(k^2 - \sin^2 \frac{\psi}{2} \right)^{1/2}, \qquad k = \sin \frac{\alpha}{2},$$

or

$$\omega \, dt = \pm \frac{d\psi}{2 \left(k^2 - \sin^2 \dfrac{\psi}{2} \right)^{1/2}}.$$

To express the right-hand side in the standard form (10.1.14) for an elliptic integral of the first kind, put

$$\sin \frac{\psi}{2} = k \sin \phi, \qquad \tfrac{1}{2} \cos \frac{\psi}{2} \, d\psi = k \cos \phi \, d\phi.$$

Then, considering only the first quarter-period of the motion, for which

$0 \leq \psi \leq \alpha$, $\dot{\psi} \geq 0$, we have

$$\omega t = \int_0^\phi \frac{d\theta}{(1 - k^2 \sin^2 \theta)^{1/2}}.$$

From equation (10.1.14),

$$\omega t = F(k, \phi) = \mathrm{sn}^{-1}(\sin \phi),$$

or

$$\sin \phi = \mathrm{sn}\,\omega t.$$

Hence

$$\sin \frac{\psi}{2} = k \sin \phi = k \,\mathrm{sn}\,\omega t,$$

or

$$\psi = 2 \sin^{-1}(k \,\mathrm{sn}\,\omega t), \qquad k = \sin \frac{\alpha}{2}.$$

The periodic property of sn ωt now shows that this solution holds throughout the motion.

The periodic time T of the oscillation is given by

$$\omega T = 4K, \qquad T = 4K(l/g)^{1/2}.$$

It is interesting to note that since K is an increasing function of k, the periodic time T increases with the amplitude α of the oscillations.

EXAMPLE 10.1.2 *Arc Length of Ellipse.* It is from this problem that elliptic integrals take their name. From elementary calculus [see Definition 6.1.3], the length of arc of the ellipse

$$x = a \cos \phi, \qquad y = b \sin \phi$$

between the points $\phi = \phi_1$ and $\phi = \phi_2 \,(>\phi_1)$ is given by

$$s = \int_{\phi_1}^{\phi_2} (a^2 \sin^2 \phi + b^2 \cos^2 \phi)^{1/2} \, d\phi$$

$$= a \int_{\phi_1}^{\phi_2} (1 - e^2 \cos^2 \phi)^{1/2} \, d\phi,$$

where e is the eccentricity of the ellipse [see V, § 1.3.2]. Let

$$t = \frac{\pi}{2} - \phi.$$

Then

$$s = a \int_{t_2}^{t_1} (1 - e^2 \sin^2 t)^{1/2} \, dt,$$

where $t = t_1$ when $\phi = \phi_1$, etc. The integral is now in the standard form (10.1.15), and hence

$$s = a[E(e, t_1) - E(e, t_2)].$$

In particular, the perimeter of the ellipse is

$$4aE\left(e, \frac{\pi}{2}\right) = 4aE(K), \qquad k = e.$$

Also, since

$$E\left(0, \frac{\pi}{2}\right) = \frac{\pi}{2},$$

the circumference of a circle of radius a is given correctly as $2\pi a$ [see V, § 1.1.8].

EXAMPLE 10.1.3.   The motion of a particle of unit mass along the x-axis is given by $x = a \operatorname{sn} \omega t$. Find the law of force.
   The force is given by

$$\frac{d^2 x}{dt^2} = \frac{d}{dt}(a\omega \operatorname{cn} \omega t \operatorname{dn} \omega t)$$

$$= -a\omega^2(\operatorname{sn} \omega t \operatorname{dn}^2 \omega t + k^2 \operatorname{sn} \omega t \operatorname{cn}^2 \omega t)$$

$$= -\omega^2\left[(1 + k^2)x - \frac{2k^2 x^3}{a^2}\right].$$

EXAMPLE 10.1.4   *Euler's Equations of Motion.*   The motion of a rigid body relative to its centre of mass is governed by Euler's equations

$$\left.\begin{array}{l} A\dot{p} - (B - C)qr = L, \\ B\dot{q} - (C - A)rp = M, \\ C\dot{r} - (A - B)pq = N, \end{array}\right\} \qquad (10.1.20)$$

where A, B, C are the principal moments of inertia of the body at the centre of mass, p, q, r are the components of angular velocity of the body about the principal axes at this point, and L, M, N are the components of the resultant applied couple about these axes. (The axes are fixed in the body and rotate with it.) We assume that $A > B > C > 0$. Equations (10.1.20) cannot be solved analytically, except in special cases. The form of the equations suggests, however, that when $L = M = N = 0$ they have a solution

$$p = \alpha \operatorname{cn} \lambda(t - t_0), \qquad q = -\beta \operatorname{sn} \lambda(t - t_0), \qquad r = \gamma \operatorname{dn} \lambda(t - t_0).$$

Substituting in equations (10.1.20), we find

$$-A\alpha\lambda + (B - C)\beta\gamma = 0,$$
$$-B\beta\lambda - (C - A)\gamma\alpha = 0,$$
$$-C\gamma k^2\lambda + (A - B)\alpha\beta = 0.$$

Multiplying these equations by α, β, γ, respectively, we obtain

$$\frac{\alpha\beta\gamma}{\lambda} = \frac{A\alpha^2}{B - C} = \frac{B\beta^2}{A - C} = \frac{C\gamma^2 k^2}{A - B} = \omega^2, \quad \text{say.}$$

Hence a solution of equations (10.1.20) when $L = M = N = 0$ is

$$
\left.
\begin{aligned}
p &= \omega \left(\frac{B - C}{A} \right)^{1/2} \operatorname{cn} \lambda (t - t_0), \\
q &= -\omega \left(\frac{A - C}{B} \right)^{1/2} \operatorname{sn} \lambda (t - t_0), \\
r &= \frac{\omega}{k} \left(\frac{A - B}{C} \right)^{1/2} \operatorname{dn} \lambda (t - t_0),
\end{aligned}
\right\}
\qquad (10.1.21)
$$

where ω, t_0 and the modulus k are arbitrary, and

$$\lambda = \frac{\alpha\beta\gamma}{\omega^2} = \frac{\omega}{k} \left[\frac{(B - C)(A - C)(A - B)}{ABC} \right]^{1/2}.$$

In the solution (10.1.21), the angular velocity r is of fixed sign, while p and q oscillate in sign. Another solution is given by

$$p = a \operatorname{dn} \mu(t - t_0), \qquad q = b \operatorname{sn} \mu(t - t_0), \qquad r = c \operatorname{cn} \mu(t - t_0),$$

where a, b, c, μ may be determined as before. In this solution, p is of fixed sign. However, there is no corresponding solution in which q is of fixed sign.

EXAMPLE 10.1.5. Find the *common volume of two circular cylinders* of radii a and b ($<a$) whose axes intersect at right angles.

Let the equations of the cylinders be [see V, § 2.1]

$$x^2 + y^2 = a^2 \quad \text{and} \quad y^2 + z^2 = b^2.$$

The common volume is [see § 6.4]

$$V = 8 \int_0^b \int_0^{(a^2 - y^2)^{1/2}} z \, dx \, dy$$

$$= 8 \int_0^b [(a^2 - y^2)(b^2 - y^2)]^{1/2} \, dy.$$

Put

$$y = b \operatorname{sn} u, \qquad dy = b \operatorname{cn} u \operatorname{dn} u \, du, \qquad k = b/a.$$

Then

$$V = 8ab^2 \int_0^K \text{cn}^2 u \, \text{dn}^2 u \, du$$

$$= 8ab^2 \int_0^K [1 - (1 + k^2) \, \text{sn}^2 u + k^4 \, \text{sn}^4 u] \, du.$$

Now

$$\int_0^K \text{sn}^2 u \, du = \int_0^K \frac{1 - \text{dn}^2 u}{k^2} \, du = \frac{K - E}{k^2},$$

and

$$\frac{d}{du} (\text{sn} \, u \, \text{cn} \, u \, \text{dn} \, u) = 1 - 2(1 + k^2) \, \text{sn}^2 u + 3k^2 \, \text{sn}^4 u,$$

giving

$$\int_0^K \text{sn}^4 u \, du = \frac{1}{3k^4} [(2 + k^2)K - 2(1 + k^2)E].$$

Hence

$$V = \frac{8ab^2}{3k^2} [(1 + k^2)E - (1 - k^2)K]$$

$$= \frac{8a}{3} [(a^2 + b^2)E - (a^2 - b^2)K].$$

10.1.4. Elliptic Integrals of the First Kind

In this section, we shall show how to reduce the elliptic integral of the first kind

$$\int \frac{dx}{X^{1/2}}$$

to the standard form (10.1.11), or its equivalent, where X is a quartic in x with real coefficients. (A cubic may be regarded as a quartic with a leading coefficient of zero.)

Real quadratic factors S_1, S_2 can always be found such that

$$X = S_1 S_2,$$

where

$$S_1 \equiv a_1 x^2 + 2b_1 x + c_1, \qquad S_2 \equiv a_2 x^2 + 2b_2 x + c_2$$

[see I, Proposition 14.5.3]. When X has four real zeros, S_1 and S_2 must be defined in such a way that the zeros of S_1 do not interlace those of S_2. This ensures that the roots of equation (10.1.22) below are always real. Now [see I, Proposition 14.5.4(iii)]

$$S_1 - \lambda S_2 \equiv (a_1 - \lambda a_2)x^2 + 2(b_1 - \lambda b_2)x + (c_1 - \lambda c_2)$$

is a perfect square if λ satisfies

$$(a_1 - \lambda a_2)(c_1 - \lambda c_2) - (b_1 - \lambda b_2)^2 = 0. \qquad (10.1.22)$$

Let λ_1, λ_2 be the (real) roots of this equation. Then

$$S_1 - \lambda_1 S_2 \equiv (a_1 - \lambda_1 a_2)(x - \alpha)^2,$$

$$S_1 - \lambda_2 S_2 \equiv (a_1 - \lambda_2 a_2)(x - \beta)^2,$$

where α and β are real. Therefore

$$S_1 \equiv A_1(x - \alpha)^2 + B_1(x - \beta)^2,$$

$$S_2 \equiv A_2(x - \alpha)^2 + B_2(x - \beta)^2,$$

where A_1, B_1, A_2, B_2 are all real.

Next, make the substitution

$$t = \frac{x - \alpha}{x - \beta}, \qquad dt = \frac{(\alpha - \beta)\,dx}{(x - \beta)^2},$$

giving

$$\int \frac{dx}{X^{1/2}} = \frac{1}{\alpha - \beta} \int \frac{dt}{[(A_1 t^2 + B_1)(A_2 t^2 + B_2)]^{1/2}}. \qquad (10.1.23)$$

For real integrals, there are only six essentially different combinations of sign, as follows.

	I	II	III	IV	V	VI
A_1	+	+	−	+	+	−
B_1	+	−	+	−	−	+
A_2	+	+	+	+	−	−
B_2	+	+	+	−	+	+

Further simple substitutions, depending on the signs of A_1 and A_2, lead to one of the following six integrals, which correspond (as indicated by the Roman numerals) to the six sign combinations in the above table.

I
$$\int \frac{dt}{[(t^2 + a^2)(t^2 + b^2)]^{1/2}}, \qquad b^2 < a^2.$$

II
$$\int \frac{dt}{[(t^2 - a^2)(t^2 + b^2)]^{1/2}}, \qquad a^2 < t^2.$$

III
$$\int \frac{dt}{[(a^2-t^2)(t^2+b^2)]^{1/2}}, \qquad t^2 < a^2.$$

IV
$$\int \frac{dt}{[(t^2-a^2)(t^2-b^2)]^{1/2}}, \qquad b^2 < a^2 < t^2.$$

V
$$\int \frac{dt}{[(t^2-a^2)(b^2-t^2)]^{1/2}}, \qquad a^2 < t^2 < b^2.$$

VI
$$\int \frac{dt}{[(a^2-t^2)(b^2-t^2)]^{1/2}}, \qquad t^2 < b^2 < a^2.$$

At this stage, it is useful to introduce *Glaisher's notation* for the reciprocals and quotients of Jacobian elliptic functions:

$$\text{ns } u = 1/\text{sn } u, \quad \text{nc } u = 1/\text{cn } u, \qquad \text{nd } u = 1/\text{dn } u,$$

$$\text{sc } u = \text{sn } u/\text{cn } u, \qquad \text{sd } u = \text{sn } u/\text{dn } u, \qquad \text{cd } u = \text{cn } u/\text{dn } u,$$

$$\text{cs } u = \text{cn } u/\text{sn } u, \qquad \text{ds } u = \text{dn } u/\text{sn } u, \qquad \text{dc } u = \text{dn } u/\text{cn } u.$$

With this notation, and naming the integrals $I_1, \ldots, I_6$, respectively, the substitution leading to the evaluation of each integral is given below.

I $t = b \text{ sc } u,$ $I_1 = \dfrac{1}{a} \text{sc}^{-1}\left(\dfrac{t}{b}\right),$ $k' = \dfrac{b}{a}.$

II $t = a \text{ nc } u,$ $I_2 = \dfrac{1}{(a^2+b^2)^{1/2}} \text{nc}^{-1}\left(\dfrac{t}{a}\right),$ $k = \dfrac{b}{(a^2+b^2)^{1/2}}.$

III $t = a \text{ cn } u,$ $I_3 = -\dfrac{1}{(a^2+b^2)^{1/2}} \text{cn}^{-1}\left(\dfrac{t}{a}\right),$ $k = \dfrac{a}{(a^2+b^2)^{1/2}}.$

IV $t = a \text{ dc } u,$ $I_4 = \dfrac{1}{a} \text{dc}^{-1}\left(\dfrac{t}{a}\right),$ $k = \dfrac{b}{a}.$

V $t = a \text{ nd } u,$ $I_5 = \dfrac{1}{b} \text{nd}^{-1}\left(\dfrac{t}{a}\right),$ $k' = \dfrac{a}{b}.$

VI $t = b \text{ sn } u,$ $I_6 = \dfrac{1}{a} \text{sn}^{-1}\left(\dfrac{t}{b}\right),$ $k = \dfrac{b}{a}.$

10.1.5.   Elliptic Integrals of the Second Kind

The elliptic integral of the second kind can be expressed in the form

$$\int \frac{t^2 \, dt}{[(A_1 t^2 + B_1)(A_2 t^2 + B_2)]^{1/2}}.$$

The same substitution is used here as in the elliptic integral of the first kind, depending on the signs of A_1, B_1, A_2, B_2. This reduces the integral to a form involving known functions and the standard elliptic integral of the second kind (10.1.18). The procedure is illustrated by the following example.

EXAMPLE 10.1.6

$$I = \int \frac{t^2 \, dt}{[(A_1 t^2 + B_1)(A_2 t^2 + B_2)]^{1/2}},$$

where A_1, B_1, A_2, B_2 are all positive.

We have

$$I = \frac{1}{(A_1 A_2)^{1/2}} \int \frac{t^2 \, dt}{[(t^2 + a^2)(t^2 + b^2)]^{1/2}},$$

where $a^2 = B_1/A_1$, $b^2 = B_2/A_2$, and we can assume without loss of generality that $b^2 < a^2$. The corresponding integral of the first kind is of Type I, and so we make the substitution

$$t = b \operatorname{sc} u, \qquad k' = b/a,$$

which gives

$$I = \frac{b^2}{(A_1 A_2)^{1/2} a} \int \frac{\operatorname{sn}^2 u}{\operatorname{cn}^2 u} \, du.$$

This integral has to be expressed in terms of

$$E(u) = \int \operatorname{dn}^2 u \, du.$$

{$E(u)$ is used here and elsewhere for the *indefinite* integral [see § 4.2] corresponding to (10.1.18). There is no ambiguity in this notation, since we always take $E(0) = 0$.}

Since

$$\frac{d}{du} (\operatorname{dn} u \operatorname{sc} u) = \operatorname{dn}^2 u - k'^2 + \frac{k'^2}{\operatorname{cn}^2 u},$$

we have

$$I = \frac{b^2}{(A_1 A_2)^{1/2} a} \int \left(\frac{1}{\operatorname{cn}^2 u} - 1 \right) du$$

$$= \frac{b^2}{(A_1 A_2)^{1/2} a} \left[\frac{1}{k'^2} \{ \operatorname{dn} u \operatorname{sc} u - E(u) + k'^2 u \} - u \right]$$

$$= \frac{a}{(A_1 A_2)^{1/2}} [\operatorname{dn} u \operatorname{sc} u - E(u)].$$

In terms of the original variable t, we find

$$I = \frac{a}{(A_1 A_2)^{1/2}} \left[\frac{t}{a} \left(\frac{t^2 + a^2}{t^2 + b^2} \right)^{1/2} - E \left\{ \operatorname{sc}^{-1} \left(\frac{t}{b} \right) \right\} \right], \qquad k' = b/a.$$

In the above example, we had to express the integral of $\text{sn}^2 u/\text{cn}^2 u$ in terms of $E(u)$. It is useful to have a list of similar integrals, and it is easy to verify, by differentiation, that the following eleven expressions are all equal to $E(u)$.

$$u - k^2 \int \text{sn}^2 u \, du,$$

$$k'^2 u + k^2 \int \text{cn}^2 u \, du,$$

$$u - \text{dn} \, u \, \text{cs} \, u - \int \text{ns}^2 u \, du,$$

$$k'^2 u + \text{dn} \, u \, \text{sc} \, u - k'^2 \int \text{nc}^2 u \, du,$$

$$k^2 \, \text{sn} \, u \, \text{cd} \, u + k'^2 \int \text{nd}^2 u \, du,$$

$$\text{dn} \, u \, \text{sc} \, u - k'^2 \int \text{sc}^2 u \, du,$$

$$k'^2 u + k^2 \, \text{sn} \, u \, \text{cd} \, u + k^2 k'^2 \int \text{sd}^2 u \, du,$$

$$u + k^2 \, \text{sn} \, u \, \text{cd} \, u - k^2 \int \text{cd}^2 u \, du,$$

$$-\text{dn} \, u \, \text{cs} \, u - \int \text{cs}^2 u \, du,$$

$$k'^2 u - \text{dn} \, u \, \text{cs} \, u - \int \text{ds}^2 u \, du,$$

$$u + \text{dn} \, u \, \text{sc} \, u - \int \text{dc}^2 u \, du.$$

The standard elliptic integral of the second kind,

$$E(u) = \int_0^u \text{dn}^2 v \, dv$$

is not periodic, but satisfies

$$E(u + 2K) = E(u) + 2E.$$

It is found convenient to introduce *Jacobi's Zeta function*

$$Z(u) = E(u) - \frac{uE}{K}, \tag{10.1.24}$$

which satisfies

$$Z(u + 2K) = Z(u),$$

and thus has $2K$ as a period. Elliptic integrals of the second kind are usually evaluated from tables of $Z(u)$. It can be shown, however, that $Z(u)$ is simply-periodic; hence it is not an elliptic function.

The above expressions for $E(u+2K)$ and $Z(u+2K)$ are special cases of the general addition formulae for $E(u)$ and $Z(u)$. These formulae are identical:

$$E(u+v) = E(u) + E(v) - k^2 \operatorname{sn} u \operatorname{sn} v \operatorname{sn}(u+v),$$

$$Z(u+v) = Z(u) + Z(v) - k^2 \operatorname{sn} u \operatorname{sn} v \operatorname{sn}(u+v).$$

10.1.6.   Elliptic Integrals of the Third Kind

The elliptic integral of the third kind can be expressed in the form

$$I = \int \frac{dt}{(1+Ct^2)[(A_1t^2+B_1)(A_2t^2+B_2)]^{1/2}}, \qquad C \neq 0.$$

Making the appropriate substitutions, as for elliptic integrals of the first and second kinds, the integral is reduced to known functions together with the standard form (10.1.19).

EXAMPLE 10.1.7

$$I = \int \frac{dt}{(1+Ct^2)[(A_1t^2+B_1)(A_2t^2+B_2)]^{1/2}}, \qquad C \neq 0,$$

where A_1, B_1, A_2, B_2 are all positive.
   We have [cf. Example 10.1.6]

$$I = \frac{1}{(A_1A_2)^{1/2}} \int \frac{dt}{(1+Ct^2)[(t^2+a^2)(t^2+b^2)]^{1/2}}, \qquad b^2 < a^2,$$

and the substitution $t = b \operatorname{sc} u$, $k' = b/a$, gives

$$I = \frac{1}{(A_1A_2)^{1/2}a} \left[u - (1+v) \int \frac{\operatorname{sn}^2 u \, du}{1+v \operatorname{sn}^2 u} \right], \qquad (10.1.25)$$

where

$$v = b^2 C - 1.$$

If $v = -k^2$, we have an elliptic integral of the second kind. For other values of v, it is convenient for computational purposes to choose a parameter h such that

$$v = -k^2 \operatorname{sn}^2 h, \qquad (10.1.26)$$

and to take

$$\Pi(u, h) = \int_0^u \frac{k^2 \operatorname{sn} h \operatorname{cn} h \operatorname{dn} h \operatorname{sn}^2 v}{1 - k^2 \operatorname{sn}^2 h \operatorname{sn}^2 v} \, dv \qquad (10.1.27)$$

as the *fundamental* elliptic integral of the third kind.

The integral (10.1.27) can be expressed in terms of *Jacobi's Theta and Zeta functions*, as follows.

$$\Pi(u, h) = \tfrac{1}{2} \log \left[\frac{\Theta(u-h)}{\Theta(u+h)} \right] + uZ(h), \qquad (10.1.28)$$

where

$$\Theta(u) = 1 + 2 \sum_{n=1}^{\infty} (-1)^n q^{n^2} \cos(n\pi u/K), \qquad (10.1.29)$$

in which $q = e^{-\pi K'/K}$. Jacobi's Zeta function is defined in (10.1.24). In practice, the series for $\Theta(u)$ is rapidly convergent [see § 1.7] and may be terminated after two or three terms.

The given integral finally becomes

$$I = \frac{1}{(A_1 A_2)^{1/2} a} \left[u - \frac{(1+v)\Pi(u, h)}{k^2 \operatorname{sn} h \operatorname{cn} h \operatorname{dn} h} \right], \qquad k' = b/a,$$

where $\Pi(u, h)$ is given by equation (10.1.28), and h is given by

$$b^2 C - 1 = -k^2 \operatorname{sn}^2 h.$$

The parameter h is purely imaginary if $b^2 C > 1$. In this case, it is necessary to use the three formulae known collectively as *Jacobi's imaginary transformation*:

$$\left. \begin{array}{l} \operatorname{sn}(iv, k) = i\operatorname{sc}(v, k'), \\[4pt] \operatorname{cn}(iv, k) = \operatorname{nc}(v, k'), \\[4pt] \operatorname{dn}(iv, k) = \operatorname{dc}(v, k'). \end{array} \right\} \qquad (10.1.30)$$

It is useful to add to these the corresponding formula for Jacobi's Zeta function:

$$Z(iv, k) = -i \left[Z(v, k') + \frac{\pi v}{2KK'} - \frac{s_1 d_1}{c_1} \right], \qquad (10.1.31)$$

where $s_1 = \operatorname{sn}(v, k')$, etc.

For completeness, we now define $\operatorname{sn} w$, $\operatorname{cn} w$ and $\operatorname{dn} w$ for all values of w, real or complex, by assuming that the addition theorems (10.1.9) remain valid when the arguments u, v are complex. Let

$$\begin{array}{ll} s = \operatorname{sn}(u, k), & s_1 = \operatorname{sn}(v, k'), \\[4pt] c = \operatorname{cn}(u, k), & c_1 = \operatorname{cn}(v, k'), \\[4pt] d = \operatorname{dn}(u, k), & d_1 = \operatorname{dn}(v, k'). \end{array}$$

Then, using Jacobi's imaginary transformation in conjunction with the addition theorems (10.1.9), we find

$$\text{sn}\,(u+iv) = (sd_1 - icds_1c_1)/(1-d^2s_1^2),$$

$$\text{cn}\,(u+iv) = (cc_1 - isds_1d_1)/(1-d^2s_1^2),$$

$$\text{dn}\,(u+iv) = (dc_1d_1 - ik^2scs_1)/(1-d^2s_1^2).$$

The formulae which involve a complex change of argument are obtained as special cases of the last three equations; these formulae are given in Appendix A.

Finally, we note that the function $\Pi(u, h)$ of equation (10.1.28) satisfies

$$\Pi(u, h) - uZ(h) = \Pi(h, u) - hZ(u),$$

since $\Theta(u)$ is an even function of u. It follows immediately that the *complete* elliptic integral of the third kind is given by

$$\Pi(K, h) = KZ(h). \tag{10.1.32}$$

10.1.7. Examples

EXAMPLE 10.1.8.   Find the length of arc of the *lemniscate*

$$r^2 = c^2 \sin 2\theta$$

between the points where $\theta = 15°$ and $\theta = 45°$.

The length of arc is [see Definition 6.1.3]

$$s = \int_{c/\sqrt{2}}^{c} \left[1 + r^2 \left(\frac{d\theta}{dr}\right)^2\right]^{1/2} dr$$

$$= c \int_{1/\sqrt{2}}^{1} \frac{dt}{(1-t^4)^{1/2}},$$

where $r = ct$. This is an elliptic integral of the first kind, of type III, with $a^2 = b^2 = 1$. Hence

$$s = -\frac{c}{\sqrt{2}} [\text{cn}^{-1} t]_{1/\sqrt{2}}^{1}, \qquad k = \frac{1}{\sqrt{2}},$$

$$= \frac{c}{\sqrt{2}} \text{cn}^{-1} \left(\frac{1}{\sqrt{2}}\right)$$

$$= 0{\cdot}70711 c\, \text{cn}^{-1}(0{\cdot}70711).$$

From Milne-Thomson's tables (1950),

$$\text{cn}\,0{\cdot}82 = 0{\cdot}71079,$$

$$\text{cn}\,0{\cdot}83 = 0{\cdot}70467,$$

and by inverse interpolation (see Appendix C) we find

$$\text{cn}^{-1}(0{\cdot}70711) = 0{\cdot}8260,$$

and hence

$$s = 0 \cdot 5841a.$$

EXAMPLE 10.1.9

$$I = \int_x^\infty \frac{dt}{[(t^2 - a^2)(t^2 - b^2)]^{1/2}}, \qquad 0 < b < a < x.$$

The integral is of the first kind, of Type IV. Hence

$$I = \frac{1}{a} \left[dc^{-1} \left(\frac{t}{a} \right) \right]_x^\infty, \qquad k = \frac{b}{a},$$

$$= \frac{1}{a} \left[dc^{-1} (\infty) - dc^{-1} \left(\frac{x}{a} \right) \right]$$

$$= \frac{1}{a} \left[K - dc^{-1} \left(\frac{x}{a} \right) \right],$$

for if $v = dc^{-1}(\infty)$ then $cn\, v = 0$, giving $v = K$. Also, from Appendix A,

$$sn\,(K - u) = cd\, u,$$

and so

$$ns\,(K - u) = dc\, u.$$

Therefore

$$I = \frac{1}{a} ns^{-1} \left(\frac{x}{a} \right), \qquad k = \frac{b}{a}.$$

Alternatively, to avoid the infinite limit of integration, and to simplify the subsequent algebra, put $s = 1/t$. Then

$$I = \frac{1}{ab} \int_0^{1/x} \frac{ds}{\left[\left(\frac{1}{a^2} - s^2 \right) \left(\frac{1}{b^2} - s^2 \right) \right]^{1/2}}, \qquad s^2 < \frac{1}{a^2} < \frac{1}{b^2},$$

which is an elliptic integral of the first kind of Type VI. Hence

$$I = \frac{1}{ab} [b\, sn^{-1}\, as]_0^{1/x}$$

$$= \frac{1}{a} ns^{-1} \left(\frac{x}{a} \right), \qquad k = \frac{b}{a},$$

as before.

EXAMPLE 10.1.10.   Evaluate

$$I = \int_1^4 \frac{dx}{[(x^2 + x - 2)(x^2 + 4x + 3)]^{1/2}}.$$

Following the method for evaluating elliptic integrals of the first kind [see § 10.1.4], we find that the zeros of S_1 and S_2 are $-2, 1$ and $-3, -1$, respectively. These pairs interlace, and so we redefine S_1 and S_2:

$$S_1 \equiv (x+2)(x+1) = x^2 + 3x + 2,$$
$$S_2 \equiv (x+3)(x-1) = x^2 + 2x - 3,$$

the zeros of which do not interlace. The quadratic (10.1.22) is

$$(1-\lambda)(2+3\lambda) - (\tfrac{3}{2}-\lambda)^2 = 0,$$

with roots

$$\lambda_1, \lambda_2 = \frac{2 \pm 3^{1/2}}{4}.$$

Hence

$$S_1 - \lambda_1 S_2 = (x^2 + 3x + 2) - \left(\frac{2+3^{1/2}}{4}\right)(x^2 + 2x - 3)$$

$$= \left(\frac{2 - 3^{1/2}}{4}\right)(x-\alpha)^2, \qquad (10.1.33)$$

which gives, on equating coefficients,

$$\alpha^2 = 71 \cdot 641016, \qquad \alpha = -8 \cdot 46410.$$

Similarly,

$$S_1 - \lambda_2 S_2 = \left(\frac{2+3^{1/2}}{4}\right)(x-\beta)^2, \qquad (10.1.34)$$

where

$$\beta^2 = 2 \cdot 358984, \qquad \beta = -1 \cdot 53590.$$

From equations (10.1.33) and (10.1.34), we find

$$S_1 = -0 \cdot 00518149 \, (x + 8 \cdot 46410)^2 + 1 \cdot 00518149 \, (x + 1 \cdot 53590)^2,$$
$$S_2 = -0 \cdot 0773503 \, (x + 8 \cdot 46410)^2 + 1 \cdot 0773503 \, (x + 1 \cdot 53590)^2.$$

Put

$$t = \frac{x - \alpha}{x - \beta}.$$

Then, using equation (10.1.23),

$$I = \frac{1}{-6\cdot92820} \int_{3\cdot73205}^{2\cdot25150}$$

$$\times \frac{dt}{[(-0\cdot005181 49t^2 + 1\cdot005 18149) \times (-0\cdot0773503t^2 + 1\cdot0773503)]^{1/2}}$$

$$= \frac{1}{0\cdot13870} \int_{2\cdot25150}^{3\cdot73205} \frac{dt}{[(193\cdot995 - t^2)(13\cdot9282 - t^2)]^{1/2}} \, ,$$

which is of Type VI with $a = 13\cdot9282$, $b = 3\cdot73205$. (Note that $a = b^2$, a special case.) Finally, put

$$t = b \operatorname{sn} u, \qquad k = b/a,$$

to give

$$I = 0\cdot51764\left[\operatorname{sn}^{-1} 1 - \operatorname{sn}^{-1}\left(\frac{2\cdot25150}{3\cdot73205}\right)\right]$$

$$= 0\cdot51764[K - \operatorname{sn}^{-1}(0\cdot60329)], \qquad k = 0\cdot26795.$$

Interpolating in the tables for k, we find $K = 1\cdot60019$. Also, since $m = k^2 = 0\cdot071797$, we evaluate $\operatorname{sn}^{-1}(0\cdot60329)$ by inverse interpolation for $m = 0, 0\cdot1$ and $0\cdot2$, and then interpolate for m. The results are as follows.

m	$\operatorname{sn}^{-1}(0\cdot60329)$
0	0·64762
0·1	0·65185
0·2	0·65623.

Hence

$$\operatorname{sn}^{-1}(0\cdot60329) = 0\cdot65064 \qquad (m = 0\cdot071797)$$

and

$$I = 0\cdot4915.$$

EXAMPLE 10.1.11.   Evaluate

$$I = \int_0^{0\cdot41} \frac{t^2 \, dt}{[(t^2 + 5)(t^2 + 2)]^{1/2}}.$$

This is an elliptic integral of the second kind, the corresponding integral of the first kind being of Type I. In the usual notation,

$$a^2 = 5, \qquad b^2 = 2, \qquad k'^2 = 0\cdot4, \qquad k^2 = 0\cdot6.$$

The appropriate substitution is therefore

$$t = 2^{1/2} \operatorname{sc} u,$$

and from example 10.1.6, the value of the integral is

$$I = 5^{1/2}[0{\cdot}28309 - E\{\text{sc}^{-1}(0{\cdot}28991)\}], \qquad k^2 = 0{\cdot}6.$$

Let

$$u = \text{sc}^{-1}(0{\cdot}28991).$$

Since

$$\text{sn}^2 u = \frac{\text{sc}^2 u}{1 + \text{sc}^2 u},$$

we find

$$\text{sn}\, u = 0{\cdot}27844, \qquad u = 0{\cdot}28443.$$

Also,

$$E(u) = Z(u) + \frac{uE}{K}$$

$$= 0{\cdot}090504 + (0{\cdot}28443)(0{\cdot}66601)$$

$$= 0{\cdot}27994,$$

where $Z(u)$ and E/K are found from the tables. Hence

$$I = 0{\cdot}00704.$$

EXAMPLE 10.1.12.   Evaluate

$$I = \int_0^{0{\cdot}77} \frac{dt}{(1 + 3t^2)[(t^2 + 5)(t^2 + 2)]^{1/2}}.$$

This is an elliptic integral of the third kind, the corresponding integral of the first kind being of Type I. Using the same substitution as in the previous example, we find

$$I = \frac{1}{5^{1/2}}\left[u_1 - 6\int_0^{u_1} \frac{\text{sn}^2 u\, du}{1 + 5\,\text{sn}^2 u}\right],$$

where

$$k'^2 = 0{\cdot}4, \qquad k^2 = 0{\cdot}6 \quad \text{and} \quad 0{\cdot}77 = 2^{1/2}\,\text{sc}\, u_1.$$

Hence

$$\text{sn}\, u_1 = 0{\cdot}47819, \qquad u_1 = 0{\cdot}51119.$$

Referring to equations (10.1.25) and (10.1.26), we have

$$\nu = 5 = -0{\cdot}6\,\text{sn}^2 h,$$

which gives

$$\text{sn}\,(h, 0{\cdot}77460) = 2{\cdot}88675i.$$

Hence

$$k^2 \operatorname{sn} h \operatorname{cn} h \operatorname{dn} h = 12 \cdot 96148i.$$

Since sn h is imaginary, we use Jacobi's imaginary transformation (10.1.30) to find h. Put $h = iv$. Then

$$\operatorname{sn}(h, 0 \cdot 77460) = \operatorname{sn}(iv, 0 \cdot 77460) = i \operatorname{sc}(v, 0 \cdot 63246),$$

and hence

$$\operatorname{sc}(v, 0 \cdot 63246) = 2 \cdot 88675,$$

giving

$$\operatorname{sn}(v, 0 \cdot 63246) = 0 \cdot 94491, \qquad v = 1 \cdot 35205, \qquad h = 1 \cdot 35205i.$$

From (10.1.27)

$$\int_0^{u_1} \frac{\operatorname{sn}^2 u \, du}{1 + v \operatorname{sn}^2 u} = \frac{\Pi(u_1, h)}{k^2 \operatorname{sn} h \operatorname{cn} h \operatorname{dn} h} = \frac{\Pi(0 \cdot 51119, 1 \cdot 35205i)}{12 \cdot 96148i},$$

where $\Pi(u_1, h)$ is given by (10.1.28).

To evaluate the appropriate Theta functions we have, for $m = k^2 = 0 \cdot 6$,

$$q = e^{-\pi K'/K} = 0 \cdot 05702026, \qquad K = 1 \cdot 9495677,$$

$$\pi(u_1 - h)/K = 0 \cdot 82375 - 2 \cdot 17873i.$$

Hence

$$\Theta(u_1 - h) = 1 - 2(0 \cdot 05702026) \cos(0 \cdot 82375 - 2 \cdot 17873i)$$

$$+ 2(0 \cdot 05702026)^4 \cos(1 \cdot 64750 - 4 \cdot 35746i)$$

$$- \text{neglected terms}$$

$$= 0 \cdot 65325 - 0 \cdot 36406i,$$

and therefore

$$\Theta(u_1 + h) = 0 \cdot 65325 + 0 \cdot 36406i,$$

giving

$$\frac{1}{2} \log \left[\frac{\Theta(u_1 - h)}{\Theta(u_1 + h)} \right] = -i \tan^{-1} \left(\frac{0 \cdot 36406}{0 \cdot 65325} \right) = -0 \cdot 50843i.$$

It remains to evaluate $Z(h) = Z(iv, k)$. We have

$$v = 1 \cdot 35205, \qquad k' = 0 \cdot 63246,$$

and, in (10.1.31),

$$Z(v, k') = 0 \cdot 073566, \qquad \pi v/2KK' = 0 \cdot 612858,$$

$$s_1 = 0 \cdot 94491, \qquad d_1 = 0 \cdot 801784, \qquad c_1 = 0 \cdot 327330,$$

giving

$$Z(h) = 1 \cdot 62810i.$$

Hence, from (10.1.28)

$$\Pi(u_1, h) = -0 \cdot 50843i + (0 \cdot 51119)(1 \cdot 62810i)$$

$$= 0 \cdot 32384i.$$

Therefore

$$\int_0^{u_1} \frac{\operatorname{sn}^2 u \, du}{1 + v \operatorname{sn}^2 u} = \frac{0 \cdot 32384i}{12 \cdot 96148i} = 0 \cdot 024985,$$

and

$$I = \frac{1}{5^{1/2}} [0 \cdot 51119 - 6(0 \cdot 024985)]$$

$$= 0 \cdot 1616.$$

10.1.8. Reduction of the General Elliptic Integral

The definition of an elliptic integral given in section 10.1.2 implies that the integrand is a rational function R_0 of x, $(1-x^2)^{1/2}$ and $(1-k^2x^2)^{1/2}$, since any quartic in x with real coefficients can be expressed as a product of real quadratic factors [cf. § 10.1.4] and a cubic may be regarded as a special case of a quartic. Equivalently, the substitution $x = \operatorname{sn} u$ shows that the integrand is a rational function of $\operatorname{sn} u$, $\operatorname{cn} u$ and $\operatorname{dn} u$.

If

$$\xi = (1-x^2)^{1/2}, \qquad \eta = (1-k^2x^2)^{1/2},$$

we can write

$$R_0(x, \xi, \eta) = \frac{a' + b'\xi + c'\eta + d'\xi\eta}{a + b\xi + c\eta + d\xi\eta},$$

where $a', a, b', b, \ldots$ are polynomials in x. To reduce this expression to the standard form for elliptic integrals, first multiply the numerator and denominator by

$$a - b\xi - c\eta + d\xi\eta.$$

Then

$$R_0(x, \xi, \eta) = \frac{A' + B'\xi + C'\eta + D'\xi\eta}{A + D\xi\eta},$$

where $A', A, \ldots$ are polynomials in x. Now multiply the numerator and denominator by

$$A - D\xi\eta.$$

Then

$$R_0(x, \xi, \eta) = R_1 + R_2\xi + R_3\eta + R_4\xi\eta,$$

where $R_1, \ldots, R_4$ are rational functions of x.

The first three terms lead to elementary integrals; the fourth can be written in the form

$$\frac{P' + xQ'}{(P + xQ)\xi\eta}$$

where P', P, Q', Q are polynomials in x^2. Now multiply the numerator and denominator of this expression by

$$P - xQ$$

to give

$$\frac{S(x^2) + xT(x^2)}{\xi\eta}$$

where $S(x^2)$ and $T(x^2)$ are rational functions of x^2. The second of these terms leads to an elementary integral; the first is reduced by expressing $S(x^2)$ in partial fractions [see I, § 14.10] and then making the substitution

$$x = \operatorname{sn}(u, k).$$

This leads to

$$\int \frac{S(x^2)}{\xi\eta}\, dx = \sum_p a_p \int \operatorname{sn}^{2p} u\, du + \sum_q b_q \int \frac{du}{(1 + \nu \operatorname{sn}^2 u)^q}$$

where p, q are non-negative integers, and a_p, b_q, ν are constants.

It remains to express the two sets of integrals on the right-hand side of the last equation in terms of standard elliptic integrals. Let

$$s = \operatorname{sn} u, \qquad c = \operatorname{cn} u, \qquad d = \operatorname{dn} u.$$

Then

$$\frac{d}{du}(s^m cd) = (m+2)k^2 s^{m+3} - (m+1)(1+k^2)s^{m+1} + ms^{m-1}$$

for all real values of m. Hence

$$G_{2p} = \int \operatorname{sn}^{2p} u\, du$$

can be expressed in terms of G_2, and [by (10.1.6) and (10.1.18)]

$$G_2 = \int \operatorname{sn}^2 u\, du = \frac{1}{k^2} \int (1 - \operatorname{dn}^2 u)\, du = \frac{u - E(u)}{k^2}.$$

Also, if

$$W = 1 + \nu s^2, \qquad \nu = \text{constant},$$

then, for all real values of m,

$$\frac{d}{du}\left(\frac{scd}{W^m}\right) = \frac{A}{W^{m+1}} + \frac{B}{W^m} + \frac{C}{W^{m-1}} + \frac{D}{W^{m-2}}$$

where A, B, C, D are constants. Hence

$$\int \frac{du}{(1 + \nu \operatorname{sn}^2 u)^q}$$

can be expressed in terms of the standard elliptic integrals u, $E(u)$ and $\Pi(u)$ of equations (10.1.17), (10.1.18) and (10.1.19), respectively. This completes the reduction of $R_0(x, \xi, \eta)$.

10.2. GAMMA AND BETA FUNCTIONS

10.2.0. Introduction

The gamma function first arose in connection with the interpolation problem for factorials (Davis, 1959). For example, if $4! = 24$ and $5! = 120$, can any meaning be assigned to $4\frac{1}{2}!$? This problem was posed by Stirling (1692–1770), Goldbach (1690–1764) and Daniel Bernoulli 1700–1784). It was solved by Euler (1707–1783) in two letters to Goldbach in 1729 and 1730, first by means of an infinite product and later as an integral. The modern notation is due to Legendre. He called the integral which Euler obtained for $n!$ the *second Eulerian integral*. Euler's derivation of this integral began with another integral which Legendre called the *first Eulerian integral*. The first and second Eulerian integrals, with slight changes in notation, are now known as the beta and gamma function, respectively.

In the 19th century, the definition of the gamma function was extended to include complex numbers, and the theory of functions of a complex variable showed that the different definitions of the gamma function for complex values were equivalent. We shall be concerned here almost entirely with real values of the variables. The gamma function is discussed from the point of view of complex function theory in (Whittaker and Watson, 1950) and (Copson, 1935).

10.2.1. Definitions and Elementary Properties

One of the definitions of the gamma function is

$$\Gamma(n) = \int_0^\infty e^{-x} x^{n-1}\, dx, \qquad n > 0. \tag{10.2.1}$$

The restriction $n > 0$ is imposed because the integral diverges at the lower limit for other values of n [see § 4.6]. We find immediately that

$$\Gamma(1) = 1.$$

Integrating equation (10.2.1) by parts [see Theorem 4.3.1], we obtain the recurrence relation

$$\Gamma(n) = (n-1)\Gamma(n-1), \qquad n > 1. \tag{10.2.2}$$

Hence [see I, § 14.13] *when n is a positive integer,*

$$\Gamma(n) = (n-1)(n-2)\ldots 3.2.1.\Gamma(1)$$

$$= (n-1)!$$

Thus the gamma function is a generalization of the factorial function [see I, (3.7.1)] to the case where n may take non-integral values.

To define $\Gamma(n)$ for negative values of n, we rewrite equation (10.2.2) as

$$\Gamma(n) = \frac{\Gamma(n+1)}{n}$$

and assume that it holds for $n < 0$. Then

$$\Gamma(n) = \frac{\Gamma(n+1)}{n} = \frac{\Gamma(n+2)}{n(n+1)} = \cdots$$

$$= \frac{\Gamma(n+r)}{n(n+1)(n+2)\ldots(n+r-1)} \tag{10.2.3}$$

where r is chosen so that $1 \le n+r < 2$. $\Gamma(x)$ is tabulated (Comrie, 1963) to six figures for $1 \le x \le 2$ in steps of 0.001, and using equation (10.2.2) or (10.2.3) any value of the argument may be brought within the range of the tables. Equation (10.2.3) shows that $\Gamma(n)$ becomes infinite if and only if n is zero or a negative integer.

The beta function is defined as

$$B(m, n) = \int_0^1 x^{m-1}(1-x)^{n-1} \, dx, \qquad m > 0, n > 0. \tag{10.2.4}$$

Making the substitution $x = 1 - y$, we find

$$B(m, n) = B(n, m).$$

The substitution

$$x = \sin^2 \theta$$

in equation (10.2.4) yields

$$B(m, n) = 2 \int_0^{\pi/2} \sin^{2m-1} \theta \cos^{2n-1} \theta \, d\theta, \tag{10.2.5}$$

or conversely,

$$\int_0^{\pi/2} \sin^p \theta \cos^q \theta \, d\theta = \tfrac{1}{2} B\left(\frac{p+1}{2}, \frac{q+1}{2}\right). \tag{10.2.6}$$

10.2.2. Relation between Beta and Gamma Functions

There is a very useful formula which expresses the beta function in terms of gamma functions. Put $x = y^2$ in equation (10.2.1) to give

$$\Gamma(n) = 2 \int_0^\infty e^{-y^2} y^{2n-1} \, dy. \tag{10.2.7}$$

Hence

$$\Gamma(m)\Gamma(n) = 4 \int_0^\infty \int_0^\infty e^{-(x^2+y^2)} x^{2m-1} y^{2n-1} \, dx \, dy.$$

Changing to polar coordinates [see V, § 1.2.6]

$$x = r \cos \theta, \qquad y = r \sin \theta,$$

we obtain [see Theorem 6.2.5]

$$\Gamma(m)\Gamma(n) = 4 \int_0^{\pi/2} \int_0^\infty e^{-r^2} r^{2m+2n-1} \cos^{2m-1} \theta \sin^{2n-1} \theta \, dr \, d\theta$$

$$= \Gamma(m+n) B(m, n),$$

using equations (10.2.5) and (10.2.7). This formula is usually written in the form

$$B(m, n) = \frac{\Gamma(m)\Gamma(n)}{\Gamma(m+n)}. \tag{10.2.8}$$

In conjunction with equation (10.2.2) it enables beta functions to be evaluated rapidly.

· Setting $p = q = 0$ in equation (10.2.6) gives

$$\frac{\pi}{2} = \tfrac{1}{2} B(\tfrac{1}{2}, \tfrac{1}{2}) = \tfrac{1}{2} [\Gamma(\tfrac{1}{2})]^2,$$

and hence we obtain the remarkable formula

$$\Gamma(\tfrac{1}{2}) = \pi^{1/2}. \tag{10.2.9}$$

Using this result together with equations (10.2.2) and (10.2.8), we can evaluate the integral of equation (10.2.6) in the case where p and q are non-negative integers:

$$\int_0^{\pi/2} \sin^p \theta \cos^q \theta \, d\theta = \frac{\tfrac{1}{2}\Gamma\left(\dfrac{p+1}{2}\right)\Gamma\left(\dfrac{q+1}{2}\right)}{\Gamma\left(\dfrac{p+q+2}{2}\right)}$$

$$= \frac{\left[(p-1)(p-3)\cdots\dfrac{4 \cdot 2}{3 \cdot 1}\right]\left[(q-1)(q-3)\cdots\dfrac{4 \cdot 2}{3 \cdot 1}\right]\left\{\dfrac{\pi}{2}\right\}}{(p+q)(p+q-2)\cdots\dfrac{4 \cdot 2}{3 \cdot 1}} \tag{10.2.10}$$

where the factor $\pi/2$ is included only if both p and q are even.

EXAMPLES 10.2.1

(i) $\displaystyle\int_0^{\pi/2} \sin^6\theta \cos^7\theta\, d\theta = \frac{5.3.1.6.4.2}{13.11.9.7.5.3.1} = \frac{16}{3003}$

(ii) $\displaystyle\int_0^{\pi/2} \sin^4\theta \cos^2\theta\, d\theta = \frac{3.1.1}{6.4.2}\frac{\pi}{2} = \frac{\pi}{32}$

(iii) $\displaystyle\int_0^{\pi/2} \sin^8\theta\, d\theta = \int_0^{\pi/2} \cos^8\theta\, d\theta = \frac{7.5.3.1}{8.6.4.2}\frac{\pi}{2} = \frac{35\pi}{256}$

(iv) $\displaystyle\int_0^{\pi} \sin^5\theta \cos^4\theta\, d\theta = 2\int_0^{\pi/2} \sin^5\theta \cos^4\theta\, d\theta$

$$= \frac{2.4.2.3.1}{9.7.5.3.1} = \frac{16}{315}$$

10.2.3. Integrals

By suitable changes of variable, a large number of integrals may be expressed in terms of gamma and beta functions.

EXAMPLES 10.2.2

(i) $\displaystyle I_1 = \int_0^a x^{m-1}(a-x)^{n-1}\, dx.$

Put $x = ay$. Then

$$I_1 = a^{m+n-1}\int_0^1 y^{m-1}(1-y)^{n-1}\, dy = a^{m+n-1}B(m, n).$$

(ii) $\displaystyle I_2 = \int_0^1 \left(\log\frac{1}{y}\right)^{n-1}\, dy.$

Put $x = \log 1/y$. Then

$$I_2 = \int_0^\infty e^{-x} x^{n-1}\, dx = \Gamma(n).$$

(iii) $\displaystyle I_3 = \int_0^\infty e^{-x} x^{n-1}(\log x)^r\, dx$

$$= \frac{d^r}{dn^r}\Gamma(n).$$

(iv) $\displaystyle I_4 = \int_0^1 x^{m-1}(1-x^p)^{n-1}\, dx.$

Put $y = x^p$. Then

$$I_4 = \frac{1}{p} \int_0^1 y^{(m/p)-1} (1-y)^{n-1} \, dy = \frac{1}{p} B\left(\frac{m}{p}, n\right).$$

(v) $I_5 = \int_0^1 \dfrac{x^{m-1}(1-x)^{n-1}}{(x+a)^{m+n}} \, dx.$

Put $y/(1+a) = x/(x+a)$, giving

$$I_5 = \frac{1}{a^n(1+a)^m} \int_0^1 y^{m-1}(1-y)^{n-1} \, dy = \frac{B(m, n)}{a^n(1+a)^m}.$$

EXAMPLE 10.2.3 *Dirichlet's integral.* Consider the integral

$$I_6 = \iiint x^p y^q z^r (1-x-y-z)^s \, dx \, dy \, dz,$$

taken throughout the interior of the tetrahedron formed by the planes

$$x = 0, \qquad y = 0, \qquad z = 0, \qquad x+y+z = 1.$$

Define new variables ξ, η, ζ by

$$\xi = x+y+z, \qquad \xi\eta = y+z, \qquad \xi\eta\zeta = z,$$

or equivalently,

$$x = \xi(1-\eta), \qquad y = \xi\eta(1-\zeta), \qquad z = \xi\eta\zeta.$$

When x, y, z are all positive and

$$x+y+z < 1,$$

ξ, η, ζ all lie between 0 and 1. Hence the tetrahedron transforms into a cube. The Jacobian of the transformation [see (5.10.3)] is

$$\frac{\partial(x, y, z)}{\partial(\xi, \eta, \zeta)} = \xi^2\eta,$$

giving [see Theorem 6.2.5]

$$I_6 = \int_0^1 d\xi \int_0^1 d\eta \int_0^1 \xi^{p+q+r+2}(1-\xi)^s \eta^{q+r+1}(1-\eta)^p \zeta^r(1-\zeta)^q \, d\zeta$$

$$= \left[\int_0^1 \xi^{p+q+r+2}(1-\xi)^s \, d\xi\right]\left[\int_0^1 \eta^{q+r+1}(1-\eta)^p \, d\eta\right]\left[\int_0^1 \zeta^r(1-\zeta)^q \, d\zeta\right]$$

$$= B(p+q+r+3, s+1)B(q+r+2, p+1)B(r+1, q+1)$$

$$= \frac{\Gamma(p+1)\Gamma(q+1)\Gamma(r+1)\Gamma(s+1)}{\Gamma(p+q+r+s+4)}.$$

This result, due to Dirichlet, can easily be extended to the case of n variables. A more general integral which can be evaluated over the same region by a similar method is

$$\int \int \int x^p y^q z^r f(x + y + z) \, dx \, dy \, dz,$$

where f is any continuous function.

EXAMPLE 10.2.4.   Consider the integral

$$I_7 = \int \int \int x^{l-1} y^{m-1} z^{n-1} \, dx \, dy \, dz,$$

taken throughout the first octant of the solid

$$0 \le \left(\frac{x}{a}\right)^p + \left(\frac{y}{b}\right)^q + \left(\frac{z}{c}\right)^r \le h, \qquad p > 0, q > 0, r > 0.$$

Put

$$\xi h = \left(\frac{x}{a}\right)^p, \qquad \eta h = \left(\frac{y}{b}\right)^q, \qquad \zeta h = \left(\frac{z}{c}\right)^r,$$

$$dx = \frac{a}{p} h^{1/p} \xi^{(1/p)-1} \, d\xi, \quad \text{etc.}$$

Then

$$I_7 = \frac{a^l b^m c^n}{pqr} h^{l/p + m/q + n/r} \int \int \int \xi^{(l/p)-1} \eta^{(m/q)-1} \zeta^{(n/r)-1} \, d\xi \, d\eta \, d\zeta,$$

where

$$0 \le \xi + \eta + \zeta \le 1.$$

Using the result of the previous example, we find

$$I_7 = \frac{a^l b^m c^n}{pqr} h^{l/p + m/q + n/r} \frac{\Gamma\left(\frac{l}{p}\right)\Gamma\left(\frac{m}{q}\right)\Gamma\left(\frac{n}{r}\right)}{\Gamma\left(\frac{l}{p} + \frac{m}{q} + \frac{n}{r} + 1\right)}.$$

The volumes of the octants are found by taking

$$l = m = n = 1,$$

and the moment of inertia about the z-axis, for example, is

$$I_z = \int \int \int (x^2 + y^2) \, dx \, dy \, dz,$$

which may be expressed as a sum of two of the given integrals.

10.2.4.   Miscellaneous Formulae

(i) An integral which occurs frequently in probability and statistics:

$$\int_0^\infty e^{-\frac{1}{2}x^2}\,dx = \left(\frac{\pi}{2}\right)^{1/2}. \qquad [\text{Put } y = \tfrac{1}{2}x^2]. \tag{10.2.11}$$

(ii) *Euler's reflection formula*:

$$\Gamma(x)\Gamma(1-x) = \frac{\pi}{\sin \pi x}. \tag{10.2.12}$$

(iii) Duplication formula:

$$\Gamma(2x) = 2^{2x-1}\pi^{-1/2}\Gamma(x)\Gamma(x + \tfrac{1}{2}). \tag{10.2.13}$$

(iv) A simple relationship between the derivative of the gamma function and *Euler's constant* $\gamma(=0\cdot5772\ldots)$:

$$\Gamma'(1) = \int_0^\infty e^{-x} \log x \, dx = -\gamma. \tag{10.2.14}$$

(v) *Stirling's asymptotic formula*:

$$\Gamma(n+1) \sim (2\pi n)^{1/2} n^n e^{-n}, \tag{10.2.15}$$

where $\sim$ means 'is approximately equal to when n is large' [see § 2.3]. The right-hand side is relatively easy to evaluate. This formula is particularly useful in statistical mechanics where the behaviour of very large numbers of particles is considered.

More generally, *Stirling's asymptotic series* for the gamma function is

$$\Gamma(n+1) \sim (2\pi n)^{1/2} n^n e^{-n}\left(1 + \frac{1}{12n} + \frac{1}{288n^2} - \frac{139}{51840n^3} - \ldots\right)$$
$$\tag{10.2.16}$$

(vi) Other definitions: Weierstrass:

$$\frac{1}{\Gamma(z)} = z\, e^{\gamma z} \prod_{n=1}^\infty \left[\left(1 + \frac{z}{n}\right) e^{-z/n}\right], \tag{10.2.17}$$

where γ is Euler's constant. Euler:

$$\Gamma(z) = \frac{1}{z} \prod_{n=1}^\infty \left[\left(1 + \frac{1}{n}\right)^z \left(1 + \frac{z}{n}\right)^{-1}\right]; \tag{10.2.18}$$

and also due to Euler:

$$\Gamma(z) = \lim_{n\to\infty} \frac{n!\,n^z}{z(z+1)\ldots(z+n)}. \tag{10.2.19}$$

These formulae are valid for all complex numbers z except zero and the negative integers.

10.3. LEGENDRE POLYNOMIALS

10.3.0. Introduction

Laplace's equation $\nabla^2 V = 0$ [see (8.1.2) and § 8.2] occurs whenever a vector field **F** satisfies

$$\mathbf{F} = -\nabla V \quad \text{and} \quad \nabla . \mathbf{F} = 0$$

[see § 17.2.3]. The first of these conditions is equivalent to

$$\nabla \times \mathbf{F} = \mathbf{0},$$

and hence **F** is said to be *irrotational*. Because of the second condition, **F** is said to be *solenoidal*. Well-known fields satisfying these conditions are

   (i) the electrostatic field due to a system of electric charges,
   (ii) the magnetostatic field in current-free regions,
   (iii) the gravitational field due to a system of masses,
   (iv) the velocity field of an irrotational, incompressible fluid.

The scalar field V is then called the electrostatic potential, magnetostatic potential, gravitational potential or velocity potential, respectively; or simply the potential if there is no ambiguity. The advantage of solving Laplace's equation rather than the two vector equations above is that Laplace's equation is a single equation in one variable. Once it has been solved for V, the components of **F** are easily found from $\mathbf{F} = -\nabla V$. Laplace's equation is also satisfied by the temperature in problems of the steady flow of heat by conduction.

The choice of coordinate system in which Laplace's equation is expressed depends on the boundary conditions which have to be satisfied. Legendre and Bessel functions arise naturally in the solution of Laplace's equation in spherical polar coordinates and cylindrical polar coordinates, respectively [see (8.2.11) and (8.2.10)]. In each case, a solution is found by the method of separation of variables [see § 8.2]. Quite general boundary conditions can be expanded as series of the functions which occur in the variables separable solution, which is itself a series of these functions. The unknown coefficients in the variables separable solution are then determined by identifying the two series on the boundary. We assume without proof that the solution of Laplace's equation for an appropriate set of initial and boundary conditions is unique. A useful introduction to Laplace's equation and its solutions is given by (Bland, 1961).

Legendre (1752–1833) introduced his polynomials into analysis in 1784. They were first used in potential theory to calculate the gravitational attraction of a near-spherical body such as the earth. This development was the direct result of more accurate astronomical calculations after the time of Newton (Bell, 1945). The properties of Legendre polynomials, with some applications, are discussed in Chapter III of (Sneddon, 1956), and in Chapter 4 of (Hildebrand, 1948).

10.3.1. Laplace's Equation in Spherical Polar Coordinates

Laplace's equation in spherical polar coordinates (r, θ, ϕ) is [cf. (8.2.11)]

$$\nabla^2 V \equiv \frac{1}{r^2}\frac{\partial}{\partial r}\left(r^2\frac{\partial V}{\partial r}\right) + \frac{1}{r^2\sin\theta}\frac{\partial}{\partial\theta}\left(\sin\theta\frac{\partial V}{\partial\theta}\right) + \frac{1}{r^2\sin^2\theta}\frac{\partial^2 V}{\partial\phi^2} = 0.$$

$$(10.3.1)$$

We assume that the dependence of V on the variable r can be separated from its dependence on θ and ϕ, that is, we seek a solution of equation (10.3.1) of the form

$$V = R(r)S(\theta, \phi).$$

This leads to

$$\frac{1}{R}\frac{d}{dr}\left(r^2\frac{dR}{dr}\right) = K,$$

$$\frac{1}{S}\left(\frac{\partial^2 S}{\partial\theta^2} + \frac{1}{\sin^2\theta}\frac{\partial^2 S}{\partial\phi^2} + \cot\theta\frac{\partial S}{\partial\theta}\right) = -K,$$

where K is a constant. Put

$$K = n(n+1),$$

where n is a non-negative integer. Then the equation for R becomes

$$r^2\frac{d^2 R}{dr^2} + 2r\frac{dR}{dr} - n(n+1)R = 0, \qquad (10.3.2)$$

and the equation for S becomes

$$\frac{\partial^2 S}{\partial\theta^2} + \frac{1}{\sin^2\theta}\frac{\partial^2 S}{\partial\phi^2} + \cot\theta\frac{\partial S}{\partial\theta} + n(n+1)S = 0. \qquad (10.3.3)$$

The general solution of equation (10.3.2) is

$$R = A_n r^n + \frac{B_n}{r^{n+1}}$$

where A_n and B_n are arbitrary constants. Denoting any solution of equation (10.3.2) by S_n, we now have

$$V = \left(A_n r^n + \frac{B_n}{r^{n+1}}\right)S_n$$

as a particular solution of equation (10.3.1). More general solutions may be obtained as a sum of such solutions for $n = 0, 1, 2, \ldots$, since equation (10.3.1) is linear and homogeneous.

When S_n is a function of θ only, that is, when there is symmetry about the polar axis $0z$, it satisfies

$$\frac{d^2 w}{d\theta^2} + \cot\theta\frac{dw}{d\theta} + n(n+1)w = 0,$$

from equation (10.3.3). This is one form of *Legendre's equation.* Alternative forms are

$$\frac{1}{\sin\theta}\frac{d}{d\theta}\left(\sin\theta\frac{dw}{d\theta}\right)+n(n+1)w=0,$$

or, putting $\cos\theta=\mu$,

$$\frac{d}{d\mu}\left[(1-\mu^2)\frac{dw}{d\mu}\right]+n(n+1)w=0,$$

or

$$(1-\mu^2)\frac{d^2w}{d\mu^2}-2\mu\frac{dw}{d\mu}+n(n+1)w=0. \tag{10.3.4}$$

We are interested in finding solutions of this equation, and in the following section we shall show that it is satisfied by a certain polynomial of degree n. This result is a consequence of the special choice of $n(n+1)$ for the constant K above.

10.3.2. Legendre Polynomials

The *Legendre polynomial* $P_n(\mu)$, of degree n in μ, may be defined as the coefficient of h^n in the expansion of the generating function

$$(1-2\mu h+h^2)^{-1/2}$$

in ascending powers of h. Thus

$$(1-2\mu h+h^2)^{-1/2}\equiv P_0(\mu)+hP_1(\mu)+h^2P_2(\mu)+\dots$$
$$+h^nP_n(\mu)+\dots . \tag{10.3.5}$$

We shall show later that $P_n(\mu)$ satisfies Legendre's equation (10.3.4).

It follows immediately from the identity (10.3.5) that

$$P_n(1)=1,\qquad P_n(-\mu)=(-1)^nP_n(\mu).$$

Carrying out the expansion in ascending powers of h [see (1.10.9)], we find

$$P_n(\mu)=\sum_{r=0}^{t}\frac{(-1)^r(2n-2r)!}{2^n r!(n-r)!(n-2r)!}\mu^{n-2r},$$

where

$$t=\tfrac{1}{2}n\qquad\text{if }n\text{ is even,}$$
$$=\tfrac{1}{2}(n-1)\quad\text{if }n\text{ is odd.}$$

The first few Legendre polynomials are

$$P_0(\mu) = 1, \qquad P_1(\mu) = \mu, \qquad P_2(\mu) = \tfrac{1}{2}(3\mu^2 - 1),$$
$$P_3(\mu) = \tfrac{1}{2}(5\mu^3 - 3\mu), \qquad P_4(\mu) = \tfrac{1}{8}(35\mu^4 - 30\mu^2 + 3),$$
$$P_5(\mu) = \tfrac{1}{8}(63\mu^5 - 70\mu^3 + 15\mu).$$

A very compact form for $P_n(\mu)$ is known as *Rodrigue's formula*:

$$P_n(\mu) = \frac{1}{2^n n!} \frac{d^n}{d\mu^n} [(\mu^2 - 1)^n].$$

To prove that $P_n(\mu)$ satisfies Legendre's equation (10.3.4), let

$$f = (\mu^2 - 1)^n.$$

Then

$$(1 - \mu^2)\frac{df}{d\mu} + 2n\mu f = 0.$$

Differentiating $n + 1$ times by Leibniz' theorem [see Theorem 3.4.1], we find

$$(1 - \mu^2)\frac{d^{n+2}f}{d\mu^{n+2}} - 2\mu \frac{d^{n+1}f}{d\mu^{n+1}} + n(n+1)\frac{d^n f}{d\mu^n} = 0,$$

and using Rodrigue's formula we see that $P_n(\mu)$ satisfies Legendre's equation (10.3.4).

The solutions of Laplace's equation (10.3.1) that we have found are therefore of the form

$$V = \sum_{n=0}^{\infty} \left(A_n r^n + \frac{B_n}{r^{n+1}} \right) P_n(\cos\theta). \tag{10.3.6}$$

10.3.3. Properties of Legendre Polynomials

It is often necessary to express an arbitrary polynomial as a sum of Legendre polynomials. The preliminary results (i) and (ii) below lead to the required formula (iii).

(i) *Integral property*

$$\int_{-1}^{1} x^r P_n(x)\, dx = 0 \qquad\qquad r < n,$$

$$= \frac{2^{n+1}(n!)^2}{(2n+1)!} \qquad r = n.$$

Hence, if

$$\phi_r(x) \equiv a_0 + a_1 x + a_2 x^2 + \ldots + a_r x^r$$

is any polynomial of degree r, then

$$\int_{-1}^{1} \phi_r(x)P_n(x)\,dx = 0 \qquad r < n,$$

$$= \frac{2^{n+1}(n!)^2 a_n}{(2n+1)!} \qquad r = n.$$

(ii) *Orthogonal property* [see § 10.5.0]
It follows from (i) that

$$\int_{-1}^{1} P_m(x)P_n(x)\,dx = 0 \qquad m \neq n,$$

$$= \frac{2}{2n+1} \qquad m = n.$$

In particular, since $P_0(x) = 1$,

$$\int_{-1}^{1} P_n(x)\,dx = 0, \qquad n \neq 0.$$

(iii) *Expression of any polynomial as a series of Legendre polynomials*
Let $\phi_n(x)$ be any polynomial of degree n in x, where x is restricted to the interval $-1 \leq x \leq 1$, and assume

$$\phi_n(x) \equiv a_0 P_0(x) + a_1 P_1(x) + a_2 P_2(x) + \ldots + a_n P_n(x).$$

Then, from (ii),

$$\int_{-1}^{1} \phi_n(x)P_m(x)\,dx = \frac{2a_m}{2m+1} \qquad m \leq n,$$

and this determines a_m uniquely. The method is analogous to that of finding Fourier coefficients [see § 20.4].

(iv) *Zeros of $P_n(x)$*
The n zeros of $P_n(x)$ are all real and distinct, and they all lie in the open interval $(-1, 1)$.

(v) *Laplace's integrals*
Laplace's first integral for $P_n(x)$:

$$P_n(x) = \frac{1}{\pi} \int_0^{\pi} [x + (x^2 - 1)^{1/2} \cos \phi]^n \, d\phi.$$

Laplace's second integral for $P_n(x)$:

$$P_n(x) = \frac{1}{\pi} \int_0^{\pi} \frac{d\psi}{[x + (x^2 - 1)^{1/2} \cos \psi]^{n+1}} \qquad x > 0,$$

$$= -\frac{1}{\pi} \int_0^{\pi} \frac{d\psi}{[x + (x^2 - 1)^{1/2} \cos \psi]^{n+1}} \qquad x < 0.$$

These integrals enable $P_n(x)$ to be evaluated numerically, and for large n will give greater accuracy than that obtained from the evaluation of a polynomial of degree n. Note that $P_n(x)$ is real when x is real, so that the imaginary parts of the integrals vanish when $|x| < 1$.

(vi) *Recurrence formulae*
These formulae are often used in numerical work on Legendre polynomials.

$$(n+1)P_{n+1}(x) - (2n+1)xP_n(x) + nP_{n-1}(x) = 0, \qquad\qquad \text{I}$$

$$nP_n(x) = xP'_n(x) - P'_{n-1}(x), \qquad\qquad \text{II}$$

$$(n+1)P_n(x) = P'_{n+1}(x) - xP'_n(x), \qquad\qquad \text{III}$$

$$(2n+1)P_n(x) = P'_{n+1}(x) - P'_{n-1}(x), \qquad\qquad \text{IV}$$

$$(x^2-1)P'_n(x) = nxP_n(x) - nP_{n-1}(x). \qquad\qquad \text{V}$$

10.3.4. Examples

EXAMPLE 10.3.1   *Isolated point charge.*   A point charge e is situated on the polar axis of a set of spherical polar coordinates, at a distance c from the origin. Find the electrostatic potential at any point in terms of Legendre polynomials.

Solution

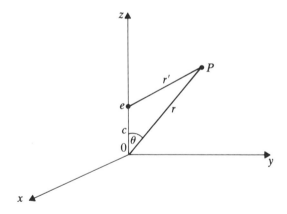

Figure 10.3.1: Point charge on polar axis.

The potential V at P is (Figure 10.3.1)

$$V = \frac{e}{r'} = \frac{e}{(r^2 - 2rc\,\cos\theta + c^2)^{1/2}}$$

$$= \frac{e}{(r^2 - 2\mu rc + c^2)^{1/2}}$$

where $\mu = \cos\theta$. There are two cases to consider.

If $r < c$

$$V = \frac{e}{c\left(1 - 2\mu\dfrac{r}{c} + \dfrac{r^2}{c^2}\right)^{1/2}}$$

$$= \frac{e}{c}\sum_{n=0}^{\infty}\left(\frac{r}{c}\right)^n P_n(\mu),$$

using the generating function (10.3.5). Similarly,

if $r > c$

$$V = \frac{e}{r\left(1 - 2\mu\dfrac{c}{r} + \dfrac{c^2}{r^2}\right)^{1/2}}$$

$$= \frac{e}{r}\sum_{n=0}^{\infty}\left(\frac{c}{r}\right)^n P_n(\mu).$$

EXAMPLE 10.3.2. *Charge outside an earthed conducting sphere.* Find the electrostatic potential at any point outside an earthed conducting sphere of radius a when a charge e is placed at a point distant c ($>a$) from the centre of the sphere.

Solution

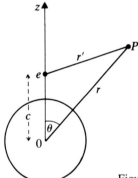

Figure 10.3.2: Point charge outside earthed conducting sphere.

Let e lie on the polar axis of a set of spherical polar coordinates (Figure 10.3.2) whose origin 0 is at the centre of the sphere. The potential at any point P outside the sphere is

$$V = \frac{e}{r'} + \text{potential due to the charge induced on the sphere}$$

$$= \frac{e}{r'} + \sum_{n=0}^{\infty}\frac{B_n}{r^{n+1}}P_n(\mu), \qquad \mu = \cos\theta, r \geq a,$$

using equation (10.3.6) with $A_n = 0$ for all n, since the terms representing the induced potential V_i must satisfy

$$\nabla^2 V_i = 0,$$

$$V_i \rightarrow 0 \quad \text{as } r \rightarrow \infty.$$

For $r < c$, the first result of Example 10.3.1 gives

$$\frac{e}{r'} = \frac{e}{c} \sum_{n=0}^{\infty} \left(\frac{r}{c}\right)^n P_n(\mu).$$

Hence the condition that $V = 0$ when $r = a$ (because the sphere is earthed) becomes

$$\frac{e}{c} \sum_{n=0}^{\infty} \left(\frac{a}{c}\right)^n P_n(\mu) + \sum_{n=0}^{\infty} \frac{B_n}{a^{n+1}} P_n(\mu) = 0.$$

It follows from the orthogonal property of section 10.3.3 that the separate coefficients of the $P_n(\mu)$ in this equation must all vanish. Thus

$$\frac{ea^n}{c^{n+1}} + \frac{B_n}{a^{n+1}} = 0 \quad \text{for all } n,$$

giving

$$B_n = -\frac{ea^{2n+1}}{c^{n+1}},$$

and hence

$$V = \frac{e}{r'} - \frac{ea}{c} \sum_{n=0}^{\infty} \left(\frac{a^2}{c}\right)^n r^{-n-1} P_n(\mu), \qquad r \geq a.$$

Comparison with the second result of Example 10.3.1 shows that the charge induced on the sphere produces the same potential outside the sphere as a point charge $-ea/c$ placed on the polar axis at a distance a^2/c from 0, that is, at the inverse point to that occupied by e. This charge is often called the *image* of e in the sphere.

EXAMPLE 10.3.3 *Conducting sphere in a uniform electric field.* A conducting sphere of radius a is placed in a uniform electric field **E**. Find the electrostatic potential at any point outside the sphere.

Solution

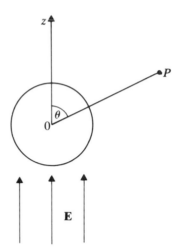

Figure 10.3.3: Conducting sphere in uniform electric field.

Let the uniform field $\mathbf{E}$ be in the direction of the polar axis of a set of spherical polar coordinates (Figure 10.3.3) whose origin 0 is at the centre of the sphere. The potential at any point P outside the sphere is

$$V = -Ez + \text{potential due to the charge induced on the sphere}$$

$$= -Er \cos \theta + \sum_{n=0}^{\infty} B_n r^{-n-1} P_n(\cos \theta),$$

since the induced potential V_i must satisfy

$$\nabla^2 V_i = 0,$$

$$V_i \to 0 \quad \text{as } r \to \infty.$$

Hence

$$V = -ErP_1(\mu) + \sum_{n=0}^{\infty} B_n r^{-n-1} P_n(\mu).$$

The potential is constant on $r = a$, giving

$$-EaP_1(\mu) + \sum_{n=0}^{\infty} B_n a^{-n-1} P_n(\mu) = \text{constant},$$

and the value of the constant determines B_0. Equating coefficients of $P_n(\mu)$ in the last equation gives:

$$n = 1 \qquad\qquad -Ea + B_1 a^{-2} = 0 \quad \Rightarrow \quad B_1 = Ea^3.$$

$$n \geq 2 \qquad\qquad B_n a^{-n-1} = 0 \quad \Rightarrow \quad B_n = 0.$$

Thus

$$V = -ErP_1(\mu) + \frac{B_0}{r} + \frac{Ea^3}{r^2} P_1(\mu)$$

$$= \frac{B_0}{r} - \frac{E}{r^2}(r^3 - a^3)\cos\theta.$$

EXAMPLE 10.3.4   *Charged circular loop of wire.*   A charge e is uniformly distributed over a thin circular loop of wire. Find the electrostatic potential at any point.

Solution

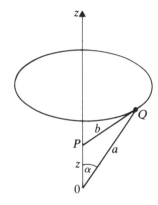

Figure 10.3.4: Charged circular loop of wire.

Take $0z$ as axis of symmetry (Figure 10.3.4); let Q be any point on the wire, P any point on the axis of symmetry, $0Q = a$, $0P = z$, $PQ = b$ and $\widehat{P0Q} = \alpha$. The potential at P is

$$V_P = \frac{e}{b} = \frac{e}{(a^2 - 2az\cos\alpha + z^2)^{1/2}}$$

$$= \frac{e}{a}\sum_{n=0}^{\infty}\left(\frac{z}{a}\right)^n P_n(\cos\alpha), \qquad z < a,$$

$$= \frac{e}{z}\sum_{n=0}^{\infty}\left(\frac{a}{z}\right)^n P_n(\cos\alpha), \qquad z > a,$$

from the results of Example 10.3.1.

At any point at a distance r from 0, not necessarily on the axis of symmetry, we must have

$$V = \sum_{n=0}^{\infty} A_n r^n P_n(\cos\theta), \qquad r < a,$$

$$= \sum_{n=0}^{\infty} B_n r^{-n-1} P_n(\cos\theta), \qquad r > a.$$

The constants A_n and B_n can be determined from the known expressions for the potential on the axis of symmetry at P. Putting $\theta = 0$ and $r = z$, we obtain

$$V_P = \sum_{n=0}^{\infty} A_n z^n = \frac{e}{a} \sum_{n=0}^{\infty} \left(\frac{z}{a}\right)^n P_n(\cos\alpha), \qquad z < a,$$

$$= \sum_{n=0}^{\infty} B_n z^{-n-1} = \frac{e}{z} \sum_{n=0}^{\infty} \left(\frac{a}{z}\right)^n P_n(\cos\alpha), \qquad z > a.$$

Hence

$$A_n = \frac{e}{a^{n+1}} P_n(\cos\alpha), \qquad B_n = ea^n P_n(\cos\alpha),$$

and therefore

$$V = \frac{e}{a} \sum_{n=0}^{\infty} \left(\frac{r}{a}\right)^n P_n(\cos\alpha) P_n(\cos\theta), \qquad r < a,$$

$$= \frac{e}{r} \sum_{n=0}^{\infty} \left(\frac{a}{r}\right)^n P_n(\cos\alpha) P_n(\cos\theta), \qquad r > a.$$

EXAMPLE 10.3.5 *Sphere moving through a liquid.* A sphere of radius a moves with constant velocity U in an incompressible, irrotational liquid which is at rest at infinity. Find the velocity of the liquid at any point.

Solution

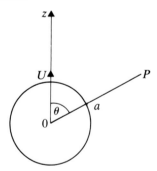

Figure 10.3.5: Sphere moving through liquid.

Suppose the sphere moves in the direction of the z-axis (Figure 10.3.5). The velocity potential V of the liquid at any point P satisfies

$$\nabla^2 V = 0,$$

$$V \to 0 \quad \text{as } r \to \infty.$$

Hence

$$V = \sum_{n=0}^{\infty} \frac{B_n}{r^{n+1}} P_n(\cos\theta).$$

At the point Q on the surface of the sphere we have the boundary condition

$$v_r = -\frac{\partial V}{\partial r} = U \cos \theta = U P_1(\cos \theta),$$

since no liquid crosses the surface. The boundary condition can be satisfied only if

$$V = \frac{B_1}{r^2} \cos \theta,$$

giving

$$-\frac{\partial V}{\partial r}\bigg|_{r=a} = \frac{2 B_1 \cos \theta}{a^3} = U \cos \theta,$$

and hence

$$B_1 = U a^3 / 2.$$

Therefore

$$V = \frac{U a^3}{2 r^2} \cos \theta,$$

and the velocity components of the liquid at any point are

$$v_r = -\frac{\partial V}{\partial r} = \frac{U a^3}{r^3} \cos \theta,$$

$$v_\theta = -\frac{1}{r}\frac{\partial V}{\partial \theta} = \frac{U a^3}{2 r^3} \sin \theta,$$

$$v_\phi = -\frac{1}{r \sin \theta}\frac{\partial V}{\partial \phi} = 0.$$

10.3.5.   Legendre Functions of the Second Kind

If the method of Frobenius (solution in series [see § 7.6.2]) is applied to Legendre's equation (10.3.4), it is found that there are two independent solutions, one of which is $P_n(\mu)$, the Legendre polynomial of degree n. The other is usually taken to be

$$Q_n(\mu) = \tfrac{1}{2} P_n(\mu) \log\left(\frac{1+\mu}{1-\mu}\right) - \sum_{r=0}^{t} \frac{2n - 4r - 1}{(2r+1)(n-r)} P_{n-2r-1}(\mu),$$

$$(10.3.7)$$

where

$$t = \tfrac{1}{2}n - 1 \qquad \text{if } n \text{ is even,}$$

$$= \tfrac{1}{2}(n-1) \quad \text{if } n \text{ is odd.}$$

The general solution of equation (10.3.4) is therefore

$$w = C_n P_n(\mu) + D_n Q_n(\mu), \tag{10.3.8}$$

where C_n and D_n are arbitrary constants. The function $Q_n(\mu)$ is called the *Legendre function of the second kind* of order n.

If $|\mu| < 1$, then [see § 2.14]

$$Q_0(\mu) = \tfrac{1}{2} \log\left(\frac{1+u}{1-\mu}\right) = \tanh^{-1}\mu,$$

$$Q_1(\mu) = \frac{\mu}{2} \log\left(\frac{1+\mu}{1-\mu}\right) - 1 = \mu Q_0(\mu) - 1.$$

Now $Q_n(\mu)$ satisfies the recurrence formulae I–IV of section 10.3.3, and using the first of these we find in succession

$$Q_2(\mu) = P_2(\mu) Q_0(\mu) - \tfrac{3}{2}\mu,$$

$$Q_3(\mu) = P_3(\mu) Q_0(\mu) - \tfrac{5}{2}\mu^2 + \tfrac{2}{3},$$

$$\text{etc., etc.}$$

In general,

$$Q_n(\mu) = P_n(\mu) Q_0(\mu) - W_{n-1}(\mu),$$

where $W_{n-1}(\mu)$ is a polynomial of degree $n-1$ in μ.

Equation (10.3.7) shows that $Q_n(\mu)$ has logarithmic singularities at $\mu = \pm 1$ [see § 9.8]. Thus, if w is to remain finite on the polar axis ($\theta = 0$ or π), the constant D_n in equation (10.3.8) must be zero. The solutions of Laplace's equation (10.3.1) corresponding to equation (10.3.8) are

$$V = \begin{pmatrix} r^n \\ r^{-n-1} \end{pmatrix} \begin{pmatrix} P_n(\cos\theta) \\ Q_n(\cos\theta) \end{pmatrix}, \tag{10.3.9}$$

the notation indicating any linear combination of the alternative functions, that is,

$$V = A_n r^n P_n(\cos\theta) + B_n r^n Q_n(\cos\theta)$$

$$+ C_n r^{-n-1} P_n(\cos\theta) + D_n r^{-n-1} Q_n(\cos\theta),$$

where A_n, B_n, C_n and D_n are arbitrary constants.

EXAMPLE 10.3.6.   A solution V of Laplace's equation is zero on the cone $\theta = \alpha$ and takes the value $\sum_{n=0}^{\infty} A_n r^n$ on the cone $\theta = \beta$. Find V when $\alpha < \theta < \beta$.

Solution.   The form of V may be taken as

$$V = \sum_{n=0}^{\infty} [C_n(r) P_n(\cos\theta) + D_n(r) Q_n(\cos\theta)],$$

where $C_n(r)$ and $D_n(r)$ are functions of r to be determined; and

$$V = 0 \qquad \text{when } \theta = \alpha,$$

$$= \sum_{n=0}^{\infty} A_n r^n \quad \text{when } \theta = \beta.$$

Hence, for any integer $n \geq 0$, the boundary conditions give

$$C_n(r)P_n(\cos \alpha) + D_n(r)Q_n(\cos \alpha) = 0,$$

$$C_n(r)P_n(\cos \beta) + D_n(r)Q_n(\cos \beta) = A_n r^n.$$

Solving for $C_n(r)$ and $D_n(r)$, we find

$$C_n(r) = \frac{A_n r^n Q_n(\cos \alpha)}{\Delta}, \qquad D_n(r) = -\frac{A_n r^n P_n(\cos \alpha)}{\Delta},$$

where

$$\Delta = P_n(\cos \beta)Q_n(\cos \alpha) - P_n(\cos \alpha)Q_n(\cos \beta),$$

and hence V is completely determined.

10.3.6. Associated Legendre Functions [see also § 8.2]

In section 10.3.1, we found a solution

$$V = \left(A_n r^n + \frac{B_n}{r^{n+1}} \right) S_n$$

of Laplace's equation (10.3.1), where S_n is a solution of equation (10.3.3). If we now suppose that in equation (10.3.3) S is of the form

$$S = w(\theta) \cos (m\phi + \varepsilon),$$

we find that $w = w(\theta)$ is a solution of

$$\frac{d^2 w}{d\theta^2} + \cot \theta \frac{dw}{d\theta} + \left[n(n+1) - \frac{m^2}{\sin^2 \theta} \right] w = 0,$$

or, with $\cos \theta = \mu$,

$$(1 - \mu^2)\frac{d^2 w}{d\mu^2} - 2\mu \frac{dw}{d\mu} + \left[n(n+1) - \frac{m^2}{1 - \mu^2} \right] w = 0. \qquad (10.3.10)$$

Equation (10.3.10) is called the *associated Legendre equation*. It reduces to Legendre's equation (10.3.4) when $m = 0$.

If m is a positive integer, and $|\mu| < 1$, the functions

$$P_n^m(\mu) = (1 - \mu^2)^{m/2} \frac{d^m}{d\mu^m} P_n(\mu),$$

$$Q_n^m(\mu) = (1 - \mu^2)^{m/2} \frac{d^m}{d\mu^m} Q_n(\mu)$$

are called *associated Legendre functions*, of degree n and order m, of the first and second kinds, respectively. These are Ferrer's definitions. Alternative definitions, due to Hobson, are often used when $|\mu| > 1$; but for most applications $\mu = \cos\theta$, and no complications arise from Ferrer's definitions in this case (Whittaker and Watson, 1950).

The associated Legendre functions satisfy equation (10.3.10). For differentiating Legendre's equation

$$(1 - \mu^2)\frac{d^2 y}{d\mu^2} - 2\mu \frac{dy}{d\mu} + n(n+1)y = 0$$

m times by Leibniz' theorem [see Theorem 3.4.1], and writing

$$v = \frac{d^m y}{d\mu^m}$$

we obtain the equation

$$(1 - \mu^2)\frac{d^2 v}{d\mu^2} - 2(m+1)\mu \frac{dv}{d\mu} + (n-m)(n+m+1)v = 0.$$

Then substituting

$$w = (1 - \mu^2)^{m/2} v$$

in this equation leads to equation (10.3.10), which is therefore the differential equation satisfied by $P_n^m(\mu)$ and $Q_n^m(\mu)$.

The solutions of Laplace's equation (10.3.1) corresponding to equation (10.3.10) are [with the notation of (10.3.9)]

$$V = \left(\begin{matrix} r^n \\ r^{-n-1} \end{matrix}\right)\left(\begin{matrix} P_n^m(\cos\theta) \\ Q_n^m(\cos\theta) \end{matrix}\right)\left(\begin{matrix} \cos m\phi \\ \sin m\phi \end{matrix}\right).$$

The function $Q_n^m(\mu)$ has singularities at $\mu = \pm 1$, and is therefore not included in any solution which must remain finite on the polar axis.

When $m = 0$, we have

$$P_n^0(\mu) = P_n(\mu), \qquad Q_n^0(\mu) = Q_n(\mu).$$

Also, since $P_n(\mu)$ is a polynomial of degree n, it follows from the definition of $P_n^m(\mu)$ that

$$P_n^m(\mu) = 0 \quad \text{if } m > n.$$

Differentiating Rodrigue's formula m times, we find

$$P_n^m(\mu) = \frac{1}{2^n n!}(1 - \mu^2)^{m/2}\frac{d^{m+n}}{d\mu^{m+n}}[(\mu^2 - 1)^n].$$

Hence, in particular,

$$P_n^m(-\mu) = (-1)^{m+n} P_n^m(\mu),$$

and

$$P_n^n(\mu) = \frac{(2n)!}{2^n n!}(1-\mu^2)^{n/2}.$$

The first few associated Legendre functions are

$$P_1^1(\mu) = (1-\mu^2)^{1/2} = \sin\theta,$$

$$P_2^1(\mu) = 3(1-\mu^2)^{1/2}\mu = \tfrac{3}{2}\sin 2\theta,$$

$$P_2^2(\mu) = 3(1-\mu^2) = \tfrac{3}{2}(1-\cos 2\theta),$$

$$P_3^1(\mu) = \tfrac{3}{2}(1-\mu^2)^{1/2}(5\mu^2 - 1) = \tfrac{3}{8}(\sin\theta + 5\sin 3\theta),$$

$$P_3^2(\mu) = 15(1-\mu^2)\mu = \tfrac{15}{4}(\cos\theta - \cos 3\theta),$$

$$P_3^3(\mu) = 15(1-\mu^2)^{3/2} = \tfrac{15}{4}(3\sin\theta - \sin 3\theta).$$

Note that if m is an even integer, and $m \le n$, then $P_n^m(\mu)$ is a polynomial of degree n.

The function $P_n^m(x)$ satisfies recurrence formulae similar to those of section 10.3.3 for $P_n(x)$. For example,

$$(n+1-m)P_{n+1}^m(x) - (2n+1)xP_n^m(x) + (n+m)P_{n-1}^m(x) = 0,$$

$$(10.3.11)$$

$$(1-x^2)P_n^{m'}(x) = (n+1)xP_n^m(x) - (n+1-m)P_{n+1}^m(x). \quad (10.3.12)$$

Finally, the orthogonal property corresponding to that of section 10.3.3 for $P_n(x)$ is

$$\int_{-1}^{1} P_n^m(x)P_r^m(x)\,dx = 0, \qquad r \ne n,$$

$$= \frac{2}{2n+1}\frac{(n+m)!}{(n-m)!} \qquad r = n, \qquad (10.3.13)$$

where $n > m$ and $r > m$.

10.4. BESSEL FUNCTIONS

10.4.0. Introduction

The earliest systematic study of Bessel functions was made by Bessel (1784–1846) in 1824 in connection with a problem in dynamical astronomy, although several special cases had appeared much earlier (Watson, 1944). For example, James Bernoulli used the series for $J_{1/3}(x)$ in 1703 in a problem on curves, Daniel Bernoulli found the function which is now called $J_0(x)$ in 1732 as the solution to the problem of the oscillations of a heavy uniform chain suspended from its upper end, and Euler found $J_n(x)$ in 1764 as the solution to the problem of the vibrations of a stretched membrane.

The introduction to the section on Legendre polynomials [see § 10.3.0], dealing with Laplace's equation, is also relevant to the present article. There are several elementary texts on the theory and applications of Bessel functions. For example, brief surveys are given in Chapter IV of (Sneddon, 1956) and in Chapter 4 of (Hildebrand, 1948), a comprehensive list of formulae and many tables appear in (McLachlan, 1955), and a short up-to-date account of Bessel functions for scientists and engineers is given by (Tranter, 1968).

10.4.1.   Laplace's Equation in Cylindrical Polar Coordinates

Laplace's equation in cylindrical polar coordinates (r, θ, z) is [cf. (8.2.10)]

$$\nabla^2 V \equiv \frac{\partial^2 V}{\partial r^2} + \frac{1}{r}\frac{\partial V}{\partial r} + \frac{1}{r^2}\frac{\partial^2 V}{\partial \theta^2} + \frac{\partial^2 V}{\partial z^2} = 0. \tag{10.4.1}$$

Assuming a solution of the form

$$V = R(r)\Theta(\theta)Z(z),$$

equation (10.4.1) yields the three equations

$$\frac{d^2 Z}{dz^2} = k^2 Z, \tag{10.4.2}$$

$$\frac{d^2 \Theta}{d\theta^2} = -n^2 \Theta, \tag{10.4.3}$$

$$r^2 \frac{d^2 R}{dr^2} + r\frac{dR}{dr} + (k^2 r^2 - n^2)R = 0, \tag{10.4.4}$$

where k^2 and n^2 are constants. The general solution of equation (10.4.2) is

$$Z = A_k e^{kz} + B_k e^{-kz},$$

where A_k and B_k are arbitrary constants, and the general solution of equation (10.4.3) is

$$\Theta = C_n \cos n\theta + D_n \sin n\theta,$$

where C_n and D_n are arbitrary constants [see Example 7.4.1].
   With the substitution $kr = u$, equation (10.4.4) becomes

$$u^2 \frac{d^2 R}{du^2} + u\frac{dR}{du} + (u^2 - n^2)R = 0, \tag{10.4.5}$$

which is *Bessel's equation* of order n [see Example 7.6.5]. Denoting a solution of this equation by

$$R = S_n(u),$$

we obtain solutions of Laplace's equation (10.4.1) of the form [with the notation of (10.3.9)]

$$V = e^{\pm kz} S_n(kr) \begin{pmatrix} \cos n\theta \\ \sin n\theta \end{pmatrix}.$$

In most physical problems, V must be a one-valued function of position in space, and so n must be an integer. If there is symmetry about the z-axis, that is, if the required solution does not depend on θ, then we take $n = 0$. We shall find some solutions of Bessel's equation (10.4.5) in the following section.

10.4.2. Bessel Coefficients and Functions

The generating function

$$\exp\left(\tfrac{1}{2}x\left(t - \frac{1}{t}\right)\right)$$

may be expanded as a convergent power series in t for all values of x and for all non-zero t [see § 2.11]. The coefficient $J_n(x)$ of t^n is called the *Bessel coefficient of the first kind* of order n. Thus

$$\exp\left(\tfrac{1}{2}x\left(t - \frac{1}{t}\right)\right) \equiv J_0(x) + t J_1(x) + t^2 J_2(x) + \dots$$

$$+ \frac{J_{-1}(x)}{t} + \frac{J_{-2}(x)}{t^2} + \dots.$$

The left-hand side is unaltered by writing $-1/t$ for t and hence

$$J_{-n}(x) = (-1)^n J_n(x), \qquad n = 0, 1, 2, \dots. \tag{10.4.6}$$

The coefficient of t^n in the expansion of the generating function is

$$J_n(x) = \frac{x^n}{2^n} \sum_{r=0}^{\infty} \frac{(-1)^r x^{2r}}{2^{2r} r! \, (n+r)!}$$

$$= \frac{x^n}{2^n n!} \left[1 - \frac{x^2}{2(2n+2)} + \frac{x^4}{2 \cdot 4(2n+2)(2n+4)} - \dots \right]. \tag{10.4.7}$$

In particular,

$$J_0(x) = 1 - \frac{x^2}{2^2} + \frac{x^4}{2^2 \cdot 4^2} - \frac{x^6}{2^2 \cdot 4^2 \cdot 6^2} + \dots. \tag{10.4.8}$$

Hence

$$J_n(0) = 0, \quad n \neq 0,$$

$$= 1, \quad n = 0.$$

The series (10.4.7) and (10.4.8) are absolutely convergent for all values of x [see § 1.9].

We shall now prove that $J_n(x)$ is a solution of Bessel's equation

$$x^2 \frac{d^2 y}{dx^2} + x \frac{dy}{dx} + (x^2 - n^2)y = 0, \tag{10.4.9}$$

which is equation (10.4.5) in a different notation. Denote the generating function by

$$F(x, t) \equiv \exp\left(\tfrac{1}{2}x\left(t - \frac{1}{t}\right)\right) = \sum_{n=-\infty}^{\infty} t^n J_n(x).$$

Then it is easily verified that

$$x^2 \frac{\partial^2 F}{\partial x^2} + x \frac{\partial F}{\partial x} + x^2 F \equiv t^2 \frac{\partial^2 F}{\partial t^2} + t \frac{\partial F}{\partial t}$$

and hence that

$$\sum_{n=-\infty}^{\infty} [x^2 J_n''(x) + x J_n'(x) + (x^2 - n^2)J_n(x)]t^n \equiv 0.$$

It follows that $J_n(x)$ satisfies Bessel's equation (10.4.9).

If n is not an integer, we define

$$J_n(x) = \frac{x^n}{2^n} \sum_{r=0}^{\infty} \frac{(-1)^r x^{2r}}{2^{2r} r! \, \Gamma(n+r+1)} \tag{10.4.10}$$

[see § 10.2.1] to be the *Bessel function of the first kind* of order n. Since Bessel's equation (10.4.9) is unaltered by writing $-n$ for n, and since it is satisfied by $J_n(x)$ of equation (10.4.10), the general solution when n is not an integer is

$$y = A J_n(x) + B J_{-n}(x),$$

where A and B are arbitrary constants [see § 7.3.1].

When n is an integer, the solutions $J_n(x)$, $J_{-n}(x)$ are not independent, because of equation (10.4.6). A second solution of Bessel's equation in this case is the limit of the expression

$$Y_n(x) = [J_n(x) \cos \pi x - J_{-n}(x)] \operatorname{cosec} \pi x \tag{10.4.11}$$

as n tends to an integral value. Explicitly, we find

$$Y_n(x) = J_n(x)(a \log x + b) + \frac{1}{x^n}(c_0 + c_1 x^2 + c_2 x^4 + \ldots),$$

where a, b, c_0, c_1, ... are constants, of which b is arbitrary (Bowman, 1958). A particular choice of b gives the generally accepted form of $Y_n(x)$, known as *Weber's Bessel function of the second kind* of order n. The general solution of Bessel's equation (10.4.9) when n is an integer is therefore taken to be

$$y = A J_n(x) + B Y_n(x),$$

where A and B are arbitrary constants. The full expression for $Y_n(x)$ is given by (McLachlan, 1955), where the function is also tabulated. It behaves like x^{-n} when x is small, and has a term in $\log x$. Because of these singularities, $Y_n(x)$ does not appear in any solution which must remain finite on the axis of symmetry.

Instead of using $J_n(x)$ and $Y_n(x)$ as independent solutions of Bessel's equation, it is sometimes advantageous to use the *Hankel functions*

$$\left.\begin{array}{l} H_n^{(1)}(x) = J_n(x) + iY_n(x), \\ H_n^{(2)}(x) = J_n(x) - iY_n(x). \end{array}\right\} \qquad (10.4.12)$$

Then the general solution of Bessel's equation (10.4.9), for any value of n, can be expressed in the form

$$y = A_1 H_n^{(1)}(x) + A_2 H_n^{(2)}(x),$$

where A_1 and A_2 are arbitrary constants. If n is not an integer, $Y_n(x)$ is defined by equation (10.4.11) as it stands.

Equation (10.4.10) may be written

$$J_n(x) = \frac{x^n}{2^n \Gamma(n+1)} \left[1 - \frac{x^2}{2(2n+2)} + \frac{x^4}{2 \cdot 4(2n+2)(2n+4)} - \cdots \right].$$

In particular,

$$J_{1/2}(x) = \left(\frac{2}{\pi x}\right)^{1/2} \left(x - \frac{x^3}{3!} + \frac{x^5}{5!} - \cdots \right)$$

$$= \left(\frac{2}{\pi x}\right)^{1/2} \sin x.$$

Similarly,

$$J_{-1/2}(x) = \left(\frac{2}{\pi x}\right)^{1/2} \cos x.$$

More generally, when the order of a Bessel function is half an odd integer, we have (Tranter, 1968)

$$J_{k+\frac{1}{2}}(x) = \left(\frac{2}{\pi}\right)^{1/2} x^{k+\frac{1}{2}} \left(-\frac{1}{x}\frac{d}{dx}\right)^k \left(\frac{\sin x}{x}\right), \qquad (10.4.13)$$

$$J_{-k-\frac{1}{2}}(x) = \left(\frac{2}{\pi}\right)^{1/2} x^{k+\frac{1}{2}} \left(\frac{1}{x}\frac{d}{dx}\right)^k \left(\frac{\cos x}{x}\right), \qquad k = 0, 1, 2, \ldots. \qquad (10.4.14)$$

10.4.3. Modified Bessel Functions

The *modified Bessel function of the first kind* of order n is defined for all values of n by

$$I_n(x) = i^{-n} J_n(ix). \qquad (10.4.15)$$

From equation (10.4.10), we find

$$I_n(x) = \frac{x^n}{2^n} \sum_{r=0}^{\infty} \frac{x^{2r}}{2^{2r} r! \, \Gamma(n+r+1)}.$$

Thus $I_n(x)$ is real when x is real. The differential equation satisfied by $I_n(x)$ is the *modified Bessel's equation* of order n:

$$x^2 \frac{d^2 y}{dx^2} + x \frac{dy}{dx} - (x^2 + n^2)y = 0, \tag{10.4.16}$$

which should be compared with Bessel's equation (10.4.9). Equation (10.4.16) may be derived from equation (10.4.4) by replacing k by ik, so that the corresponding solutions of Laplace's equation (10.4.1) are of the form [with the notation of (10.3.9)]

$$V = \begin{pmatrix} \cos kz \\ \sin kz \end{pmatrix} I_n(kr) \begin{pmatrix} \cos n\theta \\ \sin n\theta \end{pmatrix}.$$

The reason for introducing $I_n(x)$ is to express solutions of equation (10.4.16) in real instead of complex form.

The functions $I_n(x)$ and $I_{-n}(x)$ are linearly independent solutions of equation (10.4.16) except when n is an integer. The general solution of equation (10.4.16) when n is not an integer is therefore

$$y = AI_n(x) + BI_{-n}(x),$$

where A and B are arbitrary constants.

When n is an integer,

$$I_{-n}(x) = I_n(x),$$

and in this case a second solution of equation (10.4.16) is provided by the *modified Bessel function of the second kind* of order n, defined by the limit of the expression

$$K_n(x) = \frac{\pi}{2} [I_{-n}(x) - I_n(x)] \operatorname{cosec} \pi x \tag{10.4.17}$$

as n tends to an integral value. Explicitly, we find

$$K_n(x) = I_n(x)(A \log x + B) + \frac{1}{x^n}(C_0 + C_1 x^2 + C_2 x^4 + \ldots),$$

where A, B, C_0, $C_1, \ldots$ are constants [cf. the definition and properties of $Y_n(x)$ given with (10.4.11)]. The general solution of equation (10.4.16) when n is an integer is therefore taken to be

$$y = EI_n(x) + FK_n(x),$$

where E and F are arbitrary constants.

10.4.4. Properties of Bessel Functions

(i) *Bessel's integral* (for Bessel coefficients)
Let n be a positive integer or zero. Then

$$J_{-n}(x) = (-1)^n J_n(x),$$

and therefore

$$\exp\left(\tfrac{1}{2}x\left(t - \frac{1}{t}\right)\right) = J_0(x) + \left(t - \frac{1}{t}\right)J_1(x) + \left(t^2 + \frac{1}{t^2}\right)J_2(x)$$

$$+ \left(t^3 - \frac{1}{t^3}\right)J_3(x) + \dots.$$

Put $t = e^{i\phi}$. Then

$$e^{ix\sin\phi} = J_0(x) + 2iJ_1(x)\sin\phi + 2J_2(x)\cos 2\phi$$

$$+ 2iJ_3(x)\sin 3\phi + 2J_4(x)\cos 4\phi + \dots,$$

from which, incidentally, we can derive the useful formula

$$e^{ix\sin\phi} = \sum_{n=-\infty}^{\infty} J_n(x)\, e^{in\phi}. \tag{10.4.18}$$

Equating real and imaginary parts, we have

$$\cos(x\sin\phi) = J_0(x) + 2J_2(x)\cos 2\phi + 2J_4(x)\cos 4\phi + \dots,$$

$$\sin(x\sin\phi) = 2J_1(x)\sin\phi + 2J_3(x)\sin 3\phi + \dots.$$

Hence

$$\int_0^\pi \cos(x\sin\phi)\cos n\phi\, d\phi = \pi J_n(x), \quad n = 0, 2, 4, 6, \dots,$$

$$= 0, \quad n = 1, 3, 5, \dots,$$

which can be written

$$\int_0^\pi \cos(x\sin\phi)\cos n\phi\, d\phi = \frac{\pi}{2}[1 + (-1)^n]J_n(x).$$

Similarly,

$$\int_0^\pi \sin(x\sin\phi)\sin n\phi\, d\phi = \frac{\pi}{2}[1 - (-1)^n]J_n(x).$$

Adding the last two results, we find

$$J_n(x) = \frac{1}{\pi}\int_0^\pi \cos(n\phi - x\sin\phi)\, d\phi, \tag{10.4.19}$$

which is *Bessel's integral*. Bessel used this integral to determine the Fourier coefficients that are required to solve Kepler's equation of time for a planetary orbit (Bowman, 1958).

Putting $n = 0$ and $t = \sin \phi$ in Bessel's integral, we obtain a useful integral representation for $J_0(x)$:

$$J_0(x) = \frac{2}{\pi} \int_0^1 \frac{\cos xt}{(1-t^2)^{1/2}} \, dt.$$

(ii) *Recurrence formulae* (for all values of n)

$$nJ_n(x) = \tfrac{1}{2}x[J_{n-1}(x) + J_{n+1}(x)], \qquad\qquad \text{I}$$

$$J_n'(x) = \tfrac{1}{2}[J_{n-1}(x) - J_{n+1}(x)], \qquad\qquad \text{II}$$

$$J_n'(x) = J_{n-1}(x) - \frac{n}{x} J_n(x), \qquad\qquad \text{III}$$

$$J_n'(x) = \frac{n}{x} J_n(x) - J_{n+1}(x). \qquad\qquad \text{IV}$$

The last two formulae may be written as

$$\frac{d}{dx}[x^n J_n(x)] = x^n J_{n-1}(x)$$

and

$$\frac{d}{dx}[x^{-n} J_n(x)] = -x^{-n} J_{n+1}(x).$$

The functions $Y_n(x)$, $H_n^{(1)}(x)$ and $H_n^{(2)}(x)$ also satisfy equations I–IV.

Recurrence formulae for $I_n(x)$ are easily obtained from those for $J_n(x)$. Corresponding to I–IV, we find, for all values of n,

$$nI_n(x) = \tfrac{1}{2}x[I_{n-1}(x) - I_{n+1}(x)], \qquad\qquad \text{V}$$

$$I_n'(x) = \tfrac{1}{2}[I_{n-1}(x) + I_{n+1}(x)], \qquad\qquad \text{VI}$$

$$I_n'(x) = I_{n-1}(x) - \frac{n}{x} I_n(x), \qquad\qquad \text{VII}$$

$$I_n'(x) = \frac{n}{x} I_n(x) + I_{n+1}(x). \qquad\qquad \text{VIII}$$

The functions $K_n(x)$ also satisfy equations V–VIII.

(iii) *Addition formulae*

If n is *any* integer, then

$$J_n(x+y) = \sum_{r=-\infty}^{\infty} J_r(x) J_{n-r}(y).$$

If n is a *positive* integer, we can write

$$J_n(x+y) = \sum_{r=0}^{n} J_r(x)J_{n-r}(y)$$

$$+ \sum_{r=1}^{\infty} (-1)^r [J_r(x)J_{n+r}(y) + J_{n+r}(x)J_r(y)].$$

(iv) *Roots of $J_n(x) = 0$*

The following results are often used in practical applications of Bessel functions.

(a) If $J_n(\alpha) = 0$, then $J_n(-\alpha) = 0$.
(b) $J_n(x) = 0$ has an infinite number of real roots, all of which are unrepeated, except possibly the root $x = 0$. There is no other root.
(c) The positive roots of $J_n(x) = 0$ and $J_{n+1}(x) = 0$ interlace.

(v) *Integral formulae*

$$\int x^m J_n(x)\, dx = x^m J_{n+1}(x) - (m-n-1) \int x^{m-1} J_{n+1}(x)\, dx$$

$$(10.4.20)$$

is a reduction formula which yields a closed form when $m - n$ is an odd positive integer. Similarly,

$$\int x^m J_n(x)\, dx = -x^m J_{n-1}(x) + (m+n-1) \int x^{m-1} J_{n-1}(x)\, dx$$

$$(10.4.21)$$

yields a closed form when $m + n$ is an odd positive integer. When m and n are both even or both odd, use of these formulae leads to a closed form together with a multiple of

$$\int J_0(x)\, dx,$$

which is a tabulated function.

(vi) *Lommel's integrals*

If α and β are constants, then

$$(\beta^2 - \alpha^2) \int_0^1 xJ_n(\alpha x)J_n(\beta x)\, dx = \alpha J'_n(\alpha)J_n(\beta) - \beta J'_n(\beta)J_n(\alpha),$$

and

$$\int_0^1 xJ_n^2(\alpha x)\, dx = \tfrac{1}{2}[J_n'^2(\alpha) - J_{n-1}(\alpha)J_{n+1}(\alpha)].$$

If α, β $(\alpha^2 \neq \beta^2)$ are two roots of $J_n(x) = 0$, then it follows from the first of these results that

$$\int_0^1 xJ_n(\alpha x)J_n(\beta x)\, dx = 0,$$ (10.4.22)

and from the second that

$$\int_0^1 xJ_n^2(\alpha x)\, dx = \tfrac{1}{2}J_{n+1}^2(\alpha).$$ (10.4.23)

These integrals are needed to obtain the Fourier–Bessel expansion of section 10.4.5.

(vii) *A transformation of Bessel's equation*
Many useful results can be derived by giving special values to α, β, γ and n in the differential equation

$$x^2\frac{d^2y}{dx^2}+(1-2\alpha)x\frac{dy}{dx}+(\beta^2\gamma^2x^{2\gamma}+\alpha^2-n^2\gamma^2)y=0,$$

which is satisfied by

$$y = x^\alpha J_n(\beta x^\gamma).$$

(viii) *Behaviour at zero and infinity*
As $x \to 0$,

$$J_n(x) \text{ and } I_n(x) \to 0, \quad n > 0,$$
$$1, \quad n = 0,$$
$$\infty, \quad n < 0 \text{ and non-integral,}$$
$$0, \quad n \text{ a negative integer.}$$
$$Y_n(x) \to -\infty, \quad \text{all values of } n,$$
$$K_n(x) \to \infty, \quad \text{all values of } n.$$

As $x \to \infty$, and for all values of n,

$$J_n(x) \text{ and } Y_n(x) \to 0 \quad \text{like } x^{-1/2}\binom{\sin x}{\cos x},$$
$$I_n(x) \to \infty \quad \text{like } x^{-1/2}\, e^x,$$
$$K_n(x) \to 0 \quad \text{like } x^{-1/2}\, e^{-x}.$$

The behaviour of the Hankel functions $H_n^{(1)}(x)$ and $H_n^{(2)}(x)$ at zero and infinity is easily deduced from that of $J_n(x)$ and $Y_n(x)$. Asymptotic expansions of all the Bessel functions are given by (Watson, 1944).

10.4.5. Fourier–Bessel Expansion

It is possible to expand an arbitrary function $f(x)$ as an infinite series of the type

$$f(x) \equiv A_1 J_n(\alpha_1 x) + A_2 J_n(\alpha_2 x) + \dots ,$$

where $0 < x < 1$, $n \geq -\frac{1}{2}$, and $\alpha_1, \alpha_2, \dots$ are the positive roots of $J_n(x) = 0$. Sufficient conditions for the expansion to be valid are

(i)
$$\int_0^1 x^{1/2} f(x)\, dx$$

exists and (if it is an improper integral) is absolutely convergent [see § 4.6];
(ii) $f(x)$ has limited total fluctuation in (a, b), where $0 < a < x < b < 1$ (Watson, 1944).

Using Lommel's integrals [see § 10.4.4 (vi)], we find that the general coefficient in the expansion is

$$A_s = \frac{2}{J_{n+1}^2(\alpha_s)} \int_0^1 x f(x) J_n(\alpha_s x)\, dx. \tag{10.4.24}$$

10.4.6. Examples

EXAMPLE 10.4.1   *Normal modes of a circular membrane.*   A uniform circular membrane, whose circumference is fixed, vibrates symmetrically. Find the normal modes of vibration.

Solution.   Let a be the radius of the membrane, and let $u(r, t)$ be the displacement at radial distance r at time t. The differential equation governing the motion for small displacements is

$$\frac{\partial^2 u}{\partial r^2} + \frac{1}{r}\frac{\partial u}{\partial r} = \frac{1}{c^2}\frac{\partial^2 u}{\partial t^2},$$

where $c^2 = T/\sigma$, T being the uniform tension and σ the mass per unit area.

For a normal mode, in which every particle of the system vibrates at the same frequency and has the same phase,

$$u = R \cos(\omega t - \varepsilon),$$

where R depends only on r. The boundary condition is

$$u = 0 \quad \text{when } r = a \text{ for all values of } t.$$

Substituting in the differential equation, we find

$$\frac{d^2 R}{dr^2} + \frac{1}{r}\frac{dR}{dr} + \frac{\omega^2}{c^2} R = 0.$$

Comparison with equation (10.4.4) shows that this is Bessel's equation of order zero with $k = \omega/c$. The general solution is therefore

$$R = AJ_0\left(\frac{\omega r}{c}\right) + BY_0\left(\frac{\omega r}{c}\right).$$

However, since u is small when $r = 0$, but $Y_0(\omega r/c) \to -\infty$ as $r \to 0$, we must set $B = 0$, giving

$$u = AJ_0\left(\frac{\omega r}{c}\right)\cos(\omega t - \varepsilon).$$

The boundary condition is satisfied if

$$J_0\left(\frac{\omega a}{c}\right) = 0,$$

which gives

$$\frac{\omega a}{c} = \alpha_1, \alpha_2, \ldots,$$

where $\alpha_1, \alpha_2, \ldots$ are the zeros of $J_0(x)$. Hence there is an infinite number of normal modes, namely

$$u_i = A_i J_0\left(\frac{\alpha_i r}{a}\right)\cos\left(\frac{c\alpha_i t}{a} - \varepsilon_i\right), \qquad i = 1, 2, \ldots,$$

where the A_i and ε_i are arbitrary constants, the A_i being small compared with a.

EXAMPLE 10.4.2   *Cooling of a long circular cylinder.*   The surface of a long circular cylinder of radius a is kept at a temperature u_0. Initially, the interior of the cylinder is at a temperature u_1. Find the temperature distribution in the cylinder at any subsequent time.

Solution.   Assume that the cylinder is sufficiently long for the heat flow to be everywhere radial, so that the problem becomes two-dimensional. The equation of heat conduction is then

$$\frac{\partial^2 u}{\partial r^2} + \frac{1}{r}\frac{\partial u}{\partial r} = \frac{1}{K}\frac{\partial u}{\partial t}$$

where $u(r, t)$ is the temperature at radial distance r and time t, and K is the diffusivity. The initial conditions are

$$u = u_1, \qquad 0 < r < a, \qquad t = 0,$$

and the boundary conditions are

$$u = u_0, \qquad r = a, \qquad 0 \leq t < \infty.$$

For convenience, put $v = u - u_0$. Then we have to solve

$$\frac{\partial^2 v}{\partial r^2} + \frac{1}{r}\frac{\partial v}{\partial r} = \frac{1}{K}\frac{\partial v}{\partial t} \tag{10.4.25}$$

subject to

$$v = u_1 - u_0, \qquad 0 < r < a, \qquad t = 0, \tag{10.4.26}$$

$$v = 0, \qquad r = a, \qquad 0 \le t < \infty, \tag{10.4.27}$$

$$v \to 0 \quad \text{as } t \to \infty, \tag{10.4.28}$$

$$v \text{ is finite.} \tag{10.4.29}$$

If we assume a solution of the form

$$v = R(r)T(t),$$

then we obtain

$$T = A\,e^{-\mu^2 K t},$$

the exponent being chosen so that condition (10.4.28) is satisfied, and

$$\frac{d^2 R}{dr^2} + \frac{1}{r}\frac{dR}{dr} + \mu^2 R = 0,$$

which is Bessel's equation of order zero. Hence

$$R = BJ_0(\mu r),$$

there being no term in $Y_0(\mu r)$ because of condition (10.4.29).

Combining the arbitrary constants A and B, a solution of (10.4.25) is

$$v = A\,e^{-\mu^2 K t}J_0(\mu r)$$

and condition (10.4.27) gives

$$J_0(\mu a) = 0.$$

Let α_s be the sth positive root of $J_0(x) = 0$. Then a more general solution is

$$v = \sum_{s=1}^{\infty} A_s\, e^{-\alpha_s^2 K t / a^2} J_0\!\left(\frac{\alpha_s r}{a}\right),$$

and condition (10.4.26) is satisfied if

$$u_1 - u_0 = \sum_{s=1}^{\infty} A_s J_0\!\left(\frac{\alpha_s r}{a}\right).$$

To find A_s, we use (10.4.20) and (10.4.23). We find

$$A_s = \frac{2}{J_1^2(\alpha_s)} \int_0^1 x(u_1 - u_0)J_0(\alpha_s x)\,dx$$

$$= \frac{2(u_1 - u_0)}{\alpha_s J_1(\alpha_s)}$$

and hence

$$u = u_0 + 2(u_1 - u_0) \sum_{s=1}^{\infty} \frac{e^{-\alpha_s^2 Kt/a^2}}{\alpha_s J_1(\alpha_s)} J_0\left(\frac{\alpha_s r}{a}\right).$$

EXAMPLE 10.4.3 *Fourier–Bessel expansion.* Expand x^n as a Fourier–Bessel series.

Solution. Equation (10.4.24) gives

$$A_s = \frac{2}{J_{n+1}^2(\alpha_s)} \int_0^1 x^{n+1} J_n(\alpha_s x)\, dx.$$

Put $u = \alpha_s x$. Then

$$\int_0^1 x^{n+1} J_n(\alpha_s x)\, dx = \int_0^{\alpha_s} \frac{u^{n+1}}{\alpha_s^{n+2}} J_n(u)\, du$$

$$= \frac{J_{n+1}(\alpha_s)}{\alpha_s},$$

using (10.4.20). Hence

$$A_s = \frac{2}{\alpha_s J_{n+1}(\alpha_s)}$$

and

$$x^n = 2 \sum_{s=1}^{\infty} \frac{J_n(\alpha_s x)}{\alpha_s J_{n+1}(\alpha_s)}$$

where α_s, $s = 1, 2, \ldots$, are the positive roots of $J_n(x) = 0$.

EXAMPLE 10.4.4 *Eigenvalue problem for a linear differential operator.* Solve the eigenvalue problem

$$Lu = \lambda u$$

for the linear differential operator

$$L \equiv -\frac{d^2}{dx^2} - \frac{5}{x}\frac{d}{dx}, \qquad 0 < x < 1,$$

where the functions u vanish at $x = 1$ and are finite with finite derivatives at $x = 0$ [see § 7.8.1].

Solution. The differential equation to be solved for the eigenvalues λ and the corresponding eigenfunctions u is

$$\frac{d^2 u}{dx^2} + \frac{5}{x}\frac{du}{dx} + \lambda u = 0.$$

From section 10.4.4(vii), with $\alpha = -2$, $\beta = \lambda^{1/2}$, $\gamma = 1$, $n = 2$, we find

$$u = x^{-2} J_2(\lambda^{1/2} x),$$

which satisfies the boundary conditions at $x = 0$. Applying the boundary condition at $x = 1$, the values of λ are given by

$$J_2(\lambda^{1/2}) = 0.$$

Hence there is an infinite number of eigenvalues and corresponding eigenfunctions. The smallest value of $\lambda^{1/2}$ is the smallest positive root of $J_2(x) = 0$, namely $5\cdot1356$. The smallest eigenvalue is therefore $\lambda = 26\cdot37$, and the corresponding eigenfunction is

$$u = x^{-2} J_2(5\cdot1356x).$$

10.5. OTHER FUNCTIONS

10.5.0. Introduction: Orthogonal Polynomials and the Sturm–Liouville Equation

Consider a system of polynomials $p_n(x)$, $n = 0, 1, 2, \ldots$, where $p_n(x)$ is of degree n. The system is said to be *orthogonal* in the interval $a \le x \le b$ with respect to the weight function $w(x)$ (≥ 0) if

$$\int_a^b w(x) p_m(x) p_n(x)\, dx = 0, \qquad m \ne n, m, n = 0, 1, 2, \ldots. \quad (10.5.1)$$

The polynomials $p_n(x)$ are standardized by adding to equation (10.5.1) the conditions

$$\int_a^b w(x)[p_n(x)]^2\, dx = h_n, \qquad n = 0, 1, 2, \ldots, \quad (10.5.2)$$

where the h_n are given constants. If $h_n = 1$ for $n = 0, 1, 2, \ldots$, then the system is said to be *orthonormal*.

Orthogonal polynomials are used in many branches of mathematical physics and also in numerical analysis (Szegö, 1978) [see also § 20.4 and III, § 6.3]. The fundamental problem is to express an arbitrary function $F(x)$ as a weighted sum of orthogonal polynomials, that is, to find a set of coefficients a_n such that

$$F(x) = \sum_{n=0}^{\infty} a_n p_n(x), \quad (10.5.3)$$

where the polynomials $p_n(x)$ satisfy equations (10.5.1) and (10.5.2). To find a_m, multiply equation (10.5.3) throughout by $w(x) p_m(x)$ and integrate from $x = a$ to $x = b$, giving

$$\int_a^b w(x) p_m(x) F(x)\, dx = \int_a^b \sum_{n=0}^{\infty} a_n w(x) p_m(x) p_n(x)\, dx$$

$$= a_m h_m,$$

from equations (10.5.1) and (10.5.2). Hence a_m is determined uniquely by

$$a_m = \frac{1}{h_m} \int_a^b w(x) p_m(x) F(x) \, dx,$$

assuming, where necessary, that the integral converges.

The orthogonality relation (10.5.1) arises naturally from certain boundary-value problems associated with the *Sturm–Liouville equation*

$$\frac{d}{dx}\left[p(x)\frac{dy}{dx}\right] + [q(x) + \lambda w(x)]y = 0, \qquad a < x < b, \qquad (10.5.4)$$

where λ is a real parameter [see § 7.8]. For suppose that $y_m(x)$ and $y_n(x)$ are two solutions of this equation, with λ_m and λ_n as the corresponding values of λ. Then

$$\frac{d}{dx}\left[p(x)\frac{dy_m}{dx}\right] + [q(x) + \lambda_m w(x)]y_m = 0$$

and

$$\frac{d}{dx}\left[p(x)\frac{dy_n}{dx}\right] + [q(x) + \lambda_n w(x)]y_n = 0.$$

Multiply the first of these equations by y_n, the second by y_m, subtract, and integrate from $x = a$ to $x = b$, to give

$$\int_a^b \left\{ y_n \frac{d}{dx}\left[p(x)\frac{dy_m}{dx}\right] - y_m \frac{d}{dx}\left[p(x)\frac{dy_n}{dx}\right] \right\} dx$$

$$+ (\lambda_m - \lambda_n) \int_a^b w(x) y_m y_n \, dx = 0.$$

Integration by parts [see § 4.3] now yields

$$[y_n p(x) y_m' - y_m p(x) y_n']_a^b + (\lambda_m - \lambda_n) \int_a^b w(x) y_m y_n \, dx = 0.$$

If the boundary terms vanish (for example, if $y_n = y_m = 0$ at both end-points), then

$$\int_a^b w(x) y_m y_n \, dx = 0,$$

provided that $\lambda_m \neq \lambda_n$. Thus y_m and y_n are orthogonal in the sense of equation (10.5.1).

We now see that Legendre polynomials (§ 10.3) provide an example of orthogonal polynomials in the interval $-1 \leq x \leq 1$, with the weight function $w(x) = 1$ and with $h_n = 2/(2n+1)$. It is possible to derive many properties common to all orthogonal polynomials, or, more generally, orthogonal functions, but instead we shall restrict our attention to three further examples, associated with the names of Laguerre (1834–1886), Hermite (1822–1901) and Chebyshev (1821–1894), respectively.

10.5.1.   Laguerre Polynomials

The Laguerre polynomial $L_n(x)$, of degree n, is defined by

$$L_n(x) = e^x \frac{d^n}{dx^n} (x^n e^{-x}), \qquad 0 \le x < \infty.$$

The first five Laguerre polynomials are therefore

$$L_0(x) = 1,$$
$$L_1(x) = 1 - x,$$
$$L_2(x) = 2 - 4x + x^2,$$
$$L_3(x) = 6 - 18x + 9x^2 - x^3,$$
$$L_4(x) = 24 - 96x + 72x^2 - 16x^3 + x^4,$$

and in general,

$$L_n(x) = \sum_{r=0}^{n} (-1)^r \binom{n}{r}^2 (n-r)! \, x^r,$$

where $\binom{n}{r}$ is the binomial coefficient $\dfrac{n!}{r!\,(n-r)!}$ [see I, § 3.8].

Laguerre polynomials arise in quantum mechanics in the solution of Schrödinger's equation for the energy states of the hydrogen atom (Schiff, 1968) [see also III, Example 6.3.8]. The differential equation satisfied by $L_n(x)$ is *Laguerre's differential equation*

$$x \frac{d^2 y}{dx^2} + (1-x) \frac{dy}{dx} + ny = 0, \qquad 0 < x < \infty.$$

The appropriate weight function $w(x)$ in the orthogonality relation (10.5.1) may be found by expressing Laguerre's differential equation in the form of the Sturm–Liouville equation (10.5.4), giving

$$\frac{d}{dx}\left(x e^{-x} \frac{dy}{dx} \right) + n e^{-x} y = 0.$$

Hence

$$p(x) = x e^{-x}, \qquad q(x) = 0,$$

and we can take

$$w(x) = e^{-x}, \qquad \lambda = n.$$

The orthogonality relation (10.5.1) becomes

$$\int_0^\infty e^{-x}L_m(x)L_n(x)\,dx = 0, \qquad m \neq n.$$

When $m = n$, we use the fact that the coefficient of x^n in $L_n(x)$ is $(-1)^n$ to obtain

$$\int_0^\infty e^{-x}[L_n(x)]^2\,dx = (n!)^2, \qquad n = 0, 1, 2, \ldots,$$

and thus $h_n = (n!)^2$ in equation (10.5.2). It follows that the functions

$$\phi_n(x) = \frac{e^{-x/2}}{n!}L_n(x), \qquad n = 0, 1, 2, \ldots,$$

form an orthonormal system in the interval $0 \leq x < \infty$.

The following recurrence relations hold for Laguerre polynomials:

$$L_{n+1}(x) = (2n + 1 - x)L_n(x) - n^2L_{n-1}(x), \qquad\qquad \text{I}$$

$$L_n'(x) = nL_{n-1}'(x) - nL_{n-1}(x). \qquad\qquad\qquad \text{II}$$

10.5.2. Hermite Polynomials and Functions

The Hermite polynomial $H_n(x)$, of degree n, is defined for all real values of x by

$$H_n(x) = (-1)^n e^{x^2}\frac{d^n}{dx^n}(e^{-x^2}).$$

The first five Hermite polynomials are therefore

$$H_0(x) = 1,$$
$$H_1(x) = 2x,$$
$$H_2(x) = 4x^2 - 2,$$
$$H_3(x) = 8x^3 - 12x,$$
$$H_4(x) = 16x^4 - 48x^2 + 12,$$

and in general,

$$H_n(x) = n!\sum_{r=0}^{[n/2]}\frac{(-1)^r(2x)^{n-2r}}{r!\,(n-2r)!}$$

where $[n/2]$ means $n/2$ if n is even and $(n-1)/2$ if n is odd.

The differential equation satisfied by $H_n(x)$ is *Hermite's differential equation*

$$\frac{d^2y}{dx^2} - 2x\frac{dy}{dx} + 2ny = 0, \qquad -\infty < x < \infty. \qquad (10.5.5)$$

It can be shown that the only solutions of this equation which tend to infinity more slowly than $e^{x^2/2}$ as $|x| \to \infty$ are of the form $cH_n(x)$, where c is a constant. This property of $H_n(x)$ is used in the theory of the quantum oscillator (Schiff, 1968), analogous to the simple harmonic oscillator of classical mechanics.

Following the same procedure as for Laguerre's differential equation, we write equation (10.5.5) in the Sturm–Liouville form (10.5.4) giving

$$\frac{d}{dx}\left(e^{-x^2}\frac{dy}{dx}\right) + 2n\,e^{-x^2}y = 0.$$

Hence we obtain

$$p(x) = e^{-x^2}, \qquad q(x) = 0, \qquad w(x) = e^{-x^2}, \qquad \lambda = 2n,$$

and the orthogonality relation (10.5.1) becomes

$$\int_{-\infty}^{\infty} e^{-x^2} H_m(x) H_n(x)\,dx = 0, \qquad m \neq n. \tag{10.5.6}$$

When $m = n$, it can be shown that

$$\int_{-\infty}^{\infty} e^{-x^2}[H_n(x)]^2\,dx = 2^n n!\,\sqrt{\pi}, \tag{10.5.7}$$

and hence

$$h_n = 2^n n!\,\sqrt{\pi}$$

in equation (10.5.2). It follows that the functions

$$\psi_n(x) = \frac{e^{-x^2/2}}{(2^n n!\,\sqrt{\pi})^{1/2}}\,H_n(x), \qquad n = 0, 1, 2, \ldots, \tag{10.5.8}$$

form an orthonormal system in the interval $-\infty < x < \infty$.

The following recurrence relations hold for Hermite polynomials:

$$H_{n+1}(x) = 2xH_n(x) - 2nH_{n-1}(x), \qquad\qquad \text{I}$$

$$H'_n(x) = 2nH_{n-1}(x). \qquad\qquad \text{II}$$

Omitting the normalizing factor in the function $\psi_n(x)$ of equation (10.5.8), we obtain

$$\Psi_n(x) = e^{-x^2/2} H_n(x), \qquad -\infty < x < \infty,$$

which is called the *Hermite function of order n*. Differentiating twice and using Hermite's differential equation (10.5.5), we find that $\Psi_n(x)$ satisfies the differential equation

$$\frac{d^2y}{dx^2} + (1 + 2n - x^2)y = 0, \qquad -\infty < x < \infty.$$

Also, from equations (10.5.6) and (10.5.7), $\Psi_n(x)$ satisfies the orthogonality relations

$$\int_{-\infty}^{\infty} \Psi_m(x)\Psi_n(x)\,dx = 0, \qquad m \neq n,$$

$$\int_{-\infty}^{\infty} [\Psi_n(x)]^2\,dx = 2^n n!\,\sqrt{\pi}.$$

The recurrence relations for $\Psi_n(x)$, corresponding to I and II above for $H_n(x)$, are:

$$\Psi_{n+1}(x) = 2x\,\Psi_n(x) - 2n\,\Psi_{n-1}(x), \qquad\qquad \text{III}$$

$$\Psi'_n(x) = 2n\,\Psi_{n-1}(x) - x\,\Psi_n(x). \qquad\qquad \text{IV}$$

10.5.3.   Chebyshev Polynomials

Using de Moivre's theorem [see (9.5.9)], or otherwise, it is possible to expand $\cos n\theta$ as a polynomial of degree n in $\cos\theta$, and so we can write

$$\cos n\theta = \sum_{r=0}^{n} c_r \cos^r\theta. \qquad (10.5.9)$$

The *Chebyshev polynomial* $T_n(x)$, of degree n, is simply the function $\cos n\theta$ expressed in terms of the variable

$$x = \cos\theta.$$

Thus, from equation (10.5.9)

$$T_n(x) = \sum_{r=0}^{n} c_r x^r, \qquad -1 \leq x \leq 1, \qquad (10.5.10)$$

and detailed calculation shows that if $n+r$ is *even* then [see I, § 3.8]

$$c_r = (-1)^{(n-r)/2} 2^{r-1}\left[2\binom{\frac{1}{2}(n+r)}{\frac{1}{2}(n-r)} - \binom{\frac{1}{2}(n+r)-1}{\frac{1}{2}(n-r)}\right],$$

while $c_r = 0$ if $n+r$ is odd. It follows that $T_n(x)$ is an even function of x (that is, $T_n(-x) = T_n(x)$) if n is even, and is an odd function of x $(T_n(-x) = -T_n(x))$ if n is odd. The first five Chebyshev polynomials are:

$$T_0(x) = 1,$$

$$T_1(x) = x,$$

$$T_2(x) = 2x^2 - 1,$$

$$T_3(x) = 4x^3 - 3x,$$

$$T_4(x) = 8x^4 - 8x^2 + 1.$$

Chebyshev polynomials are the most efficient polynomials for approximating arbitrary functions, and are therefore much used in numerical analysis

(Lanczos, 1952 and 1957, Fox and Mayers, 1968, Stark, 1970) [see also III, § 6.3]. They combine the convergence properties and generality of Fourier series (to which they are closely related) with the analytic simplicity of polynomials.

In order to investigate the orthogonality properties of $T_n(x)$, we note that

$$T_n(x) = \cos (n \cos^{-1} x)$$

satisfies the differential equation

$$(1-x^2)\frac{d^2 y}{dx^2} - x\frac{dy}{dx} + n^2 y = 0, \qquad -1 < x < 1.$$

In Sturm–Liouville form (10.5.4), this becomes [cf. Example 7.8.1]

$$\frac{d}{dx}\left[(1-x^2)^{1/2}\frac{dy}{dx}\right] + \frac{n^2}{(1-x^2)^{1/2}} y = 0.$$

Hence we obtain

$$p(x) = (1-x^2)^{1/2}, \qquad q(x) = 0, \qquad w(x) = (1-x^2)^{-1/2}, \qquad \lambda = n^2.$$

The orthogonality relation (10.5.1) becomes

$$\int_{-1}^{1} \frac{T_m(x)T_n(x)}{(1-x^2)^{1/2}}\, dx = 0, \qquad m \neq n. \tag{10.5.11}$$

When $m = n$,

$$\int_{-1}^{1} \frac{[T_n(x)]^2}{(1-x^2)^{1/2}}\, dx = \pi, \qquad n = 0,$$

$$= \pi/2, \quad n \neq 0.$$

The substitution $x = \cos \theta$ may be used to evaluate integrals involving $T_n(x)$ [see § 4.3].

The same substitution applied to the identity [see V (1.2.19)]

$$\cos (n+1)\theta + \cos (n-1)\theta = 2 \cos \theta \cos n\theta$$

gives immediately the recurrence relation

$$T_{n+1}(x) = 2xT_n(x) - T_{n-1}(x), \qquad n \geq 1. \tag{10.5.12}$$

More generally, the identity

$$\cos (n+m)\theta + \cos (n-m)\theta = 2 \cos n\theta \cos m\theta$$

leads to

$$T_n(x)T_m(x) = \tfrac{1}{2}[T_{n+m}(x) + T_{n-m}(x)], \qquad n \geq m. \tag{10.5.13}$$

The *m*th moment about the origin of $T_n(x)$ is found from the identity [see V,

§ 1.2.3]

$$\cos^m \theta \cos n\theta = 2^{-m} \sum_{r=0}^{m} \binom{m}{r} \cos (n - m + 2r)\theta;$$

this gives

$$x^m T_n(x) = 2^{-m} \sum_{r=0}^{m} \binom{m}{r} T_{|n-m+2r|}(x). \qquad (10.5.14)$$

A formula for the indefinite integral of $T_n(x)$ for $n \geq 2$ may be found from the identity [see V (1.2.18)]

$$\cos n\theta \sin \theta = \tfrac{1}{2}[\sin (n + 1)\theta - \sin (n - 1)\theta].$$

Integrating with respect to θ and then setting $x = \cos \theta$, we find

$$\int T_n(x) \, dx = \frac{1}{2}\left[\frac{T_{n+1}(x)}{n+1} - \frac{T_{n-1}(x)}{n-1}\right], \qquad n \geq 2. \qquad (10.5.15)$$

For completeness, we also have the indefinite integrals

$$\left.\begin{aligned}\int T_0(x) \, dx &= T_1(x), \\[2mm] \int T_1(x) \, dx &= \tfrac{1}{4}T_2(x).\end{aligned}\right\} \qquad (10.5.16)$$

Using the same identity, we find

$$\left.\begin{aligned}\int_{-1}^{1} T_n(x) \, dx &= -\frac{2}{n^2 - 1}, \quad n \text{ even (including zero)}, \\[2mm] &= 0, \qquad n \text{ odd.}\end{aligned}\right\} \qquad (10.5.17)$$

The second part of this result also follows from the fact that $T_n(x)$ is an odd function of x when n is odd.

Some outstanding properties of the Chebyshev polynomial $T_n(x)$ are [cf. III, § 6.3.4]:

(i) $|T_n(x)| \leq 1$, $-1 \leq x \leq 1$.
(ii) Of all polynomials of degree n with leading coefficient 2^{n-1}, $T_n(x)$ is the only one with property (i).
(iii) $T_n(x)$ has alternate maximum and minimum values of ± 1 at the $n + 1$ points given by

$$\theta_r = \frac{r\pi}{n}, \qquad r = 0, 1, 2, \ldots, n,$$

i.e. at the points

$$x_r = \cos \frac{r\pi}{n}, \qquad r = 0, 1, 2, \ldots, n.$$

(These include the end-points $x = \pm 1$.)

In many applications it is convenient to use the range $0 \leq x \leq 1$ instead of the range $-1 \leq x \leq 1$. The *shifted Chebyshev polynomial* $T_n^*(x)$, of degree n, is again the function $\cos n\theta$, but expressed now in terms of the variable

$$x = \frac{1+\cos\theta}{2} = \cos^2\frac{\theta}{2}.$$

Thus

$$T_n^*(x) = T_n(\cos\theta) = T_n(2x - 1). \tag{10.5.18}$$

If we write $T_n^*(x)$ in the form of equation (10.5.10), namely,

$$T_n^*(x) = \sum_{r=0}^{n} c_r^* x^r,$$

we find that the coefficients c_r^* are given by

$$c_r^* = (-1)^{n+r} 2^{2r-1}\left[2\binom{n+r}{n-r} - \binom{n+r-1}{n-r} \right].$$

The first five shifted Chebyshev polynomials are:

$$T_0^*(x) = 1,$$
$$T_1^*(x) = 2x - 1,$$
$$T_2^*(x) = 8x^2 - 8x + 1,$$
$$T_3^*(x) = 32x^3 - 48x^2 + 18x - 1,$$
$$T_4^*(x) = 128x^4 - 256x^3 + 160x^2 - 32x + 1.$$

The differential equation satisfied by

$$T_n^*(x) = \cos\left[n\,\cos^{-1}(2x - 1)\right]$$

is

$$2x(1-x)\frac{d^2y}{dx^2} - (2x - 1)\frac{dy}{dx} + 2n^2 y = 0, \qquad 0 < x < 1.$$

Expressing this equation in Sturm–Liouville form, as before, or, alternatively, replacing x by $2x - 1$ in equation (10.5.11) we find that the orthogonality relation for $T_n^*(x)$ is

$$\int_0^1 \frac{T_m^*(x) T_n^*(x)}{[x(1-x)]^{1/2}}\, dx = 0, \qquad m \neq n.$$

Also, by making the substitution

$$x = \frac{1+\cos\theta}{2},$$

we find

$$\int_0^1 \frac{[T_n^*(x)]^2}{[x(1-x)]^{1/2}} dx = \pi, \qquad n = 0,$$

$$= \pi/2, \qquad n \neq 0.$$

The recurrence relation for $T_n^*(x)$ is obtained from equation (10.5.12). Replacing x by $2x - 1$ in this equation and using equation (10.5.18), we obtain

$$T_{n+1}^*(x) = 2(2x - 1) T_n^*(x) - T_{n-1}^*(x), \qquad n \geq 1.$$

The algebraic and integral relations for $T_n^*(x)$ corresponding to equations (10.5.13)–(10.5.17) for $T_n(x)$ are found to be, respectively,

$$T_n^*(x) T_m^*(x) = \tfrac{1}{2}[T_{n+m}^*(x) + T_{n-m}^*(x)], \qquad n \geq m, \qquad (10.5.19)$$

$$x^m T_n^*(x) = 2^{-2m} \sum_{r=0}^{2m} \binom{2m}{r} T_{|n-m+r|}^*(x), \qquad (10.5.20)$$

$$\int T_n^*(x) \, dx = \frac{1}{4}\left[\frac{T_{n+1}^*(x)}{n+1} - \frac{T_{n-1}^*(x)}{n-1}\right], \qquad n \geq 2, \qquad (10.5.21)$$

$$\left. \begin{array}{l} \displaystyle \int T_0^*(x) \, dx = \tfrac{1}{2} T_1^*(x), \\[2mm] \displaystyle \int T_1^*(x) \, dx = \tfrac{1}{8} T_2^*(x). \end{array} \right\} \qquad (10.5.22)$$

$$\left. \begin{array}{ll} \displaystyle \int_0^1 T_n^*(x) \, dx = -\frac{1}{n^2 - 1}, & n \text{ even (including zero),} \\[2mm] \qquad\qquad = 0, & n \text{ odd.} \end{array} \right\} \qquad (10.5.23)$$

Equations (10.5.19)–(10.5.23) may be verified by making the substitutions

$$T_n^*(x) = \cos n\theta, \qquad 2x - 1 = \cos \theta.$$

The properties of $T_n^*(x)$, $0 \leq x \leq 1$, are closely similar to those of $T_n(x)$, $-1 \leq x \leq 1$. In particular, (i)–(iii) above become

(i)* $|T_n^*(x)| \leq 1, 0 \leq x \leq 1$.

(ii)* Of all polynomials of degree n with leading coefficient 2^{2n-1}, $T_n^*(x)$ is the only one with property (i)*.

(iii)* $T_n^*(x)$ has alternate maximum and minimum values of ± 1 at the $n+1$ points given by

$$\theta_r = \frac{r\pi}{n}, \qquad r = 0, 1, 2, \ldots, n,$$

i.e. at the points

$$x_r = \frac{1}{2}\left(1 + \cos \frac{r\pi}{n}\right), \qquad r = 0, 1, 2, \ldots, n.$$

(These include the end-points $x = 0, 1$.)

REFERENCES

Bell, E. T. (1945). *Development of Mathematics*, 2nd edn., McGraw-Hill, New York.

Bland, D. R. (1961). *Solutions of Laplace's Equation*, Routledge and Kegan Paul, London.

Bowman, F. (1953). *Introduction to Elliptic Functions*, English Universities, London.

Bowman, F. (1958). *Introduction to Bessel Functions*, Dover, New York.

Copson, E. T. (1935). *An Introduction to the Theory of Functions of a Complex Variable*, Oxford University Press, London.

Davis, H. T. (1962). *Introduction to Nonlinear Differential and Integral Equations*, Dover, New York.

Davis, P. J. (1959). Leonhard Euler's Integral: A Historical Profile of the Gamma Function, *Amer. Math. Monthly* **66**, 849–869.

Dennery, P. and Krzywicki, A. (1967). *Mathematics for Physicists*, Harper and Row, New York.

Fox, L. and Mayers, D. F. (1968). *Computing Methods for Scientists and Engineers*, Oxford University Press, London.

Greenhill, A. G. (1959). *The Applications of Elliptic Functions*, Dover, New York.

Hildebrand, F. B. (1948). *Advanced Calculus for Applications*, Prentice-Hall, Englewood Cliffs, New Jersey.

Kreyszig, E. (1962). *Advanced Engineering Mathematics*, Wiley, New York.

Lanczos, C. (1957). *Applied Analysis*, Pitman, London.

McLachlan, N. W. (1955). *Bessel Functions for Engineers*, 2nd edn., Oxford University Press, London.

Neville, E. H. (1971). *Elliptic Functions: A Primer* (W. J. Langford, Editor), Pergamon, Oxford.

Schiff, L. I. (1968). *Quantum Mechanics*, 3rd edn., McGraw-Hill Kogakusha, Tokyo.

Sneddon, I. N. (1956). *Special Functions of Mathematical Physics and Chemistry*, Oliver and Boyd, Edinburgh.

Stark, P. A. (1970). *Introduction to Numerical Methods*, Collier-Macmillan, London.

Szegö, G. (1978). *Orthogonal Polynomials*, 4th edn., American Mathematical Society, Providence, R.I.

Tranter, C. J. (1968). *Bessel Functions with some Physical Applications*, English Universities, London.

Watson, G. N. (1944). *A Treatise on the Theory of Bessel Functions*, 2nd edn., Cambridge University Press, Cambridge.

Whittaker, E. T. and Watson, G. N. (1950). *Modern Analysis*, 4th edn., Cambridge University Press, Cambridge.

Theory

Hildebrand, F. B. (1948). *Advanced Calculus for Applications*, Prentice-Hall, New Jersey.

Sneddon, I. N. (1956). *Special Functions of Mathematical Physics and Chemistry*, Oliver and Boyd, Edinburgh.

Tables

Abramowitz, M. and Stegun, I. A. (Editors) (1968). *Handbook of Mathematical Functions with Formulas, Graphs and Mathematical Tables*, Dover, New York.

Comrie, L. J. (1963). *Chambers's Six-Figure Mathematical Tables*, Vol. II, W & R Chambers, Edinburgh.

Jahnke, E. and Emde, F. (1945). *Tables of Functions with Formulae and Curves*, Dover, New York.

Lanczos, C. (1952). *Tables of Chebyshev Polynomials, $S_n(x)$ and $C_n(x)$*, Nat. Bur. Standards, Appl. Math. Ser. No. 9, U.S. Government Printing Office, Washington, D.C.

Milne-Thomson, L. M. (1950). *Jacobian Elliptic Function Tables*, Dover, New York.

Spenceley, G. W. and Spenceley, R. M. (1947). *Smithsonian Elliptic Functions Tables*, Smithsonian Institution, Washington, D.C.

APPENDIX A

Change of Argument

The following table is useful in evaluating elliptic functions whose arguments fall outside the interval $[0, K]$.

Argument	sn	cn	dn
u	$\operatorname{sn} u$	$\operatorname{cn} u$	$\operatorname{dn} u$
$-u$	$-\operatorname{sn} u$	$\operatorname{cn} u$	$\operatorname{dn} u$
$u+K$	$\operatorname{cd} u$	$-k'\operatorname{sd} u$	$k'\operatorname{nd} u$
$u-K$	$-\operatorname{cd} u$	$k'\operatorname{sd} u$	$k'\operatorname{nd} u$
$K-u$	$\operatorname{cd} u$	$k'\operatorname{sd} u$	$k'\operatorname{nd} u$
$u+2K$	$-\operatorname{sn} u$	$-\operatorname{cn} u$	$\operatorname{dn} u$
$u-2K$	$-\operatorname{sn} u$	$-\operatorname{cn} u$	$\operatorname{dn} u$
$2K-u$	$\operatorname{sn} u$	$-\operatorname{cn} u$	$\operatorname{dn} u$
$u+iK'$	$k^{-1}\operatorname{ns} u$	$-ik^{-1}\operatorname{ds} u$	$-i\operatorname{cs} u$
$u+2iK'$	$\operatorname{sn} u$	$-\operatorname{cn} u$	$-\operatorname{dn} u$
$u+K+iK'$	$k^{-1}\operatorname{dc} u$	$-ik'k^{-1}\operatorname{nc} u$	$ik'\operatorname{sc} u$
$u+2K+2iK'$	$-\operatorname{sn} u$	$\operatorname{cn} u$	$-\operatorname{dn} u$
$u+4K$	$\operatorname{sn} u$	$\operatorname{cn} u$	$\operatorname{dn} u$
$u+4iK'$	$\operatorname{sn} u$	$\operatorname{cu} u$	$\operatorname{dn} u$

APPENDIX B

Change of Modulus

(i) *Reciprocal Modulus*

$$\operatorname{sn}(u, k) = k^{-1}\operatorname{sn}(ku, k^{-1})$$
$$\operatorname{cn}(u, k) = \operatorname{dn}(ku, k^{-1})$$
$$\operatorname{dn}(u, k) = \operatorname{cn}(ku, k^{-1}).$$

(ii) *Imaginary Modulus*

Let

$$k_1 = k/(1+k^2)^{1/2}.$$

Then

$$\operatorname{sn}(u, ik) = (1+k^2)^{-1/2}\operatorname{sd}[(1+k^2)^{1/2}u, k_1],$$
$$\operatorname{cn}(u, ik) = \operatorname{cd}[(1+k^2)^{1/2}u, k_1],$$
$$\operatorname{dn}(u, ik) = \operatorname{nd}[(1+k^2)^{1/2}u, k_1].$$

(iii) *Landen's Transformation to a Larger Modulus*
Let

$$k_1 = \frac{1-k'}{1+k'}.$$

Then

$$k_1 = \frac{1-k'^2}{(1+k')^2} = \frac{k}{(1+k')^2} \, k < k,$$

and

$$\text{sn}\,[(1+k')u, \, k_1] = (1+k') \, \text{sn}\,(u, \, k) \, \text{cd}\,(u, \, k),$$

$$\text{cn}\,[(1+k')u, \, k_1] = [1-(1+k') \, \text{sn}^2\,(u, \, k)] \, \text{nd}\,(u, \, k),$$

$$\text{dn}\,[(1+k')u, \, k_1] = [k'+(1-k') \, \text{cn}^2\,(u, \, k)] \, \text{nd}\,(u, \, k).$$

(iv) *Landen's Transformation to a Smaller Modulus*
Let

$$k_1 = \frac{2k^{1/2}}{1+k}.$$

Then

$$k_1 = \frac{2}{1+k} \, k^{1/2} > k^{1/2} > k,$$

and

$$\text{sn}\,[(1+k_1)u, \, k_1] = \frac{(1+k)\,\text{sn}\,(u, \, k)}{1+k\,\text{sn}^2\,(u, \, k)},$$

$$\text{cn}\,[(1+k_1)u, \, k_1] = \frac{\text{cn}\,(u, \, k)\,\text{dn}\,(u, \, k)}{1+k\,\text{sn}^2\,(u, \, k)},$$

$$\text{dn}\,[(1+k_1)u, \, k_1] = \frac{1-k\,\text{sn}^2\,(u, \, k)}{1+k\,\text{sn}^2\,(u, \, k)}.$$

APPENDIX C

Interpolation Formulae

The numerical examples in § 10.1.7 were performed with the help of Milne-Thomson's tables (1950) and a pocket calculator. The tables give the values of sn u, cn u and dn u to five places of decimals, and of K, K', E, E' and $Z(u)$ to seven places. In each case, the argument u increases in steps of 0·01 and the parameter $m = k^2$ in steps of 0·1. Values of $q = \exp(-\pi K'/K)$ and $q_1 = \exp(-\pi K/K')$ are also given, to eight places of decimals.

In using these and similar tables, interpolation formulae which use only first and second differences are recommended. For greater accuracy the fractional step length, x or y, is taken to be not greater than $0\cdot5$. This leads to the following formulae using forward differences [see III, § 1.4].

Direct Interpolation
   (i) Given

$$f(a), \Delta f(a), \Delta^2 f(a), \qquad x \leq 0\cdot5,$$

find

$$f(a + xh),$$

where

$$\Delta f(a) = f(a + h) - f(a), \quad \text{etc.}$$

We use Newton's forward formula:

$$f(a + xh) = f(a) + x\Delta f(a) + \frac{x(x-1)}{2}\Delta^2 f(a), \tag{A1}$$

ignoring third and higher-order differences. Put

$$\Delta' = \Delta f(a), \qquad \Delta'' = \Delta^2 f(a).$$

Then

$$f(a + xh) = f(a) + x\left[\Delta' - \left(\frac{1-x}{2}\right)\Delta''\right]. \tag{A2}$$

   (ii) Given

$$f(a), \Delta f(a), \Delta^2 f(a) \qquad\qquad x > 0\cdot5,$$

find

$$f(a + xh).$$

In equation (A1), put

$$y = 1 - x.$$

Then

$$f(a + xh) = f(a + h - yh) = f(a + h) - y\left[\Delta' + \left(\frac{1-y}{2}\right)\Delta''\right]. \tag{A3}$$

Inverse Interpolation
   (iii) Given

$$f(a), \Delta f(a), \Delta^2 f(a), f(a + xh),$$

find

$$a + xh, \qquad\qquad x \leq 0\cdot 5.$$

From equation (A2), a first approximation to x is

$$x_1 = \frac{f(a+xh)-f(a)}{\Delta'} = \frac{b}{\Delta'} \quad \text{say,}$$

and a second approximation is

$$x_2 = \frac{b}{\Delta' - \left(\dfrac{1-x_1}{2}\right)\Delta''}.$$

(iv) Given

$$f(a), \Delta f(a), \Delta^2 f(a), f(a+xh),$$

find

$$a + xh, \qquad\qquad x > 0\cdot 5.$$

From equation (A3), a first approximation to $y = 1 - x$ is

$$y_1 = \frac{f(a+h)-f(a+xh)}{\Delta'} = \frac{c}{\Delta'}, \quad \text{say,}$$

and a second approximation is

$$y_2 = \frac{c}{\Delta' + \left(\dfrac{1-y_1}{2}\right)\Delta''}.$$

Then $x_2 = 1 - y_2$ is the required approximation to x.

G.R.W.

CHAPTER 11

Metric Spaces

The theory of metric spaces is part of a continuing process of generalization in analysis. Once notions of continuity and limits were rigorously defined in the context of real-valued functions [see § 2.1 and § 2.3], their extension to other, similar areas was relatively straightforward: functions of several real variables [see § 5.3], complex functions [see § 9.1]. In the course of this development it was noticed that certain spaces of functions could be approached using such concepts: it is possible to speak of a continuously varying family of functions (as opposed to a single function varying continuously as its variable changes), or the limit of a sequence of functions [see § 1.12]. It was even possible to perform operations of calculus in such function spaces.

Such extensions have two purposes. One is to obtain new information in uncharted areas; the other to clarify and codify unwieldy masses of existing information. One very useful concept for clarifying analysis is that of a *metric space*: roughly, a set which possesses a sensible notion of 'distance'. Continuity is a very natural property in this setting, although even greater generality of the continuity concept can be achieved if desired (such as the theory of topological spaces [see V, Chapter 5]); and many of the basic properties of continuity can be developed easily in the context of metric spaces.

In this chapter we shall define a metric space and give several examples; discuss continuity, limits, and related concepts; and derive an important result—the *contraction mapping theorem*—with many applications in analysis (one of them is to prove the existence of solutions to systems of linear differential equations). We shall also discuss two more specialized types of metric space, which have in addition a linear structure: *Banach* and *Hilbert Spaces*. Hilbert Space has proved important in quantum mechanics, where observables cease to be represented by numbers, but instead are represented by operators on a Hilbert space. We shall not discuss such applications in any detail.

11.1. METRIC SPACES: DEFINITION AND EXAMPLES

DEFINITION 11.1.1. A *metric space* is a set X together with a *distance function* (or *metric*) $d: X \times X \to \mathbb{R}$ (that is, $d(x, y)$ is defined as a real number [see I, § 2.6.1] for all $x, y \in X$) which satisfies the following axioms for all $x, y, z \in X$:

(M1) $d(x, y) \geq 0$,   and $d(x, y) = 0$ if and only if $x = y$.

(M2) $d(x, y) = d(y, x)$.

(M3) $d(x, y) \leq d(x, z) + d(y, z)$.

Of these, (M2) is the *symmetry* property, and (M3) is called the *triangle inequality* since, in spirit, it expresses the fact that any side of a triangle is shorter than the sum of the remaining two sides (Figure 11.1.1). The variety of possible metric spaces, and their widespread occurrence 'in nature', is only hinted at by the following list of examples.

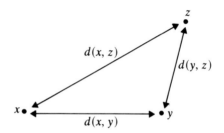

Figure 11.1.1

EXAMPLE 11.1.1.   Let $X = \mathbb{R}^n$, n-dimensional Euclidean space [see I, Example 5.2.2], and define d by the pythagorean formula

$$d(\mathbf{x}, \mathbf{y}) = [(x_1 - y_1)^2 + \ldots + (x_n - y_n)^2]^{1/2}$$

where $\mathbf{x} = (x_1, \ldots, x_n)$ and $\mathbf{y} = (y_1, \ldots, y_n)$. Axioms (M1) and (M2) are obvious; (M3) has a direct geometric interpretation in terms of a triangle (which makes it plausible) and can be proved algebraically with some care. [See I, Proposition 10.1.3 and IV, § 21.2.4.]

EXAMPLE 11.1.2.   Let X be any set whatsoever, and define d by

$$d(x, y) = \begin{cases} 0 & \text{if } x = y \\ 1 & \text{if } x \neq y. \end{cases}$$

The axioms hold trivially. We call d the *discrete* metric on X. The significance of this example is merely that *any* set can be given *some* distance function.

EXAMPLE 11.1.3.   Let $X = \mathbb{C}$, the complex plane [see I, § 2.7.2], and define

$$d(z, w) = |z - w| \qquad (z, w \in \mathbb{C}).$$

EXAMPLE 11.1.4.   Let $X = \mathbb{C}^n$, n-dimensional complex space [see I, Example 5.2.2], and define

$$d(\mathbf{z}, \mathbf{w}) = |z_1 - w_1| + \ldots + |z_n - w_n|$$

where $\mathbf{w} = (w_1, \ldots, w_n)$ and $\mathbf{z} = (z_1, \ldots, z_n)$.

EXAMPLE 11.1.5.   Let X be the set of all continuous real-valued functions defined on the unit interval $[0, 1]$ [see I (2.6.10)]; and put

$$d(f, g) = \int_0^1 |f(x) - g(x)|\, dx$$

(illustrated in Figure 11.1.2).

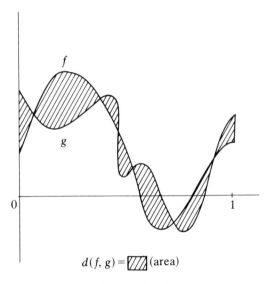

$$d(f, g) = \boxed{/\!/\!/}\,(\text{area})$$

Figure 11.1.2

EXAMPLE 11.1.6.   Let X be as in Example 11.1.5, but define a different metric d' by

$$d'(f, g) = \sup_{x \in [0,1]} |f(x) - g(x)|$$

[see I, § 2.6.3]. (Illustrated in Figure 11.1.3.)

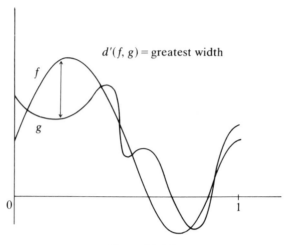

$d'(f, g) = $ greatest width

Figure 11.1.3

EXAMPLE 11.1.7.   Let X be the set of all real $n \times n$ matrices (a_{ij}) [see I, § 6.2] and define

$$d((a_{ij}), (b_{ij})) = \max_{i,j} |a_{ij} - b_{ij}|.$$

11.2.   OPEN AND CLOSED SUBSETS

The ideas of this section are preliminary to understanding continuity in a metric space setting. They are geometrically inspired, and we shall illustrate them by pictures in which the metric space X is a subset of the plane. The concepts themselves, of course, are more general.

DEFINITION 11.2.1.   Let X be a metric space, with distance function d. Let $x_0 \in X$ and let r be a real number greater than zero. We define the *open ball* radius r centre x_0 to be

$$B_r(x_0) = \{x \in X | d(x, x_0) < r\},$$

as in Figure 11.2.1, where it is the interior of a disc radius r centre x_0. Similarly there is the *closed ball*

$$B_r(x_0) = \{x \in X | d(x, x_0) \le r\},$$

which in this case would be the disc plus its boundary line.

Note that in less geometric examples these balls take on a different aspect: for instance in Example 11.1.6 above the ball around a function f consists of

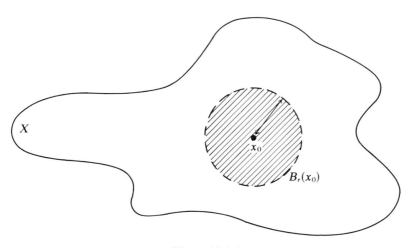

Figure 11.2.1

all functions g whose graph lies within a band about that of f, of width r either side (Figure 11.2.2).

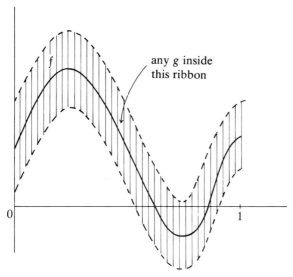

any g inside
this ribbon

Figure 11.2.2

DEFINITION 11.2.2.   A subset G of X [see I, § 1.2.3] is said to be *open* if, for every $x_0 \in G$, we can find a positive real number r such that the ball $B_r(x_0)$ is contained in G.

EXAMPLE 11.2.1.   The interior of the unit square in $\mathbb{R}^2$ is open. For here

$$G = \{(x, y) \in \mathbb{R}^2 \mid 0 < x < 1, 0 < y < 1\},$$

and for a given $(x_0, y_0) \in G$ we take the ball of radius

$$r = \tfrac{1}{2} \min (x_0, 1 - x_0, y_0, 1 - y_0).$$

Clearly $B_r(x_0, y_0)$ is contained in G (Figure 11.2.3).

Figure 11.2.3

EXAMPLE 11.2.2.   The full unit square in $\mathbb{R}^2$ is *not* open. Here the set is

$$G' = \{(x, y) \in \mathbb{R}^2 \mid 0 \le x \le 1, 0 \le y \le 1\}.$$

If we take, for example,

$$(x_0, y_0) = (1, 1)$$

it is clear that no ball $B_r(1, 1)$ lies inside G', because it will have to contain the point $(1 + r/2, 1 + r/2)$, say, lying outside G'.

Roughly speaking, an open set is one having no boundaries. We can make this notion precise by defining 'boundary'. But first we need a related idea: closed subsets.

DEFINITION 11.2.3.   A subset F of X is *closed* if and only if its complement $F^c = X \backslash F$ [see I, § 1.2.4] is open.

EXAMPLE 11.2.3.   The set $F = \{(x, y) \in \mathbb{R}^2 \mid x \le 0 \text{ or } x \ge 1 \text{ and } y \le 0 \text{ or } y \ge 1\}$ is closed.
   For its complement F^c is as in Example 11.2.1.

Sets may be neither open nor closed: for example

$$\{x \in \mathbb{R} \mid 0 \le x < 1\} \subseteq \mathbb{R}$$

is not open (no ball centre 0 lies inside it) but is not closed either (no ball centre 1 lies *out*side it).

We have the following general properties [see I, § 1.2.1]:
    The union of *any* set of open sets is open.
    The intersection of *finitely* many open sets is open.
    The union of *finitely* many closed sets is closed.
    The intersection of *any* set of closed sets is closed.
The restriction to finitely many sets, where made, is unavoidable: for example the intersection of all open intervals $(-1/n, 1/n) \subseteq \mathbb{R}$, for $n = 1, 2, 3, \ldots$ is the singleton $\{0\}$, which is not open.
    Further, both X and $\varnothing$ are both open and closed.

DEFINITION 11.2.4. Now let X be a metric space, and let A be an arbitrary subset of X. We say that $x_0 \in A$ is an *interior point* of A if $B_r(x_0) \subseteq A$ for some $r > 0$. The *interior* of A is defined to be

$$\text{Int}(A) = \{x_0 \in A \,|\, x_0 \text{ is an interior point of } A\}.$$

Equivalently, it is the largest open set contained in A, and equals the union of all open sets of X that are contained in A. Dually, the *closure* $\bar{A}$ of A is defined to be the smallest closed subset of X that contains A, which equals the intersection of all closed subsets of X that contain A.

We can specify the closure of A precisely by using the definition of closed sets in terms of open complements: the result is

$$\bar{A} = (\text{int}(A^c))^c.$$

This is a not very intuitive description, though easy to apply. We shall give another interpretation of $\bar{A}$ in terms of limits later on.

EXAMPLE 11.2.4. Let A be the subset of $X \subseteq \mathbb{R}^2$ illustrated in Figure 11.2.4a: precisely, it is

$$A = \{(x, y) \,|\, x^2 + y^2 \leq 1, \text{ and } x > 0 \text{ if } x^2 + y^2 = 1\} \cup \{(2, 2)\}$$
$$\cup \{(x, y) \,|\, (x+2)^2 + (y+3)^3 < \tfrac{3}{2} \text{ and } (x, y) \neq (-3, -3)\}.$$

Find $\text{Int}(A)$ and $\bar{A}$.
    By considering which points can be interior points, it is easy to see that

$$\text{Int}(A) = \{(x, y) \,|\, x^2 + y^2 < 1\}$$
$$\cup \{(x, y) \,|\, (x+2)^2 + (y+3)^2 < \tfrac{3}{2} \text{ and } (x, y) \neq (-3, -3)\}$$

as in Figure 11.2.4b.
    By a similar argument applied to the complement, it follows that

$$\bar{A} = \{(x, y) \,|\, x^2 + y^2 \leq 1\} \cup \{(2, 2)\} \cup \{(x, y) \,|\, (x+2)^2 + (y+3)^2 \leq \tfrac{3}{2}\},$$

shown in Figure 11.2.4c.

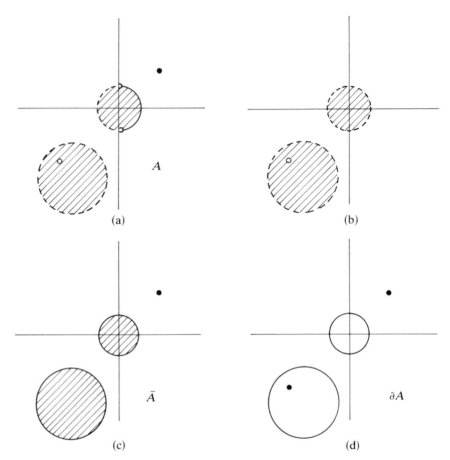

Figure 11.2.4

DEFINITION 11.2.5.   The *boundary* of a subset A of X is defined to be

$$\partial A = \bar{A}\backslash \text{Int}\,(A).$$

EXAMPLE 11.2.5.   The boundary of the set shown in Example 11.2.4 is

$$\partial A = \{(x,\,y)\,|\,x^2 + y^2 = 1\} \cup \{(2,\,2)\}$$
$$\cup \{(x,\,y)\,|\,(x+2)^2 + (y+3)^2 = \tfrac{3}{2}\} \cup \{(-3,\,-3)\}$$

as in Figure 11.2.4d.

There are many simple properties relating interiors, closures, and boundaries, and involving unions or intersections of sets. For examples, see Simmons (1963).

11.3.   SEQUENCES AND LIMITS

DEFINITION 11.3.1.   A point x in X is called a *limit point* of a subset A of X if every open ball $B_r(x)$, for $r > 0$, intersects A in at least one point other than x (Figure 11.3.1).

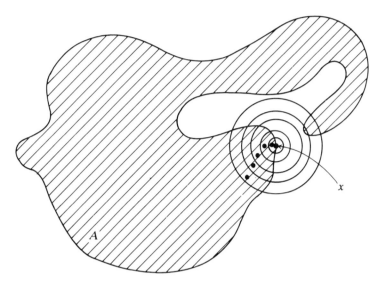

Figure 11.3.1

Roughly, we can find points in A as close as we wish to x. It is easy to see that F is a closed set if and only if it contains all of its limit points; and the closure $\bar{A}$ of A consists of A together with its limit points. This is easier to visualize than our previous definition, but technically harder to use.

The word 'limit' is also used in a slightly different sense. Consider a *sequence* (x_n) of points $x_n \in X$.

DEFINITION 11.3.2.   We say (mimicking the Definition 1.2.1 for real numbers) that the sequence *tends to a limit x* as n tends to infinity if, for all $\varepsilon > 0$, we can find N such that

$$d(x_n, x) < \varepsilon \quad \text{for all } n > N.$$

We write

$$\lim (x_n) = x$$

or

$$x_n \to x.$$

Every sequence that tends to a limit is called *convergent*; and every convergent sequence has the following property:

Given $\varepsilon > 0$ there exists N such that for all $m, n > N$

$$d(x_m, x_n) < \varepsilon. \tag{11.3.1}$$

Any sequence that satisfies (11.3.1) is called a *Cauchy sequence*.

In some metric spaces X, Cauchy sequences need not converge. For example, let $X = \mathbb{Q}$, the rationals [see I, § 2.4], and define

$$x_n = \sqrt{2} \text{ to } n \text{ decimal places.}$$

Then (x_n) is a Cauchy sequence but does not converge *in X*.

DEFINITION 11.3.3.   If every Cauchy sequence converges in X, we say that X is *complete*.

Examples of complete metric spaces include $\mathbb{R}$ and $\mathbb{C}$ with their usual metrics, and Example 11.1.6 above (whose completeness expresses the fact that a *uniformly* convergent sequence of continuous functions has a continuous limit [see § 1.12]). Completeness is an extremely useful property.

A non-complete metric space can always be embedded in a 'tight-fitting' complete one, in an essentially unique way: [see Simmons (1963), p. 84].

11.4. CONTINUITY

DEFINITION 11.4.1.   Once more mimicking the real case [cf. Definition 2.1.1], we define a function $f: X \to Y$, where X and Y are metric spaces with metrics d, e respectively, to be *continuous at* $x_0 \in X$ if for all $\varepsilon > 0$ there exists $\delta > 0$ such that $d(x, x_0) < \delta$ implies $e(f(x), f(x_0)) < \varepsilon$. And f is *continuous* if it is continuous for all $x_0 \in X$.

EXAMPLE 11.4.1.   Let $X = Y$ be as in Example 11.1.6, and define $\theta: X \to X$ by $\theta(f) = f^2$. (That is, $(\theta(f))(x) = f(x)^2$ for all $x \in X$.) Show that θ is continuous.

We have to show that if f and g are two continuous functions on $[0, 1]$ such that

$$d(f, g) = \sup_x |f(x) - g(x)| < \delta$$

then

$$d(\theta(f), \theta(g)) = \sup_x |f(x)^2 - g(x)^2| < \varepsilon$$

provided we choose δ correctly.

Now

$$f(x)^2 - g(x)^2 = (f(x) - g(x))(f(x) + g(x))$$

so we can solve the problem if we can show that, *given f* as a fixed choice, the size of $f(x) + g(x)$ is not too big. Now we know that f, being a continuous function on a closed interval, is bounded:

$$|f(x)| < M$$

for all $x \in X$, with some choice of $M \in \mathbb{R}$ [see § 2.8]. Now $|f(x) - g(x)| < \delta$ implies $|f(x) + g(x)| < 2M + \delta$. So we are finished provided we can choose δ to make $\delta(2M + \delta) < \varepsilon$. But this is easy: take

$$\delta = \min(1, \varepsilon/2M + 1).$$

Then, spelling out the argument in detail; if $d(f, g) < \delta$ then we have

$$d(\theta(f), \theta(g)) = \sup_x |f(x)^2 - g(x)^2|$$

$$= \sup_x |f(x) - g(x)| \, |f(x) + g(x)|$$

$$\leq \sup_x \delta(2M + \delta)$$

$$< \sup_x \varepsilon$$

$$= \varepsilon.$$

Therefore θ is continuous.

We give this example more to show that the definition of continuity is manageable than because the technique has specific practical significance.

We can recast the definition of continuity in a much simpler form:

DEFINITION 11.4.2. A function $f: X \to Y$ is continuous if and only if $f^{-1}(G)$ is open in X for every open set G in Y.

That is, f is continuous if the *inverse image* of an open set is always open.

There is an identical characterization with 'open' replaced by 'closed', since $f^{-1}(G^c) = (f^{-1}(G))^c$.

This characterization is elegant, easy to use in theoretical work, and leads directly to the notion of a *topological space* [Simmons (1963), p. 92; Chapter 5 of Volume V]. For example, to show that a composite of two continuous functions f and g is continuous, we merely note that $(fg)^{-1}(\text{open}) = g^{-1}(f^{-1}(\text{open})) = g^{-1}(\text{open}) = \text{open}$. (We abuse notation to convey the idea.)

DEFINITION 11.4.3. A function $f: X \to Y$ such that $e(f(x), f(y)) = d(x, y)$ for all $x, y \in X$ is called an *isometry*—it preserves distances.

It follows trivially that every isometry is continuous, although the converse is of course not true [see V, § 11.2].

11.5. THE CONTRACTION MAPPING THEOREM

DEFINITION 11.5.1. A function $f: X \to Y$ is a *contraction* if there exists a constant $k < 1$ such that

$$e(f(x), f(y)) \leq kd(x, y)$$

for all $x, y \in X$. That is, f *shrinks* distances by a definite amount k or less, where $k < 1$.

DEFINITION 11.5.2. A *fixed point* of a function $f: X \to X$ is a point x for which

$$f(x) = x.$$

Fixed points are very useful, because they are in essence solutions of equations of the form $x - f(x) = 0$, and many interesting equations can be cast in this form.

We have the following basic result.

THEOREM 11.5.1 (*The Contraction Mapping Theorem*). *If X is a complete metric space and $f: X \to X$ is a contraction, then f has a unique fixed point.*

The reason this works is illustrated in Figure 11.5.1. If we take a subset A of X and apply f to A, and if $f(A)$ is contained in A as shown, then $f(A)$ is a shrunken version of A; $f(f(A))$ is a further shrunken version lying inside $f(A)$; and so on. The successive iterations of f on A yield a nested sequence of sets, converging in on a fixed point.

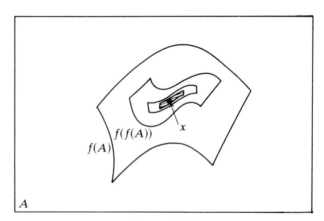

Figure 11.5.1

In practice, since the existence of such an A is not quite clear, we work differently: take any $x \in X$, form the sequence $x, f(x), f(f(x))$, and so on. This

sequence is easily seen to be a Cauchy sequence. To see this, write
$f^{(n)}(x) = f(f(\ldots f(x)))$ (n times). Let

$$M = d(x, f(x)).$$

By the contraction property,

$$d(f^{(n)}(x), f^{(n+1)}(x)) < k^n M.$$

So

$$d(f^{(m)}(x), f^{(n)}(x)) < M(k^m + \ldots + k^{n-1})$$

for $n > m$, and the right-hand side is $< k^m/(1-k)$ which can be made as small
as we wish by taking $m > N$ for some N. But this is the Cauchy property.

Now, by completeness, the sequence converges to some point, say y. Then

$$y = \lim f^{(n)}(x)$$

and since f is continuous

$$f(y) = \lim f(f^{(n)}(x)) = \lim f^{(n+1)}(x) = y.$$

So y is a fixed point.

Finally uniqueness: if also z is a fixed point then by the contraction property
$d(y, z) \le k d(y, z)$ with $k < 1$; so $d(y, z) = 0$, so $y = z$.

To illustrate the importance of this result, we shall sketch a proof of the
existence of solutions to a type of differential equation.

THEOREM 11.5.2   (*Picard's Theorem*).   *If $f(x, y)$ and $\partial f/\partial y$ are continuous
in a closed rectangle $K \subseteq \mathbb{R}^2$, and if (x_0, y_0) is interior to K, then the differential
equation*

$$\frac{dy}{dx} = f(x, y)$$

has a unique solution $y = g(x)$ passing through (x_0, y_0).

To prove this we consider instead an integral equation

$$g(x) = y_0 + \int_{x_0}^{x} f(t, g(t)) \, dt \tag{11.5.1}$$

for an unknown function g. Differentiating [see Theorem 4.7.3], we see that
any solution to (11.5.1) is also a solution to the given differential equation.

Let c be a real constant. Let X be the set of all continuous real functions
$y = g(x)$ defined on the closed interval $|x - x_0| \le a$, where a is a second constant,
and such that $|g(x) - y_0| \le ca$. This is a complete metric space under the metric
analogous to Example 11.1.6. We define $\theta: X \to X$ by

$$(\theta(g))(x) = y_0 + \int_{x_0}^{x} f(t, g(t)) \, dt.$$

Suitable choices of the constants a, c (inspired by estimates on the size of f and $\partial f/\partial y$) ensure that θ really does map X to itself, *and that θ is a contraction.*

Therefore θ has a fixed point g; and g satisfies (11.5.1) by the way we defined θ. Hence g satisfies the differential equation we began with.

Details may be found in Simmons (1963), p. 339: we are more concerned with pointing out the methodology and the way the contraction mapping theorem is involved.

11.6. NORMED, BANACH, AND HILBERT SPACES

By strengthening the axioms we can obtain a variety of more specialized structures. In particular it is often useful to consider metric spaces whose underlying set X has a linear structure—that is, is a vector space [see I, § 5.2]—nicely related to the metric. Here we briefly discuss three possibilities.

DEFINITION 11.6.1.   A *normed space* is a vector space V (over either the real or complex numbers) together with a *norm* function $V \to \mathbb{R}$, which assigns to each $\mathbf{x} \in V$ a real number $\|\mathbf{x}\|$, such that for all $\mathbf{x}, \mathbf{y} \in V$ and all scalars α,

(N1) $\|\mathbf{x}\| \geq 0, \quad \text{and} \quad \|\mathbf{x}\| = 0 \quad \text{if and only if } \mathbf{x} = \mathbf{0}.$

(N2) $\|\mathbf{x} + \mathbf{y}\| \leq \|\mathbf{x}\| + \|\mathbf{y}\|.$

(N3) $\|\alpha \mathbf{x}\| = |\alpha| \|\mathbf{x}\|.$

Every normed space is a metric space relative to the metric

$$d(\mathbf{x}, \mathbf{y}) = \|\mathbf{x} - \mathbf{y}\|.$$

Examples 11.1.1, 11.1.3, 11.1.4, 11.1.5, 11.1.6 and 11.1.7 are all metric spaces arising in this way from norms (observe the tendency for differences to appear in the definition of their metrics: the relevant norms are relatively obvious).

We speak of *real* or *complex* normed spaces, depending on the field of scalars.

DEFINITION 11.6.2.   A *Banach space* (over the real or complex numbers) is a normed space which is complete relative to the associated metric.

Because completeness is a highly useful property, it is possible to say much more about the structure of Banach spaces; in consequence much of the literature concentrates on them.

We can get a very satisfactory theory in a yet more restrictive case, by insisting not just on a norm (which measures the 'length' of a vector) but on an inner product, which allows us to talk also of angles—especially right angles. This leads to our third type of metric space:

DEFINITION 11.6.3.   A *Hilbert space* (over $\mathbb{R}$ or $\mathbb{C}$) is a vector space on which is defined an *inner product,* which assigns to any two vectors $\mathbf{x}, \mathbf{y}$ a (real or complex) scalar $(\mathbf{x}, \mathbf{y})$, and satisfies:

(H1) $\qquad\qquad (\alpha\mathbf{x} + \beta\mathbf{y}, \mathbf{z}) = \alpha(\mathbf{x}, \mathbf{z}) + \beta(\mathbf{y}, \mathbf{z}).$ $\qquad\qquad\qquad$ (11.6.1)

(H2) $\qquad\qquad \overline{(\mathbf{x}, \mathbf{y})} = (\mathbf{y}, \mathbf{x})$ $\qquad\qquad\qquad\qquad\qquad\qquad$ (11.6.2)

(H3) $\qquad\qquad (\mathbf{x}, \mathbf{x}) > 0, \quad$ and $\quad (\mathbf{x}, \mathbf{x}) = 0 \quad$ if and only if $\mathbf{x} = \mathbf{0}.$ $\qquad$ (11.6.3)

Here $\mathbf{x}, \mathbf{y}, \mathbf{z}$ are vectors and α, β are scalars. The bar denotes complex conjugation [see I § 2.7.1] (and is omitted if the underlying field is real). It follows from (H1) and (H2) that

$$(\mathbf{x}, \alpha\mathbf{y} + \beta\mathbf{z}) = \bar{\alpha}(\mathbf{x}, \mathbf{y}) + \bar{\beta}(\mathbf{x}, \mathbf{z})$$

and so

$$(\alpha\mathbf{x}, \beta\mathbf{y}) = \alpha\bar{\beta}(\mathbf{x}, \mathbf{y}). \qquad\qquad (11.6.4)$$

If we define

$$\|\mathbf{x}\| = (\mathbf{x}, \mathbf{x})^{1/2} \qquad\qquad (11.6.5)$$

we can show that this is a norm; and our final axiom is:

(H4) $\qquad\qquad$ The space is complete relative to this norm.

The above metric spaces are examples of *normed vector spaces*; a discussion of these spaces is given in I, Chapter 10.

EXAMPLE 11.6.1.   Let $H = \mathbb{R}^n$ and define an inner product by

$$(\mathbf{x}, \mathbf{y}) = x_1 y_1 + \ldots + x_n y_n$$

where $\mathbf{x} = (x_1, \ldots, x_n)$ and $\mathbf{y} = (y_1, \ldots, y_n)$. This is a Hilbert space, and its norm is the distance from $\mathbf{x}$ to the origin.

EXAMPLE 11.6.2.   Let $H = \mathbb{C}^n$ and define the inner product by

$$(\mathbf{x}, \mathbf{y}) = x_1 \overline{y_1} + \ldots + x_n \overline{y_n}.$$

This gives a complex Hilbert space.

The above examples involve *finite*-dimensional vector spaces and are not especially interesting for the general theory. In infinite dimensions there are numerous examples:

EXAMPLE 11.6.3   [see also I, Example 10.1.1].   The space $l^2(\mathbb{R})$ (or $l^2(\mathbb{C})$) of infinite sequences of real (or complex) numbers $\mathbf{x} = (x_1, \ldots, x_n, \ldots)$ such that $\sum_{i=1}^{\infty} |x_i|^2$ converges is a Hilbert space. Here the inner product is

$$(\mathbf{x}, \mathbf{y}) = \sum_{i=1}^{\infty} x_i \overline{y_i}.$$

EXAMPLE 11.6.4   [see also I, Example 10.1.2].   The space $L^2(\mathbb{C})$ of functions $f:[0, 1] \to \mathbb{C}$ which are square-integrable; that is,

$$\int_0^1 |f(x)|^2 \, dx < \infty$$

is a Hilbert space. Here the inner product is

$$(f, g) = \int_0^1 f(x) \overline{g(x)} \, dx.$$

(It is necessary to consider these integrals as *Lebesgue* integrals [see § 4.9] in the sense of measure theory to get a good structure here: the reason is the completeness axiom.)

EXAMPLE 11.6.5.   The space L_p (or L^p) of functions of class L_p (or L^p) is a Banach space [see Definition 4.9.10 and Theorem 4.9.13].

11.7.   ORTHOGONALITY PROPERTIES

From now on, let H be a complex Hilbert space (the real case is similar but not so commonly used, and at a later stage in the theory is less well behaved) with inner product $(\mathbf{x}, \mathbf{y})$.

DEFINITION 11.7.1.   We say that $\mathbf{x}$ and $\mathbf{y}$ are *orthogonal* if $(\mathbf{x}, \mathbf{y}) = \mathbf{0}$, and write

$$\mathbf{x} \perp \mathbf{y}.$$

Two subsets (especially subspaces) X and Y are orthogonal, written

$$X \perp Y,$$

if $\mathbf{x} \perp \mathbf{y}$ for all $\mathbf{x} \in X$, $\mathbf{y} \in Y$ [see also I, Definition 10.2.1]. The *orthogonal complement* of X is the set

$$X^\perp = \{\mathbf{y} \in H \,|\, \mathbf{y} \perp X\}.$$

Strictly, this is only a 'complement' if X is properly chosen. In fact, suppose that

(i) X is a linear subspace of H [see I, § 5.5],
(ii) X is *closed* as a subset of H, thought of as a metric space.

Then it may be proved (nontrivially) that $X^\perp$ is also a closed subspace, which is a linear complement to X. That is,

$$H = X + X^\perp$$

in the sense of linear algebra [see I (5.5.10)].

DEFINITION 11.7.2.   An *orthonormal set* in H is a subset $\{e_i\}$ of elements of H, such that

$$\mathbf{e}_i \perp \mathbf{e}_j \quad \text{if } i \neq j,$$
$$\|\mathbf{e}_i\| = 1 \quad \text{for all } i. \tag{11.7.1}$$

[See I, Definition 10.2.2.] For example, if $H = \mathbb{C}^n$ then, defining $\mathbf{e}_i = (0, \ldots, 0, 1, 0, \ldots, 0)$ with the 1 in the ith place, we have an orthonormal set. If we omit the second condition, the set is said to be *orthogonal*. Any orthogonal set $\{\mathbf{d}_i\}$ can be *normalized* by setting $\mathbf{e}_i = (1/\|\mathbf{d}_i\|)\mathbf{d}_i$.

A useful result is then:

THEOREM 11.7.1   (*Bessel's Inequality*).   *If* $\{\mathbf{e}_i\}$ *is an orthonormal set in a Hilbert space* H, *and* $\mathbf{x} \in H$, *then*

$$\sum_i |(\mathbf{x}, \mathbf{e}_i)|^2 \leq \|\mathbf{x}\|^2.$$

DEFINITION 11.7.3.   An *orthonormal basis* for H is an orthonormal set $\{\mathbf{e}_i\}$ with the property that no larger set is orthonormal [see also I, Definition 10.2.2].

By using Bessel's inequality and a number of other properties (see for example Simmons (1963), p. 255) we can show that every Hilbert space has an orthonormal basis $\{\mathbf{e}_i\}$ [see also I, Theorem 10.2.1] and that for every $\mathbf{x} \in H$

$$\mathbf{x} = \sum_i (\mathbf{x}, \mathbf{e}_i)\mathbf{e}_i \tag{11.7.2}$$

$$\sum_i |(\mathbf{x}, \mathbf{e}_i)|^2 = \|\mathbf{x}\|^2. \tag{11.7.3}$$

Conversely, if any orthonormal set $\{\mathbf{e}_i\}$ has either property (11.7.2) or (11.7.3) it is an orthonormal basis.

One way to construct orthonormal sets is the *Gram–Schmidt orthogonaliz-ation process*, which we shall describe in some detail. First, we need a definition:

DEFINITION 11.7.4.   A set

$$S = \{\mathbf{x}_1, \mathbf{x}_2, \ldots\} \tag{11.7.4}$$

of H is said to be *linearly independent* [see I, Definition 5.3.2] if every finite subset of S is linearly independent, that is if an equation of the form

$$a_1\mathbf{x}_{k_1} + a_2\mathbf{x}_{k_2} + \ldots + a_N\mathbf{x}_{k_n} = \mathbf{0} \tag{11.7.5}$$

implies that $a_1 = a_2 = \ldots = a_N$, where $k_1, k_2, \ldots, k_N$ are distinct positive integers.

In particular, an independent set cannot contain the zero vector; for if, say, $\mathbf{x}_1 = 0$, then the equation $1\mathbf{x}_1 = 0$ would contradict (11.7.5).

Suppose now that the set S given in (11.7.4) is linearly independent. Then the Gram–Schmidt process leads to the construction of an orthonormal set

$$E = \{\mathbf{e}_1, \mathbf{e}_2, \ldots\} \tag{11.7.6}$$

which is related to S by equations of the form

$$\mathbf{e}_i = b_{i1}\mathbf{x}_1 + b_{i2}\mathbf{x}_2 + \ldots + b_{ii}\mathbf{x}_i \tag{11.7.7}$$

$(i = 1, 2, \ldots)$, where

$$b_{ii} \neq 0. \tag{11.7.8}$$

Thus $\mathbf{B} = (b_{ij})$ is an infinite lower triangular matrix [see I, § 6.7], which on account of (11.7.8) can be inverted, that is we have that

$$\mathbf{x}_i = c_{i1}\mathbf{e}_1 + c_{i2}\mathbf{e}_2 + \ldots + c_{ii}\mathbf{e}_i$$

with suitable scalars c_{ij} and $c_{ii} \neq 0$. It is convenient to begin by solving a slightly simpler problem: we shall construct an orthogonal set

$$D = \{\mathbf{d}_1, \mathbf{d}_2, \ldots\}, \tag{11.7.9}$$

that is a set of non-zero vectors which are mutually orthogonal and satisfy equations analogous to (11.7.7), thus

$$\|\mathbf{d}_i\| \neq 0, \qquad (\mathbf{d}_i, \mathbf{d}_j) = 0 \qquad (i \neq j). \tag{11.7.10}$$

When D has been found, the vectors

$$\mathbf{e}_i = \mathbf{d}_i/\|\mathbf{d}_i\| \qquad (i = 1, 2, \ldots) \tag{11.7.11}$$

form an orthonormal set that meets all our requirements; for by (11.6.1) and (11.6.2) we have that

$$(\mathbf{e}_i, \mathbf{e}_i) = (\mathbf{d}_i, \mathbf{d}_i)/\|\mathbf{d}_i\| = 1$$

and

$$(\mathbf{e}_i, \mathbf{e}_j) = (\mathbf{d}_i, \mathbf{d}_j)/\|\mathbf{d}_i\|\|\mathbf{d}_j\| = 0$$

if $i \neq j$. To start the process we put

$$\mathbf{d}_1 = \mathbf{x}_1 \qquad (\neq 0) \tag{11.7.12}$$

and

$$\mathbf{d}_2 = \mathbf{x}_2 - \frac{(\mathbf{x}_2, \mathbf{d}_1)}{(\mathbf{d}_1, \mathbf{d}_1)}\mathbf{d}_1.$$

Using (11.6.1) we verify that

$$(\mathbf{d}_2, \mathbf{d}_1) = (\mathbf{x}_2, \mathbf{d}_1) - \frac{(\mathbf{x}_2, \mathbf{d}_1)}{(\mathbf{d}_1, \mathbf{d}_1)}(\mathbf{d}_1, \mathbf{d}_1) = 0.$$

By virtue of (11.7.12) we can write

$$\mathbf{d}_2 = \alpha_{21}\mathbf{x}_1 + \mathbf{x}_2, \tag{11.7.13}$$

where α_{21} is a suitable scalar. Since the set S is linearly independent, it follows that

$$\mathbf{d}_2 \neq \mathbf{0}.$$

Next we put

$$\mathbf{d}_3 = \mathbf{x}_3 - \frac{(\mathbf{x}_3, \mathbf{d}_1)}{(\mathbf{d}_1, \mathbf{d}_1)} \mathbf{d}_1 - \frac{(\mathbf{x}_3, \mathbf{d}_2)}{(\mathbf{d}_2, \mathbf{d}_2)} \mathbf{d}_2$$

and, as above, show that

$$(\mathbf{d}_3, \mathbf{d}_1) = (\mathbf{d}_3, \mathbf{d}_2) = 0.$$

Using (11.7.1) and (11.7.2) we can write

$$\mathbf{d}_3 = \alpha_{31}\mathbf{x}_1 + \alpha_{32}\mathbf{x}_2 + \mathbf{x}_3,$$

where α_{31} and α_{32} are suitable scalars. It follows that

$$\mathbf{d}_3 \neq \mathbf{0}.$$

The process is continued step by step. Formally pursuing the inductive argument [see I, § 2.1] we assume that the vectors

$$\mathbf{d}_1, \mathbf{d}_2, \ldots, \mathbf{d}_n$$

have already been found with the properties that

$$\mathbf{d}_i \neq \mathbf{0}, \qquad (\mathbf{d}_i, \mathbf{d}_j) = 0 \qquad (i \neq j; \, 1 \leq i, j \leq n) \qquad (11.7.14)$$

and that there exist scalars $\alpha_{i1}, \alpha_{i2}, \ldots, \alpha_{i,i-1}$ such that

$$\mathbf{d}_i = \alpha_{i1}\mathbf{x}_1 + \alpha_{i2}\mathbf{x}_2 + \ldots + \alpha_{i,i-1}\mathbf{x}_{i-1} + \mathbf{x}_i$$

$(i = 1, 2 \ldots, n)$. We then define

$$\mathbf{d}_{n+1} = \mathbf{x}_{n+1} - \frac{(\mathbf{x}_{n+1}, \mathbf{d}_1)}{(\mathbf{d}_1, \mathbf{d}_1)} - \ldots - \frac{(\mathbf{x}_{n+1}, \mathbf{d}_n)}{(\mathbf{d}_n, \mathbf{d}_n)}$$

and show that (11.7.15) remains valid when n is replaced by $n+1$. This completes the construction of D, whence E is formed by applying (11.7.11). An application of the Gram–Schmidt process is given in III, section 6.3.9.

11.8.   CONNECTIONS WITH CLASSICAL ANALYSIS

We sketch how Hilbert spaces arise in, and illuminate, classical analysis.

Consider the Hilbert space $L^2(\mathbb{C})$ analogous to Example 11.6.4, but with $[0, 1]$ replaced by $[0, 2\pi]$. We can then do Fourier analysis [see § 20.5.6]. It is easy to show that the functions

$$e^{inx} \qquad (n \in \mathbb{Z})$$

are mutually orthogonal; for [see § 20.6.3]

$$\int_0^{2\pi} e^{imx} e^{-inx} \, dx = \begin{cases} 0 & \text{if } m \neq n \\ 2\pi & \text{if } m = n. \end{cases}$$

Hence the functions

$$e_n(x) = \frac{1}{\sqrt{2\pi}} e^{inx}$$

are an orthonormal set; and if $f \in L^2(\mathbb{C})$ then the numbers

$$a_n = (f, e_n) = \frac{1}{\sqrt{2\pi}} \int_0^{2\pi} f(x) e^{-inx} dx$$

are its *Fourier coefficients* [see (20.5.45)].

The classical theory of Fourier series rests on the fact that the set $\{e_n | n \in \mathbb{Z}\}$ is not only orthonormal, but is an orthonormal *basis*. Then (11.7.2) above tells us that

$$f(x) = \frac{1}{\sqrt{2\pi}} \sum_{n=-\infty}^{\infty} a_n e^{inx},$$

that is, f has a *Fourier expansion* (converging in a measure-theoretic sense, appropriate to $L^2(\mathbb{C})$, of course: *not* pointwise). And (11.7.3) is *Parseval's Equation*

$$\sum_{n=-\infty}^{\infty} |a_n|^2 = \int_0^{2\pi} |f(x)|^2 dx.$$

The Gram–Schmidt process also has classical connections. For example, if we start with a space like $L^2(\mathbb{C})$ but for functions defined on the whole of $\mathbb{R}$, and set

$$x_n(x) = x^n e^{-x^2/2}$$

then the resulting orthonormal functions e_n are the *normalized Hermite functions*. Most of the standard sets of orthonormal functions of mathematical physics arise in this way from suitable Hilbert spaces.

Deeper connections with analysis were discovered by Hilbert. In particular he considered *linear operators* $T: H_1 \to H_2$ between Hilbert spaces H_1 and H_2 [see § 19.2.4] and proved theorems analogous to those that hold for finite-dimensional vector spaces. Further, he showed how to 'diagonalize' the equivalent of quadratic forms [cf. I, Proposition 9.1.3]. It turns out that this has important applications to the theory of integral equations (see for instance Kolmogorov and Fomin (1961), Vol. 2, p. 120). It is this deeper theory of operators that is involved in quantum mechanics, but it goes beyond our present scope.

I. S.

REFERENCES

Kolmogorov, A. N. and Fomin, S. V. (1961). *Elements of the Theory of Functions and Functional Analysis*, Graylock Press, Albany, N.J.

Simmons, G. F. (1963). *Introduction to Topology and Modern Analysis*, McGraw-Hill, New York.

CHAPTER 12

Calculus of Variations

12.1. INTRODUCTION

Extremum problems have aroused the interest and curiosity of all ages. Our walk in a straight line is the instinctive solution of an extremum problem: we want to reach the end point of our destination with as little detour as possible. The proverbial 'path of least resistance' is another acknowledgment of our instinctive desire for minimum solutions. The same interest expresses itself in the public interest attached to record achievements, to do something which 'cannot be beaten'.

Mathematically we speak of an *extremum problem* whenever the largest or smallest possible value of a quantity is involved. For example we may want to find the shortest path between two points on a surface, or the largest volume of a container cut out of a piece of metal, or the smallest possible waste in a problem of heating or lighting, or the most economic flight path for a spacecraft, and many other problems. For the solution of such problems a special branch of mathematics, called the *calculus of variations*, has been developed. This is the branch of analysis concerned with certain maximum and minimum problems, and from a modern viewpoint we can think of it as a particular division of the powerful subject of optimization theory which has been greatly developed in recent years. The title 'calculus of variations' is a purely historical one and refers to a particular method of comparison curves employed in the eighteenth century. For the most part our subject involves the optimization of integrals which depend on curves (or functions), and consequently we are dealing with an extension of the theory of maxima and minima in elementary calculus. This is a fruitful way to view the connection and we shall adopt it, even though the calculus of functions and the calculus of variations were actually developed quite separately in the early stages.

12.2. MAXIMA AND MINIMA

We shall begin by recalling some facts of elementary calculus. There we meet the idea of maxima and minima of functions $f(x)$ of a single variable x [see § 3.5].

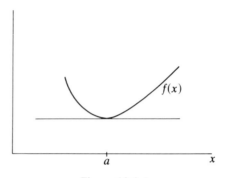

Figure 12.2.1

If $f(x)$ behaves as in Figure 12.2.1 we can say that $f(x)$ has a *minimum* at $x = a$. If the curve were the other way up, then $f(x)$ would have a *maximum* at $x = a$. In either situation it is clear that the tangent to the curve at $x = a$ is actually horizontal, so that the slope of the curve is zero at $x = a$. This means that

$$f'(x) = 0 \quad \text{at } x = a, \tag{12.2.1}$$

where $f'(x) = df/dx$ is the first derivative of the function $f(x)$. The point $x = a$, where $f'(a) = 0$, is called a *critical* or *stationary point* of $f(x)$, and the value $f(a)$ is the *stationary value* of $f(x)$.

The stationary condition (12.2.1) applies to a maximum or a minimum, and therefore by itself it is unable to distinguish which case holds. More information is needed and this is provided by looking at the difference

$$f(x) - f(a) \tag{12.2.2}$$

in a region near the point $x = a$. If this difference is always positive it follows that $f(x)$ increases as x moves away from a to right or left, and therefore $f(x)$ has a *minimum* at $x = a$ (see Figure 12.2.1). If the difference (12.2.2) is always *negative*, $f(x)$ has a *maximum* at $x = a$. From now on we concentrate on the minimum case for ease of description.

12.3. VARIATIONAL PROBLEMS

These ideas of elementary calculus can be carried over to variational problems and to see what is involved we first describe a simple example in geometry which is easily visualized.

EXAMPLE 12.3.1.   (*The shortest path problem.*)   One of the earliest variational problems was to find the shortest distance between two points in a plane. If the two points are $A(a, y_a)$ and $B(b, y_b)$, we can join them with a curve $Y = Y(x)$ (see Figure 12.3.1). An element of length ds for this curve is

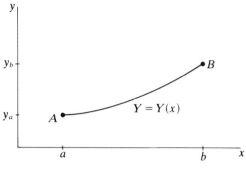

Figure 12.3.1

given by

$$ds^2 = dx^2 + dY^2, \tag{12.3.1}$$

by Pythagoras' theorem [see V, § 1.1.3], and hence

$$ds = dx\sqrt{(1 + Y'^2)}, \qquad Y' = dY/dx. \tag{12.3.2}$$

The total length of the curve $y = y(x)$ joining A and B is therefore

$$L(Y) = \int_A^B ds$$

$$= \int_a^b \sqrt{(1 + Y'^2)}\, dx. \tag{12.3.3}$$

We denote the integral in (12.3.3) by $L(Y)$ to emphasize that it is a number L (the length) depending on the curve or path taken $Y = Y(x)$. In terms of (12.3.3), the shortest distance problem consists of finding the function or curve $y = y(x)$, say, which makes $L(Y)$ a minimum. Geometric intuition suggests that the curve $y(x)$ is a straight line joining A and B (see (12.4.7)), though it is not entirely trivial to prove that this is the solution (see Young (1969)). More generally, if A and B are on some given surface, the shortest curve is a geodesic: for example, great circles on a sphere [see V, Chapter 12].

As we have seen in section 12.2, in ordinary calculus we meet the problem of finding *points* x at which a function $f(x)$ has maximum or minimum values. The above variational problem shows that in calculus of variations we try to find a *curve* or *function* y that minimizes a quantity like $L(Y)$ in equation (12.3.3). We can represent this situation by a diagram, as in Figure 12.3.2, and, as for the ordinary calculus case, we can say that the tangent to the curve $L(Y)$ at $Y = y$ is horizontal. This can be expressed by saying that (see Arthurs (1975) for details)

$$L'(y) = 0, \tag{12.3.4}$$

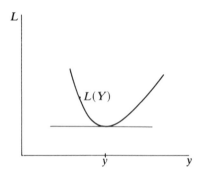

Figure 12.3.2

where $L'(y)$ is the derivative of $L(y)$ at $Y = y$. In the calculus of variations, equation (12.3.4) is called the *Euler–Lagrange equation* and its detailed form depends on the integral $L(Y)$ in question. We shall look at this more closely in a moment. Before doing so, it is important to mention that the stationary condition (12.3.4) applies to a maximum or a minimum, and therefore, as in the ordinary calculus case, by itself it is unable to distinguish which case holds. Of course it may be clear from the context that if $L(y)$ has an extremum it must be a minimum, but in general the question can be settled only by looking at the difference

$$L(Y) - L(y) \tag{12.3.5}$$

for curves Y near y. The curves y for which condition (12.3.4) holds are called *critical curves* (or *extremals*) of the integral $L(Y)$.

12.4. THE EULER–LAGRANGE EQUATION

The shortest path problem is a special case of the variational problem for the integral

$$J(Y) = \int_a^b F(x,\, Y,\, Y')\, dx. \tag{12.4.1}$$

Here the integrand F is a known function of the three variables, x, Y and Y', where in turn Y and Y' are functions of the basic independent variable x. The curves Y that may be admitted to the integral (12.4.1), called the *admissible curves* of (12.4.1), all pass through the two fixed points $A(a,\, y_a)$ and $B(b,\, y_b)$. We seek the curve y which minimizes the integral (12.4.1). From the stationary condition (12.3.4) we see that the curve y must satisfy the Euler–Lagrange equation

$$J'(y) = 0, \tag{12.4.2}$$

and for integrals of the form (12.4.1) this leads to the differential equation

$$\frac{\partial F}{\partial y} - \frac{d}{dx}\frac{\partial F}{\partial y'} = 0 \quad \text{for } a < x < b, \tag{12.4.3}$$

(see Arthurs (1975) and Gelfand and Fomin (1963), where $F = F(x, y, y')$. So the problem becomes one of solving the differential equation (12.4.3) for y, subject to the boundary conditions

$$y(a) = y_a, \qquad y(b) = y_b, \tag{12.4.4}$$

which express the fact that the curve y goes through the points A and B. In general equation (12.4.3) is a second order differential equation [see § 7.6]. For example, if $F(x, y, y') = \frac{1}{2}(y^2 + y'^2)$, we have $\partial F/\partial y = y$ and $\partial F/\partial y' = y'$, so that (12.4.3) becomes $y - y'' = 0$.

EXAMPLE 12.3.1 (continued).   The example in (12.3.3) corresponds to the choice

$$F(x, Y, Y') = \sqrt{(1 + Y'^2)}, \tag{12.4.5}$$

and we note that in this particular case x and Y do not occur explicitly in the integrand F. The Euler–Lagrange equation for the critical curve y in this example is, by (12.4.3) and (12.4.5),

$$0 = \frac{\partial F}{\partial y} - \frac{d}{dx}\frac{\partial F}{\partial y'} = 0 - \frac{d}{dx}\left\{\frac{y'}{\sqrt{(1+y'^2)}}\right\}, \tag{12.4.6}$$

which gives, on integrating once,

$$y' = \text{constant}$$

or

$$y = \alpha x + \beta. \tag{12.4.7}$$

Here α and β are constants of integration and they are determined from the boundary conditions (12.4.4). The result (12.4.7) shows that the critical curve is a straight line. This is the curve that makes $L(Y)$ stationary, and further investigation is needed to prove that it actually produces a minimum (see Young (1969)), though this is expected from the geometry of the problem.

An additional example of this method is found in example 12.5.1.

Extensions of the fundamental problem in (12.4.1) also arise, for example if the integrand F depends on Y'' as well as on x, Y, Y', or if F contains two or more independent functions as in $F(x, Y_1, Y_2, Y_1', Y_2')$. These and other extensions are dealt with in most textbooks (see Arthurs (1975) and Gelfand and Fomin (1963)) and lead to various modifications of the Euler–Lagrange differential equation (12.4.3).

The general situation is that for extremal problems involving integrals, the critical curves y are solutions of a differential equation (the Euler–Lagrange equation), subject to certain boundary conditions.

12.5. DIRECT METHODS

One difficulty with many differential equations is that they cannot be solved exactly in terms of known functions. Consequently the Euler–Lagrange equation encountered above may not be solvable for the critical curve y. This situation has led to the development of what are called *direct variational methods* in which one works directly with the integral $J(Y)$, rather than indirectly via the Euler–Lagrange differential equation.

In this context it is also of interest to add that many problems are initially formulated as boundary problems in differential equations and since these often cannot be solved it is important to develop new ways of studying them. One method is to reformulate the boundary value problem as a variational problem, if this can be done. That is, the given differential equation is regarded as the Euler–Lagrange equation of some variational problem which is then solved by using direct methods. This is the basis of variational methods for problems in differential equations.

One of the most widely used direct methods is due to *Rayleigh* and *Ritz* and this is the one that we shall now describe.

Suppose the variational problem is to find the function y which minimizes the integral

$$J(Y) = \int_a^b F(x, Y, Y')\, dx \qquad (12.5.1)$$

for curves passing through the points $A(a, 0)$ and $B(b, 0)$. The corresponding Euler–Lagrange differential equation for y is given by (12.4.3), and we shall suppose that this equation cannot be solved exactly.

In the Rayleigh–Ritz method we minimize $J(Y)$ not for the whole set Ω of admissible functions, but for a small finite subset R_n of functions. Now, unless the subset R_n contains the function y, the values of $J(Y)$ for Y chosen from R_n will all lie above $J(y)$, which is the minimum value of $J(Y)$ for Y chosen from the whole set Ω. Thus we have

$$J(y) = \min_{Y \text{ in } \Omega} J(Y) \leq \min_{Y \text{ in } R_n} J(Y). \qquad (12.5.2)$$

The idea now is to evaluate the right-hand side of (12.5.2) exactly, and hence obtain an upper bound for $J(y)$. At the same time, the minimizing function Y in R_n is an approximation to the critical curve y.

For example, we could start with $n = 1$ and take

$$Y_1 = \alpha_1(x - a)(x - b) \qquad (12.5.3)$$

which is a curve passing through the end points $A(a, 0)$ and $B(b, 0)$, where

α_1 is a parameter. We put this in $J(Y)$ which gives a function J of α_1. We then minimize $J(\alpha_1)$ by solving

$$\frac{dJ}{d\alpha_1} = 0 \tag{12.5.4}$$

for α_1 as in (12.2.1). This optimum value of α_1 is put back into $J(\alpha_1)$ and hence we obtain an upper bound for $J(y)$

$$J(y) \leq \min_{Y_1} J(Y_1), \tag{12.5.5}$$

and Y_1 in (12.5.3) is an approximation to the critical curve y.

This process can be continued, by taking for example

$$Y_2 = \alpha_1(x-a)(x-b) + \alpha_2(x-a)^2(x-b)^2 \tag{12.5.6}$$

and finding the optimum values of α_1 and α_2 from

$$\frac{\partial J}{\partial \alpha_1} = 0, \qquad \frac{\partial J}{\partial \alpha_2} = 0. \tag{12.5.7}$$

This gives

$$J(y) \leq \min_{Y_2} J(Y_2) \leq \min_{Y_1} J(Y_1). \tag{12.5.8}$$

The latter inequality in (12.5.8) arises because Y_2 cannot give a worse bound than Y_1 (since $Y_2 = Y_1$ for $\alpha_2 = 0$) and in general will give a better bound.

More generally we could take a function

$$Y_n = \sum_{k=1}^{n} \alpha_k (x-a)^k (x-b)^k \tag{12.5.9}$$

where the n parameters α_k are determined by the n equations

$$\frac{\partial J}{\partial \alpha_k} = 0, \qquad k = 1, 2, \ldots, n. \tag{12.5.10}$$

This will provide an upper bound for $J(y)$ and, as well, an approximation Y_n to the exact function y.

EXAMPLE 12.5.1.   We illustrate these ideas by considering the minimum problem for the integral

$$J(Y) = \int_0^1 (Y'^2 + Y^2 - 2Y) \, dx, \qquad Y(0) = 0 = Y(1). \tag{12.5.11}$$

We can obtain an approximation to this minimum by taking a very simple trial function

$$Y_1 = \alpha x(1-x), \tag{12.5.12}$$

following the idea of (12.5.3). This function vanishes at $x = 0$ and $x = 1$ as required, and substitution in $J(Y)$ gives

$$J(Y_1) = \tfrac{11}{30}\alpha^2 - \tfrac{1}{3}\alpha. \qquad (12.5.13)$$

The integral $J(Y_1)$ has a minimum at the solution α of

$$\frac{dJ}{d\alpha} = 0$$

that is,

$$\tfrac{11}{15}\alpha - \tfrac{1}{3} = 0,$$

or

$$\alpha = \tfrac{5}{11}. \qquad (12.5.14)$$

By (12.5.13), the corresponding value of J is

$$J = -\tfrac{10}{132} = -0 \cdot 075758. \qquad (12.5.15)$$

This is an upper bound for the exact value $J(y)$.

For such a simple example we can in fact solve the associated Euler–Lagrange equation for the critical curve y and hence find $J(y)$. Thus, since $F(x, y, y') = y'^2 + y^2 - 2y$ here, the Euler–Lagrange equation is

$$0 = \frac{\partial F}{\partial y} - \frac{d}{dx}\frac{\partial F}{\partial y'} = 2y - 2 - 2y''$$

or

$$-y'' + y = 1. \qquad (12.5.16)$$

The solution of (12.5.16) which vanishes at $x = 0$ and $x = 1$ is

$$y = 1 - \frac{1}{1+e}(e^x + e^{1-x}). \qquad (12.5.17)$$

Putting (12.5.17) in the integral $J(y)$ we then find by a simple calculation that the minimum value of J is given by

$$J(y) = -0 \cdot 075766. \qquad (12.5.18)$$

Knowledge of this exact minimum reveals how good the upper bound in (12.5.15) actually is. Usually, of course, the exact value is not known, and other estimates of the error in the upper bound are sought, for example by obtaining a lower bound as well (see Arthurs (1975)).

In practice for trial functions of the form (12.5.9) it may not be convenient to solve the equations in (12.5.10) directly for the optimum parameters α_k. Instead, the minimum value of $J(\alpha_1, \ldots, \alpha_n)$ may be determined by numerical search methods of optimization on a computer [see III, Chapter 11].

Direct methods like this have proved to be a most effective way of finding accurate solutions to variational problems and boundary value problems in differential equations arising in the physical and engineering sciences, and in biology, medicine, and economics.

EXAMPLE 12.5.2. To illustrate these ideas for boundary-value problems we consider a problem in biology, arising in the steady-state finite cable model of non-linear cell membranes (see Jack *et al.* (1975)). The problem is to solve the differential equation

$$\frac{d^2v}{dx^2} = rf(v), \qquad 0 < x < 1, \tag{12.5.19}$$

subject to the boundary conditions

$$v(0) = v_0, \tag{12.5.20}$$

$$\frac{dv}{dx}(1) = 0. \tag{12.5.21}$$

Here x measures the distance down a finite one-dimensional cable, current is injected at the end $x = 0$, and the end $x = 1$ is terminated in an open circuit, $v(x)$ is the transmembrane potential with v_0 some given constant and r is the longitudinal resistance per unit length. The membrane current per unit length $f(v)$ can be any linear or non-linear function of v which crosses the $f(v) = 0$ axis only once with positive slope. The value of $v(1)$ is not known and must be determined as part of the solution. In general such a boundary-value problem is exceedingly difficult to solve.

However, we can reformulate this problem as a variational problem which can then be used as the basis of a practical optimization procedure. If we write

$$g(V) = \int^V f(u) \, du, \tag{12.5.22}$$

then equation (12.5.19) is the Euler–Lagrange equation for the integral

$$J(V) = \int_0^1 \left\{ \frac{1}{2r} (V')^2 + g(V) \right\} dx. \tag{12.5.23}$$

In addition, if

$$\frac{df(V)}{dV} \geq 0 \quad \text{for all } V, \tag{12.5.24}$$

then the solution v of equations (12.5.19) to (12.5.21) is that function which minimizes $J(V)$ for all functions V such that $V(0) = v_0$. We can use this minimum property to obtain an approximation to the exact function v.

As an example we consider the case

$$f(v) = v + v^3 \tag{12.5.25}$$

for which (12.5.24) is satisfied. This relation is used to represent steady state outward-going rectification observed in nerve membranes (Jack *et al.* (1975),

p. 345). For the trial function we take a simple polynomial

$$V(x) = v_0 + ax + \sum_{k=1}^{6} \alpha_k x^{k+1}.$$ (12.5.26)

This satisfies the boundary condition (12.5.20) as required, and the parameter a can be chosen so that (12.5.21) is also satisfied. The parameters α_k are then optimized by minimizing J. For $r = v_0 = 1$, the optimum parameter values

$$\alpha_1 = 0\cdot7053 \qquad \alpha_4 = 0\cdot0460 \qquad a = -1\cdot0222$$

$$\alpha_2 = -0\cdot0326 \qquad \alpha_5 = 0\cdot0126 \qquad J = 0\cdot4620$$

$$\alpha_3 = -0\cdot1493 \qquad \alpha_6 = 0\cdot0001$$

were obtained in Anderson and Arthurs (1978). The corresponding variational solution (12.5.26) is very accurate, its mean square error being less than $0\cdot0001$ (see Anderson and Arthurs (1978) for details).

A. M. A.

REFERENCES

Anderson, N. and Arthurs, A. M. (1978). *Bull. Mathl. Biology*, Vol 40, pp 735–742.

Arthurs, A. M. (1975). *Calculus of Variations*, Routledge and Kegan Paul, London.

Gelfand, I. M. and Fomin, S. V. (1963). *Calculus of Variations*, Prentice-Hall, Englewood Cliffs, New Jersey.

Jack, J. J. B., Noble, D. and Tsien, R. W. (1975). *Electric Current Flow in Excitable Cells*, Clarendon Press, Oxford.

Young, L. C. (1969). *Lectures on the Calculus of Variations and Optimal Control Theory*, Saunders, Philadelphia.

CHAPTER 13

Integral Transforms

INTRODUCTION

In this chapter an account is given of the properties of the integral transforms most frequently used in applied mathematics. Since rigorous proofs of inversion theorems are long and technically rather difficult, the inversion theorems are stated without proof. Some indication is given of the field of application of each transform.

Only books are listed in the Bibliography. A useful—and very brief—account of the subject which would give the beginner an overall view of transform theory and its applications is given in Tranter (1951). A wider range of applications is covered in Sneddon (1951) and (1972). The books Davies (1978) and Miles (1971) also give a good introduction to the theory for users of mathematics.

Apart from a single reference to the Dirac delta-function in the last section, no reference is made to the theory of generalized functions or their transforms. Such an account is more appropriate to the chapter on Generalized Functions.

A good elementary account of Fourier transforms of generalized functions is given in Lighthill (1958) and in Zemanian (1965). The more sophisticated theory is developed in Zemanian (1969) but reference should also be made to Mikusinski (1959).

13.1. INTEGRAL TRANSFORMS

If we are given a function of two variables $T(x, \xi)$ defined for all values of x and ξ in a square $I \times I$ [see § 5.1] where $I = (a, b)$, that is where $a < x < b$, $a < \xi < b$, then from any prescribed function f whose domain is I we may determine a second function $\tilde{f}$ by the formula

$$\tilde{f}(\xi) = \int_I f(x) T(x, \xi) \, dx, \qquad (13.1.1)$$

provided, of course, that the integral on the right-hand side of this equation is convergent [see § 4.6]. The interval I will usually be infinite; that is [cf. I,

§ 2.6.3] will be $(-\infty, \infty)$ or $(0, \infty)$. We say that $\tilde{f}$ is the *integral transform* of f under the *kernel* $T(x, \xi)$ and we write the relation between $\tilde{f}$ and f in some such way as

$$\tilde{f} = \mathcal{T}f \tag{13.1.2}$$

This notation is adequate when f is a function of a single variable, but has to be modified if it is a function of more than one variable. For example, we write

$$\mathcal{T}[f(x, y) : x \to \xi]$$

to denote the integral

$$\int_I f(x, y) T(x, \xi)\, dx$$

and

$$\mathcal{T}[f(x, y) : y \to \eta]$$

to denote the integral

$$\int_I f(x, y) T(y, \eta)\, dy.$$

If $\tilde{f} = \mathcal{T}f$, we write $f = \mathcal{T}^{-1}\tilde{f}$. $\mathcal{T}^{-1}$ is naturally called the *inverse* of $\mathcal{T}$ and the formula which yields the form of f when $\tilde{f}$ is prescribed is called the *inversion formula* for the operator $\mathcal{T}$.

For instance we define the *Laplace transform* of a function f defined on $(0, \infty)$ by the formula

$$\bar{f}(p) = \int_0^\infty f(x)\, e^{-px}\, dx \tag{13.1.3}$$

[see § 2.11] and we write $\bar{f} = \mathcal{L}f$. Similarly the *Mellin transform* of such a function is defined by the equation

$$f^*(s) = \int_0^\infty f(x) x^{s-1}\, dx \tag{13.1.4}$$

and we write $f^* = \mathcal{M}f$. If the function f is defined on the whole real line we define its *Fourier transform* by the equation

$$\hat{f}(\xi) = \frac{1}{\sqrt{2\pi}} \int_{-\infty}^\infty f(x)\, e^{-i\xi x}\, dx \tag{13.1.5}$$

and we write $\hat{f} = \mathcal{F}f$. Similarly, *the Hankel transform of order ν* is defined by the equation

$$\bar{f}_\nu(\xi) = \int_0^\infty x f(x) J_\nu(x\xi)\, dx, \tag{13.1.6}$$

where $J_\nu(t)$ denotes the Bessel function of the first kind of order ν and argument t [see (10.4.10)]. In this case we write $\bar{f}_\nu = \mathcal{H}_\nu f$.

Closely related to the Fourier transform are the *Fourier cosine* and *Fourier sine* transforms defined, respectively, by

$$\hat{f}_c(\xi) = \sqrt{\frac{2}{\pi}} \int_0^\infty f(x) \cos (x\xi) \, dx, \qquad (13.1.7)$$

$$\hat{f}_s(\xi) = \sqrt{\frac{2}{\pi}} \int_0^\infty f(x) \sin (x\xi) \, dx, \qquad (13.1.8)$$

[see § 2.12]. The operators of the Fourier cosine and the Fourier sine transforms are denoted, respectively, by $\mathcal{F}_c$ and $\mathcal{F}_s$; that is, we write $\hat{f}_c = \mathcal{F}_c f$ and $\hat{f}_s = \mathcal{F}_s f$.

In certain cases the kernel of the transform is of the type $K(x\xi)$, that is, is such that the variables x and ξ occur only in the combination $x\xi$.

If we denote the operator of the transform in this instance by $\mathcal{H}$ we then have

$$\tilde{f}(\xi) = \mathcal{H}f(\xi) = \int_0^\infty f(x)K(x\xi) \, dx. \qquad (13.1.9)$$

If such a transform is its own inverse, that is, if

$$\mathcal{H}^{-1} = \mathcal{H} \qquad (13.1.10)$$

we say that its kernel K is a *Fourier kernel*. It is a simple matter to derive a necessary condition for a function K to be a Fourier kernel. Writing out equation (13.1.10) we see that it takes the form

$$f(x) = \int_0^\infty \tilde{f}(\xi)K(x\xi) \, d\xi. \qquad (13.1.11)$$

If we take the Mellin transform with respect to ξ of both sides of (13.1.9) and make use of the result

$$\mathcal{M}[K(x\xi); \xi \to s] = x^{-s} K^*(s),$$

(which is easily established directly from the definition (13.1.4)), where K^* is the Mellin transform of K, we see that equation (13.1.9) is equivalent to the relation

$$\tilde{f}^*(s) = K^*(s) \int_0^\infty f(x)x^{-s} \, dx.$$

Now the integral on the right-hand side of this equation is simply $f^*(1-s)$ so that (13.1.9) is equivalent to

$$\tilde{f}^*(s) = K^*(s)f^*(1-s). \qquad (13.1.12)$$

In a similar way we can show that equation (13.1.11) is equivalent to

$$f^*(s) = K^*(s)\tilde{f}^*(1-s)$$

and replacing s by $1 - s$ in this equation we obtain

$$f^*(1-s) = K^*(1-s)\tilde{f}^*(s). \qquad (13.1.13)$$

Eliminating f^* and $\tilde{f}^*$ from (13.1.12) and (13.1.13) we obtain the relation

$$K^*(s)K^*(1-s) = 1. \qquad (13.1.14)$$

It is a much more difficult thing to prove that this condition is *sufficient* as well as necessary. See Titchmarsh (1948).

Integral transforms are used in several ways in statistics and in mathematical physics but perhaps they are used most extensively in the solution of boundary value problems for partial differential equations. They arise there in two ways.

To illustrate the first case we consider the solution of the plane harmonic equation [see § 8.2]

$$\frac{\partial^2 u}{\partial x^2} + \frac{\partial^2 u}{\partial y^2} = 0, \qquad (13.1.15)$$

in the positive quadrant $x \geq 0$, $y \geq 0$, subject to the boundary conditions

(a) $u(x, 0) = f(x)$,

(b) $u(0, y) = 0$,

(c) $u(x, y) \to 0, \quad$ as $\rho \to \infty$,

where $\rho = (x^2 + y^2)^{1/2}$. Using the rule for integrating by parts (with respect to the variable x) [see Theorem 4.3.1] we see that

$$\int_0^\infty \frac{\partial^2 u}{\partial x^2} \sin(\xi x)\, dx = \left[\frac{\partial u}{\partial x}\sin(\xi x) - \xi u(x, y)\cos(\xi x)\right]_{x=0}^{x=\infty}$$

$$- \xi^2 \int_0^\infty u(x, y)\sin(\xi x)\, dx.$$

Taking into account the boundary conditions (b) and (c) we see that this reduces to the equation

$$\mathscr{F}_s[u_{xx}(x, y); x \to \xi] = -\xi^2 \hat{u}_s(\xi, y), \qquad (13.1.16)$$

where $\hat{u}_s(\xi, y) = \mathscr{F}_s[u(x, y); x \to \xi]$. Hence if we take the Fourier sine transform with respect to the variable x of both sides of equation (13.1.15) we find that the Fourier sine transform of the unknown function u satisfies the *ordinary* differential equation [see § 7.4.1]

$$\left(\frac{d^2}{dy^2} - \xi^2\right)\hat{u}_s(\xi, y) = 0.$$

Further, the boundary conditions (a) and (c) yield the following conditions on its sine transform:

$$\hat{u}_s(\xi, 0) = \hat{f}_s(\xi) = \mathscr{F}_s f(\xi)$$

$$u_s(\xi, y) \to 0 \quad \text{as } y \to \infty.$$

From these conditions we easily deduce that

$$\hat{u}_s(\xi, y) = \hat{f}_s(\xi) e^{-\xi y}.$$

Applying the inverse operator $\mathscr{F}_s^{-1}$ we then obtain the desired solution in the form

$$u(x, y) = \mathscr{F}_s^{-1}[f_s(\xi) e^{-\xi y}; \xi \to x]. \tag{13.1.17}$$

[See also § 13.2.5.]

The second approach may be illustrated by considering the problem of determining the solution $u(\rho, z)$ of the axisymmetric Laplace equation [see (8.2.10)]

$$\frac{\partial^2 u}{\partial \rho^2} + \frac{1}{\rho} \frac{\partial u}{\partial \rho} + \frac{\partial^2 u}{\partial z^2} = 0, \tag{13.1.18}$$

in the half-space $\rho \geq 0$, $z \geq 0$, satisfying the boundary conditions

(a) $\qquad\qquad u(\rho, 0) = f(\rho), \qquad \rho \geq 0;$

(b) $\qquad\qquad u(\rho, z) \to 0 \quad \text{as } r \to \infty,$

where $r = (\rho^2 + z^2)^{1/2}$. If we denote the Bessel function of the first kind of order zero and argument t by $J_0(t)$ [see (10.4.8)] we can easily show that (13.1.18) and the boundary condition (b) are satisfied by the function

$$J_0(\xi \rho) e^{-\xi z}, \qquad (\xi > 0).$$

By the principle of superposition for linear partial differential equations [see § 8.2] we deduce that

$$u(\rho, z) = \int_0^\infty \xi A(\xi) e^{-\xi z} J_0(\xi \rho) \, d\xi \tag{13.1.19}$$

is also a solution (provided that A is such that the partial derivatives of u may be calculated 'by differentiating under the integral sign' [see Theorem 4.7.2]). In terms of the Hankel transform defined by equation (13.1.6) we see that we may write this solution in the form

$$u(\rho, z) = \mathscr{H}_0[A(\xi) e^{-\xi z}; \xi \to \rho]. \tag{13.1.20}$$

Letting z tend to zero on both sides of this equation and using the condition (a) we see that the solution of the stated boundary value problem will be given by equation (13.1.20) provided that we can find a function A such that $f = \mathscr{H}_0 A$, or, in terms of the inverse transform $A = \mathscr{H}_0^{-1} f$.

From both of these examples we see that it is vitally important to be able to invert the operator of an integral transform, in other words to obtain the requisite inversion formula.

13.2.   THE FOURIER TRANSFORM

13.2.1.   General Properties

We shall begin by discussing the properties of the Fourier transform defined by equation (13.1.5).

The operator $\mathcal{F}$ defined by the relation $\hat{f} = \mathcal{F}f$ is obviously a linear operator, [see Definition 19.1.8] that is

$$\mathcal{F}(c_1 f_1 + c_2 f_2) = c_1 \mathcal{F}f_1 + c_2 \mathcal{F}f_2.$$

It can be thought of as a mapping from one space of functions onto another. The modern theory of the Fourier integral depends on the theory of the Lebesgue spaces L_p [see Example 11.6.5]. These in turn depend on the concept of the Lebesgue integral and few engineers are acquainted with it [see § 4.9]. For such readers the following may give them some feeling for the results, though what we shall say is a gross over-simplification. We say that a function belongs to the class $L_p(\Omega)$ if the integral of $|f|^p$ $(p > 1)$ taken over Ω is finite, that is if

$$\int_\Omega |f(x)|^p \, dx$$

is finite. We define the *norm* of $f \in L_p(\Omega)$ by

$$\|f\|_p = \left\{ \int_\Omega |f(x)|^p \, dx \right\}^{1/p} \qquad (p > 1).$$

When $p = \infty$ we adopt the convention that $f \in L_\infty(\Omega)$ if $|f|$ is bounded on Ω, and we write

$$\|f\|_\infty = \max_{x \in \Omega} |f(x)|.$$

In transform theory Ω is usually either $\mathbb{R}$, the set of all real numbers, or $\mathbb{R}^+$ the set of all positive real numbers [see I, § 2.6].

If the integers p and p' are related through the equation

$$\frac{1}{p} + \frac{1}{p'} = 1$$

we say that the spaces L_p and $L_{p'}$ are *conjugate*. Thus $L_2(\Omega)$ is self-conjugate and $L_1(\Omega)$ and $L_\infty(\Omega)$ are conjugate.

The theory of Fourier transforms in L_1 and in L_2 is very highly developed and represents one of the greatest triumphs of classical analysis. Full accounts are given in the books Bochner (1959) and (1949), Goldberg (1961), Titchmarsh (1948) and Wiener (1932) listed in the Bibliography at the end of this chapter; a good preliminary guide to the literature of this subject is given in section 9 of Ross (1965).

It turns out that $\mathcal{F}$ maps $L_1(\mathbb{R})$ into $L_\infty(\mathbb{R})$, that is, that if $f \in L_1$, then $\hat{f} \in L_\infty$.

As might be conjectured from the fact that $L_2(\mathbb{R})$ is self-conjugate, the L_2-theory of Fourier transforms is much more symmetrical. We find that $\mathcal{F}$ maps $L_2(\mathbb{R})$ onto itself, that is, if $f \in L_2(\mathbb{R})$, then $\hat{f} \in L_2(\mathbb{R})$.

For instance, if [see § 2.11]

$$f(x) = e^{-|x|},$$

then by (13.1.5)

$$\hat{f}(\xi) = (2\pi)^{-1/2} \int_{-\infty}^{\infty} e^{-|x|-i\xi x} \, dx$$

$$= (2/\pi)^{1/2} \int_{0}^{\infty} e^{-x} \cos(\xi x) \, dx$$

[see (9.5.8)]. The value of the latter integral is well known—it can be derived by integrating by parts twice [see Theorem 4.3.1]—and we find that

$$\hat{f}(\xi) = (2/\pi)^{1/2}(1+\xi^2)^{-1}.$$

From these expressions we can easily calculate that for this particular function f,

$$\|f\|_1 = 2, \qquad \|\hat{f}\|_\infty = (2/\pi)^{1/2}, \qquad \|f\|_2 = \|\hat{f}\|_2 = 1.$$

13.2.2.   Inversion Theorems

We shall now state (without proof) inversion theorems for the Fourier transform, that is, formulae for the inverse operator $\mathcal{F}^{-1}$. The nature of these results will vary in complexity according to how wide is the class of functions being considered. The simplest such result is:

If f is (a) *piecewise continuously differentiable, that is, if f and f' are piecewise continuous* [see Definition 2.3.5]; (b) *absolutely integrable in the Riemann sense, that is, if*

$$\int_{-\infty}^{\infty} |f(x)| \, dx$$

is finite [see § 4.6] *then $\hat{f}$ is continuous and*

$$f(x) = \frac{1}{\sqrt{(2\pi)}} \int_{-\infty}^{\infty} \hat{f}(y) \, e^{ixy} \, dy. \tag{13.2.1}$$

If we define the operator $\mathcal{F}^*$ by the equation

$$\mathcal{F}^* g(x) = \frac{1}{\sqrt{(2\pi)}} \int_{-\infty}^{\infty} g(y)\, e^{ixy}\, dy \qquad (13.2.2)$$

we may rewrite equation (13.2.1) in the form $f = \mathcal{F}^* f$. In other words

$$\mathcal{F}^{-1} = \mathcal{F}^*. \qquad (13.2.3)$$

(See § 2.2 of Sneddon (1972).)

A more sophisticated version of the Fourier inversion theorem is:

If $f \in L_1(\mathbb{R})$, then

$$f(x) = \lim_{N\to\infty} \frac{1}{\sqrt{(2\pi)}} \int_{-N}^{N} e^{ixy} \hat{f}(y)\left(1 - \frac{|y|}{N}\right) dy. \qquad (13.2.4)$$

(See § 6.C of Goldberg (1961)). The proof of this form is difficult. A simpler form is obtained by imposing an extra restriction on f. This form is:

If $f \in L_1(\mathbb{R})$ and if there exists a $\delta > 0$ such that, for each x, $t^{-1}g(t) \in L_1([0, \delta])$, where $g(t) = f(x+t) + f(x-t) - 2f(x)$, then

$$f(x) = \lim_{N\to\infty} \int_{-N}^{N} f(y)\, e^{ixy}\, dy. \qquad (13.2.5)$$

In most practical applications the function f will satisfy the conditions (a) and (b) above so that the form (13.2.1) of the inversion theorem will be adequate.

In the proofs of each of these theorems one of the basic steps in the proof is the *Riemann–Lebesgue lemma* which states that $\hat{f}(y) \to 0$ as $|y| \to \infty$.

Another important result is that if $f, g \in L_1(\mathbb{R})$ and $\hat{f} = \hat{g}$, then $f = g$, almost everywhere; in other words *a function is uniquely determined by its Fourier transform*.

13.2.3.   The Convolution Theorem

If two functions f and g are defined on $\mathbb{R}$, their *convolution* $f * g$ is defined by the equation

$$f * g(x) = \frac{1}{\sqrt{(2\pi)}} \int_{-\infty}^{\infty} f(x-y)g(y)\, dy. \qquad (13.2.6)$$

It is immediately obvious from the definition that $f * g = g * f$. It can also be shown that if f and g both belong to $L_1(\mathbb{R})$ then so does $f * g$ and

$$\mathcal{F}(f * g) = \hat{f} \cdot \hat{g}. \qquad (13.2.7)$$

This last result is known as the *convolution theorem* for Fourier transforms.

In applications it is more often used in its inverse form

$$\mathcal{F}^*(\hat{f}\hat{g}) = f * g. \tag{13.2.8}$$

13.2.4.   Useful Formulae

The Fourier transform has certain simple properties which are easily established directly from the definition but which are extremely useful in calculating the Fourier transforms of particular functions; these are

$$\mathcal{F}[f(t/a); \xi] = |a|\mathcal{F}[f(t); a\xi] \tag{13.2.9}$$

$$\mathcal{F}[f(t-a); \xi] = e^{-i\xi a}\mathcal{F}[f(t); \xi] \tag{13.2.10}$$

$$\mathcal{F}[e^{-i\lambda t}f(t); \xi] = \mathcal{F}[f(t); \xi + \lambda] \tag{13.2.11}$$

Also, if $f(t)$ is a complex-valued function of t and $\bar{f}$ denotes its complex conjugate [see I, § 2.7.1],

$$\mathcal{F}[\bar{f}(-t); \xi] = \overline{\mathcal{F}f(\xi)}. \tag{13.2.12}$$

If $f(t)$ is an *even* function of t, so that $f(-t) = f(t)$ for all real t, it follows that

$$\mathcal{F}[f(t); \xi] = \frac{1}{\sqrt{(2\pi)}}\left[\int_{-\infty}^{0} f(t)\, e^{-i\xi t}\, dt + \int_{0}^{\infty} f(t)\, e^{-i\xi t}\, dt\right]$$

$$= \frac{1}{\sqrt{(2\pi)}}\left[-\int_{\infty}^{0} f(-t)\, e^{i\xi t}\, dt + \int_{0}^{\infty} f(t)\, e^{-i\xi t}\, dt\right]$$

$$= \frac{1}{\sqrt{(2\pi)}}\int_{0}^{\infty} f(t)\{e^{i\xi t} + e^{-i\xi t}\}\, dt.$$

From this last equation and the definition (13.1.7) of the Fourier cosine transform we see that *if f is an even function*

$$\mathcal{F}[f(t); \xi] = \mathcal{F}_c[f(t); \xi]. \tag{13.2.13}$$

In a similar way we can show that *if f is an odd function*, that is, if $f(-t) = -f(t)$, for all t, then

$$\mathcal{F}[f(t); \xi] = i\mathcal{F}_s[f(t); \xi]. \tag{13.2.14}$$

By combining equations (13.2.8) and (13.2.12) we obtain an indication of an important result in the L_2-theory [see § 13.2.1]. If we write out equation (13.2.8) and replace $g(t)$ by $\bar{g}(-t)$ we obtain the equation

$$\int_{-\infty}^{\infty} f(t)\bar{g}(t-x)\, dt = \int_{-\infty}^{\infty} e^{ix\xi}\hat{f}(\xi)\mathcal{F}[\bar{g}(-t); \xi]\, d\xi.$$

Making use of equation (13.2.12) we see that this can be written in the form

$$\int_{-\infty}^{\infty} f(t)\bar{g}(t-x)\, dt = \int_{-\infty}^{\infty} e^{ix\xi}\hat{f}(\xi)\overline{\hat{g}(\xi)}\, d\xi,$$

where $\hat{g} = \mathscr{F}g$. Putting $x = 0$ and writing

$$\langle f, g \rangle = \int_{-\infty}^{\infty} f(x)\bar{g}(x)\, dx$$

we see that

$$\langle f, g \rangle = \langle \hat{f}, \hat{g} \rangle.$$

This result can be proved rigorously in the L_2-theory. $L_2(\mathbb{R})$ is in fact a Hilbert space with inner product $\langle ., . \rangle$ [see § 11.6.3]. If we put $g = f$ in this last equation we obtain *Parseval's relation*

$$\|f\|_2 = \|\hat{f}\|_2$$

which shows that $\mathscr{F}$ maps $L_2(\mathbb{R})$ onto itself *isometrically*, that is, $\mathscr{F}$ preserves 'distances' [cf. § 11.8].

Parseval's relation does not hold in the L_1-theory; there, the corresponding result is that

$$\|f\|_1 \geq \|\hat{f}\|_\infty.$$

(It will be noticed that these results were verified above in the case in which f was defined by $f(x) = e^{-|x|}$).

In applying the method of Fourier transforms to the solution of partial differential equations [see also § 13.2.5] we need to know the Fourier transform of the derivative of a function. Proceeding formally we see that

$$\mathscr{F}[f'(x); \xi] = \frac{1}{\sqrt{(2\pi)}} \int_{-\infty}^{\infty} f'(x)\, e^{-i\xi x}\, dx$$

$$= \frac{1}{\sqrt{(2\pi)}} [f(x)\, e^{-i\xi x}]_{-\infty}^{\infty} + i\xi \frac{1}{\sqrt{(2\pi)}} \int_{-\infty}^{\infty} f(x)\, e^{-i\xi x}\, dx.$$

A necessary condition for the integral defining a Fourier transform to exist is that f tend to zero at $\pm\infty$ so that the first term on the right vanishes and we have

$$\mathscr{F}[f'(x); \xi] = i\xi \mathscr{F}f(\xi). \tag{13.2.15}$$

It should be emphasized that an equation of this kind only has a meaning if both sides exist! In other words equation (13.2.15) has a meaning only for functions f of such a nature that both $\mathscr{F}f$ and $\mathscr{F}f'$ exist. Repeating the process, we see that if n is a positive integer,

$$\mathscr{F}[f^{(n)}(x); \xi] = (i\xi)^n \mathscr{F}f(\xi). \tag{13.2.16}$$

Denoting the operator d/dx by D_x we can say that under the mapping $\mathscr{F}$ the operator D_x is transformed to $(i\xi)$.

In the notation of functions of two variables, if we denote $\partial/\partial x$ by ∂_x, then the equivalent form of equation (13.2.16) is

$$\mathscr{F}[\partial_x^n u(x, y); x \to \xi] = (i\xi)^n \mathscr{F}[u(x, y); x \to \xi]. \tag{13.2.17}$$

For instance if we denote the two-dimensional Laplacian by Δ_2, then

$$\mathcal{F}[\Delta_2 u(x, y); x \to \xi] = (D_y^2 - \xi^2)\mathcal{F}[u(x, y); x \to \xi] \qquad (13.2.18)$$

$$\mathcal{F}[\Delta_2^2 u(x, y); x \to \xi] = (D_y^2 - \xi^2)^2 \mathcal{F}[u(x, y); x \to \xi]. \qquad (13.2.19)$$

Both of these formulae have been used extensively in the solution of plane strain problems in elasticity.

In a similar way we can show that

$$\mathcal{F}[x^n f(x); \xi] = i^n D_\xi^n \mathcal{F}f(\xi). \qquad (13.2.20)$$

Tables of Fourier transforms are given in Vol. 1 of Erdélyi *et al.* (1954) and in Campbell and Foster (1948). Here we shall list only a few of the more commonly occurring and useful results:

$$\mathcal{F}[e^{-a|x|}; \xi] = \left(\frac{2}{\pi}\right)^{1/2} \frac{a}{a^2 + \xi^2} \qquad (13.2.21)$$

$$\mathcal{F}[e^{-x^2/a^2}; \xi] = 2^{-1/2}|a|e^{-\frac{1}{4}\xi^2 a^2} \qquad (13.2.22)$$

$$\mathcal{F}[x^{-1}; \xi] = -i(\tfrac{1}{2}\pi)^{1/2} \operatorname{sgn} \xi, \qquad (13.2.23)$$

where sgn is defined by the equations

$$\operatorname{sgn} t = \begin{cases} +1 & \text{if } t > 0, \\ -1 & \text{if } t < 0. \end{cases}$$

Similarly, recalling the definition of the *Heaviside unit function H*:

$$H(t) = \begin{cases} 1 & t > 0, \\ 0 & t < 0, \end{cases}$$

we see that

$$\mathcal{F}[H(a - |x|); \xi] = (2/\pi)^{1/2}\xi^{-1} \sin(\xi a). \qquad (13.2.24)$$

In particular, if we take $a = \sqrt{2}$ in (13.2.22) we find that

$$\mathcal{F}[e^{-\frac{1}{2}x^2}; \xi] = e^{-\frac{1}{2}\xi^2} \qquad (13.2.25)$$

so that the function $\exp(-\tfrac{1}{2}x^2)$ is mapped into itself by the Fourier transform; such a function is said to be *self-reciprocal* under the Fourier transform.

13.2.5. Use of Fourier transforms in solving partial differential equations

To illustrate the use of Fourier transforms in the solution of boundary value problems for partial differential equations we consider the problem of determining the solution $u(x, y)$ of the two-dimensional Laplace equation (13.1.15) in the upper half-plane $y \geq 0$ satisfying the conditions

$$u(x, 0) = f(x), \qquad u(x, y) \to 0 \quad \text{as } \rho \to \infty. \qquad (13.2.26)$$

Taking the Fourier transform of both sides of equation (13.1.15) with respect to x and making use of (13.2.18) we see that $\hat{u}(\xi, y) = \mathscr{F}[u(x, y); x \to \xi]$ satisfies the ordinary differential equation [see § 7.4.1]

$$(D_y^2 - \xi^2)\hat{u}(\xi, y) = 0, \qquad y \geq 0.$$

On the other hand the boundary conditions (13.2.26) are equivalent to

$$\hat{u}(\xi, 0) = \hat{f}(\xi) = \mathscr{F}f(\xi), \qquad \hat{u}(\xi, y) \to 0 \quad \text{as } y \to \infty.$$

It is then easily shown that

$$\hat{u}(\xi, y) = \hat{f}(\xi) \, e^{-|\xi|y}.$$

Inverting this by Fourier's inversion theorem we therefore obtain the solution

$$u(x, y) = \mathscr{F}^*[\hat{f}(\xi) \, e^{-|\xi|y}; \xi \to x]. \tag{13.2.27}$$

Making use of the convolution theorem in the form (13.2.8) we see that the required solution is given by the equation

$$u(x, y) = \frac{1}{\sqrt{(2\pi)}} \int_{-\infty}^{\infty} f(t)g(x - t, y) \, dt \tag{13.2.28}$$

where $g(x, y) = \mathscr{F}^*[e^{-|\xi|y}; \xi \to x]$. Now applying the Fourier inversion theorem to the result (13.2.21) we see that

$$\frac{1}{\pi} \int_{-\infty}^{\infty} \frac{y}{x^2 + y^2} e^{ix\xi} \, dx = e^{-|\xi|y}$$

and taking the complex conjugate of both sides of this equation we obtain the equation

$$\mathscr{F}[(2/\pi)^{1/2}y(x^2 + y^2)^{-1}; x \to \xi] = e^{-|\xi|y},$$

so that $g(x, y)$ is the function on the left-hand side of this equation. Substituting this value into equation (13.2.28) we obtain the solution

$$u(x, y) = \frac{y}{\pi} \int_{-\infty}^{\infty} \frac{f(t) \, dt}{(x - t)^2 + y^2} \tag{13.2.29}$$

It should be observed that this is the solution of the stated *Dirichlet problem* only if the function f appearing in the data is such that its Fourier transform exists—that is, only if $f \in L_1(\mathbb{R})$. [*See* § 8.2].

13.2.6. Multiple Fourier Transforms

Suppose that f is a function of n real variables $(x_1, \ldots, x_n)$ defined over the whole of $\mathbb{R}^n$ [see § 5.3]. If we write $\mathbf{x} = (x_1, \ldots, x_n)$, $\mathbf{y} = (y_1, \ldots, y_n)$, $(\mathbf{x} \cdot \mathbf{y}) = x_1 y_1 + \ldots + x_n y_n$ we define the n-dimensional Fourier transform $\mathscr{F}_n$ by the equation

$$\mathscr{F}_n[f(\mathbf{x}); \mathbf{x} \to \mathbf{y}] = (2\pi)^{-n/2} \int_{\mathbb{R}^n} f(\mathbf{x}) \exp\{-i(\mathbf{x} \cdot \mathbf{y})\} \, d\mathbf{x}. \tag{13.2.30}$$

The inverse operator is given by the equation $\mathscr{F}^{-1} = \mathscr{F}^*$ where

$$\mathscr{F}^*[\hat{f}(\mathbf{y}); \mathbf{y} \rightarrow \mathbf{x}] = (2\pi)^{-n/2} \int_{\mathbb{R}^n} \hat{f}(\mathbf{y}) \exp\{i(\mathbf{x} . \mathbf{y})\} \, d\mathbf{y}. \qquad (13.2.31)$$

In $\mathbb{R}^n$ we define the convolution of two functions f and g by

$$(f * g)(\mathbf{x}) = (2\pi)^{-n/2} \int_{\mathbb{R}^n} f(\mathbf{x} - \mathbf{y}) g(\mathbf{y}) \, d\mathbf{y} \qquad (13.2.32)$$

and the convolution theorem takes the form

$$\mathscr{F}_n(f * g) = \hat{f} . \hat{g}, \qquad \hat{f} = \mathscr{F}_n f, \qquad \hat{g} = \mathscr{F}_n g. \qquad (13.2.33)$$

By a similar procedure to that outlined above for the one-dimensional case we can easily deduce that $\|\hat{f}\|_2^2 = \|f\|_2^2$ in the n-dimensional case also.

Denoting the operator of partial differentiation $\partial/\partial x_j$ by ∂_j and proceeding exactly as in the derivation of (13.2.15) we deduce the result

$$\mathscr{F}_n[\partial_j f(\mathbf{x}); \mathbf{x} \rightarrow \boldsymbol{\xi}] = i\xi_j \hat{f}(\boldsymbol{\xi}), \qquad \hat{f} = \mathscr{F}_n f. \qquad (13.2.34)$$

The extension to higher derivatives is obvious. In particular if Δ_n denotes the n-dimensional Laplacian [see § 8.2] so that

$$\Delta_n = \partial_1^2 + \partial_2^2 + \ldots + \partial_n^2,$$

we find that

$$\mathscr{F}_n[\Delta_n f(\mathbf{x}); \mathbf{x} \rightarrow \boldsymbol{\xi}] = -|\boldsymbol{\xi}|^2 \hat{f}(\boldsymbol{\xi}), \qquad (13.2.35)$$

where $\hat{f} = \mathscr{F}_n f$ and $|\boldsymbol{\xi}| = (\xi_1^2 + \xi_2^2 + \ldots + \xi_n^2)^{1/2}$.

In the special case in which f is a function only of the radial distance

$$r = (x_1^2 + x_2^2 + \ldots + x_n^2)^{1/2}$$

it turns out that

$$\mathscr{F}_n[f(r); \mathbf{x} \rightarrow \boldsymbol{\xi}] = \lambda^{-\nu} \mathscr{H}_\nu[r^\nu f(r); \lambda] \qquad (13.2.36)$$

where $\lambda = |\boldsymbol{\xi}|$, $\nu = \frac{1}{2}n - 1$ and $\mathscr{H}_\nu$ is the operator of the Hankel transform defined by (13.1.6). In particular, if $\mathbf{x} \in \mathbb{R}^2$, $\rho = |\mathbf{x}|$, then

$$\mathscr{F}_2[f(\rho); \mathbf{x} \rightarrow \boldsymbol{\xi}] = \mathscr{H}_0[f(\rho); \lambda], \qquad \lambda = (\xi_1^2 + \xi_2^2)^{1/2}, \qquad (13.2.37)$$

while if $\mathbf{x} \in \mathbb{R}^3$, $r = |\mathbf{x}|$, then

$$\mathscr{F}_3[f(r); \mathbf{x} \rightarrow \boldsymbol{\xi}] = \lambda^{-1} \mathscr{F}_s[rf(r); \lambda], \qquad \lambda = (\xi_1^2 + \xi_2^2 + \xi_3^2)^{1/2}, \qquad (13.2.38)$$

$\mathscr{F}_s$ denoting the operator of the Fourier sine transform.

As a simple illustration of the use of multiple Fourier transforms we consider the solution of the diffusion equation [see § 8.1]

$$\Delta_2 u(x, y, t) = \frac{\partial u}{\partial t}, \qquad t \geq 0, \qquad (x, y) \in \mathbb{R}^2, \qquad (13.2.39)$$

subject to the initial condition

$$u(x, y, 0) = f(x, y).$$ (13.2.40)

We shall use the double Fourier transform

$$\hat{u}_2(\xi, \eta, t) = \mathscr{F}_2[u(x, y, t); (x, y) \to (\xi, \eta)].$$ (13.2.41)

From equation (13.2.35) we deduce that (13.2.39) is equivalent to the ordinary differential equation

$$D_t \hat{u}_2 = -(\xi^2 + \eta^2)\hat{u}_2$$

while from (13.2.40) we deduce immediately the initial condition

$$\hat{u}_2(\xi, \eta, 0) = \hat{f}_2(\xi, \eta) = \mathscr{F}_2[f(x, y); (\xi, \eta)].$$

The solution of the initial value problem posed by these last two equations is readily found to be

$$\hat{u}_2(\xi, \eta, t) = \hat{f}_2(\xi, \eta) \exp\{-(\xi^2 + \eta^2)t\}.$$

Using the inversion formula, the convolution theorem (13.2.33) in its inverse form and equation (13.2.37) we find that the required solution is given by the equation

$$u(x, y, t) = \int_{-\infty}^{\infty} \int_{-\infty}^{\infty} f(x', y')K(x - x', y - y'; t) \, dx' \, dy'$$ (13.2.42)

where

$$K(x, y; t) = (2\pi)^{-1} \mathscr{H}_0[\exp(-\lambda^2 t); \lambda \to \rho], \qquad (\rho^2 = x^2 + y^2).$$

Now [see (10.4.8) for the definition of J_0]

$$\int_0^{\infty} \lambda J_0(\lambda \rho) \, e^{-\lambda^2 t} \, d\lambda = (2t)^{-1} \int_0^{\infty} J_0(\rho s^{1/2} t^{-1/2}) \, e^{-s} \, ds$$

$$= (2t)^{-1} \sum_{j=0}^{\infty} \frac{(-\rho^2/4t)^j}{j!j!} \int_0^{\infty} s^j \, e^{-s} \, ds$$

Since the value of the integral on the right side of this equation is $j!$ [see § 10.2.1] we see that

$$K(x, y, t) = (4\pi t)^{-1} \exp\{-(x^2 + y^2)/4t\}.$$ (13.2.43)

The solution of the initial value problem (13.2.39) and (13.2.40) is therefore given by equations (13.2.42) and (13.2.43).

13.3.   THE SINE AND COSINE TRANSFORMS

The Fourier sine and cosine transforms have already been defined—by equations (13.1.8) and (13.1.7) respectively. The inversion theorem for $\mathscr{F}$

[see § 13.2.2] and equation (13.2.13) gives us the inversion formula

$$\mathscr{F}_c^{-1} = \mathscr{F}_c, \tag{13.3.1}$$

that is, if $\hat{f}_c$ is defined by equation (13.1.7) then

$$f(x) = \sqrt{\frac{2}{\pi}} \int_0^\infty \hat{f}_c(\xi) \cos(x\xi)\, d\xi.$$

Similarly, from the inversion theorem for $\mathscr{F}$ and equation (13.2.14) we deduce the inversion formula

$$\mathscr{F}_s^{-1} = \mathscr{F}_s \tag{13.3.2}$$

that is, if $\hat{f}_s$ is defined by equation (13.1.8) then

$$f(x) = \sqrt{\frac{2}{\pi}} \int_0^\infty \hat{f}_s(\xi) \sin(x\xi)\, d\xi.$$

Also from the theory of Fourier transforms [see § 13.2.1] we can deduce that if $f \in L_2(\mathbb{R}^+)$, then so do $\hat{f}_c$ and $\hat{f}_s$.

The simple rules of manipulation

$$\mathscr{F}_c[f(x/a); \xi] = a\mathscr{F}_c[f(x); \xi a], \qquad (a > 0)$$
$$\mathscr{F}_s[f(x/a); \xi] = a\mathscr{F}_s[f(x); \xi a], \qquad (a > 0) \tag{13.3.3}$$

can be deduced immediately from the definitions.

From the well-known trigonometrical identity [see V (1.2.19)]

$$2 \cos(ax) \cos(\xi x) = \cos\{(a + \xi)x\} + \cos\{(a - \xi)x\}$$

we deduce the relation

$$2\mathscr{F}_c[f(x) \cos(ax); \xi] = \mathscr{F}_c[f(x); \xi + a] + \mathscr{F}_c[f(x); |\xi - a|]. \tag{13.3.4}$$

In a similar way we can prove the relations

$$2\mathscr{F}_c[f(x) \sin(ax); \xi] = \mathscr{F}_s[f(x); \xi + a] - \mathscr{F}_s[f(x); \xi - a], \tag{13.3.5}$$
$$2\mathscr{F}_s[f(x) \sin(ax); \xi] = \mathscr{F}_c[f(x); |\xi - a|] - \mathscr{F}_c[f(x); \xi + a], \tag{13.3.6}$$
$$2\mathscr{F}_s[f(x) \cos(ax); \xi] = \mathscr{F}_s[f(x); \xi + a] + \mathscr{F}_s[f(x); \xi - a]. \tag{13.3.7}$$

If we use the rule for integrating by parts [see Theorem 4.3.1] we can establish the rules

$$\mathscr{F}_c[f'(x); \xi] = -(2/\pi)^{1/2} f(0) + \xi\mathscr{F}_s[f(x); \xi] \tag{13.3.8}$$
$$\mathscr{F}_s[f'(x); \xi] = -\xi\mathscr{F}_c[f(x); \xi], \tag{13.3.9}$$

for the cosine and sine transforms of the derivative of a function.

Using (13.3.8) we see that

$$\mathscr{F}_c[f''(x); \xi] = -(2/\pi)^{1/2} f'(0) + \xi \mathscr{F}_s[f'(x); \xi]$$

and now using (13.3.9) we see that this result can be written in the form

$$\mathscr{F}_c[f''(x); \xi] = -(2/\pi)^{1/2} f'(0) - \xi^2 \mathscr{F}_c[f(x); \xi]. \tag{13.3.10}$$

Applying (13.3.9) and then (13.3.8) to f'' we can show in a similar way that

$$\mathscr{F}_s[f''(x); \xi] = (2/\pi)^{1/2} \xi f(0) - \xi^2 \mathscr{F}_s[f(x); \xi]. \tag{13.3.11}$$

For a function $u(x, y)$ of two variables, with $x \in \mathbb{R}^+$, we have the analogous formulae

$$\mathscr{F}_c[\Delta_2 u(x, y); x \to \xi] = (D_y^2 - \xi^2)\mathscr{F}_c[u(x, y); x \to \xi] - (2/\pi)^{1/2} u_x(0, y) \tag{13.3.12}$$

$$\mathscr{F}_s[\Delta_2 u(x, y); x \to \xi] = (D_y^2 - \xi^2)\mathscr{F}_s[u(x, y); x \to \xi] + (2/\pi)^{1/2} \xi u(0, y) \tag{13.3.13}$$

where D_y denotes the differential operator d/dy and Δ_2 the Laplacian operator in two dimensions [see § 8.2].

From the definitions of the transforms we see that

$$\int_0^\infty \hat{f}_c(t)\hat{g}_c(t)\, dt = \sqrt{\frac{2}{\pi}} \int_0^\infty \hat{f}_c(t)\, dt \int_0^\infty g(x) \cos(xt)\, dx$$

$$= \int_0^\infty g(x)\, dx \sqrt{\frac{2}{\pi}} \int_0^\infty \hat{f}_c(t) \cos(xt)\, dt$$

so that we obtain Parseval's relation for Fourier cosine transforms:

$$\int_0^\infty \hat{f}_c(t)\hat{g}_c(t)\, dt = \int_0^\infty f(x)g(x)\, dx. \tag{13.3.14}$$

Similarly, we can show that

$$\int_0^\infty \hat{f}_s(t)\hat{g}_s(t)\, dt = \int_0^\infty f(x)g(x)\, dx. \tag{13.3.15}$$

If we assume that $f \in L_2(\mathbb{R}^+)$, then by replacing g by $\bar{f}$, the complex conjugate of f, in (13.3.14) and (13.3.15) and introducing the norm

$$\|f\|_2^+ = \left[\int_0^\infty |f(x)|^2\, dx \right]^{1/2} \tag{13.3.16}$$

we obtain the equations

$$\|\hat{f}_c\|_2^+ = \|\hat{f}_s\|_2^+ = \|f\|_2^+. \tag{13.3.17}$$

Proceeding in a similar kind of way we have

$$2 \int_0^\infty \hat{f}_s(\xi)\hat{g}_c(\xi) \sin(\xi x)\, d\xi$$

$$= 2\sqrt{\frac{2}{\pi}} \int_0^\infty \hat{g}_c(\xi) \sin(\xi x)\, d\xi \int_0^\infty f(t) \sin(\xi t)\, dt$$

$$= \sqrt{\frac{2}{\pi}} \int_0^\infty f(t)\, dt \int_0^\infty \hat{g}_c(\xi) . 2 \sin(\xi x) \sin(\xi t)\, d\xi$$

$$= \sqrt{\frac{2}{\pi}} \int_0^\infty f(t)\, dt \int_0^\infty \hat{g}_c(\xi)[\cos\{(x-t)\xi\} - \cos\{(x+t)\xi\}]\, d\xi$$

which is equivalent to

$$2 \int_0^\infty \hat{f}_s(\xi)\hat{g}_c(\xi) \sin(\xi x)\, d\xi = \int_0^\infty f(t)\{g(|x-t|) - g(x+t)\}\, dt. \tag{13.3.18}$$

In a similar way we can show that

$$2 \int_0^\infty \hat{f}_c(\xi)\hat{g}_c(\xi) \cos(\xi x)\, d\xi = \int_0^\infty f(t)\{g(|x-t|) + g(x+t)\}\, dt. \tag{13.3.19}$$

To illustrate the use of the sine transform we shall consider the solution of Laplace's equation [see § 8.2]

$$\Delta_2 u(x, y) = 0 \tag{13.3.20}$$

in the positive quadrant ($x \geq 0$, $y \geq 0$), satisfying the boundary conditions

$$u(0, y) = 0 \tag{13.3.21}$$

$$u(x, 0) = f(x) \tag{13.3.22}$$

$$u(x, y) \to 0 \quad \text{as } (x^2 + y^2)^{1/2} \to \infty. \tag{13.3.23}$$

From equation (13.3.13) we see that the transform appropriate to the boundary condition (13.3.21) is the Fourier sine transform

$$u_s(\xi, y) = \mathscr{F}_s[u(x, y); x \to \xi].$$

From equations (13.3.13), (13.3.20) and (13.3.21) we see that $\hat{u}_s(\xi, y)$ satisfies the ordinary differential equation

$$(D_y^2 - \xi^2)\hat{u}_s = 0,$$

and from equations (13.3.22) and (13.3.23) that

$$\hat{u}_s(\xi, 0) = \hat{f}_s(\xi) = \mathscr{F}_s[f(x); \xi],$$

$$\hat{u}_s(\xi, y) \to 0 \quad \text{as } y \to \infty.$$

The function satisfying these conditions is easily seen to be

$$\hat{u}_s(\xi, y) = \hat{f}_s(\xi) e^{-\xi y}.$$

Using the Fourier sine inversion theorem we deduce immediately that

$$u(x, y) = \sqrt{\frac{2}{\pi}} \int_0^\infty \hat{f}_s(\xi) e^{-\xi y} \sin(\xi x) \, d\xi$$

$$= 2 \int_0^\infty \hat{f}_s(\xi) \hat{g}_c(\xi, y) \sin(\xi x) \, d\xi$$

where $\hat{g}_c(\xi) = (2\pi)^{-1/2} e^{-\xi y}$. Using equation (13.3.18) and the formula

$$g(x, y) = \frac{y}{\pi(x^2 + y^2)} \tag{13.3.24}$$

we see that the desired solution is

$$u(x, y) = \int_0^\infty f(t)\{g(x - t, y) - g(x + t, y)\} \, dt \tag{13.3.25}$$

with $g(x, y)$ given by (13.3.24).

On the other hand if equation (13.3.22) were replaced by the condition $u_x(x, 0) = 0$, we should use the cosine transform $\hat{u}_c(\xi, y) = \mathscr{F}_c[u(x, y); x \to \xi]$. Proceeding as before we should find that in this case

$$u(x, y) = 2 \int_0^\infty \hat{f}_c(\xi) \hat{g}_c(\xi, y) \cos(\xi x) \, d\xi$$

$$= \int_0^\infty f(t)\{g(x - t, y) + g(x + t, y)\} \, dt$$

with g being given again by (13.3.24).

13.4.   THE LAPLACE TRANSFORM

13.4.1.   Definition and Preliminary Results

The best known and most widely used integral transform is undoubtedly the Laplace transform. If the function f is defined on the positive real line $\mathbb{R}^+$ [see I, § 2.6] and if there exists a real number c such that the integral

$$\int_0^\infty e^{-px} f(x) \, dx$$

exists for Re $(p) > c$, we say that this integral defines the Laplace transform $\bar{f}(p)$ of the function $f(x)$. We write

$$\bar{f}(p) = \mathscr{L}[f(x); p]. \tag{13.4.1}$$

As in the case of the Fourier transform we adopt the notation $\mathscr{L}[u(x, y); x \to p]$

in the case of functions of more than one variable to denote that the integration is over the variable x etc.

It should be noticed that not every function has a Laplace transform. For example if $f(x) = \exp(x^2)$ there is no value of p for which the defining integral is convergent [see § 4.6] so this function does not have a Laplace transform.

The theory of the Laplace transform is developed systematically in Doetsch (1955), Widder (1941) and (1971); we shall merely state the more important results here.

If (i) $f(x)$ *is integrable over any finite interval* $[a, b]$, $0 < a < b$, (ii) *there exists a real number c such that for any positive b*

$$\int_b^N e^{-cx} f(x)\, dx$$

tends to a finite limit as $N \to \infty$ *and* (iii) *for any positive a, the integral*

$$\int_\varepsilon^a |f(x)|\, dx$$

tends to a finite limit as $\varepsilon \to 0$, *then it can be shown that* $\bar{f}(p)$ *is a holomorphic function of the complex variable p in the right half-plane* $Re(p) > c$ [*see* § 9.1].

An important result—because it asserts that a function is uniquely determined by its Laplace transform—is *Lerch's theorem*, which may be stated in the form:

If $\bar{f}_1(p) = \mathcal{L}[f_1(x); p]$ *is holomorphic in the half-plane* $Re(p) > c_1$, $\bar{f}_2(p) = \mathcal{L}[f_2(x); p]$ *is holomorphic in the half-plane* $Re(p) > c_2$ *and* $\bar{f}_1(p) = \bar{f}_2(p)$ *for* $Re(p) > c = max(c_1, c_2)$, *then* $f_1(x) = f_2(x)$.

Extensive tables of Laplace transforms of many varied types are in existence, e.g. Doetsch (1955), Erdélyi (1954), McLachlan (1950) and Oberhettinger and Badii (1973), but we shall list a few of the more elementary ones here.

The results

$$\mathcal{L}[x^\nu e^{-ax}; p] = (p-a)^{-\nu-1}\Gamma(\nu+1), \qquad Re(p) > a, \qquad \nu > -1; \tag{13.4.2}$$

$$\mathcal{L}[\sin(ax); p] = a(p^2 + a^2)^{-1}; \tag{13.4.3}$$

$$\mathcal{L}[\cos(ax); p] = p(p^2 + a^2)^{-1}; \tag{13.4.4}$$

follow immediately from the definition and elementary integrations [see § 10.2.1 for the definition of $\Gamma(n)$]. A useful pair is derived from the integral

$$\int_0^\infty e^{-a^2u^2 - (b^2/u^2)}\, du = \frac{\sqrt{\pi}}{2a} e^{-2ab}, \qquad (a > 0, b > 0) \tag{13.4.5}$$

which is not so easy to establish. See pp. 142–143 of Sneddon (1972). From this result we deduce immediately that

$$\mathcal{L}[x^{-1/2} e^{-c/x}; p] = (\pi/p)^{1/2} \exp(-2\sqrt{cp}). \tag{13.4.6}$$

Differentiating both sides of this equation with respect to c we get the second result of the pair:

$$\mathcal{L}[x^{-3/2} e^{-c/x}; p] = (\pi/c)^{1/2} \exp(-2\sqrt{cp}). \qquad (13.4.7)$$

Similarly, it is elementary to show that

$$\mathcal{L}[e^{-\frac{1}{4}x^2/a^2}; p] = \pi^{1/2} a e^{a^2 p^2} \operatorname{Erfc}(ap), \qquad (a > 0), \qquad (13.4.8)$$

where the function Erfc (z) is defined in terms of the *error function*

$$\operatorname{Erf}(z) = \frac{2}{\sqrt{\pi}} \int_0^z \exp(-s^2) \, ds,$$

by the equation

$$\operatorname{Erfc}(z) = 1 - \operatorname{Erf}(z).$$

The transform

$$\mathcal{L}[e^{ax} \operatorname{Erf}(a^{1/2} x^{1/2}); p] = a^{1/2} p^{-1/2} (p - a)^{-1} \qquad (\operatorname{Re} p > a)$$
$$(13.4.9)$$

is also useful, particularly in its inverse form

$$\mathcal{L}^{-1}[p^{-1/2}(p-a)^{-1}; x] = a^{-1/2} e^{ax} \operatorname{Erf}(a^{1/2} x^{1/2}).$$

Two simple results involving the Bessel function of the first kind [see § 10.4.2] are:

$$\mathcal{L}[x^\nu J_\nu(ax); p] = \frac{(2a)^\nu \Gamma(\nu + \frac{1}{2})}{\sqrt{\pi}(p^2 + a^2)^{\nu + \frac{1}{2}}}, \qquad (a > 0, \nu > -\tfrac{1}{2}) \quad (13.4.10)$$

$$\mathcal{L}[x^\nu J_\nu(2\sqrt{ax}); p] = a^{(1/2)\nu} p^{-\nu-1} e^{-a/p} \qquad (a > 0, \nu > -\tfrac{1}{2}) \quad (13.4.11)$$

which are easily established by substituting the series expansion for the Bessel function [see (10.4.10)] in the integral defining the Laplace transform and integrating term by term.

13.4.2.   Inversion Theorem

The inversion theorem for the Laplace transform may be derived from the Fourier inversion theorem. In its simplest form it may be stated:

If the function $\bar{f}(p)$ is holomorphic in the half-plane $\operatorname{Re}(p) > \gamma$ and if L_w denotes the line $\operatorname{Re}(p) = c > \gamma$, $-w < \operatorname{Im}(p) < w$, then as $w \to \infty$

$$\frac{1}{2\pi i} \int_{L_w} \bar{f}(p) e^{px} \, dp$$

converges to a function f such that $\mathcal{L}f = \bar{f}$. We write this result in the form

$$f(x) = \mathcal{L}^{-1} \bar{f}(x) = \frac{1}{2\pi i} \int_{c-i\infty}^{c+i\infty} \bar{f}(p) e^{px} \, dp$$

13.4.3.   **Manipulative Rules**

There are certain rules of manipulation of the Laplace transform which follow immediately from the definition but which are most useful in extending the entries in tables of transforms. The most commonly used are:

$$\mathcal{L}[f(x/a); p] = a\mathcal{L}[f(x); ap], \qquad (a > 0). \tag{13.4.12}$$

$$\mathcal{L}[e^{-ax}f(x); p] = \mathcal{L}[f(x); p+a], \qquad (\text{Re } p > -a) \tag{13.4.13}$$

$$\mathcal{L}[x^n f(x); p] = (-1)^n D_p^n \mathcal{L}[f(x); p], \tag{13.4.14}$$

with D_p denoting the differential operator d/dp, n a positive integer.

$$\mathcal{L}[f(x)H(x-a); p] = e^{pa}\mathcal{L}[f(x+a); p], \qquad (a > 0), \tag{13.4.15}$$

with H denoting Heaviside's unit function [see § 13.2.4].

$$\mathcal{L}[x^{-1}f(x); p] = \int_p^\infty \bar{f}(q)\, dq. \tag{13.4.16}$$

In applications of the Laplace transform we need to know how derivatives transform. In some cases we know that the function being transformed is only piecewise continuous, i.e. that at certain points the value of the function takes a finite 'jump' [see Definition 2.3.5]. To illustrate the procedure in such a case we now consider a function $f(x)$ defined on the positive real line which is continuous everywhere except at the point a where it has a finite discontinuity. From the definition (13.4.1) we have [see Theorem 4.1.2(iii)]

$$\mathcal{L}[f'(x); p] = \int_0^a f'(x)\, e^{-px}\, dx + \int_a^\infty f'(x)\, e^{-px}\, dx.$$

Evaluating these integrals by the rule for integrating by parts [see Theorem 4.3.1] we obtain

$$\int_0^a f'(x)\, e^{-px}\, dx = f(a-)\, e^{-pa} - f(0) + p\int_a^\infty f(x)\, e^{-px}\, dx$$

where $f(a-)$ denotes the limiting value of $f(x)$ as x tends to a from the left [see § 2.3] and

$$\int_a^\infty f'(x)\, e^{-px}\, dx = -f(a+)\, e^{-pa} + p\int_a^\infty e^{-px}f(x)\, dx,$$

where $f(a+)$ denotes the value of $f(x)$ as x approaches a from the right. Combining these integrals, we find that

$$\mathcal{L}[f'(x); p] = p\mathcal{L}[f(x); p] - f(0) - [f]_a \tag{13.4.17}$$

where

$$[f]_a = f(a+) - f(a-)$$

denotes the *saltus* or the jump in the value of $f(x)$ at $x = a$. The formula

(13.4.17) can be extended in an obvious way to the case when f has a finite number of finite discontinuities on $\mathbb{R}^+$, that is, when f is piecewise continuous on $\mathbb{R}^+$.

When f is *continuous* on $\mathbb{R}^+$, equation (13.4.17) reduces to the simple form

$$\mathscr{L}[f'(x); p] = p\mathscr{L}[f(x); p] - f(0). \qquad (13.4.18)$$

From this we deduce easily that if $f \in C^n(\mathbb{R}^+)$, that is, if f and its first n derivatives are continuous on the positive real line [see § 2.10], then

$$\mathscr{L}[f^{(n)}(x); p] = p^n \mathscr{L}[f(x); p] - \sum_{r=0}^{n-1} p^{n-r-1} f^{(r)}(0). \qquad (13.4.19)$$

In particular if $f \in C^n(\mathbb{R}^+)$ and if $f^{(r)}(0) = 0$, $(r = 0, 1, \ldots, n-1)$, then

$$\mathscr{L}[f^{(n)}(x); p] = p^n \mathscr{L}[f(x); p]. \qquad (13.4.20)$$

Also, from (13.4.19) we deduce immediately that if ϕ is the polynomial defined by

$$\phi(t) = \sum_{m=0}^{n} a_{n-m} t^m, \qquad (13.4.21)$$

and D_x denotes the differential operator d/dx, then

$$\mathscr{L}[\phi(D_x)f(x); p] = \phi(p)\bar{f}(p) - \psi(p), \qquad (13.4.22)$$

with the polynomial ψ defined by the equation

$$\psi(t) = \sum_{m=0}^{n-1} a_{n-m} \sum_{r=0}^{m} p^{m-r-1} f^{(r)}(0). \qquad (13.4.23)$$

These formulae are useful in the solution of differential equations by means of the Laplace transform as we shall see later.

13.4.4.   Convolution

For functions f and g defined on $\mathbb{R}^+$ we define *convolution* for the Laplace transform $f \circ g$ by the equation

$$(f \circ g)(x) = \int_0^x f(x-t)g(t)\, dt, \qquad (13.4.24)$$

from which definition it is obvious that $g \circ f = f \circ g$. From the definitions (13.4.1) and (13.4.24) we see that

$$\mathscr{L}[(f \circ g)(x); p] = \int_0^\infty e^{-px}\, dx \int_0^x f(x-t)g(t)\, dt.$$

Changing the order in which we perform the integrations we see that the integral on the right-hand side of this equation is equal to

$$\int_0^\infty g(t)\, dt \int_t^\infty e^{-px} f(x-t)\, dx.$$

Changing the variable of integration in the inner integral from x to y where $x = t + y$ [see Theorem 4.3.2] we find that the integral is reduced to

$$\int_0^\infty g(t)\, e^{-pt}\, dt \int_0^\infty e^{-py} f(y)\, dy,$$

so that we have shown that

$$\mathcal{L}[(f \circ g)(x); p] = \bar{f}(p)\bar{g}(p). \tag{13.4.25}$$

This is the *convolution theorem* for the Laplace transform.

If we define the nth repeated convolution of a function with itself by f^n so that

$$f^n(x) = \underbrace{(f \circ f \ldots \circ f)(x)}_{n \text{ times}}$$

it follows from repeated application of the convolution theorem that

$$\mathcal{L}f^n(p) = \{\bar{f}(p)\}^n. \tag{13.4.26}$$

In applications both (13.4.25) and (13.4.26) are most often used in their inverse forms

$$\mathcal{L}^{-1}[\bar{f}(p)\bar{g}(p); x] = (f \circ g)(x) \tag{13.4.27}$$

$$\mathcal{L}^{-1}[\{\bar{f}(p)\}^n; x] = f^n(x). \tag{13.4.28}$$

Another result which is often useful in its inverse form is

$$\mathcal{L}\left[(4\pi x^3)^{-1/2} \int_0^\infty uf(u)\, e^{-u^2/4x}\, du; p\right] = \bar{f}(p^{1/2}), \tag{13.4.29}$$

which, it can be seen, is a direct consequence of (13.4.7); its inverse form is, of course,

$$\mathcal{L}^{-1}[\bar{f}(p^{1/2}); x] = (4\pi x^3)^{-1/2} \int_0^\infty uf(u)\, e^{-u^2/4x}\, du. \tag{13.4.30}$$

13.4.5.   Watson's Lemma

We often need asymptotic formulae for the Laplace transform. It will be recalled [see § 2.15] that a series $\sum_{m=0}^\infty a_m x^{-m}$ is said to be *asymptotic to $g(x)$*, if

$$\sum_{m=0}^{n-1} a_m x^{-m} - g(x) = O(x^{-n}),$$

as $x \to \infty$, for every positive integer n and that we write $\sum a_m x^{-m} \sim g(x)$.

Watson's lemma states that *if the function $f(t)$ satisfies an inequality of the form $|f(x)| < Me^{ax}$, M a positive quantity, and, if in some neighbourhood of the origin, $f(x)$ has a Maclaurin expansion $\sum_{r=0}^{\infty} c_r x^r / r!$ [see (2.10.1)] then $\mathscr{L}f$ has the asymptotic expansion*

$$\bar{f}(p) \sim \sum_{r=0}^{\infty} c_r p^{-r-1}. \tag{13.4.31}$$

(For proof see pp. 184–189 of Sneddon (1972)). A simple consequence of this lemma—although it can be proved directly—is that if $\bar{f} = \mathscr{L}f$, then

$$\lim_{p \to \infty} p\bar{f}(p) = f(0); \tag{13.4.32}$$

this holds if f satisfies the conditions of Watson's lemma.

The inverse problem of obtaining an asymptotic expansion for $f(t)$ as $t \to \infty$, when $\bar{f}$ is known is discussed in section 13.6.4 of Davies (1978). Suppose that $\bar{f}(p)$ has a branch point at $p = a$ [see § 9.17] and that it has the asymptotic expansion

$$\bar{f}(p) \sim \sum_{m=1}^{\infty} c_m (p-a)^{-\lambda_m}, \tag{13.4.33}$$

then that branch point contributes a term

$$e^{ax} \sum_{m=1}^{\infty} \frac{c_m}{\Gamma(\lambda_m)} x^{\lambda_m - 1}. \tag{13.4.34}$$

Watson's lemma has been used in Sneddon (1972) (p. 190) to determine the asymptotic expansion of Erf (x) [see § 13.4.1] and the formulae (13.4.33) and (13.4.34) have been used in Davies (1978) (p. 85) to derive the asymptotic expansion for $J_0(x)$ [see (10.4.8)].

13.4.6. Applications of the Laplace Transform

The best known application of the Laplace transform is to the solution of ordinary differential equations with constant coefficients [see § 7.4]. If we have to solve the initial value problem

$$\left. \begin{array}{ll} \phi(D_x)y(x) = f(x), & x \geq 0, \\ y(0) = y_0, \quad y^{(r)}(0) = y_r, & (r = 1, 2, \ldots, n-1), \end{array} \right\} \tag{13.4.35}$$

with ϕ the polynomial defined by (13.4.21) and $y_0, y_1, \ldots, y_{n-1}$ prescribed constants, then taking the Laplace transform of both sides of the differential equation and making use of equation (13.4.22), we find that $\bar{y} = \mathscr{L}y$ satisfies the simple algebraic equation

$$\phi(p)\bar{y}(p) = \psi(p) + \bar{f}(p) \tag{13.4.36}$$

where $\psi(p)$ is a polynomial of degree $n-1$ in p of the form

$$\psi(p) = \sum_{m=0}^{n-1} a_{n-m} \sum_{r=0}^{m} y_r p^{m-r-1}.$$

(Cf. equation (13.4.23).)

Applying the inverse Laplace operator $\mathcal{L}^{-1}$ to both sides of equation (13.4.36) we find that the solution of the initial problem (13.4.35) is

$$y(x) = Y(x) + \eta(x) \tag{13.4.37}$$

where

$$Y(x) = \mathcal{L}^{-1}[\psi(p)/\phi(p); x], \qquad \eta(x) = \mathcal{L}^{-1}[\bar{f}(p)/\phi(p); x].$$

Since it contains the n (arbitrary) values $y_0, y_1, \ldots, y_{n-1}$, $Y(x)$ is the general solution of the homogeneous equation (that is, of the equation with $f = 0$); it is called the *complementary function* [cf. § 7.3.2]. The form of $Y(x)$ depends on the nature of the zeros of the polynomial $\phi(p)$ [see I, § 14.5]. For instance if $\phi(p)$ has n distinct zeros $a_1, a_2, \ldots, a_n$ we can resolve $\psi(p)/\phi(p)$ into partial fractions [see I, § 14.10]

$$\frac{\psi(p)}{\phi(p)} = \sum_{r=1}^{n} \frac{c_r}{p - a_r}$$

and the corresponding form of $Y(x)$ is

$$Y(x) = \sum_{r=1}^{n} c_r e^{a_r x}.$$

The procedure for multiple zeros [see I, § 14.6] is similar; for every factor of the form $(p-a)^{m+1}$ in $\phi(p)$ we have a term of the form

$$(c_1 + c_2 x + \ldots + c_m x^m) e^{ax}$$

in $Y(x)$.

The function $\eta(x)$ is called a *particular integral* of the equation [cf. § 7.3.2]; it is the solution which corresponds to the initial conditions $y(0) = y'(0) = \ldots = y^{(n-1)}(0) = 0$. If the inverse transform

$$\mathcal{L}^{-1}\left[\frac{1}{\phi(p)}; x\right] \tag{13.4.38}$$

exists—and we denote it by $K(x)$—it follows from the convolution theorem (13.4.25) that

$$\eta(x) = \int_0^x f(t) K(x - t)\, dt.$$

In certain cases the inverse transform (13.4.38) may not exist but the transform

$$M(x) = \mathcal{L}^{-1}\left[\frac{1}{p\phi(p)}; x\right]$$

does exist. Writing

$$\bar{\eta}(p) = p\bar{f}(p)\bar{M}(p),$$

and then using the convolution theorem, combined with (13.4.18) we find that

$$\eta(x) = \frac{d}{dx}\int_0^x f(t)M(x-t)\,dt.$$

As an example of this method we take the simple initial problem

$$y'' + \omega^2 y(x) = f(x), \qquad x \geq 0,$$

$$y(0) = y'(0) = 0.$$

(13.4.39)

Taking the Laplace transform of both sides of the differential equation and making use of the initial conditions we find that

$$\bar{y}(p) = \bar{K}(p)\bar{f}(p),$$

where $\bar{K}(p) = (p^2 + \omega^2)^{-1}$. From (13.4.3) and the convolution theorem (13.4.25) we deduce that the solution of (13.4.39) is

$$y(x) = \frac{1}{\omega}\int_0^x f(t)\sin\{\omega(x-t)\}\,dt. \tag{13.4.40}$$

13.4.7. Laguerre's Differential Equation

To illustrate that the Laplace transform may be used in certain cases to solve equations with variable coefficients we consider the solution of the differential equation (with n a positive integer)

$$xy''(x) + (1-x)y'(x) + ny(x) = 0; \tag{13.4.41}$$

we shall consider the solution satisfying

$$y(0) = 1. \tag{13.4.42}$$

If we denote $y'(0)$ by y_1, then, from equation (13.4.18)

$$\mathcal{L}[y'(x); p] = p\bar{y}(p) - 1$$

$$\mathcal{L}[y''(x); p] = p^2\bar{y}(p) - p - y_1.$$

From equation (13.4.14) we deduce that

$$\mathcal{L}[xy''(x); p] = -\frac{d}{dp}p^2\bar{y}(p) + 1 = -p^2\bar{y}'(p) - 2p\bar{y}(p) + 1$$

$$\mathcal{L}[xy'(x); p] = -\frac{d}{dp}p\bar{y}(p) = -p\bar{y}'(p) - \bar{y}(p).$$

Taking the Laplace transform of both sides of equation (13.4.41) and making use of these results we obtain the equation

$$p(p-1)\frac{d\bar{y}}{dp}+(p-n-1)\bar{y}(p)=0.$$

It is easily shown [see § 7.2] that the solution of this first-order equation (with 'variables separable') is

$$\bar{y}(p)=Cp^{-n-1}(p-1)^n,$$

where C is an arbitrary constant. To satisfy the initial condition we must take

$$C=\lim_{p\to\infty}p\bar{y}(p)=y(0)=1.$$

Hence the required solution may be written

$$\begin{aligned}y(x)&=\mathcal{L}^{-1}[p^{-n-1}(p-1)^n;x]\\&=\mathcal{L}^{-1}[p^{-1}(1-p^{-1})^n;x]\\&=\mathcal{L}^{-1}\sum_{r=0}^{n}\frac{n!}{r!(n-r)!}p^{-r-1}.\end{aligned}$$

Since $\mathcal{L}^{-1}p^{-r-1}=x^r/r!$ we see that the given initial value problem has the polynomial solution $y=L_n(x)$, where

$$L_n(x)=\sum_{r=0}^{n}\frac{n!x^r}{r!r!(n-r)!}. \tag{13.4.43}$$

Equation (13.4.41) is called *Laguerre's differential equation* and the polynomial $L_n(x)$ is called the *Laguerre polynomial* of degree n.

13.4.8. The Diffusion Equation

As an illustration of the use of the Laplace transform in solving partial differential equations, we consider the problem of finding the function $u(x,t)$ which satisfies the diffusion equation [see §§ 8.1 and 8.3]

(i) $$\frac{\partial^2 u}{\partial x^2}=\frac{1}{k}\frac{\partial u}{\partial t}\qquad(x\geq 0,\,t\geq 0)$$

the initial condition

(ii) $$u(x,0)=0,\qquad(x\geq 0),$$

and the boundary conditions

(iii) $$u(0,t)=f(t),\qquad(t\geq 0),\qquad u(x,t)\to 0\quad\text{as }x\to\infty.$$

If we introduce the Laplace transform

$$\bar{u}(x,p)=\mathcal{L}[u(x,t);t\to p]$$

and use the initial condition (ii) we see that

$$\mathcal{L}\left[\frac{\partial u}{\partial t}; t \to p\right] = pu(x, p),$$

so that taking the Laplace transform of both sides of the diffusion equation (i) we obtain the ordinary differential equation

$$(D_x^2 - p/k)\bar{u}(x, p) = 0,$$

for the function $\bar{u}$ which, because of (ii) also satisfies the conditions

$$\bar{u}(0, p) = \bar{f}(p), \qquad \bar{u}(x, p) \to 0 \quad \text{as } x \to \infty.$$

We therefore have

$$\bar{u}(x, p) = \bar{f}(p) e^{-x\sqrt{(p/k)}}.$$

Inverting and making use of the convolution theorem we find that the required solution is

$$u(x, t) = \int_0^x f(s)K(x - s, t) \, ds$$

where

$$K(x, t) = \mathcal{L}^{-1}[e^{-x\sqrt{(p/k)}}; p \to t].$$

Taking $c = x^2/4k$ in (13.4.7), we see that

$$K(x, t) = x(4kt^3)^{-1/2} \exp(-x^2/4kt).$$

13.4.9. An Application of the Convolution Theorem

Because of the form of the convolution theorem for the Laplace transform an integral equation of the type

$$\int_0^x f(t)K(x - t) \, dt = g(x), \qquad (x \ge 0) \tag{13.4.44}$$

can often be solved by taking the Laplace transform of both sides. With the obvious notation, we see that this leads to the relation

$$\bar{f}(p)\bar{K}(p) = \bar{g}(p),$$

connecting the Laplace transform of the unknown function f with those of the prescribed functions K and g. Solving this equation for $\bar{f}$, taking the inversion transform and making use of the convolution theorem we see that provided the inverse transform

$$\mathcal{L}^{-1}\left[\frac{1}{\bar{K}(p)}; x\right] = L(x), \tag{13.4.45}$$

exists, the solution of (13.4.44) is

$$f(x) = \int_0^x g(t)L(x-t)\,dt.$$

Also, if, as above, the inverse transform (13.4.45) does not exist but

$$\mathscr{L}^{-1}\left[\frac{1}{p\bar{K}(p)};x\right] = M(x) \tag{13.4.46}$$

does, then the solution can be written in the form

$$f(x) = \frac{d}{dx}\int_0^x g(t)M(x-t)\,dt.$$

For example, consider the integral equation

$$\int_0^x \frac{f(t)\,dt}{(x-t)^a} = g(x), \qquad (x>0, 0<a<1) \tag{13.4.47}$$

(the *generalized Abel equation*). In this case $K(x) = x^{-a}$ and it follows from equation (13.4.2) that $\bar{K}(p) = p^{a-1}\Gamma(1-a)$; it is easily seen that the corresponding inverse transform (13.4.45) does not exist, but that (13.4.46) gives in this case

$$M(x) = \mathscr{L}^{-1}[p^{-a}/\Gamma(1-a); x].$$

Again using the equation (13.4.2), in its inverse form this time, and the well-known relation $\Gamma(a)\Gamma(1-a) = \pi\,\mathrm{cosec}\,(\pi a)$ [see also (10.2.12)] we find that, in this case,

$$M(x) = \frac{\sin(\pi a)}{\pi}x^{a-1},$$

so that the required solution of (13.4.47) is

$$f(x) = \frac{\sin(\pi a)}{\pi}\frac{d}{dx}\int_0^x \frac{g(t)\,dt}{(x-t)^{1-a}}. \tag{13.4.48}$$

In cases in which $g(t)$ is given explicitly such as,

$$\int_0^x f(t)J_0(x-t)\,dt = \sin x,$$

it is easier to proceed directly by taking the Laplace transform of both sides of the equation than to make use of the general solution involving an arbitrary function g. From the above equation and equations (13.4.3) and (13.4.10) we deduce immediately that

$$\bar{f}(p)(p^2+1)^{-1/2} = (p^2+1)^{-1},$$

from which it follows that $\bar{f}(p) = (p^2+1)^{-1/2}$ and hence that $f(t) = J_0(t)$ [see (13.7.22) in a different notation].

In the space available here, we have been able to mention only a few of the many possible applications of the Laplace transform. For a fuller account of such applications, the reader should consult Churchill (1972), Davies (1978), Doetsch (1937), Sneddon (1951) and (1972) and Widder (1971).

13.4.10.   The Two-Sided Laplace Transform

It is obvious from the definition of the Laplace transform that it can be used only in the analysis of problems concerning functions defined on the positive real line. Mention should be made of a transform, again with kernel e^{-px}, which can be used to deal with functions defined on the whole real line. If $f(x)$ is defined for all real values of x, we define its *two-sided Laplace transform*

$$\bar{f}_+(p) \equiv \mathcal{L}_+[f(x); p] = \int_{-\infty}^{\infty} f(x) e^{-px} dx, \qquad (13.4.49)$$

provided, of course, that the defining integral is convergent for some range of values of p [see § 4.6]. Writing the integral on the right of equation (13.4.49) in the form

$$\int_0^{\infty} f(x) e^{-px} dx + \int_0^{\infty} f(-x) e^{px} dx,$$

[see Theorem 4.1.2] we see that we have

$$\mathcal{L}_+[f(x); p] = \mathcal{L}[f(x); p] + \mathcal{L}[f(-x); -p].$$

Suppose that the integral defining the Laplace transform of $f(x)$ is convergent only if Re $(p) > \sigma_1$ and that the integral defining the Laplace transform of $f(-x)$ is convergent only if Re $(p) > \sigma_2$. Then the two-sided transform will exist only if p lies in the strip $\sigma_1 < $ Re $(p) < -\sigma_2$. If $\sigma_1 > -\sigma_2$ the two components of $\bar{f}_+$ will have no common strip of convergence so the two-sided transform cannot be defined. For instance e^{-x} does not possess a two-sided Laplace transform since $\mathcal{L}[e^{-x}; p]$ exists if Re $(p) > -1$ but $\mathcal{L}[e^x; -p]$ exists only if Re $(p) < -1$. On the other hand $\mathcal{L}_+[e^{-x^2}; p]$ exists for all values of Re (p).

The *inversion theorem* for the two-sided Laplace transform may be stated as:

If the function $\bar{f}_+(p)$ is holomorphic in the infinite strip $k_1 < $ Re $(p) < k_2$ of the complex p-plane [see §9.1] then the function f defined by the equation

$$f(x) = \frac{1}{2\pi i} \int_{c-i\infty}^{c+i\infty} \bar{f}_+(p) e^{px} dp, \qquad k_1 < c < k_2,$$

is such that $\bar{f}_+ = \mathcal{L}_+ f$.

An operational calculus based on the two-sided Laplace transform has been devised by van der Pol and Bremmer; this, together with its applications is described in van der Pol and Bremmer (1950). A theoretical discussion of the two-sided Laplace transform is given in Chapter VI of Widder (1971).

13.4.11. The Double Laplace Transform and Other Related Transforms

The *double Laplace transform* of a function of two variables $f(x, y)$ defined in the positive quadrant of the xy-plane, is defined by the equation

$$\bar{f}(p, q) = \int_0^\infty \int_0^\infty f(x, y)\, e^{-px-qy}\, dx\, dy \qquad (13.4.50)$$

and the *iterated Laplace transform* of the same function is defined by

$$\mathcal{I}[f(x, y); p] = \bar{f}(p, p) = \int_0^\infty \int_0^\infty f(x, y)\, e^{-p(x+y)}\, dx\, dy. \qquad (13.4.51)$$

The properties of these transforms can be deduced easily from those of the Laplace transform; a full account is given on pp. 221–229 of Sneddon (1972). If $\bar{f}(p) = \mathcal{L}[f(x); p]$ we define the *Stieltjes transform* of f by the equation

$$\mathcal{S}[f(x); y] = \mathcal{L}[\bar{f}(p); y],$$

that is,

$$\mathcal{S}[f(x); y] = \int_0^\infty \frac{f(x)}{x+y}\, dx. \qquad (13.4.52)$$

Although the Stieltjes transform is a powerful tool in certain branches of analysis it has not proved to be useful in applications. The theory of the Stieltjes transform is developed fully in Chapter VIII of Widder (1941); an account of the main properties of the transform is given in section 3.20 of Sneddon (1972).

13.5. THE Z-TRANSFORM

13.5.1. Definition and Formal Properties

The transform which does for the solution of difference equations [see I, § 14.12] what the Laplace transform does for that of differential equations is called the Z-transform. It is concerned with functions defined on the set of positive integers, that is with infinite sequences $(f_0, f_1, f_2, \ldots)$ which we shall denote by $\{f_n\}_{n=0}^\infty$ or simply by $\{f_n\}$.

We define

$$\mathcal{Z}[\{f_n\}; z] = \sum_{n=0}^\infty f_n z^{-n}, \qquad (13.5.1)$$

provided that there exists a positive real number a such that the series on the right converges for $|z| > a$ [see § 1.13].

The actual relation with the Laplace transform is rather tenuous. If we define a step function f by the equation

$$f(x) = \sum_{n=0}^{\infty} f_n S_n(x),$$

where $S_n(x)$ is the *characteristic function* of the interval $(n, n+1)$, that is,

$$S_n(x) = \begin{cases} 0 & 0 \le x \le n \\ 1 & n < x < n+1 \\ 0 & x \ge n+1 \end{cases}$$

then it is a simple calculation to show that

$$\mathcal{L}[f(x); p] = (1 - e^{-p}) \mathcal{L}[\{f_n\}; e^p],$$

so that

$$\mathcal{L}[\{f_n\}; z] = z(z-1)^{-1} \mathcal{L}[f(x); \log z]. \tag{13.5.2}$$

From well-known Maclaurin expansions [see Definition 3.6.2] it is easily shown that

$$\mathcal{L}[\{k^n\}; z] = \frac{z}{z-k} \qquad |z| > k; \tag{13.5.3}$$

$$\mathcal{L}[\{n\}; z] = \frac{z}{(z-1)^2} \qquad |z| > 1; \tag{13.5.4}$$

$$\mathcal{L}[\{n^2\}; z] = \frac{z(z+1)}{(z-1)^3} \qquad |z| > 1; \tag{13.5.5}$$

$$\mathcal{L}[\{(n+1)^{-1}\}; z] = z \log \frac{z}{z-1} \qquad |z| > 1; \tag{13.5.6}$$

$$\mathcal{L}[\{k^n/n!\}; z] = e^{k/z}. \tag{13.5.7}$$

If we now denote by $\{f_{n+m}\}$ the sequence $\{f_m, f_{m+1}, f_{m+2}, \ldots\}$, then from the definition of the Z-transform we have

$$\mathcal{L}[\{f_{n+1}\}; z] = \sum_{n=0}^{\infty} f_{n+1} z^{-n} = z \sum_{n=0}^{\infty} f_{n+1} z^{-n-1}$$

so that

$$\mathcal{L}[\{f_{n+1}\}; z] = z \mathcal{L}[\{f_n\}; z] - z f_0. \tag{13.5.8}$$

Similarly, since

$$\mathcal{L}[\{f_{n+2}\}; z] = \sum_{n=0}^{\infty} f_{n+2} z^{-n},$$

it follows that

$$\mathcal{L}[\{f_{n+2}\}; z] = z^2 \mathcal{L}[\{f_n\}; z] - z^2 f_0 - z f_1, \tag{13.5.9}$$

and, in general, that

$$\mathcal{L}[\{f_{n+m}\}; z] = z^m \mathcal{L}[\{f_n\}; z] - \sum_{r=0}^{m-1} f_r z^{m-r}. \tag{13.5.10}$$

If we define the *forward difference* Δf_n [see III, § 1.4] by the equation

$$\Delta f_n = f_{n+1} - f_n \tag{13.5.11}$$

and $\Delta^m f_n$ inductively by

$$\Delta^m f_n = \Delta(\Delta^{m-1} f_n),$$

then from equation (13.5.8) we deduce that

$$\mathcal{L}[\{\Delta f_n\}; z] = (z-1)\mathcal{L}[\{f_n\}; z] - z f_0 \tag{13.5.12}$$

and hence that

$$[\mathcal{L}\{\Delta^2 f_n\}; z] = (z-1)\mathcal{L}[\{\Delta f_n\}; z] - z \Delta f_0$$
$$= (z-1)^2 [\mathcal{L}\{f_n\}; z] - z \Delta f_0 - z(z-1) f_0.$$

By induction we can then prove the general formula

$$\mathcal{L}[\{\Delta^m f_n\}; z] = (z-1)^m \mathcal{L}[\{f_n\}; z] - z \sum_{r=0}^{m-1} (z-1)^{m-r-1} \Delta^r f_0. \tag{13.5.13}$$

Similarly defining the *backward difference* ∇f_n by the equation

$$\nabla f_n = f_n - f_{n-1}, \tag{13.5.14}$$

we deduce immediately from equation (13.5.8) that

$$\mathcal{L}[\{\nabla f_n\}; z] = (1 - z^{-1})\mathcal{L}[\{f_n\}; z], \tag{13.5.15}$$

and, by induction, that

$$\mathcal{L}[\{\nabla^m f_n\}; z] = (1 - z^{-1})^m \mathcal{L}[\{f_n\}; z]. \tag{13.5.16}$$

13.5.2. Convolution

In the theory of the Z-transform the convolution $\{(f \times g)_n\}$ of two sequences $\{f_n\}$, $\{g_n\}$ is defined by the equation

$$\{f \times g\}_n = \sum_{j=0}^{n} f_j g_{n-j}. \tag{13.5.17}$$

The convolution theorem then states that

$$\mathcal{L}[\{f \times g)_n\}; z] = \mathcal{L}[\{f_n\}; z] . \mathcal{L}[\{g_n\}; z]. \tag{13.5.18}$$

13.5.3. Inversion

The inversion of a Z-transform is a remarkably simple procedure. There are two main methods. $\{f_n\} = \mathcal{Z}^{-1}[\phi(z); n]$ if f_n is the coefficient of z^{-n} in the expansion of $\phi(z)$ in descending powers of z [see § 9.7]; from this we deduce (one of the corollaries of Cauchy's integral theorem [see § 9.4]) that

$$f_n = \frac{1}{2\pi i} \int_C z^{n-1} \phi(z)\, dz, \qquad (13.5.19)$$

where the contour C encloses all the singularities of $\phi(z)$ [see § 9.8].

As an example of the use of both methods we consider the case in which

$$\mathcal{Z}[\{f_n\}; z] = z^2(z-1)^{-3}. \qquad (13.5.20)$$

Since

$$(1 - z^{-1})^{-3} = \tfrac{1}{2} \sum_{r=0}^{\infty} (r+1)(r+2)z^{-r}$$

it follows that

$$f_n = \tfrac{1}{2}n(n+1).$$

Also, by the formula (13.5.19) we have

$$f_n = \frac{1}{2\pi i} \int_C \frac{z^{n+1}}{(z-1)^3}\, dz,$$

where C is any simple closed curve enclosing the point $z = 1$, and by a well-known result in the calculus of residues [see Lemma 9.9.3 and Theorem 9.9.1] this integral has the value

$$\frac{1}{2!}\left[\frac{d^2}{dz^2} z^{n+1}\right]_{z=1} = \tfrac{1}{2}n(n+1),$$

giving the same value as the series-expansion method.

13.5.4. Examples

We shall illustrate the use of the Z-transform by solving two simple difference equations [see I, § 14.13]. The first problem is to solve the difference equation

$$a_{n+1} - ka_n = k^n, \qquad a_0 = 1. \qquad (13.5.21)$$

If we write $\mathcal{A}(z) = \mathcal{Z}[\{a_n\}; z]$, then taking the Z-transform of both sides of (13.5.20) and making use of (13.5.3) and (13.5.8) we find that $\mathcal{A}(z)$ is determined by the equation

$$z[\mathcal{A}(z) - 1] - k\mathcal{A}(z) = z(z-k)^{-1}.$$

Writing the solution in the form

$$\mathcal{A}(z) = z(z-k)^{-1} + z(z-k)^{-2}$$

and making use of equations (13.5.3) and (13.5.4) we see that the required solution is given by the equation

$$a_n = k^n + nk^{n-1}.$$

With the same notation we see that the Z-transform of the solution of the difference equation

$$a_{n+2} + 3a_{n+1} + 2a_n = 0, \qquad (a_0 = 1, \ a_1 = 0),$$

satisfies

$$\{z^2 \mathscr{A}(z) - z^2\} + 3\{z\mathscr{A}(z) - z\} + 2\mathscr{A}(z) = 0,$$

so that

$$\mathscr{A}(z) = \frac{z(z+3)}{z^2 + 3z + 2} = 2 \cdot \frac{z}{z+1} - \frac{z}{z+2}.$$

Again using (13.5.3) we find that this corresponds to the solution

$$a_n = (-2)^{n-1} + 2(-1)^n.$$

For further examples of the use of the Z-transform, the reader is referred to section 3.17 of Sneddon (1972) and to the book Jury (1964) which is devoted to the subject and is of particular interest to those interested in system theory.

13.6.   THE MELLIN TRANSFORM

13.6.1.   Definition and Formal Properties

The Mellin transform of a function defined on the positive real line is defined by the equation

$$f^*(s) = \int_0^\infty f(x) x^{s-1} \, dx, \tag{13.6.1}$$

provided that there exists a region Ω in the complex s-plane such that the integral converges for $s \in \Omega$ [see § 4.6]. To express this relationship we use the notations $f^*(s) = \mathscr{M}[f(x); s]$, $f^* = \mathscr{M}f$, $f^*(s) = \mathscr{M}f(s)$.

From the definition we deduce immediately the following results

$$\mathscr{M}[f(ax); s] = a^{-s} \mathscr{M}[f(x); s] \qquad (a > 0), \tag{13.6.2}$$

$$\mathscr{M}[x^c f(x); s] = \mathscr{M}[f(x); s + c], \tag{13.6.3}$$

$$\mathscr{M}[f(x^p); s] = p^{-1} \mathscr{M}[f(x); s/p], \qquad (p > 0). \tag{13.6.4}$$

If we recall the formula [see Example 3.2.4(ii)]

$$\frac{d}{ds} x^{s-1} = (\log x) x^{s-1}$$

we deduce that

$$\mathcal{M}[\log xf(x); s] = \frac{d}{ds} f^*(s). \tag{13.6.5}$$

There are extensive tables of the Mellin transform in Vol. 2 of Erdélyi *et al.* (1954) and in Oberhettinger (1974) but it is useful to list the following frequently occurring transforms:

$$\mathcal{M}[e^{-x}; s] = \Gamma(s), \qquad \text{Re } (s) > 0 \tag{13.6.6}$$

$$\left. \begin{aligned} \mathcal{M}[\cos x; s] &= \Gamma(s) \cos (\tfrac{1}{2}\pi s), \\ \mathcal{M}[\sin x; s] &= \Gamma(s) \sin (\tfrac{1}{2}\pi s), \end{aligned} \right\} \quad 0 < \text{Re } (s) < 1 \tag{13.6.7}$$

$$\mathcal{M}[J_\nu(x); s] = \frac{2^{s-1}\Gamma(\tfrac{1}{2}\nu + \tfrac{1}{2}s)}{\Gamma(\tfrac{1}{2}\nu - \tfrac{1}{2}s + 1)} \tag{13.6.8}$$

[for definitions of the above functions see §§ 2.11, 2.12, 10.2.1 and 10.4.2].

The results concerning the Mellin transform of derivatives of a function follow easily from the rule for integrating by parts [see Theorem 4.3.1]. If there exist real numbers s_1 and s_2 such that

$$\lim_{x \to 0} x^{s-1} f(x) = 0, \qquad \lim_{x \to \infty} x^{s-1} f(x) = 0, \qquad s_1 < \text{Re } (s) < s_2,$$

and if $f^*(s-1)$ exists in the same band of the complex s-plane, then

$$\mathcal{M}[f'(x); s] = -(s-1)f^*(s-1). \tag{13.6.9}$$

Applying this result twice we find that

$$\mathcal{M}[f''(x); s] = (s-1)(s-2)f^*(s-2)$$

and we can obviously prove by induction over n [see I, § 2.1, NA5] that, for all values of s for which $f^*(s-n)$ exists,

$$\mathcal{M}[f^{(n)}(x); s] = (-1)^n \frac{\Gamma(s)}{\Gamma(s-n)} f^*(s-n). \tag{13.6.10}$$

The operator

$$\theta = x \frac{d}{dx}$$

(sometimes called the *Euler operator*) occurs frequently in the theory of linear differential equations [cf. § 7.5]. If we combine equations (13.6.3) and (13.6.9) we obtain the equation

$$\mathcal{M}[\theta f(x); s] = -sf^*(s). \tag{13.6.11}$$

We can also prove by induction over n that

$$\mathcal{M}[\theta^n f(x); s] = (-s)^n f^*(s). \tag{13.6.12}$$

It should also be observed that if we combine (13.6.3) and (13.6.10) we obtain

$$\mathcal{M}[x^n f^{(n)}(x); s] = (-1)^n \frac{\Gamma(s+n)}{\Gamma(s)}. \tag{13.6.13}$$

In plane polar coordinates (ρ, ϕ) the two-dimensional Laplace operator Δ_2 [see § 8.2] is defined by the equation

$$\Delta_2 u(\rho, \phi) = \frac{1}{\rho} \frac{\partial}{\partial \rho} \rho \frac{\partial u}{\partial \rho} + \frac{1}{\rho^2} \frac{\partial^2 u}{\partial \phi^2}; \tag{13.6.14}$$

it follows immediately from equation (13.6.12) with $n = 2$ that

$$\mathcal{M}[\Delta_2 u(\rho, \phi); \rho \to s] = (D_\phi^2 + s^2)\mathcal{M}[u(\rho, \phi); \rho \to s]. \tag{13.6.15}$$

Also, since

$$\mathcal{M}\left[\int_0^\infty f(x/t)g(t)\frac{dt}{t}; x \to s\right] = \int_0^\infty g(t)t^{-1}\mathcal{M}[f(x/t); x \to s] dt.$$

Using equation (13.6.2) we find that the integral on the right reduces to

$$\int_0^\infty g(t)f^*(s)t^{s-1} dt,$$

where $f^* = \mathcal{M}f$. We have therefore shown that

$$\mathcal{M}\left[\int_0^\infty f(x/t)g(t)\frac{dt}{t}; x \to s\right] = f^*(s)g^*(s) \tag{13.6.16}$$

—a result which is useful in the solution of integral equations of the form

$$\int_0^\infty K(x/y)f(y) \, dy = g(x), \qquad (x > 0);$$

see, for example, p. 279 of Sneddon (1972).

In a similar way we can show that

$$\mathcal{M}\left[\int_0^\infty f(xt)g(t) \, dt; x \to s\right] = f^*(s)g^*(1-s). \tag{13.6.17}$$

13.6.2.   Inversion

The inversion theorem for the Mellin transform may be stated in the form:

Suppose that $f^*(s)$ is a holomorphic function of the complex variable $s = \sigma + i\tau$ in the infinite strip $a < \sigma < b$ [see § 9.1] and that for any arbitrary small positive number ε, $f^*(s) \to 0$ uniformly as $|\tau| \to \infty$ in the strip $a + \varepsilon < \sigma < b - \varepsilon$ [see § 1.12]. Then if the integral

$$\int_{-\infty}^\infty f^*(\sigma + i\tau) \, d\tau$$

is absolutely convergent for each value of σ in (a, b) [see § 4.6] and if for a fixed $c \in (a, b)$ we define

$$f^*(s) = \frac{1}{2\pi i} \int_{c-i\infty}^{c+i\infty} x^{-s} f(x)\, dx,$$

for positive real values of x, then $f^*(s) = \mathcal{M}[f(x); s]$ for all s in the strip $a < \mathrm{Re}\,(s) < b$.

For the proof of the theorem in this form the reader is referred to section 4.3 of Sneddon (1972).

Making use of the inversion theorem we can easily show that

$$\mathcal{M}[f(x)g(x); s] = \frac{1}{2\pi i} \int_{c-i\infty}^{c+i\infty} f^*(z)g^*(s-z)\, dz. \qquad (13.6.18)$$

The case obtained by putting $s = 1$

$$\int_0^\infty f(x)g(x)\, dx = \frac{1}{2\pi i} \int_{c-i\infty}^{c+i\infty} f^*(z)g^*(1-z)\, dz \qquad (13.6.19)$$

is of particular interest.

13.6.3.   Applications

The Mellin transform is particularly useful in the solution of boundary value problems for a wedge-shaped region. In such applications use is made of a result which was suggested by taking $e^{i\phi}$ in place of a and ρ in place of x in equation (13.6.2). Proceeding formally we obtain the equation

$$\mathcal{M}[f(\rho\, e^{i\phi}); \rho \to s] = e^{-is\phi} \mathcal{M}[f(\rho); s].$$

Equating real and imaginary parts we obtain the pair of formulae

$$\mathcal{M}^{-1}[f^*(s)\cos\,(s\phi); s \to \rho] = \mathrm{Re}\, f(\rho\, e^{i\phi})$$
$$\mathcal{M}^{-1}[f^*(s)\sin\,(s\phi); s \to \rho] = -\mathrm{Im}\, f(\rho\, e^{i\phi}). \qquad (13.6.20)$$

Of course, equation (13.6.2) was established on the assumption that a is a positive real number so the above substitution is not valid. What it does do is to suggest a theorem which can in fact be proved rigorously.

To illustrate the use of these formulae we first of all calculate $\mathcal{M}[(1+x)^{-1}; s]$. Replacing the variable x in the defining integral by $t = x/(x+1)$ [see Theorem 4.3.2] we find that

$$\mathcal{M}[(1+x)^{-1}; s] = \int_0^1 t^{s-1}(1-t)^{-s}\, dt = \Gamma(s)\Gamma(1-s),$$

$0 < \mathrm{Re}\,(s) < 1$, and by a well-known result in the theory of the gamma-function [see (10.2.12)] we have finally

$$\mathcal{M}[(1+x)^{-1}; s] = \frac{\pi}{\sin(\pi s)}, \qquad 0 < \mathrm{Re}\,(s) < 1. \qquad (13.6.21)$$

Taking $p = 2$ and $\frac{1}{2}n$ respectively in equation (13.6.4) and making use of this last result, we deduce the pair of transforms

$$\mathcal{M}[(1+x^2)^{-1}; s] = \frac{\pi}{2\sin(\frac{1}{2}\pi s)} \qquad (13.6.22)$$

$$\mathcal{M}[(1+x^{(1/2)n})^{-1}; s] = \frac{2\pi}{n\sin(2s\pi/n)}. \qquad (13.6.23)$$

From the second equation of the pair (13.6.20) and this last result, we obtain the formula

$$\mathcal{M}^{-1}\left[\frac{\sin(s\phi)}{\sin(2s\pi/n)}; s \to \rho\right] = \frac{n \cdot \rho^{(1/2)n} \sin(\frac{1}{2}n\phi)}{2\pi\{1 - 2\rho^{(1/2)n} \cos(\frac{1}{2}n\phi) + \rho^n\}}. \qquad (13.6.24)$$

Using equation (13.6.9) in the slightly different form

$$\mathcal{M}\left[\int_x^\infty f(t)\,dt; s\right] = s^{-1}f^*(s+1) \qquad (13.6.25)$$

and taking $f(x) = (1+x^2)^{-1}$, we deduce from equation (13.6.22) that

$$\mathcal{M}[\frac{1}{2} - \tan^{-1} x; s] = \frac{\pi}{2s\cos(\frac{1}{2}\pi s)}. \qquad (13.6.26)$$

To illustrate the use of the Mellin transform in the solution of partial differential equations we consider the problem of finding a solution $u(\rho, \phi)$ of Laplace's equation in the infinite wedge $-\pi/n \le \phi \le +\pi/n$, satisfying the boundary conditions

$$u(\rho, -\pi/n) = f(\rho), \qquad u(\rho, +\pi/n) = g(\rho).$$

Recalling that the Laplacian is given by equation (13.6.14) and making use of equation (13.6.15) we find that the Mellin transform $u^*(s, \phi) = \mathcal{M}[u(\rho, \phi); \rho \to s]$ of the required function is the solution of the simple boundary value problem:

$$(D_\phi^2 + s^2)u^* = 0$$

$$u^*(s, -\pi/n) = f^*(s), \qquad u^*(s, +\pi/n) = g^*(s).$$

We therefore have

$$u^*(s, \phi) = f^*(s)K^*(s, \phi + (\pi/n)) + g^*(s)K^*(s, \phi - (\pi/n))$$

where

$$K^*(s, \theta) = \frac{\sin(s\theta)}{\sin(2s\pi/n)}.$$

Using the inverse form of equation (13.6.16) we see that the required function is

$$u(\rho, \phi) = \int_0^\infty t^{-1}\{f(t)K(\rho/t, \phi + (\pi/n)) + g(t)K(\rho/t, \phi - (\pi/n))\}\, dt$$

where the kernel K is defined by the equation

$$K(\rho, \phi) = \mathcal{M}^{-1}\left[\frac{\sin (s\phi)}{\sin (2s\pi/n)}, s \to \rho\right]$$

whose form is given by equation (13.6.24).

In the particular case in which

$$f(\rho) = g(\rho) = H(1 - \rho),$$

so that

$$f^*(s) = g^*(s) = s^{-1},$$

we find that

$$u^*(s, \rho) = \frac{\cos (s\phi)}{s \cdot \cos (s\pi/n)}.$$

From equation (13.6.26) and the formula (13.6.2) we deduce that

$$\mathcal{M}[\tfrac{1}{2} - \tan^{-1} (x^{n/2}); s] = \frac{\pi}{ns \cos (\pi s/n)}.$$

From the first equation of the pair (13.6.20) we then deduce that

$$u(\rho, \phi) = \frac{n}{\pi} \operatorname{Re} [\tfrac{1}{2} - \tan^{-1} (\rho^{n/2} e^{in\phi/2})].$$

13.7.   THE HANKEL TRANSFORM

13.7.1.   Inversion and Preliminary Results

The operator of the Hankel transform was defined by (13.1.6). The proof of the inversion formula for the Hankel transform is long and difficult (see Sneddon (1972)). It turns out that $\mathcal{H}_\nu^{-1} = \mathcal{H}_\nu$, that is, that if

$$\bar{f}_\nu(\xi) = \int_0^\infty xf(x)J_\nu(\xi x)\, dx, \qquad (\nu > -\tfrac{1}{2}), \tag{13.7.1}$$

then

$$f(x) = \int_0^\infty \xi \bar{f}_\nu(\xi)J_\nu(\xi x)\, d\xi \tag{13.7.2}$$

[see (10.4.10) for the definition of $J_\nu(x)$].

It follows immediately from the definition of the Hankel transform that if $a > 0$,

$$\mathcal{H}_\nu[f(x/a); \xi] = a^2 \mathcal{H}_\nu[f(x); a\xi].$$ (13.7.3)

Using the definition and the formula for integrating by parts [see Theorem 4.3.1] we can easily show that provided f is such that

$$\lim_{x \to 0} x^{\nu+1} f(x) = 0, \qquad \lim_{x \to \infty} x^{1/2} f(x) = 0,$$

$$\mathcal{H}_\nu\left[x^{\nu-1} \frac{d}{dx} \{x^{1-\nu} f(x)\}; \xi \right] = -\xi \mathcal{H}_{\nu-1} f(\xi)$$ (13.7.4)

and

$$\mathcal{H}_\nu\left[x^{-\nu-1} \frac{d}{dx} \{x^{\nu+1} f(x)\}; \xi \right] = \xi \mathcal{H}_{\nu+1} f(\xi).$$ (13.7.5)

The particular cases, valid if $x^2 f(x) \to 0$ as $x \to 0$:

$$\mathcal{H}_1[f'(x); \xi] = -\xi \mathcal{H}_0 f(\xi)$$ (13.7.6)

$$\mathcal{H}_0\left[\frac{1}{x} \frac{d}{dx} \{xf(x)\}; \xi \right] = \xi \mathcal{H}_1 f(\xi)$$ (13.7.7)

are frequently used in applications.

By combining (13.7.4) and (13.7.5) we obtain the relation

$$\mathcal{H}_\nu[\mathcal{B}_\nu f(x); \xi] = -\xi^2 \mathcal{H}_\nu f(\xi)$$ (13.7.8)

for the *Bessel operator* $\mathcal{B}_\nu$ defined by the equation [cf. § 7.6.1]

$$\mathcal{B}_\nu f = \frac{d^2 f}{dx^2} + \frac{1}{x} \frac{df}{dx} - \frac{\nu^2}{x^2}.$$

From this last result we obtain for the Hankel transform of order ν of the three-dimensional Laplacian in cylindrical coordinates [see § 8.2]

$$\Delta_3 w = \frac{\partial^2 w}{\partial \rho^2} + \frac{1}{\rho} \frac{\partial w}{\partial \rho} + \frac{1}{\rho^2} \frac{\partial^2 w}{\partial \phi^2} + \frac{\partial^2 w}{\partial z^2}$$

the equation

$$\mathcal{H}_\nu[\Delta_3 u(\rho, z) e^{i\nu\phi}; \rho \to \xi] = (D_z^2 - \xi^2)\bar{u}_\nu(\xi, z) e^{i\nu\phi}$$ (13.7.9)

where $\bar{u}_\nu(\xi, z) = \mathcal{H}_0[u(\rho, z); \rho \to \xi]$ and

$$D_z = \frac{d}{dz}.$$

In particular for the axisymmetric operator

$$\Delta_a = \frac{\partial^2}{\partial \rho^2} + \frac{1}{\rho} \frac{\partial}{\partial \rho} + \frac{\partial^2}{\partial z^2}$$

we have

$$\mathcal{H}_0[\Delta_a u(\rho, z); \rho \rightarrow \xi] = (D_z^2 - \xi^2)\bar{u}_0(\xi, z). \tag{13.7.9*}$$

These results are of use in potential theory. In the discussion of boundary value problems in elasticity and hydrodynamics we need the corresponding expressions involving the biharmonic operators

$$\mathcal{H}_\nu[\Delta_a u(\rho, z) e^{i\nu\phi}; \rho \rightarrow \xi] = (D_z^2 - \xi^2)^2 \bar{u}_\nu(\xi, z) \tag{13.7.10}$$

$$\mathcal{H}_0[\Delta_a u(\rho, z); \rho \rightarrow \xi] = (D_z^2 - \xi^2)^2 \bar{u}_0(\xi, z). \tag{13.7.11}$$

In theoretical investigations—particularly those concerned with the solution of dual integral equations—it is convenient to use a *modified operator* of Hankel transforms $S_{\eta,\alpha}$ instead of the Hankel operator $\mathcal{H}_\nu$. We define $S_{\eta,\alpha}$ by the equation

$$S_{\eta,\alpha}[f(t); x] = 2^\alpha x^{-\alpha} \int_0^\infty t^{1-\alpha} f(t) J_{2\eta+\alpha}(xt) \, dt. \tag{13.7.12}$$

If we write $\bar{f}_{\eta,\alpha}(x)$ for the function $S_{\eta,\alpha}[f(t); x]$ we see that this equation is equivalent to the relation

$$\mathcal{H}_{2\eta+\alpha}[t^{-\alpha}f(t); x] = 2^{-\alpha} x^\alpha \bar{f}_{\eta,\alpha}(x).$$

Applying the Hankel inversion theorem we deduce from this relation that

$$f(t) = t^\alpha \mathcal{H}_{2\eta+\alpha}[2^{-\alpha} x^\alpha \bar{f}_{\eta,\alpha}(x); t].$$

Comparing the right-hand side of this equation with the definition (13.7.12) we see that it can be interpreted to give the inversion formula

$$S_{\eta,\alpha}^{-1} = S_{\eta+\alpha,-\alpha} \tag{13.7.13}$$

for the modified operator of the Hankel transform.

13.7.2.   The Macauley–Owen Formula

There is no simple theorem of convolution type for the Hankel transform but Macauley–Owen (1939) proved the following theorem of Parseval type

$$\int_0^\infty x\bar{f}_\nu(x)\bar{g}_\nu(x) \, dx = \int_0^\infty xf(x)g(x) \, dx, \tag{13.7.14}$$

where $\bar{f}_\nu = \mathcal{H}_\nu f$, $\bar{g}_\nu = \mathcal{H}_\nu g$ ($\nu > -\frac{1}{2}$). This is often useful in the evaluation of integrals; for examples see Sneddon (1972).

Extensive tables of Hankel transforms are given in Vol. 2 of the set Erdélyi *et al.* (1954) and in Oberhettinger (1972). We shall list here some simple results which are used frequently:

$$\mathcal{H}_\nu[x^\nu(a^2 - x^2)^{\mu-\nu-1}H(a-x); \xi] = 2^{\mu-\nu-1}\Gamma(\mu-\nu)a^\mu\xi^{\nu-\mu}J_\mu(\xi a) \tag{13.7.15}$$

($a > 0$, $\mu > \nu \geq 0$). Making use of the Hankel inversion theorem we deduce immediately from this result the equation

$$2^{\mu-\nu-1}\Gamma(\mu-\nu)a^\mu \mathscr{H}_\nu[\xi^{\nu-\mu}J_\mu(\xi a); x]=x^\nu(a^2-x^2)^{\mu-\nu-1}H(a-x). \quad (13.7.16)$$

If we expand the Bessel function occurring in the definition of the Hankel transform [see (10.4.10)] and integrate term by term we obtain the equations

$$\mathscr{H}_\nu[x^{\nu-1}e^{-ax}; \xi]=2^\nu\pi^{-1/2}\Gamma(\nu+\tfrac{1}{2})(a^2+\xi^2)^{-\nu-(1/2)}\xi^\nu \quad (13.7.17)$$

$$\mathscr{H}_\nu[x^\nu e^{-ax}; \xi]=2^{\nu+1}\pi^{-1/2}a\Gamma(\nu+\tfrac{3}{2})(a^2+\xi^2)^{-\nu-(3/2)}\xi^\nu \quad (13.7.18)$$

$$\mathscr{H}_\nu[x^\nu e^{-ax^2/2}; \xi]=2^{\nu+1}a^{-\nu-1}\xi^\nu e^{-\xi^2/a}. \quad (13.7.19)$$

In each of these equations we have the conditions $a>0$, $\nu\geq0$.
Taking $a=\tfrac{1}{2}$ in (13.7.19) we find that

$$\mathscr{H}_\nu[x^\nu e^{-x^2/2}; \xi]=\xi^\nu e^{-\xi^2/2}$$

showing that the function $x^\nu e^{-x^2/2}$ is self-reciprocal under the Hankel transform of order ν [cf. § 13.2.4].

If we interchange the roles of x and ξ in equation (13.7.15) and then write out the left-hand side in accordance with the definition of the Hankel transform we obtain the result

$$\int_0^\infty x^{\nu-\mu+1}J_\nu(x\xi)J_\mu(xa)\,dx=\frac{\xi^\nu(a^2-\xi^2)^{\mu-\nu-1}}{2^{\mu-\nu-1}\Gamma(\mu-\nu)a^\mu}H(a-\xi),$$
$$(13.7.20)$$

$(a>0, \xi>0, \mu>\nu\geq0)$.

Special cases of these results are of particular interest. For example, if we take $\nu=0$ in (13.7.18) and (13.7.17) respectively we obtain the equations

$$\mathscr{H}_0[e^{-ax}; \xi]=a(a^2+\xi^2)^{-3/2}, \quad (13.7.21)$$

$$\mathscr{H}_0[x^{-1}e^{-ax}; \xi]=(a^2+\xi^2)^{-1/2}. \quad (13.7.22)$$

If in the last equation we put $a=p+it$ and proceed to the limit $p\to0$ we obtain the pair of equations $(t>0)$

$$\mathscr{H}_0[x^{-1}\cos(xt); x\to\xi]=\frac{H(\xi-t)}{\sqrt{\xi^2-t^2)}} \quad (13.7.23)$$

$$\mathscr{H}_0[x^{-1}\sin(xt); x\to\xi]=\frac{H(t-\xi)}{\sqrt{(t^2-\xi^2)}}. \quad (13.7.24)$$

Applying the Hankel inversion theorem to this pair of equations we derive the pair

$$\mathscr{H}_0\left[\frac{H(x-t)}{\sqrt{(x^2-t^2)}}; x\to\xi\right]=\xi^{-1}\cos(\xi t), \quad (13.7.25)$$

$$\mathscr{H}_0\left[\frac{H(t-x)}{\sqrt{(t^2-x^2)}}; x\to\xi\right]=\xi^{-1}\sin(\xi t), \quad (13.7.26)$$

$(t > 0)$. Another important result is obtained by noting that

$$\mathcal{H}_\nu[x^{s-1}; \xi] = \mathcal{M}[J_\nu(\xi x); x \to s + 1]$$

and then evaluating the Mellin transform by means of equation (13.6.8) we find that

$$\mathcal{H}_\nu[x^{s-1}; \xi] = \frac{2^s \Gamma(\tfrac{1}{2}s + \tfrac{1}{2}\nu + \tfrac{1}{2})}{\xi^{s+1} \Gamma(\tfrac{1}{2}\nu - \tfrac{1}{2}s + \tfrac{1}{2})}. \qquad (13.7.27)$$

13.7.3.   The Abel Transforms

Equations (13.7.25) and (13.7.26) suggest that there is a relationship between the Hankel transform and the Fourier transform. To express this relationship it has been found convenient to introduce two new operators $\mathcal{A}_1$ and $\mathcal{A}_2$ defined by the equations

$$\mathcal{A}_1[f(t); x] = \sqrt{\frac{2}{\pi}} \int_0^x \frac{f(t)\,dt}{\sqrt{(x^2 - t^2)}} \qquad (13.7.28)$$

$$\mathcal{A}_2[f(t); x] = \sqrt{\frac{2}{\pi}} \int_x^\infty \frac{f(t)\,dt}{\sqrt{(t^2 - x^2)}}. \qquad (13.7.29)$$

Since the determination of the inverse operators $\mathcal{A}_1^{-1}$ and $\mathcal{A}_2^{-1}$ is, in each case, equivalent to solving an integral equation of Abel type [see III, § 10.4.1], we call $\mathcal{A}_1$ the *Abel transform of the first kind* and $\mathcal{A}_2$ the *Abel transform of the second kind*. Proceeding by the method suggested we find that

$$\mathcal{A}_1^{-1}[g(x); t] = D_t \mathcal{A}_1[xg(x); t] \qquad (13.7.30)$$

$$\mathcal{A}_2^{-1}[g(x); t] = -D_t \mathcal{A}_2[xg(x); t] \qquad (13.7.31)$$

$(D_t = d/dt)$.

We can now interpret the well-known integrals [cf. § 10.4.4(i)]

$$\int_0^x \frac{\cos(\xi t)\,dt}{\sqrt{(x^2 - t^2)}} = \tfrac{1}{2}\pi J_0(\xi x), \qquad (x > 0, \, \xi > 0) \qquad (13.7.32)$$

$$\int_x^\infty \frac{\sin(\xi t)\,dt}{\sqrt{(t^2 - x^2)}} = \tfrac{1}{2}\pi J_0(\xi x), \qquad (x > 0, \, \xi > 0), \qquad (13.7.33)$$

in the transform form

$$\mathcal{A}_1[\cos(\xi t); x] = \sqrt{(\tfrac{1}{2}\pi)} J_0(\xi x) \qquad (13.7.34)$$

$$\mathcal{A}_2[\sin(\xi t); x] = \sqrt{(\tfrac{1}{2}\pi)} J_0(\xi x). \qquad (13.7.35)$$

Suppose now that $K(\xi, x)$ is an integrable kernel and that $\hat{f}_1 = \mathcal{A}_1 f$; then from the definition (13.7.28) and interchanging the order of the integrations we obtain the relation

$$\int_0^\infty K(t, x)\hat{f}_1(x)\, dx = \int_0^\infty f(x)\hat{K}_2(t, x)\, dx, \tag{13.7.36}$$

where

$$\hat{K}_2(t, x) = \mathcal{A}_2[K(t, y); y \to x]. \tag{13.7.37}$$

Similarly, we can show that

$$\int_0^\infty K(t, x)\hat{f}_2(x)\, dx = \int_0^\infty f(x)\hat{K}_1(t, x)\, dx, \tag{13.7.38}$$

where

$$K_1(t, x) = \mathcal{A}_1[K(t, y); y \to x]. \tag{13.7.39}$$

If we take $K(t, x) = (2/\pi)^{1/2} \sin(tx)$, then it follows from (13.7.35) that $\hat{K}_2(t, x) = J_0(tx)$ and from (13.7.36) that [see § 13.3]

$$\mathscr{F}_s\mathcal{A}_1 f(t) = \mathscr{H}_0[x^{-1}f(x); t]. \tag{13.7.40}$$

In a similar way we can use (13.7.38) and (13.7.34) to prove that

$$\mathscr{F}_c\mathcal{A}_2 f(t) = \mathscr{H}_0[x^{-1}f(x); t]. \tag{13.7.41}$$

Since $\mathscr{F}_s^{-1} = \mathscr{F}_s$ and $\mathscr{F}_c^{-1} = \mathscr{F}_c$, it follows from these last two equations that

$$\mathscr{F}_s\mathscr{H}_0 f(x) = \mathcal{A}_1[tf(t); x], \tag{13.7.42}$$

$$\mathscr{F}_c\mathscr{H}_0 f(x) = \mathcal{A}_2[tf(t); x]. \tag{13.7.43}$$

From the definitions of the operators involved we see that

$$\mathscr{H}_0[t^{-1}\hat{f}_c(t); x] = (2/\pi)^{1/2} \int_0^\infty J_0(tx)\, dt \int_0^\infty \cos(ty)f(y)\, dy$$

$$= (2/\pi)^{1/2} \int_0^\infty f(y)\, dy \int_0^\infty J_0(tx)\cos(ty)\, dt$$

$$= (2/\pi)^{1/2} \int_0^\infty f(y)\frac{H(x-y)}{\sqrt{(x^2-y^2)}}\, dy,$$

[by (13.7.23)] which is equivalent to the relation

$$\mathscr{H}_0[t^{-1}f_c(t); x] = \mathcal{A}_1 f(x). \tag{13.7.44}$$

In a precisely similar fashion we can establish the companion relation

$$\mathscr{H}_0[t^{-1}f_s(t); x] = \mathcal{A}_2 f(x). \tag{13.7.45}$$

If we replace f by f' in this equation and use $\mathscr{F}_s f'(t) = -t\hat{f}_c(t)$—see equation (13.3.9)—we have the relation

$$\mathscr{H}_0[\hat{f}_c(t); x] = -\mathcal{A}_2 f'(x). \tag{13.7.46}$$

Similarly, from equation (13.7.44) and using (13.3.8) we deduce that if $f(0) = 0$,

$$\mathcal{H}_0[\hat{f}_s(t); x] = \mathcal{A}_1 f'(x). \tag{13.7.47}$$

13.7.4.   Applications

To illustrate the use of the Hankel transform in the solution of mixed boundary value problems for a half-space, we consider the problem of finding the function $u(\rho, z)$ which satisfies the axisymmetric Laplace equation [see (8.2.10)]

$$\Delta_a u(\rho, z) = 0$$

in the half-space $z \geq 0$, when the plane boundary $z = 0$ is subjected to the mixed conditions

$$u(\rho, 0) = f(\rho), \qquad 0 \leq \rho < a, \tag{13.7.48}$$

$$\frac{\partial u(\rho, 0)}{\partial z} = 0, \qquad \rho > a, \tag{13.7.49}$$

and $u(\rho, z) \to 0$ as $(\rho^2 + z^2)^{1/2} \to \infty$. From equation (13.7.9*) we see that

$$\bar{u}(\xi, z) = \mathcal{H}_0[u(\rho, z); \rho \to \xi]$$

satisfies the ordinary differential equation

$$(D_z^2 - \xi^2)\bar{u}(\xi, z) = 0,$$

with the limiting condition $\bar{u}(\xi, z) \to 0$ as $z \to \infty$. We may therefore take

$$\bar{u}(\xi, z) = \xi^{-1} A(\xi) e^{-\xi z},$$

where the function A is arbitrary. Inverting this solution by means of the Hankel inversion theorem we find that

$$u(\rho, z) = \mathcal{H}_0[\xi^{-1} A(\xi) e^{-\xi z}; \xi \to \rho]. \tag{13.7.50}$$

From equations (13.7.48) and (13.7.49) we deduce that (13.7.50) is the solution of the stated problem if the function A satisfies the equations

$$\begin{aligned} \mathcal{H}_0[\xi^{-1} A(\xi); \rho] &= f(\rho), \qquad 0 \leq \rho < a; \\ \mathcal{H}_0[A(\xi); \rho] &= 0, \qquad \rho > a. \end{aligned} \tag{13.7.51}$$

Equations of this kind are known as *dual integral equations*.

To solve them we assume a representation of the form [see § 13.3]

$$A(\xi) = \mathcal{F}_c[g(t)H(t - a); \xi] \tag{13.7.52}$$

for the unknown function A. Substituting this expression into equations

(13.7.44) and (13.7.46) we obtain

$$\mathcal{H}_0[\xi^{-1}A(\xi); \rho] = (2/\pi)^{1/2} \int_0^{\min(\rho,a)} \frac{g(t)\, dt}{\sqrt{(\rho^2 - t^2)}}$$

$$\mathcal{H}_0[A(\xi); \rho] = -(2/\pi)^{1/2} H(a - \rho) \int_\rho^a \frac{g'(t)\, dt}{\sqrt{(t^2 - \rho^2)}}.$$

We see from this second formula that the form (13.7.52) automatically satisfies the second of the dual integral equations (13.7.51) and from the first one that the remaining equation of the pair is satisfied if g is the solution of

$$\mathcal{A}_1 g(\rho) = f(\rho), \qquad (0 \leq \rho < a)$$

that is, if

$$g(t) = D_t \mathcal{A}_1[\rho f(\rho); t], \qquad (0 \leq t < a).$$

13.8. FRACTIONAL INTEGRATION

We shall now consider generalizations of the Abel transforms defined by equations (13.7.28) and (13.7.29). The operators of these transforms are called the *Erdélyi–Kober operators*; they have proved to play a central role in the theory of dual and triple integral equations which itself is of great use in the discussion of mixed boundary value problems—see § 8.2 and, for instance, Sneddon (1966).

The operators $I_{\eta,\alpha}$ and $K_{\eta,\alpha}$ are defined for $\alpha > 0$, $\eta > -\frac{1}{2}$, by the equations

$$I_{\eta,\alpha} f(x) = \frac{2x^{-2\alpha - 2\eta}}{\Gamma(\alpha)} \int_0^x (x^2 - u^2)^{\alpha - 1} u^{2\eta + 1} f(u)\, du, \qquad (13.8.1)$$

$$K_{\eta,\alpha} f(x) = \frac{2x^{2\eta}}{\Gamma(\alpha)} \int_x^\infty (u^2 - x^2)^{\alpha - 1} u^{-2\alpha - 2\eta + 1} f(u)\, du, \qquad (13.8.2)$$

[see § 10.2.1] and for $\alpha < 0$, $\eta > -\frac{1}{2}$, by the equations

$$I_{\eta,\alpha} f(x) = x^{-2\alpha - 2\eta - 1} \mathcal{D}_x^n x^{2n + 2\alpha + 2\eta + 1} I_{\eta,\alpha + n} f(x), \qquad (13.8.3)$$

$$K_{\eta,\alpha} f(x) = (-1)^n x^{2\eta - 1} \mathcal{D}_x^n x^{2n - 2\eta + 1} K_{\eta - n, \alpha + n} f(x), \qquad (13.8.4)$$

where $\mathcal{D}_x$ denotes the differential operator defined by the equation

$$\mathcal{D}_x f(x) = \frac{1}{2} \frac{d}{dx} [x^{-1} f(x)] \qquad (13.8.5)$$

and n is a positive integer such that $0 \leq \alpha + n < 1$.

Changing the variable of integration in (13.8.1) from u to t where $u = xt^{1/2}$ [see Theorem 4.3.2] we see that we may write

$$I_{\eta,\alpha} f(x) = \frac{1}{\Gamma(\alpha)} \int_0^1 (1 - t)^{\alpha - 1} t^\eta f(xt^{1/2})\, dt.$$

Taking the Mellin transform of both sides of this equation and making use of (13.6.2) we find that

$$\mathcal{M}I_{\eta,\alpha}f(s) = \frac{1}{\Gamma(\alpha)} \int_0^1 (1-t)^{\alpha-1} t^{\eta-\frac{1}{2}s} f^*(s)\, dt$$

where $f^* = \mathcal{M}f$. The evaluation of the integral is now trivial [see (10.2.4)] and we find that [see (10.2.8)]

$$\mathcal{M}I_{\eta,\alpha}f(s) = \frac{\Gamma(\eta+1-\frac{1}{2}s)}{\Gamma(\alpha+\eta+1-\frac{1}{2}s)} f^*(s) \qquad (\alpha>0,\ \eta>-\tfrac{1}{2}). \qquad (13.8.6)$$

Similarly by changing the variable of integration in the integral on the right of (13.8.2) to t where $u = xt^{-1/2}$ we obtain the equation

$$K_{\eta,\alpha}f(x) = \frac{1}{\Gamma(\alpha)} \int_0^1 (1-t)^{\alpha-1} t^{\eta-1} f(xt^{-1/2})\, dt.$$

Again using (13.6.2) we find that

$$\mathcal{M}K_{\eta,\alpha}f(s) = \frac{\Gamma(\eta+\frac{1}{2}s)}{\Gamma(\alpha+\eta+\frac{1}{2}s)} f^*(s) \qquad (\alpha>0,\ \eta>-\tfrac{1}{2}). \qquad (13.8.7)$$

These results have been derived on the basis that $\alpha>0$. If we take the Mellin transforms of equations (13.8.3) and (13.8.4) and make use of the result

$$\mathcal{M}[\mathcal{D}_x f(x);\, s] = -\tfrac{1}{2}(s-1)f^*(s-2)$$

[see (13.6.3) and (13.6.9)] we find that equations (13.8.6) and (13.8.7) are valid for $\alpha<0$ also.

We need also the Mellin transform of $S_{\eta,\alpha}f(x)$ where $S_{\eta,\alpha}$ is the operator of the generalized Hankel transform defined by equation (13.7.12).

Making use of equations (13.6.8) and (13.6.2) we find on taking the Mellin transform of both sides of equation (13.7.12) that

$$\mathcal{M}S_{\eta,\alpha}f(s) = 2^{s-1} \frac{\Gamma(\eta+\frac{1}{2}s)}{\Gamma(1+\eta+\alpha-\frac{1}{2}s)} f^*(2-s). \qquad (13.8.8)$$

Making use of (13.8.6), (13.8.7) and (13.8.8) we can establish important relations connecting the Erdélyi–Kober operators with the generalized operator of the Hankel transform. For instance from equation (13.8.6) we have

$$\mathcal{M}I_{\eta,\alpha}I_{\eta+\alpha,-\alpha}f(s) = f^*(s),$$

showing that $I_{\eta,\alpha}I_{\eta+\alpha,-\alpha}$ is the identity operator. Hence the inversion theorem for $I_{\eta,\alpha}$ is

$$I_{\eta,\alpha}^{-1} = I_{\eta+\alpha,-\alpha}. \qquad (13.8.9)$$

Similarly, from equation (13.8.7) we can deduce that

$$K_{\eta,\alpha}^{-1} = K_{\eta+\alpha,-\alpha}. \qquad (13.8.10)$$

These results can of course be derived by using the solution of integral equations of generalized-Abel type. The inversion theorem for the generalized operator of the Hankel transform can be derived in a similar way from (13.8.8); the result, of course, has already been derived by means of the Hankel inversion theorem—see (13.7.13).

From (13.8.6) and (13.6.3) we have

$$\mathcal{M}[I_{\eta,\alpha}\{x^{2\beta}f(x)\}; s] = \frac{\Gamma(1+\eta-\tfrac{1}{2}s)}{\Gamma(1+\eta+\alpha-\tfrac{1}{2}s)} \mathcal{M}[x^{2\beta}f(x); s]$$

$$= \frac{\Gamma(1+\eta-\tfrac{1}{2}s)}{\Gamma(1+\eta+\alpha-\tfrac{1}{2}s)} f^*(s+2\beta),$$

and

$$\mathcal{M}[x^{2\beta}I_{\eta+\beta,\alpha}f(x); s] = \mathcal{M}[I_{\eta+\beta,\alpha}f(x); s+2\beta]$$

$$= \frac{\Gamma(1+\eta-\tfrac{1}{2}s)}{\Gamma(1+\eta+\alpha-\tfrac{1}{2}s)} f^*(s+2\beta),$$

showing that

$$I_{\eta,\alpha}[x^{2\beta}f(x)] = x^{2\beta}I_{\eta+\beta,\alpha}f(x). \tag{13.8.11}$$

In a similar way we can show that

$$K_{\eta,\alpha}[x^{2\beta}f(x)] = x^{2\beta}K_{\eta-\beta,\alpha}f(x). \tag{13.8.12}$$

From equations (13.8.6) and (13.8.8) we deduce that

$$\mathcal{M}I_{\eta+\alpha,\beta}S_{\eta,\alpha}f(s) = \frac{\Gamma(1+\eta+\alpha-\tfrac{1}{2}s)}{\Gamma(1+\eta+\alpha+\beta-\tfrac{1}{2}s)} \mathcal{M}S_{\eta,\alpha}f(s)$$

$$= 2^{s-1}\frac{\Gamma(\eta+\tfrac{1}{2}s)}{\Gamma(1+\eta+\alpha+\beta-\tfrac{1}{2}s)} f^*(2-s)$$

$$= \mathcal{M}S_{\eta,\alpha+\beta}f(s),$$

and hence that

$$I_{\eta+\alpha,\beta}S_{\eta,\alpha} = S_{\eta,\alpha+\beta}. \tag{13.8.13}$$

By the same method we can show that

$$S_{\eta+\alpha,\beta}S_{\eta,\alpha} = I_{\eta,\alpha+\beta} \tag{13.8.14}$$

$$K_{\eta,\alpha}S_{\eta+\alpha,\beta} = S_{\eta,\alpha+\beta} \tag{13.8.15}$$

$$S_{\eta,\alpha}S_{\eta+\alpha,\beta} = K_{\eta,\alpha+\beta} \tag{13.8.16}$$

$$S_{\eta+\alpha,\beta}I_{\eta,\alpha} = S_{\eta,\alpha+\beta} \tag{13.8.17}$$

$$S_{\eta,\alpha}K_{\eta+\alpha,\beta} = S_{\eta,\alpha+\beta}. \tag{13.8.18}$$

These relations may be established by the use of standard results in the theory of Bessel functions [see § 10.4]. For example, to prove (13.8.13) we write $I_{\eta+\alpha,\beta}S_{\eta,\alpha}f$ as a repeated integral, change the order of the integrations and use Sonine's first integral [see Watson (1944) Chapter 13] to evaluate the inner integral; we then identify the result as $S_{\eta,\alpha+\beta}f$.

By the same method we can derive the product rules

$$I_{\eta,\alpha}I_{\eta+\alpha,\beta} = I_{\eta,\alpha+\beta} \tag{13.8.19}$$

$$K_{\eta,\alpha}K_{\eta+\alpha,\beta} = K_{\eta,\alpha+\beta}. \tag{13.8.20}$$

A result due to Erdélyi—and called by him the rule for *fractional integration by parts*—should also be noticed. It states that

$$\int_0^\infty xf(x)I_{\eta,\alpha}g(x)\,dx = \int_0^\infty xg(x)K_{\eta,\alpha}f(x)\,dx. \tag{13.8.21}$$

This result can be looked at in a different way. In a certain sense it shows that $K_{\eta,\alpha}$ is the adjoint of the operator $I_{\eta,\alpha}$ [see Definition 19.2.22].

There is also a connection between the Erdélyi–Kober operators and the operator L defined by the equation

$$L_\nu = \frac{\partial^2}{\partial\rho^2} + \frac{2\nu+1}{\rho}\frac{\partial}{\partial\rho} \tag{13.8.22}$$

which occurs in generalized axisymmetric potential theory (GASPT). The axisymmetric Laplace equation in $2\nu+3$ dimensions is [see § 8.2]

$$\Lambda_\nu u(\rho, z) = 0 \tag{13.8.23}$$

where

$$\Lambda_\nu = L_\nu + D_z^2. \tag{13.8.24}$$

It can be shown that under very wide circumstances

$$I_{\nu,\alpha}L_\nu = L_{\nu+\alpha}I_{\eta,\alpha}. \tag{13.8.25}$$

If we call a solution of (13.8.23) a $(2\nu+3)$-dimensional *symmetric potential* then equation (13.8.25) leads to the interesting result that *if u is a $(2\nu+3)$-dimensional potential then $I_{\eta,\alpha}u$ exists and is a $(2\nu+2\alpha+3)$-dimensional symmetric potential.*

This last result enables us to generate symmetric potentials from plane harmonic functions [see § 9.13]. For instance, if $h(x, y)$ is a two-dimensional harmonic function, that is, is a harmonic function which is an even function of x, then if $\nu > -1$,

$$u(\rho, z) = I_{-(1/2),\nu+(1/2)}h(\rho, z) \tag{13.8.26}$$

exists and is a $(2\nu+3)$-dimensional symmetric potential.

For example, the function

$$h(x, y) = \tfrac{1}{2}\Gamma(\nu+\tfrac{1}{2})[f(y+ix)+f(y-ix)]$$

is harmonic for every twice-differentiable arbitrary function f and is even in x so that

$$u(\rho, z) = \tfrac{1}{2}\Gamma(\nu + \tfrac{1}{2})I_{-(1/2),\nu+(1/2)}[\{f(z + is) + f(z - is)\}; s \to \rho]$$

is a $(2\nu + 3)$-dimensional symmetric potential. Writing out the integral on the right-hand side of this equation we obtain

$$u(\rho, z) = \rho^{-2\nu} \int_0^\infty (\rho^2 - s^2)^{\nu-(1/2)}[f(z + is) + f(z - is)]\, ds$$

which is easily transformed to

$$u(\rho, z) = \int_0^\pi \sin(2\nu\theta)f(z + i\rho \cos \theta)\, d\theta. \qquad (13.8.27)$$

13.9. OTHER TRANSFORMS

The integral transforms listed above are the ones which occur with the greatest frequency in applications but there are a few others which are used occasionally and which therefore should be mentioned.

The Y-transform is defined in terms of $Y_\nu(x)$, the Bessel function of the second kind of order ν, by the formula

$$\mathcal{Y}_\nu[f(x); t] = \int_0^\infty (xt)^{1/2} Y_\nu(xt)f(x)\, dx, \qquad (13.9.1)$$

and the **H**-transform is defined in terms of *Struve's function* $\mathbf{H}_\nu(x)$ by

$$\mathcal{H}_\nu^*[f(x); t] = \int_0^\infty (xt)^{1/2}\mathbf{H}_\nu(xt)f(x)\, dx. \qquad (13.9.2)$$

Tables of the Y- and **H**-transforms are given in Oberhettinger (1972). It turns out that

$$\mathcal{Y}_\nu^{-1} = \mathcal{H}_\nu^*,$$

that is that the equation $g = \mathcal{Y}_\nu f$ implies that $f = \mathcal{H}_\nu^* g$.

The *Hilbert transform* is defined by the equation

$$\mathcal{H}f(x) = \frac{1}{\pi} \int_{-\infty}^\infty \frac{f(t)\, dt}{t - x}, \qquad (13.9.3)$$

where the Cauchy principal value of the integral [see (9.10.4)] is taken, or, alternatively, by the equation

$$\mathcal{H}f(x) = -\frac{1}{\pi} \frac{d}{dx} \int_{-\infty}^\infty f(t) \log |1 - (x/t)|\, dt. \qquad (13.9.4)$$

It can be shown that, if $f \in L_2(\mathbb{R})$, then $\mathcal{H}f \in L_2(\mathbb{R})$ [see § 13.2.1] and that

$$\|f\|_2 = \|\bar{f}_H\|_2, \qquad \bar{f}_H = \mathcal{H}f. \qquad (13.9.5)$$

The inversion theorem for the Hilbert transform takes the simple form

$$\mathscr{H}^{-1} = -\mathscr{H}, \tag{13.9.5}$$

that is, $\bar{f}_H = \mathscr{H}f$ implies that $f = -\mathscr{H}\bar{f}_H$.

An account of the main properties of the Hilbert transform is given on pp. 233–238 of Sneddon (1972); tables of the Hilbert transform are given in Vol. 2 of Erdélyi *et al.* (1954).

A transform which is useful in the analysis of boundary value problems in a wedge-shaped region described by cylindrical coordinates by the inequalities $\rho \geq 0$, $0 \leq \phi \leq \alpha$, $0 \leq z \leq a$, is the *Kantorovich–Lebedev transform* defined by the equation

$$\mathscr{H}f(\tau) = \int_0^\infty x^{-1}f(x)K_{i\tau}(x)\,dx, \tag{13.9.6}$$

where $K_{i\tau}(x)$ is the *Macdonald function* defined by the equation [see § 2.13]

$$K_{i\tau}(x) = \int_0^\infty \exp\left(-x\cosh t\right)\cos\left(\tau t\right)\,dt. \tag{13.9.7}$$

For this transform, the inversion theorem is: *If $\tilde{f}(\tau) = \mathscr{H}f(\tau)$, then*

$$f(x) = \frac{2}{\pi^2}\int_0^\infty K_{i\tau}(x)\sinh\left(\pi\tau\right)\tilde{f}(\tau)\,d\tau. \tag{13.9.8}$$

The properties of the Kantorovich–Lebedev transform are discussed in Chapter 6 of Sneddon (1972), there an account is given of some applications. The reader should also consult section 17.4 of Davies (1978). Tables of the Kantorovich–Lebedev transform are contained in Oberhettinger and Higgins (1961).

The *Mehler–Fock transform* is useful in deriving the solution of problems formulated in a system of toroidal coordinates. The simplest transform—that of order zero—is defined by the equation

$$f_0^*(\tau) = \int_0^\infty f(x)P_{-(1/2)+i\tau}(\cosh x)\sinh x\,dx, \tag{13.9.9}$$

where $P_{-(1/2)+i\tau}(\cosh x)$ is the *Legendre function of complex order* defined by the equation

$$P_{-(1/2)+i\tau}(\cosh x) = \frac{\sqrt{2}}{\pi}\int_0^x \frac{\cos\left(\tau t\right)\,dt}{\sqrt{(\cosh x - \cosh t)}}. \tag{13.9.10}$$

For this transform the inversion formula is

$$f(x) = \int_0^\infty \tau\tanh\left(\pi\tau\right)P_{-(1/2)+i\tau}(\cosh x)f^*(\tau)\,d\tau. \tag{13.9.11}$$

An account of the main properties of Mehler–Kock transforms and of some of their simpler applications is given in Chapter 7 of Sneddon (1972). Tables of this transform are to be found in Oberhettinger and Higgins (1961).

A transform called the *Radon transform* has recently been used in the discussion of wave problems in theoretical physics. It is perhaps best approached in what, at first sight, seems a rather roundabout way. We consider two functions $f(\mathbf{n}, t)$ and $\tilde{f}(\mathbf{x})$ where $t \in \mathbb{R}$, $\mathbf{x} \in \mathbb{R}^3$ and $\mathbf{n} \in S^2 = \{\mathbf{x} \in \mathbb{R}^3 : |\mathbf{x}| = 1\}$. We say that $\tilde{f}$ is the $\mathcal{R}$-transform of f if [see § 13.2.6]

$$\mathcal{F}_3[\tilde{f}(\mathbf{x}); \mathbf{x} \to \boldsymbol{\xi}] = \mathcal{F}[f(|\boldsymbol{\xi}|^{-1}\boldsymbol{\xi}; t); t \to |\boldsymbol{\xi}|] \qquad (13.9.12)$$

and write $\tilde{f} = \mathcal{R}f$. $f = \mathcal{R}^{-1}\tilde{f}$, the inverse transform of f is called the *Radon transform of f.*

If we write

$$\hat{f}(\mathbf{n}, \eta) = \mathcal{F}[f(\mathbf{n}, t); t \to \eta] \qquad (13.9.13)$$

then

$$\mathcal{F}_3[\tilde{f}(\mathbf{x}); \mathbf{x} \to \boldsymbol{\xi}] = \hat{f}(|\boldsymbol{\xi}|^{-1}\boldsymbol{\xi}, |\boldsymbol{\xi}|), \qquad (13.9.14)$$

and by the Fourier inversion theorem [see § 13.2.2]

$$\tilde{f}(\mathbf{x}) = (2\pi)^{-3/2} \int_{\mathbb{R}^3} \hat{f}(|\boldsymbol{\xi}|^{-1}\boldsymbol{\xi}, |\boldsymbol{\xi}|) \exp\{i(\boldsymbol{\xi} \cdot \mathbf{x})\} \, d\boldsymbol{\xi}.$$

Writing $\boldsymbol{\xi} = k\mathbf{k}$, where $\mathbf{k} \in S^2$, we reduce this to

$$\tilde{f}(\mathbf{x}) = (2\pi)^{-3/2} \int_{S^2} d\mathbf{k} \int_0^\infty \hat{f}(\mathbf{k}, k) \exp\{ik(\mathbf{k} \cdot \mathbf{x})\}k^2 \, dk$$

$$= -\Delta_3(2\pi)^{-3/2} \int_{S^2} d\mathbf{k} \int_0^\infty \hat{f}(\mathbf{k}, k) \exp\{ik(\mathbf{k} \cdot \mathbf{x})\} \, dk.$$

If, in addition, we require of the function f that it have the symmetry property

$$f(-\mathbf{n}, -t) = f(\mathbf{n}, t), \qquad (13.9.15)$$

then

$$\hat{f}(-\mathbf{n}, -\eta) = \hat{f}(\mathbf{n}, \eta)$$

and so

$$\tilde{f}(\mathbf{x}) = -\Delta_3(2\pi)^{-3/2}\tfrac{1}{2} \int_{S^2} d\mathbf{k} \int_{-\infty}^\infty \hat{f}(\mathbf{k}, k) \exp\{ik(\mathbf{k} \cdot \mathbf{x})\} \, dk$$

$$= -\Delta_3 \frac{1}{4\pi} \int_{S^2} f\{\mathbf{k}, (\mathbf{k} \cdot \mathbf{x})\} \, d\mathbf{k}$$

$$= -\frac{1}{4\pi} \int_{S^2} f''\{\mathbf{k}, (\mathbf{k} \cdot \mathbf{x})\} \, d\mathbf{k} \qquad (13.9.16)$$

where $f''(\mathbf{n}, t) = D_t^2 f(\mathbf{n}, t)$. Equation (13.9.16) can be regarded as the definition of the $\mathcal{R}$-transform rather than the relation (13.9.12).

To find the form of the operator $\mathcal{R}^{-1}$ we write $\boldsymbol{\xi} = k\mathbf{k}$, $(\mathbf{k} \in S^2)$ in (13.9.12) and invert to find that

$$f(\mathbf{k}, t) = \mathcal{F}^* \left[(2\pi)^{-3/2} \int_{\mathbb{R}^3} \tilde{f}(\mathbf{x}) \, e^{-ik(\mathbf{k}.\mathbf{x})} \, d\mathbf{x}; \, k \to t \right].$$

Now

$$\mathcal{F}^* [e^{-ikp}; \, k \to t] = (2\pi)^{1/2} \delta(t - p)$$

so

$$f(\mathbf{k}, t) = \frac{1}{2\pi} \int_{\mathbb{R}^3} f(\mathbf{x}) \delta(t - \mathbf{k}.\mathbf{x}) \, d\mathbf{x}. \qquad (13.9.17)$$

Equation (13.9.17) defines the Radon transform. The detailed properties of the Radon transform are discussed in Gel'fand, Graev and Vilenkin (1966).

I. N. S.

REFERENCES

Bochner, S. (1959). Lectures on Fourier Integrals, *Ann. Math. Studies No. 42*, Princeton University Press.

Bochner, S. and Chandrasekharan, K. (1949). Fourier Transforms, *Annals of Mathematics Studies, No. 19*, Princeton University Press.

Campbell, G. A. and Foster, R. M. (1948). *Fourier Integrals for Practical Application*, John Wiley & Sons, New York.

Carslaw, H. S. (1930). *Fourier Series and Integrals*, 3rd edn., Macmillan, London.

Churchill, R. V. (1972). *Operational Mathematics*, 3rd edn., McGraw-Hill, New York.

Davies, B. (1978). *Integral Transforms and their Applications*, Springer-Verlag, New York.

Doetsch, G. (1937). *Theorie und Anwendung der Laplace-Transformation*, Springer-Verlag, Berlin.

Doetsch, G. (1950, 1955). *Handbuch der Laplace-Transformation*, Birkhauser, Basel, Bd. 1, Bd. 2.

Erdélyi, A. *et al.* (1954). *Tables of Integral Transforms*, 2 Vols., McGraw-Hill, New York.

Goldberg, R. R. (1961). *Fourier Transforms*, Cambridge University Press.

Jury, E. I. (1964). *Theory and Application of the Z-Transform*, John Wiley & Sons, New York.

Lighthill, M. J. (1958). *Introduction to Fourier Analysis and Generalised Functions*, Cambridge University Press.

McLachlan, N. W. and Humbert, P. (1950). *Formulaire pour le Calcul Symbolique*, Gauthier-Villars, Paris.

Mikusinski, J. G. (1959). *Rachunek Operatow*, Polskie Towarzystwo, Warsaw, 1953: Engl. Trans., *The Calculus of Operators*, Pergamon Press, Oxford.

Miles, J. W. (1971). *Integral Transforms and their Applications*, Cambridge University Press.

Oberhettinger, F. (1972). *Tables of Bessel Transforms*, Springer-Verlag, Berlin.

Oberhettinger, F. (1974). *Tables of Mellin Transforms*, Springer-Verlag, Berlin.

Oberhettinger, F. and Badii, L. (1973). *Tables of Laplace Transforms*, Springer-Verlag, Berlin.

Oberhettinger, F. and Higgins, T. P. (1961). *Tables of Lebedev, Mehler and Generalized Mehler Transforms*, Boeing Sci. Res. Lab., Seattle.

Paley, R. E. A. C. and Wiener, N. (1934). *Fourier Transforms in the Complex Domain*, Amer. Math. Soc. Colloquium Publ., XV, New York.

Papoulis, A. (1963). *The Fourier Integral and its Applications*, McGraw-Hill, New York.

van der Pol, B. and Bremmer, H. (1950). *Operational Calculus based on the Two-Sided Laplace Transform*, Cambridge University Press.

Ross, K. A. (1965). *Fourier Series and Integrals*, (mimeographed lecture notes), Yale University.

Sneddon, I. N. (1951). *Fourier Transforms*, McGraw-Hill, New York.

Sneddon, I. N. (1966). *Mixed Boundary Value Problems in Potential Theory*, North Holland Publ. Co., Amsterdam.

Sneddon, I. N. (1972). *The Use of Integral Transforms*, McGraw-Hill, New York.

Titchmarsh, E. C. (1948). *An Introduction to the Theory of Fourier Integrals*, 2nd edn., Oxford University Press.

Tranter, C. J. (1951). *Integral Transforms in Mathematical Physics*, Methuen, London.

Watson, G. N. (1944). *The Theory of Bessel Functions*, Cambridge University Press.

Widder, D. V. (1941). *The Laplace Transform*, Princeton University Press.

Widder, D. V. (1971). *An Introduction to Transform Theory*, Academic Press, New York.

Wiener, N. (1932). *The Fourier Integral*, Cambridge University Press.

Zemanian, A. H. (1965). *Distribution Theory and Transform Analysis*, McGraw-Hill, New York.

Zemanian, A. H. (1969). *Generalized Integral Transforms*, John Wiley & Sons, New York.

Gel'fand, I. M., Graev, M. I. and Vilenkin, N. Ya. (1966). *Generalized Functions*, Vol. 5, Academic Press, New York.

CHAPTER 14

Mathematical Modelling

"Nature, and Nature's Laws lay hid in night.
God said 'Let Newton be!' and all was light."
Alexander Pope.

"It did not last. The Devil, howling 'Ho!
Let Einstein be!' restored the status quo."
Sir John Collings Squire.

14.1. MATHEMATICS AND SCIENCE

It is a historical fact that mathematics is a useful and important scientific tool. Indeed there has developed a strong tendency to measure the progress of a science by the extent to which it has been placed on a mathematical basis. That mathematics, in essence a mental construct, tells us so much about the real world, is in some ways a surprise. The most straightforward explanation—that mathematics has this property because it was designed to have it—seems not to take account of the full subtlety of the rich and complex interplay between mathematics and science. While many applications of mathematics arise 'by design', there are many that do not.

The *power* of mathematics may be traced to its ability to draw far-reaching conclusions from simple hypotheses. Its logical precision—not perfect, but refined to a high degree—permits lengthy and involved deductions to be carried out with confidence in the result. A feature whose importance is becoming increasingly apparent is the *unity* of mathematics, whereby a problem posed in one area (differential equations, say) may be transformed into a more tractable one in a totally different area (such as algebraic geometry) where it is finally solved. This unity—which is not apparent to the casual observer—casts doubt on snap judgements of the relative importance, for applications, of different areas. The theory of differential equations, both by its nature and by its evolution, has obvious importance for applied science. Algebraic geometry, largely an internal invention of pure mathematics, does not. If mathematical ideas were *only* relevant to the purposes that brought them about, it would be sensible to dismiss algebraic geometry as an applicable

591

subject. But once a connection with another area is established, the argument loses its force. And almost every branch of 'mainstream' mathematics impinges on almost every other. Thus almost every branch is *potentially* applicable. This potential may never be realized, or it may prove illusory—but attempts to assess it in advance are likely to prove ill-advised.

Why the power of mathematics can actually be brought to bear on real-world problems is much less clear. It seems (and at best this observation merely reformulates the problem) that mathematical structures have the ability to mimic events in the real world. This is again an observed fact, not fully accounted for by the 'argument from design', and it is an almost total mystery. Possibly it has more to do with the way the human mind perceives organized structure than with the real behaviour of the natural world: be that as it may, we do not understand it at all well.

This *Handbook* is about 'applicable' mathematics, which differs from 'applied' mathematics in the same way that a scalpel differs from an appendectomy. The adjective 'applicable' reflects a growing realization that the customary (and rather stereotyped) classification of mathematics into 'pure' and 'applied' territories leaves much to be desired. Previous chapters have concentrated on presenting the *mathematics*, as a body of ideas and techniques. In this chapter we take a look at the process by which mathematics *becomes* applicable: mathematical modelling.

14.2. MODELS

Felix Klein once published a paper which included a 'proof' that a certain partial differential equation, defined on a Riemann surface, possessed a solution. The proof was to imagine the surface to be made of tin, and to apply an electric current. The resulting equilibrium charge distribution would afford a solution to the equation.

Klein seems to have seen nothing peculiar about this argument, though it is patently absurd. Even if we could guarantee that the physical reality really did correspond to the equation—which we cannot—no amount of experimentation can show that *every* electrical input leads to a stable charge distribution. (In particular, it could oscillate.)

Throughout the eighteenth and nineteenth centuries mathematicians seem to have made no serious distinction between mathematical equations, and the phenomena those equations described. Thus Fourier, by a fundamentally fallacious argument involving power series, obtained the correct formulae [see (20.5.1) and (20.5.2)] for the coefficients of a Fourier expansion (in fact they had already been found by Euler), and then asserted on physical grounds that *every* function possessed a convergent Fourier series. The upshot of almost a century of chaos was that (a) Fourier was dead wrong, and the precise circumstances for possession of a convergent series [see Theorem 20.5.1] were remarkably elusive, (b) Fourier had had an absolutely first-class idea of enormous importance because he was right often enough for the method to be useful, and it was a very powerful method. This illustrates vividly both the dangers and the benefits of purely 'physical' reasoning about mathematics.

The attitude generally adopted today is that any mathematical description of the real world is a *model*. By manipulating the model we hope to comprehend something of the reality. And we no longer ask whether the model is *true*: we ask merely whether its implications check out experimentally.

The shift of emphasis permits a caveat of some importance. The real world is 'true' in all circumstances—that is, the universe behaves however the universe behaves, and we don't imagine that it can somehow 'go wrong'. A model, as now understood, is essentially an imperfect idealization, and it always goes wrong somewhere. We expect it to be useful only within a certain 'field of competence'.

Continuum mechanics, for example, models a fluid as a continuous density distribution, and treats time and space and matter as being infinitely divisible. This is nonsense on the level of atomic structure, so in that sense the model is simply false. However, it remains highly successful and useful; first at the macroscopic level where it is a good approximation (even if we can't prove that mathematically from the Schrödinger equation for a system of 10^{25} atoms), and more paradoxically at the molecular level too, in certain circumstances. There are in addition cases where 'atomic' phenomena are amplified into macroscopic effects inconsistent with a continuum model, though apparently within its field of competence. It is hard to explain a laser, or a pocket calculator, using macroscopic continuum mechanics.

Mathematical descriptions of reality, in practice, tend to form hierarchies of models whose fields of competence overlap in a fairly comfortable way without too much disagreement on the overlap. The theorist selects whichever model looks most appropriate for the problem in hand.

The situation is even more fluid when—as is commonly the case—the theorist is working in a field where there are no well-established models at all. This is equally the case in the 'hard' sciences and the 'soft'. The 'laws' of particle physics are essentially a mystery, although physicists have innumerable ideas as to what they might resemble. The 'laws' of psychology are also a mystery, and there is far less evidence that, in any useful sense, they exist at all.

Applications of mathematics, therefore, may arise either as the tentative advancement of new models, or as more or less rigorously argued consequences of existing models. In the case of a well-established model whose field of competence is well understood, there is a justifiable (but essentially pragmatic) tendency to expect the results to correspond closely to reality. In more tentative cases, initial confidence will be less. The failure of a well-established model is more interesting than a success; and the success of a tentative one more interesting than a failure, even if only partial. Models are most interesting when they tell us something we did not expect. The interest has no necessary connection with the utility, in this sense, because they are often most useful when they tell us something as a matter of routine.

14.3.   THE MODELLING PROCESS

A mathematical model is designed for a purpose. This may be precisely defined, for example, to find the optimum level of production for an industrial process; or it may be much vaguer, say to gain some understanding of the formation of raindrops. The success of the model, in the first instance, should be assessed in terms of the intended purpose. Further, the purpose conditions the choice of model. As Aris (1978) remarks: 'The purpose for which a model is constructed should not be taken for granted but, at any rate initially, needs to be made explicit'.

Whatever the purpose, however, the model will generally be used to make some deductions about the phenomenon to which it relates. Since the model is not the phenomenon, this deduction must take place in (at least) three steps:

(1) *Formulate* a model,
(2) *Deduce* mathematical implications within the model,
(3) *Interpret* these in real terms.

In a living science actual practice is often more complicated. The passage from model to reality or back may be made several times, the model may be modified in various ways, and so on.

For the scientist who is fortunate enough to be working in a thoroughly established field, the above steps may boil down to a rather simpler process: (1) write down 'the laws', (2) solve 'the equations', (3) read off 'the answer'. Here the transition between model and reality, being a standard step, is usually performed in a perfunctory way. It is however worth noting that it is still there.

EXAMPLE 14.3.1   (*Epidemics.*)   To illustrate the three steps we take a model from population theory, known as the Reed–Frost model, which describes an epidemic. We are concerned not with its correctness, but with its construction and analysis. Suppose we have a population of N individuals, affected by a disease which remains infectious in a given individual for a certain period of time. How does the pattern of contact between individuals affect the spread of the disease?

Step 1: Formulation

We must accept at the outset that our model will be grossly oversimplified: if it were made sufficiently complicated to take even reasonable variations into account, it would be impossible to analyse. We shall assume

(1) The infectious period is constant.

This is a mild simplification, but the next is not. We may scale time so that the infectious period is 1 unit: we now *discretize* time by assuming the whole process proceeds in steps of 1 unit of time. Contacts between individuals are only considered to happen at whole-number times. This assumption is wildly unreasonable by comparison with a real-life epidemic: unabashed, we make it anyway. Let's put it on record:

(2) The process occurs in discrete time-steps.

Next we make the simplest reasonable assumptions about contacts: *homogeneous mixing*. That is,

(3) During each discrete time-step contacts between individuals are independent random events [see II, § 3.5]. The probability of a given pair making contact is a constant p.

We let $q = 1 - p$, the probability of no contact. We have $0 \le p, q \le 1$.

(4) There are three types of individual: *susceptibles*, open to the disease; *infectives*, who suffer from it and can pass it on by contact to susceptibles during the single time-step that the infection persists; and *immunes* who can neither catch nor pass on the disease, having passed the end of their infective period. Transitions between types occur *only* according to the scheme

$$\text{susceptible} \rightarrow \text{infective} \rightarrow \text{immune}$$

and the *only* way to catch the disease is by contact with an infective.

We let $s(t)$, $i(t)$, and $r(t)$ denote the numbers of individuals in each type, respectively, at time t. (If the disease is possibly fatal, 'immune' includes 'dead'.) At the start of the epidemic the immunes already present play no further part, and we may assume that $r(0) = 0$ by excluding them from the total population N; so $N = s(0) + i(0)$ and we have

$$s(t) + i(t) + r(t) = N$$

for all t. Since we may solve this for $r(t)$ in terms of $s(t)$ and $i(t)$ we may ignore $r(t)$ from now on.

Imagine we are at time t, and seeking the behaviour at $t + 1$.

The probability of a given susceptible contacting *no* infectives by time $t + 1$ is $q^{i(t)}$. The probability of contacting at least one infective is therefore $1 - q^{i(t)}$, and the *expected* number of infectives at time $t + 1$ is the product of this with the number of susceptibles, namely $s(t)[1 - q^{i(t)}]$.

Still in search of simplicity, we make the model *deterministic* by assuming

(5) The expected happens.

Then we have the equation

$$i(t + 1) = s(t)[1 - q^{i(t)}]. \tag{14.3.1}$$

The present generation of infectives becomes immune. The only susceptibles left are the original ones, minus those who have just become infective, so

$$s(t + 1) = s(t) - i(t + 1)$$

$$= s(t)q^{i(t)}. \tag{14.3.2}$$

We take the system of equations (14.3.1), (14.3.2) as our model, whose formulation is now complete.

Step 2: Deduction

The heuristic reasoning employed in formulating the model is now inappropriate: we must take the model as given and study its mathematical properties.

We can, in a sense, solve it exactly. The equations (14.3.1), (14.3.2) form a *recurrence equation* in the following sense [see I, §§ 14.12, 14.13]. Starting with the known initial conditions $s(0)$ and $i(0)$ we may use (14.3.1), (14.3.2) with $t = 0$ to deduce $s(1)$ and $i(1)$; then with $t = 2$ to deduce $s(2)$ and $i(2)$, and so on. In fact:

$$t = 1: \quad s(1) = s(0)q^{i(0)}$$

$$i(1) = s(0) - s(1).$$

$$t = 2: \quad s(2) = s(0)q^{i(0)+i(1)}$$

$$i(2) = s(1) - s(2),$$

and so on.

If, however, we try to substitute back to get everything in terms of $s(0)$ and $i(0)$, the resulting expressions are cumbersome to say the least. To say something more useful we must set our sights on the right target.

We concentrate here on the *long-term* behaviour: the way $s(t)$ and $i(t)$ vary as t tends to infinity. It is easy to prove that in general

$$s(t+1) = s(0)q^{i(0)+i(1)+\ldots+i(t)}$$

$$i(t+1) = s(t) - s(t+1).$$

Since the $i(t)$ are positive and $q \leq 1$ it follows that the $s(t)$ form a decreasing sequence, bounded below (by 0). Hence $s(t)$ tends to a limit s_∞ as $t \to \infty$ [see § 1.5]. The second equation then implies that $i(t) \to 0$ as $t \to \infty$. So the problem is, to find s_∞.

Now

$$i(0) + i(1) + \ldots + i(t) = i(0) + s(0) - s(1) + s(1) - s(2) + \ldots + s(t-1) - s(t)$$

$$= i(0) + s(0) - s(t)$$

$$= N - s(t).$$

Therefore we have

$$s(t+1) = s(0)q^{N-s(t)}.$$

Letting $t \to \infty$ we get

$$s_\infty = s(0)q^{N-s_\infty}.$$

This is a (transcendental) equation for s_∞, given N and $s(0)$. It may be solved graphically by plotting the graphs of the left and right hand sides against s_∞. Before we do this more carefully, we can make life easier by putting the equation into a 'dimensionless' form. To do this, introduce new variables

$$\sigma(t) = s(t)/N$$

$$\sigma_\infty = s_\infty/N.$$

We further let

$$q^N = e^{-k}$$

where $k \geq 0$ is given by $k = -N \log q$ [see (2.11.6)].
    Now we have

$$\sigma_\infty = \sigma_0 e^{k(\sigma_\infty - 1)}. \tag{14.3.3}$$

Graphically, the solutions of (14.3.3) look like Figure 14.3.1. The exponential curve on the right-hand side can cross the diagonal line, the graph of the

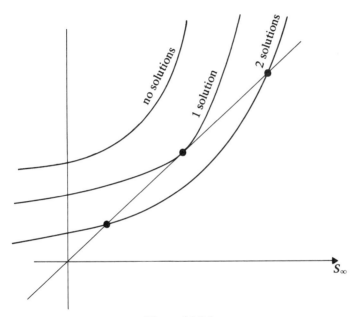

Figure 14.3.1

left-hand side, in 0, 1, or 2 points. But we must also bear in mind the restrictions imposed by our definition of the variables, namely

$$0 \leq \sigma_0 \leq 1$$

$$0 \leq \sigma_\infty \leq 1.$$

For certain values of k, these restrictions cut down the number of solutions further.
    To be more specific, suppose first that $\sigma_0 = 1$ (no infectives initially). Then $\sigma_\infty = 1$ is a solution of (14.3.3). For $k < 1$ the other solution for σ_∞ is bigger than 1 and may be ignored. For $k > 1$ there is a second solution $\sigma_\infty = \rho$, where ρ is a good bit smaller than 1. See Figure 14.3.2.

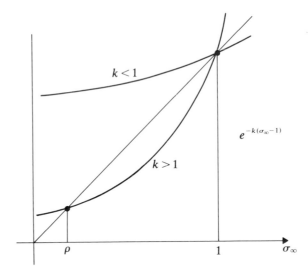

Figure 14.3.2

Now in practice $\sigma_0 = 1$ is not very interesting: no infectives means no possibility of an epidemic! But the number of initial infectives is usually a small proportion of the total population (or else we already *have* an epidemic) so we can assume that σ_0 is *near* 1. This pushes the exponential curve a little lower, so that when $k > 1$ we still get two solutions for σ_∞: one of them near 1, one near ρ. See Figure 14.3.3.

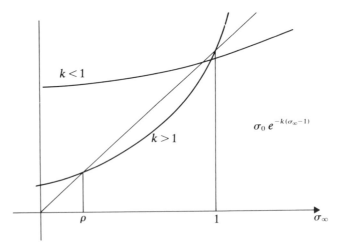

Figure 14.3.3

It can be shown by further analysis (which is not difficult, but which I do not wish to embark upon here) that the solution near ρ, when $k > 1$, is a *stable* one. By this I mean that, if we start with a proportion of susceptibles σ near to ρ, then it will stay near to ρ as time tends to infinity (and in fact approach ρ). The solution near 1, on the other hand, is *unstable*: if we displace the system slightly, and let it run, the value of σ will move *away* as time increases (and in fact generally will tend to ρ instead).

This is exactly analogous to stable and unstable equilibria in, say, dynamics. A rigid pendulum will hang vertically downwards, and in principle will balance pointing vertically upwards. The downwards position is *stable*, the upwards one *unstable*. We do not expect to encounter an upward-pointing, balanced pendulum, in practice! And for the same reason we may reject unstable solutions of the equations in our model of epidemics. Figure 14.3.4 illustrates schematically the instability of the solution near 1. At successive time-steps, the value of σ follows the 'staircase' and moves towards ρ.

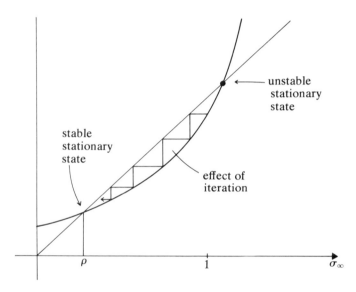

Figure 14.3.4

Step 3: Interpretation

What do these results mean in terms of epidemics?

If $k < 1$ there is a unique possible value of s_∞, which is also near 1. In other words, almost the whole population remains susceptible but does *not* (by definition!) catch the disease.

If $k > 1$ the long-term behaviour takes the system to a value of s_∞ much less than 1. We may choose to interpret this as representing an epidemic, with a proportion ρ of survivors. In fact for k near 1 we find ρ near 1: the severity of the epidemic lessens.

As for the parameter k which is crucial to the distinction: we can interpret that too. We have

$$(1-p)^N = q^N = e^{-k}$$

so [see (2.11.1)]

$$1 - p = e^{-k/N} = 1 - \frac{k}{N} + \dots .$$

If k is reasonably near 1 and N is large, we may truncate the power series expansion of $e^{-k/N}$ at this point, and we get $p \sim k/N$. Hence k is the expected number of contacts involving a given infective. This is a very reasonable conclusion: if each infective makes contact with more than one susceptible, the epidemic grows initially; and if he makes contact with less than one susceptible, it dies away.

Note that in this interpretation we have taken a *mathematical* distinction— whose appearance is in some sense 'natural'—namely whether $k > 1$ or $k < 1$; and *interpreted* this as the distinction epidemic/no epidemic. In view of the behaviour this is reasonable; but it is *not* part of the mathematics, and it is not even part of the assumptions of the model: it is an additional step in the interpretation of the results, and as such is as open to criticism as the initial modelling assumptions are. (In particular, would $k = 1 \cdot 0000001$ *really* count as an epidemic, given that the proportion of survivors would be, say, $0 \cdot 99999$ or thereabouts?)

EXAMPLE 14.3.2   (*The Pendulum.*)   As a further example of this process, consider one of the standard models of mechanics: the simple pendulum. We start with the physical observation that if a heavy bob is suspended by a thread, and set swinging, then it undergoes repeated and regular oscillations.

To formulate our model, we make the usual simplifying assumptions that

    (i) the bob is a point mass;
    (ii) the thread is weightless and inelastic;
    (iii) the pendulum swings in a vertical plane;
    (iv) effects of friction, air-resistance, etc. may be ignored.

This leads us to the following figure, in which θ denotes the angle through which the bob swings, m is its mass, and l its length. We are now in a very fortunate position: we can work within a standard theoretical framework, namely, Newton's laws of motion, to obtain a differential equation describing the system. For present purposes all we need to know is that the mass of a body multiplied by its acceleration in a given direction is equal to the force acting in that direction.

Considering a direction at right angles to the thread, we obtain

$$m \cdot \frac{d^2}{dt^2}(l\theta) = -mg \sin \theta \tag{14.3.4}$$

Figure 14.3.5

where g is the acceleration due to gravity. *This equation is our model.* Its formulation here is on the basis of general principles, rather than some *ad hoc* procedure: in less standard circumstances we might not have this luxury.

Now for some mathematical deductions.

First, note that there is a factor m on each side of (14.3.4) which may be cancelled out, to get (after some rearrangement of terms)

$$\frac{d^2\theta}{dt^2} + \frac{g}{l}\sin\theta = 0. \tag{14.3.5}$$

The solution of this differential equation is not easy, and the usual trick is to bend the rules and assume

(v) the angle θ is small.

Then it is plausible (though please note we have not proved this rigorously) that we may replace $\sin\theta$ in (14.3.5) by θ, to get the new equation

$$\frac{d^2\theta}{dt^2} + \frac{g}{l}\theta = 0. \tag{14.3.6}$$

This has the general solution (obtainable by standard methods [see Example 7.4.1])

$$\theta = A\sin(\rho t) + B\cos(\rho t) \tag{14.3.7}$$

where $\rho^2 = g/l$.

This opens the way to step 3 in the process: interpreting the results. Equation (14.3.7) describes a steady sinusoidal oscillation with period $2\pi/\rho = 2\pi\sqrt{l/g}$. Hence our model predicts steady oscillations with this period. Note that the period is independent of the mass m (a fact that is already apparent by equation (14.3.5) above) and of the coefficients A and B (related to the phase and amplitude of the oscillation) in (14.3.7).

This model for the pendulum differs from that for epidemics in many ways. In particular it is based on (more or less) sound physical principles of a more general nature. Nevertheless it exhibits the same three basic steps.

It also exhibits the slightly schizophrenic tendency of even the best-formulated model: to solve the mathematical problem exactly may not be feasible, and we may have to settle for approximate answers. Nobody—not even a pure mathematician—should be upset by an approximation to a solution, provided it is a good one! In particular the fact that one has approximated the *problem* may not imply that the result approximates a *solution*—as we discuss in the next section. Thus the change from equation (14.3.5) to (14.3.6) above, ideally, requires justification—and moderately sophisticated justification, at that. So does the approximation 'σ_0 near 1' in the epidemic analysis. In these two cases the approximations may be proved valid, and all is well. If no justification is known, but the modeller has good reason to believe his procedure is valid, he is of course welcome to press ahead regardless: no doubt time will tell. *Experimental* confirmation that the results of his calculations appear accurate is encouraging, and may often be sufficient; but it is no more *proof* that the approximations are mathematically valid than Klein's electric current was a proof that his theorem was mathematically valid.

With these two examples behind us, we shall now look in turn at some general principles behind each of the three modelling steps.

14.4.   FORMULATION OF A MODEL

No model can take into account every conceivable aspect of the phenomenon it describes. This remains true even when the model is to be formulated in terms of well-established physical 'laws' such as conservation of energy, the second law of thermodynamics, or the Schrödinger equation. All of these 'laws' are in some sense idealizations of reality (although energy conservation is so fundamental that its apparent failures have always been circumvented by defining new types of energy to make the equation balance again). These laws are of course extremely important, and scientists quite rightly prefer to base models on them whenever this is (a) possible and (b) comprehensible. William Kingdon Clifford (1956) put it this way:

'But between this inconceivably small error and no error at all, there is fixed an enormous gulf; the gulf between practical and theoretical exactness, and, what is even more important, the gulf between what is practically universal and what is theoretically universal. I say that a law is practically universal which is more exact than experiment for all cases that might be got at by such experiments as we can make. We assume this kind of universality, and we find that it pays us to assume it. But a law would be theoretically universal if it were true of all cases whatever; and this is what we do not know of any law at all.'

There are two basic criteria that a model should fulfil, and they are somewhat contradictory both in spirit and in substance. The first is that the model should reflect as accurately as possible the established features of the phenomenon. The second is that the model must be amenable to mathematical analysis with the available techniques.

The first is reasonably sound, philosophically speaking: it tells us to use the best theories that we have, to set up the model. The second is really rather objectionable: it constrains our choice of model not in terms of what we know of Nature, but in terms of a purely artificial construct: mathematics. But in practical terms a model is of very little use if we cannot handle the calculations that it involves.

Thus it is probably true that the main properties of bulk matter—say, the laws of elasticity—are derivable (at least as excellent approximations) from the Schrödinger equation for systems of, say, 10^{25} atoms. If macroscopic physics is indeed a consequence of microscopic, as is generally believed, this ought to be true. However, such an equation would be far too complicated to handle directly: it would not even be accessible to numerical calculation on a computer. (In fact the theoretical calculations for single atoms are of doubtful accuracy for transuranic elements, having of the order of 300 nucleons and 300 electrons: what hope is there for a steel bar?) Elasticians instead resort to a continuum model, in which the material properties are held to vary continuously through the material, and where quantum effects are neglected altogether [see § 8.6.4]. This gives a very successful model, but one that is based only tenuously on *fundamental* physical laws.

The practitioner of mathematical modelling, ideally, must have a working knowledge of two distinct areas. The first: the known physical properties of the system being modelled, and the type of model already known to provide useful insight. The second: those areas of mathematics from which new types of model might be drawn, if required.

The range of models currently applied is broad; but the vast majority fall into a fairly narrow range sanctioned by past successes. Paramount among these are systems of partial differential equations. (Indeed, to many scientists, 'model' *means* a system of partial differential equations.) Also noteworthy are statistical models, and discrete models such as finite difference equations (and modern 'finite element' models). More conceptual models, based on recent mathematical discoveries, are beginning to emerge: notably topological models and abstract algebraic structures. (Group theory has had a long record of success in physics, from crystallography to fundamental particles: as the natural formalization of symmetry concepts [see V, Chapter 11], this is not entirely surprising.)

The modeller must choose which features of the system under study are felt to be sufficiently important to include, and which may safely be neglected. This choice is normally dictated by experience and criteria such as mathematical rigour do not apply. A model that is known to be 'robust' in some sense, that is, insensitive to certain types of perturbations, carries with it a partial guarantee of its conclusions: all else being equal, a 'robust' model is to be preferred to a 'fragile' one. *But* the type of robustness present must be appropriate to the perturbations of interest, and again the best guide is experience. 'Fragile' models may prove misleading, unless it is known that the perturbations that destroy them are irrelevant to the purpose at hand.

For example, the casual neglect of 'small' terms from systems of differential equations may have more drastic effects than the modeller expects. A case in point is the Navier–Stokes equations for the flow of a viscous fluid [see § 8.6.1]: the zero-viscosity limit of solutions may be quite different from the exact solutions to the zero-viscosity equations. (This happens, for instance, in two-dimensional flow round a circular disc.) So neglecting the viscosity terms, even when the viscosity is very small, can be dangerous.

I am going to concoct a simple example to show how tricky this kind of thing can be. It is of course *too* simple, in that it can be solved easily and hence would be unlikely to trap anyone; but it is quite revealing and should serve as an adequate warning.

The problem is to solve the differential equation

$$\frac{dx}{dt} = x^2 + a^2$$

for small a, subject to the initial conditions

$$x = -1 \quad \text{when } t = 0.$$

(1) Suppose we neglect the (small) term a^2. Then we get

$$\frac{dx}{dt} = x^2$$

which can be solved by separation of variables [see § 7.2]

$$\int \frac{dx}{x^2} = \int dt.$$

This gives

$$\frac{-1}{x} = t + c$$

where c is constant: the initial conditions lead to $c = 1$. Thus the solution is

$$x = \frac{-1}{1 + t}. \tag{14.4.1}$$

(2) Now we solve exactly, with the a^2 term included. Separating the variables gives

$$\int \frac{dx}{x^2 + a^2} = \int dt$$

so that [cf. Table 4.2.1]

$$\frac{1}{a} \tan^{-1}\left(\frac{x}{a}\right) = t + c.$$

Putting $t = 0$, $x = -1$ we find that

$$c = \frac{1}{a} \tan^{-1}\left(\frac{-1}{a}\right) = -\frac{1}{a} \tan^{-1}\left(\frac{1}{a}\right).$$

Solving for x, we get

$$x = a \tan\left(at - \tan^{-1}\left(\frac{1}{a}\right)\right). \qquad (14.4.2)$$

(3) Now compare the behaviour of the solutions (14.4.1) and (14.4.2). The 'approximate' solution (14.4.1) represents behaviour in which x increases gradually from -1 towards 0, becoming closer and closer to 0 as time t tends to infinity.

However, equation (14.4.2), the exact solution, has x tending to infinity so fast that it gets there in finite time, namely when

$$at - \tan^{-1}\left(\frac{1}{a}\right) = \frac{\pi}{2}$$

that is,

$$t = \frac{\pi}{2a} + \frac{1}{a} \tan^{-1}\left(\frac{1}{a}\right).$$

This is quite different behaviour (see Figure 14.4.1).

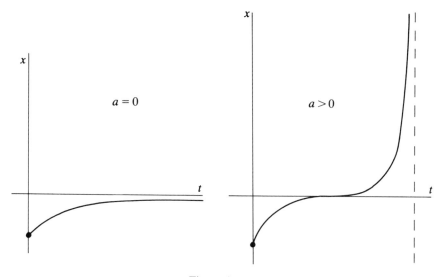

$a = 0$ $a > 0$

Figure 14.4.1

There are various ways of seeing why this happens without going through the solution itself. For example, if we seek stationary solutions $dx/dt = 0$ we see that these require $x^2 + a^2 = 0$, which is not possible when $a > 0$, although we can take $x = 0$ when $a = 0$. This warns us that the behaviour *near* zero values of a, and that *at* zero values, is different. In fact, what happens is that the zero-a approximation starts off quite well at $x = -1$, $t = 0$, but suddenly starts to go badly wrong as x nears the origin. The solution x drifts very slowly past the origin (and the smaller a is, the slower is the drift) but eventually it succeeds in getting past, and head off towards infinity at ever-increasing speed as the x^2 term starts to dominate the right-hand side of the equation.

It follows that one's experience of what terms may be omitted is best guided, not just by physical knowledge of what is small, but by an understanding of the mathematical effects of small changes to the model.

However, the problem is often more subtle than simply a choice of important or negligible terms.

For example, in early work on numerical weather prediction (see Thompson (1978)), it was found that the numerical methods used could 'resonate' and produce grossly inaccurate solutions. (This again is a numerical point: approximation of a differential equation by a finite difference equation can behave strangely when periodic solutions, whose periods take certain values related to the size of the finite differences chosen, can occur.) In this case, the periodic solutions were sound-waves. Unfortunately, sound-waves of arbitrary frequency are possible as solutions: resonances occur with *any* choice of finite difference. The way out, it transpired, was not to add in extra terms to the equations, making them more accurate: it was to *modify* the equations and make them more *coarse*. Sound-wave solutions were deliberately excluded by a kind of 'filtering' process. The resulting equations were *inferior* to the originals, as a description of reality, since phenomena known to occur had been deliberately excluded. But, for reasons more to do with the technicalities of *solving* them, the new equations gave a *better* model of the atmospheric waves (Rossby waves) relevant to weather-forecasting.

This example also illustrates the need to bear in mind the *purpose* of the model.

14.5.   DEDUCTIONS WITHIN A MODEL

The overriding principle here (though it is often broken: see below) is that once a model is formulated, deductions within it should be made as rigorously as possible. The ability to check the final answer against experiment should *not* be used as an excuse for sloppy mathematics.

The reason for this is simple. Suppose you set up a model, make a deduction, and test it—and it fails. You would like to conclude that the model is faulty. But, if your deductions have not been rigorous, no such conclusion may be drawn: the error may be merely an artefact of bad mathematics.

In practice, the ideal of rigour is attainable only by judicious interpretation of the qualifying phrase 'as possible'. Many computational methods of undoubted importance have yet to be justified in full rigour (renormalization group, numerous types of perturbation expansion, finite element methods, ...). While one might hope that mathematicians would be encouraged to investigate such methods with a view to establishing them rigorously, it seems unduly heavy-handed to insist that nobody use them until this has been done. (The obvious example is calculus, invented in the mid-seventeenth century and used widely thereafter, but not given a rigorous justification until the middle of the nineteenth century. Another is Fourier analysis, which took about a century to be properly understood.) It is equally heavy-handed to insist that the detailed justification be exhibited every time the method is used. Thus the basic theorems of calculus generally require the functions involved to be continuous [see, for example, Theorems 3.1.1 and 4.1.1]: an engineer who pays lip-service to this by remarking (correctly) that the function he is about to integrate is 'obviously continuous' has given ample notice that he appreciates the point and may be trusted to proceed sensibly. (He has also done far more than any self-respecting engineer would care to.)

If it transpires that the method can be rigorized in its entirety, or under hypotheses generally known to hold, then such lip-service is satisfactory. If the method is known to work, it may be used without apology: the precise form of words chosen is immaterial. Having spent much of their undergraduate careers learning not to evaluate a limit by putting $n = \infty$, most mathematicians find themselves doing precisely that, safe in the knowledge of when it works [see § 1.2].

Within the calculus, a more specific example is the use of certain formal and apparently unjustified manipulations to solve differential equations, introduced by Heaviside and often referred to as the 'operational calculus'.

EXAMPLE 14.5.1.   To solve the differential equation

$$2\frac{dy}{dx} + y = x^2$$

one introduces an 'operator' $D = d/dx$ [see also § 7.4.2] and proceeds in the following manner:

$$(2D + 1)y = x^2$$

so

$$y = \frac{1}{2D + 1}x^2$$

$$= (1 - 2D + 4D^2 - 8D^3 + \ldots)x^2$$

by binomial expansion [see (1.10.9)]

$$= x^2 - 2D(x^2) + 4D^2(x^2) - 8D^3(x^2) + \ldots.$$

Now if $D = d/dx$ then $D^2 = d/dx(d/dx) = d^2/dx^2$, etc. So this is

$$x^2 - 2(2x) + 4(2) - 8(0) + \ldots$$

and the remaining terms vanish, leaving

$$y = x^2 - 4x + 8.$$

Ignoring the niceties, we can check this:

$$2\frac{dy}{dx} + y = 2(2x - 4) + (x^2 - 4x + 8) = x^2.$$

The same sort of thing can be made to work for integral equations too. One of my favourites is the equation

$$\int y \, dx = y.$$

To solve this, write it as

$$\left(1 - \int\right) y = 0.$$

Then

$$y = \frac{1}{1 - \int} \, 0$$

$$= \left(1 + \int + \int^2 + \int^3 + \ldots\right) 0$$

$$= 0 + \int 0 + \int\int 0 + \int\int\int 0 + \ldots$$

$$= 0 + 1 + x + \frac{x^2}{2} + \frac{x^3}{6} + \ldots$$

$$= e^x \qquad [\text{see } (2.11.1)].$$

However unhappy one may feel with this, it is a fact that $y = e^x$ *does* satisfy $\int y \, dx = y$.

Such methods were first introduced without much attention either to their rigour, their meaning, or the conditions under which they were valid. It was always possible to check the answer at the end, you see.

Further, it is an 'experimental' fact that the methods generally work rather well. To ignore such evidence would be unimaginative: to leave it unexplained would be unmathematical. And of course there are good explanations (involving, in sophisticated versions, linear operators on a Banach space) that make sense of the steps used *and* let us keep control of when they are valid. The advantage, apart from professional pride, is that the tedious *a posteriori* verifications may be dispensed with. And the rigorous setting suggests more elaborate or powerful methods.

The tricky cases are (a) when the method *can* be made rigorous, but only at the expense of tedious computations, and (b) when it is known that the method *can* fail, 'often' works, and no good test is known to decide which. Here a certain pragmatism is in order, but if rigour is not used, the results are best presented as heuristic. Case (a) is exacerbated if (as is sometimes the case) it is known that the necessary conditions for rigour 'almost always' hold. (For example, 'almost all' zeros of a function are simple [see § 9.8], although it is often quite hard to check this explicitly because second derivatives can become very complicated.)

Avoidance of rigour is least justifiable when a rigorous argument is perfectly feasible, but (either through ignorance or plain sloppiness) it is not given. Worse, a certain amount of 'hand-waving' may be used to give the impression that an argument is rigorous when in truth it is nothing of the kind. (The standard applied mathematics text's 'proof' of Stokes's Theorem [see Theorem 6.3.1] springs to mind. Heuristic motivation, certainly; valuable as motivation to a student who neither needs, nor will be able to follow, a rigorous proof. But don't *call* it a proof.)

For a simple example, let us consider the bread-and-butter Taylor series approximation, which pervades applications of mathematics. It is often used as follows. Given an equation

$$f(x, y) = 0 \tag{14.5.1}$$

to solve, one expands f in a Taylor series up to some order [see § 5.8], say

$$f(x, y) = \bar{f}(x, y) + \text{higher terms},$$

and argues that solutions to (14.5.1) will be approximated by solutions to the polynomial equation

$$\bar{f}(x, y) = 0. \tag{14.5.2}$$

As icing on the cake, one further remarks that this approximation is justified because f is analytic [see Definition 5.8.1].

But suppose that

$$f(x, y) = (x^2 - y^2)^2 + \varepsilon y^6$$

where ε is small. The Taylor series, up to order 4, gives

$$\bar{f}(x, y) = (x^2 - y^2)^2.$$

Solutions to (14.5.2) are the lines $y = \pm x$ (Figure 14.5.1b). On the other hand, solutions to (14.5.1) consist of a curve having eight branches at the origin when $\varepsilon < 0$, and just the origin when $\varepsilon > 0$ (Figures 14.5.1a, 14.5.1c). Inasmuch as ε is small, Taylor truncation at order 4 might seem reasonable: the fact is that it does not work. (To see that there is an essential difficulty here, note that y^6 can equally well be replaced by y^{2n} for arbitrarily large n: the same problem arises. If you knew that f had Taylor series $(x^2 - y^2)^2$ for

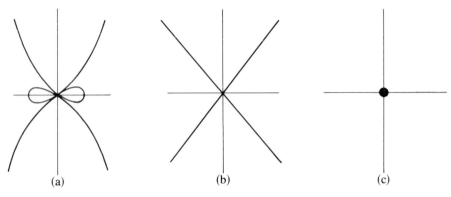

Figure 14.5.1

the first million terms, the zeros of f might still look quite different from $y = \pm x$.)

This difficulty arises here for a function f which is *analytic* (indeed polynomial): hence the much-cited analyticity is a red herring. We need to be able to deduce that the set of *zeros* of f is near to that of $\bar{f}$: analyticity implies that the *values* of $f(x, y)$ are near to those of $\bar{f}(x, y)$, which simply is not the same thing.

The true situation is quite complex. For functions of one variable [see § 2.10], it is sufficient to truncate at the first nonvanishing term. (This works for smooth functions, not just analytic, provided there *is* a first nonvanishing term.) This rule-of-thumb fails dismally for two or more variables. When symmetries are involved, the whole thing becomes desperately counter-intuitive.

Cavalier rule-of-thumb (which is what one generally finds) is unlikely to prove helpful in such questions. It was justified until very recently by the absence of rigorous tests. Such tests now exist, and are quite simple to apply (see catastrophe theory, Volume V, Chapter 8): it is to be hoped that their use will increase. (Of course, in very simple cases, rule-of-thumb works perfectly, so the danger of error is minimal: even so, one finds $x^2 y + y^4$ incorrectly truncated to $x^2 y$ in the literature.)

It is of course often the case that in the course of a calculation one encounters insuperable difficulties, as far as rigour goes. While it is thereafter hard to maintain that anything has been *proved*, much useful information may be derived heuristically. Thus innumerable approximations to the Navier–Stokes equations have been shown to provide good models of fluid flow under certain circumstances. But in fact, what can be claimed in the absence of rigorous verification that *solutions* to these equations approximate those to the Navier–Stokes equation, is that certain *modified* models appear to test out experimentally. Rigorous verifications remain as an intriguing challenge to the mathematical community, and the importance of such heuristics in suggesting lines of investigation should not be underestimated. In addition, the rigorous theory of some of the modified equations leads to some remarkable results of enormous interest.

14.6. ARIS'S MAXIMS

At this stage it is appropriate to reproduce a list of maxims for modellers, suggested by the chemical engineer Rutherford Aris and reproduced in his book (1978) (which is highly recommended to readers of this chapter who wish to pursue matters more deeply).

(1) Cast the problem in as elegant a form as possible.
(2) Choose a sympathetic notation, but don't become too attached to it.
(3) Make the variables dimensionless [cf. Example 14.3.1], since this is the only way in which their magnitudes take on general significance, but do not lose sight of the quantities which have to be varied later on in the problem nor forget the physical origin of each part.
(4) Use *a priori* bounds of physical or mathematical origin to keep all variables of the same order of magnitude, letting the dimensionless parameters show the relative size of the several terms.
(5) Think geometrically. See when you can reduce the number of variables (even at the expense of first treating an oversimplified problem), but keep in mind the needs of the general case.
(6) Use rough and ready methods, but don't carry them beyond their point of usefulness (e.g. isoclines in the phase plane).
(7) Find critical points [see § 5.6] and how the system behaves near them, or what is asymptotic behaviour at long or short times.
(8) Check limiting cases and see how they tie in with simpler problems that can be solved explicitly.
(9) Use crude approximations. Trade on the analogies they suggest, but remember their limitations.
(10) Rearrange the problem. Don't get fixed ideas on what are the knowns and what the unknowns. Be prepared to work with implicit solutions [see § 5.13].
(11) Neglect small terms, but distinguish between regular and singular perturbations.

(A few words of explanation are in order here. To solve a problem by perturbation methods one takes certain terms to be 'small'. Ignoring them altogether gives an equation which is solved exactly. It is then assumed that the solution to the true equation may be obtained by adding small terms to this solution. Such terms are substituted in, retaining only 'first order perturbations'—that is neglecting products of two or more 'small' terms.

Even when this process appears to work well, rigorous justification is often lacking. But there are well-understood cases in which it will *not* work well: when the problem involves some 'degeneracy'. Such a problem is said to be *singular*: recognizing this in advance warns us to be extra careful. The Taylor series example $(x^2 - y^2)^2 + \varepsilon y^6 = 0$ above conveys the spirit, if not the technicalities: one reason why it behaves badly is the 'degeneracy' in $(x^2 - y^2)^2 = 0$, whose solution is that of $x^2 - y^2 = 0$, but 'repeated twice'.

(A *regular* problem, of course, is one that is not singular.)

(12) Use partial insights and despise them not (e.g. Descartes' rule of signs [see I, Theorem 14.8.1]).

(13) These maxims will self-destruct. Make your own!

Aris's Maxims are most appropriate to differential equation models, but they have plenty to say elsewhere. Inasmuch as they come from one who (unlike myself) is actively involved in formulating and analysing mathematical models of real things, the reader should pay more attention to them than to the rest of the chapter.

Maxim (5) deserves comment. As a rough-and-ready rule, some 90% of mathematicians like to think geometrically. The other 10% think symbolically. There is an interesting discussion in Chapters 6 and 7 of Hadamard (1954) on the thought-processes of scientists (mostly mathematicians). Those who think symbolically are unlikely to benefit from being told to think geometrically.

That geometric intuitions can prove useful is well illustrated by the following (true) tale [refer to § 11.2 and V, § 5.2 for mathematical definitions].

Consider incompressible inviscid fluid flow in an open subset of the plane, subject to the constraint of being stationary at infinity. Let F be the set of all such flows, and let E be the subset comprising those of compact support (i.e. where the velocity is zero outside some bounded set). It is plausible that anything in F is approximable by things in E, that is, that the closure $\bar{E}$ of E is F. A sketch proof would go like this. Things in F are zero at infinity, hence small outside some compact set K. Restrict to K and chop down to zero outside a compact neighbourhood K' of K (using a bump function to keep things smooth). The result is in E and close to the original flow.

The literature of fluid dynamics contains a somewhat different, analytic argument, proving that $\bar{E} = F$. It is fairly technical and complicated. It seems to have been accepted for quite a long time.

It has recently been proved false.

Consider a flow through a hole in the y-axis (Figure 14.6.1). By incompressibility, flows in F are area-preserving. On any compact set, invariant under the flow, net flow through the hole must be zero. Hence net flow through the hole for anything in E is zero: by continuity the same goes for $\bar{E}$. But the flow illustrated lies in F and has nonzero net flow through the hole.

We can now see the fallacy in the 'sketch proof' above: if net flow through the hole is non-zero, say a quantity q of fluid per second, then the error involved in chopping down to a flow on a compact set is of the order of magnitude of q, no matter what set we use. There is simply no way to dispose of this net flow except by draining it away to infinity. What the fallacy in the analytic argument was I have no idea, but there is always room for error in

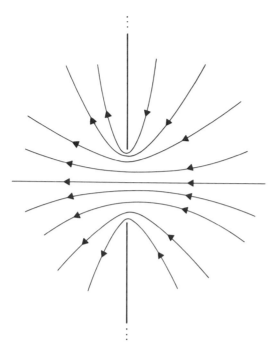

Figure 14.6.1

complicated and delicate calculation, especially if it is of a formal, non-intuitive nature.

The key insight in this problem is geometric (in fact, physical) and elementary. We are often told that a great advantage of 'applied' mathematics is that since it deals with 'real' things it offers useful physical insights. And that is true here: incompressibility implies area-preservation. Unfortunately, the fluid dynamicists' insight seems to have been working erratically in this case. At a guess, they got trapped by their own analytic paradigm: ignoring intuition, they turned the handle of the only machine they knew.

Comparable blunders by geometric intuition, shown up by analysis, are easy enough to find too. (The belief that continuous functions are almost everywhere differentiable was long held until Weierstrass wrecked it.) Perhaps the real lesson here is one about narrowness of conceptual vision. Concentrating too hard on specific, limited objectives may be conducive to intellectual blinkers. It is a pure-mathematical reflex to play around with a problem, reformulate it, try related or analogous problems, and so on. The best applied mathematicians have a similar reflex, but it is not so firmly embedded in the general consciousness. Applied mathematics could make good use of a little more awareness of the deformability of mathematical concepts: its thinking is often too rigid.

14.7. INTERPRETING PREDICTIONS

The guiding principle here is trite. *Be careful.* No matter how rigorously
mathematical conclusions have been derived from the model, they are worthless if incorrectly interpreted.

How *can* conclusions be incorrectly interpreted? It ought never to happen.
Anyone who understands the relation between the model and the reality,
well enough to set the model up, and who understands the mathematics well
enough to derive conclusions within the model, is hardly likely to make errors
of interpretation.

On the whole, in fact, this is so. But these stages are not always carried
out by the same person. When one is using somebody else's model, or
mathematical theorem, or technique, there is room for confusion.

A startling case reported in a paper of Gonzalez and Byrd (1978) concerns
the interpretation of statistical data about the effects of drugs on the central
nervous system.

In a typical experiment, a baboon was trained to push a lever in response
to food as a stimulus. Without the administration of some drug, its response
rate was R_c (control rate). With the drug, the rate was R_d. The drug rate,
expressed as a percentage of the corresponding control rate (at a given period
during the experiment), was plotted against control rate, using logarithmic
scales on both axes. A best-fit line of regression was found (Figure 14.7.1).

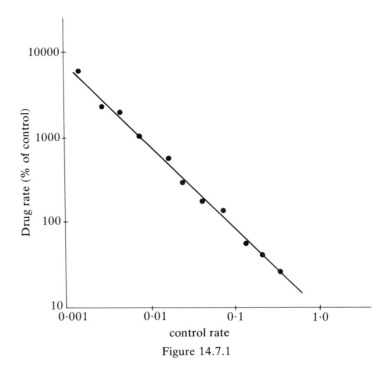

Figure 14.7.1

This line has a noticeable slope, close to -1. Such results, which occur with fair regularity, have often been interpreted in the literature as indicating that the drug effects are dependent on the control rates of responding.

However, as Gonzalez and Byrd (1978) pointed out, this interpretation is incorrect. The drug rate, as percentage of control rate, is given by $100 R_d / R_c$. If the slope of the regression line is j (and if the line fits well enough to indicate the trend) then its equation (given logarithmic scaling) will be of the form

$$\log (100 R_d / R_c) = j . \log R_c + k$$

for some constant k. That is,

$$R_d = m R_c^{j+1}$$

for a constant m.

When the slope $j = -1$, this gives $R_d = m$. That is, the drug rate is constant (on the regression line, hence approximately constant if the fit is good). The *effect* of the drug, which is what R_d is supposed to be measuring, is independent of the control rate.

A *post mortem* analysis might trace the error to two sources. One is a simple failure to understand the mathematical basis of the technique. The other is a failure to remember the purpose of the model, which is to discover the *effect* of the drug. Using a variable R_d / R_c to measure this is unwise, since it builds in a possibility of artefactual control-dependence (which happened regularly). Gonzalez and Byrd put it this way.

'*The relation between R_d and R_c can be unclear when the ratio R_d / R_c is plotted as a function of R_c . . . it can be made more apparent and less subject to misinterpretation by plotting R_d as a function of R_c.*'

Thus the error of interpretation, more properly, is an error in the choice of model (which variables to plot—I'm assuming that the model is the graph, since there's nothing else around). And it results from a failure to remember the purpose of the model.

The tale is also a monument to the blind pursuit of standard methods. Did nobody *notice* that R_d was roughly constant?

A second, more subtle misinterpretation occurs in the famous work of Alexander Thom (1967) on ancient megalithic structures, in connection with the 'megalithic yard'. This alleged standard unit of length was supposedly used by megalithic man to measure out his numerous stone circles. These circles are found over a wide area of Europe: such a widespread use of a standard unit argues for a well-developed system of communications, and a whole tower of speculation about megalithic man rests on this foundation.

The evidence for the existence of the unit is statistical. If measurements are taken on actual megalithic structures, and subjected to a certain statistical test (described in Thom's book), it transpires that there is an overwhelming probability that the measurements are integer multiples of a common length. The length is around three feet, hence the name *megalithic yard*.

The statistical analysis is impressive and looks convincing, especially to anyone unfamiliar with statistics. But it is also flawed.

The error in interpretation was found by Porteous (1973). What happened was that the statistical test used compared two hypotheses. One was integer multiples of a common length; the other was *uniform* distribution of lengths.

There is, however, no reason to suppose that the distribution might be uniform. A reasonable alternative would be that megalithic man paced out his measurements. Since different individuals have different strides, the distribution of lengths would be quite complicated; but hardly uniform.

A more appropriate statistical test would compare integer multiples of a common length with integer multiples of a *variable* length, distributed (perhaps normally [see II, § 11.4]) around a fixed average. Such a test is hard to devise, and Porteous resorted to a more direct approach. He got a number of different people to pace out various numbers of strides, measured the lengths, and plugged the results into a statistical test used by Thom.

The result, according to the test used, was an even more overwhelming probability that the measurements were integer multiples of a *fixed* length. Of course the conclusion is that the test works badly if the alternatives are not uniform.

It is therefore much more likely that megalithic man paced out his structures: the alleged unit of length is probably the average stride. Of course we haven't *proved* this conclusively—but the megalithic yard looks pretty shaky.

I'm not sure whether the archaeologists recognize this now. Some years after Porteous's paper I recall seeing an article which accepted the megalithic yard as proven fact. But it was in a newspaper. . . .

It is no accident that both the misinterpretations quoted above involved statistics. This is almost certainly the most abused of mathematical techniques. One reason for this is that it is widely used in the social and human sciences, whose practitioners are in general less mathematically minded than the physical scientists. Another is the easy availability of computerized packages for statistical analysis: it's much easier to play with a computer than it is to understand what it is doing.

The way to avoid errors of interpretation is to make sure that you fully understand what you are doing. This does not just mean that you must understand the mathematics. It might seem that, in the megalithic yard débâcle, it was the archaeologist who was at fault. But Thom consulted a mathematician to find his statistical test, and he can hardly be blamed for believing what the expert told him. So there's a moral for mathematicians who offer advice to other scientists: make sure you understand in some detail what the chap really wants to know. (And there's a moral to everybody: don't place too much trust in what the expert says.)

So, when interpreting your conclusions, keep in mind the purpose of the model, the real meaning of the variables, the distinction between raw data and smoothed data, the scales on which the data are plotted . . . . And remember the type of model. Is it dynamic or static (or quasistatic)?* What is the timescale? What variables are assumed irrelevant? Does it apply to an individual, a population, or an average? What sort of average?

And above all, don't forget that *all* conclusions rest on the hypothesis that the model is appropriate. However well-founded it seems, it may not be.

14.8. NEW TYPES OF MODEL

New discoveries in mathematics suggest new models for the applied scientist to use. The applied mathematician may rightly chide his purer brethren for their lack of attention to problems from the world of nature—but his brethren, the tool-makers of the profession, may with equal justice berate his failure to make good use of the tools they have so diligently provided. Modern 'pure' mathematics has produced an enormous range of potentially applicable ideas: it is questionable how effectively this potential has yet been realized. The problem, of course, is *communication* between disciplines whose thinking and working habits are not entirely in tune with one another. There have been enough improvements in such communication over the past few decades for the general prospect to be encouraging, however, and there is every reason to expect this to continue.

As an example of a burgeoning body of new ideas, with its origins both in the pure and the applied zones, and immeasurably enriched by the collaboration of both, I shall mention what is fashionably being called *chaos*. This may be described as the discovery that deterministic systems may exhibit random behaviour. Its precise implications are probably not yet understood in detail, but here is one possibility. It is commonly assumed, by engineers for example, that stochastic *effects* should be modelled by incorporating stochastic *terms* in the equations. 'Chaos' tells us that this may not be as necessary as it appears. But this does not mean that it is not useful, and any general assessment would be premature at this stage. However, those in the business of mathematical modelling would be well advised to keep a weather-eye open, because chaotic models look likely to become important in the future.

The archetypal example of a chaotic system is the nonlinear difference equation

$$x_{n+1} = kx_n(1 - x_n).$$

---

* In a *dynamic* model quantities vary with time. In a *static* model one deals only with steady-state or equilibrium solutions. In a *quasistatic* model one assumes the system is always in equilibrium, but that the actual equilibria may vary with time. Effectively this involves two independent timescales: a 'fast' one for changes in the position of the equilibria, and a much slower one for the variations that move the equilibria around. For example, an arch buckling under slowly increasing load may be modelled quasistatically, assuming that the arch responds so quickly to variations in load that it stays in equilibrium. If it collapses, this approximation tends to break down. If you are interested in vibration of the arch about equilibria, it is not an appropriate model anyway.

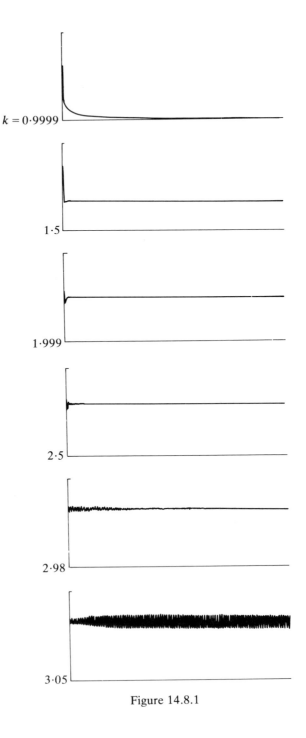

Figure 14.8.1

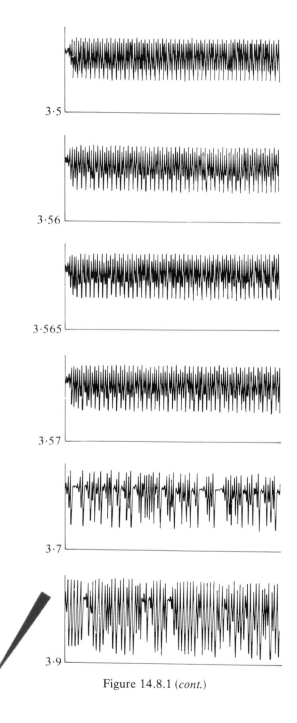

Figure 14.8.1 (*cont.*)

Here $0 \le x_n \le 1$, and k is a constant between 1 and 4. From an initial value x_0, successive x_n are found by iteration. What is the long-term behaviour?

A detailed treatment is given in a beautiful expository paper by Guckenheimer, Oster, and Ipaktchi (1977), to which the reader is strongly directed. I shall merely record some of the results. Figure 14.8.1 shows graphs of x_n against n for a representative selection of values of k.

For $k < 1$, x_n tends to zero. For $1 < k < 2$ it tends to a fixed non-zero value. For $2 < k < 3$ it does the same, but the convergence is oscillatory, alternately over- and under-shooting the limiting value. For k a little larger than 3 a steady oscillation of period 2 appears. As k increases this becomes an oscillation of period 4, 8, 16, ... doubling successively. By the time k has reached about 3·6 all trace of regularity seems to have vanished; and for $k = 4$ the wild behaviour is extremely marked.

This wild behaviour is 'chaos'. It gives every appearance of being random, stochastic—unless you graph x_{n+1} against x_n, of course—and in fact the 'statistical dynamics' of the system can be written down. So, paradoxically, not only does this deterministic system exhibit stochastic behaviour, but the statistical features may be computed precisely.

There are a number of lessons to be learned. One is that a simple model may give more complicated behaviour than the modeller bargained for. The model knows more than its creator. Another is that these complexities, even when they arise in such a simple way, may be hard to manage on anything other than a statistical level. For example, if numerical solutions are worked out, even to high precision, on two different computers, they have a habit of disagreeing entirely after the first 20 or 30 iterations. The chaotic system amplifies the rounding errors in the arithmetic to such an extent that they swamp the actual behaviour.

But more subtly, many important features of the behaviour survive rounding errors unscathed—for example, the chaotic appearance itself. Many statistical features seem similarly insensitive.

Chaotic behaviour is not unusual. In fact, it is beginning to be realized that in many ways it is far more common than non-chaotic behaviour! If you could pick a system of differential equations at random, it would not be a surprise if it were operating chaotically. (Though it would not be too much of a surprise if it were not: both appear reasonable in the light of current knowledge.)

At the time of writing, one of the most actively pursued problems to which chaotic dynamics is felt potentially applicable is the study of turbulence in fluids. Here we have precisely the features that chaotic systems like to exhibit: 'nice' behaviour suddenly becoming perversely complicated and random.

Applied scientists encountered special cases of chaos in various models. Pure mathematicians first began to recognize its wide occurrence in very general studies of differential equations which had no particular practical application in mind. And the combination of both kinds of insight gives every promise of enriching the entire subject area.

This development is, I hope, typical of the way that mathematicians and applied scientists will collaborate in future. Not only must the toolmakers—the mathematicians—pay closer attention to the demands of the tool-users: the users must take more trouble to understand how to use the tools that have already been made.

This article has largely concentrated on two great traditional types of mathematical model: differential and finite difference equations. Even in these traditional areas, new ideas and methods are becoming prominent. Many areas of modelling have not been discussed—for example, probabilistic models, simulations, numerical models, combinatorial models, integral equations, equations with delay, functional equations, topological models—it would be hard to catalogue them in detail. A hint of the variety of modelling viewpoints available may be gleaned from the fascinating survey edited by Levin (1978) of mathematical models in biology.

Every scientist knows the types of model prevalent in his own field—and, I hope, in nearby fields as well. But models from entirely different fields may be ripe for transplant, and may prove fertile in new ground. Insights from other disciplines should not be despised merely because they are unfamiliar, or couched in uncouth language. Nor should their value be judged too hastily: it will take time and effort to formulate them correctly and develop their potential. Many attempts will fail altogether. But the effort is well worth making: we have far more to fear from narrowness of vision than we can ever have from breadth.

I. S.

REFERENCES

Aris, R. (1978). Mathematical Modelling Techniques, *Research Notes in Mathematics* 24, Pitman, Boston and London.

Clifford, W. K. (1956). The Exactness of Mathematical Laws, in *The World of Mathematics*, ed. J. R. Newman, Simon & Schuster, New York, pp. 548–551.

Gonzalez, F. A. and Byrd, L. D. (1978). Mathematics Underlying the Rate-Dependency Hypothesis, *Science* **195**, 546–550.

Guckenheimer, J., Oster, G. and Ipaktchi, A. (1977). The Dynamics of Density-Dependent Population Models, *J. Math. Biology* **4**, 101–147.

Hadamard, J. (1954). *The Psychology of Invention in the Mathematical Field*, Dover Publications Inc., New York.

Levin, S. (ed.) (1978). *Studies in Mathematical Biology*, 2 vols., MAA *Studies in Mathematics Vols. 15 and 16*, Mathematical Association of America.

Porteous, H. L. (1973). Megalithic Yard or Megalithic Myth? *J. History of Astronomy* **4**, 22–24.

Thom, A. (1967). *Megalithic Sites in Britain*, Oxford University Press, Oxford.

Thompson, P. D. (1978). The Mathematics of Meteorology, in *Mathematics Today*, ed. L. A. Steen, Springer, Berlin and New York, pp. 127–152.

CHAPTER 15

Non-linear Programming

15.1. INTRODUCTION

15.1.1. Unconstrained Optimization

We shall take the standard unconstrained optimization problem as the choice of a variable x in a set $\mathcal{X}$ so as to maximize a scalar function $f(x)$. Typically x could represent an allocation of one's resources and $f(x)$ the consequent economic return. Alternatively x might represent design of an engine or a process and $f(x)$ some measure of its efficiency. One usually refers to $f(x)$ as the *objective* or *criterion* function.

The problem of minimizing an objective (e.g. a cost $C(x)$) can be rephrased as the maximization of $f(x) = -C(x)$.

In some cases there are several objectives which one would wish to maximize simultaneously, in some sense. This aim of *multi-criterion planning* raises new conceptual and mathematical issues. However, we can say that, if there is effectively a single optimizer (so that there is no question of conflicting interests, or of different degrees of controllability or observability for optimization of the different objectives), then, under plausible assumptions on behaviour, the optimizer should behave as if he were maximizing some non-negatively weighted combination $\sum_i \alpha_i f_i(x)$ of the individual objective functions $f_i(x)$ [see Whittle (1971) p. 211].

From the purely mathematical point of view, $f(x)$ may not possess a maximizing value in $\mathcal{X}$: This situation is variously expressed by saying that '$f(x)$ does not attain its supremum in $\mathcal{X}$', or 'the optimization problem does not possess a solution'. Suppose, for example, that Cinderella experiences a total pleasure proportional to the time x that she spends at the ball, but that this pleasure diminishes to zero if she stays until midnight, represented by $x = 1$, say, when she expects to be exposed. She then wishes to maximize

$$f(x) = \begin{cases} x, & 0 \le x < 1 \\ 0, & x \ge 1 \end{cases}$$

in $x \ge 0$. She can bring f arbitrarily near to 1 by postponing the moment of departure almost to midnight, but there is no value of x for which $f(x) = 1$.

Problems showing such behaviour usually neglect some point of physical realism or realizability, and are in fact ill-posed. We shall assume problems so posed that they *do* possess a solution, but the point may require checking in particular cases. A sufficient condition is provided by

THEOREM 15.1.1. *If $\mathscr{X}$ is a compact set in a finite-dimensional Euclidean space [see V, § 5.2], and f is continuous [see Definition 5.3.1], then f attains its extreme values in $\mathscr{X}$.*

15.1.2. Optimality Conditions

A point $\bar{x}$ of $\mathscr{X}$ will solve the optimization problem if and only if

$$f(\bar{x}) \geq f(x) \qquad (x \in \mathscr{X}). \tag{15.1.1}$$

We shall suppose henceforth that such an $\bar{x}$ exists (although it may not be unique) and that $\mathscr{X}$ is a set in a finite-dimensional Euclidean space, $\mathbb{R}^n$. The argument x can thus be regarded as a column vector

$$\mathbf{x} = \begin{pmatrix} x_1 \\ x_2 \\ \vdots \\ x_n \end{pmatrix} \tag{15.1.2}$$

with Cartesian components x_k [see I, Example 5.2.2].

Let us say that a direction $\mathbf{s}$ (an n-dimensional vector or 'n-vector') is *interior to $\mathscr{X}$ at $\bar{x}$* if the point $\bar{x} + \varepsilon \mathbf{s}$ lies in $\mathscr{X}$ for all non-negative ε smaller than some positive value. That is, if one moves from $\bar{x}$ in the direction $\mathbf{s}$, then one remains in $\mathscr{X}$, at least initially. Suppose also that f is differentiable at $\bar{x}$, with row vector of derivatives $f_\mathbf{x}$ [see Definition 5.9.1]. It follows then, from (15.1.1), that

$$f_\mathbf{x}\mathbf{s} \leq 0. \tag{15.1.3}$$

for interior directions $\mathbf{s}$. That is, all directions in which one *can* move from $\bar{x}$ are 'down-hill'.

We shall say that $\bar{x}$ is an *interior point* of $\mathscr{X}$ if an ε neighbourhood of $\bar{x}$ lies wholly in $\mathscr{X}$. Otherwise it is a *boundary point* [see V, § 5.2].

If $\bar{x}$ is an interior point, then all directions are interior, so that (15.1.3) holds for all $\mathbf{s}$. This implies that

$$f_\mathbf{x} = \mathbf{0} \tag{15.1.4}$$

at $\bar{x}$, i.e. that $\bar{x}$ is a *stationary point* [see § 5.6].

Summing up, we see that a maximizing point must fall into one of three categories:

(i) $\bar{\mathbf{x}}$ is a stationary point;
(ii) $\bar{\mathbf{x}}$ is a boundary point;
(iii) $\bar{\mathbf{x}}$ is a point at which f is not differentiable.

These categories are not completely mutually exclusive, but $\bar{\mathbf{x}}$ must fall in at least one of them. To some extent categories (ii) and (iii) are rather composite. For example, one will still expect partial stationarity at a boundary point, in that (15.1.3) must still hold for interior directions $\mathbf{s}$ if f is differentiable. For another example, the functions f may have directional derivatives [see § 15.2.7] without being differentiable, when (15.1.3) will still possess an analogue.

However, the condition of stationarity is of great help in locating interior maximizing points when f is prescribed analytically. If the problem of maximization has to be tackled numerically then the notion of trying to find 'up-hill' directions is basic to a large class of algorithms [see § 15.9.5].

15.1.3. Constrained Optimization

Very often the maximization of $f(\mathbf{x})$ is subject, not only to the requirement that $\mathbf{x}$ be confined to some set $\mathscr{X}$, but also to a number of subsidiary constraints

$$g_j(\mathbf{x}) = b_j \qquad (j = 1, 2, \ldots, m) \tag{15.1.5}$$

which one should sum up in the m-vector equation

$$\mathbf{g}(\mathbf{x}) = \mathbf{b}. \tag{15.1.6}$$

The resource allocation problem is the prototype of this situation. If x_k represents the rate at which activity k is undertaken, then $\mathbf{x}$ will be confined to that set of values $\mathscr{X}$ which is technically feasible (which will be in many cases taken just as the set $\mathbf{x} \geq \mathbf{0}$). However, activities consume various resources (labour, energy, raw materials, etc.) whose supply may be restricted. Constraint (15.1.5) would then represent the fact that $g_j(\mathbf{x})$, the amount of resource j consumed with activity vector $\mathbf{x}$, must equal the supply b_j of that resource. Actually, for the allocation problem a more reasonable form of constraint is

$$\mathbf{g}(\mathbf{x}) \leq \mathbf{b} \tag{15.1.7}$$

since the compulsion is to use no more resources than are available, rather than to use exactly what is available. Constraints (15.1.5) are *equality constraints*; constraints (15.1.7) are *inequality constraints*. A formulation of the constraint that includes both cases is to stipulate that

$$\mathbf{b} - \mathbf{g}(\mathbf{x}) \in \mathscr{C} \tag{15.1.8}$$

where $\mathscr{C}$ is some set in $\mathbb{R}^m$ which in case (15.1.6) consists purely of the origin, $\mathbf{0}$; in case (15.1.7), of the non-negative orthant $\mathbb{R}^m_+$ [see Example 19.4.1].

This last version of the problem can be reduced to the case of equality constraints by introducing a vector $\mathbf{z}$ in $\mathbb{R}^m$, the *slack vector*, and regarding the problem as the choice of $(\mathbf{x}, \mathbf{z})$ in $(\mathcal{X}, \mathcal{C})$ to maximize $f(\mathbf{x})$ subject to the equality constraint

$$\mathbf{g}(\mathbf{x}) + \mathbf{z} = \mathbf{b}. \tag{15.1.9}$$

For example, in the allocation case with constraint (15.1.7), $\mathbf{z}$ would be subject to $\mathbf{z} \geq \mathbf{0}$ and would be interpretable as the difference between the availability vector $\mathbf{b}$ and the consumption vector $\mathbf{g}(\mathbf{x})$, i.e. the vector of amounts of resources left unused. The appropriateness of the term 'slack' is clear; it is not until $\mathbf{z}$ is on the boundary $\mathcal{C}$ that some of the constraints become 'active' and actually constrain.

Since problem (15.1.8) can be reduced to the case of equality constraint, there might seem to be no point in considering anything but equality constraints. This is not quite true; the status of the 'structural variables' $\mathbf{x}$ and the 'slack variables' $\mathbf{z}$ is different, and it is as well to preserve the distinction between them. For any of the constraints (15.1.6), (15.1.7) and (15.1.8) it turns out to be natural to regard the constrained optimization problem as one of a family, generated by varying $\mathbf{b}$. For example, in the allocation problem it is natural to consider how both the optimal solution and the optimal return vary as the vector of availabilities $\mathbf{b}$ is varied. That is, one embeds the given problem in a family of problems.

We shall denote the problem of maximizing $f(\mathbf{x})$ in $\mathcal{X}$ subject to (15.1.8) by $P_{\mathcal{C}}(\mathbf{b})$ to emphasize the dependence of the problem upon $\mathcal{C}$ and $\mathbf{b}$. The equality constraint problem (15.1.6) will then be written $P_0(\mathbf{b})$.

We shall say that a value of $\mathbf{x}$ is *feasible* for $P_{\mathcal{C}}(\mathbf{b})$ if it satisfies all constraints, i.e. lies in $\mathcal{X}$ and satisfies (19.1.8). The set of such values is the *feasible set*. If the feasible set is empty, then the constraints are mutually incompatible, and no solution to the problem exists (in a different sense to the type of failure mentioned in section 15.1.1). By a *solution* to the problem we usually mean a feasible value of $\mathbf{x}$ which is also optimizing, although in some parts of the literature (e.g. that on linear programming [see I, § 11.1.2]) the term implies no more than feasibility.

In general $\mathbf{b}$ cannot be assigned an arbitrary value if the problem is to possess a solution. Let $\mathcal{B}$ denote the *attainable set*; the set of values of $\mathbf{b}$ for which the set of feasible $\mathbf{x}$ is not empty. In the case of $P_0(\mathbf{b})$ this is the set of values of $\mathbf{b}$ for which $\mathbf{b} = \mathbf{g}(\mathbf{x})$ for some $\mathbf{x}$ in $\mathcal{X}$; we shall write this as

$$\mathcal{B} = \mathbf{g}(\mathcal{X}). \tag{15.1.10}$$

In the case of $P_{\mathcal{C}}(\mathbf{b})$ it is the set of values of $\mathbf{b}$ for which $\mathbf{b} = \mathbf{g}(\mathbf{x}) + \mathbf{z}$ for some $\mathbf{x}$ in $\mathcal{X}$ and some $\mathbf{z}$ in $\mathcal{C}$; we shall write this as

$$\mathcal{B} = \mathbf{g}(\mathcal{X}) + \mathcal{C}. \tag{15.1.11}$$

15.1.4. Lagrange Multipliers

Historically, the Lagrange multiplier technique has provided an elegant way of generalizing the stationarity criteria (15.1.3), (15.1.4) to the case of constrained optimization. In fact, the technique has stronger formulations and deeper bases which have only become apparent in the last few decades.

Like most fundamental subjects, it can be approached in many ways. To form a first acquaintance we follow a local approach.

Consider first the case of equality constraints. Let $U(\mathbf{b})$ be the optimal return for $P_0(\mathbf{b})$; i.e. the maximal value of $f(\mathbf{x})$ under the constraint of $P_0(\mathbf{b})$. Then U is a function defined on $\mathcal{B}$, and characterized exactly by the requirement that

$$U(\mathbf{g}(\mathbf{x})) \geq f(\mathbf{x}) \qquad (\mathbf{x} \in \mathcal{X}) \tag{15.1.12}$$

with equality for some $\mathbf{x}$ for every value of $\mathbf{g}$ in $\mathcal{B}$. Let $\bar{\mathbf{x}}$ be a solution to $P_0(\mathbf{b})$, so that

$$U(\mathbf{b}) = U(\mathbf{g}(\bar{\mathbf{x}})) = f(\bar{\mathbf{x}}), \tag{15.1.13}$$

and

$$f(\mathbf{x}) - f(\bar{\mathbf{x}}) \leq U(\mathbf{g}(\mathbf{x})) - U(\mathbf{g}(\bar{\mathbf{x}})). \tag{15.1.14}$$

Suppose that $\mathbf{s}$ is an interior direction from $\bar{\mathbf{x}}$ and that $f, \mathbf{g}$ are differentiable at $\bar{\mathbf{x}}$ and U at $\mathbf{b}$. Setting $\mathbf{x} = \bar{\mathbf{x}} + \varepsilon \mathbf{s}$ we deduce then from (15.1.4) that

$$(f_{\mathbf{x}} - U_{\mathbf{b}} \mathbf{g}_{\mathbf{x}}) \mathbf{s} \leq 0 \tag{15.1.15}$$

or

$$(f_{\mathbf{x}} - \mathbf{y}' \mathbf{g}_{\mathbf{x}}) \mathbf{s} \leq 0 \tag{15.1.16}$$

where

$$\mathbf{y}' = U_{\mathbf{b}}. \tag{15.1.17}$$

If $\bar{\mathbf{x}}$ is interior to $\mathcal{X}$ then (15.1.16) implies that

$$f_{\mathbf{x}} - \mathbf{y}' \mathbf{g}_{\mathbf{x}} = \mathbf{0}. \tag{15.1.18}$$

Relations (15.1.16), (15.1.18) are modified forms of the stationarity conditions (15.1.3) and (15.1.4). Stationarity of f itself at $\bar{\mathbf{x}}$ is replaced in the constrained case (under the regularity condition assumed) by stationarity of the *Lagrangian form* $f - \mathbf{y}' \mathbf{g}$, where $\mathbf{y}$ is the vector of *Lagrangian multipliers*.

One could view the multipliers in the following way. In the constrained case $f(\mathbf{x})$ is still decreased by perturbations of $\mathbf{x}$ from $\bar{\mathbf{x}}$ in directions which are consistent with the constraints, i.e. in directions which are *feasible*. It may not however, be decreasing in directions which are infeasible. The multipliers $\mathbf{y}$ in the form $f - \mathbf{y}' \mathbf{g}$ make a correction for the non-zero gradient in these infeasible directions, so that the form satisfies a stationarity condition (15.1.18) for *all* interior directions.

This last interpretation hints at, but does not fully yield, the interpretation of the multipliers given in (15.1.17). Written componentwise, this is

$$y_j = \frac{\partial U(b)}{\partial b_j}. \tag{15.1.17'}$$

That is, y_j is the rate of change of optimal return with the value of b_j. This implies the interpretation of y_j as a 'shadow price', the marginal worth (in the currency in which f is measured) of a unit change in b_j. In the allocation problem this would be the marginal value of additional supplies of resource j: a shadow price for resource j.

We summarize the versions of those assertions for $P_{\mathscr{C}}(\mathbf{b})$ in

THEOREM 15.1.2

(i) *For $P_{\mathscr{C}}(\mathbf{b})$ the function $U(\mathbf{b})$ is defined on $B = \mathbf{g}(\mathscr{X}) + \mathscr{C}$, and is characterized by the fact that*

$$U(\mathbf{g}(\mathbf{x}) + \mathbf{z}) \geq f(\mathbf{x}) \tag{15.1.19}$$

for all $(\mathbf{x}, \mathbf{z})$ in $(\mathscr{X}, \mathscr{C})$ with equality for some $(\mathbf{x}, \mathbf{z})$ for every attainable value of the argument of U.

(ii) *Suppose $P_{\mathscr{C}}(\mathbf{b})$ has a solution at $(\bar{\mathbf{x}}, \bar{\mathbf{z}})$, and that f and $\mathbf{g}$ are differentiable at $\bar{\mathbf{x}}$ and U at $\mathbf{b}$. Then*

$$(f_{\mathbf{x}} - \mathbf{y}'\mathbf{g}_{\mathbf{x}})\mathbf{s} \leq 0 \tag{15.1.20}$$

$$\mathbf{y}'\mathbf{t} \geq 0 \tag{15.1.21}$$

for all directions $\mathbf{s}$ from $\bar{\mathbf{x}}$ interior to $\mathscr{X}$, and for all directions $\mathbf{t}$ from $\bar{\mathbf{z}}$ interior to $\mathscr{C}$. Here

$$\mathbf{y}' = U_{\mathbf{b}}. \tag{15.1.22}$$

We see from (15.1.20), (15.1.21) that one can in fact make stronger statements when the constraint is relaxed. Relation (15.1.20) states that a Lagrangian stationarity condition still holds. Relation (15.1.21) places constraints as the value of the multiplier y. For example, in the case of the inequality constraint (15.1.7) we shall see later [see §§ 15.2.4, 15.3.1, 15.4.1] that (15.1.21) implies that $\mathbf{y} \geq \mathbf{0}$ and that $y_j = 0$ if $\bar{z}_j > 0$. That is, that shadow prices are non-negative (for an increase in resource can never make matters worse) and that the shadow price is zero for a resource not fully utilized in the optimal allocation (when an increase in that resource would simply increase the surplus without affecting the allocation).

To develop fully the implications of (15.1.21) one needs to examine the nature of $\mathscr{C}$ rather more closely. Such a study leads one to stronger formulations of the Lagrangian principle [see § 15.4.1].

A situation where Lagrange multipliers are particularly helpful is that of a *decomposable* or *separable problem* (which is not quite the same as the 'separable programming' of I, § 12.1.5. This is the situation where f and $\mathbf{g}$ decompose additively:

$$f(\mathbf{x}) = \sum f_k(x_k)$$

$$\mathbf{g}(\mathbf{x}) = \sum \mathbf{g}_k(x_k),$$

where x_k is a component of the variable associated with the 'kth subsystem' of the total system. In this case, in a Lagrangian approach, x_k will be chosen to maximize $f_k(x_k) - \mathbf{y}'\mathbf{g}_k(x_k)$. The optimization thus reduces to a separate optimization for each subsystem, these being linked only by the multiplier values which must later be determined (probably by duality; see § 15.3.3). If $\mathbf{b} - \sum \mathbf{g}_k(x_k) \in \mathscr{C}$ is still regarded as a resource constraint, then $\mathbf{y}$ is an appropriate price vector, and in the Lagrangian approach optimizations are coupled by the existence of an effective price rather than by an explicit constraint.

15.1.5. Notation

We have already established several conventions on notation. In this section we shall summarize conventions already made and to come.

A vector such as $\mathbf{x}$ is always a column vector with elements x_k. We use $\mathbf{A}'$ to denote the transpose of a matrix $\mathbf{A}$, so that the row vector corresponding to $\mathbf{x}$ is $\mathbf{x}'$ [see I, § 6.5].

A Euclidean space of n dimensions will be denoted $\mathbb{R}^n$. The non-negative orthant in $\mathbb{R}^n$ will be denoted $\mathbb{R}^n_+$. If $\mathbf{x} \in \mathbb{R}^n$ then we shall refer to $\mathbf{x}$ as an *n-vector*. In fact, in this treatment $\mathbf{x}$ is consistently an n-vector, and $\mathbf{g}(\mathbf{x})$ an m-vector [see Example 19.4.1 and I, Example 5.2.2].

If $f(\mathbf{x})$ is a scalar-valued function of $\mathbf{x}$, then $f_\mathbf{x}$ will denote the *row* vector of first derivatives $\partial f / \partial x_k$ [see § 5.3]. Correspondingly, $\mathbf{g}_\mathbf{x}$ is the $m \times n$ matrix of derivatives $\partial g_i / \partial x_k$. Simple matrix notation does not in general extend to higher order derivatives. However, we shall use $f_{\mathbf{xx}}$ to denote the Hessian of f: the $n \times n$ matrix of second derivatives $\partial^2 f / \partial x_j \, \partial x_k$ [see § 5.7].

If $\mathscr{X}$ and $\mathscr{C}$ are two sets, then we shall usually indicate the product set by $(\mathscr{X}, \mathscr{C})$. That is, $(\mathscr{X}, \mathscr{C})$ is the set of pairs $(\mathbf{x}, \mathbf{z})$ such that $\mathbf{x} \in \mathscr{X}$ and $\mathbf{z} \in \mathscr{C}$ [see I, § 1.2.6].

We also employ the useful functional notation for sets. So $\mathbf{g}(\mathscr{X})$ is the set of values generated by $\mathbf{g}(\mathbf{x})$ as $\mathbf{x}$ traverses $\mathscr{X}$. Correspondingly, if $\mathbf{x}$, $\mathbf{y}$ take values in $\mathscr{X}$, $\mathscr{Y}$ respectively then $h(\mathscr{X}, \mathscr{Y})$ will denote the set of values generated by $h(\mathbf{x}, \mathbf{y})$ as $(\mathbf{x}, \mathbf{y})$ traverses $(\mathscr{X}, \mathscr{Y})$. So, $\mathscr{X} - \mathscr{Y}$ would denote the set generated by $\mathbf{x} - \mathbf{y}$ (and is not to be confused with $\mathscr{X} \backslash \mathscr{Y}$: see section 15.2.1 for a definition of this and other notations associated with sets). For another example which will occur, $\mathbf{g}(\mathscr{X}) + \mathscr{C}$ denotes the set of values generated by $\mathbf{g}(\mathbf{x}) + \mathbf{z}$ for $(\mathbf{x}, \mathbf{z})$ in $(\mathscr{X}, \mathscr{C})$.

There are several grades of inequality for vectors. We shall write $\mathbf{x} \geq \mathbf{0}$ if $x_k \geq 0$ for all k; $\mathbf{x} > \mathbf{0}$ if $x_k \geq 0$ for all k with strict inequality for some k, and $\mathbf{x} \gg \mathbf{0}$ if $x_k > 0$ for all k. If $\mathbf{x}$ is an n-vector such that $\mathbf{x} \geq \mathbf{0}$ and $\sum x_k = 1$ we shall write $\mathbf{x} \in P_n$ (so that P_n is the set of distributions on the integers $1, 2, \ldots, n$ [see II (4.3.5))].

15.2.   CONVEXITY

15.2.1.   Sets, Interiors and Boundaries

One of the most important ideas in mathematics, and certainly one of the most important to Lagrangian theory, is that of convexity. This is a property which can be associated with point sets in topological vector spaces. For simplicity, we shall consider only sets in a finite-dimensional Euclidean space $\mathbb{R}^n$, although the whole development transfers, with some precautions to the more general case. A full treatment of the subject is given in V, Chapter 4. In this section we just clarify a few definitions and key concepts. We use the usual notation and definitions [see I, § 1.2]; if $\mathcal{A}$, $\mathcal{B}$ are sets of points in $\mathbb{R}^n$ then we define:

Union, $\mathcal{A} \cup \mathcal{B}$; the set of all points contained in at least one of the two sets.
Intersection, $\mathcal{A} \cap \mathcal{B}$; the set of all points contained in both sets.
Difference, $\mathcal{A} \backslash \mathcal{B}$; the set of all points contained in $\mathcal{A}$ but not also in $\mathcal{B}$.
Inclusion, $\mathcal{A} \subset \mathcal{B}$; the statement that all points of $\mathcal{A}$ belong to $\mathcal{B}$.
The *empty set* is denoted $\varnothing$. Sets $\mathcal{A}$, $\mathcal{B}$ are *disjoint* if they have no points in common, that is $A \cap B = \varnothing$.

A set $\mathcal{A}$ is *closed* if all limit points of $\mathcal{A}$ belong to $\mathcal{A}$ [see § 11.3]. The *closure* of a set $\mathcal{A}$ is the smallest closed set containing $\mathcal{A}$, denoted $\bar{\mathcal{A}}$. A set is *bounded* if it is contained within a sphere of finite radius. It is *compact* if it is closed and bounded [see V, § 5.2].

A point x is *interior* to $\mathcal{A}$ if some ε-neighbourhood of x belongs to $\mathcal{A}$ [see V, § 5.2]. The set of such points constitute the interior of $\mathcal{A}$, denoted $\mathcal{A}^0$. The set of points of $\bar{\mathcal{A}}$ which do not belong to $\mathcal{A}^0$ is termed the *boundary* of A, denoted

$$\partial \mathcal{A} = \bar{\mathcal{A}} \backslash \mathcal{A}^0.$$

'Boundary' and 'interior' are evidently formalizations of the intuitive notions these names convey.

Often a set $\mathcal{A}$ in $\mathbb{R}^n$ is in fact confined to a subspace of lower dimensionality (its *affine hull*). In such a case $\mathcal{A}$ has empty interior, because $\mathcal{A}$ is 'flat'. However, it would have an interior if one considered it as a set in the subspace, rather than in $\mathbb{R}^n$. The 'interior' then defined is termed the *relative interior*, denoted ri $\mathcal{A}$. The *relative boundary* is then defined as

$$\text{rebd } \mathcal{A} = \bar{\mathcal{A}} \backslash \text{ri } \mathcal{A}.$$

15.2.2.   Convex Sets

A set $\mathcal{A}$ is said to be *convex* if, for any $\mathbf{x}^{(1)}$ and $\mathbf{x}^{(2)}$ in $\mathcal{A}$, the averaged combination

$$\boldsymbol{\xi} = p_1 \mathbf{x}^{(1)} + p_2 \mathbf{x}^{(2)} \tag{15.2.1}$$

$(p_1, p_2 \geq 0; p_1 + p_2 = 1)$ also belongs to $\mathcal{A}$. One could express this property by saying that $\mathcal{A}$ is *closed under averaging*. Alternatively, one could characterize the property geometrically by saying that all points on the line segment joining $\mathbf{x}^{(1)}$ and $\mathbf{x}^{(2)}$ belong to $\mathcal{A}$. So, in Figure 15.2.1 we give examples of convex and non-convex sets.

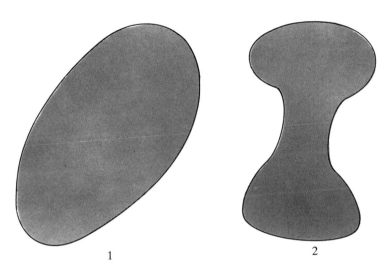

1
2

Figure 15.2.1: 1 Convex set, 2 Non-convex set.

A corollary of property (15.2.1) is that an average of any finite number of points $\mathbf{x}^{(i)}$ of $\mathcal{A}$

$$\boldsymbol{\xi} = \sum_{1}^{r} p_i \mathbf{x}^{(i)} \tag{15.2.2}$$

(the p_i non-negative and summing to unity) belongs to $\mathcal{A}$. Geometrically, the smallest polygon containing $\mathbf{x}^{(1)}, \mathbf{x}^{(2)} \ldots$ belongs to $\mathcal{A}$.

If $\mathcal{A}$ is also closed, then one can assert quite simply that $\boldsymbol{\xi} = E(\mathbf{x})$ belongs to $\mathcal{A}$, where E is the expectation operator corresponding to a distribution with support confined to $\mathcal{A}$ [see II, Definition 8.1.1]. The smallest convex set containing $\mathcal{A}$ is termed the *convex hull* of $\mathcal{A}$, denoted $[\mathcal{A}]$. This is the set of points generated by $\boldsymbol{\xi} = E(\mathbf{x})$ as the distribution varies.

THEOREM 15.2.1   (*Carathéodory's theorem*).   *Let $\mathcal{A}$ be a set in $\mathbb{R}^n$. Then a point $\mathbf{x}$ of $[\mathcal{A}]$ can be expressed as an average of at most $n + 1$ points of $\mathcal{A}$.*

The *extreme points* of a set are its 'vertices': those points $\boldsymbol{\xi}$ for which the only possible representations $\boldsymbol{\xi} = E(\mathbf{x})$ are those for which the distribution assigns weight only to a single point of the set.

THEOREM 15.2.2. (*Minkowski's theorem*).   *If $\mathscr{A}$ is a non-empty compact convex set, then it is the convex hull of its extreme points.*

A convex set with a finite number of extreme points is termed a *convex polytope* [see V, § 4.2.3].

15.2.3. Separating and Supporting Hyperplane Theorems

A *hyperplane* $\mathscr{H}$ in $\mathbb{R}^n$ is the set of $\mathbf{x}$ in $\mathbb{R}^n$ satisfying a single linear constraint

$$\mathbf{a}'\mathbf{x} = a_0. \tag{15.2.3}$$

One says that $\mathscr{H}$ *separates* the sets $\mathscr{A}$ and $\mathscr{B}$ if

$$\mathbf{a}'\mathbf{x} \geq a_0 \qquad (\mathbf{x} \in \mathscr{A})$$
$$\mathbf{a}'\mathbf{x} \leq a_0 \qquad (\mathbf{x} \in \mathscr{B}) \tag{15.2.4}$$

and equality does not hold for all $\mathbf{x}$ in both $\mathscr{A}$ and $\mathscr{B}$. (The directions of inequality in (15.2.4) have been chosen for definiteness; all that is really required is that they should be different for the two sets.) The hyperplane is *strictly separating* if the inequalities in (15.2.4) are strict.

So, $\mathscr{H}$ separates $\mathscr{A}$ and $\mathscr{B}$ strictly if $\mathscr{A}$ lies strictly to one side of $\mathscr{H}$, and $\mathscr{B}$ lies strictly to the other. In dispensing with strictness one allows both $\mathscr{A}$ and $\mathscr{B}$ to touch $\mathscr{H}$, but not both to lie wholly in $\mathscr{H}$, when the characterization would be so degenerate as not to embody any ideas of separation at all.

The celebrated *separating hyperplane theorem* states, roughly, that a separating hyperplane exists if $\mathscr{A}$ and $\mathscr{B}$ are convex and disjoint. We indicate the situation in Figure 15.2.2, and also a case where a separating hyperplane does not exist, because of absence of convexity.

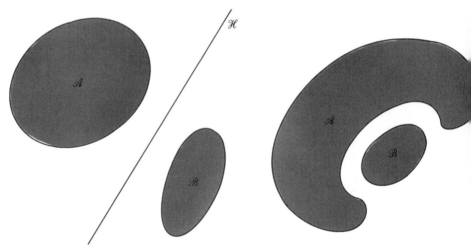

Figure 15.2.2: Cases where sets $\mathscr{A}$ and $\mathscr{B}$ can and cannot be separated by a hyperplane $\mathscr{H}$.

The following two theorems express this statement rather more circumspectly.

THEOREM 15.2.3 (*The strict separating hyperplane theorem*). *Let $\mathcal{A}$, $\mathcal{B}$ be non-empty disjoint convex sets, such that $\mathcal{A}$ is closed and $\mathcal{B}$ is compact. Then a hyperplane exists which strictly separates $\mathcal{A}$ and $\mathcal{B}$.*

THEOREM 15.2.4 (*The separating hyperplane theorem*). *Let $\mathcal{A}$, $\mathcal{B}$ be non-empty convex sets with disjoint relative interiors. Then a hyperplane separating $\mathcal{A}$ and $\mathcal{B}$ exists.*

We shall say that $\mathcal{H}$ is a *supporting hyperplane* to $\mathcal{A}$ if

$$\mathbf{a}'\mathbf{x} \le a_0 \qquad (\mathbf{x} \in \mathcal{A}) \tag{15.2.5}$$

with equality for some $\mathbf{x}$ in $\bar{\mathcal{A}}$. (Again, the direction of inequality is immaterial; all that is required is that there be a consistent inequality.) That is, $\mathcal{A}$ lies totally to one side of $\mathcal{H}$, but $\mathcal{H}$ meets $\mathcal{A}$ at least insofar as it meets $\bar{\mathcal{A}}$. The points in which it meets $\bar{\mathcal{A}}$ must be boundary points of $\bar{\mathcal{A}}$. If $\mathbf{x}_0$ is a boundary point of $\mathcal{A}$ for which equality holds in (15.2.5) we shall say that $\mathcal{H}$ is a supporting hyperplane to $\mathcal{A}$ at $\mathbf{x}_0$. Clearly, $\mathbf{x}_0$ is thus a value of $\mathbf{x}$ maximizing the linear form $\mathbf{a}'\mathbf{x}$ in $\mathcal{A}$. We shall say that $\mathcal{H}$ is a *non-trivial supporting hyperplane* to $\mathcal{A}$ if it is supporting but does not contain $\mathcal{A}$; i.e. if equality does not hold in (15.2.5) for all $\mathbf{x}$ of $\mathcal{A}$.

The *supporting hyperplane theorem* states, roughly, that a convex set has a supporting hyperplane at all boundary points. Another way to express this is to say that any boundary point of a convex set $\mathcal{A}$ maximizes some linear form $\mathbf{a}'\mathbf{x}$ in $\mathcal{A}$. In Figure 15.2.3 we illustrate existence and non-existence of a supporting hyperplane at a boundary point $\mathbf{x}_0$.

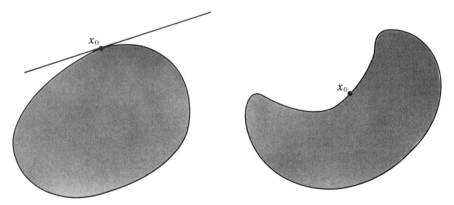

Figure 15.2.3: Cases where sets do and do not possess a supporting hyperplane at a boundary point x_0.

THEOREM 15.2.5   (*The supporting hyperplane theorem*).   *Let $\mathcal{A}$ be a convex set. A supporting hyperplane to $\mathcal{A}$ exists at each point of the boundary at $\mathcal{A}$, and a non-trivial supporting hyperplane at each point of the relative boundary of $\mathcal{A}$.*

The supporting and separating hyperplane theorems are closely related. If, in the separating hyperplane theorem, one allows $\mathcal{B}$ to shrink to becoming a boundary point $\mathbf{x}_0$ of $\mathcal{A}$, then the separating hyperplane becomes a supporting hyperplane to $\mathcal{A}$ at $\mathbf{x}_0$ (the argument requires care!). Going the other way, a supporting hyperplane to $\mathcal{A}\backslash\mathcal{B}$ separates $\mathcal{A}$ and $\mathcal{B}$.

A hyperplane $\mathcal{H}$ is *tangent* to the set $\mathcal{A}$ at the boundary point $\mathbf{x}_0$ if it is tangent to the boundary; that is if the perpendicular distance to $\mathcal{H}$ from any other boundary point $\mathbf{x}$ is $o(|\mathbf{x}-\mathbf{x}_0|)$. A set can show all combinations of behaviour as to whether it does or does not possess tangent or supporting hyperplanes at a given boundary point, see Figure 15.2.4. However, if a convex set possesses a tangent hyperplane at $\mathbf{x}_0$, then this is also the unique supporting hyperplane at $\mathbf{x}_0$.

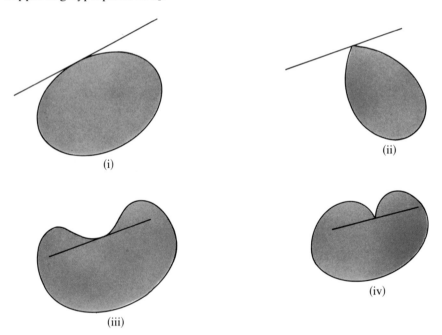

Figure 15.2.4: Cases where a set admits at a given boundary point a hyperplane which is (i) both supporting and tangent, (ii) supporting but not tangent, (iii) tangent but not supporting, and (iv) neither.

15.2.4.   Cones [see also § 19.4.2]

A *cone with vertex at the origin* is a set $\mathcal{C}$ closed under dilations. That is, if $\mathbf{x} \in \mathcal{C}$ then $\lambda\mathbf{x} \in \mathcal{C}$ where λ is a positive scalar. If $\mathcal{C}$ is closed then the origin

also evidently belongs to $\mathscr{C}$. A cone with vertex elsewhere than at the origin
is just a translated version.

A convex cone is closed under both averaging and dilation, and so is closed
under *positive linear operations*. That is, if $\mathbf{x}^{(1)}. \mathbf{x}^{(2)}, \ldots, \mathbf{x}^{(r)}$ belong to $\mathscr{C}$ then
so does

$$\boldsymbol{\xi} = \sum_{1}^{r} \lambda_i \mathbf{x}^{(i)} \quad \text{for } \lambda_i > 0.$$

If $\mathscr{C}$ is a convex cone and $\mathbf{x}_0$ a point external to $\bar{\mathscr{C}}$ then $\mathscr{C}$ will meet the ray
$\lambda \mathbf{x}_0$ only at the origin, and one can separate $\lambda \mathbf{x}_0$ and $\mathscr{C}$ by a hyperplane
passing through the origin.

THEOREM 15.2.6. *Let* $\mathbf{b}$ *be a point external to a closed convex cone* $\mathscr{C}$.
Then $\{\mathbf{b}\}$ *can be separated strictly from* $\mathscr{C}$ *by a hyperplane passing through the
origin.*

The following development of this result will prove useful in the sequel.

THEOREM 15.2.7. *Suppose that* $\mathbf{c}'\mathbf{x} \leq 0$ *for all* $\mathbf{x}$ *such that* $\mathbf{A}\mathbf{x} = \mathbf{b}$ *and*
$\mathbf{x} \in \mathscr{H}$, *where* $\mathscr{H}$ *is a closed convex cone. Then there exists a vector* $\mathbf{y}$ *such that*

$$(\mathbf{c}' - \mathbf{y}'\mathbf{A})\mathbf{x} \leq 0$$

for $\mathbf{x} \in \mathscr{H}$.

To establish the result, one considers the convex cone $\mathscr{L} = (\mathbf{A}\mathscr{H}, \mathbf{c}'\mathscr{H})$ in $\mathbb{R}^{m+1}$
(we suppose $\mathbf{A}$ to be $m \times n$, as usual). The initial hypothesis of the theorem
implies that $(0, \delta)$ is external to $\mathscr{L}$, for $\delta > 0$. There is then a hyperplane
through the origin which separates $(0, \delta)$ strictly from $\mathscr{L}$; this assertion proves
the result.

Let $\mathscr{C}$ be a cone in $\mathbb{R}^n$. Then it emerges (from conditions such as (15.1.21))
that an important concept is the *conjugate* of $\mathscr{C}$: the set $\mathscr{C}^*$ in $\mathbb{R}^n$ of points
$\mathbf{y}$ such that

$$\mathbf{y}'\mathbf{x} \geq 0$$

for all $\mathbf{x}$ in $\mathscr{C}$. It is plain that $\mathscr{C}^*$ is also a cone, and that $\mathscr{C}^*$ decreases (from
$\mathbb{R}^n$ to 0) as $\mathscr{C}$ increases (from 0 to $\mathbb{R}^n$). One can say more:

THEOREM 15.2.8
   (i) $\mathscr{C}^*$ *is a closed convex cone*
   (ii) $\mathscr{C}^{**}$ *is the smallest convex cone containing* $\mathscr{C}$
   (iii) $\mathscr{C}^{**} = \mathscr{C}$ *if and only if* $\mathscr{C}$ *is a closed convex cone.*

The set $-\mathscr{C}^*$, whose elements $\mathbf{y}$ satisfy

$$\mathbf{y}'\mathbf{x} \leq 0$$

for $\mathbf{x}$ in $\mathscr{C}$, is sometimes called the *polar of* $\mathscr{C}$. The polar is rather more natural than the conjugate for minimization problems, but corresponds to a trivial change of convention.

A *polyhedral cone* is one generated from a finite number of rays.

THEOREM 15.2.9.   *A cone* $\mathscr{C}$ *in* $\mathbb{R}^m$ *is polyhedral if and only if it can be written* $\{\mathbf{Ax}: \mathbf{x} \geq 0\}$ *for some finite* $m \times n$ *matrix* $\mathbf{A}$. *The conjugate* $\mathscr{C}^*$ *is* $\{\mathbf{y}: \mathbf{A'y} \leq 0\}$.

15.2.5.   The Farkas Lemmas

The Farkas lemmas are often appealed to in treatments of linear and convex programming. They amount to somewhat special and disguised versions of the separating hyperplane theorem. In general, we prefer to refer back directly to this theorem. However, for the sake of reference, we record the relationship.

The lemmas all amount to a statement that one of two mutually exclusive events must be true; a given vector $\mathbf{b}$ in $\mathbb{R}^m$ either belongs to a given closed convex set $\mathscr{K}$ in $\mathbb{R}^m$ or it does not. In the first case it is representable as an average of extreme points of $\mathscr{K}$; in the second, there is a hyperplane strictly separating $\mathbf{b}$ and $\mathscr{K}$.

LEMMA.   *Let* $\mathbf{A}$ *be an* $m \times n$ *matrix,* $\mathbf{b}$ *a non-zero vector in* $\mathbb{R}^m$. *Then just one of the two following problems possesses a solution:*
  (i) *Find* $\mathbf{x} > 0$ *such that* $\mathbf{Ax} = \mathbf{b}$,
  (ii) *Find* $\mathbf{y}$ *in* $\mathbb{R}^m$ *such that* $\mathbf{y'A} > 0$, $\mathbf{y'b} < 0$.

The set $\mathscr{K}$ is the convex cone in $\mathbb{R}^m$ generated from the points represented by the columns of $\mathbf{A}$, and with generic point $\mathbf{Ax}$ $(x > 0)$. We interpret $\mathbf{y}$ as the vector of coefficients for the hyperplane through the origin separating $\mathbf{b}$ from $\mathscr{K}$.

Modifications of the problem lead to the following variants of the Lemma.

LEMMA.   *Just one of the two following problems possesses a solution:*
  (i) *Find* $\mathbf{x} \geq 0$ *such that* $\mathbf{Ax} \leq \mathbf{b}$
  (ii) *Find* $\mathbf{y} > 0$ *such that* $\mathbf{y'A} \geq 0$, $\mathbf{y'b} < 0$.

LEMMA.   *Just one of the two following problems possesses a solution:*
  (i) *Find* $\mathbf{x} > 0$ *such that* $\mathbf{Ax} \leq 0$.
  (ii) *Find* $\mathbf{y} > 0$ *such that* $\mathbf{y'A} \gg 0$.

15.2.6.   Convex and Concave Functions

Consider a scalar-valued function $f(\mathbf{x})$ with natural domain of definition $\mathscr{X}$, a set in $\mathbb{R}^n$. Then the *function f* is said to be *convex* if the set of $(z, \mathbf{x})$ in $\mathbb{R}^{n+1}$ satisfying $z \geq f(\mathbf{x})$ and $\mathbf{x} \in \mathscr{X}$ is convex. This is the set of points above the graph of $f(\mathbf{x})$, sometimes called the *epigraph* of $f(\mathbf{x})$, and alternatively express-ible as $(f(\mathcal{X})+\mathbb{R}_+, \mathcal{X})$ (see Figure 15.2.5).

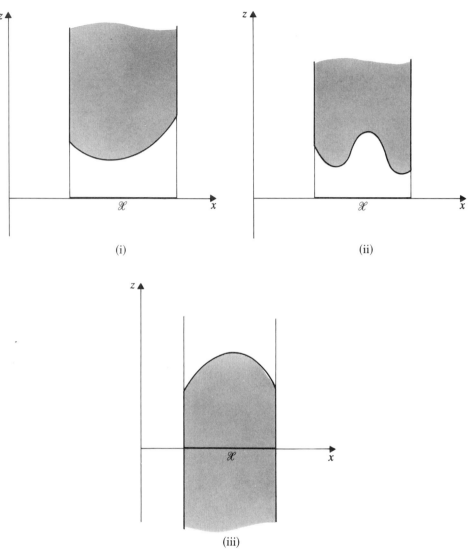

Figure 15.2.5: (i) The epigraph of a convex function. (ii) The epigraph of a non-convex function. (iii) The hypograph of a concave function.

The definition implies that the set $\mathcal{X}$ is itself convex, and that

$$f(p_1\mathbf{x}^{(1)}+p_2\mathbf{x}^{(2)}) \leq p_1 f(\mathbf{x}^{(1)})+p_2 f(\mathbf{x}^{(2)}) \tag{15.2.6}$$

for $\mathbf{x}^{(1)}, \mathbf{x}^{(2)}$ in $\mathcal{X}$ and (p_1, p_2) a distribution [see II, § 4.3].

The more natural property in maximization problems is concavity. The function $f(\mathbf{x})$ is concave if $-f(\mathbf{x})$ is convex. This means that the set *below* its graph is convex. That is, the set $(f(\mathcal{X}) - \mathbb{R}_+, \mathcal{X})$. This set is sometimes called the *hypograph* of $f(\mathbf{x})$, abbreviated to hyp f. A concave function obeys the reverse inequality to (15.2.6):

$$f(p_1\mathbf{x}^{(1)} + p_2\mathbf{x}^{(2)}) \geq p_1 f(\mathbf{x}^{(1)}) + p_2 f(\mathbf{x}^{(2)}). \tag{15.2.7}$$

If one were interpreting $f(\mathbf{x})$ as the return from an allocation $\mathbf{x}$, then relation (15.2.7) can be expressed as saying that the return from an averaged allocation is not more than the averaged return from different allocations. That is, that uncertainty is in general costly, which is often a natural property of objective functions.

The function f will be *strictly* convex (concave) if inequality (15.2.6) (resp. (15.2.7)) is strict.

A function f can be both concave and convex if and only if it is linear, although neither property is then strict. An archetypal form of concave function is the minimum of several linear forms:

$$f(\mathbf{x}) = \min_i (\mathbf{c}_i'\mathbf{x}).$$

In general, the pointwise minimum (maximum) of several concave (convex) functions is concave (convex).

A concave (convex) function can be formally extended outside $\mathcal{X}$, consistent with concavity (convexity) by setting $f = -\infty$ $(+\infty)$ outside $\mathcal{X}$. *Jensen's inequality* is the version of (15.2.6) for general distributions over $\mathcal{X}$:

$$f(E\mathbf{x}) \leq Ef(\mathbf{x}) \tag{15.2.8}$$

[see also § 21.3].

15.2.7. Differentiability Properties: Sub-gradients

We shall in most cases quote results only for concave functions, the analogue for convex functions being immediate. The reason for our preference is that, as stated, concavity is the natural property in the maximization context.

If f is twice-differentiable at a point, then the $n \times n$ matrix of second derivatives (the *Hessian*) will be denoted $f_{\mathbf{xx}}$ [see § 5.7]. A fairly immediate consequence of (15.2.7) is

THEOREM 15.2.10. *If f is concave (strictly concave), and is twice-differentiable at a point, then $f_{\mathbf{xx}}$ there is negative semi-definite (negative definite)* [*see* I, § 9.2].

In fact, a concave function need not even be differentiable, although it can be shown to be differentiable almost everywhere.

Suppose that the expression

$$\frac{f(\mathbf{x}+\varepsilon \mathbf{s})-f(\mathbf{x})}{\varepsilon} \tag{15.2.9}$$

has a limit as $\varepsilon \downarrow 0$ (that is, as ε converges to zero monotonically from above—see § 1.4). We shall denote the limit by $D_\mathbf{s} f(\mathbf{x})$, and term it the *derivative of f at* $\mathbf{x}$ *in direction* $\mathbf{s}$. If f is differentiable in the ordinary sense at $\mathbf{x}$ then we have, of course

$$D_\mathbf{s} f = f_\mathbf{x} \mathbf{s} \tag{15.2.10}$$

but the directional derivatives may exist even if $f_\mathbf{x}$ does not. This is in fact the case for concave functions, because (15.2.7) implies that expression (15.2.9) is monotonic in ε.

THEOREM 15.2.11.   *If f is concave, and if f is finite at* $\mathbf{x}$ *then* $D_\mathbf{s} f(\mathbf{x})$ *exists for all interior directions* $\mathbf{s}$. *Moreover,* $D_\mathbf{s} f(\mathbf{x})$ *is concave in* $\mathbf{s}$.

Figure 15.2.6 illustrates how f can be non-differentiable at a point, and yet possess directional derivatives there. The second assertion of the theorem supplies a partial analogue of Theorem 15.2.10.

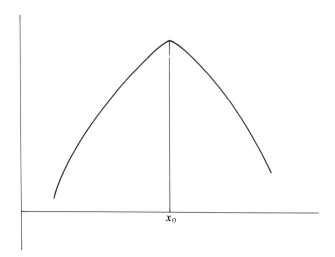

Figure 15.2.6: A concave function, non-differentiable at x_0, but with directional derivatives.

The supporting hyperplane theorem (15.2.5) must have implications for convex and concave functions.

THEOREM 15.2.12.   *Suppose* $\mathcal{X}$ *convex,* $f(\mathbf{x})$ *concave in* $\mathcal{X}$, *and finite at the finite point* $\bar{\mathbf{x}}$ *of* $\mathcal{X}$. *Then*

(i) *A vector* **y** *and a non-negative scalar w exist such that* $wf(\mathbf{x}) - \mathbf{y}'\mathbf{x}$ *is maximal in* $\mathscr{X}$ *at* $\mathbf{x} = \bar{\mathbf{x}}$.

(ii) *The directional derivatives of f at* $\bar{\mathbf{x}}$ *satisfy*

$$\mathbf{y}'\mathbf{s} \geq wD_\mathbf{s}f \qquad (15.2.11)$$

for direction **s** *interior to* $\mathscr{X}$ *from* $\bar{\mathbf{x}}$, *and any* (**y**, *w*) *satisfying* (15.2.11) *have the property enunciated in* (i).

(iii) *The constant w may be normalized to unity if either*: (*a*) $\bar{\mathbf{x}}$ *is interior to* $\mathscr{X}$ *and* $f(\mathbf{x})$ *finite in a neighbourhood* $N(\bar{\mathbf{x}})$ *of* $\bar{\mathbf{x}}$, *or* (*b*) $f(\mathbf{x})$ *is piecewise linear and finite in* $N(\bar{\mathbf{x}}) \cap \mathscr{X}$, *or* (*c*) $D_\mathbf{s}f$ *is finite for finite interior* **s**.

(iv) *Suppose that* $\bar{\mathbf{x}}$ *is interior to* $\mathscr{X}$, *and that* $f_\mathbf{x}$ *exists there, and is finite. If w is normalized to unity* (*as is possible, by* (iii)), *then* **y** *necessarily equals* $f_\mathbf{x}'$.

Assertion (i) is just a statement of the existence of a supporting hyperplane to hyp f at the boundary point $(\bar{\mathbf{x}}, f(\bar{\mathbf{x}}))$. The coefficient w of $f(\mathbf{x})$ is non-negative because $(\bar{\mathbf{x}}, f(\bar{\mathbf{x}}) + \delta)$ lies outside the hypograph (and will be strictly separated from it by the hyperplane if $w > 0$). The case one would wish to exclude is $w = 0$, when the hyperplane is vertical. Virtually the only case in which the hyperplane *must* be vertical is that where f has infinite derivative at $\bar{\mathbf{x}}$, and this is excluded in various ways by conditions (a), (b), (c).

In the case where w can be normalized to unity then (15.2.11) becomes

$$D_\mathbf{s}f \leq \mathbf{y}'\mathbf{s} \qquad (15.2.12)$$

or

$$D_\mathbf{s}f \geq \mathbf{y}'\mathbf{s} \qquad (15.2.13)$$

in the case of a convex function. A vector satisfying (15.2.13) is termed a *sub-gradient* of f at $\bar{\mathbf{x}}$; comparison of (15.2.10) and (15.2.13) makes appropriateness of the term plain. A **y** satisfying (15.2.12) is presumably a *super-gradient*. In either case, **y** is the gradient of a supporting hyperplane to the graph of $f(\mathbf{x})$ at $\bar{\mathbf{x}}$. If f is differentiable at $\bar{\mathbf{x}}$, an interior point of $\mathscr{X}$, then **y** is unique and equal to the gradient of f there.

15.2.8.   Conjugate Functions

The original variable **x** of our optimization problem is a *point variable*; the multiplier vector **y** is essentially a *slope variable*, as one sees from (15.1.22). The duality between these two types of variable is a fundamental one. The material of this section prepares us for a formal study of the duality.

Let $f(\mathbf{x})$ be a function, not necessarily concave, defined on a set $\mathscr{X}$ of $\mathbb{R}^n$, not necessarily convex. If values of **x** outside $\mathscr{X}$ are regarded as in some sense inadmissible, we shall extend the definition of f to the whole space by setting $f = -\infty$ outside $\mathscr{X}$. We now define the *maximum transform* or *convex conjugate* of f as

$$f^*(\mathbf{y}) = \sup_\mathbf{x} [f(\mathbf{x}) - \mathbf{y}'\mathbf{x}] \qquad (15.2.14)$$

where the supremum is taken over the whole space. Correspondingly, we define the *minimum transform* or *concave conjugate* of f^* as

$$f^{**}(\mathbf{x}) = \inf_{\mathbf{y}} [f^*(\mathbf{y}) + \mathbf{y}'\mathbf{x}].$$ (15.2.15)

The appropriateness of the term 'convex conjugate' follows from the readily proved

THEOREM 15.2.13. $f^*(\mathbf{y})$ *is convex.*

In a later context [§ 15.3.3] $\mathbf{y}$ will emerge as the Lagrange multiplier vector for a constrained maximization problem, the actual values of the multiplier being determined from a minimization problem analogous to (15.2.15), the problem *dual* to the original, or *primal* problem. In this context, the following theorem is important.

THEOREM 15.2.14
(i) f^{**} *is concave.*
(ii) $f^{**} \geq f$.
(iii) f^{**} *is the smallest function with properties* (i) *and* (ii), *that is the smallest* concave majorant *to f.*
(iv) $f^{**}(\mathbf{x}) = f(\mathbf{x})$ *if there is a non-vertical supporting hyperplane to the hypo-* graph of f *at* $\mathbf{x}$.
(v) $f^{**} = f$ if f *is concave.*

Assertion (v) obviously follows from (iii). There is an illuminating graphical argument which makes assertions (ii) and (iv) very plain, and which emphasizes the role of $\mathbf{y}$ as a slope variable.

For a given fixed finite $\mathbf{y}$ suppose that the supremum in (15.2.14) is attained at $\mathbf{x} = \mathbf{x}_0$. That is, $(\mathbf{x}_0, f(\mathbf{x}_0))$ is the point on the graph of $f(\mathbf{x})$ at which a hyperplane with slope $\mathbf{y}$ supports the hypograph of f. It will be the tangent-plane to the graph of f at this point if there is one. We have then

$$f^*(\mathbf{y}) + \mathbf{y}'\mathbf{x} = f(\mathbf{x}_0) + \mathbf{y}'(\mathbf{x} - \mathbf{x}_0)$$

$$= PR$$ (15.2.16)

where by PR we mean the length of this line segment in Figure 15.2.7, the height at $\mathbf{x}$ of the 'lowest' hyperplane of slope $\mathbf{y}$ which lies above the graph of $f(\mathbf{x})$. Thus

$$(f^*(\mathbf{y}) + \mathbf{y}'\mathbf{x}) - f(\mathbf{x}) = QR$$ (15.2.17)

which is plainly positive. Minimizing with respect to $\mathbf{y}$ we deduce assertion (ii) of the theorem. We shall achieve a minimal value of QR equal to zero if we can change $\mathbf{y}$ (that is roll the supporting hyperplane over the graph of f) until it meets the graph at the very point $R = (\mathbf{x}, f(\mathbf{x}))$. That is, if a supporting hyperplane to the hypograph of $\mathbf{f}$ exists at $\mathbf{x}$—this is just assertion (iv).

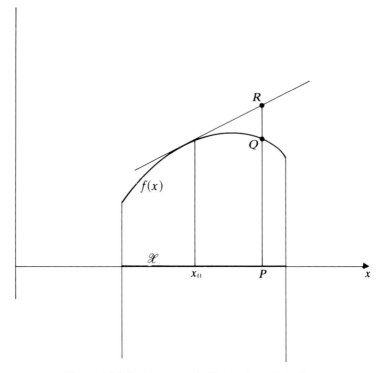

Figure 15.2.7: A geometric illustration of duality.

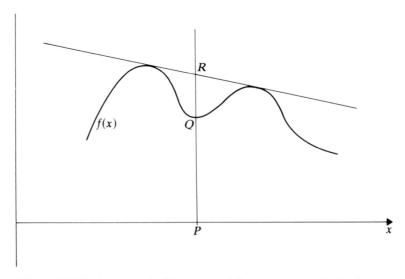

Figure 15.2.8: A geometric illustration of the occurrence of a duality gap.

The case in which QR *cannot* be decreased to zero, so that $f^{**}(\mathbf{x}) > f(\mathbf{x})$, is just that in which f demonstrates non-concavity at $\mathbf{x}$, and so has no supporting hyperplane to its hypograph there. In Figure 15.2.8 we have chosen the supporting hyperplane which minimizes QR. But since Q lies in a 'valley', the distance QR cannot be reduced to zero. This is the case of the so-called 'duality gap', for which the notion of a point solution must be generalized to the idea of a 'mixed solution' [see § 15.5].

15.2.9. $\mathscr{C}$-convexity

This is an extension of the concept of convexity which we should need later. Let $\mathbf{g}(\mathbf{x})$ be a function from $\mathscr{X}$ to $\mathbb{R}^m$, and $\mathscr{C}$ a set in $\mathbb{R}^m$. Then $\mathbf{g}$ is said to be $\mathscr{C}$-*convex* if

$$p_1\mathbf{g}(\mathbf{x}^{(1)}) + p_2\mathbf{g}(\mathbf{x}^{(2)}) - \mathbf{g}(p_1\mathbf{x}^{(1)} + p_2\mathbf{x}^{(2)}) \in \mathscr{C}$$

for any two-point distribution (p_1, p_2) and any two points $\mathbf{x}^{(1)}$, $\mathbf{x}^{(2)}$ of $\mathscr{X}$.

This amounts to the ordinary definition of convexity if $\mathscr{C} = \mathbb{R}_+$. If $\mathscr{C} = \mathbb{R}^m$ then it is no constraint at all. If $\mathscr{C} = \{0\}$ then $\mathbf{g}$ is constrained to be linear.

15.2.10. Conditions For Global Maximality

Note that, under appropriate convexity conditions, local maximality conditions are sufficient for global maximality.

THEOREM 15.2.15. *Suppose $\mathscr{X}$ convex, $f(\mathbf{x})$ concave in $\mathscr{X}$. If at $\bar{\mathbf{x}}$ we have $D_\mathbf{s}f \leq 0$ for all directions $\mathbf{s}$ interior to $\mathscr{X}$, then $\bar{\mathbf{x}}$ maximizes f in $\mathscr{X}$.*

15.3. STRONG LAGRANGIAN METHODS

15.3.1. The Strong Lagrangian Principle

The introductory approach of section 15.1.4 led one to expect that, under certain conditions, a multiplier vector $\mathbf{y}$ would exist such that the solution $\bar{\mathbf{x}}$ of the constrained maximization problem $P(\mathbf{b})$ lay at a stationary value of the form $f(\mathbf{x}) - \mathbf{y}'\mathbf{g}(\mathbf{x})$. It turns out that, under other conditions, one can make a stronger assertion: that a $\mathbf{y}$ exists such that a solution $\bar{\mathbf{x}}$ of $P(\mathbf{b})$ *maximizes* $f(\mathbf{x}) - \mathbf{y}'\mathbf{g}(\mathbf{x})$ *globally in* $\mathscr{X}$. We shall refer to the two types of assertion as *weak* and *strong* Lagrangian principles respectively. The weak principle asserts a local property (stationarity) and requires local conditions (differentiability, etc.) for its validity. The strong principle asserts a global property (maximization in $\mathscr{X}$) and requires global conditions (convexity etc.) for its validity. The weak principle is historically the earlier; the strong principle (when valid) endows the problem with more structure. The strong version is also distinctly the simpler, so we shall take it first.

Note first one simple sufficient condition for the validity of the strong principle: 'if it works, it is valid'.

THEOREM 15.3.1.   *Suppose that a* $\mathbf{y}$ *exists such that a value* $\bar{\mathbf{x}}$ *of* $\mathbf{x}$ *maximiz-* *ing* $f(\mathbf{x}) - \mathbf{y}'\mathbf{g}(\mathbf{x})$ *in* $\mathcal{X}$ *is feasible for* $P_0(\mathbf{b})$. *Then* $\bar{\mathbf{x}}$ *solves* $P_0(\mathbf{b})$.

The proof is immediate. Let us denote, as before [see § 15.1.4] the con-strained maximum of f by $U(\mathbf{b})$. Let us also denote the feasible set for $P_0(\mathbf{b})$ by $\mathcal{X}(\mathbf{b})$. Then, since $\bar{\mathbf{x}}$ is feasible, we have

$$f(\bar{\mathbf{x}}) \le U(\mathbf{b}) = \max_{\mathcal{X}(\mathbf{b})} f$$

$$= \max_{\mathcal{X}(\mathbf{b})} [f + \mathbf{y}'(\mathbf{b} - \mathbf{g})]$$

$$\le \max_{\mathcal{X}} [f + \mathbf{y}'(\mathbf{b} - \mathbf{g})]$$

$$= f(\bar{\mathbf{x}}) + \mathbf{y}'(\mathbf{b} - \mathbf{g}(\bar{\mathbf{x}}))$$

$$= f(\bar{\mathbf{x}}). \tag{15.3.1}$$

We then have $f(\bar{\mathbf{x}}) = U(\mathbf{b})$, so that $\bar{\mathbf{x}}$ solves $P(\mathbf{b})$.

The analogous assertion for $P_{\mathscr{C}}(\mathbf{b})$ is plain; we record it together with a more specific alternative sufficient condition.

THEOREM 15.3.2
   (i) *Suppose that a* $\mathbf{y}$ *exists such that a value* $(\bar{\mathbf{x}}, \bar{\mathbf{z}})$ *maximizing* $f(\mathbf{x}) - \mathbf{y}'(\mathbf{g}(\mathbf{x}) + \mathbf{z})$ *is feasible for* $P_{\mathscr{C}}(\mathbf{b})$. *Then* $\bar{\mathbf{x}}$ *solves* $P_{\mathscr{C}}(\mathbf{b})$.
   (ii) *Suppose* $\mathscr{C}$ *a cone, and that* $\mathbf{y}$ *in* $\mathscr{C}^*$ *exists such that a value* $\bar{\mathbf{x}}$ *maximizing* $f(\mathbf{x}) - \mathbf{y}'\mathbf{g}(\mathbf{x})$ *in* $\mathcal{X}$ *is feasible for* $P_{\mathscr{C}}(\mathbf{b})$, *and*

$$\mathbf{y}'(\mathbf{b} - \mathbf{g}(\bar{\mathbf{x}})) = \mathbf{0}. \tag{15.3.2}$$

Then $\bar{\mathbf{x}}$ *solves* $P_{\mathscr{C}}(\mathbf{b})$. *This condition is equivalent to the maximization condition* *of* (i) *in the case when* $\mathscr{C}$ *is a convex cone*.

Statement (ii) implies that the assertions $\mathbf{y} \in \mathscr{C}^*$ and the *complementary slackness* condition (15.3.2), otherwise expressible $\mathbf{y}'\bar{\mathbf{z}} = 0$, are *sufficient* to ensure that $\bar{\mathbf{z}}$ maximizes $-\mathbf{y}'\mathbf{z}$ in $\mathscr{C}$. They are also *necessary* if $\mathscr{C}$ is a convex cone.
   The most familiar case is that of inequality conditions (15.1.7), when $\mathscr{C} = \mathbb{R}_+^m$, and also $\mathscr{C}^* = \mathbb{R}_+^m$. This is the situation pertaining for the familiar allocation problem. If there is an appropriate multiplier vector then we must have $\mathbf{y} \ge \mathbf{0}$, while the complementary slackness condition (15.3.2) implies that it cannot be true that $b_j - g_j(\bar{\mathbf{x}})$ and y_j are simultaneously positive for any j. Anticipating the price interpretation of the multiplier, we can say that prices are non-negative if dispersal is costless, and that a resource in surplus cannot command a positive price.

We now continue with the determination of necessary conditions for the validity of the strong Lagrangian principle. So, if $\bar{\mathbf{x}}$ solves $P_{\mathscr{C}}(\mathbf{b})$, we require conditions which assume the existence of a $\mathbf{y}$ such that $f(\mathbf{x}) - \mathbf{y}'\mathbf{g}(\mathbf{x})$ is maximal in $\mathscr{X}$ at $\bar{\mathbf{x}}$. We shall also admit a somewhat weakened *homogeneous form* of the principle: there exists a non-zero $(m+1)$-vector $(w, \mathbf{y})$ such that $wf(\mathbf{x}) - \mathbf{y}'\mathbf{g}(\mathbf{x})$ is maximal in $\mathscr{X}$ at $\bar{\mathbf{x}}$. It will necessarily be true that $w \geq 0$. If $w > 0$ then we can normalize to $w = 1$, and are back in the previous form of the principle. If $w = 0$, as occasionally can happen, then one is in a situation where the conventional multiplier vector $\mathbf{y}$ is effectively infinite. This is a somewhat bothersome point, which corresponds to an extreme form of problem, but which does need some attention.

15.3.2. Geometrical Characterization of the Strong Lagrangian Principle

For the purposes of this section we need only consider the problem with equality constraints, $P_0(\mathbf{b})$. The fact that others are reducible to it means that the argument covers all cases. The special features of $P_{\mathscr{C}}(\mathbf{b})$ will be considered in the next section.

Consider the set of values generated by

$$\boldsymbol{\xi} = \mathbf{g}(\mathbf{x})$$
$$\eta = f(\mathbf{x}) \tag{15.3.3}$$

as $\mathbf{x}$ traverses $\mathscr{X}$. The point with co-ordinates $(\boldsymbol{\xi}, \eta)$ takes values in $\mathbb{R}^{m+1}$; the set thus generated can be denoted $\mathscr{D} = (\mathbf{g}(\mathscr{X}), f(\mathscr{X}))$. The 'upper boundary' of this set is a surface described by

$$\eta = U(\boldsymbol{\xi}) \qquad (\boldsymbol{\xi} \in \mathscr{B}) \tag{15.3.4}$$

where $\mathscr{B}$ is the attainable set and U the optimal return function, defined in section 1.4. (By 'upper surface' we mean the set of points such that $(\boldsymbol{\xi}, \eta)$ belongs to $\mathscr{D}$, but $(\boldsymbol{\xi}, \eta + \delta)$ does not, for any $\delta > 0$.)

By considering the set $\mathscr{D}$ rather than behaviour of f and $\mathbf{g}$ on $\mathscr{X}$ one submerges some aspects of the original problem, in that one considers merely what combinations of f and $\mathbf{g}$ values are possible, without concerning oneself as to what value of $\mathbf{x}$ realizes these. One knows simply that any $(\boldsymbol{\xi}, \eta)$ value in $\mathscr{D}$ can be realized for $(\mathbf{g}, f)$ for some $\mathbf{x}$ in $\mathscr{X}$. This subordination of some aspects of the problem leaves us with those which are most important structurally.

Note now that maximization of a form $wf(\mathbf{x}) - \mathbf{y}'\mathbf{g}(\mathbf{x})$ in $\mathscr{X}$ is equivalent to maximization of $w\eta - \mathbf{y}'\boldsymbol{\xi}$ in $\mathscr{D}$. So, the fact that $wf(\mathbf{x}) - \mathbf{y}'\mathbf{g}(\mathbf{x})$ is maximal at $\mathbf{x} = \bar{\mathbf{x}}$ (say) is equivalent to the fact that $\mathscr{D}$ has a supporting hyperplane $w\eta - \mathbf{y}'\boldsymbol{\xi} = \text{const.}$, which meets it at the point $(\mathbf{g}(\bar{\mathbf{x}}), f(\bar{\mathbf{x}}))$. This must be a boundary point of $\mathscr{D}$. Furthermore, if $w > 0$ (when we may suppose that $w = 1$)

then the point must be on the upper boundary, so that $f(\bar{\mathbf{x}}) = U(\mathbf{g}(\bar{\mathbf{x}}))$. Finally, if $\bar{\mathbf{x}}$ is to solve $P_0(\boldsymbol{b})$, then one must have $g(\bar{\mathbf{x}}) = \mathbf{b}$. We derive then

THEOREM 15.3.3

(i) $P_0(\mathbf{b})$ *is soluble by strong Lagrangian methods if and only if* $\mathcal{D}$ *possesses a supporting hyperplane at* $(\mathbf{b}, U(\mathbf{b}))$. *That is, if the hypograph of* $U(\boldsymbol{\xi})$ *possesses a supporting hyperplane at* $\boldsymbol{\xi} = \mathbf{b}$. *The case when one may set* $w = 1$ *corresponds to the case when this hyperplane may be chosen non-vertical.*

(ii) *If* $w = 1$ *then the multiplier vector* $\mathbf{y}$ *is identifiable as a super-gradient of* U *at* $\mathbf{b}$. *This must be the gradient of* U *at* $\mathbf{b}$, *if* U *is differentiable and* $\mathbf{b} \in \mathrm{ri}\,(\mathcal{B})$.

The necessary and sufficient condition for the strong Lagrangian principle to be valid (possibly in the homogeneous form) *for all attainable* $\mathbf{b}$ is then simply that U be concave, because this is the necessary and sufficient condition for hyp U to have a supporting hyperplane for all $\mathbf{b}$ in $\mathcal{B}$. In the next section we shall develop conditions of f, $\mathbf{g}$, $\mathcal{X}$ and $\mathcal{C}$ which will ensure this. If we wish the strong principle to be valid in the classic inhomogeneous form (with $w = 1$) then we must add further conditions, derived from Theorem 15.2.11.

THEOREM 15.3.4.   *Suppose* U *concave. Then* $P_0(\mathbf{b})$ *is soluble by strong Lagrangian methods with* $w = 1$ *for all* $\mathbf{b}$ *for which* U *possesses finite supergradient. Sufficient conditions for this latter are* (a) *that* $\mathbf{b} \in \mathrm{ri}\,(\mathcal{B})$ *and* U *be finite in* $N(\mathbf{b}) \cap \mathrm{ri}\,(\mathcal{B})$ *for some neighbourhood* $N(\mathbf{b})$ *of* $\mathbf{b}$, *or* (b) *that* U *be piecewise linear and finite in* $N(\mathbf{b}) \cap \mathcal{B}$, *or* (c) *that* $D_\mathbf{s}U$ *be finite for all* $\mathbf{s}$ *interior to* $\mathcal{B}$ *from* $\mathbf{b}$.

All the assertions of Theorems 15.3.3 and 15.3.4 hold for $P_\mathcal{C}(\mathbf{b})$ with the only modification that now

$$\mathcal{D} = (\mathbf{g}(\mathcal{X}) + \mathcal{C}, f(\mathcal{X})),$$

and that the maximization of $wf - \mathbf{y}'\mathbf{g}$ in $\mathcal{X}$ is to be replaced by the maximization of $wf - \mathbf{y}'(\mathbf{g}+z)$ in $(\mathcal{X}, \mathcal{C})$.

15.3.3.   Conditions for Concavity of U

Conditions on f, $\mathbf{g}$, $\mathcal{X}$ and $\mathcal{C}$ which imply concavity of U on $\mathcal{B}$ are supplied by

THEOREM 15.3.5.   *Suppose that* $\mathcal{X}$, $\mathcal{C}$ *are convex, that* f *is concave, and* $\mathbf{g}$ *is* $\mathcal{C}$-*convex. Then* $\mathcal{B}$ *is convex, and* U *is concave on* $\mathcal{B}$.

The reader is reminded of the definition of $\mathcal{C}$-convexity in section 15.2.9. The theorem is easily proved. The conditions of the theorem are sufficient for the concavity of $\mathcal{U}$. However, they are not necessary, as is fortunate, because the conditions can be irksomely restrictive. For example, if the constraints are all equalities ($\mathscr{C} = \{0\}$) then **g** is restricted to be linear. More relaxed conditions are given in Mangasarian (1969), p. 161.

Note that if $\mathscr{C}$ is a convex cone then $\bar{\mathbf{x}}, \bar{\mathbf{z}}$ will maximize $f(\mathbf{x}) - \mathbf{y}'(\mathbf{g}(\mathbf{x}) + \mathbf{z})$ in $(\mathscr{X}, \mathscr{C})$ if and only if (i) $\bar{\mathbf{x}}$ maximizes $f(\mathbf{x}) - \mathbf{y}'\mathbf{g}(\mathbf{x})$ in $\mathscr{X}$, (ii) $\mathbf{y} \in \mathscr{C}^*$, and (iii) $\mathbf{y}$ and $\bar{\mathbf{z}}$ obey the *complementary slackness principle*:

$$\mathbf{y}'\bar{\mathbf{z}} = \mathbf{0}.$$

So, as we have before observed, a relaxation of the constraint to $\mathbf{b} - \mathbf{g}(\mathbf{x}) \in \mathscr{C}$ imposes constraints on the possible values of the multiplier vector $\mathbf{y}$. As we observed at the end of section 15.3.1, these constraints have a natural interpretation.

15.3.4. Duality

Let us define, for $P_{\mathscr{C}}(\mathbf{b})$,

$$F(\mathbf{y}, \mathbf{b}) = \max_{\substack{\mathbf{x} \in \mathscr{X} \\ \mathbf{z} \in \mathscr{C}}} [f(\mathbf{x}) + \mathbf{y}'(\mathbf{b} - \mathbf{g}(\mathbf{x}) - \mathbf{z})] \left.\vphantom{\begin{matrix}1\\1\\1\end{matrix}}\right\}$$

$$= \max_{\mathbf{\xi} \in \mathscr{B}} [U(\mathbf{\xi}) + \mathbf{y}'(\mathbf{b} - \mathbf{\xi})] \qquad (15.3.5)$$

and

$$\hat{U}(\mathbf{b}) = \min_{\mathbf{y}} F(\mathbf{y}, \mathbf{b}). \qquad (15.3.6)$$

It is plain, then, from the second expression of (15.3.5), that

$$F(\mathbf{y}, \mathbf{b}) = U^*(\mathbf{y}) + \mathbf{y}'\mathbf{b} \qquad (15.3.7)$$

and that

$$\hat{U}(\mathbf{b}) = U^{**}(\mathbf{b}) \qquad (15.3.8)$$

where U^* is the convex conjugate of U, defined in section 15.2.8, and U^{**} the concave conjugate of U^*. We have then, from Theorem 15.2.14,

THEOREM 15.3.6
(i) $\hat{U}$ *is the smallest concave majorant to U, and $\hat{U}(\mathbf{b}) = U(\mathbf{b})$ if there is a non-vertical supporting hyperplane to the hypograph of U at $\mathbf{b}$.*

(ii) *The slope vector $\mathbf{y}$ of such a supporting hyperplane (if it exists) minimizes $F(\mathbf{y}, \mathbf{b})$, and supplies a value of the multiplier for the strong Lagrangian solution of $P_{\mathscr{C}}(\mathbf{b})$.*

It is the second part of the assertion which is of immediate interest: that if a multiplier vector $\mathbf{y}$ exists which supplies a strong Lagrangian solution to the problem (true if and only if the non-vertical supporting hyperplane to hyp $U(\mathbf{b})$ exists, by Theorem 15.3.3), then this value minimizes $F(\mathbf{y}, \mathbf{b})$. That is, the multiplier vector solves a *minimization* problem. This problem is referred to as the *dual* problem, the original problem $P_{\mathscr{C}}(\mathbf{b})$ being referred to as the *primal*.

Given the price interpretation of the vectors $\mathbf{y}$, the dual problem often has an interesting economic interpretation. Linear programmes have the conjugacy property, that the dual of the dual is the primal [see § 11.2.1]; this is not true in general. We shall find it useful to formulate the dual in several cases; see §§ 15.6.1, 15.6.6, 15.7.

In section 15.5 we examine the significance of the dual in the case when $U(\mathbf{b}) < \hat{U}(\mathbf{b})$, and $P_{\mathscr{C}}(\mathbf{b})$ is not soluble by strong Lagrangian methods.

The coupling between primal and dual problems is related to the ideas of the saddlepoint of a function and the *min–max theorem*: A function $K(\mathbf{x}, \mathbf{y})$ of two sets of variables $\mathbf{x}$ and $\mathbf{y}$ has a *saddlepoint* at $(\bar{\mathbf{x}}, \bar{\mathbf{y}})$ if

$$K(\mathbf{x}, \bar{\mathbf{y}}) \le K(\bar{\mathbf{x}}, \bar{\mathbf{y}}) \le K(\bar{\mathbf{x}}, \mathbf{y}) \tag{15.3.9}$$

for all admissible $\mathbf{x}, \mathbf{y}$ [see Figure 5.6.4]. If such a saddlepoint exists then

$$\min_{\mathbf{y}} \max_{\mathbf{x}} K(\mathbf{x}, \mathbf{y}) = \max_{\mathbf{x}} \min_{\mathbf{y}} K(\mathbf{x}, \mathbf{y}) \tag{15.3.10}$$

i.e. the operations $\max_{\mathbf{x}}$ and $\min_{\mathbf{y}}$ commute when applied to K. If we define

$$K(\mathbf{x}, \mathbf{y}) = f(\mathbf{x}) + \mathbf{y}'(\mathbf{b} - \mathbf{g}(\mathbf{x})) \tag{15.3.11}$$

(for $P_0(\mathbf{b})$) then we can say that the primal problem is the maximization of $\min_{\mathbf{y}} K(\mathbf{x}, \mathbf{y})$ and the dual problem is the minimization of $\max_{\mathbf{x}} K(\mathbf{x}, \mathbf{y})$; furthermore, that

$$\left.\begin{aligned}U(\mathbf{b}) &= \max_{\mathbf{x}} \min_{\mathbf{y}} K(\mathbf{x}, \mathbf{y}) \\[2mm] \hat{U}(\mathbf{b}) &= \min_{\mathbf{y}} \max_{\mathbf{x}} K(\mathbf{x}, \mathbf{y}).\end{aligned}\right\} \tag{15.3.12}$$

Primal and dual are always related to a min–max problem in this way. Moreover, they yield the same extremum (i.e. the two expressions (15.2.12) are equal) just exactly when K possesses a saddlepoint. The various convexity assumptions of Theorem 15.3.5 assure the existence of such a saddlepoint for all attainable $\mathbf{b}$.

15.4.   WEAK LAGRANGIAN METHODS

15.4.1.   The Weak Lagrangian Principle; the Kuhn–Tucker Conditions

The Lagrangian principles state that, under suitable conditions, there is a $\mathbf{y}$ such that the Lagrangian form $f(\mathbf{x}) - \mathbf{y}'\mathbf{g}(\mathbf{x})$ is maximized, in some sense, at the solution $\bar{\mathbf{x}}$ of $P_0(\mathbf{b})$. The strong principle asserts that the form finds its global maximum at $\bar{\mathbf{x}}$. The weak principle asserts that the form obeys stationarity conditions, demonstrating that it has a local maximum at $\bar{\mathbf{x}}$.

We have demonstrated the weak principle in section 15.1.4 under the assumption that U is differentiable at $\mathbf{b}$ (as well as f and $\mathbf{g}$ at $\bar{\mathbf{x}}$). We see, then, that the weak and strong principles correspond respectively to the existence of *tangent* and *supporting* hyperplanes to hyp U at f. These hyperplanes must be non-vertical if the multipliers are to be finite.

Part (ii) of Theorem 15.1.1 gives an exact statement of the weak Lagrangian principle, under the assumption of the existence of this tangent hyperplane. With stronger assumptions on the form of U, the stationarity conditions (15.1.20), (15.1.21) yield the so-called *Kuhn–Tucker necessary conditions*, which we summarize in the following

THEOREM 15.4.1.   *Suppose that* $P_\mathscr{C}(\mathbf{b})$ *has a solution at* $\bar{\mathbf{x}}, \bar{\mathbf{z}}$; *that f and $\mathbf{g}$ are differentiable at* $\bar{\mathbf{x}}$, *and that U possesses finite gradient* $\mathbf{y}'$ *at* $\mathbf{b}$. *Then*
  (KT1) *For all directions* $\mathbf{s}$ *interior to* $\mathscr{X}$ *from* $\bar{\mathbf{x}}$

$$(f_\mathbf{x} - \mathbf{y}'\mathbf{g}_\mathbf{x})\mathbf{s} \le 0. \tag{15.4.1}$$

  (KT2) *For all directions* $\mathbf{t}$ *interior to* $\mathscr{C}$ *from* $\bar{\mathbf{z}}$

$$\mathbf{y}'\mathbf{t} \ge 0. \tag{15.4.2}$$

  (KT3) *If $\mathscr{X}$ is a cone then*

$$(f_\mathbf{x} - \mathbf{y}'\mathbf{g}_\mathbf{x})\bar{\mathbf{x}} = 0. \tag{15.4.3}$$

  (KT4) *If $\mathscr{C}$ is a cone then* $\mathbf{y}'\bar{\mathbf{z}} = 0$, *or*

$$\mathbf{y}'(\mathbf{b} - \mathbf{g}(\bar{\mathbf{x}})) = 0. \tag{15.4.4}$$

To these can be added conditions on $\mathbf{y}$ implied by (15.4.2).

THEOREM 15.4.2.   *Under the conditions of Theorem* 15.4.1
  (i) *If $\mathscr{X}$ is a convex cone then*

$$\mathbf{g}'_\mathbf{x}\mathbf{y} - f'_\mathbf{x} \in \mathscr{X}^*$$

  (ii) *If $\mathscr{C}$ is a convex cone then* $\mathbf{y} \in \mathscr{C}^*$.
The weak principle demands differentiability rather than convexity (although there may be local convexity demands: see section 15.4.3). However, we may note

THEOREM 15.4.3.   *Suppose f, $\mathbf{g}$, $\mathscr{X}$ and $\mathscr{C}$ obey the convexity conditions of Theorem* 15.3.5, *as well as the differentiability conditions of Theorem* 15.4.1. *Then conditions* (15.4.1), (15.4.2) *of Theorem* 15.4.1 *are sufficient as well as necessary for a feasible pair* $(\bar{\mathbf{x}}, \bar{\mathbf{y}})$ *to solve* $P_\mathscr{C}(\mathbf{b})$.

15.4.2. Failure of the Weak Principle

The weak principle will presumably fail if U is non-differentiable at $\mathbf{b}$. However, it (or the usual statement of it) can also fail if the gradient $U_\mathbf{b}$ exists, but is infinite, otherwise expressed, the weak principle will apply to the homogeneous form $wf(\mathbf{x}) - \mathbf{y}'\mathbf{g}(\mathbf{x})$ but with $w = 0$.

There is a standard example illustrating this point. It is set in the plane with typical point $\mathbf{x} = (x_1, x_2)$, and $\mathscr{X}$ taken as the upper half-plane $x_2 \geq 0$. One wishes to maximize $f(\mathbf{x}) = x_1$, subject to

$$x_2 + x_1^3 = 0. \tag{15.4.5}$$

The solution is plainly at $\bar{\mathbf{x}} = (0, 0)$. If we try to find this solution by Lagrangian methods, we fail. The (inhomogeneous) Lagrangian form is

$$L = x_1 - y(x_2 + x_1^3)$$

and at $\bar{\mathbf{x}}$ we have

$$\frac{\partial L}{\partial x_1} = 1 - 3yx_1^3 = 1$$

so that L is not stationary with respect to x_1 variations at $\bar{\mathbf{x}}$.

To explain the trouble, let us generalize the constraint (15.4.5) to

$$x_2 + x_1^3 = b.$$

Thus $x_1 = (b - x_2)^{1/3}$ (the sign of x_1 being that of $b - x_2$), whence

$$U(b) = b^{1/3}$$

(where we take the real root, and give $b^{1/3}$ the sign of b).

So, $U(b)$ is exceptional only insofar as its derivative is infinite at $b = 0$. This does mean that the tangent to U at $b = 0$ is vertical and the corresponding y infinite (or $w = 0$ in the homogeneous formulation). In fact, the stationarity conditions of the Lagrangian $wx_1 - y(x_2 + x_1^3)$ at $(0, 0)$ are

$$w = 0, \qquad -y \leq 0.$$

This is the behaviour we have excluded for the strong principle, by various conditions in Theorem 15.3.4. In the case of the weak principle, one attempts to exclude it by a so-called *constraint qualification*.

15.4.3.   Constraint Qualifications

The condition that U_b not merely exist, but also be finite, is sufficient to exclude the anomaly just discussed. However, one would require conditions more immediately verifiable.

Let us define first the notion of a *contained path*. This is a path in $\mathbb{R}^n$ described by $\mathbf{x} = \mathbf{h}(t)$, where t is a scalar parameter, and such that $\mathbf{h}(t)$ varies continuously with t and is feasible for $P_{\mathscr{C}}(\mathbf{b})$ for all $t \geq 0$. Suppose a contained path exists such that $\mathbf{h}(0) = \bar{\mathbf{x}}$ and $\mathbf{h}'(0) = \mathbf{s}$. Then the direction $\mathbf{s}$ is said to be *attainable* from $\bar{\mathbf{x}}$.

Suppose for the moment that we are dealing with $P_0(\mathbf{b})$ and the equality constraint

$$\mathbf{g}(\mathbf{x}) = \mathbf{b}; \tag{15.4.6}$$

all cases are reducible to this. Suppose also that $\mathbf{g_x}$ exists at $\bar{\mathbf{x}}$. Then attainable directions $\mathbf{s}$ must satisfy

$$\mathbf{g_x s} = 0 \qquad (15.4.7)$$

that is, they must lie in the *linearizing cone* or *tangent cone* to the hypersurface (15.4.6) at $\bar{\mathbf{x}}$. They must also be interior to $\mathscr{X}$ at $\bar{\mathbf{x}}$. Directions $\mathbf{s}$ satisfying these two constraints are called *locally constrained* and form a convex cone $\mathscr{G}$. The set of attainable directions forms a cone, not necessarily convex, which is contained in $\mathscr{G}$.

The converse is not necessarily true: that all directions in $\mathscr{G}$ must be attainable. Consider the example of the last section for which, at the solution point $\bar{\mathbf{x}} = (0, 0)$, the cone $\mathscr{G}$ is specified by $\mathbf{g_z} = 0$. The direction $\mathbf{s} = (1, 0)$ lies in $\mathscr{G}$, and yet is not attainable.

For $P_{\mathscr{C}}(\mathbf{b})$ the constraint (15.4.7) on $\mathbf{s}$ is replaced by the requirement that

$$\mathbf{g} + \varepsilon \mathbf{g_x s} \in \mathscr{C} \qquad (15.4.8)$$

for all sufficiently small positive ε, where $\mathbf{g}$ and $\mathbf{g_x}$ are evaluated at $\bar{\mathbf{x}}$. The cone of locally constrained directions then consists of all directions satisfying this constraint which are also interior to $\mathscr{X}$.

The *Kuhn–Tucker constraint qualification* (KTCQ) is the requirement that all locally constrained directions be attainable.

THEOREM 15.4.4.   *Suppose $\bar{\mathbf{x}}$ is a solution of $P_{\mathscr{C}}(\mathbf{b})$ at which f and $\mathbf{g}$ are differentiable, and at which the Kuhn–Tucker constraint qualification is satisfied.*

Then there exists a multiplier vector $\mathbf{y}$ satisfying the Kuhn–Tucker conditions of Theorem 15.4.1.

The proof follows by appeal to Theorem 15.2.7.

In the case of $P_0(\mathbf{b})$ the KTCQ is certainly satisfied if the constraints are linear, or if $\bar{\mathbf{x}}$ is interior to and $\mathbf{g_x}$ is of full rank [see I, § 5.6].

There are other constraint qualifications which also assure existence and finiteness of $\mathbf{y}$: see Mangasarian (1969) p. 102.

15.5.   MIXED SOLUTIONS

15.5.1.   The Non-convex Case; Duality Gaps

The function $\hat{U}(\mathbf{b})$ defined in section 15.3.4 can be written

$$\hat{U}(\mathbf{b}) = \min_{\mathbf{y}} \max_{\boldsymbol{\xi}} [U(\boldsymbol{\xi}) + \mathbf{y}'(\mathbf{b} - \boldsymbol{\xi})]. \qquad (15.5.1)$$

We know from Theorem 15.3.6 that $\hat{U}$ is the smallest concave majorant to U, and that $\hat{U}(\mathbf{b}) = U(\mathbf{b})$ if there is a non-vertical supporting hyperplane to hyp U at $\mathbf{b}$. Suppose there is not, as can happen if U is non-convex. This is

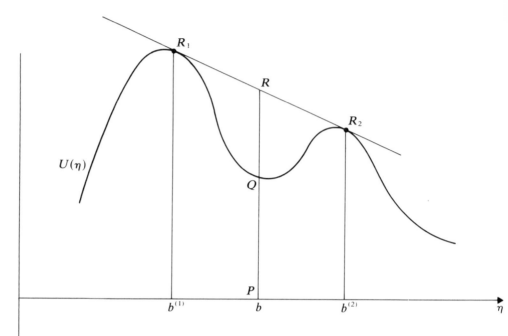

Figure 15.5.1: Geometric interpretation of a duality gap and of a randomised solution.

the case of a *duality gap*, illustrated for the case $m = 1$ in Figure 15.5.1, the analogue of Figure 15.2.8 in the present case. The quantity $U(b)$ has the value PQ, the quantity $\hat{U}(b)$ yielded by (15.5.1) has the value PR. The line R_1RR_2 is the nearest thing to a supporting hyperplane to hyp U at b which can be achieved. It is generated by the min–max operations of (15.5.1), in that the y-value at the min–max is the slope of this hyperplane, and the ξ value(s) are the co-ordinates $b^{(1)}$, $b^{(2)}$ at which $U(\xi) - y'\xi$ is maximal.

Note now that, because R_1RR_2 is a straight line, we can find positive numbers p, q such that

$$1 = p + q \tag{15.5.2}$$

$$b = pb^{(1)} + qb^{(2)} \tag{15.5.3}$$

$$\hat{U}(b) = pU(b^{(1)}) + qU(b^{(2)}) \tag{15.5.4}$$

[see V, § 2.2.2]. That is, it is as though in the constraint

$$b - g(x) \in \mathscr{C} \tag{15.5.5}$$

the value b were replaced by $b^{(1)}$ and $b^{(2)}$ with probabilities p and q. In virtue of (15.5.3) the constraint (15.5.3) is still satisfied *on average*, but the average return (15.5.4) is greater than would be possible if one held the single value b.

If a solution to $P_\mathscr{C}(b)$ is denoted $\bar{x}(b)$, then it is as though we had chosen $\bar{x}(b^{(1)})$ and $\bar{x}(b^{(2)})$ with respective probabilities p and q, instead of the single solution $\bar{x}(b)$. Such a solution is known as a *mixed* or *randomized* solution.

Generally, if one allows mixed solutions, then the attainable set $\mathscr{B}$ is expanded to its convex hull $[\mathscr{B}]$, and $U(\mathbf{b})$ is replaced by its concave majorant $\hat{U}(\mathbf{b})$, defined on $[\mathscr{B}]$. Randomization thus has the effect of introducing appropriate convexities throughout. Indeed, if one allows randomized solutions, then one is effectively reducing the constrained optimization problem to a linear programme (albeit an infinite-dimensional one in general). For example, in the extended version of $P_0(\mathbf{b})$, instead of looking for an $\mathbf{x}$ in $\mathscr{X}$ maximizing $f(\mathbf{x})$ subject to $\mathbf{g}(\mathbf{x}) = \mathbf{b}$ one looks for a probability measure μ on $\mathscr{X}$ maximizing $\int f(\mathbf{x})\mu(\mathbf{dx})$ subject to

$$\int \mathbf{g}(\mathbf{x})\mu(\mathbf{dx}) = \mathbf{b}.$$

This could be regarded as a linear programme in μ [see I, Chapter 11].

15.5.2. Randomized Solutions

We can collect the principal assertions of interest in

THEOREM 15.5.1
   (i) *The upper bound*

$$\hat{U}(\mathbf{b}) = \min_{\mathbf{y}} \max_{\mathbf{x},\mathbf{z}} [f(\mathbf{x}) + \mathbf{y}'(\mathbf{b} - \mathbf{z} - \mathbf{g}(\mathbf{x}))] \qquad (15.5.6)$$

to $U(\mathbf{b})$ is the average return obtainable with an optimal randomized solution to $P_\mathscr{C}(\mathbf{b})$, at least if the minimizing $\mathbf{y}$ in (15.5.6) is finite.
   (ii) *The minimizing $\mathbf{y}$ is a shadow price, in that it is a super-gradient to $\hat{U}(\mathbf{b})$.*
   (iii) *The support of the optimal mixed solution is restricted to the set of x achieving the maximum in* (i).
   (iv) *If this set is bounded, then an optimal randomized solution with support on at most $m + 1$ x-values exists.*

These results follow by direct appeal to the supporting-hyperplane arguments, and Carathéodory's theorem [see § 15.2.3 and Theorem 15.2.1].

15.5.3. Physical Realization of Randomized Solutions

The equality constraint $\mathbf{g}(\mathbf{x}) = \mathbf{b}$ will often represent the fact that supply of a given resource-bundle is maintained at rate $\mathbf{b}$ per unit time, or may represent a conservational balance of flows within a system. If a mixture of rates $\mathbf{b}^{(1)}$

and $\mathbf{b}^{(2)}$ in the proportion $p:q$ turns out to be advantageous, as in the case envisaged in section 15.5.1, then one can vary the rate between these two values over time. If the time allocated to the two rates is held to the ratio $p:q$, then there will be no net depletion or accumulation of resources, by (15.5.3). However, if transient imbalances in flow are to be accommodated, then buffer stores of adequate size must be inserted on lines through which the flow is being varied. It was pointed out by Simmons and White (1977) that it is the existence of such buffer stores that will permit one to realize a randomized solution.

15.6. SPECIAL ALGORITHMS

15.6.1. Recollections of the Simplex Algorithm

The simplex algorithm [see I, § 11.4] or variants of it, continue to play a role even for non-linear optimization problems. We therefore recall some appropriate points for reference.

The standard problem is the maximization of $f(\mathbf{x}) = \mathbf{c}'\mathbf{x}$ in $\mathbf{x} \geq \mathbf{0}$ subject to

$$\mathbf{Ax} = \mathbf{b}. \tag{15.6.1}$$

If there are m effective constraints then at most m elements of $\mathbf{x}$ need be positive in an optimal solution. The m values of j thus favoured are termed the *basis*, denoted B, and constraints (15.6.1) are rewritten as

$$x_j = \bar{x}_j - \sum_{k \in N} \alpha_{jk} x_k \qquad (j \in B) \tag{15.6.2}$$

where N, the set of non-basic subscripts, is the complement of B in $\{1, 2, \ldots, n\}$. In (15.6.2) $\bar{x}_j$ is the value of the basic variable x_j when the non-basic variables x_k are set equal to zero. One also writes

$$f(\mathbf{x}) = f(\bar{\mathbf{x}}) - \sum_N \gamma_k x_k. \tag{15.6.3}$$

The quantities $f(\bar{\mathbf{x}})$, γ_k and $\bar{x}_j$ are sometimes written α_{00}, α_{0k} and α_{j0} respectively, because all elements of the matrix

$$\begin{pmatrix} f(\bar{\mathbf{x}}) & \boldsymbol{\gamma}' \\ \bar{\mathbf{x}} & \boldsymbol{\alpha} \end{pmatrix} \tag{15.6.4}$$

follow the same tableau transformation rules when the basis is changed [see I, § 5.12].

If there is no obvious basic solution which is primal feasible with which to start the simplex algorithm, one can use the method of *artificial variables*. That is, one rewrites (15.6.1) as

$$(\mathbf{Ax})_j + \operatorname{sgn}(b_j) z_j = b_j \tag{15.6.5}$$

and considers the preliminary problem of minimizing the *sum of infeasibilities*

$$-f(\mathbf{x}, \mathbf{z}) = \sum z_j \tag{15.6.6}$$

subject to $(\mathbf{x}, \mathbf{z}) \geq \mathbf{0}$ and constraints (15.6.5). A feasible basic solution for this modified problem is plainly $\mathbf{x} = \mathbf{0}$, $z_j = |b_j|$. Iteration of the simplex method will lead to a feasible basic solution of the original problem, if one exists.

In the *primal simplex method* one starts from a basis which is primal feasible, and continues with changes of basis (in such a direction as to increase f) until $\boldsymbol{\gamma} \geq \mathbf{0}$, that is until the solution is dual feasible. In the *dual simplex method* one starts with a basic solution which is dual feasible, that is for which $\boldsymbol{\gamma} \geq \mathbf{0}$, and continues with changes of basis (in such a direction as to decrease f) until $\bar{\mathbf{x}} \geq \mathbf{0}$, and the solution is also primal feasible. This method is advantageous if one wishes to study the effect of changes in $\mathbf{b}$; a solution which is dual feasible will remain so under such changes, and so serves as a starting point for application of the dual simplex method.

15.6.2.   Non-linear Problems Reducible to Linear Form

Before we consider problems which are inescapably non-linear, let us note that some problems can be exactly linearized by an appropriate transformation.

Suppose, for example, that the objective function

$$f(\mathbf{x}) = \min_i \left(\sum_k c_{ik} x_k \right) \tag{15.6.7}$$

is to be maximized subject to linear constraints on $\mathbf{x}$. This problem can be replaced by the maximization of the objective function ξ with respect to $(\mathbf{x}, \xi)$ subject to the constraints

$$\sum_k c_{ik} x_k \geq \xi \tag{15.6.8}$$

on $(\mathbf{x}, \xi)$ plus the linear constraints originally imposed upon $\mathbf{x}$. This is a complete linearization of the problem.

Note that expression (15.6.7) is the typical finitely-generated concave function, and that the procedure we have just described has some affinity with the cutting-plane method of section 15.10.7 [see also I, § 12.3.1(b)].

Another reducible problem is that of *fractional programming*, in which the ratio of linear forms

$$f(\mathbf{x}) = \frac{c_0 + \mathbf{c}'\mathbf{x}}{d_0 + \mathbf{d}'\mathbf{x}} \tag{15.6.9}$$

is to be maximized subject to linear constraints on $\mathbf{x}$.

Suppose these are the standard ones

$$\mathbf{x} \geq \mathbf{0}$$
$$\mathbf{A}\mathbf{x} = \mathbf{b}. \tag{15.6.10}$$

Suppose also that it is clear that $d_0 + \mathbf{d}'\mathbf{x}$ is positive at the optimizing value (as would certainly be true if the expression were positive for all possible $\mathbf{x}$). We can then define the transformed variables

$$\xi = (d_0 + \mathbf{d}'\mathbf{x})^{-1}$$

$$\tilde{\mathbf{x}} = \mathbf{x}\xi$$

and rewrite the problem as the maximization of

$$\tilde{f} = c_0\xi + \mathbf{c}'\tilde{\mathbf{x}}$$

with respect to $(\tilde{\mathbf{x}}, \xi)$ subject to

$$\tilde{\mathbf{x}} \geq \mathbf{0}, \qquad \xi \geq 0$$

$$\mathbf{A}\tilde{\mathbf{x}} = \mathbf{b}\xi$$

$$d_0\xi + \mathbf{d}'\tilde{\mathbf{x}} = 1.$$

This is a linear problem. If the sign of $\xi^{-1} = d_0 + \mathbf{d}'\mathbf{x}$ at the optimum is not obvious, then one considers two problems, one constraining ξ to be non-negative, the other constraining ξ to be non-positive. One problem will have a solution and the other will not.

15.6.3. Quadratic Programming

One standard form of the quadratic programme is: to maximize

$$f(\mathbf{x}) = \mathbf{c}'\mathbf{x} + \tfrac{1}{2}\mathbf{x}'\mathbf{D}\mathbf{x} \tag{15.6.11}$$

in $\mathbf{x} \geq \mathbf{0}$ subject to

$$\mathbf{A}\mathbf{x} = \mathbf{b}. \tag{15.6.12}$$

An alternative form prescribes inequality constraints

$$\mathbf{A}\mathbf{x} \leq \mathbf{b}. \tag{15.6.13}$$

The objective function (15.6.11) will be concave if and only if $\mathbf{D} \leq \mathbf{0}$ (that is $\mathbf{D}$ is non-positive definite—see I, § 9.2); this condition may or may not be imposed.

A quadratic programme can be regarded as an improved approximation over a linear programme to the case when the objective function is non-linear. (For example the 'portfolio selection' problem.) However, sometimes the quadratic form of f is itself quite natural.

From the Lagrangian form $f(\mathbf{x}) + \mathbf{y}'(\mathbf{b} - \mathbf{A}\mathbf{x})$ we derive the Kuhn–Tucker conditions

$$\mathbf{c} + \mathbf{D}\mathbf{x} - \mathbf{A}'\mathbf{y} \leq \mathbf{0} \tag{15.6.14}$$

$$\mathbf{x}'(\mathbf{c} + \mathbf{D}\mathbf{x} - \mathbf{A}'\mathbf{y}) = 0 \tag{15.6.15}$$

for the constraint (15.6.12). For the inequality constraint (15.6.13) one has, in addition,

$$\mathbf{y} \ge \mathbf{0} \tag{15.6.16}$$

$$\mathbf{y}'(\mathbf{b} - \mathbf{Ax}) = 0. \tag{15.6.17}$$

THEOREM 15.6.1.   *Suppose* $\mathbf{D} \le \mathbf{0}$. *Then the necessary and sufficient condition for* $\mathbf{x}$ *to maximize expression* (15.6.11) *in* $\mathbf{x} \ge \mathbf{0}$ *subject to* (15.6.12) *is that* $\mathbf{x}$ *be feasible and satisfy* (15.6.14), (15.6.15) *for some* $\mathbf{y}$. *The necessary and sufficient condition for* $\mathbf{x}$ *to maximize expression* (15.6.11) *in* $\mathbf{x} \ge \mathbf{0}$ *subject to* (15.6.13) *is that* $\mathbf{x}$ *be feasible and satisfy* (15.6.14), (15.6.15), (15.6.17) *for some non-negative* $\mathbf{y}$.

15.6.4.   Quadratic Programming; Wolfe's Method

For the equality constraint case (15.6.12) we can collect all the constraint and optimality conditions into the system

$$\mathbf{Ax} = \mathbf{b} \tag{15.6.18}$$

$$\mathbf{c} + \mathbf{Dx} - \mathbf{A}'\mathbf{y} + \mathbf{w} = \mathbf{0} \tag{15.6.19}$$

$$(\mathbf{x}, \mathbf{w}) \ge \mathbf{0} \tag{15.6.20}$$

$$\mathbf{x}'\mathbf{w} = 0 \tag{15.6.21}$$

where $\mathbf{w}$ is an n-vector of slack-variables. The non-negative vector $(\mathbf{x}, \mathbf{w})$ is thus subject to the equality constraints (15.6.18), (15.6.19) and the complementarity condition (15.6.21).

Wolfe's method uses the artificial variable method to find a non-negative solution of (15.6.18), (15.6.19), with observance of (15.6.21). Artificial variables of required sign are included in relations (15.6.18) and (15.6.19), and one uses the simplex method to minimize the sum of infeasibilities [see § 15.6.1]. However, at each step in the simplex method one avoids introduction of x_j if w_j is already in the basis, or of w_j if x_j is already in the basis, thus preserving the complementarity property (15.6.21). Let us, for future reference, refer to such a constraint on the simplex method as a *complementarity-preserving constraint*.

If $\mathbf{D} \le \mathbf{0}$ then the Wolfe method converges to a solution of the quadratic programme in a finite number of steps.

15.6.5.   Quadratic Programming; Lemke's Method

Consider the case of an inequality constraint (15.6.13), and define the $(m + n)$-vectors

$$\mathbf{z} = \begin{pmatrix} \mathbf{x} \\ \mathbf{y} \end{pmatrix}, \qquad \mathbf{w} = \begin{pmatrix} -\mathbf{c} - \mathbf{Dx} + \mathbf{A}'\mathbf{y} \\ \mathbf{b} - \mathbf{Ax} \end{pmatrix}. \tag{15.6.22}$$

The constraint and optimality conditions can then be written

$$(\mathbf{w}, \mathbf{z}) \geq 0 \tag{15.6.23}$$

$$\mathbf{w} = \mathbf{q} + \mathbf{Mz} \tag{15.6.24}$$

$$\mathbf{w}'\mathbf{z} = 0 \tag{15.6.25}$$

where

$$\mathbf{q} = \begin{pmatrix} -\mathbf{c} \\ \mathbf{b} \end{pmatrix}, \qquad \mathbf{M} = \begin{pmatrix} -\mathbf{D} & \mathbf{A}' \\ -\mathbf{A} & \mathbf{0} \end{pmatrix}. \tag{15.6.26}$$

Relations (15.6.23)–(15.6.25) can thus be summarized as a positivity condition, a linear constraint and a complementarity condition on the $(m+n)$-vector $(\mathbf{w}, \mathbf{z})$. Solution of these relations amounts to solution of the quadratic programme, by Theorem 15.6.1.

Wolfe's method could obviously be applied to (15.6.23)–(15.6.25), by introducing $(m+n)$ artificial variables into (15.6.24). Lemke's method is a more economic version, which needs only a single artificial variable.

Conditions (15.6.23), (15.6.25) imply that at most $(m+n)$ elements of $(\mathbf{w}, \mathbf{z})$ will be non-zero. Term such a solution *basic*. If $\mathbf{q} \geq 0$ then one has a solution $\mathbf{z} = 0$, $\mathbf{w} = \mathbf{q}$ to (15.6.23)–(15.6.25). Assume then that $\mathbf{q} \not\geq 0$.

Lemke defines a vector $\mathbf{h}$ by

$$h_j = \begin{cases} 0 & \text{if } q_j \geq 0 \\ 1 & \text{if } q_j < 0 \end{cases} \tag{15.6.27}$$

and relaxes (15.6.24) to

$$\mathbf{w} - \mathbf{Mz} - \theta\mathbf{h} = \mathbf{q}, \tag{15.6.28}$$

where θ is a non-negative scalar just large enough to ensure that the solution

$$\mathbf{z} = 0$$
$$\mathbf{w} = \mathbf{q} + \theta\mathbf{h} \tag{15.6.29}$$

is non-negative, that is

$$\theta = -\min_j q_j. \tag{15.6.30}$$

This solution satisfies (15.6.23), (15.6.25), (15.6.28), and θ is a measure of its infeasibility, in that it does not satisfy (15.6.24).

One now continues with the simplex method [see § 15.6.1] for the set of variables $(\mathbf{w}, \mathbf{z}, \theta)$ so as to minimize θ, but observing the complementarity-preserving constraint between $\mathbf{w}$ and $\mathbf{z}$ at every step. Relations (15.6.29), (15.6.30) determine the initial basic solution. Relations (15.6.28) give the representation of basic in terms of non-basic variables and so the first tableau. More explicitly, if the minimum in (15.6.30) is attained for $j = i$ (so that $w_i = 0$, and does not belong to the basis) then the ith relation of (15.6.28)

$$\theta = -q_i + w_i - (\mathbf{Mz})_i \tag{15.6.28'}$$

yields the relevant representation of θ. Substituting this expression for θ into the other relations of (15.6.28) one obtains the representation of the other basic variables w_j.

The simplex method is applied until $\theta = 0$, when one has a solution to the original system (15.6.23)–(15.6.25). As before, one reaches the solution in a finite number of steps if $\mathbf{D} \le \mathbf{0}$.

The system (15.6.23)–(15.6.25) is of interest in contexts other than quadratic programming, e.g. the solution of two-person zero-sum games [see I, § 13.2]. Lemke (1968) proves more generally that if $\mathbf{M}$ is such that $\mathbf{z} \ge \mathbf{0}$ and $\mathbf{z}'\mathbf{Mz} = \mathbf{0}$ imply that $(\mathbf{M} + \mathbf{M}')\mathbf{z} = \mathbf{0}$ then his method terminates either in a solution or in a demonstration of infeasibility.

15.6.6. Geometric Programming

Consider a cost function with m non-negative arguments $t_1, t_2, \ldots, t_m$ and n components, of the form

$$C(\mathbf{t}) = \sum_{k=1}^{n} c_k \prod_{j=1}^{m} t_j^{a_{jk}}, \tag{15.6.31}$$

where the c_k are non-negative, and the exponents a_{jk} of such value that $C(\mathbf{t})$ becomes infinite as any t_j approaches 0 or $+\infty$. Such functions occur frequently in engineering applications, with t_j typically being the jth dimension of the object being optimized. A function $C(\mathbf{t})$ of this type is sometimes termed a *posynomial*. This problem can be related (by duality) to another, which is often simpler.

THEOREM 15.6.2. *Consider the maximization of*

$$f(\mathbf{x}) = \sum_{k=1}^{n} x_k \log (c_k/x_k) \tag{15.6.32}$$

in $\mathbb{R}^n_+$ *subject to conditions*

$$\sum x_k = 1$$
$$\sum_k a_{jk} x_k = 0 \qquad (j = 1, 2, \ldots, m). \tag{15.6.33}$$

Suppose the c_k positive and the a_{jk} such that the system (15.6.33) is of full rank and possesses solutions in the interior of $\mathbb{R}^n_+$. *Then a modified form of the dual of this problem is: minimize expression (15.6.31) in* $\mathbb{R}^m_+$. *If the solutions and extrema reached in these respective problems are denoted* $\bar{\mathbf{x}}, \bar{\mathbf{t}}$ *and* $\bar{f}, \bar{C}$ *then*

$$\bar{x}_k = \left(c_k \prod_j \bar{t}_j^{a_{jk}} \right) \Big/ \bar{C} \tag{15.6.34}$$

$$\bar{C} = e^{\bar{f}} = \prod_k (\bar{x}_k/c_k)^{\bar{x}_k}. \tag{15.6.35}$$

Equations (15.6.34), (15.6.35) relate the solutions of the two problems. If $n - m$ is small, then the first problem will be easier to solve. Indeed, if $n = m + 1$ then there is only the one **x**-value which is feasible.

Note the interpretation of $\bar{x}_k$ that follows from (15.6.33): it is the proportion of the minimal cost $\bar{C}$ attributable to the kth component.

15.6.7.   Separable Programming

Suppose that the objective function separates into a sum of non-linear functions of the *single* components of **x**:

$$f(\mathbf{x}) = \sum_1^n \phi_k(x_k).$$
(15.6.36)

We assume the usual positivity and linear constraints

$$\mathbf{x} \geq \mathbf{0}$$
(15.6.37)

$$\mathbf{A}\mathbf{x} = \mathbf{b}$$
(15.6.38)

although these can also be generalized.

On the x_k-axis one now selects a set of values x_{ki} such that ϕ_k can be sufficiently well represented by linear interpolation between consecutive values. One assigns non-negative weights λ_{ki} to these values

$$\lambda_{ki} \geq 0$$

$$\sum_i \lambda_{ki} = 1 \qquad (k = 1, 2, \ldots, m)$$
(15.6.39)

and defines an 'average' x_k value

$$\bar{x}_k = \sum_i \lambda_{ki} x_{ki}.$$
(15.6.40)

The problem is now modified to the linear one: maximize

$$\Phi(\boldsymbol{\lambda}) = \sum_k \sum_i \lambda_{ki} \phi(x_{ki})$$
(15.6.41)

with respect to the λ_{ki} subject to (15.6.39) and

$$\mathbf{A}\mathbf{x} = \mathbf{b}.$$
(15.6.42)

Put this way, the effective linearization of the problem appears as nothing but a randomized solution [cf. § 15.5] restricted for simplicity to the set of points with component values x_{ki}. Because of the separable nature of objective function and constraints, one does not have to specify the probability weighting associated with an n-tuple $(x_{1i_1}, x_{2i_2}, \ldots, x_{ni_n})$ but simply that, λ_{ki}, associated with a given component value x_{ki}.

However, one departs somewhat from the full randomized solution in that one normally imposes an *adjacency condition*: that for a given k not more than two λ_{ki} should be non-zero, and that these should correspond to consecutive values of i. This means that one realizes the linear interpolation between the values $\phi_k(x_{ki})$ $(i = 1, 2, \ldots)$ rather than the concave majorant to these values.

The simplex method is followed as usual, but the adjacency condition observed in changes of basis.

One can deal with separable constraints, in that

$$\sum_k \psi_k(x_k) \in \mathscr{C}$$

would be modified to

$$\sum_k \sum_i \lambda_{ki} \psi_k(x_{ki}) \in \mathscr{C}.$$

15.7. NETWORK OPTIMIZATION

15.7.1. Maximal Flow Between Two Points of a Network

There are many optimization problems on networks and graphs [see V, Chapter 6]. In this section we consider one particular important case: the maximization of flow from one given node to another (the 'source' and the 'sink') in a network carrying a single type of traffic, subject to capacity constraints on the arcs [see V, § 6.4].

Let the nodes of the network be labelled $j = 1, 2, \ldots, n$, with source and sink corresponding to $j = 1$ and $j = n$ respectively. Let x_{jk} denote the rate of direct traffic flow from node j to node k. We distinguish between traffic in the two directions, at rates x_{jk} and x_{kj} respectively, and assume each non-negative, and subject to a capacity constraint

$$0 \le x_{jk} \le b_{jk}. \tag{15.7.1}$$

We are thus essentially considering a *directed* network [see V § 6.1] with the jk link having capacity b_{jk}.

Let v denote the net flow through the network, from source to sink. The flow is subject to the nodal balance equations

$$\sum_k (x_{jk} - x_{kj}) = r_j \tag{15.7.2}$$

where

$$r_j = \begin{cases} v & \text{if } j = 1 \\ -v & \text{if } j = n \\ 0 & \text{otherwise.} \end{cases} \tag{15.7.3}$$

The aim is to maximize v subject to constraints (15.7.1)–(15.7.3).

The dual form of the problem is interesting, insofar as the multiplier y_j can be regarded as a 'potential', defined for every node rather than for every arc' of the network.

THEOREM 15.7.1.   *The dual problem is: minimize*

$$D(\mathbf{y}) = \sum_j \sum_k b_{jk}(y_k - y_j)_+ \qquad (15\!:\!7.4)$$

subject to

$$y_n - y_1 = 1. \qquad (15.7.5)$$

The optimal flow in arc jk is b_{jk} *if* $y_k > y_j$, *zero if* $y_k < y_j$, *and undetermined if* $y_k = y_j$.

The implications of the dual and a workable algorithm can be developed if one introduces the idea of a *cut* [see V, § 6.4.1]. Suppose the nodes of the network are grouped into two complementary sets: a set s containing the source, and its complement $\bar{s}$ containing the sink. The *cut* corresponding to this division is the set of arcs jk with j in s and k in $\bar{s}$. *The value of the cut* is the sum of the capacities of these arcs:

$$V = \sum_{j \in s} \sum_{k \in \bar{s}} b_{jk}. \qquad (15.7.6)$$

This is certainly an upper bound to the maximal flow from source to sink, since this flow must traverse any cut. One can say more.

THEOREM 15.7.2   (*The maximal-flow minimal-cut theorem*).
  (i) *The value of a cut is not less than the value of the flow.*
  (ii) *The maximal flow equals the minimal cut.*
  (iii) *If* s, $\bar{s}$ *is the division of nodes corresponding to a minimal cut, then a solution of the dual problem is*

$$y_j = \begin{cases} 0 & \text{if } j \in s \\ 1 & \text{if } j \in \bar{s}. \end{cases} \qquad (15.7.7)$$

By 'minimal cut' we mean the 'minimal value of cut'. The minimal cut is a bottleneck for the flow problem.

In the *Ford–Fulkerson algorithm* one starts from a trial flow, and recursively 'marks' all nodes which can be reached from the source by a chain of arcs along which there is either a sub-capacity flow out from the source, or a flow back to the source. Let the marked set be denoted s^*.

THEOREM 15.7.3
  (i) *If* s^* *includes the sink, then the flow can be improved. It can be improved by an integral amount if flows and capacities are integral.*
  (ii) *If* s^* *does not include the sink then* $s^*, \bar{s}^*$ *defines a minimal cut, and the flow is maximal. This stage can be reached in a finite number of improvement steps if capacities are integral, and maximal flow finite.*

For, if s^* includes the sink, then there is at least one chain of sub-capacity or reverse-flow arcs from source to sink. The direct flow along this chain can be increased until either direct flow has been brought up to the full capacity of some arc of the chain, or reverse flow is reduced to zero in some arc of the chain.

15.8. MISCELLANEOUS METHODS

15.8.1. Branch-and-bound Methods

This is a method of constrained optimization, initially proposed heuristically which has been developed to a relatively sophisticated state, and which can prove effective where special-purpose algorithms are inapplicable. For a general discussion, see Mitten (1970). One particular application has been to integer programming (IP), that is to an LP in which at least some of the elements of $\mathbf{x}$ are required to be integral [see I, § 12.3.1]. We shall take this as a concrete example when explaining the method.

It is required to maximize the objective $f(\mathbf{x})$ in some feasible set $\mathcal{X}$. Specification of $\mathcal{X}$ thus implies specification of all constraints in this section. It is supposed that one can solve the problem in a set $\mathcal{X}' \supset \mathcal{X}$, that is in a relaxed version. For the LP case, $\mathcal{X}'$ could initially be the feasible set for the corresponding LP, in which integrality constraints are neglected. The solution in this relaxed version will in general be infeasible, and one must endeavour to pinch off these infeasibilities one at a time.

Let f^* be $\sup_x f$, the optimal value of the objective. At any point in the calculation the maximizing value in the smallest $\mathcal{X}' \supset \mathcal{X}$ hitherto used supplies an upper bound $\bar{f}$ on f^*. The largest value of $f(\mathbf{x})$ for feasible $\mathbf{x}$ hitherto examined provides a lower bound $\underline{f}$. The aim is to continue until $\bar{f} = \underline{f}$.

If the current solution $\bar{\mathbf{x}}$ in $\mathcal{X}'$ is infeasible, then some element of infeasibility i can be isolated, and it can be stated that for feasibility $\mathbf{x}$ must lie in one of a discrete number of sets $\mathcal{A}_{ij}$ ($j = 1, 2, \ldots$). For example, in the IP case i could be the label of one of the variables x_i which should be integral, but for which $\bar{x}_i$ is non-integral. Then j could label the two alternatives

$$x_i \leq [\bar{x}_i]$$

$$x_i \geq [\bar{x}_i] + 1$$

one of which must be true ($[\cdot]$ denoting 'integral part of'). This is the step known as 'branching' in that the problem is broken up into alternative sub-problems, corresponding to a free maximization in each of the $\mathcal{A}_{ij}$. However, sub-problems are stored and examined in reverse order, in that one examines a particular $\mathcal{A}_{ij}$ and (if necessary) branches on that before returning to examine the other sub-problems. More specifically, the procedure is

(i) Maximize freely in some $\mathcal{A}_{ij}$.
(ii) If the maximum found is not greater than $\underline{f}$, revert to another problem.

(iii) If the maximum exceeds $\underline{f}$, one tests for feasibility. If feasible, one revises $\underline{f}$, and reverts to another problem. If infeasible, one branches within $\overline{\mathscr{A}}_{ij}$, possibly with a revised $\overline{f}$.

(iv) One continues then until $\overline{f} = \underline{f}$.

There is considerable latitude in the algorithm in that, as one branches, one has a choice of the *branching node i* and the *branch j*. That is, of the particular infeasibility one wishes to examine, and the direction in which one moves to resolve this infeasibility. Let $\bar{\mathbf{x}}_{ij}$ be a value in $\mathscr{A}_{ij}$ near to the current (infeasible) maximizer $\bar{\mathbf{x}}$. Consider the 'penalty'

$$\Delta_{ij} = f(\bar{\mathbf{x}}) - f(\bar{\mathbf{x}}_{ij});$$

the change in upper bound on f^* as one moves from $\bar{\mathbf{x}}$ to $\bar{\mathbf{x}}_{ij}$. Then a strategy recommended by Beale and Driebeek is: choose an i maximizing $\max_j \Delta_{ij}$ and then choose a j minimizing Δ_{ij}. That is, choose a node (mode of infeasibility) for which the objective shows greatest sensitivity to infeasibility, and then a branch (resolution of the infeasibility) for which the drop in objective is least.

15.9.   NUMERICAL METHODS FOR UNCONSTRAINED OPTIMIZATION

15.9.1.   Types of Method

Lagrangian theory does not in itself provide a numerical method of optimization, but only a method of in some sense reducing the constrained case to the unconstrained case. Numerical methods must be developed as a separate issue.

We already have optimization algorithms for particular cases in the various simplex-type algorithms for linear and quadratic programming [see § 15.6]. These solve the problem of optimization in the positive orthant subject to linear constraints with linear and quadratic objective function, respectively. The methods will still be of some use for more general objective functions, if these can be regarded as locally linear or quadratic.

Then there are problems of very special structure, such as the network-flow problems of section 15.7 best solved by specially-tailored algorithms.

What we are concerned with in this part are general algorithms, for numerically maximizing an objective function $f(\mathbf{x})$ under minimal assumptions on the form of f. We shall consider unconstrained maximization in this chapter, constrained maximization in the next.

Consider the unconstrained problem: that of maximizing $f(\mathbf{x})$ in $\mathscr{X}$. The assumption is that the function f is in principle known numerically, in that the value of $f(\mathbf{x})$ can be determined for any desired $\mathbf{x}$ in $\mathscr{X}$. It may or may not be true that one can also evaluate the gradient vector $f_{\mathbf{x}}$ numerically. The aim is to carry out a sequence of evaluations at points $\mathbf{x}_1, \mathbf{x}_2, \ldots$, these points being determined sequentially from evaluations already made, in such a way that the sequence $\{\mathbf{x}_i\}$ converges as quickly as possible to a maximizing value of f. (Note that in this chapter and the next $\mathbf{x}_i$ denotes the ith value of a vector, that is the co-ordinates at an ith point, rather than the ith component of a vector.)

The commonest type of method is a *gradient* or *hill-climbing* method, in which the new $\mathbf{x}$-value is chosen in a direction in which f would seem to be increasing, on the basis of previous evaluations. We shall consider such methods in sections 15.9.6–15.9.8. However, at least for the case of one-dimensional or *line search*, there is the alternative of 'quartering' methods such as Fibonacci search, described in sections 15.9.4, 15.9.5.

15.9.2. Rates of Convergence

Suppose that a sequence of vectors $\{\mathbf{x}_i\}$ converges to $\bar{\mathbf{x}}$ [see § 1.2]. The *order of convergence* r of the sequence is defined as the supremum of non-negative numbers p satisfying [see Definition 1.2.3 and I, (9.1.2)]

$$\overline{\lim} \frac{\|\mathbf{x}_{i+1} - \bar{\mathbf{x}}\|}{\|\mathbf{x}_i - \bar{\mathbf{x}}\|^p} < \infty. \tag{15.9.1}$$

So, very roughly, if $\Delta_i = \|\mathbf{x}_i - \bar{\mathbf{x}}\|$, then $\Delta_{i+1} = O(\Delta_i^p)$, or $\Delta_i = O(\Delta^{p^i})$ for some small quantity Δ [see Definition 2.3.2].

Within the class of order one convergences one distinguishes *linear convergence* for which

$$\overline{\lim} \frac{\|\mathbf{x}_{i+1} - \bar{\mathbf{x}}\|}{\|\mathbf{x}_i - \bar{\mathbf{x}}\|} = \beta < 1. \tag{15.9.2}$$

Here β is the *convergence ratio*, and one can say that $\Delta_i = O(\beta^i)$. So linear convergence corresponds to convergence at an exponential rate (or geometric rate), somewhat confusingly. If in (15.9.2) $\beta = 0$ then one says that convergence is *superlinear*. Convergence of any order greater than unity is superlinear, although the converse is not true.

15.9.3. Line Search: The Newton–Raphson Method

Suppose one is solving an equation $h(x) = 0$ for a scalar x. Let the last evaluation of h and its gradient be at x_i. Then, if h were linear, the root of the equation would be at

$$x_{i+1} = x_i - \frac{h(x_i)}{h'(x_i)}. \tag{15.9.3}$$

If h is almost linear, then x_{i+1} will be an improved evaluation of the root, and one can hope that the sequence $\{x_i\}$ will converge to the root of the equation.

This is the Newton–Raphson method [see III, § 5.4.1] also applicable to maximization of a function f if this is regular enough. If the maximum is attained at a stationary point then it satisfies $f'(\bar{x}) = 0$, and the analogue of (15.9.3) is

$$x_{i+1} = x_i - \frac{f'(x_i)}{f''(x_i)}. \tag{15.9.4}$$

This is as much of a hill-climbing approach as is possible in one dimension. The process will converge to a maximizing value in one step if f is quadratic, and converge superlinearly if f is concave [see III, § 5.4.2].

15.9.4. Golden Section Search

This is an efficient method of one-dimensional maximization which does not use gradient evaluations. The method is based on the assumption that f is *unimodal* in the interval of interest (that is monotonic on either side of the maximum—see § 2.7).

Suppose it is known that the maximum lies in an interval AB, and that f has been evaluated at the extreme points A and B. To locate the maximum within a sub-interval one needs function evaluations at two internal points of the interval, C and D, say. One can then locate the maximum within AD or CB according as f is larger at C or D. It now seems reasonable to choose the points C and D so that the lengths of the intervals satisfy

$$AC = DB \tag{15.9.5}$$

and

$$\frac{AC}{AD} = \frac{AD}{AB}. \tag{15.9.6}$$

see Figure 15.9.1. In this way, if the maximum is found to lie in AD, say, then the point C at which one still has an interval evaluation divides the new interval AD in the same ratio as D divided AB. One has then simply to add another evaluation at the point symmetrically placed to C in AD, and one has a scaled-down version of the initial situation. Correspondingly if the maximum is found to be in CB.

Figure 15.9.1: Evaluation points for golden section search.

In fact, if $p = AD/AB$ then the interval of interest is scaled down by a factor of p at each step. Relations (15.9.5), (15.9.6) imply the equation

$$\frac{1-p}{p} = p \tag{15.9.7}$$

whence

$$p = \frac{\sqrt{5}-1}{2} \sim 0 \cdot 618, \tag{15.9.8}$$

the so-called 'golden section'. (A rectangle with sides in ratio $1 : p$ preserves these proportions if the square erected on the shorter side is removed).

So, suppose one starts with an initial interval of length L, with f evaluated at each end-point, and is permitted r further function evaluations. Golden section search will then reduce the interval of uncertainty to one of length Lp^{r-1}, where p is given by (15.9.8).

15.9.5. Fibonacci Search [see also III, Algorithm 11.1.2].

One may ask whether golden section search is optimal, in that the length of the interval within which the maximum is located after r function evaluations is minimal. Actually, it is almost optimal, but the truly optimal method in the above sense is that of *Fibonacci search*, as proved by Kiefer (1953). The Fibonacci numbers F_r are determined by $F_0 = F_1 = 1$ and

$$F_r = F_{r-1} + F_{r-2} \qquad (r \ge 2) \tag{15.9.9}$$

so that [see I, Example 14.13.1] the sequence is $(1, 1, 2, 3, 5, 8, 13, 21, 34, \ldots)$ and for large r

$$F_r \sim \frac{p^{-r}}{1+p^2} \tag{15.9.10}$$

where p has the evaluation (15.9.8).

THEOREM 15.9.1. *Suppose that f is evaluated at the ends of the interval, and one is permitted r further function evaluations. Then the optimal method is to evaluate f at points dividing the interval in the ratios $F_{r-1}:F_{r-2}$ and $F_{r-2}:F_{r-1}$. By this method r internal evaluations will reduce an interval of length L to one of length L/F_r.*

The rule is self-consistent in that, because of (15.9.9), the surviving interval evaluation at each step divides the remaining sub-interval in one of the two desired proportions. So, referring back to diagram (Figure 15.9.1), if C and D divide AB in ratios $F_{r-2}:F_{r-1}$ and $F_{r-1}:F_{r-2}$, then C divides AD in ratio $F_{r-2}:F_{r-3}$, as would be required at the next step.

With two evaluations remaining the interval evaluations should divide the interval in ratios $F_0:F_1$ and $F_1:F_0$, that is, both at the mid-point. This should be construed as an evaluation infinitesimally to each side of the mid-point, so that one can determine in which half-interval the maximum should be.

The size of the final interval for golden section search exceeds that for the optimal method by a factor of

$$\frac{p^{r-1}}{(1+p^2)p^r} = \frac{1}{p(1+p^2)} \sim 1\cdot171.$$

So, the golden section method reduces intervals in the same ratio as the optimal method does asymptotically, but gives a final interval about 17% larger. On the other hand, its invariant nature makes it simpler to apply.

15.9.6.   Gradient Methods; Newton and Quasi-Newton Methods [see also III, § 11.3]

Consider the problem of maximizing a function f without further constraint than that $\mathbf{x}$ lies in a subset $\mathscr{X}$ of $\mathbb{R}^n$ which is usually connected [see V, § 5.2], even convex. We shall assume that both f and its vector of first derivatives $\mathbf{d} = f'_\mathbf{x}$ can be evaluated at any desired point.

The general technique adopted is to choose a fixed direction of search, $\mathbf{s}$, and to consider the one-dimensional problem of maximizing $f(\mathbf{x})$ in this direction. That is, if one is currently at point $\mathbf{x}_i$ then one searches in a direction $\mathbf{s}_i$, by maximizing $f(\mathbf{x}_i + \alpha\mathbf{s}_i)$ with respect to the scalar α. One thus arrives at a new point

$$\mathbf{x}_{i+1} = \mathbf{x}_i + \alpha_i\mathbf{s}_i$$

from which one searches again in a new direction etc.

The maximization of $f(\mathbf{x}_i + \alpha\mathbf{s}_i)$ with respect to α is known as a *line search* in the direction $\mathbf{s}_i$. If one indeed locates the maximum, then the search is termed *perfect*. If one merely determines an α_i, and so an $\mathbf{x}_{i+1}$, such that

$$f(\mathbf{x}_{i+1}) \geq f(\mathbf{x}_i) + 0\cdot01\mathbf{d}'_i\mathbf{s}_i \qquad (15.9.11)$$

$$\mathbf{d}'_{j+1}\mathbf{s}_i \leq 0\cdot5\mathbf{d}'_i\mathbf{s}_i \qquad (15.9.12)$$

then the search is termed *acceptable*. (Here $\mathbf{d}'_i = f_\mathbf{x}(\mathbf{x}_i)$). These conditions state, respectively, that the move from $\mathbf{x}_i$ to $\mathbf{x}_{i+1}$ has increased f by at least 1% of what a linear approximation to f at $\mathbf{x}_i$ would have predicted, and that the move has reduced the gradient in the search direction $\mathbf{s}_i$ by at least 50%.

The natural direction of search to take is that in which f increases fastest, the direction of local gradient. That is, one chooses $\mathbf{s}_i = \mathbf{d}_i = f'_\mathbf{x}(\mathbf{x}_i)$. This is the *method of steepest ascents*.

However, the steepest ascent method can work badly. For example, if the graph of f has a curved ridge, then ascent of the ridge can take place in a tight zig-zag, whose many steps make progress slow. A variant of steepest ascents which virtually eliminates such behaviour is to adapt the Newton–Raphson method of section 15.9.3, and choose

$$\mathbf{x}_{i+1} = \mathbf{x}_i - f_{\mathbf{xx}}^{-1}f'_\mathbf{x}$$

$$= \mathbf{x}_i + \mathbf{G}_i\mathbf{d}_i. \qquad (15.9.13)$$

Here f_x and f_{xx} are the *gradient vector* and the *Hessian matrix* of f at x_i (the row vector and matrix of first and second derivatives at x_i respectively—see § 5.3 and § 5.7).

If f were a strictly concave quadratic function then evaluation (15.9.13) would find the maximum in a single step. In other cases it is a great improvement on steepest ascents, often giving superlinear convergence to the maximum. Effectively, one is making a second-order rather than a first-order local approximation to f. Variants of this approach are called *quasi-Newton methods*: also *variable metric* methods, because premultiplication of f'_x by f_{xx}^{-1} (or by an estimate of it) implies a use of local curvature to standardize local gradient.

However, one generally needs to vary the method from the simple prescription (15.9.13) for at least two reasons. For the first, it is usually necessary to regard (15.9.13) as recommending a *search direction* rather than the actual displacement taken, and replace (15.9.13) by

$$\mathbf{x}_{i+1} = \mathbf{x}_i + \alpha_i \mathbf{G}_i \mathbf{d}_i \qquad (15.9.14)$$

where α_i is an appropriately chosen scalar. The raw choice $\alpha_i = 1$ can lead to numerical instability. For the second, it is not in general realistic to assume that second derivatives f_{xx} can be calculated. The variable metric methods use only function and gradient evaluations, and from those build up a sequence of matrices $\mathbf{G}_i$ which can be regarded as estimates of $-f_{xx}^{-1}$ at the current x_i. We return to this point in section 15.9.8.

15.9.7. Conjugate Gradients [see also III, § 11.3]

One way of choosing search directions efficiently without calculation of second-order derivatives is the so-called method of conjugate gradients. This method requires that one chooses $\mathbf{s}_1 = \mathbf{d}_1$ but that for $i > 1$

$$(\mathbf{d}_j - \mathbf{d}_{j-1})'\mathbf{s}_i = 0 \qquad (j \le i) \qquad (15.9.15)$$

that is that the search direction from x_i be orthogonal to the change in gradient at all previous steps. The point of this method is expressed in

THEOREM 15.9.2. *If f is concave quadratic then the method of conjugate gradients (with perfect line searches) will locate the maximum of f in at most n steps.*

So the number of steps needed is bounded if f is quadratic; a property sometimes referred to as *quadratic termination*. The conjugacy requirement (15.9.15) effectively utilises information on second derivatives, without explicit calculation of these. One can generate a sequence of search directions satisfying (15.9.15) by the recursive relation

$$\mathbf{s}_i = \mathbf{d}_i - \frac{(\mathbf{d}_i - \mathbf{d}_{i-1})'\mathbf{d}_i}{(\mathbf{d}_i - \mathbf{d}_{i-1})'\mathbf{s}_{i-1}} \mathbf{s}_{i-1} \qquad (15.9.16)$$

and so achieve a method requiring little more computer storage than the simple method of steepest ascent. For non-quadratic functions one needs more than n steps, of course, but convergence is greatly accelerated as compared with steepest ascents, and will often be superlinear. However, one obviously cannot require that (15.9.15) holds over more than n steps, and so one has to apply a *restart* procedure. That is, after so many iterations one should scrap the accumulated conjugacy conditions, and make a fresh start from the current value of $\mathbf{x}_i$.

15.9.8. Variable-metric Methods

For the variable-metric methods we shall write

$$\mathbf{x}_{i+1} = \mathbf{x}_i + \alpha_i \mathbf{G}_i \mathbf{d}_i$$

$$= \mathbf{x}_i + \alpha_i \mathbf{B}_i^{-1} \mathbf{d}_i \qquad (15.9.17)$$

so that the recommended search direction is $\mathbf{s}_i = \mathbf{G}_i \mathbf{d}_i = \mathbf{B}_i^{-1} \mathbf{d}_i$. The sequence of matrices $\mathbf{B}_i = \mathbf{G}_i^{-1}$ is to be regarded as a sequence of estimates of the negative Hessian $-f_{\mathbf{xx}}$ at $\mathbf{x}_i$, although the method can still work well if there is considerable departure from this expectation. The matrices are generated by a simple recursion, or 'update'.

We shall write

$$\boldsymbol{\sigma}_i = \mathbf{x}_{i+1} - \mathbf{x}_i$$
$$\boldsymbol{\gamma}_i = \mathbf{d}_i - \mathbf{d}_{i+1}, \qquad (15.9.18)$$

the increments in co-ordinate and *negative* gradient (in the maximization case) respectively.

One of the most celebrated of the variable-metric methods is the *Davidon–Fletcher–Powell method* (DFP method [see also III, (11.3.8) and (11.3.9)]) for which the recursion is

$$\mathbf{G} \rightarrow \mathbf{G} - \frac{\mathbf{G}\boldsymbol{\gamma}\boldsymbol{\gamma}'\mathbf{G}}{\boldsymbol{\gamma}'\mathbf{G}\boldsymbol{\gamma}} + \frac{\boldsymbol{\sigma}\boldsymbol{\sigma}'}{\boldsymbol{\sigma}'\boldsymbol{\gamma}}. \qquad (15.9.19)$$

That is, if one attaches subscript i to all quantities in the right-hand member of (15.9.19), then the resultant expression defines $\mathbf{G}_{i+1}$.

THEOREM 15.9.3.   *With perfect line searches and quadratic (concave) f the DFP method has the properties:*
   (i) *The search directions $\mathbf{s}_i$ have the conjugate gradient property*
   (ii) *$\mathbf{x}_i$ converges to a maximizing value in at most n steps*
   (iii) *$\mathbf{G}_i$ converges to $-f_{\mathbf{xx}}^{-1}$ in at most n steps.*

For more general functions (but perfect line searches) one can add

   (iv) *The method is superlinearly convergent if f is concave*
   (v) *Apart from the choice of starting point, the method is invariant to a rescaling of* $\mathbf{x}$.

In practice one starts from $\mathbf{G}_0 = \mathbf{I}$ and a convenient $\mathbf{x}_0$.

The rationale of the recursion (15.9.19) is that the final correction term brings $\mathbf{G}$ nearer what $-f_{\mathbf{xx}}^{-1}$ would be if the search directions indeed had the conjugate gradient property, while the correction before that maintains the identity $\boldsymbol{\gamma}_i = \mathbf{G}_i \boldsymbol{\sigma}_i$ which one would expect in the quadratic case.

The DFP method has later been seen as one of a continuum of variable-metric methods. The method which has found widest recent acceptance is the *Broyden–Fletcher–Goldfarb–Shanno* (BFGS) method [see also III (11.3.10)] which appears superior when line-searches are not perfect. This gives a recursion for $\mathbf{B}$:

$$\mathbf{B} \to \mathbf{B} - \frac{\mathbf{B}\boldsymbol{\sigma}\boldsymbol{\sigma}'\mathbf{B}}{\boldsymbol{\sigma}'\mathbf{B}\boldsymbol{\sigma}} + \frac{\boldsymbol{\gamma}\boldsymbol{\gamma}'}{\boldsymbol{\gamma}'\boldsymbol{\sigma}} \tag{15.9.20}$$

which is *complementary* to the DFP formula in that one is obtained from the other by interchange of $\mathbf{G}$ and $\mathbf{B}$ plus interchange of $\boldsymbol{\sigma}$ and $\boldsymbol{\gamma}$. The BFGS formula (15.9.20) implies the update

$$\mathbf{G} \to \left(\mathbf{I} - \frac{\boldsymbol{\sigma}\boldsymbol{\gamma}'}{\boldsymbol{\gamma}'\boldsymbol{\sigma}}\right)\mathbf{G}\left(\mathbf{I} - \frac{\boldsymbol{\gamma}\boldsymbol{\sigma}'}{\boldsymbol{\sigma}'\boldsymbol{\gamma}}\right) + \frac{\boldsymbol{\sigma}\boldsymbol{\sigma}'}{\boldsymbol{\sigma}'\boldsymbol{\gamma}} \tag{15.9.21}$$

for $\mathbf{G}$. The complementary version of this gives the DFP update for $\mathbf{B}$.

The BFGS gives superlinear convergence for concave f with acceptable (but not necessarily perfect) line searches.

15.10. NUMERICAL METHODS FOR CONSTRAINED OPTIMIZATION

15.10.1. Constrained Maximization

We shall consider first methods suitable for linear constraints, and, first of all, the case in which the criterion function is also quadratic. There are adaptations of the simplex method [§ 15.6.1] which are natural for such cases.

For the case of more general constraints we then consider penalty function methods, and, their development, augmented Lagrangian methods. It will be noted that both these and the simplex-variants we describe first make quite explicit use of Lagrangian multipliers [see § 15.1.4].

15.10.2. A Feasible Direction Method, Especially Suited for Quadratic Programming

We formulate a direct method of constrained maximization in general terms. The method turns out to be directly applicable and effective for quadratic programmes (that is the maximization of a quadratic function subject to linear constraints—see § 15.6.3) but to need development for more general cases.

Consider the maximization of $f(\mathbf{x})$ in $\mathbb{R}^n$ subject only to the constraints

$$g_j(\mathbf{x}) = 0 \qquad (j = 1, 2, \ldots, m')$$
$$g_j(\mathbf{x}) \geq 0 \qquad (j = m'+1, \ldots, m).$$

The mixed equality/inequality specification of constraints is general enough to cover most cases. A *feasible* point $\mathbf{x}$ is one that satisfies the constraints. The *active set* I_i at stage i is the subset of integers j of $\{1, 2, \ldots, m\}$ for which $g_j(\mathbf{x}) = 0$, that is for which the jth constraint is binding. This then always includes the integers $j = 1, 2, \ldots, m'$.

The algorithm goes through the following sequence of steps.

(1) One chooses a feasible starting vector $\mathbf{x}_0$, with strict inequality in all the inequality constraints, so that $I_1 = \{1, 2, \ldots, m'\}$.

(2) At step i one chooses the step to $\mathbf{x}_{i+1} = \mathbf{x}_i + \boldsymbol{\delta}_i$ so as to maximize $f(\mathbf{x}_{i+1})$ subject to the constraints in the active set $\mathbf{I}_i$, that is so that $g_j(\mathbf{x}_{i+1}) = 0$ $(j \in I_i)$.

(3) If $\mathbf{x}_{i+1}$ thus determined is feasible one goes to step 4, otherwise one determines $\mathbf{x}_{i+1}$ from

$$\mathbf{x}_{i+1} = \mathbf{x}_i + \alpha_i \boldsymbol{\delta}_i$$

where $\boldsymbol{\delta}_i$ is the increment determined in (2), and α_i the largest value for which $\mathbf{x}_{i+1}$ is feasible.

Modify I_i to I_{i+1} by adding to it the labels of the new equality constraint(s) thus introduced. Return to (2).

(4) This step can be entered only from (2). Determine multipliers y_i $(j \in I_i)$ from the equations

$$f_{\mathbf{x}} - \mathbf{y}' \mathbf{g}_{\mathbf{x}} = \mathbf{0} \tag{15.10.1}$$

at $\mathbf{x}_{i+1}$. Terminate if $y_j \leq 0$ for $j > m'$. Otherwise let $I_{i+1} = I_i - \{t\}$ where t is an integer in I_i such that $t > m'$ and $y_i > 0$. Return to (2).

The technique is similar to the simplex algorithm [see § 15.6.1] insofar as one 'changes basis': one can acquire new equality constraints in step (3) as the attempt at minimization brings one against a constraint which was not previously binding; one removes equality constraints in step (4) by identifying constraints which it would be profitable to relax. The y_i determined in step (4) are just Lagrangian multipliers [see § 15.1.4] indicating how the value of f would change if the assigned value of g_i were changed.

Step (2) itself involves a constrained maximization, which can be carried out explicitly if f is quadratic and the g_i are linear. If $\mathbf{d}$ is the gradient [see (5.10.2)] and $-\mathbf{H}$ the second-derivative matrix of f at $\mathbf{x}_i$ [see § 5.7], and $\mathbf{A}$ the matrix of the active linear constraints, then one is choosing the perturbation vector $\boldsymbol{\delta}$ to maximize

$$\mathbf{d}'\boldsymbol{\delta} - \tfrac{1}{2}\boldsymbol{\delta}'\mathbf{H}\boldsymbol{\delta} \tag{15.10.2}$$

subject to

$$\mathbf{A}\boldsymbol{\delta} = \mathbf{0}.$$

Using Lagrangian methods, one finds the optimal δ to be

$$\delta = [\mathbf{H}^{-1} - \mathbf{H}^{-1}\mathbf{A}'(\mathbf{A}\mathbf{H}^{-1}\mathbf{A}')^{-1}\mathbf{A}\mathbf{H}^{-1}]\mathbf{d}. \tag{15.10.3}$$

Step (4) involves determination of the Lagrangian multipliers; these are determined from the equation system (15.10.1).

For quadratic f and linear g the process terminates, and, when it terminates, solves the original problem.

15.10.3. Maximization of a General Function Subject to Linear Constraints

One can adapt the treatment of the previous section to the case of a general objective function $f(\mathbf{x})$. The principal difference now is that the maximization in step 2 cannot be carried out explicitly; one must rather carry out a sequence of line searches, with the step δ given by (15.10.3) being regarded as the recommended search direction. If this relation is written $\delta = \mathbf{J}\mathbf{d}$, so that

$$\mathbf{J} = \mathbf{H}^{-1} - \mathbf{H}^{-1}\mathbf{A}'(\mathbf{A}\mathbf{H}^{-1}\mathbf{A}')^{-1}\mathbf{A}\mathbf{H}^{-1}$$

then there is an approximate up dating formula for $\mathbf{J}$ which saves complete recalculation at each state. This updating is based on the observations that $\mathbf{J}$ is a positive semi-definite matrix [see I, § 9.2] satisfying

$$\mathbf{J}\mathbf{A}' = \mathbf{0}$$

and

$$\mathbf{A}\mathbf{z} = \mathbf{0} \quad \Rightarrow \quad \mathbf{J}\mathbf{H}\mathbf{z} = \mathbf{z}.$$

The updating formula is [cf. (15.9.19) and (15.9.20)]

$$\mathbf{J} \to \mathbf{J} - \frac{\mathbf{J}\gamma\gamma'\mathbf{J}}{\gamma'\mathbf{J}\gamma} + \frac{\sigma\sigma'}{\sigma'\gamma}$$

(although the sequence this generates only approximates to a sequence of true $\mathbf{J}$ values).

Special updatings are needed when a constraint is added to, or dropped from the active set. If constraint $g_t(\mathbf{x}) \geq 0$ is added, then

$$\mathbf{J} \to \mathbf{J} - \frac{\mathbf{J}\mathbf{a}_t\mathbf{a}_t'\mathbf{J}}{\mathbf{a}_t'\mathbf{J}\mathbf{a}_t}$$

where $\mathbf{a}_t = \nabla g_t$. If $g_s(\mathbf{x}) \geq 0$ is removed from the active set, then

$$\mathbf{J} \to \mathbf{J} + \frac{(\hat{\mathbf{a}}_s - \mathbf{J}\mathbf{H}\hat{\mathbf{a}}_s)(\hat{\mathbf{a}}_s - \mathbf{J}\mathbf{H}\hat{\mathbf{a}}_s)'}{\mathbf{a}_s'\mathbf{a}_s}$$

where $\hat{\mathbf{a}}_s$ is the *orthogonal complement* of $\mathbf{a}_s$ in the set of gradient vectors of the new active set, that is

$$|\hat{\mathbf{a}}_s|^2 = \min_{\theta} \left| \mathbf{a}_s - \sum_I \theta_j \mathbf{a}_j \right|^2.$$

15.10.4.   Penalty Function and Augmented Lagrangian Methods

Consider the maximization of f subject to the equality constraints

$$g_j = 0 \qquad (j = 1, 2, \ldots, m). \tag{15.10.4}$$

One way of approximating the solution to this problem is to maximize the form

$$\Phi(\mathbf{x}, r) = f(\mathbf{x}) - \frac{r}{2} \sum_{1}^{m} (g_j(\mathbf{x}))^2 \tag{15.10.5}$$

freely, where r is a finite but large constant. The final term in this expression is a *penalty function*, measuring deviation from the constraints (15.10.4). In the full procedure ('sequential unconstrained maximization') one increases r indefinitely in discrete stages, taking the maximizing value of $\mathbf{x}$, $\bar{\mathbf{x}}_r$ say, from the previous stage as an initial value for an iterative free maximization at the new value of r. One can prove, under quite weak assumptions, that the minimizing value tends, with increasing r, to a solution of the original constrained problem.

However, $\bar{\mathbf{x}}_r$ will not solve the original problem for any finite r. A more efficient procedure turns out to be a combination of Lagrangian and penalty function techniques. One chooses $\mathbf{x}$ to maximize

$$\Phi(\mathbf{x}, \mathbf{y}, r) = f(\mathbf{x}) - \mathbf{y}'\mathbf{g}(\mathbf{x}) - \frac{r}{2} \sum (g_j(\mathbf{x}))^2$$

where the $\mathbf{y}$ is so chosen that the form has zero gradient (in $\mathbf{x}$) at the minimizing value $\bar{\mathbf{x}}$. Let this value be denoted $\bar{\mathbf{y}}$. Denote by $\nabla_{\mathbf{x}}$ the gradient operator [see (5.10.2) or (17.3.14)]. One can then assert the

THEOREM 15.10.1.   *If there exists a positive constant ε such that*

$$\mathbf{z}'\nabla_{\mathbf{x}}^2[f(\bar{\mathbf{x}}) - \bar{\mathbf{y}}'g(\bar{\mathbf{x}})]\mathbf{z} \le -\varepsilon |\mathbf{z}|^2$$

for all vectors $\mathbf{z}$ satisfying $\mathbf{z}'\nabla_{\mathbf{x}}\mathbf{g}(\bar{\mathbf{x}}) = 0$, then there exists a finite $\bar{r}$ such that $\bar{\mathbf{x}}$ minimizes $\Phi(\mathbf{x}, \bar{\mathbf{y}}, r)$ for all $r > \bar{r}$.

That is, under appropriate *local* convexity assumptions the penalty function will enforce the exact constraint even with finite r.

The problem is, as ever, to find $\bar{\mathbf{y}}$, and here one appeals to a type of duality again: choosing the value of $\mathbf{y}$ that minimizes

$$\phi(\mathbf{y}, r) \triangleq \max_{\mathbf{x}} \Phi(\mathbf{x}, \mathbf{y}, r).$$

One alternates $\mathbf{x}$-maximization and $\mathbf{y}$-minimization, and obtains excellent convergence with fixed finite r.

Of course $\phi(\mathbf{y}, r)$ is not known as an explicit function of $\mathbf{y}$: one uses a quadratic approximation to it based upon

$$\phi(\mathbf{y}, r) = \Phi(\mathbf{x}, \mathbf{y}, r)$$

$$\nabla_{\mathbf{y}}\phi(\mathbf{y}, r) = -\mathbf{g}(\mathbf{x})$$
$$\nabla_{\mathbf{y}}^2\phi(\mathbf{y}, r) = (\nabla_{\mathbf{x}}\mathbf{g})'(\nabla_{\mathbf{x}}^2 f)^{-1}\nabla_{\mathbf{x}}\mathbf{g},$$

where $\mathbf{x}$ has its maximizing value for the given value of $\mathbf{y}$.

In dealing with inequality constraints, $g_i(\mathbf{x}) \geq 0$, one can replace the penalty term $-(r/2)g_i^2$ by one such as $-\exp(-rg_i)$.

Alternatively, one introduces a slack variable z_i to make the constraint

$$g_i - z_i = 0$$

$(z_i \geq 0)$. On maximizing

$$-y_i(g_i - z_i) - \frac{r}{2}(g_i - z_i)^2$$

one obtains a term

$$\frac{y_i^2}{2r} \quad \text{if } g_i + \frac{y_i}{r} \geq 0$$

$$-g_i y_i + \frac{r}{2}g_i^2 \quad \text{otherwise.}$$

As a function of $\mathbf{y}$, $\phi(\mathbf{y}, r)$ now shows discontinuities in its second derivatives [see Definition 5.1.3].

15.10.5. Variable-metric Methods for Non-linear Constraints; Outline

The very recent adaptation of the variable metric method to constrained problems by Biggs (1975) Han (1977) and Powell (1978) has produced an order-of-magnitude improvement in computational efficiency. Numerical experience indicates that these methods may require only a fifth of the number of functions and gradient evaluations required by other methods of constrained optimization. It is true that each cycle of computation is heavier than for the augmented Lagrangian method, say, but overall the new methods still constitute a dramatic advance.

The Biggs–Han–Powell (BHP) method, as developed finally by Powell, essentially applies the variable metric method to the Lagrangian function

$$L(\mathbf{x}, \mathbf{y}) = f(\mathbf{x}) - \mathbf{y}'\mathbf{g}(\mathbf{x}). \tag{15.10.6}$$

Suppose, to begin with, that the constraints consist of an m-vector equality constraint

$$\mathbf{g}(\mathbf{x}) = \mathbf{0}. \tag{15.10.7}$$

In the unconstrained case the search direction $\mathbf{s}_i = \mathbf{B}_i^{-1}\mathbf{d}_i$ recommended by the variable metric method at stage i can be regarded as that which maximizes a quadratic approximation to f:

$$f(\mathbf{x}_i + \mathbf{s}_i) = f(\mathbf{x}_i) + \mathbf{d}_i'\mathbf{s}_i - \tfrac{1}{2}\mathbf{s}_i'\mathbf{B}_i\mathbf{s}_i. \tag{15.10.8}$$

In the BHP method one chooses the **s** into the interior of $\mathscr{X}$ which maximizes expression (15.10.8) subject to

$$\mathbf{g}(\mathbf{x}_i) + \mathbf{g}_x(\mathbf{x}_i)\mathbf{s}_i = \mathbf{0}; \tag{15.10.9}$$

a linearized version of the constraint (15.10.7). The search direction $\mathbf{s}_i$ is thus determined by a quadratic programming problem [see § 15.6.3] whose solution is immediate if (15.10.9) is the only constraint [see § 15.6.4]. In solving this quadratic programming problem, one simultaneously determines a Lagrangian multiplier vector $\mathbf{y}_i$ associated with constraints (15.10.9).

One then carries out a line search in the direction of $\mathbf{s}_i$, as usual, with some precautions which we shall outline in the next section. The matrix $\mathbf{B}_i$ is updated to the next stage by one of the usual algorithms (the BFGS method is recommended by Powell) except that the decrement $\boldsymbol{\gamma}_i$ in the gradient of f is replaced by the decrement in the gradient of the Lagrangian function

$$\boldsymbol{\gamma}_i^* = \boldsymbol{\nabla}_x L(\mathbf{x}_i, \mathbf{y}_i) - \boldsymbol{\nabla}_x L(\mathbf{x}_{i+1}, \mathbf{y}_i). \tag{15.10.10}$$

In the next section we shall describe the implementation of this method in the case of somewhat more general constraints, and when a number of necessary precautions are incorporated. These precautions are due to the facts that (i) the linearized constraints (15.10.9) can be inadequate as a local approximation to (15.10.7), (ii) the line search in direction $\mathbf{s}_i$ must still take account of constraint preservation as well as objective maximization, and (iii) precautions must be taken in the up-date of $\mathbf{B}$ to ensure that $\mathbf{B}$ remains positive definite [see I, § 9.2].

In the unconstrained variable-metric methods, a convergent method will also be superlinearly convergent if $\mathbf{B}$ converges to $-f_{\mathbf{xx}}$ at the limit point. In the constrained case, Powell shows that a convergent method will be super-linearly convergent if $\mathbf{B}$ converges to $-\boldsymbol{\nabla}_x^2 L(\mathbf{x}, \mathbf{y})$ at the limit $\mathbf{x}, \mathbf{y}$ for all directions which are locally constraint-consistent from the limit $\mathbf{x}$. So, even though $-\boldsymbol{\nabla}_x^2 L$ may not be positive-definite, one can still preserve positive-definiteness of $\mathbf{B}$.

15.10.6. Variable-metric Methods for Non-linear Constraints; Description of the Biggs–Han–Powell Method

As a case of moderate but useful generality let us suppose that the constraints are of the mixed equality/inequality type

$$g_j(\mathbf{x}) \circ 0 \tag{15.10.11}$$

where $\circ$ represents equality $(=)$ for $j \in J$ and inequality $(\leq)$ for $j \in J'$, with the $g_j(\mathbf{x})$ scalar. The linear approximation to this set, analogous to (15.10.9) is

$$g_j(\mathbf{x}_i) + \mathbf{s}_i' \boldsymbol{\nabla}_x g_j(\mathbf{x}_i) \circ 0. \tag{15.10.12}$$

However, it may occasionally happen that the linear equation system (15.10.12) for $\mathbf{s}_i$ is inconsistent and has no solution. This can happen even if the system (15.10.11) is consistent. In this case (15.10.12) is modified to

$$\xi_i g_i(\mathbf{x}_i) + \mathbf{s}_i' \nabla_{\mathbf{x}} g_i(\mathbf{x}_i) \circ 0 \qquad (15.10.13)$$

where ξ_i has the value 1 if (15.10.11) is satisfied at $\mathbf{x}_i$. Otherwise ξ_i is given the value ξ, and one takes the largest value of ξ in $[0, 1]$ which assures consistency of the system (15.10.12).

One now maximizes expression (15.10.8) subject to (15.10.12) (or (15.10.13), if necessary), and to the fact that $\mathbf{s}_i$ must be a direction internal to $\mathscr{X}$ from $\mathbf{x}_i$. This is a quadratic problem, which generates a solution $\mathbf{s}_i$ and a vector $\mathbf{y}_i$ of Lagrangian multipliers associated with constraints (15.10.12) (or (15.10.13)).

The $\mathbf{s}_i$ thus determined is taken as direction for a line search, so that one has to determine the scalar α_i in

$$\mathbf{x}_{i+1} = \mathbf{x}_i + \alpha_i \mathbf{s}_i. \qquad (15.10.14)$$

However, the search can not be by simple maximization of the objective; in taking the finite perturbation (15.10.14) one must continue to be sensitive to constraints. One therefore maximizes rather the penalised function

$$f(\mathbf{x}) - \sum_J \mu_j |g_j(\mathbf{x})| - \sum_{J'} \mu_j [g_j(\mathbf{x})]_+ \qquad (15.10.15)$$

where the μ_j are appropriate non-negative weighting coefficients. Powell recommends the choices

$$\mu_{j0} = |y_{j0}|$$

$$\mu_{ji} = \max\left[|y_{ji}|, \tfrac{1}{2}(\mu_{j,i-1} + |y_{ji}|) \right] \qquad i > 0$$

where μ_{ji} and y_{ji} are the values of the jth coefficient and jth multiplier at the ith stage.

The use of (15.10.15) may seem like a reversion to penalty-function methods, which are the very methods one is hoping to improve upon. However, these methods are used only for determination of the step-length parameter α, not for a full n-dimensional search.

Finally, in the updating of $\mathbf{B}$, one must preserve positive-definiteness at all stages. Recall that $\mathbf{B}$ is updated by the BFGS formula (15.9.20) with γ replaced by the γ of (15.10.10). However, it may happen that $\sigma'\gamma$ is negative, in which case one loses positive-definiteness in the up-date. To avoid this, Powell recommends that one replaces γ in formula (15.9.20) by

$$\boldsymbol{\eta} = \theta\boldsymbol{\gamma}^* + (1 - \theta)\mathbf{B}\boldsymbol{\sigma}$$

where θ is a scalar in $[0, 1]$. The value of θ is clearly as large as possible consistent with the empirically determined condition

$$\boldsymbol{\sigma}'\boldsymbol{\eta} \geq 0 \cdot 2\boldsymbol{\sigma}'\mathbf{B}\boldsymbol{\sigma}.$$

Thus

$$
\theta = \begin{cases} 1 & \text{if } \boldsymbol{\sigma}'\boldsymbol{\gamma}^* \geq 0\cdot 2\boldsymbol{\sigma}'\mathbf{B}\boldsymbol{\sigma} \\[2ex] \dfrac{0\cdot 8\boldsymbol{\sigma}'\mathbf{B}\boldsymbol{\sigma}}{\boldsymbol{\sigma}'\mathbf{B}\boldsymbol{\sigma} - \boldsymbol{\sigma}'\boldsymbol{\gamma}^*} & \text{otherwise.} \end{cases}
$$

15.10.7.   The Cutting-plane Method [see also I, § 12.3.1(b)]

Consider the problem of maximizing $f(\mathbf{x})$ subject to

$$g_j(\mathbf{x}) \leq 0 \qquad (j = 1, 2, \ldots, m) \tag{15.10.16}$$

where $f(\mathbf{x})$ is concave and the $g_j(\mathbf{x})$ convex. We imagine that constraints (15.10.16) embody the specification of both the set $\mathcal{X}$ and of any additional constraints; the feasible set F thus specified is clearly convex.

We can normalize the problem to the case where f is linear:

$$f(\mathbf{x}) = \mathbf{c}'\mathbf{x} \tag{15.10.17}$$

say, so that a solution exists on the boundary of F. Quite simply, we replace the problem for general concave f by the maximization of ξ subject to

$$\xi - f(\mathbf{x}) \leq 0 \tag{15.10.18}$$

and constraints (15.10.17). This modified problem is plainly of the standardized form in the pair of variables $(\mathbf{x}, \xi)$.

In the cutting-plane method the feasible set F is approximated at the ith stage by a polytope P_i, which contains F. That is, constraints (15.10.16) are approximated by a number of linear inequalities, which are also a relaxed version of (15.10.16).

One maximizes the objective function $\mathbf{c}'\mathbf{x}$ over P_i: a linear programming problem. If the solution $\bar{\mathbf{x}}_i$ lies in F then one has solved the original problem. If not, then one adds an additional linear constraint, that is a new face to P_i, in such a way as to improve the approximation to F in the neighbourhood of $\bar{\mathbf{x}}_i$. The additional face will usually separate $\bar{\mathbf{x}}_i$ from F, and so is a *cutting-plane*. The new polytope is denoted P_{i+1}, and the sequence repeated. The sequence $\{P_i\}$ is clearly monotonic decreasing [see § 1.4], as also is $\{f(\bar{\mathbf{x}}_i)\}$. If the cutting-plane is chosen as a supporting hyperplane to F at some point $\tilde{\mathbf{x}}_i$ of F which approximates $\bar{\mathbf{x}}_i$ then clearly

$$f(\tilde{\mathbf{x}}_i) \leq f(\bar{\mathbf{x}}) \leq f(\bar{\mathbf{x}}_i) \tag{15.10.19}$$

where $\bar{\mathbf{x}}$ is a solution to the original problem.

The method is clearly a sequence of linear approximations to the original problem. Variants of the method differ in the way the new cutting-plane is chosen. We shall describe only the method due to Kelley (1960). In this one looks for the constraint (15.10.16) which is most violated at $\bar{\mathbf{x}}_i$, that is for the value of j which maximizes $g_j(\bar{\mathbf{x}}_i)$. Suppose this is at $j = t$. One then adds the

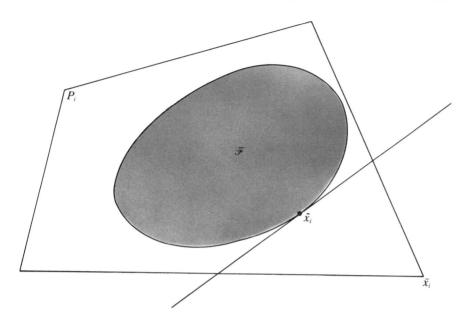

Figure 15.10.1: The principle of the cutting-plane method.

new constraint to the linear system

$$g_t(\bar{\mathbf{x}}_i) + (\mathbf{x} - \bar{\mathbf{x}}_i)' \nabla_{\mathbf{x}} g_t(\bar{\mathbf{x}}_i) \le 0. \qquad (15.10.20)$$

One can in general discard non-binding constraints from the linear constraint set without affecting the convergence of the method.

It can be shown that any limit point of the sequence $\{\bar{\mathbf{x}}_i\}$ is a solution to the original problem if the g_j are continuously differentiable [see Definition 5.9.2]. However, convergence can be quite slow.

P. W.

REFERENCES

Biggs, M. C. (1975). Constrained Minimization Using Recursive Quadratic Programming: Some Alternative Subproblem Formulations. *Towards global optimization*, Eds. Dixon and Szegö, pp. 341–349.

Busacker, R. G. and Saaty, T. L. (1969). *Finite Graphs and Networks*, McGraw-Hill.

Berge, C. and Ghouila-Houri, A. (1969). *Programming, Games and Transportation Networks*, Wiley.

Christofides, N. (1973). *Graph Theory: An Algorithmic Approach*, Academic Press.

Dantzig, G. B. and Veinott, A. F. (1968). *Mathematics of the Decision Sciences*, Part I, American Mathematical Society.

Dennis, I. E. and Moré, J. (1977). Quasi-Newton Methods, Motivation and Theory, *SIAM Review* **19**, 46–89.

Duffin, R. J., Peterson, E. C. and Zener, C. (1967). *Geometric programming*, Wiley.

Farkas, J. (1901). Über die Theorie der einfachen Ungleichungen, *J. Reine Angew. Math.* **124**, 1–27.

Fletcher, R. (1972). Minimizing General Functions Subject to Linear Constraints. *Numerical Methods for Nonlinear Optimization*, Ed. F. A. Lootsma, Academic Press.

Fletcher, R. (1980). *Practical Methods of Optimization. Vol. 1. Unconstrained Optimization*, Wiley.

Fletcher, R. and Powell, M. J. D. (1963). A Rapidly Convergent Descent Method for Minimization, *Computer J.* **6**, 163–168.

Ford, L. R. and Fulkerson, D. R. (1962). *Flows in networks*, Princeton U.P.

Han, S. P. (1977). A Globally Convergent Method for Nonlinear Programming. *J. Opt. Th. Appls.*, **22**, 297–309.

Hu, T. C. (1969). *Integer Programming and Networks Flows*, Addison-Wesley.

Kelley, J. E. (1960). The Cutting-Plane Method for Solving Convex Programs, *J. Soc. Ind. Appl. Math.* **8**, 703–712.

Kiefer, J. (1953). Sequential Minimax Search for a Maximum, *Proc. Ann. Math. Soc.* **4**, 502–506.

Lemke, C. E. (1968). On Complementary Pivot Theory, *Mathematics of the Decision Sciences, Part 1*, Eds. Dantzig and Veinott), pp. 95–136.

Luenberger, D. G. (1973). *Introduction to Linear and Nonlinear Programming*, Addison-Wesley.

Mangasarian, D. C. (1969). *Nonlinear Programming*, McGraw-Hill.

Mitten, L. C. (1970). Branch-and-Bound Methods; General Formulation and Properties, *Operations Research* **18**, 24–34.

Murray, W. (1972). *Numerical Methods for Unconstrained Optimization*, Academic Press.

Potts, R. B. and Oliver, R. M. (1972). *Flows in Transportation Networks*, Academic Press.

Powell, M. J. D. (1978). A Fast Algorithm for Nonlinearly Constrained Optimization Problems. *Numerical Analysis Dundee. Lecture Notes in Mathematics 630*, Ed. G. A. Watson, pp. 144–157, Springer.

Rockafellar, R. T. (1970). *Convex Analysis*, Princeton U.P.

Simmons, M. D. and White, G. W. T. (1977). The Decentralized Profit Maximization of Interconnected Productions Systems, *Int. J. Control* **25**, 705–744.

Varaiya, P. P. (1972). *Notes on Optimization*, Van Nostrand.

Walsh, G. R. (1979). *Methods of Optimization*, Wiley.

Whittle, P. (1971). *Optimization under Constraints*, Wiley.

CHAPTER 16

Dynamic Programming

16.1. INTRODUCTION

The term 'dynamic programming' was coined by R. Bellman (1957) to describe the mathematical techniques which he had brought together to study a class of optimization problems involving sequences of decisions. There have been many applications and further developments of the method since that time, but we shall not attempt to cover all of these. The aim of this chapter is to describe the method and indicate the range of possible applications by means of examples.

Our subject is sequential or multi-stage decision problems and, since the time variable is used in organizing the sequence, it is helpful to classify problems according to whether time is discrete or continuous. The next section is concerned with discrete dynamic programming and the basic theory will be developed there. Although models in continuous time require greater mathematical sophistication, the effort is often worthwhile because we can make use of powerful tools from the calculus. Continuous models will be discussed in section 16.3. In particular, it will be shown how dynamic programming provides an alternative approach to classical problems in the calculus of variations: see Chapter 12. The final section is concerned with mathematical models involving random behaviour and depends on some elementary notions of probability theory, such as expectations [see Volume II, Chapter 8]. In practice, decisions must be taken in the face of uncertainty about the future and the theory of Markov decision processes has been developed to deal with systems in which the state variable changes according to a known probability distribution for each possible decision [see II, Chapter 19].

Progress has also been made by applying similar methods to statistical problems where the changes of state depend on unknown probabilities. Such models are more complicated and beyond the scope of this chapter, but it is worth mentioning that some of the techniques of dynamic programming first emerged in the investigation of sequential decisions in statistics by Wald and others. A comprehensive account of these developments can be found in the book by De Groot (1970).

Bellman certainly deserves credit for giving a clear statement of the principles of dynamic programming and for demonstrating a wide range of applications, but it would be misleading to suggest that the subject is entirely modern. The central idea is that of mathematical induction which is very old indeed. We conclude this introduction with a very informal illustration of how this idea can be applied to multi-stage decision problems.

Consider a restaurant in which the same menu is repeated each day and imagine a customer who attempts to maximize his personal satisfaction from eating lunch there on five successive days. He knows the cost of the items on the menu and he begins by assigning to each of them another positive number called the *utility* which measures the satisfaction he would obtain by choosing the item. For one day, he then has a simple decision problem provided that he restricts himself to one or two courses: he must choose in order to maximize his utility, subject to any limit imposed on the cost of the meal. At first sight, there is no reason why he should not treat different days quite separately. However, there is a connection between the five decisions if he sets a limit on the total cost: obviously, his choice on the last day will depend on what money he has left. It would also be sensible to impose a constraint on successive choices in order to achieve a balanced diet; otherwise, he might repeat the same meal every day. Such constraints mean that he has a genuine multi-stage decision problem and he cannot maximize the total utility by separating the decisions. However, by working backwards in time, it is possible to reduce the problem. Roughly speaking, the situation on the last day requires a simple choice, whatever happens previously, and by expressing the utility of this final choice in a convenient way, the problem of reaching a decision on the last day but one can be simplified. The inductive argument eventually leads to a coherent policy for the whole sequence of decisions.

In general, the method treats the sequence of decisions in reverse order and, for this reason, it is called *backwards induction*. This is more natural than it seems, but our restaurant customer could be forgiven for abandoning his attempt to use a mathematical model. Unfortunately, in order to maximize his total utility, he has to analyze the whole set of decisions before ordering his lunch on the first day.

16.2. MULTI-STAGE DECISION PROBLEMS

Before we attempt to construct a general model and give a statement of Bellman's principle of optimality, it will be helpful to consider two examples.

EXAMPLE 16.2.1. The entries in the matrix

$$\begin{pmatrix} 3 & 5 & 1 & 6 & 8 \\ 6 & 9 & 2 & 1 & 4 \\ 3 & 6 & 2 & 5 & 9 \\ 0 & 7 & 8 & 5 & 0 \end{pmatrix}$$

represent costs associated with the positions in the rectangle. It is required to find an optimal route from the top left-hand corner to the bottom right-hand corner which consists of steps, either to the right or downwards at each stage. The cost of following any particular route is the sum of all the entries encountered on the way. For example, the cost of moving down the first column and along the bottom row is $3+6+3+0+7+8+5+0=32$. An optimal route is one for which the total cost is a minimum.

In order to find such a route, we shall construct another matrix in which each entry represents the minimum cost, for the corresponding position, of reaching the bottom right-hand corner. The complete minimum-cost matrix is

$$\begin{pmatrix} 22 \rightarrow 19 \rightarrow 14 & 17 & 21 \\ \downarrow \quad \downarrow \quad \downarrow \\ 27 & 22 \rightarrow 13 \rightarrow 11 & 13 \\ \downarrow \qquad\qquad \downarrow \quad \downarrow \\ 21 \rightarrow 18 \rightarrow 12 \rightarrow 10 & 9 \\ \downarrow \quad \downarrow \\ 20 \rightarrow 20 \rightarrow 13 \rightarrow 5 \rightarrow 0 \end{pmatrix}$$

and the construction proceeds by examining the columns in reverse order. Thus, for the position in row 1, column 5, there is only one admissible path, as indicated by the arrows, and the minimum cost must be $8+4+9+0=21$. We can now deal with the positions in column 4, starting at the bottom where the minimum cost is simply $5+0=5$. The next entry in row 3, column 4, is obtained by noting that it is preferable to make the initial move downwards, giving a total cost $5+5=10$, rather than moving to the right. It does not take long to work backwards through the matrix in this way; the minimum cost for any position can be determined by comparison as soon as the two neighbouring entries on the right and below are known. The last entry to be calculated is the one in the top left-hand corner and this is based on observing that a total cost of $3+19=22$ can be achieved by moving to the right. Notice that the optimal route through the matrix is determined by following the arrows and, finally, we can verify that the total cost is correctly determined by adding up the appropriate entries in the original matrix: $3+5+1+2+1+5+5+0=22$.

EXAMPLE 16.2.2. This example is concerned with maximizing utilities and it depends on a given utility function U, which represents the satisfaction obtained by a particular individual from spending his money. When he spends an amount a, the corresponding utility is $U(a)$ and we shall investigate how to maximize the total utility obtained over a number of decisions to spend. We assume that

$$U(0) = 0$$

and that the function has a derivative U' with

$$U'(a) > 0,$$

so that $U(a)$ is positive and strictly increasing in a [see § 3.1.1].

Consider a man with initial capital x_0 and let $x_1, x_2, \ldots$ be his capital at the times $1, 2, \ldots$. At time t, suppose he decides to spend an amount a_t within the range

$$0 \leq a_t \leq x_t. \tag{16.2.1}$$

The rest of his money is invested until time $t + 1$. More precisely, we assume that

$$x_{t+1} = \lambda(x_t - a_t), \tag{16.2.2}$$

where $\lambda > 1$ is a constant determined by the rate of interest [see I, § 15.1]. For example, if the unit of time is one year and interest is at 5%, $\lambda = 1 \cdot 05$. Over a period of length T, the total utility obtained is

$$U(a_0) + U(a_1) + \ldots + U(a_{T-1}) \tag{16.2.3}$$

and the aim is to maximize this quantity, subject to the constraints imposed by relations (16.2.1) and (16.2.2). Thus, we are interested in a function of T variables, $a_0, a_1, \ldots, a_{T-1}$, which must be chosen under rather complicated conditions determined by the fact that x_t must remain positive for $t = 1, 2, \ldots, T$. Notice that any capital x_T left over at the end of the period makes no contribution because no further spending is considered.

It is useful to define a new time variable

$$n = T - t, \tag{16.2.4}$$

which represents the time to go. Consider the situation at time $t = T - 1$ when x_{T-1} becomes known. In this case $n = 1$ and this is the number of decisions remaining to be taken. It is clear that the maximization problem is relatively simple if n is small and, for this reason, the method of dynamic programming proceeds by induction on n.

In the first place, we turn our attention to the special case of *linear utilities*, taking

$$U(a) = a \tag{16.2.5}$$

for all $a \geq 0$. It is not difficult to guess the form of the optimal policy in this case, but it will be instructive to work through the backwards induction in detail. We begin by reformulating the maximization problem in terms of the time to go n and the initial capital x_0. It is important to realize that the maximum total utility obtainable from a sequence of spending decisions depends only on the appropriate values of n and x_0. This means that, for any positive integer n and any real $x \geq 0$,

$$f_n(x) = \text{maximum total utility} \tag{16.2.6}$$

is properly defined.

Let us calculate $f_1(x)$ for an arbitrary value of x. Since there is only a single amount a to be chosen in the case $n = 1$, we have

$$f_1(x) = \max_{0 \le a \le x} a.$$

As might be expected, the amount spent should be chosen as large as possible: $a = x$, so that

$$f_1(x) = x.$$

Thus, we have a very simple formula for the function f_1. In order to record the dependence of the optimal decision on n and x, we introduce a *decision function* d_n and write

$$a = d_n(x). \tag{16.2.7}$$

In particular, we have just worked out that

$$d_1(x) = x.$$

The case $n = 2$ is more typical of the induction process. We observe that

$$f_2(x) = a + f_1(y),$$

where a now represents the amount spent in the first of two stages, x is the capital available at the start of the first stage and y is the capital for the second stage. Thus, a must be chosen within the range

$$0 \le a \le x$$

and the choice determines

$$y = \lambda(x - a),$$

by using equation (16.2.2). Since we already know how to attain the maximum utility $f_1(y)$ at the final stage, we can evaluate

$$f_2(x) = \max_{0 \le a \le x} \{a + \lambda(x - a)\}.$$

Here, the expression

$$a + \lambda(x - a) = \lambda x - (\lambda - 1)a$$

and, since $\lambda > 1$, the coefficient of a is negative. Hence, the maximum is attained by choosing a as small as possible: $a = 0$. The results from our investigation of the case $n = 2$ can be summarized as follows:

$$f_2(x) = \lambda x, \qquad d_2(x) = 0,$$

for all $x \ge 0$.

The same argument can be used to determine the function f_3 now that we know f_2, and so on. In general, let us suppose that f_{n-1} is a known function, for some $n \ge 2$. Then, by considering the capital x available at the start of n

stages and the amount a spent in the first of these stages, we obtain

$$f_n(x) = \max_{0 \le a \le x} \{a + f_{n-1}(\lambda(x-a))\}. \tag{16.2.8}$$

This relation determines f_n in terms of f_{n-1}. In fact, we can use the relation to show that

$$f_n(x) = \lambda^{n-1}x, \tag{16.2.9}$$

for all n and x.

It is easy to prove the general formula (16.2.9) by induction. In the special cases $n = 1$ and $n = 2$, our previous calculations show that the formula is correct. We now assume that, for some $n \ge 2$ and any amount $y \ge 0$,

$$f_{n-1}(y) = \lambda^{n-2}y.$$

On substituting this expression into (16.2.8) with $y = \lambda(x-a)$, we obtain

$$f_n(x) = \max_{0 \le a \le x} \{a + \lambda^{n-1}(x-a)\}.$$

Since $\lambda^{n-1} > 1$, the coefficient of a is negative in the expression on the right and it follows that the maximum is obtained by setting $a = 0$, which leads immediately to the required formula (16.2.9).

Notice that, in view of the choice $a = 0$, the decision function corresponding to (16.2.9) is simply

$$d_n(x) = 0. \tag{16.2.10}$$

This holds for every $n \ge 2$ but, as we remarked earlier,

$$d_1(x) = x.$$

The optimal policy is implicit in these decision functions and it only remains to interpret the results. We need to examine the sequence of values $x_0, x_1, \ldots, x_T$ determined by (16.2.2) when we substitute the optimal amounts $a_0, a_1, \ldots, a_{T-1}$. Here, it is natural to work forwards in time so we rely on relation (16.2.4). Initially, $t = 0$, $n = T$ and the capital is x_0. Assuming that $T \ge 2$, the formula (16.2.10) shows that

$$a_0 = d_T(x_0) = 0$$

and then we can apply (16.2.2), obtaining

$$x_1 = \lambda x_0.$$

Similarly, we determine a_1 in terms of x_1 and then find x_2,

$$a_1 = d_{T-1}(x_1) = 0, \qquad x_2 = \lambda x_1 = \lambda^2 x_0.$$

In fact, all the quantities $a_0, a_1, \ldots, a_{T-2}$ must be set equal to zero and we obtain, in succession,

$$x_3 = \lambda^3 x_0, \ldots, x_{T-1} = \lambda^{T-1}x_0.$$

Finally, the choice of a_{T-1} is determined by the decision function d_1:

$$a_{T-1} = d_1(x_{T-1}) = x_{T-1},$$

which means that $x_T = 0$. For the linear utility function, the optimal policy is to spend nothing until the final period and then spend all the accumulated funds. The utility obtained is equal to the final amount spent:

$$a_{T-1} = \lambda^{T-1} x_0.$$

This agrees with our general formula (16.2.9) which shows that the maximum total utility over the whole sequence of decisions is

$$f_T(x_0) = \lambda^{T-1} x_0.$$

Of course, the results obtained here depend on the special form of utility (16.2.5). As we shall see later, more interesting policies are obtained when the utility function is non-linear. We are now in a position to develop a much more general model, but the approach and some of the notation will be the same as in Example 16.2.2.

16.2.1. A General Model

Imagine a dynamic system which moves through a sequence of states $x_0, x_1, x_2, \ldots$ at the times $0, 1, 2, \ldots$. We suppose that the motion is controlled by choosing a sequence of actions $a_0, a_1, a_2, \ldots$. If there is a cost associated with each transition from one state to the next, then it is natural to seek a policy which minimizes the sum of these costs. A wide range of optimization problems can be specified in this way.

The state x_t and the corresponding action a_t at time t could be vectors of given dimensions [see I, § 5.4] and the method is sometimes applied to systems with more general state and action spaces. However, in what follows, both x_t and a_t will always be real numbers. The essential elements of the model are that, given x_t at time t, the choice of a_t determines both the next state x_{t+1} and the cost c_t of the transition from x_t to x_{t+1}. We suppose that two functions K and L are prescribed and that, in general,

$$c_t = K(x_t, a_t, t), \tag{16.2.11}$$

$$x_{t+1} = L(x_t, a_t, t). \tag{16.2.12}$$

Thus, K determines the immediate *cost* of any particular action, whereas L defines the *law of motion*. The problem is to choose the sequence of actions in order to minimize the total cost over a given period. More precisely, we shall restrict attention to a period of length T and our criterion will be the sum

$$c_0 + c_1 + \ldots + c_{T-1}. \tag{16.2.13}$$

The aim is to minimize this quantity, for a given initial state x_0, by choosing $a_0, a_1, \ldots, a_{T-1}$. Since we are neglecting costs incurred at time T or later, it is clear that the actions $a_T, a_{T-1}, \ldots$ are irrelevant.

A key idea underlying the method of dynamic programming is the recognition that at any time t, during the period of control, we need only concern ourselves with minimizing the total future cost:

$$c_t + c_{t+1} + \ldots + c_{T-1},$$

since the choice of present and future actions cannot possibly affect the past. This reduced problem involves the actions $a_t, a_{t+1}, \ldots, a_{T-1}$, so that the effective number of decision variables is

$$n = T - t. \tag{16.2.14}$$

Clearly, the minimization problem is simpler if n is small. The idea of separating past and future costs reveals that it is natural to work backwards in time, considering the cases $n = 1, 2, 3, \ldots$ in that order. Roughly speaking, we are able to choose the actions one at a time, but in order to do this we shall need to rely on the following principle.

PRINCIPLE OF OPTIMALITY.   *An optimal policy has the property that, whatever the initial state and initial decision are, the remaining decisions must constitute an optimal policy with regard to the state resulting from the first decision.*

Bellman's book (1957) gives only a brief comment by way of justifying this claim. It is true that 'a proof by contradiction is immediate', but some further explanation might be helpful. Suppose the remaining decisions $a_1, a_2, \ldots, a_{T-1}$ do not constitute an optimal policy as the principle asserts. Then, starting at x_0, the policy determines a_0 and hence x_1, but the total future cost

$$c_1 + c_2 + \ldots + c_{T-1}$$

is not minimized. In other words, by changing the remaining decisions, the future cost can be reduced. However, any such changes can be included in the policy $a_0, a_1, \ldots, a_{T-1}$ for the whole period, which shows that the original policy cannot be optimal and provides the contradiction.

Another point worth noting is that an optimal policy may not exist. It may be that the required minimum cost cannot be attained exactly. This is why we refer to a principle rather than a theorem, but the essential idea still works even when it is necessary to deal with an infimum [see I, § 2.6.3] rather than an exact minimum. In fact, we need to deal with many different infima; one for each possible state x and index n. It is useful to regard these quantities as the values of a sequence of functions known as the *minimum future cost functions*:

DEFINITION 16.2.1.   For any state x and positive integer $n = T - t$ we define

$$f_n(x) = \inf_{a_t, a_{t+1}, \ldots, a_{T-1}} \{c_t + c_{t+1} + \ldots c_{T-1}\},$$

where $x_t = x$, $x_{t+1} = L(x, a_t, T - n)$ and so on.

This emphasizes that the relevant variables are n and x. The notation is an abbreviation for the result of a rather complicated minimization process and it provides a very convenient summary of the future possibilities at each stage. We regard x as the initial state and then apply the principle of optimality to the choice of the initial action. Of course, we cannot explicitly calculate each of the functions $f_1, f_2, \ldots$ for the general model defined by (16.2.11) and (16.2.12), but we are now in a position to see how the backwards induction works.

At each stage in the argument, we consider a typical transition from x to

$$y = L(x, a, T - n) \qquad (16.2.15)$$

generated by choosing the action a. For simplicity, let us assume that there is an optimal action in every case and denote this by

$$a = d_n(x). \qquad (16.2.16)$$

Consider the case $n = 1$ when $t = T - 1$. Then there is only a single cost term

$$c_{T-1} = K(x, a, T - 1)$$

and we have

$$f_1(x) = \inf_a K(x, a, T - 1). \qquad (16.2.17)$$

This determines the function f_1 and we can proceed to the case $n = 2$. Here, we are concerned with two actions and, by using the principle of optimality, we may assume that the second of these is optimal when we consider choosing the first. Thus, for any initial state x and action a, the total cost is

$$K(x, a, T - 2) + f_1(y),$$

where

$$y = L(x, a, T - 2).$$

Hence

$$f_2(x) = \inf_a \{K(x, a, T - 2) + f_1(L(x, a, T - 2))\}.$$

In general, we can determine f_n in terms of f_{n-1} by using exactly the same argument:

$$f_n(x) = \inf_a \{K(x, a, T - n) + f_{n-1}(L(x, a, T - n))\}. \qquad (16.2.18)$$

The optimal choice of a serves to define the corresponding *decision function*:

$$a = d_n(x).$$

However, it is important to recognize that the method is based on the sequence of minimum future cost functions; the decision functions are a useful biproduct in most, but not all, applications.

Sometimes the optimal actions do not exist and the corresponding decision functions are not properly defined. Strictly speaking, in such cases there is no optimal policy, but policies can be found which approximate the required infimum of the total cost. In other cases when an optimal policy exists, it may not be unique. However, these difficulties are not very important in practice. For example, the first cannot occur when the set of possible actions at each stage is finite; the infimum over a finite set is always a minimum. Most applications involve computations of the minimum future cost functions, rather than exact mathematical formulae for f_1, f_2, etc. This means that both states and actions must be restricted to a finite number of possibilities, depending on the capacity of the computer. Then the computations always determine a sequence of decision functions and an optimal policy.

16.2.2.   Applications

The next example is a problem of maximizing utilities and is a special case of the model described earlier in Example 16.2.2, with a non-linear utility function. This will be followed by a simple numerical application which illustrates the possibility of discovering policies which are optimal over an infinite period of time.

EXAMPLE 16.2.3.   We return to the model specified by relations (16.2.1) and (16.2.2), where x_t represents the capital owned by an individual at time t and a_t is the amount he chooses to spend. It is natural to impose the constraint

$$0 \le a_t \le x_t \qquad\qquad (16.2.19)$$

on the choice of a_t. As before, we suppose that a utility function U is prescribed and consider the problem of maximizing the total utility

$$U(a_0) + U(a_1) + \ldots + U(a_{T-1})$$

for a given value of the initial capital x_0.

The general model of subsection 16.2.1 can be applied directly by defining the functions K and L as follows:

$$K(x_t, a_t, t) = -U(a_t), \qquad\qquad (16.2.20)$$

$$L(x_t, a_t, t) = \lambda (x_t - a_t), \qquad\qquad (16.2.21)$$

where $\lambda > 1$ is a constant. This determines the law of motion

$$x_{t+1} = \lambda (x_t - a_t) \qquad\qquad (16.2.22)$$

exactly as before, and the total cost we are required to minimize is

$$-\{U(a_0) + U(a_1) + \ldots + U(a_{T-1})\}.$$

This is obviously equivalent to maximizing the total utility and we can avoid the repetition of negative signs, simply by defining the functions f_1, f_2, etc. as maximum future rewards, rather than minimum costs. Thus,

$$f_1(x) = \max_{0 \le a \le x} U(a) \tag{16.2.23}$$

and, for each $n \ge 2$,

$$f_n(x) = \max_{0 \le a \le x} \{U(a) + f_{n-1}(\lambda (x - a))\}. \tag{16.2.24}$$

These equations are special cases of (16.2.17) and (16.2.18), apart from the convenient device of changing the signs.

The utility function U may be specified in many different ways, leading to different types of optimal policy. In most cases, the task of carrying out the maximization on the right of (16.2.24) becomes complicated after the first few stages, but the case we shall consider is relatively straightforward. Let

$$U(a) = a^{1/2}. \tag{16.2.25}$$

Notice that the utility is positive and increasing in $a > 0$, but its derivative is a decreasing function of a which means that U is a concave function [see § 15.2.6].

For $n = 1$, we have

$$f_1(x) = \max_{0 \le a \le x} a^{1/2} = x^{1/2} \tag{16.2.26}$$

and the corresponding decision function is

$$d_1(x) = x. \tag{16.2.27}$$

Then, by taking $n = 2$ in (16.2.24) and by using the formula we have obtained for f_1, we find that

$$f_2(x) = \max_{0 \le a \le x} \{a^{1/2} + (\lambda (x - a))^{1/2}\}.$$

It is a simple exercise to maximize the expression on the right. The partial derivative with respect to a [see § 5.3] is

$$\tfrac{1}{2}a^{-1/2} - \tfrac{1}{2}\lambda^{1/2}(x - a)^{-1/2}.$$

This changes sign from positive to negative as a increases from 0 to x and it follows that the required maximum is uniquely determined by setting the partial derivative equal to zero [see § 5.6]. Hence,

$$\lambda a = x - a, \qquad a = x/(1 + \lambda).$$

The decision function determined by this result is

$$d_2(x) = x/(1 + \lambda).$$

When a is replaced by $x/(1 + \lambda)$ in the above formula for f_2, the terms can be rearranged to give

$$f_2(x) = (1+\lambda)^{1/2}x^{1/2}.$$

It is perhaps unwise to guess the general form of $f_n(x)$ without carrying out another iteration, but let us anticipate the result. In fact,

$$f_n(x) = (1+\lambda+\lambda^2+\ldots+\lambda^{n-1})^{1/2}x^{1/2}, \qquad (16.2.28)$$

$$d_n(x) = x/(1+\lambda+\lambda^2+\ldots+\lambda^{n-1}) \qquad (16.2.29)$$

and both these formulae are valid for all $n \geq 1$ and $x \geq 0$. The results are not difficult to prove by induction from the general recurrence relation (16.2.24). The argument is based on a calculation very similar to the one we have just carried out to determine $f_2(x)$, so the details will be omitted.

Finally, we can examine the optimal policy over T stages, starting with the capital x_0. It follows from (16.2.29) with $n = T$ that

$$a_0 = d_T(x_0) = x_0/(1+\lambda+\lambda^2+\ldots+\lambda^{T-1}).$$

Then we find that

$$x_1 = \lambda(x_0 - a_0)$$

and a short calculation shows that

$$a_1 = d_{T-1}(x_1) = \lambda^2 x_0/(1+\lambda+\lambda^2+\ldots+\lambda^{T-1}).$$

Thus,

$$a_1 = \lambda^2 a_0$$

and it is not very difficult to verify that

$$a_2 = \lambda^4 a_0$$

and so on. This kind of spending pattern, in which the amounts increase by a factor of λ^2 at each stage, seems more reasonable than the optimal policy we obtained for the linear utility function in Example 16.2.2.

It is interesting to examine the behaviour of (16.2.28) and (16.2.29) when n becomes large. We recall the assumption that $\lambda > 1$, which implies that the geometric series

$$1+\lambda+\lambda^2+\ldots$$

diverges [see Example 1.7.1]. Hence, $f_n(x)$ becomes infinite as $n \to \infty$, but $d_n(x)$ converges to zero. At first sight, this suggests that a policy corresponding to the decision function

$$d(x) = 0$$

might be a good one, in the long run. But the implication is that no capital would ever be spent, so the total utility would be zero. However, in some applications, a more useful long-term policy can be found by letting $n \to \infty$.

EXAMPLE 16.2.4. As a final example of solving a multi-stage decision problem, we consider a *replacement or repair* model. This is a simple illustration of the kind of problem which occurs when we try to answer the question whether to continue operating a machine or renew it in some way. More realistic models can be developed by representing the gradual deterioration of the machine as a random process, which leads into the field of Markov decision processes [see II, Chapter 19]. The present model, however, is deterministic.

Imagine a system in which the state x_t represents the performance level of a machine. It changes according to the rule

$$x_{t+1} = x_t + 1,$$

unless an 'overhaul' takes place, and an overhaul simply means that $x_{t+1} = 0$, which is the ideal performance level. We measure the deterioration in the operation of the machine by taking the cost of a transition from x_t to x_{t+1} as x_t. In other words, the state represents the net cost of running the machine for the next unit of time. For an overhaul, we suppose that a fixed cost of 7 units is incurred. Of course, the relative magnitudes of the running and repair costs determine how frequently we should overhaul the machine, but a similar pattern emerges for other values of the overhaul cost.

We define $f_n(x)$ as the minimum cost of maintaining the system for n units of time, where $x = 0, 1, 2, \ldots$ is the initial state of the machine. In this application of dynamic programming, it is hardly necessary to refer to the general model and equations (16.2.17) and (16.2.18) before starting the construction of the minimum cost functions; the relations between them are very simple. We have

$$f_1(x) = \min \{x, 7\}, \tag{16.2.30}$$

since the two alternative actions are to continue operation at cost x or to overhaul it at cost 7. In general

$$f_n(x) = \min \{x + f_{n-1}(x+1), 7 + f_{n-1}(0)\} \tag{16.2.31}$$

and the choice between the two terms on the right indicates whether it is optimal to continue or to carry out an overhaul, respectively. The results of the construction are shown in Table 16.2.1.

Equation (16.2.30) yields all the values of f_1 in the table and then, for the next row, we have $f_1(0) = 0$ and

$$f_2(x) = \min \{x + f_1(x+1), 7\}.$$

Each entry depends on a comparison between two of the entries in the previous row, so the calculation is very simple. It soon becomes clear that $f_n(x)$ is constant when x is sufficiently large, but the critical value of x corresponding to the switch from one action to the other is not uniquely determined. For

x	0	1	2	3	4	5	6	7	8	9	$\cdot$
f_1	0	1	2	3	4	5	6	7	7	7	$\cdot$
f_2	1	3	5	7	7	7	$\cdot$	$\cdot$	$\cdot$	$\cdot$	$\cdot$
f_3	3	6	8	8	8	8	$\cdot$	$\cdot$		
f_4	6	9	10	10	10	10	$\cdot$	$\cdot$		
f_5	9	11	12	13	13	13	$\cdot$	$\cdot$		
f_6	11	13	15	16	16	16	$\cdot$	$\cdot$		
f_7	13	16	18	18	18	18	$\cdot$	$\cdot$		
f_8	16	19	20	20	20	20	$\cdot$	$\cdot$		
f_9	19	21	22	23	23	23	$\cdot$	$\cdot$		
$\cdot$	$\cdot$	$\cdot$	$\cdot$								
$\cdot$	$\cdot$	$\cdot$									

Table 16.2.1

example, when $n = 5$ and $x = 3$, we have

$$f_5(3) = \min\{3 + 10, 7 + 6\} = 13.$$

Since both alternatives lead to the same cost, it is immaterial which action is used.

The optimal policy here is not unique, but it exhibits a definite pattern as n increases. In order to see this, notice that

$$f_7(x) = f_3(x) + 10$$

for all x. The function f_8 is similarly related to f_4 and so on.

In fact

$$f_n(x) = f_{n-4}(x) + 10, \tag{16.2.32}$$

for all x and every $n \geq 7$. This pattern has two useful implications. In the first place, it shows that the average cost of an optimal maintenance policy over a long period is $10/4 = 2 \cdot 5$. For example,

$$f_{100}(0) = f_{96}(0) + 10 = \ldots = f_4(0) + 240 = 246,$$

so the average cost is $2 \cdot 46$ in this case. More precisely, it can be shown that the ratio $f_n(x)/n$ converges to $2 \cdot 5$ as $n \to \infty$ for every x [see § 1.2]. The other consequence of (16.2.32) is the existence of a policy, not depending on n, which attains the minimum average cost, in the long run. When we examine the values of f_4, f_5, f_6, and f_7, we notice that the rule: 'overhaul if and only if $x \geq 3$', is optimal for these values of n. But in view of (16.2.32), the same is true for all $n \geq 4$.

16.3. CONTINUOUS MODELS

The calculus of variations is concerned with the problem of finding a smooth curve which is optimal in some sense. For example, we might wish to determine a path joining two points and lying in a given surface such that its length is a minimum. In general, the optimality criterion is specified by means of a functional which associates a value with each possible curve and the problem is to maximize or minimize this value. The classical approach to this type of problem is based on extending the notion of differentiation from functions to functionals, as described in Chapter 12. The idea of differentiation with respect to local perturbations of a whole curve is not an easy one and the main aim of this section is to describe a simpler point of view, using the technique of backwards induction [see § 16.1].

For discrete models of the type studied earlier in this chapter, optimal policies can often be justified directly from the inductive approach. We are not concerned here with rigorous proofs, but it should be mentioned that there are genuine mathematical difficulties in attempting to carry out induction with respect to a continuous time variable. This leaves plenty of scope for mathematical ingenuity in most applications. On the other hand, the method of dynamic programming extends in a natural way which is intuitively revealing. We shall restrict our attention to a standard form of problem in the calculus of variations and give a heuristic derivation of the partial differential equation known as the Euler–Lagrange equation [see § 12.4]. This will be followed by two examples giving solutions of the equation and the corresponding optimal paths.

As in section 16.2.1 we suppose there is a given cost function $K(x, a, t)$, but here we shall interpret the control variable a as the slope of a smooth curve:

$$\frac{dx}{dt} = a. \tag{16.3.1}$$

The total cost associated with the whole curve is

$$\int_0^T K(x, a, t)\, dt = \int_0^T K\left(x(t), \frac{dx}{dt}(t), t\right) dt \tag{16.3.2}$$

and it is required to minimize this integral, subject to the conditions that the curve must start at $x_0 = x(0)$ and finish at $x_T = x(T)$. Let us imagine that the path reaches a state x at some intermediate time t. Then we can express the minimum cost incurred over the rest of the path in terms of x and t. Define

$$f(x, t) = \min \int_t^T K(x(s), a(s), s)\, ds, \tag{16.3.3}$$

where the state $x = x(t)$ affects the cost because it determines the starting point for the remainder of the path.

The idea is to consider an initial step from this point to a new state $x + \delta x$ at time $t + \delta t$. The immediate cost is $K(x, a, t)\delta t$ and $a = \delta x/\delta t$ is the slope of the curve, apart from smaller terms which become negligible when δt tends to zero. Hence, we can apply the optimality principle by noting that the minimum future cost, from the new state, is $f(x + \delta x, t + \delta t)$, where $\delta x = a\delta t$.

Thus,

$$f(x, t) = \min_a \{K(x, a, t)\delta t + f(x + a\delta t, t + \delta t)\}. \qquad (16.3.4)$$

This is similar to the basic relation (16.2.18) used in discrete dynamic programming, but in the continuous time model we can make use of the fact that δt is arbitrarily small. We have

$$f(x + a\delta t, t + \delta t) = f(x, t) + \frac{\partial f}{\partial x}(x, t)a\delta t + \frac{\partial f}{\partial t}(x, t)\delta t$$

and, when the terms of order δt in (16.3.4) are collected together, it turns out that

$$\min_a \left\{ K(x, a, t) + a\frac{\partial f}{\partial x}(x, t) + \frac{\partial f}{\partial t}(x, t) \right\} = 0. \qquad (16.3.5)$$

This is the *dynamic programming equation* for our problem.

Now suppose that the minimum on the left of (16.3.5) is obtained by setting the partial derivative with respect to a equal to zero. This leads to the *optimality equation*

$$\frac{\partial K}{\partial a}(x, a, t) + \frac{\partial f}{\partial x}(x, t) = 0. \qquad (16.3.6)$$

Another equation called the *conservation equation* is also implicit in (16.3.5) because the choice of $a = dx/dt$ must satisfy the condition

$$K(x, a, t) + a\frac{\partial f}{\partial x}(x, t) + \frac{\partial f}{\partial t}(x, t) = 0. \qquad (16.3.7)$$

In applications, it is often convenient to solve the last two equations for the unknown function f. However, in order to see how this approach is related to the classical one, we must eliminate f. This can be done by further differentiations. We first obtain the time derivative of (16.3.6) using the fact that $dx/dt = a$ [see §§ 5.4, 5.5].

$$\frac{d}{dt}\frac{\partial K}{\partial a} + a\frac{\partial^2 f}{\partial x^2} + \frac{\partial^2 f}{\partial x \, \partial t} = 0. \qquad (16.3.8)$$

On the other hand, the partial derivative of (16.3.7) with respect to x is

$$\frac{\partial K}{\partial x} + \left\{ \frac{\partial K}{\partial a} + \frac{\partial f}{\partial x} \right\}\frac{\partial a}{\partial x} + a\frac{\partial^2 f}{\partial x^2} + \frac{\partial^2 f}{\partial x \, \partial t} = 0. \qquad (16.3.9)$$

The expression in brackets must vanish, by (16.3.6), and then (16.3.8) and (16.3.9) together show that

$$\frac{d}{dt}\left\{ \frac{\partial K}{\partial a}(x, a, t) \right\} = \frac{\partial K}{\partial x}(x, a, t). \qquad (16.3.10)$$

This is known as the *Euler–Lagrange equation* in the calculus of variations: see section 12.4.

EXAMPLE 16.3.1. Consider the case when

$$K(x, a, t) = \lambda^2 + a^2,$$ (16.3.11)

where λ is a given positive constant. Notice that the state variable x and the time t do not appear in this cost function. The aim here is to minimize the total cost of reaching a target state, taken to be $x_T = 0$, from an arbitrary initial state x_0. The total cost is

$$\int_0^T \left\{ \lambda^2 + \left(\frac{dx}{dt} \right)^2 \right\} dt$$ (16.3.12)

and the two terms in the integrand represent costs associated with the time taken and the speed, respectively. Thus, if λ is large, the main objective is to minimize the time to reach the target, but there is a relatively small additional cost depending on the choice of $a = dx/dt$.

The first step is to realize that the minimum cost of reaching state zero from any given state x depends on x, but not on the particular time t. When we substitute (16.3.11) into (16.3.5) and write $f(x, t) = f(x)$, the dynamic programming equation reduces to

$$\min_a \{ \lambda^2 + a^2 + af'(x) \} = 0,$$ (16.3.13)

where $f'(x)$ is an ordinary derivative. In this case, it is not difficult to see that the minimum is always determined by differentiating (16.3.13) with respect to a. The optimality equation is

$$2a + f'(x) = 0$$ (16.3.14)

and the conservation equation is

$$\lambda^2 + a^2 + af'(x) = 0.$$ (16.3.15)

We can easily solve this pair of equations for a and $f'(x)$. In fact,

$$f'(x) = -2a, \qquad a^2 = \lambda^2.$$ (16.3.16)

It only remains to interpret these results; there is no need to use Euler's equation directly.

According to (16.3.16) there are two possible forms of the derivative $f'(x)$, corresponding to the choices $a = \pm\lambda$. However, the definition of $f(x)$ provides us with further information. Since $f(x)$ is the minimum of an integral of strictly positive costs, it follows that $f(x) > 0$, except that the total cost reduces to zero when the initial state x coincides with the target state, i.e. $f(0) = 0$. The only appropriate solution of (16.3.16) is obtained by setting

$$a = -\lambda, \qquad f'(x) = 2\lambda,$$

whenever $x > 0$ and

$$a = \lambda, \qquad f'(x) = -2\lambda,$$

whenever $x < 0$. Hence

$$f(x) = 2\lambda x, \qquad f(x) = -2\lambda x$$

according as x is positive or negative. In other words

$$f(x) = 2\lambda |x| \tag{16.3.17}$$

and this formula holds for all x. The optimal control policy for this example is to maintain a constant speed $a = \lambda$, always choosing the direction so as to reduce $|x|$ and move towards the target. In particular, if $x_0 = x(0) > 0$, we set $a = dx/dt = -\lambda$ and the motion of the state is described by

$$x(t) = x_0 - \lambda t. \tag{16.3.18}$$

This holds for $0 < t < T$ and the final time is determined by the condition that $x(T) = x_T = 0$. This means that $T = x_0/\lambda$.

EXAMPLE 16.3.2.   We now turn to a more realistic application. The basic cost function is again quadratic, but not so simple as in the previous example. This will illustrate how changes in the costs made in designing a model for a particular situation can have a substantial effect on the solution.

It is required to increase the temperature x of a chemical reactor towards a prescribed level $\theta > 0$, by applying heat at a controlled rate $u \geq 0$. For temperatures $x \leq \theta$, any choice of u determines the rate of increase

$$\frac{dx}{dt} = \beta(\theta - x) + u \tag{16.3.19}$$

and the corresponding cost per unit time is given by

$$\gamma(\theta - x)^2 + u^2, \tag{16.3.20}$$

where β and γ are positive constants. Suppose that, at time $t = 0$, the temperature $x = 0$ and consider the problem of choosing a control function $u = u(t)$ which determines the behaviour of $x(t)$ for $t > 0$ in such a way that the total cost is minimized.

In the first place, we need a slight change of notation in order to reduce the equation (16.3.19) to the standard form (16.3.1). The control variable u can be replaced by

$$a = \beta(\theta - x) + u \tag{16.3.21}$$

so that the cost function (16.3.20) becomes

$$K(x, a, t) = \{a - \beta(\theta - x)\}^2 + \gamma(\theta - x)^2. \tag{16.3.22}$$

This depends on the state x and the new control variable a, but not on the time t. Hence, as in the previous example, the minimum cost of raising the temperature from any level $x < \theta$ can be expressed as a function of x alone; we write $f(x, t) = f(x)$ in the dynamic programming equation (16.3.5) and find that

$$\min_{a} \{(a - \beta(\theta - x))^2 + \gamma(\theta - x)^2 + af'(x)\} = 0. \qquad (16.3.23)$$

Strictly speaking, we should impose a constraint on the choice of a here: it follows from the condition $u \geq 0$ that $a \geq \beta(\theta - x)$ must always hold. However, it turns out that the unconstrained solution of (16.3.23) does satisfy this inequality, so it makes sense from a practical point of view.

Since the expression on the left of (16.3.23) is quadratic in a, the required minimum must occur where the partial derivative with respect to a is zero [see § 5.6]. Hence,

$$2\{a - \beta(\theta - x)\} + f'(x) = 0 \qquad (16.3.24)$$

and equation (16.3.23) also implies that

$$\{a - \beta(\theta - x)\}^2 + \gamma(\theta - x)^2 + af'(x) = 0. \qquad (16.3.25)$$

We can eliminate the unknown function $f'(x)$ between the last two relations and obtain an expression for the optimal policy in terms of x:

$$a^2 = (\beta^2 + \gamma)(\theta - x)^2. \qquad (16.3.26)$$

This means that

$$a = \lambda(\theta - x), \qquad \lambda = (\beta^2 + \gamma)^{1/2}, \qquad (16.3.27)$$

where we have introduced a new constant $\lambda > 0$, for convenience of notation. Since $\lambda > \beta$, (16.3.27) implies that $a \geq \beta(\theta - x)$ whenever $x < \theta$. This shows that the rate of heat supplied $u \geq 0$ and it is worth noting that the alternative solution of (16.3.26), in which $a = -\lambda(\theta - x)$, can be ruled out because it does not satisfy this constraint.

In order to examine what happens to the temperature under the optimal policy (16.3.27), we have to solve the differential equation

$$\frac{dx}{dt} = a = \lambda(\theta - x), \qquad (16.3.28)$$

starting with $x(0) = 0$. The solution [see § 7.2] is

$$x(t) = \theta(1 - e^{-\lambda t}). \qquad (16.3.29)$$

Then, by using relation (16.3.21), we can determine the rate u at which heat must be supplied in order to achieve this increase in temperature:

$$u(t) = (\lambda - \beta)\theta e^{-\lambda t}. \qquad (16.3.30)$$

We remark that the prescribed temperature θ is never actually attained except as a limit as $t \to \infty$. Finally, let us evaluate the minimum total cost, which can be done in two different ways. One method is to use equations (16.3.24) and

(16.3.27) to find $f'(x)$ and then to integrate the resulting expression, with $f(0) = 0$. This approach shows that

$$f(x) = (\lambda - \beta)(\theta - x)^2. \tag{16.3.31}$$

The formula applies for all $x \leq \theta$, although we are mainly interested in the total cost, starting with $x = 0$. The other method is to evaluate the total cost directly by using (16.3.20), (16.3.29) and (16.3.30). A useful check on the detailed arguments given above is to verify that the two methods yield the same result:

$$f(0) = \int_0^\infty \{\gamma(\theta - x(t))^2 + u(t)^2\} \, dt = (\lambda - \beta)\theta^2. \tag{16.3.32}$$

This is a straightforward exercise.

16.4.   DECISIONS INVOLVING UNCERTAINTY

In principle, the techniques of dynamic programming can be applied to problems where the gains and losses are uncertain, simply by considering expectations. We try to maximize the expected gain or minimize the expected loss by introducing probabilities for events in the future. Many of the most interesting applications are concerned with this type of problem. However, we shall not attempt to give a comprehensive account. The particular application described below is known as the 'marriage' or 'secretary' problem and it will serve to illustrate the basic ideas. Further examples can be found in the books by Bellman (1957) and De Groot (1970) or in the elementary treatment of Markov decision processes by Howard (1960).

We shall need some basic ideas of probability, including conditional probabilities and expectations, as described in Volume II, sections 3.9.1 and 8.1, but only at an elementary level. For example, consider the case of a man deciding whether or not to take his umbrella to work. He assesses the cost of carrying the umbrella as 1 unit and the cost of getting wet if it rains as 10 units. The decision should also depend on the probability of rain and we might consider too the possibility of losing the umbrella. Suppose the probability of rain is 0·4 and the probability of losing the umbrella, if he takes it, is 0·1; the cost of a replacement being 25 units. With all these assumptions, he calculates the expected cost of a decision to take the umbrella as $1 + 0·1 \times 25 = 3·5$ units and the expected cost of not taking it is $0·4 \times 10 = 4$, so he should carry the umbrella.

16.4.1.   The Secretary Problem

Suppose that our decision-maker has the task of choosing a new secretary from a number of candidates interviewed at different times. He wishes to select the best candidate, according to his own personal preferences, but he must decide immediately after each interview whether or not to offer the job to the candidate he has just seen. There is no problem of selection if he is allowed to wait until all the interviews are completed before making a decision, since he will then know which candidate is best. However, let us assume that he may offer the job at any stage, if no previous offer has been made, but only to the candidate just interviewed. He may not recall any previous candidate and it is assumed that when the job is offered, it must be accepted. These conditions are perhaps more appropriate to a young lady receiving a sequence of offers of marriage; her decision problem is similar. For either interpretation, if the number of candidates is known in advance, there is a decision procedure which maximizes the probability of selecting the best candidate, as we shall see.

In the first place, consider the case of 3 candidates who present themselves for interview in random order. This means that, if their relative ranks are numbered 1, 2 and 3 in decreasing order of preference, each of the 6 possible arrangements of the interviews [see I, § 3.7] has probability $\frac{1}{6}$ [see II, § 3.3]. There are essentially three different decision rules for consideration: offer the job to the first candidate, make an offer to the second candidate if she is better than the first and otherwise wait for the third candidate, or reject the first two candidates and wait for the third. Let p_1, p_2 and p_3 denote the probabilities of success for each of these rules, where 'success' is defined to be the event that the rule selects the best candidate. The 6 possible arrangements of the interviews can be represented as follows:

$$(1, 2, 3), \quad (1, 3, 2), \quad (3, 1, 2), \quad (2, 1, 3), \quad (2, 3, 1), \quad (3, 2, 1).$$

Since the first rule succeeds in two of these cases, $p_1 = \frac{1}{3}$ and, similarly, $p_3 = \frac{1}{3}$. But the second rule succeeds for each of the arrangements:

$$(3, 1, 2), \quad (2, 1, 3), \quad (2, 3, 1),$$

so $p_2 = \frac{1}{2}$, which is the maximum probability of success. Notice that the optimal decision rule can be expressed in the form: wait until the kth candidate is interviewed and then offer the job to the next candidate who turns out to be the best so far. In general, the optimal policy has the same form, where the critical number k depends on the total number of candidates.

Suppose now that there are T candidates and imagine, for convenience, that they are interviewed at the times $t = 1, 2, \ldots, T$. In order to apply the method of backwards induction [see § 16.1], we must consider the state of information available to the decision-maker after interviewing the tth candidate, assuming that no previous offer has been made. In effect, there are only two states, depending on whether the tth candidate is the best so far, or not. Instead of using the standard dynamic programming notation of section 16.2, let us denote by u_t the maximum probability of success, given that the tth candidate is the best so far and that the job is still available, and let v_t denote the corresponding maximum probability when the tth candidate is not the best so far. Notice that the first candidate must certainly be the best so far, so we interpret v_1 as the probability of success, conditional on rejecting the first candidate. In the case of the final candidate, we know that

$$u_T = 1, \qquad v_T = 0. \tag{16.4.1}$$

Roughly speaking, we must work our way backwards through the candidates until we can evaluate the required maximum probability $p = u_1$. The inductive process will be based on the recurrence relations [see I, § 14.13]

$$u_t = \max\left\{\frac{t}{T}, v_t\right\}, \tag{16.4.2}$$

$$v_t = \frac{1}{(t+1)} u_{t+1} + \frac{t}{(t+1)} v_{t+1}. \tag{16.4.3}$$

The first of these is obtained by considering whether it is advantageous to offer the job to the tth candidate or not: the ratio t/T is the probability that the best of the first t candidates interviewed is also the best of all. The expression on the right of (16.4.3) is obtained by considering whether the next candidate will turn out to be better than any of the first t [see II, Theorem 16.2.1]. According to our assumption that all the candidates are arranged in random order, this event occurs with probability $1/(t+1)$, in which case the conditional probability of success is u_{t+1}. Otherwise with probability $t/(t+1)$, the $(t+1)$th candidate will not be better than all previous ones and the conditional probability of success is v_{t+1}. The relations (16.4.2) and (16.4.3) summarize the effects of probability on our problem. It only remains to solve them, using (16.4.1).

The crucial step in the investigation is to see that, for a suitable integer k,

$$u_t = v_t > \frac{t}{T}, \tag{16.4.4}$$

for $t = 1, 2, \ldots, k-1$, and

$$u_t = \frac{t}{T}, \tag{16.4.5}$$

for $t = k, k+1, \ldots, T$. We can prove this by showing that, if

$$u_{t+1} > \frac{t+1}{T}, \tag{16.4.6}$$

then $u_t > t/T$. By using (16.4.2), with t replaced by $(t+1)$, we note that (16.4.6) implies the equation $u_{t+1} = v_{t+1}$ and then (16.4.3) shows that $v_t = v_{t+1}$. But (16.4.2) implies that $u_t \geq v_t$, so we now have the sequence of relations:

$$u_t \geq v_t = v_{t+1} = u_{t+1} > \frac{t+1}{T} > \frac{t}{T},$$

which shows that $u_t > t/T$, as required. Thus, (16.4.4) and (16.4.5) hold for some integer k in the range $1, 2, \ldots, T$, and since (16.4.5) means that the optimal decision is to offer the job if the latest candidate is best so far, the form of the decision rule is now clear. We must determine k and then the optimal policy is to offer the job to the first of candidates $k, k+1, \ldots, T$ who turn out to be the best so far interviewed.

It is not difficult to find the critical number k by using the relations we have established. Equation (16.4.4) means that $u_{t+1} = v_{t+1}$ for $t = 1, 2, \ldots, k-2$, and (16.4.3) then shows that $v_t = v_{t+1}$, so we obtain

$$u_1 = v_1 = u_2 = v_2 = \ldots = u_{k-1} = v_{k-1}. \tag{16.4.7}$$

Next, we can apply (16.4.3) and (16.4.5) to express v_{k-1} in terms of v_k:

$$v_{k-1} = \frac{1}{T} + \frac{(k-1)}{k} v_k. \tag{16.4.8}$$

Similarly, v_k can be expressed in terms of v_{k+1} and so on. The result of combining these equations together and using the fact that $v_T = 0$ is

$$v_{k-1} = \frac{(k-1)}{T} \left\{ \frac{1}{k-1} + \frac{1}{k} + \ldots + \frac{1}{T-1} \right\}. \tag{16.4.9}$$

We also need the conditions that $v_{k-1} > (k-1)/T$, which arises from (16.4.4), and that $v_k \le k/T$, which is implicit in (16.4.2). The first of these conditions, together with (16.4.9), means that

$$\frac{1}{T-1} + \frac{1}{T-2} + \ldots + \frac{1}{k-1} > 1 \tag{16.4.10}$$

and the second can be combined with (16.4.8) and (16.4.9) to show that

$$\frac{1}{T-1} + \frac{1}{T-2} + \ldots + \frac{1}{k} \le 1. \tag{16.4.11}$$

It follows from these two inequalities that k must be chosen as the smallest integer for which (16.4.11) holds. Thus, for any given integer T, we can determine $k = k(T)$ by calculating

$$\frac{1}{T-1}, \frac{1}{T-1} + \frac{1}{T-2}, \ldots$$

in succession, until the total reaches 1. Table 16.4.1 below gives several values of k, together with the corresponding maximum probability of success $p = u_1$, obtained from (16.4.7) and (16.4.9).

T	1	2	3	4	5	10	15	20	30
k	1	1	2	2	3	4	6	8	12
p	1	0·5	0·5	0·458	0·433	0·399	0·389	0·384	0·379

Table 16.4.1

Finally it can be shown that when T is large, a good approximation for k is T/e, where $e = 2 \cdot 718$. The probability $p = p(T)$, using the optimal decision procedure, decreases to the limit $1/e = 0 \cdot 3679$ when T becomes infinite. It is perhaps surprising that such a high probability of finding the best candidate can be achieved, no matter how many candidates there are.

J. A. B.

REFERENCES

Bellman, R. (1957). *Dynamic Programming*, Princeton University Press, Princeton, New Jersey.

De Groot, M. H. (1970). *Optimal Statistical Decisions*, McGraw-Hill.

Howard, R. (1960). *Dynamic Programming and Markov Processes*, M.I.T. Press and Wiley.

CHAPTER 17

Classical Mechanics

17.1. INTRODUCTION

Classical mechanics is the study of the differential equations that arise from *Newton's laws of motion*:

Law 1. Every particle moves with constant velocity, unless acted on by a force.

Law 2. The rate of change of the momentum of a particle with time is proportional to the force acting on it, and is in the direction of the force.

Law 3. The forces exerted by two particles on each other are always equal in magnitude and opposite in direction.

Since Newton's day there have been criticisms of the logical presentation of his definitions and laws. However, for a suitably restricted class of problem, the basic correctness of his laws has not been in doubt. These provide an accurate description of the motion of massive bodies, provided that they are of sufficiently large size for quantum effects to be small and that they are moving sufficiently slowly for relativistic effects to be negligible. (The adjective 'classical' in classical mechanics is meant to differentiate it from quantum mechanics and relativistic mechanics.)*

A full account of classical mechanics would involve a discussion of the physical concepts involved in Newton's laws; space, time, mass and force. However, in the following it will be assumed that the meaning of these terms is familiar to the reader. Further assumptions and idealizations will also have to be made from time to time in constructing a mathematical model of physical systems [see Chapter 14]. For example, it is customary to introduce the concept of a 'particle' as an approximation to a massive body; a particle is a body having mass but no appreciable size, so that its state can be described

---

* It must also be remembered that the third law only applies to simple interactions, for example the collisional impact of idealized bodies and gravitational or electrostatic interactions; electromagnetic interactions of moving charged particles will in general violate the third law.

completely by its point position. It will also be necessary to make assumptions about the nature of the forces of interaction between particles.

In the historical development of classical mechanics two different mathematical descriptions have been used, leading to two distinct branches of the subject—vectorial mechanics and analytical mechanics. In the first of these the quantities of mechanics (position, velocity, force) are denoted by vectors [see I, § 5.1]. One considers separately the individual particles of the system and obtains vector equations that describe the motion of each of them. This has the advantage that results can be interpreted easily since the mathematical variables are just the quantities of physical interest. It is also possible to deal with quite general kinds of forces, like friction. Difficulties arise, however, when constraints are imposed on a system. For example, if a particle is constrained to move along a curved wire, then there will be a force of reaction between the particle and the wire that is not known initially. These forces must, of course, be included in the vector equations of motion, a complication that is avoided when using the methods of analytical mechanics.

In analytical mechanics generalized co-ordinates are introduced to define the state of the system as a whole [see § 17.4.1]. They are chosen primarily to simplify the imposition of constraints. Also, provided that the forces are of a sufficiently simple kind, the system is defined completely when the kinetic and potential energies are expressed in terms of the generalized co-ordinates. The equations of motion then follow simply from a variational principle. Analytical mechanics also allows a deeper insight into the mathematical structure of classical mechanics and provides powerful methods for investigating problems.

In the following there are sections on vectorial mechanics and analytical mechanics. Both concentrate on systems that contain a finite number of particles. Two topics that involve systems containing an infinite number of particles are the motion of rigid bodies and the motion of continuous systems, like the particles that comprise an elastic vibrating string; these lie outside the scope of this article.

17.2.   VECTORS

(As a preliminary to the section on vectorial mechanics an account is given of vector algebra and some vector calculus [see also V, Chapter 13].)

17.2.1.   Definitions

A *scalar* is a quantity that is completely specified when its magnitude in some appropriate system of units is given. (Time t, mass m are scalars.)

A *vector* is a quantity that is specified by a magnitude and a direction in space (which for our purposes is 3-dimensional Euclidean space [see I, Example 5.2.2]). It can be represented geometrically by a straight line with an arrow attached to it (Figure 17.2.1). The length of the line is equal to the

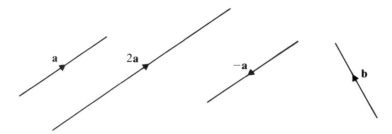

Figure 17.2.1

magnitude of the vector (in some appropriate system of units); the orientation of the line and the direction of the arrow gives the *direction* of the vector. A vector is usually denoted by a letter printed in **Clarendon** type, such as **a**. It can also be represented by a symbol $\overrightarrow{OP}$, say, which means the vector represented by the straight line joining the point O to the point P [see also I, § 5.1].

A vector of zero magnitude is denoted by **0** and has no direction associated with it. A vector of unit magnitude in the direction of the vector **a** is called a *unit vector* and denoted **â**.

Two vectors are equal if and only if they have the same magnitude and direction.

A vector must also satisfy the laws of vector algebra given below.

EXAMPLE 17.2.1.   Velocity **v** and force **F** are vectors. If a particle is moved from point O to point P then its displacement is the vector $\overrightarrow{OP}$.

EXAMPLE 17.2.2.   *Ox, Oy, Oz* are co-ordinate axes of a cartesian co-ordinate system with origin O [see V, § 2.1.4]. P is a point in space. The vector $\overrightarrow{OP}$ is called *the position vector of P*, usually denoted by $\overrightarrow{OP} = \mathbf{r}$.

17.2.2. Vector Algebra

(i) *Addition of vectors.*   Vector addition is defined by the *parallelogram rule*. If vectors **a** and **b** lie along two sides of a parallelogram as shown in Figure 17.2.2a, then $\mathbf{c} = \mathbf{a} + \mathbf{b}$ is the vector represented by the diagonal. In

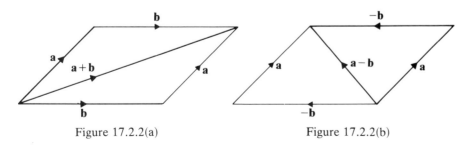

Figure 17.2.2(a)   Figure 17.2.2(b)

terms of displacements we can write $\overrightarrow{OQ} = \overrightarrow{OP} + \overrightarrow{PQ}$. It follows from the definition that

$$\mathbf{a} + \mathbf{b} = \mathbf{b} + \mathbf{a} \qquad (17.2.1)$$

and it can be shown that

$$\mathbf{a} + (\mathbf{b} + \mathbf{c}) = (\mathbf{a} + \mathbf{b}) + \mathbf{c}. \qquad (17.2.2)$$

(This definition corresponds to our notion of the addition of displacements and velocities. It follows from Newton's second law that forces add in this way.)

The negative of a vector $\mathbf{a}$ is denoted by $-\mathbf{a}$ and is the vector equal in magnitude to $\mathbf{a}$ but in the opposite direction (Figure 17.2.1). The difference of two vectors $\mathbf{a}$ and $\mathbf{b}$ can now be defined as

$$\mathbf{a} - \mathbf{b} = \mathbf{a} + (-\mathbf{b}). \quad \text{(Figure 17.2.2 b)} \qquad (17.2.3)$$

(ii) *Multiplication of a vector by a scalar.* If s is a positive scalar $s\mathbf{a}$ denotes the vector in the direction of $\mathbf{a}$ and with magnitude equal to s times the magnitude of $\mathbf{a}$ (Figure 17.2.1). The product of $\mathbf{a}$ and the negative scalar $(-s)$ is equal in magnitude but opposite in direction to $s\mathbf{a}$. From this definition it follows that, for any scalars s and t,

$$s(t\mathbf{a}) = st\mathbf{a}, \qquad (17.2.4)$$

$$(s + t)\mathbf{a} = s\mathbf{a} + t\mathbf{a}, \qquad (17.2.5)$$

and

$$s(\mathbf{a} + \mathbf{b}) = s\mathbf{a} + s\mathbf{b}. \qquad (17.2.6)$$

(iii) *The magnitude of a vector.* The magnitude of $\mathbf{a}$ is denoted by $|\mathbf{a}|$. If $\hat{\mathbf{a}}$ is a unit vector in the direction of $\mathbf{a}$ then we can write

$$\mathbf{a} = |\mathbf{a}|\hat{\mathbf{a}}. \qquad (17.2.7)$$

Note that

$$|\mathbf{a} + \mathbf{b}| \le |\mathbf{a}| + |\mathbf{b}|, \qquad (17.2.8)$$

which states that the length of any side of a triangle is less than or equal to the sum of the lengths of the other two sides.

(iv) *Components of a vector.* Let $\mathbf{i}$, $\mathbf{j}$ and $\mathbf{k}$ be unit vectors along the x, y and z axes of a Cartesian co-ordinate system [see V, § 2.1.4]. P is a point in space with co-ordinates (x, y, z); the position vector of P with respect to the origin O is $\overrightarrow{OP} = \mathbf{r}$. The vector $\overrightarrow{OP}$ can be obtained by adding displacements of magnitude x, y and z in the directions of the vectors $\mathbf{i}$, $\mathbf{j}$ and $\mathbf{k}$ respectively. That is

$$\mathbf{r} = x\mathbf{i} + y\mathbf{j} + z\mathbf{k}. \qquad (17.2.9)$$

The scalars x, y and z are the projections of the vector $\mathbf{r}$ onto Ox, Oy and Oz and are called the *components* of $\mathbf{r}$ with respect to $\mathbf{i}$, $\mathbf{j}$ and $\mathbf{k}$. In fact, if a is any vector with projections a_x, a_y and a_z onto the co-ordinate axes, then

$$\mathbf{a} = a_x\mathbf{i} + a_y\mathbf{j} + a_z\mathbf{k} \tag{17.2.10}$$

and a_x, a_y and a_z are called the components of $\mathbf{a}$ with respect to $\mathbf{i}$, $\mathbf{j}$, $\mathbf{k}$ (Figure 17.2.3).

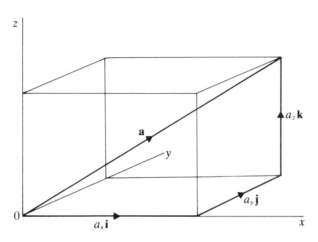

Figure 17.2.3

The vectors $\mathbf{i}$, $\mathbf{j}$ and $\mathbf{k}$ are called a set of *basis vectors*. Any set of non-coplanar vectors can be used as a set of basis vectors in terms of which any other vector can be expressed. If $\mathbf{e}_1$, $\mathbf{e}_2$ and $\mathbf{e}_3$ are such a set, not necessarily at right angles to each other or unit vectors, we can write any vector $\mathbf{a}$ uniquely as

$$\mathbf{a} = a_1\mathbf{e}_1 + a_2\mathbf{e}_2 + a_3\mathbf{e}_3,$$

where the scalars a_1, a_2 and a_3 are called the components of $\mathbf{a}$ with respect to the basis vectors $\mathbf{e}_1$, $\mathbf{e}_2$ and $\mathbf{e}_3$. When there is no ambiguity as to which set of basis vectors is being used, the vector $\mathbf{a}$ is sometimes written merely as the set of its components:

$$\mathbf{a} = (a_x, a_y, a_z). \tag{17.2.11}$$

If

$$\mathbf{a} = a_x\mathbf{i} + a_y\mathbf{j} + a_z\mathbf{k}$$

and

$$\mathbf{b} = b_x\mathbf{i} + b_y\mathbf{j} + b_z\mathbf{k},$$

then

$$\mathbf{a} + \mathbf{b} = (a_x + b_x)\mathbf{i} + (a_y + b_y)\mathbf{j} + (a_z + b_z)\mathbf{k}.$$

That is, vectors are added (or subtracted) by adding (or subtracting) components. Similarly, if s is a scalar

$$sa = sa_x\mathbf{i} + sa_y\mathbf{j} + sa_z\mathbf{k}.$$

That is, scalar multiplication is accomplished by multiplying each component by the scalar.

17.2.3.   Vector Products

Before discussing products of vectors it is necessary to define a *right-handed system* of vectors. The thumb, first finger and second finger of the right hand can be held so that they are at right angles to each other. If the vectors $\mathbf{i}$, $\mathbf{j}$, $\mathbf{k}$, *in that order*, can be aligned with these directions of the thumb, first finger and second finger of the right hand, then $\mathbf{i}$, $\mathbf{j}$, $\mathbf{k}$ is called a right-handed system of vectors (R.H.S.). In a similar way a co-ordinate system with axes Ox, Oy, Oz can be described as right-handed. Only two orientations of the three basis vectors are possible and if a set of vectors is not right-handed it is called a *left-handed system* (L.H.S.). Note that if $\mathbf{i}$, $\mathbf{j}$, $\mathbf{k}$ is a R.H.S. then so are $\mathbf{k}$, $\mathbf{i}$, $\mathbf{j}$ and $\mathbf{j}$, $\mathbf{k}$, $\mathbf{i}$, obtained by permuting the vectors cyclically. All other orders of the vectors produce a L.H.S. In the following $\mathbf{i}$, $\mathbf{j}$, $\mathbf{k}$ will always be taken to be a R.H.S.

(i) *The scalar product.*   Given any two vectors $\mathbf{a}$ and $\mathbf{b}$, it is possible to construct from them a scalar quantity that is useful in many mathematical applications. This quantity is called the *scalar product* of $\mathbf{a}$ and $\mathbf{b}$ and is written $\mathbf{a} \cdot \mathbf{b}$. It is defined to be

$$\mathbf{a} \cdot \mathbf{b} = |\mathbf{a}|\,|\mathbf{b}|\cos\theta \tag{17.2.12}$$

Figure 17.2.4

where θ is the angle between $\mathbf{a}$ and $\mathbf{b}$, $0 \le \theta \le \pi$. (See Figure 17.2.4). It can be shown that

$$\mathbf{a} \cdot \mathbf{b} = \mathbf{b} \cdot \mathbf{a}, \tag{17.2.13}$$

$$\mathbf{a} \cdot (\mathbf{b} + \mathbf{c}) = \mathbf{a} \cdot \mathbf{b} + \mathbf{a} \cdot \mathbf{c} \tag{17.2.14}$$

and that

$$(s\mathbf{a}) \cdot (t\mathbf{b}) = (st)\mathbf{a} \cdot \mathbf{b}. \tag{17.2.15}$$

Note that if $\mathbf{a} \neq \mathbf{0}$ and $\mathbf{b} \neq \mathbf{0}$ then $\mathbf{a} \cdot \mathbf{b} = 0$ implies that $\cos \theta = 0$; that is $\theta = \pi/2$ [see § 2.12] and $\mathbf{a} \perp \mathbf{b}$.

EXAMPLE 17.2.3.   If $\mathbf{F}$ is a force and $\hat{\mathbf{a}}$ a unit vector then $\hat{\mathbf{a}} \cdot \mathbf{F} = |\mathbf{F}| \cos \theta$, where θ is the angle between $\hat{\mathbf{a}}$ and $\mathbf{F}$. That is $\hat{\mathbf{a}} \cdot \mathbf{F}$ is the component of the force $\mathbf{F}$ in the direction $\hat{\mathbf{a}}$.

EXAMPLE 17.2.4.   The quantity $\mathbf{a} \cdot \mathbf{a}$ is often written as $\mathbf{a}^2$ and $|\mathbf{a}|$ is often written as a. Note that

$$\mathbf{a}^2 = a^2 \cos 0 = a^2.$$

EXAMPLE 17.2.5.   The work W done by a force $\mathbf{F}$ moving through a displacement $\mathbf{x}$ is equal to the distance moved times the component of the force in the direction of the displacement. This can be expressed as the scalar product

$$W = \mathbf{F} \cdot \mathbf{x} = |\mathbf{F}||\mathbf{x}| \cos \theta$$

where θ is the angle between $\mathbf{F}$ and $\mathbf{x}$.

(ii) *The vector product.*   Given any two vectors $\mathbf{a}$ and $\mathbf{b}$ it is possible to construct from them a vector quantity that is useful in many mathematical applications. This quantity is called the vector product of $\mathbf{a}$ and $\mathbf{b}$ and is written $\mathbf{a} \wedge \mathbf{b}$ (or $\mathbf{a} \times \mathbf{b}$). It is defined by

$$\mathbf{a} \wedge \mathbf{b} = |\mathbf{a}||\mathbf{b}| \sin \theta \, \mathbf{n}, \qquad (17.2.16)$$

where θ is defined as above and $\mathbf{n}$ is a unit vector perpendicular to both $\mathbf{a}$ and $\mathbf{b}$ and such that $\mathbf{a}, \mathbf{b}, \mathbf{n}$ is a R.H.S. It can be shown that

$$\mathbf{a} \wedge \mathbf{b} = -\mathbf{b} \wedge \mathbf{a} \qquad (17.2.17)$$

$$\mathbf{a} \wedge (\mathbf{b} + \mathbf{c}) = \mathbf{a} \wedge \mathbf{b} + \mathbf{a} \wedge \mathbf{c} \qquad (17.2.18)$$

and

$$(s\mathbf{a}) \wedge (t\mathbf{b}) = (st)\mathbf{a} \wedge \mathbf{b}. \qquad (17.2.19)$$

Note that if $a \neq 0$ and $b \neq 0$ then $\mathbf{a} \wedge \mathbf{b} = \mathbf{0}$ implies that $\sin \theta = 0$; that is $\theta = 0$ and $\mathbf{a} \| \mathbf{b}$. It also follows that for any vector $\mathbf{a}$

$$\mathbf{a} \wedge \mathbf{a} = \mathbf{0}. \qquad (17.2.20)$$

EXAMPLE 17.2.6.   If $\mathbf{F}$ is a force acting through a point P with position vector $\mathbf{r}$ with respect to an origin O, then the *moment of* $\mathbf{F}$ about O is equal to the magnitude of $\mathbf{F}$ multiplied by the perpendicular distance from O to the line through which the force acts. The moment of $\mathbf{F}$ about O can be expressed in terms of the vector product as

$$\mathbf{G} = \mathbf{r} \wedge \mathbf{F} = |\mathbf{F}| r \cos \theta \, \mathbf{n}$$

where **n** is a unit vector perpendicular to **r** and **F** and such that **r**, **F** and **n** is a R.H.S.

EXAMPLE 17.2.7. A particle with position vector **r** rotates about a line with angular speed ω (Figure 17.2.5). We can define an angular velocity vector

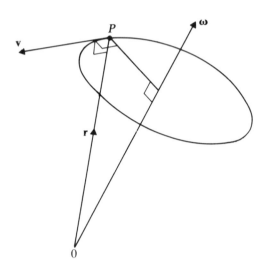

Figure 17.2.5

ω as the vector of magnitude ω with the orientation of the given line. The direction of ω along this line is determined from the sense of rotation by the corkscrew rule. (Rotate a corkscrew in the specified direction and ω points in the direction that the corkscrew would penetrate.) We can now write the velocity of the particle as $\mathbf{v} = \omega \wedge \mathbf{r}$. Since velocities are vectors it follows that angular velocities can be added like vectors.

EXAMPLE 17.2.8. The vector product of **a** and **b** is equal in magnitude to the area A of the parallelogram with sides **a** and **b**;

$$\mathbf{a} \wedge \mathbf{b} = A\mathbf{n},$$

where **n** is a unit vector such that **a**, **b**, **n** is a R.H.S. This quantity is called the *vector area* of the figure.

(iii) *Scalar and vector product in component form.* Using the definitions of scalar and vector products one can deduce the values of products of the basis vectors **i**, **j** and **k**:

$$\mathbf{i}.\mathbf{i} = \mathbf{j}.\mathbf{j} = \mathbf{k}.\mathbf{k} = 1, \qquad \mathbf{i}.\mathbf{j} = \mathbf{j}.\mathbf{k} = \mathbf{k}.\mathbf{i} = 0,$$

$$\mathbf{i} \wedge \mathbf{i} = \mathbf{j} \wedge \mathbf{j} = \mathbf{k} \wedge \mathbf{k} = 0, \qquad \mathbf{i} \wedge \mathbf{j} = \mathbf{k}, \qquad \mathbf{k} \wedge \mathbf{i} = \mathbf{j}, \qquad \mathbf{j} \wedge \mathbf{k} = \mathbf{i}.$$

Then

$$\mathbf{a}.\mathbf{b} = (a_x\mathbf{i} + a_y\mathbf{j} + a_z\mathbf{k}).(b_x\mathbf{i} + b_y\mathbf{j} + b_z\mathbf{k})$$
$$= a_xb_x\mathbf{i}.\mathbf{i} + a_xb_y\mathbf{i}.\mathbf{j}... + a_zb_z\mathbf{k}.\mathbf{k}$$
$$= a_xb_x + a_yb_y + a_zb_z. \tag{17.2.21}$$

Similarly

$$\mathbf{a} \wedge \mathbf{b} = (a_x\mathbf{i} + a_y\mathbf{j} + a_z\mathbf{k}) \wedge (b_x\mathbf{i} + b_y\mathbf{j} + b_z\mathbf{k})$$
$$= a_xb_x\mathbf{i} \wedge \mathbf{i} + a_xb_y\mathbf{i} \wedge \mathbf{j}... + a_zb_z\mathbf{k} \wedge \mathbf{k}$$
$$= (a_yb_z - a_zb_y)\mathbf{i} + (a_zb_x - a_xb_z)\mathbf{j} + (a_xb_y - a_yb_x)\mathbf{k}. \tag{17.2.22}$$

A convenient way of remembering this formula is to write

$$\mathbf{a} \wedge \mathbf{b} = \begin{vmatrix} \mathbf{i} & \mathbf{j} & \mathbf{k} \\ a_x & a_y & a_z \\ b_x & b_y & b_z \end{vmatrix} \tag{17.2.23}$$

and expand the right-hand side as one would a determinant [see I, § 6.9].

EXAMPLE 17.2.9.   If $\mathbf{a} = (2, 1, 3)$, $\mathbf{b} = (5, 1, 4)$ then

$$\mathbf{a}.\mathbf{b} = 2.5 + 1.1 + 3.4 = 23$$

and

$$\mathbf{a} \wedge \mathbf{b} = \begin{vmatrix} \mathbf{i} & \mathbf{j} & \mathbf{k} \\ 2 & 1 & 3 \\ 5 & 1 & 4 \end{vmatrix} = \mathbf{i} + 7\mathbf{j} - 3\mathbf{k}.$$

17.2.4.   Products of Three Vectors

(i) *The scalar triple product*

$$\mathbf{a}.(\mathbf{b} \wedge \mathbf{c}) = (a_x\mathbf{i} + a_y\mathbf{j} + a_z\mathbf{k}).\begin{vmatrix} \mathbf{i} & \mathbf{j} & \mathbf{k} \\ b_x & b_y & b_z \\ c_x & c_y & c_z \end{vmatrix}$$

$$= \begin{vmatrix} a_x & a_y & a_z \\ b_x & b_y & b_z \\ c_x & c_y & c_z \end{vmatrix}. \tag{17.2.24}$$

Using the property of determinants of odd order that the value is unchanged if the rows are permuted cyclically [see I, § 6.10(III)] it follows that

$$\mathbf{a}.(\mathbf{b} \wedge \mathbf{c}) = \mathbf{c}.(\mathbf{a} \wedge \mathbf{b}) = \mathbf{b}.(\mathbf{c} \wedge \mathbf{a}).$$

Also since the scalar product is symmetric [see (17.2.13)]

$$\mathbf{a} \cdot (\mathbf{b} \wedge \mathbf{c}) = (\mathbf{b} \wedge \mathbf{c}) \cdot \mathbf{a} \quad \text{etc.}$$

In fact, in the scalar triple product, the product symbols can be interchanged and the vectors permuted cyclically without altering the value.

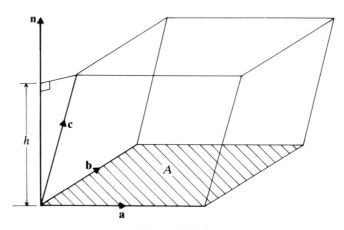

Figure 17.2.6

If $\mathbf{a}, \mathbf{b}$ and $\mathbf{c}$ form three sides of a parallelepiped (Figure 17.2.6) then a simple geometrical interpretation can be given for the scalar triple product. The vector product $\mathbf{a} \wedge \mathbf{b}$ represents the vector area of the base, $A\mathbf{n}$ [see Example 17.2.8]. The scalar product $\mathbf{n} \cdot \mathbf{c}$ is the projection of $\mathbf{c}$ onto the vector $\mathbf{n}$ and so equals $\pm h$ where h is the perpendicular height of the parallelepiped. As a result $(\mathbf{a} \wedge \mathbf{b}) \cdot \mathbf{c}$ equals $\pm V$, where V is the volume of the parallelepiped, the plus sign applying when $\mathbf{a}, \mathbf{b}, \mathbf{c}$ is a R.H.S., the minus sign applying otherwise. Note that $(\mathbf{a} \wedge \mathbf{b}) \cdot \mathbf{c} = 0$ if the three vectors are coplanar.

(ii) *The vector triple product.*   Given three vectors $\mathbf{a}, \mathbf{b}$ and $\mathbf{c}$ then the vector triple product $\mathbf{a} \wedge (\mathbf{b} \wedge \mathbf{c})$ is well defined. It can be checked, using the formulae for vector and scalar products in component form, that

$$\mathbf{a} \wedge (\mathbf{b} \wedge \mathbf{c}) = (\mathbf{a} \cdot \mathbf{c})\mathbf{b} - (\mathbf{a} \cdot \mathbf{b})\mathbf{c}. \tag{17.2.25}$$

Note that, in writing the vector triple product, the position of the brackets must be specified. In general

$$\mathbf{a} \wedge (\mathbf{b} \wedge \mathbf{c}) \neq (\mathbf{a} \wedge \mathbf{b}) \wedge \mathbf{c}.$$

17.2.5.   Vector Calculus

(i) *Differentiation of vectors.*   If $\mathbf{r}$ is the position vector of a point P and if P traces out a path in space as time t varies, then the co-ordinates of the point vary with time and, by (17.2.9), so does $\mathbf{r}$. It is, therefore, possible to define the derivative of the vector $\mathbf{r}(t)$ with respect to the variable t [cf. Definition 3.1.1]:

$$\frac{d\mathbf{r}}{dt} = \lim_{\delta t \to 0} \frac{\mathbf{r}(t+\delta t) - \mathbf{r}(t)}{\delta t} = \lim_{\delta t \to 0} \frac{\delta \mathbf{r}}{\delta t} \qquad \text{(Figure 17.2.7)}.$$

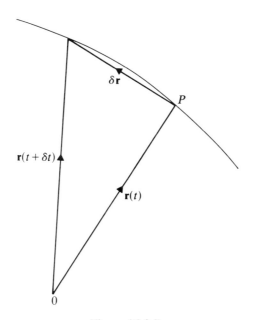

Figure 17.2.7

Note that, in the limit as $\delta t \to 0$, $\delta \mathbf{r}$ is tangent to the path followed by P and $|\delta \mathbf{r}|$ is the distance travelled by P in time δt [see § 3.1.1]. It follows that $d\mathbf{r}/dt$ is in the *direction of motion* of P and its magnitude is equal to the *speed* of P. That is $d\mathbf{r}/dt = \mathbf{v}$, the velocity of the point P. In terms of components, since $\mathbf{r} = x\mathbf{i} + y\mathbf{j} + z\mathbf{k}$ and since the basis vectors do not vary along the path (having the same magnitude and direction at each point of space and time) one obtains

$$\frac{d\mathbf{r}}{dt} = \dot{\mathbf{r}} = \dot{x}\mathbf{i} + \dot{y}\mathbf{j} + \dot{z}\mathbf{k}, \qquad (17.2.26)$$

where the dot above a quantity denotes differentiation with respect to t.

It then follows, from a consideration of vectors in component form, that

$$\frac{d}{dt}(f\mathbf{a}) = \dot{f}\mathbf{a} + f\dot{\mathbf{a}}, \qquad (17.2.27)$$

where f is a scalar function of t,

$$\frac{d}{dt}(\mathbf{a} \cdot \mathbf{b}) = \dot{\mathbf{a}} \cdot \mathbf{b} + \mathbf{a} \cdot \dot{\mathbf{b}}, \qquad (17.2.28)$$

and

$$\frac{d}{dt}(\mathbf{a} \wedge \mathbf{b}) = \dot{\mathbf{a}} \wedge \mathbf{b} + \mathbf{a} \wedge \dot{\mathbf{b}}. \tag{17.2.29}$$

In the above we have considered the vector $\mathbf{r}(t)$. In a similar way, if a vector $\mathbf{a}$ varies with some parameter s, then $\mathbf{a} = \mathbf{a}(s)$ and quantities like $d\mathbf{a}/ds$ can be defined.

EXAMPLE 17.2.10. If $\mathbf{a}$ is a vector of constant length then $\mathbf{a} \cdot \mathbf{a} = a^2 = $ constant. Differentiating with respect to time one obtains $\mathbf{a} \cdot \dot{\mathbf{a}} = 0$. That is either $\dot{\mathbf{a}} = 0$ or $\dot{\mathbf{a}}$ is perpendicular to $\mathbf{a}$.

EXAMPLE 17.2.11.   If $\mathbf{a}$ is a vector of constant length rotating with angular velocity $\boldsymbol{\omega}$ about one end then the velocities of the other end is [see Example 17.2.7]

$$\dot{\mathbf{a}} = \boldsymbol{\omega} \wedge \mathbf{a}.$$

EXAMPLE 17.2.12

$$\frac{d}{dt}\mathbf{r}^2 = 2\mathbf{r} \cdot \dot{\mathbf{r}}.$$

$$\frac{d}{dt}\dot{\mathbf{r}}^2 = 2\dot{\mathbf{r}} \cdot \ddot{\mathbf{r}}.$$

$$\frac{d}{dt}(\mathbf{r} \wedge \dot{\mathbf{r}}) = \dot{\mathbf{r}} \wedge \dot{\mathbf{r}} + \mathbf{r} \wedge \ddot{\mathbf{r}} = \mathbf{r} \wedge \ddot{\mathbf{r}},$$

since $\dot{\mathbf{r}} \wedge \dot{\mathbf{r}} = \mathbf{0}$ [see (17.2.20)].

(ii) *Integration of vectors.*   If $\mathbf{a}$ depends upon t then we can define the integral $\int \mathbf{a} \, dt$ to be a vector $\mathbf{b}$ whose derivative is $\mathbf{a}$.

In terms of components

$$\mathbf{b} = \mathbf{i} \int a_x \, dt + \mathbf{j} \int a_y \, dt + \mathbf{k} \int a_z \, dt, \tag{17.2.30}$$

where each integral is just the usual integral of a scalar function [see § 4.1].

17.3.   VECTORIAL MECHANICS

17.3.1.   Velocity and Acceleration

A particle moves in space so that at time t it is at the point P with position vector $\mathbf{r}(t)$. As explained above [see § 17.2.5(i)] the velocity of P is $\mathbf{v} = \dot{\mathbf{r}}$. The derivative of $\mathbf{v}$ with time gives the (vector) acceleration $\mathbf{a} = \dot{\mathbf{v}} = \ddot{\mathbf{r}}$.

In Cartesian co-ordinates the components of velocity and acceleration are

$$\mathbf{v} = (\dot{x}, \dot{y}, \dot{z}) \quad \text{and} \quad \mathbf{a} = (\ddot{x}, \ddot{y}, \ddot{z}). \qquad (17.3.1)$$

In cylindrical co-ordinates (r, θ, z), with unit basis vectors $\mathbf{e}_1, \mathbf{e}_2, \mathbf{e}_3$ in the co-ordinate directions at any point [see § 17.2.2(iv) and V, § 2.1.5] the components of $\mathbf{v}$ and $\mathbf{a}$ can be found as follows. Using the relations:

$$x = r \cos \theta, \qquad y = r \sin \theta,$$

$$\mathbf{i} = \cos \theta \mathbf{e}_1 - \sin \theta \mathbf{e}_2, \qquad \mathbf{j} = \sin \theta \mathbf{e}_1 + \cos \theta \mathbf{e}_2, \qquad \mathbf{k} = \mathbf{e}_3$$

then

$$\dot{\mathbf{r}} = \dot{x}\mathbf{i} + \dot{y}\mathbf{j} + \dot{z}\mathbf{k}$$

$$= (r \cos \theta)(\cos \theta \mathbf{e}_1 - \sin \theta \mathbf{e}_2) + (r \sin \theta)(\sin \theta \mathbf{e}_1 + \cos \theta \mathbf{e}_2) + \dot{z}\mathbf{e}_3$$

$$= \dot{r}\mathbf{e}_1 + r\dot{\theta}\mathbf{e}_2 + \dot{z}\mathbf{e}_3.$$

Differentiating again one can find $\ddot{\mathbf{r}}$, and obtain finally for the r, θ and z components of velocity and acceleration:

and

$$\left.\begin{array}{l} \mathbf{v} = (\dot{r}, r\dot{\theta}, \dot{z}) \\ \mathbf{a} = (\ddot{r} - r\dot{\theta}^2, r\ddot{\theta} + 2\dot{r}\dot{\theta}, \ddot{z}) \end{array}\right\}. \qquad (17.3.2)$$

Similarly, in spherical polar co-ordinates (r, θ, ϕ) with

$$x = r \sin \theta \cos \phi, \qquad y = r \sin \theta \sin \phi, \qquad z = r \cos \theta,$$

[see V, § 2.1.5] the r, θ and ϕ components of velocity and acceleration are found to be

$$\mathbf{v} = (\dot{r}, r\dot{\theta}, r \sin \theta \, \dot{\phi}) \qquad (17.3.3)$$

and

$$\left.\begin{array}{l} a_r = \ddot{r} - r\dot{\theta}^2 - r \sin^2 \theta \, \dot{\phi}^2, \\ a_\theta = r\ddot{\theta} + 2\dot{r}\dot{\theta} - r \sin \theta \cos \theta \, \dot{\phi}^2, \\ a_\phi = r \sin \theta \ddot{\phi} + 2 \sin \theta \dot{r}\dot{\phi} + 2r \cos \theta \, \dot{\theta}\dot{\phi} \end{array}\right\}. \qquad (17.3.4)$$

17.3.2. Rotating Frames of Reference

A *frame of reference* in space is determined when the origin and the co-ordinate axes are specified. Let Ox, Oy and Oz be one set of Cartesian co-ordinate axes that defines a frame of reference [see V, § 2.1.4] with unit vectors $\mathbf{i}$, $\mathbf{j}$ and $\mathbf{k}$ in the co-ordinate directions. Another set of Cartesian co-ordinate axes OX, OY, OZ with the same origin O is rotating with respect to the first frame of reference with angular velocity $\boldsymbol{\omega}$. If $\mathbf{i}'$, $\mathbf{j}'$ and $\mathbf{k}'$ are unit vectors along OX, OY and OZ, then we can express a vector $\mathbf{a}$ in terms of either set of basis vectors [see § 17.2.2(iv)]:

$$\mathbf{a} = a_x\mathbf{i} + a_y\mathbf{j} + a_z\mathbf{k} = a_X\mathbf{i}' + a_Y\mathbf{j}' + a_Z\mathbf{k}'.$$

However, the time derivative of $\mathbf{a}$ [see § 17.2.5(i)] will be different for observers fixed in the different reference frames. Writing d/dt and $\delta/\delta t$ to be the rate of change with time in the first and second frames of reference respectively, then

$$\frac{d\mathbf{a}}{dt} = \frac{d}{dt}(a_X\mathbf{i}' + a_Y\mathbf{j}' + a_Z\mathbf{k}')$$

$$= (\dot{a}_X\mathbf{i}' + \dot{a}_Y\mathbf{j}' + \dot{a}_Z\mathbf{k}') + \left(a_X\frac{d\mathbf{i}'}{dt} + a_Y\frac{d\mathbf{j}'}{dt} + a_Z\frac{d\mathbf{k}'}{dt}\right).$$

Now $d\mathbf{i}'/dt = \boldsymbol{\omega} \wedge \mathbf{i}'$, $d\mathbf{j}'/dt = \boldsymbol{\omega} \wedge \mathbf{j}'$ etc., [see Example 17.2.7] therefore

$$\frac{d\mathbf{a}}{dt} = \frac{\delta\mathbf{a}}{\delta t} + \boldsymbol{\omega} \wedge \mathbf{a}. \qquad (17.3.5)$$

That is

$$\frac{d}{dt} = \frac{\delta}{\delta t} + \boldsymbol{\omega} \wedge \quad . \qquad (17.3.6)$$

(This relation between the two frames is symmetric since the angular velocity of the first with respect to the second is $-\boldsymbol{\omega}$.)

By substituting $\mathbf{r}$ for $\mathbf{a}$ in the above result, expressions can be obtained for the velocity and acceleration in the two frames:

$$\frac{d\mathbf{r}}{dt} = \frac{\delta\mathbf{r}}{\delta t} + \boldsymbol{\omega} \wedge \mathbf{r}; \qquad (17.3.7)$$

$$\frac{d^2\mathbf{r}}{dt^2} = \left(\frac{\delta}{\delta t} + \boldsymbol{\omega} \wedge \right)\left(\frac{\delta\mathbf{r}}{\delta t} + \boldsymbol{\omega} \wedge \mathbf{r}\right)$$

$$= \frac{\delta^2\mathbf{r}}{\delta t^2} + 2\boldsymbol{\omega} \wedge \frac{\delta\mathbf{r}}{\delta t} + \boldsymbol{\omega} \wedge (\boldsymbol{\omega} \wedge \mathbf{r}) + \frac{\delta\boldsymbol{\omega}}{\delta t} \wedge \mathbf{r}. \qquad (17.3.8)$$

The first term on the right-hand side of (17.3.8) is the acceleration of the point P with respect to the second reference frame. The second term is called the *Coriolis acceleration*, the third term the *centripetal acceleration* and the last term involves the rate of change of $\boldsymbol{\omega}$ with respect to the second frame [see also (17.3.27)].

Furthermore, if the second frame is moving so that its origin has position vector $\mathbf{s}$ with respect to the origin of the first frame, then quantities $d\mathbf{s}/dt$ and $d^2\mathbf{s}/dt^2$ must be added to the right-hand sides of the formulae (17.3.7) and (17.3.8) respectively.

17.3.3. Motion of a Single Particle

The *linear momentum* $\mathbf{p}$ of a particle of mass m is defined as

$$\mathbf{p} = m\mathbf{v}. \qquad (17.3.9)$$

Newton's second law of motion states that

$$\frac{d}{dt}\mathbf{p} = \mathbf{F} \qquad\qquad (17.3.10)$$

where $\mathbf{F}$ is the total force acting on the particle. For a particle of constant mass the law takes the form

$$m\ddot{\mathbf{r}} = \mathbf{F}. \qquad\qquad (17.3.11)$$

When $\mathbf{F} = \mathbf{0}$ the equation $\ddot{\mathbf{r}} = \mathbf{0}$ can be integrated once to give $\mathbf{v} = $ constant. This result, that in the absence of forces a particle moves with constant velocity, is Newton's first law of motion. The reader may well be puzzled at this point since it appears that one fundamental law has been deduced from another. The explanation is that Newton's laws of motion only apply in special reference frames, the so called 'inertial reference frames' in which bodies in the absence of forces move with constant velocity. For all practical purposes such inertial frames are those in which the distant stars appear to be fixed. In a rotating frame Newton's laws do not apply and extra terms (like the Coriolis acceleration) appear in the equations. Newton's first law is, therefore, not redundant but can be regarded as a statement about inertial frames.

For some forces $\mathbf{F}$ equation (17.3.11) can be solved in terms of known functions. Most cases are not soluble in this way but it is possible to obtain some understanding of the particle motion by finding quantities that remain constant as the particle moves along its trajectory. These are known by a variety of terms: constants or integrals of the motion, invariants, conserved quantities.

The *angular momentum* $\mathbf{h}$ of a particle about the origin is defined to be:

$$\mathbf{h} = \mathbf{r} \wedge \mathbf{p} = m\mathbf{r} \wedge \dot{\mathbf{r}}. \qquad\qquad (17.3.12)$$

(If $\mathbf{r}'$ is the position vector of the particle with respect to a point A, then $\mathbf{r}' \wedge \mathbf{p}$ is the angular momentum about A.) Taking the vector product of (17.3.11) with $\mathbf{r}$ leads to the equation for $\mathbf{h}$:

$$\mathbf{h} = \mathbf{G}, \qquad\qquad (17.3.13)$$

where $\mathbf{G} = \mathbf{r} \wedge \mathbf{F}$ is the moment of $\mathbf{F}$ about O.

EXAMPLE 17.3.1.   *Constant gravitational force,* $\mathbf{F} = -\mathbf{g}$. The equation of motion can be integrated to give

$$\mathbf{r} = -\tfrac{1}{2}\mathbf{g}t^2 + \mathbf{u}t + \mathbf{r}_0$$

as the position at time t of the particle that at time $t = 0$ was at $\mathbf{r}_0$ with velocity $\mathbf{u}$.

EXAMPLE 17.3.2.   *Motion under a central force,* $\mathbf{F} = f(\mathbf{r}, t)\mathbf{r}$. That is, the force is directed along the radius vector. The equation of motion

$$m\ddot{\mathbf{r}} = f\mathbf{r}$$

is not, in general, soluble in terms of simple functions. Because $\mathbf{G} = \mathbf{r} \wedge \mathbf{F} = \mathbf{0}$, $\dot{\mathbf{h}} = \mathbf{0}$ and the angular momentum about the origin is a (vector) constant of the motion (the law of conservation of angular momentum for motion under a central force).

EXAMPLE 17.3.3.   *Force* $\mathbf{F}$ *derivable from a potential.* If $V(\mathbf{r})$ is a scalar function of position then the vector with x, y and z components $(\partial V/\partial x, \partial V/\partial y, \partial V/\partial z)$ is called the *gradient* of V and written grad V or ∇V, where

$$\nabla = \mathbf{i}\frac{\partial}{\partial x} + \mathbf{j}\frac{\partial}{\partial y} + \mathbf{k}\frac{\partial}{\partial z} \qquad (17.3.14)$$

[see (5.10.27)]. When the force $\mathbf{F} = -\nabla V$ for some function V then V is called a *potential function* and $\mathbf{F}$ is said to be derived from a potential. Since the work done against $\mathbf{F}$ in a displacement $d\mathbf{r}$ is

$$-\mathbf{F} \cdot d\mathbf{r} = \left(\frac{\partial V}{\partial x}dx + \frac{\partial V}{\partial y}dy + \frac{\partial V}{\partial z}dz\right) = dV, \qquad (17.3.15)$$

the work done in moving a particle to any point $\mathbf{r}$ is, to within a constant, equal to $V(\mathbf{r})$. The particle is then said to have *potential energy* V. If V is a function of position only and not of time then the force $\mathbf{F} = -\nabla V$ is called *conservative* as the work done against it in completing a closed circuit is zero.

The equation of motion is

$$m\ddot{\mathbf{r}} = -\nabla V.$$

The scalar product of this equation with $\dot{\mathbf{r}}$ gives

$$m\ddot{\mathbf{r}} \cdot \dot{\mathbf{r}} = -\nabla V \cdot \dot{\mathbf{r}},$$

the left-hand side of which is $(d/dt)(\tfrac{1}{2}m\dot{\mathbf{r}}^2)$ and the right-hand side is $-dV/dt$. Therefore

$$\frac{d}{dt}(\tfrac{1}{2}m\mathbf{v}^2 + V) = 0. \qquad (17.3.16)$$

The quantity $T = \tfrac{1}{2}m\mathbf{v}^2$ is the *kinetic energy* of the particle. The *total* energy is $E = T + V$. Equation (17.3.16) states that E is a constant of the motion, (the law of conservation of energy for a particle under a conservative force).

17.3.4.   Motion of a System of Particles

The equations of motion for a system of n particles with masses m_i and position vectors $\mathbf{r}_i$ are

$$m_i\ddot{\mathbf{r}}_i = \mathbf{F}_i, \qquad i = 1, 2, \ldots, n. \qquad (17.3.17)$$

The vector $\mathbf{F}_i$ is the force acting on the ith particle and is made up of forces of interaction due to the other particles plus external forces, $\mathbf{F}_i'$. In this section it will be assumed that the particles interact through two-body forces; the force on the ith particle due to interaction with the jth particle is written $\mathbf{F}_{ij}$. It will also be assumed in this section that the forces of interaction satisfy $\mathbf{F}_{ij} = -\mathbf{F}_{ji}$ and act along the line of centres of the two particles. (Note that $\mathbf{F}_{ii} = 0$.) Equation (17.3.17) can then be written as

$$m_i \ddot{\mathbf{r}}_i = \mathbf{F}'_i + \sum_{j=1}^{n} \mathbf{F}_{ij}, \qquad i = 1, 2, \ldots, n. \tag{17.3.18}$$

The *centre of mass* of a system of particles is the point with position vector

$$\mathbf{R} = \left(\sum_{i=1}^{n} m_i \mathbf{r}_i \right) \Big/ m \tag{17.3.19}$$

where the total mass is $m = \sum_{i=1}^{n} m_i$.

The total linear momentum of the system is [compare (17.3.9)]

$$\mathbf{P} = \sum_{i=1}^{n} m_i \dot{\mathbf{r}}_i. \tag{17.3.20}$$

Adding together the n equations (17.3.18) gives

$$\frac{d\mathbf{P}}{dt} = m\ddot{\mathbf{R}} = \sum_{i=1}^{n} \left(\mathbf{F}'_i + \sum_{j=1}^{n} \mathbf{F}_{ij} \right) = \sum_{i=1}^{n} \mathbf{F}'_i,$$

since the interaction forces cancel. Thus

$$\frac{d\mathbf{P}}{dt} = m\ddot{\mathbf{R}} = \sum_{i=1}^{n} \mathbf{F}'_i = \mathbf{F}', \tag{17.3.21}$$

where $\mathbf{F}'$ is the total of the external forces.

We now consider the total angular momentum $\mathbf{H}$ of the system of particles about the origin [compare (17.3.12)]:

$$\mathbf{H} = \sum_{i=1}^{n} m_i \mathbf{r}_i \wedge \dot{\mathbf{r}}_i \tag{17.3.22}$$

and therefore

$$\frac{d\mathbf{H}}{dt} = \sum_{i=1}^{n} m_i \mathbf{r}_i \wedge \ddot{\mathbf{r}}_i = \sum_{i=1}^{n} \sum_{j=1}^{n} \mathbf{r}_i \wedge \mathbf{F}_{ij} + \sum_{i=1}^{n} \mathbf{r}_i \wedge \mathbf{F}'_i. \tag{17.3.23}$$

Consider the forces of interaction between the particles labelled 1 and 2. The contribution they make to (17.3.23) is

$$\mathbf{r}_1 \wedge \mathbf{F}_{12} + \mathbf{r}_2 \wedge \mathbf{F}_{21} = (\mathbf{r}_1 - \mathbf{r}_2) \wedge \mathbf{F}_{12}.$$

This will vanish [see § 17.2.3(ii)] since the force of interaction $\mathbf{F}_{12}$ is along the line of centres and so is parallel to $(\mathbf{r}_1 - \mathbf{r}_2)$. The double sum on the right of (7.3.23) vanishes and

$$\frac{d\mathbf{H}}{dt} = \sum_{i=1}^{n} \mathbf{r}_i \wedge \mathbf{F}'_i = \mathbf{G}, \tag{17.3.24}$$

where **G** is the total moment of the external forces about the origin [compare Example 17.2.6].

EXAMPLE 17.3.4.   The sum of the external forces is zero, $\mathbf{F}' = 0$. From (17.3.21) we deduce that **P** is a constant of the motion. It is also evident that $\dot{\mathbf{R}}$ is a constant so that the centre of mass of the system moves uniformly.

In these circumstances and with the assumptions made at the beginning of this section it also follows that $\mathbf{G} = 0$ and **H** is a constant.

17.3.5.   Motion in a Rotating Frame of Reference

The equation of motion of mass m in a frame of reference rotating with respect to an inertial frame [see § 17.3.2] is,

$$m\ddot{\mathbf{r}} + 2m\boldsymbol{\omega} \wedge \dot{\mathbf{r}} + m\boldsymbol{\omega} \wedge (\boldsymbol{\omega} \wedge \mathbf{r}) + m\dot{\boldsymbol{\omega}} \wedge \mathbf{r} = \mathbf{F}, \qquad (17.3.25)$$

where dots denote rate of change with time measured with respect to the moving frame. This can be written

$$m\ddot{\mathbf{r}} = \mathbf{F}', \qquad (17.3.26)$$

where

$$\mathbf{F}' = \mathbf{F} - 2m\boldsymbol{\omega} \wedge \dot{\mathbf{r}} - m\boldsymbol{\omega} \wedge (\boldsymbol{\omega} \wedge \mathbf{r}) - m\dot{\boldsymbol{\omega}} \wedge \mathbf{r}. \qquad (17.3.27)$$

Equation (17.3.26) has the appearance of the equation of motion with respect to an inertial frame but with extra terms included in the force. The second and third terms in the expression for **F**' are called the *Coriolis force* and *centrifugal force* [compare (17.3.8)].

If the position vector of the origin of the second frame with respect to the origin of the first is **s** and **s** changes with time, then a term $-\ddot{\mathbf{s}}$ must also be included in the expression for **F**'.

EXAMPLE 17.3.5. O is the origin of an inertial frame. A sphere centred at O rotates with constant angular velocity $\boldsymbol{\omega}$. A second frame is fixed on the surface of the sphere with its origin O' having position vector **s** with respect to O. Find the equation of motion referred to the frame fixed in the sphere.

The term $\ddot{\mathbf{s}} = \boldsymbol{\omega} \wedge (\boldsymbol{\omega} \wedge \mathbf{s})$ and $\dot{\boldsymbol{\omega}} = 0$ so the equation of motion is

$$m\ddot{\mathbf{r}} = \mathbf{F} - 2m\boldsymbol{\omega} \wedge \dot{\mathbf{r}} - m\boldsymbol{\omega} \wedge (\boldsymbol{\omega} \wedge \mathbf{r}) - m\boldsymbol{\omega} \wedge (\boldsymbol{\omega} \wedge \mathbf{s}).$$

If the terms involving ω^2 can be neglected in comparison with the other terms one obtains

$$m\ddot{\mathbf{r}} = \mathbf{F} - 2m\boldsymbol{\omega} \wedge \dot{\mathbf{r}}.$$

Solutions of this equation give good approximations to the motion of projectiles and pendulums with respect to the rotating earth.

17.4. ANALYTICAL MECHANICS

17.4.1. Constraints

The equations of motion for a system of n particles, in the language of vectorial mechanics [see § 17.3.4] are:

$$m_i \ddot{\mathbf{r}}_i = \mathbf{F}_i, \qquad i = 1, 2, \ldots, n. \tag{17.4.1}$$

These are $3n$ equations for the $3n$ co-ordinates of the particles. However, there may exist constraints on the possible configurations of the systems. For example, a particle moving under gravity in a smooth, spherical bowl of radius a has the values of its co-ordinates (x, y, z) related by $x^2 + y^2 + z^2 = a^2$; the co-ordinates cannot be chosen independently. Note also that, besides reducing the number of independent co-ordinates, the constraints introduce extra forces (so called *forces of constraint*). In the example quoted, the normal reaction of the sphere on the particle must be included among the forces on the particle. This force is not known *a priori* and finding its magnitude is part of the problem.

The simplest form of constraint is one which gives a relation between the particle co-ordinates of the form:

$$f(\mathbf{r}_1, \mathbf{r}_2, \ldots, \mathbf{r}_n) = 0.$$

Such constraints are called *holonomic*. (Although not written explicitly in the above formula, the function f can also depend on time t.) A more general kind that involves not only co-ordinates $\mathbf{r}_i$ but velocities $\dot{\mathbf{r}}_i$ is called non-holonomic. In the following only holonomic constraints are considered.

If k independent holonomic constraints are specified,

$$f_m(\mathbf{r}_1, \mathbf{r}_2, \ldots, \mathbf{r}_n) = 0, \qquad m = 1, 2, \ldots, k,$$

then it is possible to choose new coordinates $(q_1, q_2, \ldots, q_{3n})$ to describe the position of the system with the last k of these co-ordinates chosen to be of the form

$$q_{3n-m+1} = f_m, \qquad m = 1, 2, \ldots, k.$$

The imposition of the constraints then merely involves setting the last k co-ordinates equal to zero. The position of the particles can be written in terms of the remaining $N = 3n - k$ co-ordinates:

$$\mathbf{r}_i = \mathbf{r}_i(q_1, q_2, \ldots, q_N). \tag{17.4.2}$$

These co-ordinates $(q_1, q_2, \ldots, q_N)$ are called *generalized co-ordinates*, the space spanned by them is called *configuration space*, and N is called the number of *degrees of freedom* of the system. The set of co-ordinates $(q_1, q_2, \ldots, q_N)$ is frequently referred to below by the symbol $\mathbf{q}$. Note that

$$\dot{\mathbf{r}}_i = \sum_{k=1}^{N} \frac{\partial \mathbf{r}_i}{\partial q_k} \dot{q}_k$$

is a function of both the generalized co-ordinates and their time derivatives,
$\dot{\mathbf{r}}_i = \dot{\mathbf{r}}_i(\mathbf{q}, \dot{\mathbf{q}})$.

17.4.2. Lagrange's Equations

Consider a displacement of the system with each particle displaced from $\mathbf{r}_i$
to $\mathbf{r}_i + d\mathbf{r}_i$, the only restriction on the $d\mathbf{r}_i$ being that they are consistent with
the constraints on the system. The work done by the forces in making this
displacement is [compare Example 17.2.5]

$$W = \sum_{i=1}^{n} \mathbf{F}_i \cdot d\mathbf{r}_i, \tag{17.4.3}$$

where $\mathbf{F}_i$ contains all forces on the ith particle, external forces, forces of
interaction and forces of constraint. Using (17.4.1) the work can be written

$$\sum_{i=1}^{n} \mathbf{F}_i \cdot d\mathbf{r}_i = \sum_{i=1}^{n} m_i \ddot{\mathbf{r}}_i \cdot d\mathbf{r}_i. \tag{17.4.4}$$

The extra assumption will now be made that the forces of constraint are such
that they do no work in a displacement consistent with the constraints. (This
is true for forces of constraint that are perpendicular to the particle displacements, e.g. the reactions at smooth surfaces and pivots. The main forces of
reaction that do work in such a displacement are those involving friction.)
With this assumption the forces of constraint therefore disappear from
(17.4.4).

Introducing generalized co-ordinates, the right-hand side of (17.4.4) can,
after some manipulation, be written as [see § 5.3]

$$\sum_{k=1}^{N} \left\{ \frac{d}{dt} \left(\frac{\partial T}{\partial \dot{q}_k} \right) - \frac{\partial T}{\partial q_k} \right\} dq_k, \tag{17.4.5}$$

where T is the kinetic energy [see § 17.3.3] expressed in terms of the generalized co-ordinates $\mathbf{q}$ and velocities $\dot{\mathbf{q}}$. The left-hand side of (17.4.4) can be
written

$$\sum_{k=1}^{N} \sum_{i=1}^{n} \mathbf{F}_i \cdot \frac{\partial \mathbf{r}_i}{\partial q_k} dq_k = \sum_{k=1}^{N} Q_k \, dq_k, \tag{17.4.6}$$

where Q_k is called the kth *generalized force component*. Equating (17.4.5) and
(17.4.6), and noting that the dq_k can be chosen independently of each other,
one obtains as the equations of motion of the system

$$\frac{d}{dt} \left(\frac{\partial T}{\partial \dot{q}_k} \right) - \frac{\partial T}{\partial q_k} = Q_k, \qquad k = 1, 2, \ldots, N. \tag{17.4.7}$$

These are Lagrange's form of the equations of motion. When the forces on
each particle are potential forces ($F_i = -\nabla V_i$) [see Example 17.3.3]) then one

obtains, using (17.3.15)

$$Q_k = -\frac{\partial V}{\partial q_k} \qquad (17.4.8)$$

where $V = \sum_{i=1}^{n} V_i$ is the total potential energy of the system. Introducing the *Lagrangian function* $L = T - V$, Lagrange's equations can be written

$$\frac{d}{dt}\left(\frac{\partial L}{\partial \dot{q}_k}\right) - \frac{\partial L}{\partial \dot{q}_k} = 0, \qquad k = 1, 2, \ldots, N. \qquad (17.4.9)$$

To sum up, provided that a system is subject to holonomic constraints and that the forces of constraint do no work in a displacement consistent with the constraints, then generalized co-ordinates $(q_1, q_2, \ldots, q_N)$ can be introduced such that the equations of motion of the system are given by (17.4.7) or (17.4.9). (If Q_k is of the form

$$Q_k = \frac{d}{dt}\left(\frac{\partial M}{\partial \dot{q}_k}\right) - \frac{\partial M}{\partial \dot{q}_k} \qquad (17.4.10)$$

with M a function of both co-ordinates q_k and velocities $\dot{q}_k$, then one can write $L = T - M$ and the equations of motion are again of the form (17.4.9). Electromagnetic forces on charged particles are of this nature.)

EXAMPLE 17.4.1.   A particle of mass m moves in 3 dimensions under the action of a potential force.

The position of the particle can be specified by its Cartesian co-ordinates (x, y, z). There are no constraints on the system. We take as the generalized co-ordinates $q_1 = x$, $q_2 = y$, $q_3 = z$. The velocity of the particle has components $(\dot{x}, \dot{y}, \dot{z})$. The kinetic energy $T = \frac{1}{2}m(\dot{x}^2 + \dot{y}^2 + \dot{z}^2)$, the potential energy $V = V(x, y, z)$. The Lagrangian function

$$L = T - V = \tfrac{1}{2}m(\dot{x}^2 + \dot{y}^2 + \dot{z}^2) - V(x, y, z).$$

Lagrange's equation of motion for the x-co-ordinate is

$$\frac{d}{dt}\left(\frac{\partial L}{\partial \dot{x}}\right) - \frac{\partial L}{\partial x} = 0$$

that is

$$\frac{d}{dt}(m\dot{x}) + \frac{\partial V}{\partial x} = 0 \quad \text{or} \quad m\ddot{x} = -\frac{\partial V}{\partial x}.$$

There are similar equations for the y and z co-ordinates which, when combined, agree with the corresponding equation of vectorial mechanics

$$m\ddot{\mathbf{r}} = -\nabla V.$$

EXAMPLE 17.4.2.   *Motion of a particle of mass m under the action of a central force* $\mathbf{F} = f(r)\hat{\mathbf{r}}$. There is in this case a potential $V(r) = -\int^r f(r')\, dr'$. Motion is in a plane so we can describe the particle position in terms of plane polar co-ordinates (r, θ) [see V, § 1.2.5]. There are no constraints. We take as generalized co-ordinates $q_1 = r$, $q_2 = \theta$. Then

$$T = \tfrac{1}{2}mv^2 = \tfrac{1}{2}m(\dot{r}^2 + r^2\dot{\theta}^2).$$

The *r*-equation of motion is

$$\frac{d}{dt}\left(\frac{\partial L}{\partial \dot{r}}\right) - \frac{\partial L}{\partial r} = 0,$$

that is

$$m\ddot{r} - mr\dot{\theta}^2 - f = 0,$$

or

$$m(\ddot{r} - r\dot{\theta}^2) = f.$$

The θ-equation is similarly,

$$\frac{d}{dt}(mr^2\dot{\theta}) = mr^2\ddot{\theta} + 2mr\dot{r}\dot{\theta} = 0.$$

Note that it is easier to obtain the equations of motion this way than to calculate the accelerations in polar co-ordinates (17.3.2).

EXAMPLE 17.4.3.   *The spherical pendulum.* (Motion of a particle of mass *m* under gravity in a smooth spherical bowl of radius *a*.) If the position of the particle is given by its Cartesian co-ordinates (x, y, z) then there is a single holonomic constraint between the co-ordinates, $x^2 + y^2 + z^2 - a^2 = 0$. In terms of spherical polar co-ordinates (r, θ, ϕ) [see V, § 2.1.5] the constraint is just $r - a = 0$. The force of constraint is the reaction of the bowl on the particle, normal to the spherical surface; the force of constraint, therefore, does no work during the motion. We introduce $q_1 = \theta$, $q_2 = \phi$, $q_3 = r - a$ and the imposition of the constraint merely involves setting $q_3 = 0$. There are two generalized co-ordinates (q_1, q_2) and, therefore, two degrees of freedom. The kinetic energy $T = \tfrac{1}{2}m(a^2\dot{\theta}^2 + a^2 \sin^2 \theta \dot{\phi}^2)$ and $V = -mga \cos \theta$, for suitably chosen θ. The Lagrangian function can then be written and the θ- and ϕ-equations obtained.

17.4.3.   Hamilton's Principle

Instead of deriving Lagrange's equations from Newton's equations, it is also possible to formulate analytical mechanics in another way. Define the *Action functional S* to be

$$S = \int_{t_0}^{t_1} L(\mathbf{q}, \dot{\mathbf{q}}, t)\, dt, \qquad\qquad (17.4.11)$$

where $\mathbf{q} = \mathbf{q}(t)$ defines a curve in configuration space [see § 17.4.1]. Then Hamilton's principle states that the trajectories of a mechanical system with

Lagrangian L are just the extremals of the functional S. The equations of motion are just the Euler–Lagrange equations of the functional S. [See § 12.4.]

An extremal of S in configuration space is obviously independent of the particular co-ordinates chosen to specify it; any set of generalized co-ordinates would be suitable. The Lagrangian could be expressed, say, in terms of other generalized co-ordinates $\mathbf{Q}$ and velocities $\dot{\mathbf{Q}}$:

$$L(\mathbf{q}, \dot{\mathbf{q}}, t) = \mathscr{L}(\mathbf{Q}, \dot{\mathbf{Q}}, t). \tag{17.4.12}$$

Hamilton's principle then shows that in terms of the new co-ordinates the equations of motion are

$$\frac{d}{dt}\left(\frac{\partial \mathscr{L}}{\partial \dot{Q}_l}\right) - \frac{\partial \mathscr{L}}{\partial Q_l} = 0, \qquad l = 1, 2, \ldots, N. \tag{17.4.13}$$

That is, Lagrange's equations are invariant in form under point transformations:

$$Q_l = Q_l(q_1, q_2, \ldots, q_N), \qquad l = 1, 2, \ldots, N. \tag{17.4.14}$$

EXAMPLE 17.4.4.   *Change of variables.* The Lagrangian for a particle moving under a potential force is in terms of Cartesian components:

$$L = \tfrac{1}{2}m(\dot{x}^2 + \dot{y}^2 + \dot{z}^2) - V(x, y, z)$$

giving as the Lagrange equations:

$$m\ddot{x} = -\partial V/\partial x, \qquad m\ddot{y} = -\partial V/\partial y, \qquad m\ddot{z} = -\partial V/\partial z$$

[see Example 17.4.1]. In spherical polar co-ordinates:

$$L = \tfrac{1}{2}m(\dot{r}^2 + r^2\dot{\theta}^2 + r^2 \sin^2 \theta \, \dot{\phi}^2) - V(r, \theta, \phi)$$

giving as the Lagrange equations:

$$m\ddot{r} - mr\dot{\theta}^2 - mr \sin^2 \theta \, \dot{\phi}^2 = -\partial V/\partial r$$

$$(d/dt)(mr^2\dot{\theta}) - mr^2 \sin \theta \cos \theta \, \dot{\phi}^2 = -\partial V/\partial \theta$$

$$(d/dt)(mr^2 \sin^2 \theta \, \dot{\phi}) = -\partial V/\partial \phi.$$

These equations are equivalent to the equations for (x, y, z) and could have been obtained from them, after a tedious calculation, by changing co-ordinates.

17.4.4.   Ignorable Co-ordinates and Invariants

If the Lagrangian of a system is independent of a generalized co-ordinate q_k, then q_k is said to be *ignorable* and Lagrange's equation for q_k reduces to

$$\frac{d}{dt}\left(\frac{\partial L}{\partial \dot{q}_k}\right) = 0.$$

That is, $\partial L/\partial \dot{q}_k$ is an invariant.

It is also possible to show that if L is independent of explicit dependence on the time t (that is $\partial L/\partial t = 0$), then

$$E = \sum_{k=1}^{N} \frac{\partial L}{\partial \dot{q}_k} \dot{q}_k - L$$

is an invariant known as the *Jacobian integral*.

EXAMPLE 17.4.5.   *Constants of the motion.*
   (i) Motion under the action of a potential force [see Example 17.3.3]. The Lagrangian is independent of t and therefore the Jacobian integral E is a constant of the motion. For *natural* systems (L independent of t, T a homogeneous quadratic function of the $\dot{q}_k$ [see § 5.1] and V a function of $\mathbf{q}$ only) the Jacobian integral is just $T + V$, the total energy.
   (ii) Motion under a central force [see Example 17.3.2]

$$L = \tfrac{1}{2}m(\dot{r}^2 + r^2\dot{\theta}^2) - V(r).$$

The generalized co-ordinate θ is absent from L so $\partial L/\partial\dot{\theta} = mr^2\dot{\theta}$ is a constant along an orbit of the system. (This is the angular momentum about the origin.) Also L is independent of t so $E = T + V$ is a constant.
   (iii) The spherical pendulum [see Example 17.4.3]. The co-ordinate ϕ is ignorable so $\partial L/\partial\dot{\phi}$ is a constant (angular momentum about the vertical axis). L is independent of t so the total energy of the particle remains constant.

17.4.5.   Small Oscillations About a Position of Equilibrium

Lagrangian theory is useful when considering the problem of small oscillations. For a natural system with N degrees of freedom, the kinetic energy can be written

$$T = \tfrac{1}{2} \sum_{i=1}^{N} \sum_{j=1}^{N} g_{ij}\dot{q}_i\dot{q}_j,$$

where the g_{ij} are the elements of a positive-definite, real, symmetric matrix [see I, § 6.7(v)] and are functions of the co-ordinates $\mathbf{q}$. The potential energy $V = V(\mathbf{q})$ is a function only of position. A *point of equilibrium* of the system is one where all the forces vanish and the system remains at rest, that is a point at which $\partial V/\partial\mathbf{q} = 0$. If we take this point to be at the origin, then $\mathbf{q}$ is a measure of the displacement from equilibrium. For small $\mathbf{q}$ the dominant terms in the Taylor expansions of T and V [see § 5.8] give as the Lagrangian of the system:

$$L = \tfrac{1}{2} \sum_{i=1}^{N} \sum_{j=1}^{N} (a_{ij}\dot{q}_i\dot{q}_j - b_{ij}q_iq_j), \tag{17.4.15}$$

where a_{ij} and b_{ij} are the values at $\mathbf{q} = \mathbf{0}$ of g_{ij} and $\partial^2 V/\partial q_i\, \partial q_j$. This can be written in matrix form as

$$L = \tfrac{1}{2}\dot{\mathbf{q}}'\mathbf{A}\dot{\mathbf{q}} - \tfrac{1}{2}\mathbf{q}'\mathbf{B}\mathbf{q},$$

where **A** and **B** are the matrices with elements a_{ij} and b_{ij} respectively and **q** is the column vector with elements $q_1, q_2, \ldots, q_N$. The problem now becomes an exercise in Linear Algebra. Lagrange's equations are

$$\mathbf{A\ddot{q}} + \mathbf{Bq} = \mathbf{0},$$

a set of coupled, linear equations with constant coefficients with solutions of the form $\mathbf{q} = \boldsymbol{\eta} \exp{(i\omega t)}$ [see § 7.4.1]. Substitution in Lagrange's equations shows that

$$(\mathbf{A}\omega^2 - \mathbf{B})\boldsymbol{\eta} = \mathbf{0}$$

and for non-trivial solutions we must have [see I, § 5.8(ii) and Theorem 5.9.1]

$$|\mathbf{A}\omega^2 - \mathbf{B}| = 0.$$

This is a polynomial equation of degree N for ω^2 which gives N real values for ω^2. The set of vectors $\boldsymbol{\eta}$ corresponding to the set of values of ω^2 are the columns of a matrix **S** and can be chosen so that

$$\mathbf{S'AS} = \mathbf{I} \quad \text{and} \quad \mathbf{S'BS} = \mathbf{D}$$

where **I** is the identity matrix and **D** is the diagonal matrix with diagonal elements equal to the N values of ω^2 [see I, § 7.12]. If new co-ordinates **Q** are introduced with $\mathbf{q} = \mathbf{SQ}$ then

$$L = \tfrac{1}{2}\mathbf{\dot{Q}'\dot{Q}} - \tfrac{1}{2}\mathbf{Q'DQ}.$$

Lagrange's equations are then

$$\mathbf{\ddot{Q}} + \mathbf{DQ} = \mathbf{0},$$

N uncoupled equations that are simply solved [see § 7.4.1]. The values of ω are called the *natural frequencies* of the system, the components of **Q** are the *normal co-ordinates* of the system and their separate oscillations are the *normal modes of oscillation*.

EXAMPLE 17.4.6.   A spring length $3l$, with spring constant k, has its two ends fixed. Particles of equal mass m are fixed to it at $x = l$ and $x = 2l$ and the system is in equilibrium. The two masses are now displaced small distances from their equilibrium positions and released. Describe the subsequent motion.

Let the positions of the two masses be x_1 and x_2, then

$$L = \tfrac{1}{2}m(\dot{x}_1^2 + \dot{x}_2^2) - \tfrac{1}{2}kx_1^2 - \tfrac{1}{2}k(x_2 - x_1)^2 - \tfrac{1}{2}kx_2^2.$$

Lagrange's equations of motion are (writing $k/m = \alpha$)

$$\ddot{x}_1 + 2\alpha x_1 - \alpha x_2 = 0,$$

$$\ddot{x}_2 - \alpha x_1 + 2\alpha x_2 = 0,$$

that is

$$A = \begin{pmatrix} 1 & 0 \\ 0 & 1 \end{pmatrix}, \quad B = \begin{pmatrix} 2\alpha & -\alpha \\ -\alpha & 2\alpha \end{pmatrix}.$$

$$|A\omega^2 - B| = (3\alpha - \omega^2)(\alpha - \omega^2) = 0 \quad \text{so } \omega^2 = \alpha \text{ or } 3\alpha.$$

The matrix S is found to be

$$(1/\sqrt{2}) \begin{pmatrix} 1 & 1 \\ 1 & -1 \end{pmatrix}.$$

Write

$$Q = \begin{pmatrix} Q_1 \\ Q_2 \end{pmatrix} = S \begin{pmatrix} x_1 \\ x_2 \end{pmatrix} = (1/\sqrt{2}) \begin{pmatrix} x_1 + x_2 \\ x_1 - x_2 \end{pmatrix}.$$

Q_1 and Q_2 are normal co-ordinates which oscillate with frequencies $\sqrt{\alpha}$ and $\sqrt{3\alpha}$ respectively. The first normal mode occurs when $Q_2 = 0$, $Q_1 \neq 0$; that is $x_1 = x_2$; both particles move in the same direction with equal amplitudes. In the second normal mode $x_1 = -x_2$ and the particles move in opposite directions with equal amplitudes.

17.4.6. Hamilton's Equations

Lagrange's equations are equivalent to Newton's equations of motion. Their derivation using Hamilton's principle [see § 17.4.3] indicates that the form of the equations is invariant under co-ordinate transformations (17.4.14). From Lagrange's equations can be formed a different but equivalent set, Hamilton's equations. These are important because they are invariant under a larger class of transformations, the canonical transformations [see § 17.4.7] and because of the way they lead to further developments in mechanics.

Lagrange's equations are N second-order equations for the N generalized co-ordinates q_k which span the configuration space of the system [see § 17.4.1]. In order to derive Hamilton's equations we first define new variables

$$p_k = \partial L / \partial \dot{q}_k, \quad k = 1, 2, \ldots, N, \tag{17.4.16}$$

the *generalized momenta* of the system; p_k is said to be *conjugate* to q_k, q_k and p_k being called *conjugate variables*. As before we denote the set $(p_1, p_2, \ldots, p_N)$ by $\mathbf{p}$. We will assume that the relations (17.4.16) that define the p's in terms of the q's and $\dot{q}$'s can be inverted to give the $\dot{q}$'s in terms of the q's and p's.

We now form the *Hamiltonian function* $H(\mathbf{q}, \mathbf{p}, t)$:

$$H = \sum_{k=1}^{N} p_k \dot{q}_k - L, \tag{17.4.17}$$

where, on the right-hand side, wherever $\dot{q}_k$ appears it is expressed in terms of the q's and p's.

If the conjugate variables $\mathbf{q}$ and $\mathbf{p}$ and also t are increased by $d\mathbf{q}$, $d\mathbf{p}$ and dt, then the increase in H is

$$dH = \sum_{k=1}^{N} \left(\frac{\partial H}{\partial p_k} dp_k + \frac{\partial H}{\partial q_k} dq_k \right) + \frac{\partial H}{\partial t} dt. \qquad (17.4.18)$$

But from (17.4.17) we also see that

$$dH = \sum_{k=1}^{N} \left(dp_k \, \dot{q}_k + p_k \, d\dot{q}_k - \frac{\partial L}{\partial \dot{q}_k} d\dot{q}_k - \frac{\partial L}{\partial q_k} dq_k \right) - \frac{\partial L}{\partial t} dt. \qquad (17.4.19)$$

Equating coefficients of the $d\mathbf{p}$, $d\mathbf{q}$ and dt we obtain

$$\left. \begin{aligned} \frac{dq_k}{dt} &= \frac{\partial H}{\partial p_k} \\[2mm] \frac{\partial L}{\partial q_k} &= -\frac{\partial H}{\partial q_k} \end{aligned} \right\} \quad k = 1, 2, \ldots, N, \qquad (17.4.20)$$

and

$$\frac{\partial L}{\partial t} = -\frac{\partial H}{\partial t}.$$

Now if Lagrange's equations hold [see (17.4.9)] then (17.4.20) become

$$\left. \begin{aligned} \frac{dq_k}{dt} &= \frac{\partial H}{\partial p_k} \\[2mm] \frac{dp_k}{dt} &= -\frac{\partial H}{\partial q_k} \end{aligned} \right\} \quad k = 1, 2, \ldots, N. \qquad (17.4.21)$$

These are Hamilton's Equations, $2N$ first-order equations for the conjugate variables $(\mathbf{q}, \mathbf{p})$, and are equivalent to Lagrange's equations, N second-order equations for the variables $\mathbf{q}$. The solutions $\mathbf{q} = \mathbf{q}(t)$, $\mathbf{p} = \mathbf{p}(t)$ define curves in *phase space*, the space spanned by co-ordinates $\mathbf{q}$ and $\mathbf{p}$. If, for brevity, we write

$$\frac{\partial}{\partial \mathbf{q}} = \left(\frac{\partial}{\partial q_1}, \frac{\partial}{\partial q_2}, \ldots, \frac{\partial}{\partial q_N} \right),$$

we can write Hamilton's equations as

$$\dot{\mathbf{p}} = -\frac{\partial H}{\partial \mathbf{q}}, \qquad \dot{\mathbf{q}} = \frac{\partial H}{\partial \mathbf{p}}. \qquad (17.4.22)$$

It is also possible to derive Hamilton's equations from Hamilton's variational principle [see § 17.4.3]. Writing

$$S = \int_{t_0}^{t_1} L \, dt = \int_{t_0}^{t_1} \left(\sum_{k=1}^{N} p_k \dot{q}_k - H \right) dt, \qquad (17.4.23)$$

then Hamilton's principle requires that S be stationary with respect to variations in the variables $\mathbf{q}$. Since the variables $\mathbf{p}$ are given functions of $\mathbf{q}$, $\dot{\mathbf{q}}$ and

t, the variations in $\mathbf{p}$ are not independent of the variations in $\mathbf{q}$. However, because of the definition of the Hamiltonian, it can be seen that the variables $\mathbf{p}$ and $\mathbf{q}$ can be regarded as independent in the variational principle. S is then a functional of the two sets of functions $\mathbf{p}$ and $\mathbf{q}$. The equations for the extremals are

$$\frac{d}{dt}\left(\frac{\partial S}{\partial \dot{q}_k}\right) - \frac{\partial S}{\partial q_k} = 0$$

and

$$\frac{d}{dt}\left(\frac{\partial S}{\partial \dot{p}_k}\right) - \frac{\partial S}{\partial p_k} = 0.$$

That is

$$\dot{p}_k + \frac{\partial H}{\partial q_k} = 0$$

and

$$\dot{q}_k - \frac{\partial H}{\partial p_k} = 0,$$

which are just Hamilton's equations.

With a dynamical system written in Hamiltonian form it is possible to find invariants whenever a co-ordinate is absent from the Hamiltonian. If H is independent of a canonical co-ordinate then the conjugate co-ordinate is a constant of the motion. If H is independent of t $(\partial H/\partial t = 0)$ then H itself is a constant for

$$\frac{dH}{dt} = \frac{\partial H}{\partial t} + \sum_{k=1}^{N}\left(\frac{\partial H}{\partial q_k}\dot{q}_k + \frac{\partial H}{\partial p_k}\dot{p}_k\right) = 0$$

from (17.4.21). If f and g are functions of the canonical variables $\mathbf{q}$ and $\mathbf{p}$, then

$$[f, g] = \sum_{k=1}^{N}\left(\frac{\partial f}{\partial q_k}\frac{\partial g}{\partial p_k} - \frac{\partial f}{\partial p_k}\frac{\partial g}{\partial q_k}\right) \tag{17.4.24}$$

is called the *Poisson bracket* of f and g with respect to $\mathbf{q}$ and $\mathbf{p}$. In terms of the Poisson bracket, Hamilton's equations can be written

$$\frac{dq_k}{dt} = [q_k, H],$$

$$\frac{dp_k}{dt} = [p_k, H]. \tag{17.4.25}$$

In fact, for a function $f(\mathbf{q}, \mathbf{p}, t)$

$$\frac{df}{dt} = \frac{\partial f}{\partial t} + [f, H].$$

The function f is an invariant if and only if $df/dt = 0$. The Poisson bracket is anti-symmetric

$$[f, g] = -[g, f]$$

and also satisfies *Jacobi's identity*

$$[f, [g, h]] + [h, [f, g]] + [g, [h, f]] = 0.$$

EXAMPLE 17.4.7. *Simple harmonic motion*. Given a particle of mass m at position x with restoring force $-kx$ then

$$L = \tfrac{1}{2}m\dot{x}^2 - \tfrac{1}{2}kx^2$$

with the generalized co-ordinate $q = x$.

The first step in finding the Hamiltonian is to obtain $\mathbf{p} = \partial L/\partial \dot{q}$. That is

$$p_x = \partial L/\partial \dot{x} = m\dot{x}.$$

This can be solved to give $\dot{x}$ in terms of p_x:

$$\dot{x} = p_x/m.$$

We now form

$$H = p_x\dot{x} - L$$
$$= p_x^2/m - p_x^2/2m + \tfrac{1}{2}kx^2$$
$$= \tfrac{1}{2}(p_x^2/m + kx^2).$$

The equations of motion are

$$\dot{x} = \partial H/\partial p_x = p_x/m, \qquad \dot{p}_x = -\partial H/\partial x = -kx$$

which together are equivalent to Lagrange's equation $m\ddot{x} + kx = 0$.

EXAMPLE 17.4.8. *Motion of a particle due to a central force* [see Examples 17.3.2 and 17.4.2].

$$L = \tfrac{1}{2}m(\dot{r}^2 + r^2\dot{\theta}^2) - V(r)$$

with generalized co-ordinates $q_1 = r$, $q_2 = \theta$. Then $p_r = \partial L/\partial \dot{r} = m\dot{r}$ and $p_\theta = \partial L/\partial \dot{\theta} = mr^2\dot{\theta}$. Forming

$$H = \sum_{k=1}^{2} p_k\dot{q}_k - L$$

one obtains

$$H = (1/2m)(p_r^2 + p_\theta^2/r^2) + V(r),$$

from which Hamilton's equations can be obtained. Since H is independent of θ, θ is ignorable and $p_\theta = mr^2\dot{\theta}$ is a constant of the motion (the angular momentum about the origin).

17.4.7.   Canonical Transformations

Any transformation from canonical co-ordinates $(\mathbf{q}, \mathbf{p})$ to new co-ordinates $(\mathbf{Q}, \mathbf{P})$ that leaves the form of the equations of motion invariant is called a canonical transformation; that is, if a dynamical system has Hamiltonian $H(\mathbf{q}, \mathbf{p}, t)$ with equations of motion

$$\dot{\mathbf{q}} = \frac{\partial H}{\partial \mathbf{p}}, \qquad \dot{\mathbf{p}} = -\frac{\partial H}{\partial \mathbf{q}}$$

then the transformation $(\mathbf{q}, \mathbf{p}) \rightarrow (\mathbf{Q}, \mathbf{P})$ is *canonical* if there exists a function $K(\mathbf{Q}, \mathbf{P}, t)$ such that the equations of motion in the new co-ordinates are

$$\dot{\mathbf{Q}} = \frac{\partial K}{\partial \mathbf{P}}, \qquad \dot{\mathbf{P}} = -\frac{\partial K}{\partial \mathbf{Q}}.$$

Hamilton's principle in canonical form provides a means of generating canonical transformations. The functionals

$$\int_{t_0}^{t_1} \left(\sum_{k=1}^{N} p_k \dot{q}_k - H \right) dt \quad \text{and} \quad \int_{t_0}^{t_1} \left(\sum_{k=1}^{N} P_k \dot{Q}_k - K \right) dt$$

have the same extremals (i.e. represent the same dynamical system) provided that the integrands differ by no more than the time derivative of a function, $d\phi/dt$. (This ensures that the functionals differ only by a constant, which makes no difference to the extremal equations [see § 12.3]). The condition for the functionals to be equivalent in this way is for

$$\sum_{k=1}^{N} p_k \, dq_k - H \, dt = \sum_{k=1}^{N} P_k \, dQ_k - K \, dt + d\phi. \qquad (17.4.26)$$

This equation contains $4N$ variables, $\mathbf{q}, \mathbf{p}, \mathbf{Q}, \mathbf{P}$ of which $2N$ can be regarded as independent and $2N$ as dependent on them through the canonical transformation.

In the first case let us take the variables $\mathbf{q}, \mathbf{Q}$ as independent and express ϕ as a function of $\mathbf{q}, \mathbf{Q}$ and t, so that

$$d\phi = \sum_{k=1}^{N} \left(\frac{\partial \phi}{\partial q_k} \, dq_k + \frac{\partial \phi}{\partial Q_k} \, dQ_k \right) + \frac{\partial \phi}{\partial t} \, dt. \qquad (17.4.27)$$

Equating coefficients of the differentials dq_k, dQ_k in (17.4.26) then gives

$$p_k = \frac{\partial \phi}{\partial q_k}, \qquad P_k = -\frac{\partial \phi}{\partial Q_k}, \qquad (17.4.28)$$

which define the canonical transformation relating $(\mathbf{q}, \mathbf{p})$ and $(\mathbf{Q}, \mathbf{P})$.

Equating to zero the coefficient of the differential dt leads to

$$K = H + \frac{\partial \phi}{\partial t} \qquad (17.4.29)$$

which gives the new Hamiltonian, K, in terms of the old Hamiltonian H. Other forms of generating function are as follows:

$\phi = \phi(\mathbf{q}, \mathbf{P}, t)$ giving

$$p_k = \frac{\partial \phi}{\partial q_k}, \qquad Q_k = \frac{\partial \phi}{\partial P_k}; \qquad (17.4.30)$$

$\phi = \phi(\mathbf{Q}, \mathbf{p}, t)$ giving

$$P_k = -\frac{\partial \phi}{\partial Q_k}, \qquad q_k = -\frac{\partial \phi}{\partial p_k}; \qquad (17.4.31)$$

$\phi = \phi(\mathbf{p}, \mathbf{P}, t)$ giving

$$q_k = -\frac{\partial \phi}{\partial p_k}, \qquad Q_k = \frac{\partial \phi}{\partial P_k}. \qquad (17.4.32)$$

In all cases $K = H + \partial \phi / \partial t$.

By their name it is clear that generating functions *generate* transformations. A set of conditions that *test* whether a given transformation is canonical or not is the following:

A transformation $(\mathbf{q}, \mathbf{p}) \to (\mathbf{Q}, \mathbf{P})$ is canonical if and only if

$$[Q_i, Q_j] = 0, \qquad [P_i, P_j] = 0,$$
$$[Q_i, P_j] = \delta_{ij} \qquad (17.4.33)$$

for all $i, j = 1, 2, \ldots, N$ [see (17.4.24)].

EXAMPLE 17.4.9. *Canonical transformations*

(i)
$$\phi = \sum_{k=1}^{N} q_k Q_k;$$

$$p_k = \partial \phi / \partial q_k = Q_k, \qquad P_k = -\partial \phi / \partial Q_k = -q_k.$$

This simple canonical transformation interchanges the role of the p's and q's showing that the original nomenclature of 'co-ordinate' and 'momentum' [see § 17.4.6] is not necessarily meaningful.

(ii)
$$\phi = \sum_{k=1}^{N} q_k P_k;$$

$$p_k = \partial \phi / \partial q_k = P_k, \qquad Q_k = \partial \phi / \partial P_k = q_k.$$

This is just the identity transformation.

(iii)
$$\phi = \tfrac{1}{2} \omega q^2 \cot Q;$$

$$p = \partial \phi / \partial q = \omega q \cot Q, \qquad P = -\partial \phi / \partial Q = \tfrac{1}{2} \omega q^2 \operatorname{cosec}^2 Q.$$

Solving for p and q in terms of P and Q gives [see § 2.12]

$$q = \sqrt{2P/\omega} \sin Q, \qquad p = \sqrt{2P\omega} \cos Q. \qquad (17.4.34)$$

This canonical transformation changes the Hamiltonian for simple harmonic motion, $H = \frac{1}{2}(p^2 + \omega^2 q^2)$ [see Example 17.4.7] into the new Hamiltonian $K = \omega P$. The equations of motion are then $\dot{P} = 0$, $\dot{Q} = \omega$ which have the immediate solution $P = \alpha$, $Q = \omega t + \beta$ where α and β are constants.

One can also check that (17.4.34) is a canonical transformation by showing that $[q, p] = 1$.

17.4.8. Hamilton–Jacobi Theory

One use of canonical transformations is to introduce new canonical co-ordinates in terms of which the new Hamiltonian takes a simple form; it may, for example, exhibit ignorable co-ordinates. In particular one can look for a generating function ϕ that makes the new Hamiltonian K equal to a constant, independent of the canonical co-ordinates and time. (This is a simplification that can be achieved at least locally at a general point in phase space.) If we take the constant value of the Hamiltonian to be zero and $\phi(\mathbf{q}, \mathbf{P}, t)$ to be the generating function, then the canonical transformation is defined by

$$Q_k = \frac{\partial \phi}{\partial P_k}, \qquad p_k = \frac{\partial \phi}{\partial q_k}$$

and the new Hamiltonian

$$K = H\left(\mathbf{q}, \frac{\partial \phi}{\partial \mathbf{q}}, t\right) + \frac{\partial \phi}{\partial t} = 0. \qquad (17.4.35)$$

Equation (17.4.35) is called the Hamilton–Jacobi (H–J) equation. A complete integral of the equation will depend on $\mathbf{q}$ and on N independent constants of integration which can be taken, without loss of generality, to be the new constant momenta $P_1, P_2, \ldots, P_N$. The generating function ϕ is called *Hamilton's Principal Function*.

When H is independent of t, and is therefore a constant of the motion, the H–J equation can be solved as far as its dependence on t is concerned by separation of variables [see § 8.2]. One obtains $\phi = S - Et$ where E is the constant value of H and $S(\mathbf{q}, \mathbf{P})$ is called *Hamilton's Characteristic Function*. In terms of S the canonical transformation is

$$Q_k = \frac{\partial S}{\partial P_k}, \qquad p_k = \frac{\partial S}{\partial q_k}$$

and the new Hamiltonian is equal to H since S is independent of t. In this solution of the H–J equation there are $N + 1$ constants, the P_k and E. Only N of them are independent and usually E is taken to depend on the P_k. As a result the Hamiltonian $k = H(\mathbf{P})$ and every co-ordinate is ignorable. Writing

$\partial k/\partial P_k = \nu_k$ then the equations of motion become

$$\dot{P}_k = 0, \qquad \dot{Q}_k = \nu_k$$

with solution

$$P_k = \gamma_k, \qquad Q_k = \nu_k t + \beta_k,$$

where γ_k and β_k are constants. A system such as this with N independent constants of the motion is called an integrable system.

Finding Hamilton's principal or characteristic functions is the aim of H–J theory since then the equations of motion of the dynamical system are trivially solved. In practice this can only be done with ease when the H–J equation can be solved by separation of variables. For example, it sometimes happens that Hamilton's characteristic function can be written

$$S = \sum_{i=1}^{N} S_i(q_i, \mathbf{P})$$

and the H–J equation splits into N equations of the form

$$H_i(q_i, \partial S_i/\partial q_i, \mathbf{P}) = P_i, \qquad i = 1, 2, \ldots, N.$$

Each equation contains only one co-ordinate q_i and can be written as an equation for $\partial S/\partial q_i$ that can be solved by integration. Many of the Hamiltonians of interest in physics are of this form.

Another case of particular interest is when the Hamiltonian is time independent and the motion is periodic in time. As above the characteristic function defines a canonical transformation with the new momenta P_k constant and the new co-ordinates Q_k increasing linearly with time. By a suitable rescaling one can define a new co-ordinate J_k in place of P_k with

$$J_k = \oint P_k \, dQ_k = \gamma_k \nu_k T_k$$

where the integral is taken over a complete period of the co-ordinate, Q_k, and T_k is the period of oscillation. The corresponding conjugate co-ordinate is

$$w_k = t/T_k + \delta_k.$$

Note that w_k changes by unity over a period as t increases by T_k. The variables J_k have the dimension of *action* and the variables w_k can be interpreted as *angles*. The combined set $(\mathbf{J}, \mathbf{w})$ are known as the *action-angle* variables for the periodic system. In the space with co-ordinates (J_k, w_k) the dynamical system trajectory is a circle of radius J_k and w_k is the polar angle. With several action and angle variables the orbit lies on a torus [see Example 6.3.3] in a higher dimensional space.

EXAMPLE 17.4.10. *Simple harmonic motion,* $H = \tfrac{1}{2}(p^2 + \omega^2 q^2)$ [see Example 17.4.7]. H is independent of t so we can find Hamilton's characteristic function $S(q, p)$ by solving

$$\frac{1}{2}\left(\frac{\partial S}{\partial q}\right)^2 + \tfrac{1}{2}\omega^2 q^2 = E,$$

that is

$$\frac{\partial S}{\partial q} = \sqrt{2E - \omega^2 q^2}.$$

This can be integrated to give

$$S = \tfrac{1}{2}q\sqrt{2E - \omega^2 q^2} + (E/\omega)\sin^{-1}(\omega q/\sqrt{2E}).$$

If we equate the constant E to the new momentum P we get

$$p = \partial S/\partial q = \sqrt{2P - \omega^2 q^2}, \qquad Q = \partial S/\partial P = (1/\omega)\sin^{-1}(\omega q/\sqrt{2P}.$$

That is

$$q = \sqrt{2P/\omega^2}\sin \omega Q, \qquad P = \sqrt{2P}\cos \omega Q.$$

This canonical transformation changes $H(q, P)$ into the new Hamiltonian $K(Q, P) = P$. The equations of motion are $\dot{Q} = 1$, $\dot{P} = 0$ giving $P = \alpha$, $Q = t + \beta$ with α, β constant.

The motion is periodic (frequency ω). The above canonical co-ordinates are not action-angle variables as Q does not change by unity as t increases by $(1/\omega)$. However, if we equate the constant E to ωP, then the canonical transformation becomes

$$q = \sqrt{2P/\omega}\sin Q, \qquad p = \sqrt{2P\omega}\cos Q$$

and the new Hamiltonian $K = \omega P$. Now the solution of the equations of motion is $P = \alpha$, $Q = \omega t + \beta$ so that Q is a proper angle variable. The pair (P, Q) are now the action-angle variables of the dynamical system.

K. J. W.

REFERENCES

Arnold, V. I. (1978). *Mathematical Methods of Classical Mechanics*, Springer.
Goldstein, H. (1980). *Classical Mechanics* 2nd edn. Addison-Wesley.

CHAPTER 18

Stochastic Differential Equations

18.1. INTRODUCTION

The mathematical theory of stochastic differential equations is a highly technical and abstract analysis of the effects of continuous random disturbances on dynamical systems. The complex nature of the mathematical theory is necessary to explain these effects, many of which are counter-intuitive. It is the purpose of this chapter to present the mathematical results and to indicate their areas of application in a form which will be readable to someone interested in those applications rather than in the mathematical theory. Proofs of the results are not given. However an intuitive basis for non-intuitive results is given where possible. The major area of application is in control engineering and applications are also indicated in biological, chemical and financial modelling, etc.

The reader is referred to Chapters 18–22 of Volume II for the general theory of random (stochastic) processes.

18.1.1. Examples of Random Systems

The need for stochastic differential equations arises from a desire to give a rigorous representation of random disturbances in continuous dynamical systems. Consider, for instance, the measurement and communications systems illustrated in Figure 18.1.1. Random disturbances of various types affect all parts of the systems continuously. In these examples of measurement and communication, the random effects are properly called disturbances since they interfere with the smooth running of the system, and our implied aim is to reduce their effect or to find means of ignoring them if we possibly can. There are other systems in which random processes are not something to be eliminated but are a more intrinsic part of our system whose effects we desire to predict, for instance the biological and financial systems illustrated in Figure 18.1.2.

Our ways of looking at the random processes in Figures 18.1.1 and 18.1.2 are different, but our analysis of the random effects will be the same. These

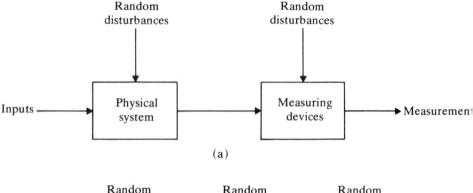

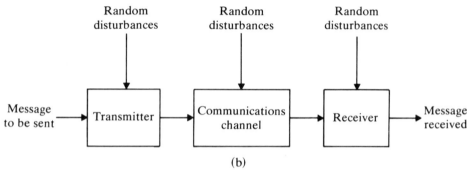

Figure 18.1.1: (a) Random disturbances affecting measurements of a physical system. (b) Random disturbances affecting a communications system.

examples can be used to bring out a point about random processes which will be important later on. This concerns the variance of the process. Is the variance of the random process constant, or does it depend on the magnitude of other variables in the system? The disturbance on the communications channel in Figure 18.1.1(b) may well be independent of the signal being sent, but the random disturbance affecting the measurements in Figure 18.1.1(a) may well depend on the magnitude of the variable being measured, for example, if the tolerance on the measuring device is expressed as a percentage of the measurement. Similarly, in Figure 18.1.2(a), the number of individuals reproducing or dying is dependent on the total number in the population and in Figure 18.1.2(b) research has shown [see Cootner (1964)] that the random fluctuations in stock market prices are proportional to the prices themselves. The counter-intuitive results of the theory of stochastic differential equations apply to the cases where the variance of the random process depends on other variables of the system. Mathematically this can be illustrated with a stationary random process $S(t, \omega)$, where t is time, and the randomness derives from different choices of ω from a sample space [see II, § 18.0.5]. If S is stationary then its probability distribution at t is the same as that at any other time. If x is another variable in the system under consideration then S has constant

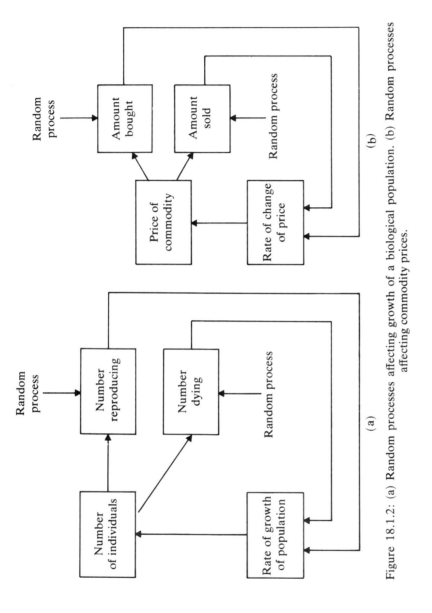

Figure 18.1.2: (a) Random processes affecting growth of a biological population. (b) Random processes affecting commodity prices.

variance, but for a given value of x at time t, xS has variance depending on that given value of x.

18.1.2.   Dynamic Equations for Random Systems

Work in control engineering and later in the dynamic modelling of, for example, economic and biological systems, has led to fairly general use of the 'state space' representation of dynamical systems where certain key variables, known as state variables, $\mathbf{x} = (x_1, \ldots, x_n)$, are chosen to represent the dynamics of the system by means of a system of differential equations:

$$\frac{dx_i}{dt} = f_i(\mathbf{x}, t) \qquad i = 1, \ldots, n \tag{18.1.1}$$

where t is time [see § 7.9 and § 7.11].

To introduce the random processes into the equations let us simplify matters and take the one-dimensional case:

$$\frac{dx}{dt} = f(x, t) + g(x, t) W(t, \omega) \tag{18.1.2}$$

where $W(t, \omega)$ is a random function, the randomness being determined by different choices of points, ω, in a sample space.

$x = x(t, \omega)$ is now also a random function.

$g(x, t)$ is a function indicating how the variance of the random disturbance depends on a given value of the state variable x at time t.

dx/dt is a function of x, t, ω and is thus also random. It gives the rate of change of x with t for given x, t, ω.

Without loss of generality, we can assume that the mean, or expected value, $E[W(t, \omega)]$, of $W(t, \omega)$ (averaged over all points, ω, of the sample space) is zero, since any non-zero mean can be absorbed into the function $f(x, t)$ [see II, Chapter 8].

At any time, t, (18.1.1) indicates the rate of change of $\mathbf{x}$ given its value at that time. Equation (18.1.2) serves a corresponding purpose in the stochastic case. If we know the state, x, of the system at time, t, then $f(x, t)$, $g(x, t)$ are known, so that (18.1.2) can be written in terms of conditional probabilities [see II, §§ 3.9.1 and 6.5]:

$$P\left[\alpha < \frac{1}{g(x, t)}\left(\frac{dx}{dt} - f(x, t)\right) < \beta \,\middle|\, x, t\right] = P[\alpha < W(t, \omega) < \beta | x, t].$$

This is the basis of any numerical procedure for integrating (18.1.1) or (18.1.2) forwards in time, in order to simulate it on a digital computer. Given a starting state x_0 at a starting time t_0 and the magnitude of the random function $W(t_0, \omega)$, we can calculate the rate of change, dx/dt for that x_0, t_0, ω. This rate of change allows us to get an approximation to the value of x a small increment of time afterwards, e.g. at $t_0 + \delta t$, for small δt. (The simplest such approximation is $x_0 + (dx/dt)\delta t$, but see section 18.4 below.) Then the procedure is repeated between $t_0 + \delta t$ and $t_0 + 2\delta t$.

Equation (18.1.2) can be applied to the example in Figure 18.1.2(b) as follows. The amount, B, of the commodity bought is inversely related to price, p, but is also subject to random fluctuations, for example $B = (b/p)(1 + W_1)$. Similarly, suppose the amount, S, sold is given by $S = sp(1 + W_2)$; where b, s are constants and W_1, W_2 are random variables with zero means. Then, suppose the rate of change of price is proportional to $B - S$, we have

$$\frac{dp}{dt} = k\left(\frac{b}{p} - sp\right) + \frac{kb}{p} W_1 - ksp W_2$$

where k is a constant of proportionality.

This is an example extending (18.1.2) to the case of two random variables W_1, W_2 on the right-hand side. In section 18.3 below we shall consider extensions with several state variables, x, and several random variables.

18.1.3. Use of the Wiener Process to Represent Random Effects

Equation (18.1.2) can be written in integral form:

$$x(t, \omega) - x(a, \omega) = \int_a^t f(x, s)\, ds + \int_a^t g(x, s)\, dw\,(s, \omega) \qquad (18.1.3)$$

where $w(t, \omega)$ is the integral of $W(t, \omega)$ with respect to t. Wiener was the first person to use this representation where the second integral on the right-hand side is called a *stochastic integral*. He considered the most natural form of random function, viz. a normally distributed one [see II, § 11.4], and this is very important in applications today. Consider a random function $W(t, \omega)$ which is normally distributed at any time, t, but where $W(t_1, \omega)$, $W(t_2, \omega)$ are uncorrelated for different times t_1, t_2 [see II, § 9.6.3]. Let us call this function *white noise*. It is, in fact, a special case of a continuous time Markov process [see Volume II, Chapter 20]. Such a function is a mathematical abstraction since it is not possible to have a physical realization of a variable whose value at t_1 is uncorrelated with that at t_2 no matter how close t_1 and t_2 are. Interference on radio reception (Figure 18.1.1(b)) is a good approximation to white noise, and many other examples of random disturbance in physical systems are too. It may seem strange to consider a mathematical function which is not physically realizable in a paper on applications, but we are faced with a choice. Either we develop a theory for our unrealizable function and use it to approximate the physical world; or we develop a theory for realizable functions. The latter course turns out to be extremely difficult and the former course is the theory of stochastic differential equations. It can be shown [e.g. Wong (1971)] that if $W(t, \omega)$ is of this normally distributed but unrealizable form then $w(t, \omega)$

is also normally distributed with:

$$E[w(t, \omega)] = 0 \tag{18.1.4(a)}$$

$$E[(w(t, \omega) - w(s, \omega))^2] = \sigma^2 |t - s| \tag{18.1.4(b)}$$

$$E[w(t, \omega) w(s, \omega)] = 0 \qquad t \neq s \tag{18.1.4(c)}$$

$$E[w(u, \omega)(w(t, \omega) - w(s, \omega))] = 0 \qquad t \geq s \geq u \tag{18.1.4(d)}$$

for some constant σ^2. We call $w(t, \omega)$ a *Wiener Process* [cf. II, Examples 22.1.1 and 22.2.1]. Like $W(t, \omega)$ it is a Markov Process.

If we concern ourselves with small increments, δw, in w from time t to time $t + \delta t$, then these properties lead to:

$$E[\delta w] = 0 \tag{18.1.5(a)}$$

$$E[(\delta w)^2] = \sigma^2 \delta t \tag{18.1.5(b)}$$

where the expectation is taken over all points, ω, in the sample space.

In the second of these equations, we have a first-order term, δt, in a small quantity related to a second-order term $(\delta w)^2$ in another small quantity. This unusual circumstance gives rise to many of the counter-intuitive properties of stochastic differential equations described above.

Before we can proceed further, we must make precise the definition of the second (stochastic) integral in (18.1.3). This is done in the next section, for the Wiener process, $w(t, \omega)$, and also for the other random function that is important in applications, viz. the Poisson process [see II, § 20.1].

18.2. STOCHASTIC INTEGRALS

18.2.1. Definitions of Stochastic Integrals

In this section we concern ourselves with the definition of integrals such as the second in (18.1.3). The normal procedure in the case of non-random functions is to divide the interval $[a, t]$ into subintervals by $a = t_0 < t_1 < \ldots < t_n = t$ and to approximate the integral by a sequence of finite sums in which the lengths of the subintervals tend to zero:

$$\int_a^t g(x, s) \, dw(s, \omega) = \lim_{\delta \to 0} \sum_{r=0}^{n-1} g(x(\tau_r, \omega), \tau_r)\{w(t_{r+1}, \omega) - w(t_r, \omega)\}$$

where $\delta = \max_r (t_{r+1} - t_r)$ and the point of evaluation of the integrand, g, is somewhere in the appropriate subinterval: $\tau_r \in [t_r, t_{r+1}]$ [see § 4.8].

However, Doob (1953) p. 395 shows that the Wiener process is of unbounded variation on any interval (i.e. the length of the graph of $w(t, \omega)$ against t is unboundedly large on any interval of time [see also Definition 2.7.3]) and this results in the above definition giving rise to different values of the integral for different positions of τ_r in the interval $[t_r, t_{r+1}]$.

In order to make the definition of stochastic integrals self-consistent, different authors have proposed different values of τ_r. Ito takes $\tau_r = t_r$, see, for example, Gihman and Skorohod (1972); Stratonovich (1966) takes $\tau_r = \frac{1}{2}(t_r + t_{r+1})$ and any other point could have been chosen. Another notation for (18.1.3) is:

$$dx = f(x, t)\, dt + g(x, t)\, dw. \tag{18.2.1(a)}$$

This is a stochastic differential equation and it is important to note that it is not an equation in small increments dx, dt, dw but instead is a simplified notation for (18.1.3). To be meaningful it requires the specification of the stochastic integral in (18.1.3).

18.2.2.   Desirable Properties of the Ito Integral

Equations (18.2.1(a)) and (18.1.3) are simply different notations for the same thing, and, in choosing a notation, it is convenient to choose one which has some reasonable intuitive basis. The intuitive basis for (18.2.1(a)) can be seen if we *formally* rewrite it in terms of small increments. It should be emphasized that this has no strict mathematical basis but is done simply to provide an intuitive meaning for (18.2.1(a)). *Formally* in terms of small increments (18.2.1(a)) corresponds to:

$$\delta x = f(x, t)\, \delta t + g(x, t)\, \delta w. \tag{18.2.1(b)}$$

We now look for desirable properties of this equation (18.2.1(b)) and then stipulate that those properties should also apply to (18.2.1(a)). This will enable us to choose a definition of the stochastic integral in (18.1.3) which gives those properties to (18.2.1(a)). This definition of the stochastic integral can therefore be said to be intuitively desirable, although other definitions are mathematically possible.

Before indicating the desirable property that we can derive from (18.2.1(b)) we first specify a detail about the small increments, so as to make them correspond with the numerical procedure mentioned in section 18.1.2 for integrating the equation forward in time. This procedure was based on starting at a given value of x_0 at time t_0, moving forward a small time step, say to $t_0 + \delta t$, and then repeating in small steps over whatever time interval we are interested in. In order to do this on (18.2.1(b)) it is important that δx, δt, δw should be increments starting at time t. In other words we are interpreting (18.2.1(b)) to imply:

$$x(t + \delta t, \omega) = x(t, \omega) + f(x(t, \omega), t)\, \delta t + g(x(t, \omega), t)(w(t + \delta t, \omega) - w(t, \omega)).$$

Here, in the last term on the right-hand side, $x(t, \omega)$ depends on $w(s, \omega)$ for $s \leq t$, so that· (18.1.4(d)) implies that $g(x(t, \omega), t)$ is independent of $w(t + \delta t, \omega) - w(t, \omega) = \delta w$.

Now we can obtain our desirable property by taking the expectation of (18.2.1(b)), averaging over all points ω in the sample space [cf. II, Theorem

8.4.1]:

$$E[\delta x] = E[f(x, t)]\delta t + E[g(x, t)\delta w].$$

The last term on the right-hand side is zero because of the independence just mentioned and property (18.1.5(a)) [cf. II, Theorem 8.6.1]. Thus

$$E[\delta x] = E[f(x, t)]\,\delta t.$$

Applying this over a sequence of consecutive small intervals δt, and then letting $\delta t \to 0$, we have:

$$E[x(t, \omega) - x(a, \omega)] = E\int_a^t f(x, t)\,dt.$$

Comparing this with (18.1.3) we obtain:

$$E\left[\int_a^t g(x, s)\,dw(s, \omega)\right] = 0 \tag{18.2.2}$$

Processes satisfying relations of the type (18.2.2) are called *Martingales* (that is, processes $y(t, \omega)$ satisfying $E[y(t, \omega) - y(a, \omega)] = 0$).

It is worth repeating that we have not proved (18.2.2) to be true for the stochastic integral in (18.1.3) but rather we have shown that if we are to give (18.2.1(a)) the intuitive meaning implied by (18.2.1(b)) then we would like (18.2.2) to be true. Indeed (18.2.2) is a very natural property since the integral, $\int_a^t g(x, s)\,dw(s, \omega)$, is the limit of a sum of terms each of which has zero expectation. So it is natural to adopt a definition of the integral in which its own expected value is zero.

This is precisely what Ito's definition (with $\tau_r = t_r$) does for us. It can be shown that his integral is a Martingale, see, for instance, Gihman and Skorohod (1972). Also Ito's integral has a very natural appeal in the numerical integration of (18.2.1(a)) which we shall discuss in sections 18.4 and 18.5 below. Stratonovich's definition (1966) (with $\tau_r = \frac{1}{2}(t_r + t_{r+1})$) does not have either of these desirable properties.

If the integrand $g(x, t)$ is independent of $x(t, \omega)$, that is, if g is a non-random function, then we have the trivial case in which all the integrals agree, and none of the interesting considerations described below apply. This is the case when the disturbance to our dynamical system has variance independent of the state variables of the system. This dependence of the variance of the disturbance on the state variables was mentioned in section 18.1 as determining the relevance of stochastic calculus to the system.

18.2.3.   Example of Stochastic Integral Involving Wiener Process

The difference between Ito's integral and the Riemann integral of ordinary calculus [see § 4.1] can be illustrated by the following example. Consider evaluating the integral of the Wiener process itself:

$$\int_a^t w(s, \omega)\, dw(s, \omega).$$

Under Ito's definition this is approximated by the sum:

$$\sum_{r=0}^{n-1} w_r\{w_{r+1} - w_r\}$$

where we have simplified the notation: $w_r = w(t_r, \omega)$.

From (18.1.5(b)) we have approximately:

$$\sum_{r=0}^{n-1} \{w_{r+1} - w_r\}^2 = \sigma^2(t - a). \tag{18.2.3}$$

This is made precise by Wong (1971, p. 53). And by algebraic manipulation we find that:

$$\sum_{r=0}^{n-1} \{w_{r+1} + w_r\}\{w_{r+1} - w_r\} = w(t, \omega)^2 - w(a, \omega)^2. \tag{18.2.4}$$

Subtracting (18.2.3) from (18.2.4) and taking the limit $\delta \to 0$, we obtain, under Ito's interpretation:

$$I = \int_a^t w(s, \omega)\, dw(s, \omega) = \tfrac{1}{2}\{w(t, \omega)^2 - w(a, \omega)^2 - \sigma^2(t - a)\} \tag{18.2.5}$$

whereas under ordinary calculus the term $\sigma^2(t - a)$ is not present. This term has arisen from second-order and first-order terms in small quantities being related in (18.2.3), corresponding to the comment we made in section 18.1 about (18.1.5(b)).

The Martingale property (18.2.2) of Ito integrals can easily be verified for (18.2.5) by using (18.1.4(b)) and (18.1.4(c)). However, the price we must pay for this desirable property is a departure from the rules of ordinary calculus.

On the other hand, under the Stratonovich interpretation, our integral is approximated by the sum:

$$\sum_{r=0}^{n-1} w_{r+\frac{1}{2}}\{w_{r+1} - w_r\}$$

$$= \sum_{r=0}^{n-1} w_{r+\frac{1}{2}}\{(w_{r+1} - w_{r+\frac{1}{2}}) + (w_{r+\frac{1}{2}} - w_r)\}$$

$$= \sum_{r=0}^{n-1} w_{r+\frac{1}{2}}(w_{r+1} - w_{r+\frac{1}{2}}) + \sum_{r=0}^{n-1} (w_{r+\frac{1}{2}} - w_r + w_r)(w_{r+\frac{1}{2}} - w_r)$$

$$= \sum_{r=0}^{n-1} \{w_{r+\frac{1}{2}}(w_{r+1} - w_{r+\frac{1}{2}}) + w_r(w_{r+\frac{1}{2}} - w_r)\} + \sum_{r=0}^{n-1} (w_{r+\frac{1}{2}} - w_r)^2.$$

Taking the limit as $\delta \to 0$, the first of these sums is the Ito integral, I, and the second is $\frac{1}{2}\sigma^2 (t - a)$ by (18.1.4(b)). Thus the *Stratonovich* integral, S, is given by:

$$S = I + \tfrac{1}{2}\sigma^2(t - a)$$

$$= \tfrac{1}{2}\{w(t, \omega)^2 - w(a, \omega)^2\}$$

as in ordinary calculus. Stratonovich (1966) shows that, in fact, it is a general rule that his integral obeys the rules of ordinary calculus [see Theorem 4.1.2]. However, it does not have the desirable Martingale property possessed by the Ito integral.

Those properties of Ito and Stratonovich integrals are summarized in Table 18.2.1.

	Martingale property	Satisfies rules of ordinary calculus
Ito	Yes	No
Stratonovich	No	Yes

Table 18.2.1: Properties of Ito and Stratonovich Integrals.

18.2.4. Stochastic Integrals with respect to the Poisson Process

It is possible to have stochastic integrals with respect to processes other than the Wiener process. Ito (1951a) extended his definition to discrete processes, the main practical advantage of the extension being that Poisson processes are catered for [see II, § 20.1]. These often occur in applications where we wish to simulate discrete events such as the arrival of customers for service, or the reproduction of animals, see Figure 18.1.2(a). The two desirable properties mentioned above for the Wiener case still apply. Another stochastic integral applying to Poisson processes has been given by Marcus (1978). It obeys the rules of ordinary calculus.

A third stochastic integral has been defined by McShane (1969a, b). He allows τ_r to go outside the interval $[t_r, t_{r+1}]$ and, specifically, to take any value less than or equal to t_r. He then takes the limit $t_{r+1} - \tau_r \to 0$. This definition allows a wider range of stochastic processes $w(t, \omega)$ than Ito's but is more restrictive on the allowable integrands, $g(x, t)$. However, where the regions of definition overlap, Ito's and McShane's integrals agree.

18.3. STOCHASTIC CALCULUS

18.3.1. Differential Rules

The price we must pay for the desirable properties in section 18.2, above, is that, under Ito's definition, the rules of ordinary calculus break down. Under

Stratonovich's definition, the rules of ordinary calculus still hold, but here we give the modifications required by Ito.

(a) *Wiener Case*

Consider a function, $Y(x, t)$, where $x(t, \omega)$ satisfies the stochastic differential equation (18.2.1(a)) for a Wiener process $w(t, \omega)$. We now ask what stochastic differential equation Y satisfies. Before quoting the result, we give an intuitive basis for it in terms of small increments, δx, δt:

$$\delta Y = \frac{\partial Y}{\partial t} \delta t + \frac{\partial Y}{\partial x} \delta x + \tfrac{1}{2} \frac{\partial^2 Y}{\partial x^2} (\delta x)^2 + \frac{\partial^2 Y}{\partial x\, \partial t} \delta x\, \delta t + \tfrac{1}{2} \frac{\partial^2 Y}{\partial t^2} (\delta t)^2 + \ldots$$

$$= \frac{\partial Y}{\partial t} \delta t + \frac{\partial Y}{\partial x} (f\, \delta t + g\, \delta w) + \tfrac{1}{2} \frac{\partial^2 Y}{\partial x^2} (f\, \delta t + g\, \delta w)$$

$$+ \frac{\partial^2 Y}{\partial x\, \partial t} (f\, \delta t + g\, \delta w)\, \delta t + \tfrac{1}{2} \frac{\partial^2 Y}{\partial t^2} (\delta t)^2 + \ldots.$$

Normally, we would ignore second-order terms, but here we cannot since, from (18.1.4(b)), $E[(\delta w)^2] = \sigma^2 \delta t$. Ito (1950, 1951b) showed that $(\delta w)^2$ can effectively be replaced by its expectation so that, to first order in δw, δt, we have:

$$\delta Y = \frac{\partial Y}{\partial t} \delta t + \frac{\partial Y}{\partial x} (f\, \delta t + g\, \delta w) + \tfrac{1}{2} \sigma^2 \frac{\partial^2 Y}{\partial x^2} g^2\, \delta t.$$

This corresponds directly with the *Ito differential rule* which we now quote to be:

$$dY = \frac{\partial Y}{\partial t} dt + \frac{\partial Y}{\partial x} (f\, dt + g\, dw) + \tfrac{1}{2} \sigma^2 \frac{\partial^2 Y}{\partial x^2} g^2\, dt. \qquad (18.3.1)$$

If Y is a function of $w(t)$ and t, rather than of $x(t)$, t, we have:

$$dY = \frac{\partial Y}{\partial t} dt + \frac{\partial Y}{\partial w} dw + \tfrac{1}{2} \sigma^2 \frac{\partial^2 Y}{\partial w^2} dt. \qquad (18.3.2)$$

The above derivation in terms of small increments δx, δt is not rigorous, but has been included as an intuitive basis for a non-intuitive result. The rigorous proof is given by Ito (1950, 1951b). In future we shall manipulate dx, dt as though they were small increments δx, δt with the same motivation of indicating the plausibility of the result.

We illustrate the Ito differential rule on the important example $Y = Y_0 \exp(w - \tfrac{1}{2}\sigma^2 t)$. Applying (18.3.2) to this, we get:

$$dY = -\tfrac{1}{2}\sigma^2 Y\, dt + Y\, dw + \tfrac{1}{2}\sigma^2 Y\, dt$$

$$= Y\, dw.$$

Thus, under the Ito interpretation, the solution of the stochastic differential equation, $dY = Y\,dw$, is:

$$Y = Y_0 \exp\left(w - \tfrac{1}{2}\sigma^2 t\right). \tag{18.3.3}$$

We can verify the Martingale property (18.2.2) for this example as follows. For simplicity, let us assume that $w(0) = 0$. Then:

$$E\left[\int_0^t Y\,dw\right] = E[Y(t) - Y(0)]$$

$$= Y_0 E[\exp w]\exp\left(-\tfrac{1}{2}\sigma^2 t\right) - Y_0$$

$$= 0.$$

(This uses the result: $E[\exp w] = (2\pi\sigma^2 t)^{-1/2}\int_{-\infty}^{\infty} \exp w \exp\left(-\tfrac{1}{2}w^2/\sigma^2 t\right)dw = \exp\left(\tfrac{1}{2}\sigma^2 t\right)$.)

The above scalar Ito differential rules (18.3.1), (18.3.2) can be extended to the multidimensional case in two ways corresponding to two possible multidimensional extensions of (18.2.1(a)), viz.

$$dx_i = f_i(\mathbf{x}, t)\,dt + \sum_p G_{ip}(\mathbf{x}, t)\,dw_p \tag{18.3.4}$$

and

$$dx_i = f_i(\mathbf{x}, t)\,dt + \sum_p h_p(\mathbf{x}, t)\,dw_{ip} \tag{18.3.5}$$

where $\mathbf{f}, \mathbf{G}, \mathbf{h}, \mathbf{x}, \mathbf{w}$ are now vectors or matrices of the appropriate (finite) dimensions.

Equation (18.3.4) is the most natural extension of (18.2.1(a)) but (18.3.5) is also useful, see sections 18.5.3 and 18.5.4 below. The Ito differential rules for a function $Y(\mathbf{x}, t)$ in these two cases are:

$$dY = \frac{\partial Y}{\partial t}\,dt + \sum_i \frac{\partial Y}{\partial x_i}\left(f_i\,dt + \sum_p G_{ip}\,dw_p\right) + \tfrac{1}{2}\sum_{i,j,p,q}\frac{\partial^2 Y}{\partial x_i\,\partial x_j}G_{ip}G_{jq}R_{pq}\,dt \tag{18.3.6}$$

and

$$dY = \frac{\partial Y}{\partial t}\,dt + \sum_i \frac{\partial Y}{\partial x_i}\left(f_i\,dt + \sum_p h_p\,dw_{ip}\right) + \tfrac{1}{2}\sum_{i,j,p,q}\frac{\partial^2 Y}{\partial x_i\,\partial x_j}h_p h_q R_{ipjq}\,dt \tag{18.3.7}$$

where (18.3.5(b)) is now extended to:

$$E[\delta w_p\,\delta w_q] = R_{pq}\,\delta t$$

and

$$E[\delta w_{ip}\,\delta w_{jq}] = R_{ipjq}\,\delta t.$$

From (18.3.6) can be derived the rules for the derivative of a function of a function and of a product [cf. (3.2.8) and (3.2.6)]:

$$dY(Z(\mathbf{x}, t)) = \frac{dY}{dZ}\,dZ + \tfrac{1}{2}\frac{d^2 Y}{dZ^2}\sum_{i,j,p,q}\frac{\partial Z}{\partial x_i}\frac{\partial Z}{\partial x_j}G_{ip}G_{jq}R_{pq}\,dt$$

and

$$d(Y(\mathbf{x}, t)Z(\mathbf{x}, t)) = Y\,dZ + Z\,dY + \sum_{i,j,p,q} \frac{\partial Y}{\partial x_i} \frac{\partial Z}{\partial x_j} G_{ip} G_{jq} R_{pq}\, dt$$

with analogous results for (18.3.5) from (18.3.7).

(b) *Poisson Case*

The point was made in section 18.2.4 that Ito's definition of stochastic integration is not restricted to the case of Wiener processes $w(t)$. Whilst the latter are the most commonly found in practical examples, we shall also have occasion to use (in § 18.5.5) stochastic differential equations driven by Poisson processes $p(t)$. These are processes consisting of discrete jumps with probability distribution:

$$P[\delta p = k] = \frac{\exp(-c\,\delta t)(c\,\delta t)^k}{k!}$$

where $\delta p = p(t + \delta t) - p(t)$ is the jump occurring in a small time interval δt, c is a constant determining the mean jump: $E[\delta p] = c\,\delta t$, and k is an integer [cf. II (20.1.6)].

The differential rule in this case has been given by Gihman and Skorohod (1972, p. 269). Consider the scalar stochastic differential equation:

$$dx = f(x, t)\, dt + g(x, t)\, dp \qquad (18.3.8)$$

where $p(t)$ is a Poisson process with $E[\delta p] = c\,\delta t$. Then for a function $Y(x,(t), t)$, the differential rule is:

$$dY = \frac{\partial Y}{\partial t}\,dt + \frac{\partial Y}{\partial x} f(x, t)\, dt + \{Y(x + G, t) - Y(x, t)\}\, dp. \qquad (18.3.9)$$

Gihman and Skorohod give only the trivial multidimensional extension of this result where $\mathbf{x}, \mathbf{f}, \mathbf{g}$ are vectors but p remains a scalar. They do, however, show the more interesting result that independent Wiener and Poisson processes can be combined additively in a single stochastic differential equation (for example, combining (18.2.1(a)) and (18.3.8)) giving the corresponding combination in the differential rule (for example, a combination of (18.3.1) and (18.3.9)).

18.3.2. Relationship between Ordinary and Stochastic Differential Equations

(a) *Wiener Case*

Since the Wiener process is not physically realizable (being of unbounded variation on any non-degenerate interval, see section 18.2.1), and since stochastic differential equations driven by it do not satisfy the rules of ordinary calculus, the question arises as to whether we can approximate these equations by sequences of other equations which are driven by physically realizable approximations to the Wiener process and which, therefore, do obey the rules of ordinary calculus. This question is very important in practical applications of stochastic differential equations, see section 18.5, and in their simulation on computers, see section 18.4. It was answered by Wong and Zakai (1965a, b) in the scalar, Wiener case as follows. Let $w''(t)$, for $n = 1, 2, 3, \ldots$ be a sequence of physically realizable (that is, continuous [see Definition 2.1.2] and of bounded variation [see Definition 2.7.3]) approximations (with probability one) to the Wiener process, $w(t)$. Then the sequence of solutions, $x''(t)$, to the ordinary differential equations:

$$dx''(t) = f(x'', t) \, dt + g(x'', t) \, dw''$$

converges (in the mean) to the solution, $x(t)$, of a stochastic differential equation containing an 'additional term':

$$dx = \left\{ f(x, t) + \tfrac{1}{2}\sigma^2 g(x, t) \frac{\partial g}{\partial x} \right\} dt + g(x, t) \, dw. \qquad (18.3.10)$$

An intuitive basis for this result can be obtained by considering the stochastic differential equation for the function $Y(x, t) = \int^x (1/g(\mu, t)) \, d\mu$. From (18.3.1) and (18.3.10) this is:

$$dY(x, t) = \frac{\partial Y}{\partial t} \, dt + \frac{1}{g} \left\{ f \, dt + \tfrac{1}{2}\sigma^2 g \frac{\partial g}{\partial x} \, dt + g \, dw \right\} - \tfrac{1}{2}\sigma^2 \frac{1}{g^2} \frac{\partial g}{\partial x} g^2 \, dt$$

$$= \frac{\partial Y}{\partial t} \, dt + \frac{f(x, t)}{g(x, t)} \, dt + dw. \qquad (18.3.11)$$

The ordinary differential equation for $Y(x'', t)$ is seen by ordinary calculus to correspond precisely with this:

$$dY(x'', t) = \frac{\partial Y}{\partial t} \, dt + \frac{f(x'', t)}{g(x'', t)} \, dt + dw''. \qquad (18.3.12)$$

The function Y has been chosen so that the coefficients of dw and dw'' in these equations do not depend on x, so that the variance of the random effect is constant. In this case there is no difference between the corresponding stochastic and ordinary integrals [see §§ 18.1.1 and 18.2.2]. The correspondence between these equations is thus a necessary condition on (18.3.10) for it to be the equation for the limit $x(t)$ of the sequence $x''(t)$.

The extension of this result to the multidimensional case (18.3.4) is given by McShane (1971) who shows that the 'additional term' needed is:

$$\tfrac{1}{2} \sum_{j,p,q} G_{jp} \frac{\partial G_{iq}}{\partial x_j} R_{pq}. \qquad (18.3.13)$$

The corresponding additional term for (18.3.5) is:

$$\tfrac{1}{2} \sum_{j,p,q} h_q \frac{\partial h_p}{\partial x_j} R_{ipjq}. \qquad (18.3.14)$$

The intuitive basis of these multidimensional results is similar to the scalar case.

These results (18.3.13), (18.3.14), indicate a qualitative difference between the two multidimensional stochastic differential equations (18.3.4), (18.3.5), which pertains even in the simple case when all the Wiener processes are independent: $R_{pq} = \delta_{pq}$, $R_{ipjq} = \delta_{ij} \delta_{pq}$. In this case (18.3.14) is simply a function of $\mathbf{h'h}$, but (18.3.13) is not simply a function of $\mathbf{G'G}$ [' denoting the transpose—see I, § 6.5]. A direct transformation of (18.3.5) into (18.3.4), using

$$dw_{j+n(i-1)} = dw_{ij}, \tag{18.3.15}$$

(where n is the dimension of $\mathbf{h}$) indicates that the corresponding $\mathbf{G}$ is related to $\mathbf{h}$ by: $\mathbf{G'G} = \mathbf{h'hI}$ [where $\mathbf{I}$ denotes the unit matrix—see I (6.2.7)]. Thus equations of the type (18.3.5) can be transformed into equations of the type (18.3.4) which have 'additional terms' (18.3.13) depending only on $\mathbf{G'G}$ whereas, in general, equations of the type (18.3.4) have 'additional terms' which depend directly on $\mathbf{G}$.

(b) *Poisson Case*

The Poisson process, unlike the Wiener one, is physically realizable so the question of approximations to it does not arise. However, it is often useful in simulation [see § 18.4.2] to be able to transform from an Ito Poisson differential equation (obeying the differential rule (18.3.9)) to an ordinary Poisson differential equation (obeying the rules of ordinary calculus). This enables a given system to be modelled by either type of equation. The necessary transformation is given by Wright (1980) as follows. The Ito equation:

$$dx = f(x, t)\, dt + g(x, t)\, dp \tag{18.3.16}$$

corresponds to the 'ordinary' equation:

$$dx^* = f(x^*, t)\, dt + g(x^*, t)\, dp \tag{18.3.17}$$

where

$$\int_x^{x+\gamma} \frac{1}{g(\mu, t)}\, d\mu = 1. \tag{18.3.18}$$

The stochastic integral implied in (18.3.17) which obeys the rules of ordinary calculus is that defined by Marcus (1978) see section 18.2.4. We now give an intuitive basis for the result (18.3.18). As in the Wiener case, we consider the function $Y(x, t) = \int^x (1/g(\mu, t))\, d\mu$, with the same motivation of obtaining equations for Y where the coefficients of the random terms are unity. From (18.3.16), using the differential rule (18.3.9) we obtain:

$$dY(x, t) = \frac{\partial Y}{\partial t}\, dt + \frac{f(x, t)}{g(x, t)}\, dt + \int_x^{x+\gamma} \frac{1}{g(\mu, t)}\, d\mu\, dp. \tag{18.3.19}$$

From (18.3.17), using ordinary calculus:

$$dY(x^*, t) = \frac{\partial Y}{\partial t} dt + \frac{f(x^*, t)}{g(x^*, t)} dt + dp. \tag{18.3.20}$$

A necessary condition for $x(t)$ and $x^*(t)$ to model the same system is that (18.3.19) and (18.3.20) correspond for this particular choice of Y. Hence the relationship (18.3.18) between $\gamma(x, t)$ and $g(x, t)$ follows. The usefulness of this result is described in section 18.4.2.

18.3.3.   Analytic Solutions of Stochastic Differential Equations

Stochastic differential equations with analytic solutions are few and far between. The most important example is a slight extension of (18.3.3) above: the solution of the Ito equation:

$$dx = \alpha(t)x\, dw \tag{18.3.21}$$

is

$$x(t) = \exp(M - \tfrac{1}{2}N), \tag{18.3.22}$$

where

$$M(w, t) = \int^t \alpha(s)\, dw(s) \tag{18.3.23}$$

and

$$N(t) = \int^t \sigma^2 \alpha(s)\, ds.$$

This can easily be verified from the differential rule (18.3.2).

McKean (1969, pp. 36–38) shows that this solution can be expressed as a series of Hermite polynomials, H_n [see § 10.5.2]:

$$x(t) = \sum_{n=0}^{\infty} \frac{1}{n!} Z_n \tag{18.3.24}$$

where $Z_n = N^{(1/2)n} H_n(M/N^{1/2})$.

If (18.3.21) had been an ordinary differential equation, its solution would have been [using (2.11.1)]

$$x(t) = \exp\left\{\int^t \alpha(s)\, dw(s)\right\} = \sum_{n=0}^{\infty} \frac{1}{n!} \left\{\int^t \alpha(s)\, dw(s)\right\}^n.$$

Thus the functions Z_n in stochastic calculus play the role of the functions $\{\int^t \alpha(s)\, dw(s)\}^n$ in ordinary calculus.

As in ordinary calculus, the explicit solution (18.3.22) to the time dependent linear equation (18.3.21) cannot be extended to the multidimensional case. However, the series solution (18.3.24) has been so extended by Wright (1975) to apply to the equation:

$$dx_i = \alpha(t) \sum_j x_j \, dw_{ij}. \qquad (18.3.25)$$

Using (18.3.15) and a natural multidimensional extension of (18.3.23), we define

$$Z^n_{j_1 \cdots j_n} = N^{\frac{n}{2}} H^n_{j_1 \cdots j_n} (M/N^{1/2})$$

in terms of the multidimensional Hermite polynomials $H_{j_1 \cdots j_n}$.
   The *multidimensional Hermite polynomials* $H^n_{j_1 \cdots j_n}$ are defined by

$$H^n_{j_1 \cdots j_n} = \frac{(-1)^n}{\omega} \frac{\partial^n \omega}{\partial x_{j_1} \cdots \partial x_{j_n}}$$

where   (18.3.26)

$$\omega = (2\pi)^{-n/2} \exp\left(-\tfrac{1}{2} \sum_{i=1}^n x_i^2\right).$$

The first few such polynomials are

$$H^0 = 1$$

$$H^1_j = x_j$$

$$H^2_{j_1 j_2} = x_{j_1 j_2} - \delta_{j_1 j_2}$$

$$H^3_{j_1 j_2 j_3} = x_{j_1} x_{j_2} x_{j_3} - (x_{j_1} \delta_{j_2 j_3} + x_{j_2} \delta_{j_1 j_3} + x_3 \delta_{j_1 j_2}).$$

Grad (1949) has shown that the multidimensional Hermite polynomials defined by (18.3.26) satisfy the recurrence relations

$$\left.\begin{array}{l} \dfrac{\partial}{\partial x_i} H^n_{j_1 \cdots j_n} = \delta_{ij_1} H^{n-1}_{j_2 \cdots j_n} + \delta_{ij_2} H^{n-1}_{j_1 j_3 \cdots j_n} + \ldots + \delta_{ij_n} H^{n-1}_{j_1 \cdots j_{n-1}} \\[3mm] H^{n+1}_{ij_1 \cdots j_n} = x_i H^n_{j_1 \cdots j_n} - \delta_{ij_1} H^{n-1}_{j_2 \cdots j_n} - \delta_{ij_2} H^{n-1}_{j_1 j_3 \cdots j_n} - \ldots - \delta_{ij_n} H^{n-1}_{j_1 \cdots j_{n-1}} \end{array}\right\} \quad (18.3.27)$$

Also they are orthogonal in the sense that the following integral taken over all space $-\infty < x_i < \infty$ $(i = 1, 2, \ldots, n)$

$$\int w H^m_{i_1 \cdots i_m} H^n_{j_1 \cdots j_n} \, d\Omega \qquad (18.3.28)$$

is zero

   (i) *whenever* $m \neq n$, and
   (ii) *also whenever* $m = n$ *and* $(i_1, \ldots, i_n)$ *is not a permutation of* $(j_1, \ldots, j_n)$;
        otherwise the integral equals unity.

Thus polynomials of different degrees are orthogonal, as are different polynomials of the same degree. The formulae (18.3.26) to (18.3.28) can be seen to correspond to the formulae for ordinary Hermite polynomials given in section 10.5.2.

The solution of (18.3.25) is then:

$$x_i = \sum_{n=0}^{\infty} \frac{1}{n!} a^n_{j_1 \ldots j_n i} Z^n_{j_1 \ldots j_n}$$

for any cyclically symmetric tensors a [see V, § 7.3.3] satisfying:

$$a^n_{j_1 \ldots j_{n-1} ki} = a^{n-1}_{j_1 \ldots j_{n-1} k - p(i-1)} \quad \text{for } 0 < k - p(i-1) \le p$$

$$= 0 \qquad\qquad\qquad \text{otherwise,}$$

where p is the dimension of $\mathbf{x}$.

18.4.   COMPUTER SIMULATION

In section 18.5 below, we see that certain practical systems can be represented by the Ito interpretation of a stochastic differential equation with the modified rules of calculus given in section 18.3.1, rather than by the Stratonovich or Marcus interpretations which obey ordinary calculus. In simulating such systems it is therefore important to know whether the standard numerical integration procedures are appropriate or not. In this section we shall discuss which numerical methods are appropriate for the Ito interpretation and which for the Stratonovich and Marcus interpretations.

18.4.1.   Wiener Case

The two most commonly used numerical integration methods are the Runge–Kutta methods [see III, § 8.2.2] and the Predictor Corrector methods [see III, § 8.3.1]. See, for instance Wilkes (1966, pp. 52–58). The simplest of the latter is the Euler method where the differential equation is regarded as discrete over the integration step interval, that is, the equation:

$$dx = f(x, t)\, dt + g(x, t)\, dw$$

is approximated by:

$$x_{r+1} = x_r + f(x_r, t_r)h + g(x_r, t_r)\{w(t_{r+1}) - w(t_r)\}$$

where the time interval over which the equation is to be integrated is divided into equal subintervals $[t_r, t_{r+1}]$ with $t_{r+1} - t_r = h$ [see III, § 8.2.1].

It is clear that the last term corresponds precisely with the definition of a stochastic integral under the Ito interpretation: $\tau_r = t_r$ given in section 18.2.1. On the other hand, the Runge–Kutta methods which involve successive approximations to the integrand at various points within the interval $[t_r, t_{r+1}]$ averaging near the centre, correspond with the Stratonovich interpretation: $\tau_r = \frac{1}{2}(t_r + t_{r+1})$. Thus one can expect intuitively that the Euler method will produce the Ito solution of a stochastic differential equation and that the Runge–Kutta methods will produce the Stratonovich solution. This hypothesis was verified for a specific example by Wright (1974). In extending the Euler method to Predictor and Predictor Corrector methods, he found that the Predictor method produces the Ito solution, but that when the Corrector and Modifier are also applied it is unclear what is the corresponding interpretation of the stochastic differential equation.

Later work has pursued the simulation of Ito equations. Harris (1977) has proposed a more complex simulation technique which converges more rapidly. To simulate the Ito solution of (18.3.4):

$$dx_i = f_i(\mathbf{x}, t)\, dt + \sum_p G_{ip}(\mathbf{x}, t)\, dw_p$$

he shows that one can use the numerical procedure:

$$x_i^{(r+1)} = x_i^{(r)} + h f_i^{(r)} + \sum_p G_{ip}^{(r)} Z_p^{(r)} + \tfrac{1}{2} \sum_{p,k} \left(\frac{\partial G_{ip}}{\partial x_k} \right)^{(r)} G_{kp}^{(r)} ((Z_p^{(r)})^2 - h)$$

$$+ \tfrac{1}{2} \sum_{p,q,k} \left(\frac{\partial G_{ip}}{\partial x_k} \right)^{(r)} G_{kp}^{(r)} Z_q^{(r)} Z_p^{(r)} \tag{18.4.1}$$

where a superscript in brackets indicates the time step and without loss of generality, the covariance matrix R_{pq} of the Wiener increments [see II, Definition 13.3.1] is diagonal. The Z_p are independent random variables: $Z_p \sim N(0, h R_{pp})$ [see II, Definition 6.6.1 and § 11.4.3].

18.4.2. Poisson Case

A numerical procedure of the type (18.4.1) is not available in the Poisson case. However, we now show intuitively that we can expect the Euler method to give the Ito solution and the Runge–Kutta method to give the Marcus solution. This hypothesis is substantiated by the following argument. (For simplicity we take $f = 0$ in (18.3.16) and (18.3.17) since this term is not involved in the transformation in any case.)

The Euler solution of the Ito equation (18.3.16) is:

$$x_{r+1} = x_r + \gamma(x_r, t_r)\{p_{r+1} - p_r\}. \tag{18.4.2}$$

The Runge–Kutta solution of (18.3.17) can be assumed to approximate it continuously:

$$\frac{dx^*}{dt} = g(x^*, t) \frac{p_{r+1} - p_r}{t_{r+1} - t_r}$$

that is

$$\int_{x_r^*}^{x_{r+1}^*} \frac{1}{g(\mu, t)} \, d\mu = \frac{p_{r+1} - p_r}{t_{r+1} - t_r} \int_{t_r}^{t_{r+1}} dt$$

$$= p_{r+1} - p_r. \tag{18.4.3}$$

Now $P[p_{r+1} - p_r > 1] = O((t_{r+1} - t_r)^2)$, so, as the integration interval $t_{r+1} - t_r \to 0$, it is sufficient to consider only two possible values of $p_{r+1} - p_r$, viz. 0 and

1. For each of these values (18.3.18) combined with (18.4.3) implies:

$$x^*_{r+1} = x^*_r + \gamma(x^*_r, t_r)\{p_{r+1} - p_r\}. \tag{18.4.4}$$

Thus as $t_{r+1} - t_r \to 0$ the Euler simulation (18.4.2) of (18.3.16) converges to the same value as the Runge–Kutta simulation (18.4.4) of (18.3.17).

These results were verified on numerical examples by Wright (1980). He found that better accuracy was sometimes obtained by one of these methods and sometimes by the other. Equation (18.3.18) can be used to transform between them to find which is better for a specific system.

18.5.   APPLICATIONS

18.5.1.   Stochastic Control

One of the first areas of application of stochastic calculus was in missile guidance, that is, the automatic optimal control of a system subject to continuous random disturbances. Optimal control is a major application of stochastic calculus; however it will not be dealt with here since it will be dealt with in another part of this Handbook [see the guidebook for engineers].

18.5.2.   Non-linear Filtering

Let us consider a system following the Ito stochastic differential equation (18.3.4) which we rewrite as:

$$dx_i = f_i(\mathbf{x}, t)\, dt + \sum_p G_{ip}(\mathbf{x}, t)\, dw_p. \tag{18.5.1}$$

Suppose noisy observations, $\mathbf{y}$, are made of the state variables, $\mathbf{x}$, according to:

$$dy_i = h_i(\mathbf{x}, t)\, dt + dq_i \tag{18.5.2}$$

where $q_i(t)$ are Wiener processes uncorrelated with the Wiener processes $w_p(t)$. Moreover, without loss of generality, let us assume that the cross-correlation matrix of the w_p is the identity, that is [see I (6.2.8)] $R_{pq} = \delta_{pq}$, and that the cross-correlation matrix of the q_i is $\mathbf{V}$.

The filtering problem is to estimate the x_i, given the observation y_i. Such a problem is an important preliminary to the automatic control problem of section 18.5.1 and has applications in all types of process control. The basic result, due to Kushner (1967a) gives the Ito equation for the conditional probability density, $P(\mathbf{x}, t)$, of the state variables $\mathbf{x}$, given the observations $\mathbf{y}$ up to time t [see II, § 6.5]:

$$dP = (L^*P)\, dt + P \sum_{i,p} (dy_i - E[h_i]\, dt)\, V_{ip}^{-1} (h_p - E[h_p]) \tag{18.5.3}$$

where

$$L^*P = -\sum_i \frac{\partial}{\partial x_i}(f_i P) + \tfrac{1}{2} \sum_{i,j,k} \frac{\partial^2}{\partial x_i \, \partial x_j}(G_{ik}G_{jk}P)$$

and V_{ip}^{-1} is the i, pth component of $\mathbf{V}^{-1}$.

This equation (18.5.3) can be used to obtain estimates of $\mathbf{x}(t)$ given the observations $\mathbf{y}$.

One such estimate is the least squares estimate [see VI, Chapter 11] which can be shown to be given by the mean of the distribution. The equation for the expected value of any function $\phi(\mathbf{x})$ is obtained from (18.5.3) by integration by parts:

$$d(E\phi(\mathbf{x})) = E[L\phi(\mathbf{x})]\, dt + \sum_{i,p}(dy_i - Eh_i \, dt)\, V_{ip}^{-1}\,(E[\phi h_p] - E[\phi]E[h_p])$$

$$(18.5.4)$$

where L is the adjoint of the operator L^* [see Definition 19.2.22].

As an example of the use of (18.5.4) we give the equations for the mean, m_1, of $\mathbf{x}$ and the variance m_2, in the scalar case [see II, § 9.11]:

$$dm_1 = Ef \, dt + (dy - m_1 \, dt)m_2/V$$

$$dm_2 = \{2E[(x - m_1)f] + E[G^2] - m_2^2/V\}\, dt + (dy - m_1 \, dt)m_3/V.$$

It can be seen that the right-hand sides of these equations involve higher moments $m_3, m_4, \ldots$. Thus the optimal filter is infinite dimensional. Approximations to it have been proposed by Nakamiyo (1970) and Bass *et al.* (1966). The innovations approach of Frost and Kailath (1971) has also been used. We cannot deal with these in detail here.

Another estimate of the state $\mathbf{x}(t)$ given the observations, is the maximum likelihood estimate [see VI, Chapter 6] given by the mode of the distribution. Again an infinite system of equations is obtained by Kushner (1967b) whose algebraic form is more complex than those for the moments given above. We therefore state only the first one, for the mode x^* in the scalar case:

$$dx^* = -P_1 \left\{ \frac{1}{P}\frac{\partial}{\partial x}(L^*P)\, dt + (dy - h \, dt)\frac{1}{V}\frac{\partial h}{\partial x} \right\}$$

$$+ \frac{P_1^2}{V}\frac{\partial h}{\partial x}\frac{\partial^2 h}{\partial x^2}\, dt + \frac{1}{2}\frac{P_2}{V}\left(\frac{\partial h}{\partial x}\right)^2 dt$$

where $P_1, P_2, \ldots$ are given by other equations in the system.

These non-linear filters and the finite dimensional approximations to them involve random terms (involving dy) whose variance depends on other variables in the system. They are derived using the rules of Ito stochastic calculus, e.g. (18.3.1), and should therefore be given the Ito interpretation. According to section 18.4.1 they should be simulated, either using the Euler method or using (18.4.1). Alternatively they could be transformed according to (18.3.13) and the Runge–Kutta method used.

18.5.3. Linear Systems with Random Parameters

Ariaratnam and Graefe (1965) have studied the behaviour of linear systems
with randomly varying parameters:

$$dx_i = \sum_j A_{ij}x_j\, dt + \sum_j x_j\, dw_{ij} + dw_{i0} \qquad (18.5.5)$$

where w_{ij} are Wiener Processes with variance $E[dw_{ip}\, dw_{jq}] = R_{ipjq}\, dt$.

Without using stochastic calculus, they obtained the equations for the
moments of the distribution of x [see II, § 9.11], and observed that there are
two possible results. The equations for the moments can be obtained much
more quickly using the differential rule (18.3.7) to give the equation for
$Y = x_1^{k_1} x_2^{k_2} \ldots x_n^{k_n}$, and then taking the expectation to give the equation for
the Nth moment $m_N(k_1, \ldots, k_n) = E[Y]$, where $\sum k_i = N$. The two possible
results correspond to the Ito and Stratonovich interpretations of (18.5.5)
which are related by the additional term (18.3.14), which, for (18.5.5) is:

$$\tfrac{1}{2} \sum_{j,p,q} x_q \delta_{pj} R_{ipjq} = \tfrac{1}{2} \sum_{j,q} x_q R_{ijjq}.$$

This means that a difference in interpretation results in a difference in the
deterministic part, A_{ij}, of the equation.

The equations for the moments were used by Pell and Aris (1969) to study
a sequence of chemical reactions $A \rightarrow B \rightarrow C$ with randomly fluctuating rate
constants. Clearly care is necessary in the interpretation of equations used to
model such a system. More recently Harris (1978) has used a specifically Ito
representation of randomly varying rates of the biochemical processes in
waste-water treatment systems.

18.5.4. Financial Modelling

The commercial gains possible from useful models of commodity or stock
market prices are considerable and attempts have been made dating from the
turn of the last century, by Bachelier [see Cootner (1964)]. More recently
Osborne (1959) has proposed a model in which the proportional change in
price, $p(t)$ is a Wiener increment, that is

$$\frac{dp}{p} = dw. \qquad (18.5.6)$$

This is the equation considered as an example in section 18.3.1 with solution
(18.3.3) under the Ito interpretation. This interpretation also leads to there
being no rise or fall in price on average over time, because of the Martingale
property (18.2.2). This is desirable intuitively and is a reason for choosing
the Ito interpretation. The observed rise in price can then be represented
deterministically in the equation:

$$dp = ap\, dt + p\, dw.$$

Another reason for choosing the Ito interpretation of (18.5.6) is that the jobber (U.K. terminology) has to state his prices before knowing how much he is being asked to buy or sell. Since this amount is the random factor dw, this means that p in (18.5.6) is evaluated at a time before dw. This corresponds to the McShane integral, with $\tau_r \leq t_r$, which, as stated in section 18.2.4, agrees with the Ito integral when both exist.

The multidimensional case of (18.5.6) representing the interactions between the prices of several stocks cannot be solved by an extension of (18.3.3) and the solution is given instead by (18.3.25).

There is a considerable literature on the use of stochastic differential equations with the Ito interpretation in modelling stock market prices. Merton (1971) studies the optimum portfolio problem and Smith (1976) reviews various approaches to option pricing. Other financial modelling applications are given in Merton (1973), (1974).

18.5.5. Population Dynamics

Consider a population of simple individuals which reproduce by dividing in two. Let the probability of an individual reproducing in time δt be $\eta_n \delta t$. The equation for the probability $P_n(t)$ of there being n individuals in the population at time t can be derived as:

$$\frac{dP_n}{dt} = -n\eta_n P_n(t) + (n-1)\eta_{n-1} P_{n-1}(t).$$

Using a moment generating function [see II, § 12.2], the equations for the moments, m_i, of the distribution of n can be found. For instance if we include a crowding effect by taking $\eta_n = \lambda - n\mu$, then we get

$$\frac{dm_1}{dt} = \lambda m_1 - \mu m_2 \qquad (18.5.7)$$

for $n < \lambda/\mu$.

This representation of the behaviour of a system by its moments is similar to the approach in section 18.5.3. Here again we can derive the same results more quickly using stochastic differential equations. However, in section 18.5.3 we had a Wiener disturbance and here we have a Poisson one. The stochastic differential equation for n is

$$dn = (\lambda - n\mu)n\, dt + dp$$

where $E[dp] = 0$.

The equation (18.5.7) for the first moment follows directly from this by taking the expectation. The equations for higher moments are obtained using the differential rule (18.3.9). For instance

$$d(n^2) = \{(n+1)^2 - n^2\}\{(\lambda - n\mu)n\, dt + dp\}$$
$$= \{n\lambda + n^2(2\lambda - \mu) - 2n^3\mu\}\, dt + (2n+1)\, dp.$$

Taking expectations, we obtain the equation for the second moment:

$$\frac{dm_2}{dt} = \lambda m_1 + (2\lambda - \mu)m_2 - 2\mu m_3.$$

This use of stochastic differential rules to obtain equations for the moments of random processes is a major advantage of using stochastic differential equations to model those processes.

18.5.6.   Other Applications

Possible applications that have not been explored in detail in the literature occur in particle physics. The velocity, $\mathbf{v}$, of a particle of charge, e, in a randomly varying electromagnetic field, $\mathbf{E}$, $\mathbf{B}$ (for instance that due to black body radiation inside a heated cavity) is given by:

$$m\dot{\mathbf{v}} = e\mathbf{E} + \frac{e}{c}\mathbf{v} \wedge \mathbf{B}$$

where m, c are constants depending on the units used.

This equation gives different results under the Ito and Stratonovich interpretations and is complicated by the fact that $\mathbf{E}$, $\mathbf{B}$ are related by the Maxwell equations.

The quantum mechanical representation using the Schrödinger equation:

$$i\hbar\dot{\psi} = (T + V)\psi$$

also has different Ito and Stratonovich interpretations when the potential energy V is random.

18.6.   CONCLUSION

Above we have given an intuitive basis wherever possible for some of the results of the theory of stochastic differential equations that are most useful in practical applications. We have then indicated the diversity of those areas of application and how the results can be used in them.

D. J. W.

REFERENCES

Ariaratnam, S. T. and Graefe, P. W. U. (1965). Linear Systems with Stochastic Coefficients, *Int. J. Contr. Pt. I* **1** (3), 239–250; *Pt. II* **2** (2), 161–169; *Pt. III* **2** (3), 205–210.

Bass, R. W., Norum, V. D. and Schwartz, L. (1966). Optimal Multichannel Non-Linear Filtering, *J. Math. Anal. Appl.* **16**, 152–164.

Cootner, P. H. (1964). *The Random Character of Stock Market Prices*, MIT Press.

Doob, J. L. (1953). *Stochastic Processes*, John Wiley.

Frost, P. A. and Kailath, T. (1971). An Innovations Approach to Least Squares Estimation—Part III: Non-Linear Estimation in White Gaussian Noise, *IEEE Trans. A.C.* **16** (3), 217–226.

Gihman, I. I. and Skorohod, A. V. (1972). *Stochastic Differential Equations*, Springer-Verlag.

Grad, H. (1949). Note on n-Dimensional Hermite Polynomials, *Comm. Pure Appl. Maths.* **2**, 325–330.

Harris, C. J. (1977). Modelling, Simulation and Control of Stochastic Systems, *Int. J. Syst. Sci.* **8** (4), 393–411.

Harris, C. J. (1978). Minimum Variance Control, Estimation and Simulation of Non-Linear Models of Wastewater Treatment Systems, *Proc. IEEE* **125** (5), 441–446.

Ito, K. (1950). Stochastic Differential Equations on a Differentiable Manifold, *Nagoya Math. J.* **1**, 35–47.

Ito, K. (1951a). Stochastic Differential Equations, *Mem. Amer. Math. Soc.* **4**.

Ito, K. (1951b). On a Formula Concerning Stochastic Differentials, *Nagoya Math. J.* **3**, 55–65.

Kushner, H. J. (1967a). Dynamical Equations for Optimal Non-Linear Filtering, *J. Diff. Equns.* **3**, 179–190.

Kushner, H. J. (1967b). Non Linear Filtering: The Exact Dynamical Equations Satisfied by the Conditional Mode, *IEEE Trans. A.C.* **12**, 262–267.

Marcus, S. I. (1978). Modelling and Analysis of Stochastic Differential Equations Driven by Point Processes, *IEEE Trans. Inf. Theory* **24**, 164–172.

McKean, H. P. Jnr. (1969). *Stochastic Integrals*, Academic Press.

McShane, E. J. (1969a). Towards a Stochastic Calculus, *Proc. Nat. Acad. Sci.* **63** Pt. I, 275–280, Pt. II, 1084–1087.

McShane, E. J. (1969b). Stochastic Integrals and Stochastic Functional Equations, *SIAM J. App. Maths.* **17** (2), 287–306.

McShane, E. J. (1971). *Stochastic Differential Equations and Models of Random Processes*, Proc. 6th Berkeley Symposium.

Merton, R. C. (1971). Optimum Consumption and Portfolio Rules in a Continuous Time Model, *J. Ec. Thy.* **3** (4), 373–413.

Merton, R. C. (1973). An Intertemporal Capital Asset Pricing Model, *Econometrica* **41** (5), 867–887.

Merton, R. C. (1974). On the Pricing of Corporate Debt: The Risk Structure of Interest Rates, *J. Finance* **29** (2), 449–470.

Nakamizo, T. (1970). On State Estimation for Non-Linear Dynamical Systems, *Int. J. Contr.* **11**, 683–695.

Osborne, M. F. M. (1959). Brownian Motion in the Stock Market, *Operations Res.* **7**, 145–173.

Pell, T. M. and Aris, R. (1969). Some Problems in Chemical Reactor Analysis with Stochastic Features, *Ind. Eng. Chem. Fund.* **8**, 339–345.

Smith, C. W. Jr. (1976). Option Pricing: a Review, *J. Fin. Econs.* **3**, 3–51.

Stratonovich, R. L. (1966). A New Representation for Stochastic Integrals and Equations, *J. SIAM Contr.* **4** (2), 362–371.

Wilkes, M. V. (1966). *A Short Introduction to Numerical Analysis*, Cambridge University Press.

Wong, E. (1971). *Stochastic Processes in Information and Dynamical Systems*, McGraw Hill.

Wong, E. and Zakai, M. (1965a). On the Convergence of Ordinary Integrals to Stochastic Integrals, *Annals Math. Stat.* **36**, 1560–1564.

Wong, E. and Zakai, M. (1965b). On the Relation Between Ordinary and Stochastic Differential Equations, *Int. J. Engng. Sci.* **3**, 213–229.

Wright, D. J. (1974). Digital Simulation of Stochastic Differential Equations, *IEEE Trans. A.C.* **19**, 75–76.

Wright, D. J. (1975). Multidimensional Hermite Polynomials and Ito Equations, *J. SIAM App. Maths.* **28** (3), 555–558.

Wright, D. J. (1980). Digital Simulation of Poisson Stochastic Differential Equations, *Int. J. Syst. Sci.* **11** (6), 781–785.

CHAPTER 19

Functional Analysis

In Rudin's monograph (Rudin (1973)), functional analysis is described as the study of certain topological-algebraic structures and of the methods by which knowledge of these structures can be applied to analytic problems. In this chapter a choice of the methods suitable for applications will be presented. In the light of the ideas mentioned in the introductory section of Volume I, Chapter 1, we emphasize the concepts and methods which are of a more computational nature than those of conceptional character. This in fact implies that the topological aspects of our explanation are based on the simplest but most important concepts and actually only the norm and metric topologies occur in this chapter.

Many problems of analysis are concerned not only with single objects such as a function [see I, § 1.4] or a measure [see, for example, Definition 4.9.9], but with classes of such objects. In this way there occur linear vector spaces and their maps—i.e. operators [see I, § 5.2 and § 5.11] either with real scalars or with complex ones. An interplay between various objects is reflected by concepts through which some features such as closeness, position and the corresponding quantities such as distances, angles etc. can be expressed. This leads to a necessity of supplying the spaces with metrics or topologies [see § 11.1]. The simplest but most fruitful way of doing that is to introduce a norm. The resulting structure is called normed linear vector space, or shortly, normed space [see § 11.6].

Because of the paramount importance of the field of real numbers $\mathbb{R}$ and the field of complex numbers $\mathbb{C}$ [see I, § 2.6 and § 2.7] we consider only vector spaces over these fields. If it is not necessary to distinguish between them we use the symbol $\mathbb{F}$ for either of the two fields $\mathbb{R}$ or $\mathbb{C}$ and in such situations we frequently omit mentioning the field at all.

19.1. REMARKS ON VECTOR SPACES AND LINEAR OPERATORS

We briefly summarize some basic properties of two very important tools of modern mathematics: linear vector spaces and linear maps on such spaces.

765

These topics are discussed in Chapters 5, 6, 7 and 10 of Volume I with an explicit emphasis on finite dimensionality. The reader can find some examples and motivation for examining such concepts there.

19.1.1. Vector Spaces

DEFINITION 19.1.1. A *vector space* or *linear space* X over the field $\mathbb{F}$ consists of a set of objects called vectors with the following properties:

  (i) any two vectors $\mathbf{a}$ and $\mathbf{b}$ have a sum $\mathbf{a} + \mathbf{b}$ which is a uniquely defined vector;
  (ii) if $\mathbf{a}$ is a vector and α is a scalar then a unique vector $\alpha\mathbf{a}$ (or $\mathbf{a}\alpha$) is defined.

The compositions referred to in (i) and (ii) satisfy certain rules, a complete list of which is in Volume I, § 5.1.

DEFINITION 19.1.2. A non-void subset M of a vector space X is called a *subspace* of X if $\mathbf{a} + \mathbf{b}$ and $\alpha\mathbf{a}$ are in M for arbitrary $\alpha \in \mathbb{F}$ and $\mathbf{a}$ and $\mathbf{b}$ in M. A subspace M is called *proper* if $M \neq X$ and *non-trivial* if M is proper and M contains at least two different vectors.

EXAMPLE 19.1.1. Let $\Omega \subset \mathbb{R}^n$ $n \geq 1$ be an open region and $\bar{\Omega}$ its closure [see § 11.2]. The set of all scalar functions u continuous on $\bar{\Omega}$ is denoted by $C^0(\bar{\Omega})$. Obviously, $C^0(\bar{\Omega})$ is a linear space the operations being such that $u + v = w$ and $\alpha u = z$ mean $w(s) = u(s) + v(s)$ and $z(s) = \alpha u(s)$ for $s \in \bar{\Omega}$. [See also Example 11.1.5.]
  We write $C^0([a, b])$ instead of $C^0(\bar{\Omega})$ if $n = 1$ and $\Omega = (a, b)$. We also use this notation in other similar situations.
  Let $n \geq 1$ and $\mathbf{a} = (a_1, \ldots, a_n)$ be a vector whose components are non-negative integers. We let $|\mathbf{a}| = \sum a_j$.
  Let u be a scalar function on $\Omega \subset \mathbb{R}^n$. We denote by $D^{\mathbf{a}}u$ the partial derivative

$$\frac{\partial^{|\mathbf{a}|} u}{\partial^{a_1} x_1, \ldots, \partial^{a_n} x_n}$$

[see § 5.5 and cf. § 8.1].

EXAMPLE 19.1.2. Let $\Omega \subset \mathbb{R}^n$, $n \geq 1$ be an open region [see § 11.2]. By $C^k(\Omega)$ we denote the space of scalar-valued functions u on Ω such that all the derivatives $D^{\mathbf{a}}u$ of orders $|\mathbf{a}| \leq k$ are continuous on Ω [see Definition 5.9.2]. We write $C(\Omega)$ instead of $C^0(\Omega)$ and $C(\bar{\Omega})$ instead of $C^0(\bar{\Omega})$.
  The space $C^k(\Omega)$ contains a subspace, the space $C^k(\bar{\Omega})$ consisting of functions u being such that $D^{\mathbf{a}}u(\mathbf{x}) = v_{\mathbf{a}}(\mathbf{x})$ for $\mathbf{x} \in \Omega$ and $v_{\mathbf{a}} \in C(\bar{\Omega})$.

DEFINITION 19.1.3. Let u be a scalar function on $\Omega \subset \mathbb{R}^n$. The closure of the set $M = \{\mathbf{x} \in \Omega: u(\mathbf{x}) \neq 0\}$ is called the *support* of u and is denoted by $\operatorname{supp} u$.

We introduce some further spaces defined as follows; $C_0^k(\Omega)$ is the space of functions $u \in C^k(\Omega)$ such that $\text{supp}\, u \subset \Omega$. Furthermore, let

$$C_0^\infty(\Omega) = \bigcap_{k=0}^{\infty} C^k(\Omega) \quad \text{and} \quad C_0^\infty(\bar{\Omega}) = \bigcap_{k=0}^{\infty} C^k(\bar{\Omega}).$$

DEFINITION 19.1.4. Let S be an arbitrary subset of a vector space X. The collection of all finite linear combinations of vectors in S forms a subspace called the *linear hull* of S. The linear hull of a set $S \subset X$ is the smallest (with respect to the set inclusion) vector subspace of X containing S.

DEFINITION 19.1.5. Let $r \geq 2$ be a positive integer and $M_1, \ldots, M_r$ be vector sub-spaces of X. Let $\mathbf{x}_j \in M_j$ and let $\alpha \mathbf{x}_1 + \ldots + \alpha \mathbf{x}_r = \mathbf{0}$, $\alpha_j \in \mathbb{F}$, $\alpha_j \neq 0$, $j = 1, \ldots, r$, imply that $\mathbf{x}_j = \mathbf{0}$. Then the linear hull of the union $M_1 \cup \ldots \cup M_r$ is called the *direct sum* of the subspaces $M_1, \ldots, M_r$ and is denoted by $M_1 \oplus \ldots \oplus M_r$ [see also I, § 5.5].

DEFINITION 19.1.6. Let M be a subspace of X. We call two vectors $\mathbf{x}, \mathbf{y} \in X$ *equivalent modulo* M if $\mathbf{x} - \mathbf{y} \in M$; we write $\mathbf{x} \equiv \mathbf{y} (\text{mod } M)$. This relation is reflexive, symmetric and transitive, and hence the space X is divided into mutually disjoint equivalence classes [see I, § 1.3.3] two vectors being in the same equivalence class if and only if they are equivalent modulo M. The set of all these equivalence classes is denoted by X/M. Let $[\mathbf{x}]$ denote the equivalence class of the vector $\mathbf{x}$, thus $[\mathbf{x}] = [\mathbf{y}]$ if and only if $\mathbf{x} \equiv \mathbf{y} (\text{mod } M)$. Defining $[\mathbf{x}] + [\mathbf{y}] = [\mathbf{x} + \mathbf{y}]$ and $\alpha[\mathbf{x}] = [\alpha \mathbf{x}]$ for $\mathbf{x}, \mathbf{y} \in X$ and $\alpha \in \mathbb{F}$, X/M becomes a vector space, It is called the *quotient space of* X *modulo* M. The mapping ϕ of X onto X/M defined by $\phi(\mathbf{x}) = [\mathbf{x}]$ is called the *canonical mapping* of X onto the quotient X/M.

DEFINITION 19.1.7. A set A is called an *algebra over* $\mathbb{F}$ if (1) A is a vector space over $\mathbb{F}$ and (2) a binary operation [see I, § 1.5] called *multiplication* of vectors is defined on A such that the following rules hold:

(i) $\alpha(\mathbf{xy}) = (\alpha \mathbf{x})\mathbf{y}$;
(ii) $\mathbf{x}(\alpha \mathbf{y}) = \alpha(\mathbf{xy})$;
(iii) $(\mathbf{x} + \mathbf{y})\mathbf{z} = \mathbf{xz} + \mathbf{yz}$;
(iv) $\mathbf{x}(\mathbf{y} + \mathbf{z}) = \mathbf{xy} + \mathbf{xz}$;
(v) there is a vector $\mathbf{e} \in A$ called the *unit element of* A such that $\mathbf{xe} = \mathbf{ex} = \mathbf{x}$, for arbitrary $\mathbf{x}, \mathbf{y}, \mathbf{z} \in A$ and $\alpha \in \mathbb{F}$.
(vi) If $\mathbf{xy} = \mathbf{yx}$ for $\mathbf{x}, \mathbf{y} \in A$ then A is called *commutative*.

EXAMPLE 19.1.3. Denote by M_n the set of all square matrices of order $n \geq 1$. If we define addition and componentwise multiplication by scalars, M_n constitutes a vector space [see I, § 6.2]. The multiplication of matrices (rows by columns) may be accepted as the second operation. Thus, M_n is an algebra. For $n > 1$, M_n is non-commutative; the subalgebra D_n of diagonal matrices is obviously commutative [see I, § 6.7(iv)].

19.1.2. Linear Operators

A linear operator is a kind of function whose domain is a linear space and whose range is contained in another linear space (possibly the same as the first one). If **T** is a linear operator we generally omit parentheses and write **Tx** instead of **T(x)** whenever it seems convenient.

DEFINITION 19.1.8.   Let X and Y be vector spaces over the same field $\mathbb{F}$. Let **T** be a function with domain X and range contained in Y. Then **T** is called a *linear operator* or, more explicitly, a linear operator on X into Y, if the following two conditions are satisfied: $\mathbf{T}(\mathbf{x}+\mathbf{y}) = \mathbf{Tx}+\mathbf{Ty}$ and $\mathbf{T}(\alpha\mathbf{x}) = \alpha\mathbf{Tx}$ for arbitrary $\mathbf{x}, \mathbf{y} \in X$ and $\alpha \in \mathbb{F}$. If the range of a linear operator **T** is contained in the field $\mathbb{F}$, we call **T** a *linear functional* on X.

Sometimes we consider a linear operator **T** whose domain is a proper subset of a given vector space X. The domain is itself a vector space (by definition) [see I, § 5.5]. We denote this domain of **T** by $\mathscr{D}(\mathbf{T})$ and the range of **T** by $\mathscr{R}(\mathbf{T})$.

EXAMPLE 19.1.4.   Every linear operator **T** mapping $\mathbb{R}^n$ into $\mathbb{R}^n$ can be represented by a matrix; let us denote it also by **T** [see I, § 5.11].

EXAMPLE 19.1.5.   Let $X = C([0, 1])$ be the space defined in Example 19.1.1. Let $\mathscr{D}$ be a subspace consisting of those functions u which have first and second order derivatives (denoted by primes) continuous on $[0, 1]$ and which, in addition, are such that $u(0) = u'(1) = 0$. Let the functions p and q be in X. We define T by setting $v = Tu$, where $v(s) = u''(s) + p(s)u'(s) + q(s)u(s)$. Then T is a linear operator on $\mathscr{D}$ into X.

DEFINITION 19.1.9.   Suppose X and Y are vector spaces and **T** is an operator on X into Y. The set $\mathscr{N}(\mathbf{T}) = \{\mathbf{x} \in X : \mathbf{Tx} = \mathbf{0}\}$ is called the *kernel* or the *null space* of **T**.

DEFINITION 19.1.10.   A linear operator **T** on X into Y is called *injective* if $\mathscr{N}(\mathbf{T})$ consists only of the zero vector; and **T** is called *surjective* if its range $\mathscr{R}(\mathbf{T}) = Y$. A linear operator **T** which is both injective and surjective is called an *isomorphism* and the corresponding spaces X and Y *isomorphic*.

DEFINITION 19.1.11.   A linear mapping **H** of an algebra A into an algebra B is called a *homomorphism* if $\mathbf{H}(\mathbf{xy}) = \mathbf{H}(\mathbf{x})\mathbf{H}(\mathbf{y})$ for arbitrary $\mathbf{x}, \mathbf{y} \in A$.

19.2. NORMED VECTOR SPACES AND THEIR MAPS

19.2.1. Normed Vector Spaces

A most useful tool of functional analysis is the concept of a normed vector space [see § 11.6].

DEFINITION 19.2.1.   A *norm* ν on a vector space X is a real-valued function on X such that

(i) $0 \leq \nu(\mathbf{x}) < \infty$ for $\mathbf{x} \in X$,
(ii) $\nu(\alpha \mathbf{x}) = |\alpha|(\mathbf{x})$ for $\mathbf{x} \in X$ and $\alpha \in \mathbb{F}$,
(iii) $\nu(\mathbf{x} + \mathbf{y}) \leq \nu(\mathbf{x}) + \nu(\mathbf{y})$ for $\mathbf{x}, \mathbf{y} \in X$ and
(iv) $\nu(\mathbf{x}) = 0$ if and only if $\mathbf{x} = \mathbf{0}$.

If ν satisfies (i), (ii) and (iii) but not (iv) it is called a *pseudonorm* or a *semi-norm*.

A vector space X equipped with a norm ν is called a *normed vector space* over $\mathbb{F}$. More explicitly we write (X, ν) in order to distinguish between possible different norms on X. However, usually there is only one single norm on the space considered and hence we use the symbol X not only to denote a vector space but also a normed space. We follow this usage and denote the norms in various spaces using the same single symbol $\|.\|$. We also call $\|\mathbf{x}\|$ the norm of $\mathbf{x} \in X$. Obviously, a subspace of a normed vector space is a normed vector space.

EXAMPLE 19.2.1.   Let $\Omega \subset \mathbb{R}^n$, $n \geq 1$, be an open and bounded region. Then the formula

$$\|\mathbf{x}\| = \sup \{|\mathbf{x}(\mathbf{s})| : \mathbf{s} \in \bar{\Omega}\}$$

[see I, § 2.6.3] defines a norm on the space $C(\bar{\Omega})$. Usually the symbol $C(\bar{\Omega})$ refers to the normed vector space equipped with the above *sup-norm*.

EXAMPLE 19.2.2.   Let l^p be the space of all sequences $\mathbf{x} = (\xi_1, \ldots, \xi_k, \ldots)$ such that $\xi_k \in \mathbb{C}$ and $\sum_{k=1}^{\infty} |\xi_k|^p$ converges [see § 1.7]; in l^p the linear vector space operations are the standard componentwise ones. Then $\|\mathbf{x}\| = (\sum |\xi_k|^p)^{1/p}$ is a norm on l^p [cf. Example 11.6.3].

EXAMPLE 19.2.3.   Let p be a positive integer. Let $L^p(\Omega)$ denote the collection of equivalence classes of scalar functions, representatives of which are Lebesgue integrable on Ω with the power p [see Definition 4.9.6]. If $[u] \in L^p(\Omega)$ and u is a representative of $[u]$, we define

$$\|[u]\| = \left\{ \int_{\Omega} |u(s)|^p \, ds \right\}^{1/p}. \tag{19.2.1}$$

For $p \geq 1$ every $L^p(\Omega)$ is a normed vector space. But for $0 < p < 1$ formula (19.2.1) is not a norm. Usually one does not distinguish between an element (a class) $[u] \in L^p(\Omega)$ and its representative u (a function).

Let X and Y be normed vector spaces and $Z = X \times Y$ be the Cartesian product of X and Y [see I (1.2.18)]. We let

$$\|\mathbf{z}\| = \|\mathbf{x}\| + \|\mathbf{y}\|, \tag{19.2.2}$$

where $(\mathbf{x}, \mathbf{y}) = \mathbf{z} \in Z$, $\mathbf{x} \in X$ and $\mathbf{y} \in Y$. Hence the Cartesian product $X \times Y$ is a normed vector space.

A norm on a vector space X induces a 'metric' in X by setting $d(\mathbf{x}, \mathbf{y}) = \|\mathbf{x} - \mathbf{y}\|$ [see § 11.1 and I, § 10.1]. More generally we have:

DEFINITION 19.2.2.   A *metric* or *distance function* d on a vector space X is a real valued function on the Cartesian product $X \times X$ such that

(a) $0 \le d(\mathbf{x}, \mathbf{y}) < \infty$ for $\mathbf{x}, \mathbf{y} \in X$ and $d(\mathbf{x}, \mathbf{y}) = 0$ if and only if $\mathbf{x} = \mathbf{y}$;

(b) $d(\mathbf{x}, \mathbf{y}) = d(\mathbf{y}, \mathbf{x})$ for $\mathbf{x}, \mathbf{y} \in X$;

(c) $d(\mathbf{x}, \mathbf{z}) \le d(\mathbf{x}, \mathbf{y}) + d(\mathbf{y}, \mathbf{z})$ for $\mathbf{x}, \mathbf{y}, \mathbf{z} \in X$. A distance function d is called *invariant* if

(d) $d(\mathbf{x} + \mathbf{z}, \mathbf{y} + \mathbf{z}) = d(\mathbf{x}, \mathbf{y})$.

A vector space X equipped with a distance function d is called a *metric space*, denoted (X, d) or simply by X if no confusion can arise about the nature of the distance function.

We see that a normed vector space is a metric space with an invariant metric. Any metric on a vector space X induces a notion of convergence in X:

DEFINITION 19.2.3.   A sequence $\{\mathbf{x}_k\}$, $\mathbf{x}_k \in X$, is *convergent* if there is a vector $\mathbf{x} \in X$, called the *limit*, such that $\lim d(\mathbf{x}_k, \mathbf{x}) = 0$ [see § 1.2]. If the metric is defined by $d(\mathbf{x}, \mathbf{y}) = \|\mathbf{x} - \mathbf{y}\|$, where $\|.\|$ is some norm, such convergence is called *norm convergence* or *strong convergence*.

DEFINITION 19.2.4.   A subset M of a metric space (X, d) is said to be *open* if $\mathbf{x} \in M$ implies the existence of a real $\rho > 0$ such that $\mathbf{y} \in M$ whenever $d(\mathbf{x}, \mathbf{y}) < \rho$. A subset Q of a metric space X is called *closed* if there is an open set $M \subset X$ such that Q is the complement of M, i.e. $Q = X \backslash M$. Recall that if M is closed and $\{\mathbf{x}_k\}$ is convergent, where $\mathbf{x}_k \in M$, then $\mathbf{x} \in M$. [See § 11.3.]

Let X be a normed vector space and Q a closed subspace. We let $[\mathbf{x}] \in X/M$ [see Definition 19.1.6] and put

$$\|[\mathbf{x}]\| = \inf \{\|\mathbf{x}\|: \mathbf{x} \in [\mathbf{x}]\}$$

[see I, § 2.6.3]. In this way the quotient space X/Q is a normed vector space.

DEFINITION 19.2.5.   Let M be any subset of X, where X is a metric vector space, and let S be any closed set containing M. The intersection of all these closed sets is the *closure* of M and it is denoted by $\overline{M}$. A subset M of a metric space X is called *dense* in X if $\overline{M} = X$. A metric space X is called *separable* if there is a countable dense subset $S \subset X$ [see I, § 1.6].

Most of the function spaces occurring in functional analysis are separable, e.g. the spaces $C(\Omega)$, $L^p(\Omega)$, l^p etc. discussed in Examples 19.2.1 to 19.2.3 are separable. On the other hand, the space $BV(0, 1)$ consisting of all functions of bounded variation on $[0, 1]$ is not separable [see Definition 2.7.3].

DEFINITION 19.2.6. A sequence $\{\mathbf{x}_k\}$, $\mathbf{x}_k \in X$, where X is a metric space, is called a *Cauchy sequence*, if $d(\mathbf{x}_m, \mathbf{x}_n) \to 0$ as $m \to \infty$ and $n \to \infty$. A metric space X is called *complete* if every Cauchy sequence $\{\mathbf{x}_k\}$, $\mathbf{x}_k \in X$, is convergent in X. A complete normed vector space X is called a *Banach space*. [See § 11.3 and § 11.6.]

DEFINITION 19.2.7. An algebra A is called a *normed algebra* if the multiplication of vectors in A satisfies the relation $\|\mathbf{xy}\| \le \|\mathbf{x}\| \|\mathbf{y}\|$ for every pair $\mathbf{x}, \mathbf{y} \in A$. If A is complete as a normed vector space, A is called a *Banach algebra*.

Let $A = C([0, 1])$ with the standard pointwise operations and the sup-norm [see Example 19.2.1]. Then A is a Banach algebra.

THEOREM 19.2.1. *Let V be a normed vector space and μ a norm on V. There exists a complete normed space X with the norm ν such that $V \subset X$ and V is ν-dense in X, where $\nu(\mathbf{x}) = \mu(\mathbf{x})$ for $\mathbf{x} \in V$. We call the space X* a completion of the space V in the norm μ.

A well-known example for an illustration of Theorem 19.2.1 is the field of real numbers $\mathbb{R}$. The space $\mathbb{R}$ with the absolute value as norm is a completion of the space $\mathbb{Q}$ of rational numbers [see § 11.3 or I, § 2.6.2].

EXAMPLE 19.2.4. Let k and p be integers, $k \ge 0$ and $p \ge 1$. Define a norm

$$\|u\|_{k,p} = \sum_{|\mathbf{b}| \le k} \left[\int_\Omega |D^\mathbf{b} u(\mathbf{s})|^p \, d\mathbf{s} \right]^{1/p} \qquad (19.2.3)$$

where $\Omega \subset \mathbb{R}^n$, $n \ge 1$, $\mathbf{s} = (s_1, \ldots, s_n)$, and u is a scalar-valued function on Ω. A completion of the space $C^k(\Omega)$ in the norm (19.2.3) constitutes a Banach space. It is called a *Sobolev space* and is denoted by $W^{k,p}(\Omega)$. If $u \in W^{k,p}(\Omega)$ then the functions $D^\mathbf{a} u$ with $|\mathbf{a}| \ge 1$ are called *generalized derivatives* of u. We denote by $W_0^{k,p}(\Omega)$ the completion of the space $C_0^\infty(\Omega)$ [see Example 19.1.2] in the norm (19.2.3) [cf. Definition 8.5.8].

It is a natural property of the generalized derivatives that they coincide with the classical derivatives if the latter exist.

We observe that $W^{0,p}(\Omega) = L^p(\Omega)$ [see Example 19.2.3].

DEFINITION 19.2.8. Let $\{\mathbf{x}_k\}$ be a sequence in a metric space X. We say that the series $\sum_{k=1}^\infty \mathbf{x}_k$ is *convergent* (in X) if the sequence $\{\mathbf{s}_n\}$, where $\mathbf{s}_n = \sum_{k=1}^n \mathbf{x}_k$, converges in X [see Definition 19.2.3]; its limit $\mathbf{x}$ is said to be the *sum* of the series $\sum_{k=1}^\infty \mathbf{x}_k$.

It is easy to see that if X is a Banach space and the scalar series $\sum_{k=1}^\infty \|\mathbf{x}_k\| < \infty$, then the series $\sum_{k=1}^\infty \mathbf{x}_k$ is convergent in X.

A remarkable refinement of the concept of a norm is an *inner product function*. Some of the properties of inner product functions are discussed in Chapter 10 of Volume I.

DEFINITION 19.2.9.   An *inner product function*, or briefly *inner product*, on a vector space X is a scalar valued function φ on the Cartesian product $X \times X$ having the following properties:

(i) $\varphi(\lambda \mathbf{x} + \mu \mathbf{y}, \mathbf{z}) = \lambda \varphi(\mathbf{x}, \mathbf{z}) + \mu \varphi(\mathbf{y}, \mathbf{z})$ for arbitrary $\mathbf{x}, \mathbf{y}, \mathbf{z} \in X$ and $\lambda, \mu \in \mathbb{F}$;

(ii) $\varphi(\mathbf{x}, \mathbf{y}) = \overline{\varphi(\mathbf{y}, \mathbf{x})}$ for $\mathbf{x}, \mathbf{y} \in X$;

(iii) $0 \le \varphi(\mathbf{x}, \mathbf{x}) < \infty$ for all $\mathbf{x} \in X$ and $\varphi(\mathbf{x}, \mathbf{x}) = 0$ if and only if $\mathbf{x} = \mathbf{0}$.

The value $\varphi(\mathbf{x}, \mathbf{y})$ is also called an inner product of the vectors $\mathbf{x}$ and $\mathbf{y}$. A vector space X equipped with an inner product function φ is called an *inner product space*. When there is no danger of misunderstanding we use the same symbol $(.,.)$ to denote any inner product on various inner product spaces under consideration.

A direct verification shows that the function defined by setting $\nu(\mathbf{x}) = (\mathbf{x}, \mathbf{x})^{1/2}$ for $\mathbf{x} \in X$, is a norm on X. This norm is said to be *induced* by the inner product φ.

DEFINITION 19.2.10.   A complete inner product space X is called a *Hilbert space*. The term *pre-Hilbert space* sometimes occurs in the literature to denote an incomplete inner product space [see § 11.6].

Note that the vector spaces l^p, $L^2(\Omega)$ and $W^{k,2}(\Omega)$ are Hilbert spaces, as also are the Euclidean spaces $\mathbb{R}^n$ and $\mathbb{C}^n$. The inner product on $L^2(\Omega)$ is given by

$$(u, v) = \int_\Omega u(\mathbf{s}) \overline{v(\mathbf{s})} \, d\mathbf{s}.$$

DEFINITION 19.2.11.   Let $\mathbf{x}$ and $\mathbf{y}$ be two vectors in an inner product space X. If $(\mathbf{x}, \mathbf{y}) = 0$ we call those vectors *orthogonal*. We write $\mathbf{x} \perp \mathbf{y}$ or $\mathbf{y} \perp \mathbf{x}$. Let $M < X$. The set $\{\mathbf{x} \in X : (\mathbf{x}, \mathbf{y}) = 0 \text{ for all } \mathbf{y} \in M\}$ is called the *orthogonal comple-ment of M* and is denoted by $M^\perp$.

If M is a closed subspace of a Hilbert space X, then $X = M \oplus M^\perp$.

DEFINITION 19.2.12.   A set S of vectors in an inner product space X is called *orthogonal* if $\mathbf{x} \perp \mathbf{y}$ for every pair $\mathbf{x}, \mathbf{y} \in S$ and $\mathbf{x} \ne \mathbf{y}$. If, in addition, $(\mathbf{x}, \mathbf{x}) = 1$ for every $\mathbf{x} \in S$, the set S is called *orthonormal*. An orthonormal set $S \subset X$ is called *complete* if there is no orthonormal set in X such that S is a proper subset of it.

Note that the completeness of orthogonal sets plays an important role in the theory of Fourier series, and in other fields of mathematics [see § 20.4].

Up to now we have considered all vector spaces over either of the fields $\mathbb{R}$ or $\mathbb{C}$. However, there are some concepts in functional analysis which are meaningful only in vector spaces over $\mathbb{R}$. In more advanced work it is essential to use the field of complex numbers. There is a simple procedure for embedding a vector space over the real numbers into what is called its 'complexification':

DEFINITION 19.2.13. Let Y be a normed vector space over $\mathbb{R}$. We call the direct sum $Y \oplus iY = X$ a *complexification of* Y, where $i^2 = -1$. A norm on X is defined as follows: if $\mathbf{z} = \mathbf{x} + i\mathbf{y}$,

$$\|\mathbf{z}\| = \sup\{\|\mathbf{x}\cos\theta + \mathbf{y}\sin\theta\|: 0 \le \theta \le \pi\}.$$

If Y is an inner product space then, if $\mathbf{w} = \mathbf{u} + i\mathbf{v}$,

$$(\mathbf{z}, \mathbf{w}) = (\mathbf{x}, \mathbf{u}) - (\mathbf{y}, \mathbf{v}) + [(\mathbf{x}, \mathbf{v}) + (\mathbf{y}, \mathbf{u})]i$$

is an inner product on the complexification X. A complexification X of the space Y is complete whenever Y is complete.

19.2.2. Spaces of Linear Operators

Let X and Y denote Banach or Hilbert spaces. Let $\mathbf{T}$ be a linear operator mapping X into Y. Let

$$\|\mathbf{T}\| = \sup\{\|\mathbf{Tx}\|: \mathbf{x} \in X, \|\mathbf{x}\| \le 1\}. \qquad (19.2.4)$$

Here $\|\mathbf{Tx}\|$ denotes the norm in Y while $\|\mathbf{x}\|$ is the norm in X.

DEFINITION 19.2.14. Let $\mathbf{T}$ be a linear operator on X into Y. We call $\mathbf{T}$ *bounded* if $\|\mathbf{T}\| < \infty$, and $\|\mathbf{T}\|$ is called the *norm* of $\mathbf{T}$.

It can be shown that a linear operator $\mathbf{T}$ is continuous at some vector $\mathbf{x}$ [see Definition 5.3.1] if and only if it is continuous at $\mathbf{0}$ and that a linear operator $\mathbf{T}$ is continuous if and only if it is bounded.

If $\mathbf{T}$ and $\mathbf{S}$ are linear operators on X into Y we define a *sum* and a *scalar multiple* as follows: $\mathbf{U} = \mathbf{T} + \mathbf{S}$ and $\mathbf{V} = \alpha\mathbf{T}$ means that $\mathbf{Ux} = \mathbf{Tx} + \mathbf{Sx}$ and $\mathbf{Vx} = \alpha\mathbf{Tx}$ for all $\mathbf{x} \in X$ and $\alpha \in \mathbb{F}$.

The collection of all bounded linear operators mapping X into Y forms a normed vector space, the norm of which is given by (19.2.4). This space is denoted by $\mathscr{B}(X, Y)$; it is a Banach space whenever Y is complete. If $Y = X$ we write $\mathscr{B}(X)$ instead of $\mathscr{B}(X, X)$ and if $Y = \mathbb{F}$ we denote $\mathscr{B}(X, \mathbb{F})$ by the symbol X' and call it the *dual space of* X; the terms *adjoint* or *conjugate* are also used. The elements of X' are called *continuous* or *bounded linear functionals*.

THEOREM 19.2.2. *If X is a Hilbert space, then there is a conjugate linear isometry $\mathbf{x} \to \mathbf{x}'$ of X onto X' [see § 11.4] given by*

$$\mathbf{x}'(\mathbf{x}) = (\mathbf{x}, \mathbf{y}_{\mathbf{x}'}), \qquad \mathbf{x}, \mathbf{y}_{\mathbf{x}'} \in X, \mathbf{x}' \in X', \qquad (19.2.5)$$

where $\mathbf{y}_{\mathbf{x}'}$ is a suitable element.

This fact is a restatement of the famous *Riesz representation theorem* (Taylor (1958)).

DEFINITION 19.2.15.   If $\mathbf{T} \in \mathcal{B}(Y)$, where Y is a normed vector space over $\mathbb{F}$, then $\tilde{\mathbf{T}}$ defined by setting $\tilde{\mathbf{T}}\mathbf{z} = \mathbf{T}\mathbf{x} + i\mathbf{T}\mathbf{y}$ for $\mathbf{z} = \mathbf{x} + i\mathbf{y}$, where $\mathbf{x}, \mathbf{y} \in Y$ (i.e. $\mathbf{z}$ belongs to the complexification $X = Y \oplus iY$) is called a *complex extension of* $\mathbf{T}$.

This definition offers an effective way of investigating spectral properties of operators on vector spaces over $\mathbb{R}$. As usual, an operator $\mathbf{S}$ mapping its domain $\mathcal{D}(\mathbf{S}) \subset X$ into Y is called an *extension* of another linear operator $\mathbf{T}$ on $\mathcal{D}(\mathbf{T})$ into Y, if $\mathbf{S}\mathbf{x} = \mathbf{T}\mathbf{x}$ for $\mathbf{x} \in \mathcal{D}(\mathbf{T}) \subset \mathcal{D}(\mathbf{S})$.

If $\mathbf{T}$ is a bounded linear operator mapping a dense domain $\mathcal{D}(\mathbf{T})$ into Y, then the operator $\mathbf{S}$, where $\mathbf{S}\mathbf{x} = \lim \mathbf{T}\mathbf{x}_k$ for $\mathbf{x} \in X$, $\mathbf{x} = \lim \mathbf{x}_k$, is an extension of $\mathbf{T}$ such that $\overline{\mathcal{D}(\mathbf{T})} = X$ and $\mathbf{S} \in \mathcal{B}(X, Y)$.

PROPOSITION 19.2.3.   *Let X be a Banach space and $\mathbf{T}$ and $\mathbf{S}$ operators in* $\mathcal{B}(X)$. *The following relation holds for every pair of elements in* $\mathcal{B}(X)$:

$$\|\mathbf{TS}\| = \|\mathbf{ST}\| \le \|\mathbf{T}\| \|\mathbf{S}\|.$$

Also

$$\|\mathbf{I}\| = 1,$$

where $\mathbf{I}$ denotes the identity operator. Thus $\mathcal{B}(X)$ is a Banach algebra.

The latter result has a decisive impact on operator theory and its applications.

DEFINITION 19.2.16.   Let $\mathbf{T} \in \mathcal{B}(X, Y)$. We define the *dual* or *conjugate* $\mathbf{T}'$ of the operator $\mathbf{T}$ by setting $\mathbf{T}'\mathbf{y}' = \mathbf{x}'$ if and only if $\mathbf{x}'(\mathbf{x}) = \mathbf{y}'(\mathbf{T}\mathbf{x})$ for all $\mathbf{x} \in X$ and $\mathbf{y}' \in Y'$.

Evidently, $\mathbf{T}' \in \mathcal{B}(Y', X')$ and $\|\mathbf{T}'\| = \|\mathbf{T}\|$.

DEFINITION 19.2.17.   An operator $\mathbf{P}$ mapping a normed vector space X into itself is called a *projection* if $\mathbf{P}^2\mathbf{x} = \mathbf{P}\mathbf{x}$ for every $\mathbf{x} \in X$.

We see that the zero operator $\mathbf{0}$ and the identity operator $\mathbf{I}$ are projections and that $\|\mathbf{P}\| \ge 1$ whenever $\mathbf{P} \ne \mathbf{0}$. Obviously $\mathbf{I} - \mathbf{P}$ is also a projection.

DEFINITION 19.2.18.   Let $\mathbf{T}$ be a linear operator mapping its domain $\mathcal{D}(\mathbf{T}) \subset X$ into Y. The operator $\mathbf{T}$ is said to be *closed*, if its *graph* $\mathbf{G}(\mathbf{T}) = \{[\mathbf{x}, \mathbf{T}\mathbf{x}]: \mathbf{x} \in \mathcal{D}(\mathbf{T})\}$ is closed in $X \times Y$.

It is easy to see that a continuous linear operator is closed, but closed linear operators which are discontinuous do exist (see Taylor (1958)).

EXAMPLE 19.2.5.   Suppose p and q are real positive numbers, $p > 1$ and such that $1/p + 1/q = 1$. Let $X = L^p(0, 1)$ and $t = t(s, s')$ be a scalar-valued function on $(0, 1) \times (0, 1)$ such that

$$\int_0^1 \int_0^1 |t(s, s')|^q \, ds \, ds' < \infty.$$

Then T, defined by $Tx = y$ where $y(s) = \int_0^1 t(s, s') x(s') \, ds'$, is a linear operator and $T \in \mathcal{B}(L^p(0, 1), L^q(0, 1))$.

19.2.3. Three Fundamental Principles

Among the many deep ideas and results of functional analysis there are three which stand out as being most useful and applicable. These are the 'Hahn–Banach theorem', the 'closed graph theorem' and the 'uniform boundedness principle'. In today's mathematics they are considered as principles. In this section only some elementary versions of these principles are presented.

THEOREM 19.2.4 (*Hahn–Banach Theorem*). *Let X be a normed vector space, M a subspace of X, p a seminorm on X and f a linear functional such that $|f(\mathbf{x})| \leq p(\mathbf{x})$, for $\mathbf{x} \in M$. Then there exists an extension F of f such that $|F(\mathbf{x})| \leq p(\mathbf{x})$ for $\mathbf{x} \in X$; that is $F \in X'$.*

This is an immensely important fact telling us, among other things, that the dual space X' does contain a sufficiently large number of elements.

COROLLARY 19.2.5. *Let X be a normed vector space and $\mathbf{x}_0 \in X$, $\mathbf{x}_0 \neq \mathbf{0}$. Then there is an $\mathbf{x}' \in X'$ such that $\mathbf{x}'(\mathbf{x}_0) = \|\mathbf{x}_0\|$ and $\|\mathbf{x}'\| = 1$.*

Another consequence is

PROPOSITION 19.2.6. *For every $\mathbf{x} \in X$, where X is a normed vector space, we have that $\|\mathbf{x}\| = \sup \{|\mathbf{x}'(\mathbf{x})| : \mathbf{x}' \in X'\}$.*

Hence $\mathbf{x}'(\mathbf{x}) = 0$ for all $\mathbf{x}' \in X'$ implies $\mathbf{x} = \mathbf{0}$. In particular, if X is an inner product space, then $(\mathbf{x}, \mathbf{u}) = 0$ for every $\mathbf{x} \in X$ implies that $\mathbf{u} = \mathbf{0}$: The only vector orthogonal to all other vectors is the zero vector.

DEFINITION 19.2.19. A sequence $\{\mathbf{x}_k\}$, $\mathbf{x}_k \in X$, is called *weakly convergent* if the sequence of scalars $\{\mathbf{x}'(\mathbf{x}_k)\}$ is convergent for every $\mathbf{x}' \in X'$.

Obviously a strongly convergent sequence is also weakly convergent, the limit being the same, but the converse is not true.

EXAMPLE 19.2.6. Let $X = l^2$ [see Example 19.2.2] and $\mathbf{e}_k = (0, \ldots, 1, 0, \ldots)$, $k = 1, 2, \ldots$. The sequence (of sequences) $\{\mathbf{e}_k\}$ converges weakly to the zero vector. This is because of the linear conjugate isometry of X and

X' [see (19.2.5)] according to which $\mathbf{x}'(\mathbf{e}_k) = (\mathbf{e}_k, \mathbf{y}_{x'})$ for some $\mathbf{y}'_{x'} \in l^2$; but $(\mathbf{e}_k, \mathbf{y}_{x'}) = \eta_k$, where $\mathbf{y}_{x'} = (\eta_1, \ldots, \eta_k, \ldots)$ and $\eta_k \to 0$. On the other hand, $\|\mathbf{e}_k\| = 1$.

DEFINITION 19.2.20.   Let X'' denote the dual of X'; X'' is called the *second conjugate* or *bidual* of X. Let $\mathbf{J}$ be a map of X'' into X'' defined by $\mathbf{Jx} = \mathbf{x}''$ if and only if $\mathbf{x}''(\mathbf{x}') = \mathbf{x}'(\mathbf{x})$, where $\mathbf{x} \in X$, $\mathbf{x}' \in X'$ and $\mathbf{x}'' \in X''$. The map $\mathbf{J}$ is called the *canonical map of X into X''*: $\mathbf{J}$ is an isometry on X'': $\|\mathbf{Jx}\| = \|\mathbf{x}\|$ [see § 11.4]. If the range of J is the whole space X'' then the space X is called *reflexive*.

If X is reflexive then so are X' and every closed subspace $M \subset X$.

Note that Hilbert spaces are obviously reflexive and so also are the spaces $L^p(\Omega)$ and $W^{k,p}(\Omega)$, $1 < p < \infty$. On the other hand, the spaces $C^k(\bar{\Omega})$ are not reflexive.

It is the reflexivity of a space X that implies some useful properties of certain subsets of vector spaces, e.g. 'bounded sequences in a reflexive space contain weakly convergent subsequences' etc.

The remaining two principles are based on the completeness of the spaces considered.

THEOREM 19.2.7   (*Closed graph theorem*).   *Let X and Y be complete metric linear spaces and suppose $\mathbf{T}$ is a closed linear operator mapping X into Y. Then $\mathbf{T}$ is continuous.*

THEOREM 19.2.8.   *Let X be a Banach space and let S be a non-empty subset of X. Suppose that $\sup \{|\mathbf{x}'(\mathbf{x})|: \mathbf{x} \in S\} < \infty$ for each $\mathbf{x}' \in X'$. Then $\sup \{\|\mathbf{x}\|: \mathbf{x} \in S\} < \infty$, that is, S is bounded.*

One more theorem explains the term '*uniform boundedness*':

THEOREM 19.2.9.   *Let X be a normed vector space and Y a Banach space. Let $\mathcal{T}$ be a family of operators of $\mathcal{B}(X, Y)$ such that for a given $\mathbf{x} \in X$, $\|\mathbf{Tx}\| \le \mathcal{H}(\mathbf{x})$ with some $\mathcal{H}(\mathbf{x})$ for each $\mathbf{T} \in \mathcal{T}$. Then there is a constant $\mathcal{H}$ independent of $\mathbf{x} \in X$ such that $\|\mathbf{T}\| \le \mathcal{H}$ for all $\mathbf{T} \in \mathcal{T}$; that is, the set $\mathcal{T}$ is* uniformly bounded.

DEFINITION 19.2.21.   Let X be a complex Banach space and Δ an open set in the complex plane. We call a function $\mathbf{f}$ defined on Δ with values in X, *locally analytic* in Δ if $\mathbf{f}$ is differentiable with respect to the complex variable λ at each point λ_0 of Δ; that is there is an $\mathbf{f}'(\lambda_0) \in X$ such that

$$\lim_{\lambda \to \lambda_0} \left\| \frac{\mathbf{f}(\lambda) - \mathbf{f}(\lambda_0)}{\lambda - \lambda_0} - \mathbf{f}'(\lambda_0) \right\| = 0$$

[see also § 9.1].

An interesting application of the uniform boundedness principle, used in section 19.3.2, is the following theorem.

THEOREM 19.2.10. *Let X and Y be complex Banach spaces and let $\mathbf{F}$ be a function with values in $\mathscr{B}(X, Y)$, defined on an open set Δ in the complex plane. Suppose that $\mathbf{x}'(\mathbf{F}(\lambda)\mathbf{x})$ is differentiable at each point of Δ and every $\mathbf{x} \in X$ and $\mathbf{x}' \in Y'$. Then $\mathbf{F} = \mathbf{F}(\lambda)$ is locally analytic on Δ.*

19.2.4. Operators on Hilbert Spaces

In this section let X denote a complex Hilbert space.

DEFINITION 19.2.22. Let $\mathbf{T}$ be a linear operator mapping its dense domain $\mathscr{D}(\mathbf{T})$ into X. Let $\mathscr{D} \subset X$ be the set of all $\mathbf{y} \in X$ for which there exists a $\mathbf{y}^* \in X$ such that

$$(\mathbf{Tx}, \mathbf{y}) = (\mathbf{x}, \mathbf{y}^*) \tag{19.2.6}$$

holds for all $\mathbf{x} \in \mathscr{D}(\mathbf{T})$. By setting $\mathbf{y}^* = \mathbf{T}^*\mathbf{y}$ an operator $\mathbf{T}^*$ is defined; it maps $\mathscr{D}$ into X and is called the *adjoint* or *conjugate* of $\mathbf{T}$.

If $\mathbf{T} \in \mathscr{B}(X)$ then $\mathbf{T}^* \in \mathscr{B}(X)$ and $\|\mathbf{T}^*\| = \|\mathbf{T}\|$.

DEFINITION 19.2.23. A densely defined linear operator $\mathbf{T}$ on $\mathscr{D}(\mathbf{T})$ into X is called *symmetric* if $(\mathbf{Tx}, \mathbf{y}) = (\mathbf{x}, \mathbf{Ty})$ holds for all $\mathbf{x}, \mathbf{y} \in \mathscr{D}(\mathbf{T})$. A symmetric operator $\mathbf{T}$ is said to be *self-adjoint* if $\mathbf{T}^* = \mathbf{T}$.
It can be shown that a self-adjoint operator is closed and hence, if a symmetric operator $\mathbf{T}$ is not closed, $\mathbf{T}^* \neq \mathbf{T}$.
Obviously, every $\mathbf{T} \in \mathscr{B}(X)$ is self-adjoint if and only if it is symmetric. Self-adjoint operators $\mathbf{T} \in \mathscr{B}(X)$ are called *Hermitian operators*.

DEFINITION 19.2.24. An operator $\mathbf{T} \in \mathscr{B}(X)$ is called *skew symmetric* if $\mathbf{T}^* = -\mathbf{T}$ and *unitary* if $\mathbf{T}^*\mathbf{T} = \mathbf{TT}^* = \mathbf{I}$. More generally, an operator $\mathbf{T} \in \mathscr{B}(X)$ is called *normal* if $\mathbf{T}^*\mathbf{T} = \mathbf{TT}^*$.

EXAMPLE 19.2.7. Operators on $\mathbb{C}^n$ which are represented by symmetric (skew symmetric) matrices [see I, § 6.7] are symmetric (skew symmetric) operators. If $p = 2$ in Example 19.2.5 and the kernel $t = t(s, s')$ satisfies the relation $t(s, s') = \overline{t(s', s)}$ for $s, s' \in [0, 1]$, then the operator defined by the kernel T is symmetric in $L^2(0, 1)$.

DEFINITION 19.2.25. An operator $\mathbf{T}$ mapping a dense domain $\mathscr{D}(T) \subset X$ into X is called *bounded below* (*above*) if there is a constant $\gamma \in \mathbb{R}$ independent of $\mathbf{x}$ such that $(\mathbf{Tx}, \mathbf{x}) \geq \gamma(\mathbf{x}, \mathbf{x})$ $((\mathbf{Tx}, \mathbf{x}) \leq \gamma(\mathbf{x}, \mathbf{x}))$. A quantity γ_0 is called a *lower bound* of a bounded below operator $\mathbf{T}$, if γ_0 is the supremum of γ's for which the above relation holds. If γ_0 is positive, the operator **T** is called *positive definite*. In a similar way one defines *negative definite* operators.

PROPOSITION 19.2.11.   *A positive definite operator* **T** *is symmetric.*

Let $X = \mathbb{R}^2$ and let **T** be represented by the matrix

$$\mathbf{T} = \begin{bmatrix} 1 & \varepsilon \\ 0 & 1 \end{bmatrix}, 0 < \varepsilon < 1.$$

For every vector $\mathbf{x} = (\xi_1, \xi_2)'$ in $\mathbb{R}^2$ we have $(\mathbf{Tx}, \mathbf{x}) = \xi_1^2 + \varepsilon \xi_1 \xi_2 + \xi_2^2 \geq (1 - \varepsilon)(\mathbf{x}, \mathbf{x})$. Here a standard inner product is used in $\mathbb{R}^2$ [see I (9.1.1)]. We see that Proposition 19.2.11 may fail in Hilbert spaces over $\mathbb{R}$.

DEFINITION 19.2.26.   A projection $\mathbf{P} \in \mathcal{B}(X)$ is called *orthogonal* if $\mathbf{P}^2 = \mathbf{P}$.

The space X can be split into a direct sum $\mathcal{R}(\mathbf{P}) \oplus [\mathcal{R}(\mathbf{P})]^\perp$ and $\mathcal{R}(\mathbf{P}) \oplus \mathcal{R}(\mathbf{I} - \mathbf{P})$. This explains why **P** is called orthogonal. For every $\mathbf{x} \in X$ we thus have $\mathbf{x} = \mathbf{x}_1 + \mathbf{x}_2$, where $\mathbf{x}_1 \in \mathcal{R}(\mathbf{P})$ and $\mathbf{x}_2 \in \mathcal{R}(\mathbf{I} - \mathbf{P})$ and $\|\mathbf{x}\|^2 = \|\mathbf{x}_1\|^2 + \|\mathbf{x}_2\|^2$.

19.3.   SPECTRAL PROPERTIES OF LINEAR OPERATORS

19.3.1.   Motivation

EXAMPLE 19.3.1.   Let $\Omega \subset \mathbb{R}^n$, $n \geq 1$, be a bounded open region whose boundary $\partial \Omega$ is smooth [see § 11.2 and Definition 6.3.2]. Let us consider the following boundary value problem: Let $\sigma \geq 0$ and **b** be a piecewise continuous function on Ω and $D_j \geq d > 0, j = 1, \ldots, n$, be piecewise differentiable functions in Ω [see Definition 20.2.1]. We seek a function **u** such that

$$\sum_{j=1}^n \frac{\partial}{\partial s_j} \left(D_j(\mathbf{s}) \frac{\partial \mathbf{u}}{\partial s_j} \right) - \sigma(\mathbf{s}) \mathbf{u} + \lambda \mathbf{u} = \mathbf{b} \quad \text{in } \Omega \tag{19.3.1}$$

and

$$\mathbf{u}(\mathbf{s}) = \mathbf{0} \quad \text{on } \partial \Omega. \tag{19.3.2}$$

Here λ is scalar.

Note that problems of this kind appear in various branches of mathematical physics, mechanics, engineering etc.

We let $X = L^2(\Omega)$ [see Example 19.2.3]

$$\mathscr{D}(\mathbf{T}) = \left\{ \mathbf{u} \in X : \frac{\partial}{\partial s_j} \left(D_j \frac{\partial \mathbf{u}}{\partial s_j} \right) \in X \quad \text{and} \quad \mathbf{u}(\mathbf{s}) = \mathbf{0} \text{ on } \partial \Omega \right\} \tag{19.3.3}$$

and define **T** such that **Tu = v**, if

$$\mathbf{v}(s_1, \ldots, s_n) = -\sum_{j=1}^{n} \frac{\partial}{\partial s_j} \left(D_j(s_1, \ldots, s_n) \frac{\partial \mathbf{u}}{\partial s_j}(s_1, \ldots, s_n) \right.$$

$$\left. + \sigma(s_1, \ldots, s_n)\mathbf{u}(s_1, \ldots, s_n) \right)$$

for $\mathbf{u} \in \mathscr{D}(\mathbf{T})$.

The problem (19.3.1)–(19.3.2) can be reformulated: To find **u** such that $\lambda \mathbf{u} - \mathbf{Tu} = \mathbf{b}$.

Formally, $\mathbf{u} = (\lambda \mathbf{I} - \mathbf{T})^{-1}\mathbf{b}$ is the required solution [see I, § 5.8]. To analyse the problem we have to find conditions which guarantee the existence of the inverse $(\lambda \mathbf{I} - \mathbf{T})^{-1}$ depending on λ. Here a non-trivial solution of the homogeneous equation $\lambda \mathbf{u} = \mathbf{Tu}$ is of great importance; those λ's for which the inverse of $\lambda \mathbf{I} - \mathbf{T}$ does not exist belong to the *spectrum* (the set of eigenvalues) of **T** [see I, Theorem 7.2.3]. The concept of spectrum has its origin in physics. The questions just mentioned belong to a mathematical discipline called *spectral analysis*.

In Example (19.3.1) the operator **T** is linear if the coefficients D_j, σ and **b** do not depend upon the required solution **u**. Otherwise we call them non-linear operators. For such operators the spectral theory is not as rich as for the linear ones. Non-linear spectral theory is now an important topic of research.

If the boundary $\partial \Omega$ is not smooth enough we must generalize the concept of a solution. This has led to the notion of a weak solution.

A bilinear form $B(\mathbf{u}, \mathbf{v})$ [see I, § 9.1] on the Sobolev space $W_0^{1,2}(\Omega) \times W_0^{1,2}(\Omega)$ [see Example 19.2.4] is associated with (19.3.1)–(19.3.2). Thus:

$$B(\mathbf{u}, \mathbf{v}) = \int_\Omega \left(D_j \frac{\partial \mathbf{u}}{\partial x_j} \frac{\partial \bar{\mathbf{v}}}{\partial s_j} + \sigma \mathbf{u}\bar{\mathbf{v}} \right) d\mathbf{s},$$

where the symbols $\partial \mathbf{u}/\partial x_j$, $\partial \bar{\mathbf{v}}/\partial s_j$ denote the generalized derivatives. Then $\mathbf{u} \in X = W_0^{1,2}(\Omega)$ is a weak solution to (19.3.1)–(19.3.2) if the relations

$$B(\mathbf{u}, \mathbf{v}) = \lambda \int_\Omega \mathbf{u}\bar{\mathbf{v}} \, d\mathbf{s} - \int_\Omega \mathbf{b}\bar{\mathbf{v}} \, d\mathbf{s}$$

hold for all $\mathbf{v} \in X$.

We see that a weak solution may exist whilst the classical solution does not. The form B contains first-order derivatives only. This is very important, in particular for numerical computation. The computation can be performed by discretizing the problem as follows: taking a suitable finite-dimensional subspace $X_h \subset X$ we consider instead of (19.3.1)–(19.3.2) the following problem. To find $\mathbf{u}_h \in X_h$ such that

$$B(\mathbf{u}_h, \mathbf{v}) = \lambda \int_\Omega \mathbf{u}\bar{\mathbf{v}} \, d\mathbf{s} - \int_\Omega \mathbf{b}\bar{\mathbf{v}} \, d\mathbf{s} \qquad (19.3.4)$$

is valid for all $\mathbf{v} \in X_h$. Such a $\mathbf{u}_h$ is an approximate solution to the original problem. It has been shown that this is a very powerful method. Its convergence (i.e. $\|\mathbf{u}_h - \mathbf{u}\| \to 0$—see Definition 19.2.3) depends upon an approximation of X by its finite dimensional subspaces X_h [see III, § 6.2.4]. Note that (19.3.4) represents a system of algebraic equations for the unknown coefficients of $\mathbf{u}_h$ in a basis of X_h.

19.3.2.   Spectrum of a Linear Operator

Consider a Banach space X over the field of complex numbers. If Y is a vector space over $\mathbb{R}$ we embed it into its complexification $Y \oplus iY$ [see Definition 19.2.13]. This remark also applies to other sections where concepts of the spectral theory occur. We assume that $\mathbf{T}$ is a linear operator with a dense domain $\mathcal{D}(\mathbf{T})$ in X mapping $\mathcal{D}(\mathbf{T})$ into X. We write $\mathbf{T}_\lambda$ instead of $\lambda \mathbf{I} - \mathbf{T}$ and $\mathbf{R}_\lambda$ instead of $\mathbf{T}_\lambda^{-1} = (\lambda \mathbf{I} - \mathbf{T})^{-1}$. The range of $\mathbf{T}_\lambda$ is denoted by $\mathcal{R}_\lambda$. The operator function $\mathbf{R}_\lambda = \mathbf{R}(\lambda, \mathbf{T})$ is called the *resolvent operator of* $\mathbf{T}$.

DEFINITION 19.3.1.   If λ is such that the range of $\mathbf{T}_\lambda$ is dense in X and $\mathbf{T}_\lambda$ has a bounded inverse (i.e. there is a linear operator $\mathbf{S}$ mapping the range $\mathcal{R}_\lambda$ into X such that $\mathbf{S}\mathbf{T}_\lambda \mathbf{x} = \mathbf{x}$ for $\mathbf{x} \in \mathcal{D}(\mathbf{T})$ and $\mathbf{T}_\lambda \mathbf{S}\mathbf{y} = \mathbf{y}$ for $\mathbf{y} \in \mathcal{R}_\lambda$) we say that λ is in the *resolvent set of* $\mathbf{T}$; this set of values λ is denoted by $\rho(\mathbf{T})$. All scalars λ not in $\rho(\mathbf{T})$ comprise the set called the *spectrum* of $\mathbf{T}$; it is denoted by $\sigma(\mathbf{T})$.

EXAMPLE 19.3.2.   Let $X = \mathbb{R}^n$ and $\mathbf{T}$ be an $n \times n$ matrix with complex entries. Then the spectrum of the corresponding operator, also denoted by $\mathbf{T}$, consists of eigenvalues $\lambda_1, \ldots, \lambda_s$ and these are the roots of the characteristic polynomial $\det(\lambda \mathbf{I} - \mathbf{T}) = (\lambda - \lambda_1)^{r_1} \ldots (\lambda - \lambda_s)^{r_s}$ and $r_1 + \ldots + r_s = n$ [see I, Theorem 7.2.3].

EXAMPLE 19.3.3.   (a) Let $X = C([0, 1])$. Define T by setting $Tx = y$ if $y(s) = sx(s)$, $x \in X$, $s \in [0, 1]$. Then $\sigma(T) = [0, 1]$.
   (b) Let $X = L^2(0, 1)$ and T defined by the same formula as in (a) but in $L^2(0, 1)$. Then again $\sigma(T) = [0, 1]$.

If no specific assumptions are made concerning the operator $\mathbf{T}$, the spectrum $\sigma(\mathbf{T})$ may happen to be empty or coincide with the whole complex plane. We assume explicitly that the resolvent set $\rho(\mathbf{T})$ is *not empty*. If $\mathbf{T}$ is closed then $\mathcal{R}_\lambda = X$ for $\lambda \in \rho(\mathbf{T})$; this property emphasizes the importance of closed operators. Obviously, every operator $\mathbf{T} \in \mathcal{B}(X)$ is closed.

PROPOSITION 19.3.1.   *The resolvent set* $\rho(\mathbf{T})$ *is open and hence the spectrum* $\sigma(\mathbf{T})$ *is closed.*

THEOREM 19.3.2.   *Suppose* $\mathbf{T}$ *is such that* $\mathcal{R}_\lambda = X$ *if* $\lambda \in \rho(\mathbf{T})$. *Then if* λ *and* μ *are any two points in* $\rho(\mathbf{T})$, $\mathbf{R}_\lambda$ *and* $\mathbf{R}_\mu$ *satisfy the relations*

$$\mathbf{R}_\lambda - \mathbf{R}_\mu = (\mu - \lambda)\mathbf{R}_\lambda \mathbf{R}_\mu \qquad (19.3.5)$$

and

$$\mathbf{R}_\lambda \mathbf{R}_\mu = \mathbf{R}_\mu \mathbf{R}_\lambda. \qquad (19.3.6)$$

The resolvent operator $\mathbf{R}_\lambda$ *is locally analytic as a function on* $\rho(\mathbf{T})$ *with values in* $\mathcal{B}(X)$.

For bounded operators we have an important result.

THEOREM 19.3.3.   *Let* $\mathbf{T} \in \mathcal{B}(X)$. *Then* $\sigma(\mathbf{T})$ *is not empty.*

DEFINITION 19.3.2.   Supposing $\sigma(\mathbf{T})$ is non-empty and bounded we define

$$r(\mathbf{T}) = \sup \{|\lambda| : \lambda \in \sigma(\mathbf{T})\}$$

and call $r(\mathbf{T})$ the *spectral radius of* $\mathbf{T}$.

THEOREM 19.3.4   (*Spectral radius formula*).   *Suppose* $\mathbf{T} \in \mathcal{B}(X)$. *Then*

$$r(\mathbf{T}) = \lim_{n \to \infty} \|\mathbf{T}^n\|^{1/n} = \inf_n \|\mathbf{T}^n\|^{1/n}$$

and hence

$$r(\mathbf{T}) \le \|\mathbf{T}\|.$$

DEFINITION 19.3.3.   Let $\mathbf{T}$ be a linear operator mapping $\mathcal{D}(\mathbf{T}) \subset X$ into X. Let $\rho(\mathbf{T})$ be non-empty. We say that $\lambda \in \sigma(\mathbf{T})$ belongs to the *point spectrum* $P\sigma(\mathbf{T})$ of $\mathbf{T}$ if the inverse of $\mathbf{T}_\lambda$ does not exist; λ belongs to the *continuous spectrum* $C\sigma(\mathbf{T})$ if $\mathcal{R}_\lambda$ is dense in X and the inverse $\mathbf{T}_\lambda^{-1}$ exists but is not bounded; and λ belongs to the *residual spectrum* $R\sigma(\mathbf{T})$ if $\mathbf{T}_\lambda^{-1}$ exists but the range $\mathcal{R}_\lambda$ is not dense in X. A point $\lambda \in \sigma(\mathbf{T})$ is an *eigenvalue* of $\mathbf{T}$ if there is a vector $\mathbf{x}_\lambda \neq \mathbf{0}$ such that $\mathbf{T}\mathbf{x}_\lambda = \lambda \mathbf{x}_\lambda$ [see I, Definition 7.1.1].

EXAMPLE 19.3.2 (continued).   We see that $\sigma(\mathbf{T}) = P\sigma(\mathbf{T})$ and it consists only of eigenvalues of $\mathbf{T}$.

EXAMPLE 19.3.3 (continued).   There is no eigenvalue in $\sigma(T)$ in either case (a) or (b); however, $\sigma(T) = R\sigma(T)$ for T as an operator on $C([0, 1])$, while $\sigma(T) = C\sigma(T)$ for T as an operator on $L^2(0, 1)$.

PROPOSITION 19.3.5.   *The spectrum of an operator* $\mathbf{T}$ *is divided into three mutually exclusive parts*:

$$\sigma(\mathbf{T}) = P\sigma(\mathbf{T}) \cup C\sigma(\mathbf{T}) \cup R\sigma(\mathbf{T}).$$

PROPOSITION 19.3.6.   *If* $\mathbf{T} \in \mathcal{B}(X)$ *then* $\mathbf{T}$ *and its dual* $\mathbf{T}'$ *have the same spectrum and the same resolvent set.*

If X is a Hilbert space and $\mathbf{T}^*$ is the adjoint of $\mathbf{T}$, then $\lambda \in \sigma(\mathbf{T})$ implies that $\bar{\lambda} \in \sigma(\mathbf{T}^*)$.

PROPOSITION 19.3.7 [see I, § 10.3].   *Let X be a complex Hilbert space and* $\mathbf{T} \in \mathcal{B}(X)$. *Then the following statements hold*: (i) *if* $\mathbf{T}$ *is Hermitian then* $\lambda \in \sigma(\mathbf{T})$ *is real*; (ii) *if* $\mathbf{T}$ *is unitary then* $\lambda \in \sigma(\mathbf{T})$ *implies that* $|\lambda| = 1$.

19.3.3.   An Operational Calculus

THEOREM 19.3.8.   *Let* $\mathbf{T} \in \mathcal{B}(X)$ *and let* $|\lambda| > r(\mathbf{T})$. *Then*

$$\mathbf{R}(\lambda, \mathbf{T}) = \sum_{k=1}^{\infty} \lambda^{-k} \mathbf{T}^{k-1}. \tag{19.3.7}$$

Choose $\lambda \neq 0$ and consider the function f defined by

$$f(z) = \sum_{k=1}^{\infty} \lambda^{-k} z^{k-1} = \frac{1}{\lambda} \sum_{k=0}^{\infty} (z/\lambda)^k$$

for z such that $|\lambda^{-1} z| < 1$. We see that $f(z) = (\lambda - z)^{-1}$. According to (19.3.7) we may write $\mathbf{R}(\lambda, \mathbf{T}) = f(\mathbf{T})$ as a function of $\mathbf{T}$.

EXAMPLE 19.3.4.   Let $X = \mathbb{R}^n$ and $\mathbf{T}$ be represented by a matrix also denoted by $\mathbf{T}$. We want to solve a *Cauchy problem*

$$\frac{d\mathbf{u}}{dt} = \mathbf{T}\mathbf{u}, \qquad \mathbf{u}(0) = \mathbf{u}_0, \qquad t \in [0, 1], \qquad 0 < \|\mathbf{T}\| < \infty,$$

where $\mathbf{u}_0 \in \mathbb{R}^n$.

It is known that the unique solution $\mathbf{u}$ is given by

$$\mathbf{u}(t) = \exp(t\mathbf{T})\mathbf{u}_0,$$

where

$$\exp(\mathbf{T}) = \sum_{k=0}^{\infty} \frac{1}{k!} \mathbf{T}^k \tag{19.3.8}$$

[see § 7.9.2]. The computation of all the powers of $\mathbf{T}$ may be tedious and perhaps impossible even with a computer.

In this section we shall show another way of expressing functions whose argument is a linear operator, not necessarily bounded.

Let $\mathbf{T}$ be a closed linear operator with a non-empty resolvent $\rho(\mathbf{T})$. Since $\mathbf{R}_\lambda$ is a locally analytic function in $\rho(\mathbf{T})$ we may obtain some important results by the use of contour integrals in the complex plane. These integrals with values in $\mathcal{B}(X)$ may be defined just as in classical analysis [see § 9.3]. The convergence is guaranteed by the completeness of $\mathcal{B}(X)$.

DEFINITION 19.3.4. A set D in the complex plane is called a *Cauchy domain* if (1) D is open; (2) D has a finite number of components, with mutually disjoint closures; (3) the boundary of D is composed of a finite positive number of closed rectifiable curves, no two of which intersect [see § 9.4]. The positively oriented boundary of D is denoted by $B(D)$.

DEFINITION 19.3.5. Suppose $\mathbf{T}$ is a closed linear operator mapping a dense domain $\mathcal{D}(\mathbf{T}) \subset X$ into X and having a non-empty resolvent set. By $\mathcal{F}_\infty(\mathbf{T})$ we denote the set of complex-valued functions f such that (1) the domain $\Delta(f)$ is an open set in the complex plane which contains $\sigma(\mathbf{T})$ and is such that (1a) the complement of $\Delta(f)$ is closed and bounded; (2) f is differentiable in $\Delta(f)$ [see § 9.1] and (2a) $f(\lambda)$ is bounded as $\lambda \to \infty$. The set of those f satisfying (1) and (2) but possibly not (1a) and (2a) is denoted by $\mathcal{F}(\mathbf{T})$; obviously $\mathcal{F}(\mathbf{T}) \supset \mathcal{F}_\infty(\mathbf{T})$.

DEFINITION 19.3.6. Let D be an unbounded Cauchy domain of a closed linear operator $\mathbf{T}$ with non-empty $\rho(\mathbf{T})$. Let $\sigma(\mathbf{T}) \subset D \subset \bar{D} \subset \Delta(f)$. For $f \in \mathcal{F}_\infty(\mathbf{T})$ we let

$$f(\mathbf{T}) = f(\infty)\mathbf{I} + \frac{1}{2\pi i} \int_{B(D)} f(\lambda)\mathbf{R}(\lambda, \mathbf{T}) \, d\lambda. \qquad (19.3.9)$$

DEFINITION 19.3.7. We call two functions $f, g \in \mathcal{F}(\mathbf{T})$ (or $\in \mathcal{F}_\infty(\mathbf{T})$) *equivalent* if $f(\lambda) = g(\lambda)$ on a neighbourhood of $\sigma(\mathbf{T})$ (and at ∞).

THEOREM 19.3.9. *The mapping $f \to f(\mathbf{T})$ defined in (19.3.9) is an algebraic homomorphism of equivalence classes of $\mathcal{F}_\infty(\mathbf{T})$ into the algebra $\mathcal{B}(X)$ and this homomorphism is such that the image of the function $f(\lambda) \equiv 1$ is the identity operator.*

If $\mathbf{T} \in \mathcal{B}(X)$ then (19.3.9) reduces to

$$f(\mathbf{T}) = \frac{1}{2\pi i} \int_{B(D_1)} f(\lambda)\mathbf{R}(\lambda, \mathbf{T}) \, d\lambda \qquad (19.3.10)$$

where D_1 is a bounded Cauchy domain such that $\sigma(\mathbf{T}) \subset D_1 \subset \bar{D}_1 \subset \Delta(f)$. In addition we have

THEOREM 19.3.10. *Let $\mathbf{T} \in \mathcal{B}(X)$. The mapping $f \to f(\mathbf{T})$ defined in (19.3.10) is an algebraic homomorphism of the algebra of equivalence classes of $\mathcal{F}(\mathbf{T})$ into the algebra $\mathcal{B}(X)$. This carries the function $f(\lambda) \equiv 1$ into $\mathbf{I}$ and $f(\lambda) \equiv \lambda$ into $\mathbf{T}$.*

EXAMPLE 19.3.4 (continued). According to (19.3.10),

$$\exp\,(t\mathbf{T}) = \frac{1}{2\pi i}\int_{B(D)} e^{\lambda t}\mathbf{R}(\lambda,\mathbf{T})\,d\lambda, \tag{19.3.11}$$

where $D = \{\lambda : |\lambda| \le \|\mathbf{T}\| + 1\}$. The contour integral in (19.3.11) can easily be evaluated and we obtain:

$$\exp\,(t\mathbf{T}) = \sum_{j=1}^{s}\sum_{k=1}^{q_j}\frac{e^{\lambda_j t}t^{k-1}}{(k-1)!}\mathbf{B}_{j,k}, \tag{19.3.12}$$

where $\sigma(\mathbf{T}) = \{\lambda_1, \ldots, \lambda_s\}$ and $q_1, \ldots, q_s$ are the multiplicities of $\lambda_1, \ldots, \lambda_s$ in the minimal polynomial of $\mathbf{T}$ [see I, § 7.5]; the operators $\mathbf{B}_{j,k}$ are defined as follows:

$$\mathbf{B}_{j,1} = \frac{1}{2\pi i}\int_{B(D_j)}\mathbf{R}(\lambda,\mathbf{T})\,d\lambda,$$

where $D_j = \{\lambda : |\lambda - \lambda_j| \le \rho_j, \rho_j > 0\}$, $D_j \cap D_k = \varnothing$ for $j \ne k$ and

$$\mathbf{B}_{j,k+1} = (\mathbf{T} - \lambda_j\mathbf{I})\mathbf{B}_{j,k}\qquad j = 1, \ldots, s,\qquad k = 1, 2, \ldots .$$

Although formula (19.3.12) is also of little use in practical computations of the exponential of a matrix, it gives much more information than formula (19.3.8) does.

The following interesting relations are valid

$$\mathbf{T} = \sum_{j=1}^{s}(\lambda_j\mathbf{B}_{j,1} + \mathbf{B}_{j,2})$$

and

$$\mathbf{I} = \sum_{j=1}^{s}\mathbf{B}_{j,1},\qquad \mathbf{B}_{j,1}^2 = \mathbf{B}_{j,1},\qquad j = 1, \ldots, s.$$

We observe that the operational calculus introduced by (19.3.9) or (19.3.10) covers only functions of operators belonging to $\mathcal{B}(X)$. A more general operational calculus, including unbounded functions of an unbounded operator, would need some more subtle conditions concerning the operators under consideration. The very important case of self-adjoint operators is discussed in the next section.

19.3.4. Spectral Representation of a Self-adjoint Operator

Let X be a complex Hilbert space and let $\mathbf{T}$ be a self-adjoint operator mapping $\mathcal{D}(\mathbf{T}) \subset X$ into X.

DEFINITION 19.3.8.   We say that an operator $\mathbf{B} \in \mathcal{B}(X)$ *commutes* with the operator $\mathbf{T}$ if $\mathcal{D}(\mathbf{TB}) \supset \mathcal{D}(\mathbf{BT})$ and $\mathbf{TBx} = \mathbf{BTx}$ for $\mathbf{x} \in \mathcal{D}(\mathbf{BT})$; we write $\mathbf{TB} \supset \mathbf{BT}$.

THEOREM 19.3.11.   *There exists a uniquely determined operator function* $\mathbf{E} = \mathbf{E}(\lambda)$, $-\infty < \lambda < +\infty$ *and this function has the following properties*:

  (i) $[\mathbf{E}(\lambda)]^2 = \mathbf{E}(\lambda) = [\mathbf{E}(\lambda)]^*$ *for* $\lambda \in (-\infty, +\infty)$;
  (ii) $\mathbf{E}(\lambda)\mathbf{E}(\mu) = \mathbf{E}(\lambda)$ *for* $\lambda \leq \mu$;
  (iii) $\mathbf{E}(\lambda)\mathbf{B} = \mathbf{B}\mathbf{E}(\lambda)$ *for all* $\lambda \in (-\infty, +\infty)$ *and for every* $\mathbf{B} \in \mathscr{B}(X)$ *commuting with* $\mathbf{T}$;
  (iv) $\lim_{\lambda \to -\infty} \mathbf{E}(\lambda)\mathbf{x} = 0$; $\lim_{\lambda \to +\infty} \mathbf{E}(\lambda)\mathbf{x} = \mathbf{x}$ *for every* $\mathbf{x} \in X$;
  (v) $\lim_{\lambda \to \lambda_0} \mathbf{E}(\lambda)\mathbf{x} = \mathbf{E}(\lambda_0)\mathbf{x}$, *for* $\lambda_0 \in (-\infty, +\infty)$, $\mathbf{x} \in X$;
  (vi) $\mathbf{x} \in \mathscr{D}(\mathbf{T})$ *if and only if*

$$\int_{-\infty}^{\infty} |\lambda|^2 \, d(\mathbf{E}(\lambda)\mathbf{x}, \mathbf{x})$$

and then

$$(\mathbf{T}\mathbf{x}, \mathbf{y}) = \int_{-\infty}^{\infty} \lambda \, d(\mathbf{E}(\lambda)\mathbf{x}, \mathbf{y}) \quad \textit{for } \mathbf{x} \in \mathscr{D}(\mathbf{T}), \mathbf{y} \in X. \qquad (19.3.13)$$

Note that the integrals considered in Theorem 19.3.11 are of Riemann–Stieltjes type [see § 4.8].

DEFINITION 19.3.9.   A collection of orthogonal projections $\mathbf{E}(\lambda)$ described in Theorem 19.3.11 is called a *spectral resolution of identity* with respect to the operator $\mathbf{T}$ and the formula (19.3.13) determines the spectral representation of the operator $\mathbf{T}$.

DEFINITION 19.3.10.   Let $\mathbf{E}(\lambda)$ be the spectral resolution of identity with respect to a self-adjoint operator $\mathbf{T}$. Let f be a scalar-valued function continuous on $(-\infty, +\infty)$. Further, let $f_n(\lambda) = f(\lambda)$ for $|\lambda| \leq n$ and $f_n(\lambda) = 0$ for $|\lambda| > n$, $n = 1, 2, \ldots$, and

$$(f_n(\mathbf{T})\mathbf{x}, \mathbf{y}) = \int_{-n}^{n} f_n(\lambda) \, d(\mathbf{E}(\lambda)\mathbf{x}, \mathbf{y}) \quad \textit{for } \mathbf{x} \in \mathscr{D}(\mathbf{T}) \textit{ and } \mathbf{y} \in X. \quad (19.3.14)$$

Define

$$\mathscr{D}(f(\mathbf{T})) = \{\mathbf{x} \in X : \lim_{n \to \infty} f_n(\mathbf{T})\mathbf{x} \in X\}$$

and

$$f(\mathbf{T})\mathbf{x} = \lim_{n \to \infty} f_n(\mathbf{T})\mathbf{x}, \qquad \mathbf{x} \in \mathscr{D}(f(\mathbf{T})). \qquad (19.3.15)$$

THEOREM 19.3.12.   *The operational calculus introduced in* (19.3.14)–(19.3.15) *has the following properties*:

  (i) $\mathscr{D}(f(\mathbf{T})) = \{\mathbf{x} \in X : \int |f(\lambda)|^2 \, d(E(\lambda)\mathbf{x}, \mathbf{x}) < \infty\}$
  (ii) $(f(\mathbf{T})\mathbf{x}, \mathbf{y}) = \int f(\lambda) \, d(\mathbf{E}(\lambda)\mathbf{x}, \mathbf{y})$, $\mathbf{x} \in \mathscr{D}(f(\mathbf{T})), \mathbf{y} \in X$,
  (iii) $\|f(\mathbf{T})\mathbf{x}\|^2 = \int |f(\lambda)|^2 \, d(\mathbf{E}(\lambda)\mathbf{x}, \mathbf{x})$, $\mathbf{x} \in \mathscr{D}(f(\mathbf{T}))$,
  (iv) $[f(\mathbf{T})]^* = \bar{f}(\mathbf{T})$,

where $\bar{f}(\lambda) = \overline{f(\lambda)}$, $\lambda \in (-\infty, +\infty)$.

In a way similar to that shown for self-adjoint operators we can obtain a spectral representation of other classes of operators in Hilbert spaces, such as unitary and, more generally, normal operators. There is also a remarkable generalization of the spectral representation theory for essentially non-symmetric operators. This general theory has been developed by N. Dunford and is known as *the theory of spectral operators* (see Dunford and Schwartz (1971)).

Let us turn again to Example 19.3.1. It can be shown that the operator **T** defined there is symmetric. However, the domain determined in (19.3.3) may not be large enough for **T** to becomes self-adjoint in $L^2(\Omega)$. It is an easy matter to observe that the operator **T** is positive definite. The question of extending a symmetric operator to a self-adjoint one is discussed in section 19.3.5.

19.3.5. A Self-adjoint Extension of a Symmetric Operator

In this section we assume that X is a complex Hilbert space and **T** a linear operator mapping a dense domain $\mathscr{D}(\mathbf{T}) \subset X$ into X.

DEFINITION 19.3.11. The dimensions of the subspaces $\mathscr{R}(\mathbf{T} - i\mathbf{I})^{\perp}$ and $\mathscr{R}(\mathbf{T} + i\mathbf{I})^{\perp}$, where $i^2 = -1$, are called *deficiency indices* of a symmetric operator **T**.

THEOREM 19.3.13. *Let **T** be a symmetric operator densely defined on $\mathscr{D}(\mathbf{T})$. Let $(\mathscr{M}, \mathscr{N})$ be its deficiency indices. Then the operator **T** has a self-adjoint extension if and only if $\mathscr{M} = \mathscr{N}$.*

A subspace of a separable Hilbert space is either finite-dimensional or its dimension is a countable cardinal number [see I, §§ 5.4, 5.5 and 1.6].

A very important strengthening of the previous result is concerned with symmetric operators bounded from below.

DEFINITION 19.3.12. Let **T** be a symmetric operator mapping $\mathscr{D}(\mathbf{T})$ into X. A point λ of the complex plane is called a *point of regularity* of **T** if there is a real $\rho > 0$ such that $\|(\mathbf{T} - \lambda \mathbf{I})\mathbf{x}\| \geq \rho\|\mathbf{x}\|$, $\mathbf{x} \in \mathscr{D}(\mathbf{T})$.

PROPOSITION 19.3.14. *If a symmetric operator **T** has a real point of regularity then its deficiency indices coincide.*

THEOREM 19.3.15. *Every symmetric bounded-below operator **T** with a dense domain $\mathscr{D}(\mathbf{T})$ has at least one self-adjoint extension which is also bounded below and the bounds of both of the operators are the same.*

A complete description of all self-adjoint extensions of a symmetric bounded-below operator is given by M. G. Krein (Achieser and Glasmann (1971)).

Since the operator of Example 19.3.1 is positive definite we see that it has a self-adjoint extension and the results of this section apply.

19.3.6.   Compact Linear Operators

A well-known representative of a compact operator is a *Fredholm integral operator* [see III, § 10.1.1].

EXAMPLE 19.3.5.   Let $t = t(s, s')$ be a scalar-valued function continuous on $[a, b] \times [c, d]$. Define a linear operator T by setting $Tu = v$ if $v(s) = \int_c^d t(s, s') u(s') \, ds'$. Thus T is a continuous linear operator mapping $C([c, d])$ into $C([a, b])$. T is an example of a Fredholm integral operator.

DEFINITION 19.3.13.   Let X and Y be normed vector spaces over $\mathbb{F}$. Suppose T is a linear operator mapping X into Y. We say that T is *compact* if for each bounded sequence $\{x_k\}$ in X the sequence $\{Tx_k\}$ contains a subsequence convergent to some limit vector in Y. Note that the term *completely continuous* is also used for a compact linear operator.

THEOREM 19.3.16.   *Suppose* $T \in \mathcal{B}(X)$ *and* T *is compact. Then the spectrum* $\sigma(T)$ *contains at most a countable set of points and they have no limit point except possibly* $\lambda = 0$ [see § 11.3]. *Every element* $\lambda \in \sigma(T)$, $\lambda \neq 0$, *is an eigenvalue of* T, *i.e. there is a vector* $x_\lambda \in X$, $x_\lambda \neq 0$, *such that* $Tx_\lambda = \lambda x_\lambda$. *If* $\lambda \neq 0$ *then the kernels of* T_λ^n *and* $(T_\lambda')^n$, *have the same (finite) dimension. The dual operator* T' *is also compact.*

THEOREM 19.3.17   (*Fredholm alternative*).   *Suppose* $T \in \mathcal{B}(X)$ *is a compact operator and let us consider the equation* (19.3.14)

$$\lambda x - Tx = b, \qquad b \in X. \tag{19.3.16}$$

Then exactly one of the two following possibilities occurs: Either

(1) *there is a unique solution* x *of* (19.3.16) *for every* $b \in X$ *and then the inverse of* $\lambda I - T$ *exists and belongs to* $\mathcal{B}(X)$, *or*
(2) *the homogeneous equations*

$$\lambda u = Tu \quad and \quad \lambda u' = T'u'$$

both have non-trivial solutions and $\dim \mathcal{N}(T_\lambda) = \dim \mathcal{N}(T_\lambda')$; *the equation* (19.3.14) *possesses a solution if and only if* $u'(b) = 0$ *for every* $u' \in \mathcal{N}(T_\lambda')$.

We conclude this section by noticing that the theory presented here may be applied to problems governed by operators T, some function $f(T)$ of which is compact, and where T itself is not necessarily compact. This is the case of the operator in Example 19.3.1. It can be shown that the operator T in Example 19.3.1 belongs to the class $\mathcal{C}_s$ defined as follows:

DEFINITION 19.3.14.   Let X be a normed vector space over $\mathbb{F}$. An operator $\mathbf{T} \in \mathcal{B}(X)$ is said to belong to the *class* $\mathscr{C}_s$ if there is a positive integer $s \geq 1$ such that $\mathbf{T}^s$ is compact.

19.4.   PARTIALLY ORDERED SPACES

19.4.1.   Motivation

Let us consider the operator $\mathbf{T}$ of Example 19.3.1. If $(\mathbf{Tu})(\mathbf{s}) \geq \mathbf{0}$ for all $\mathbf{s} \in \Omega$ then also $\mathbf{u}(\mathbf{s}) = \int G(\mathbf{s}, \mathbf{s}')(\mathbf{Tu})(\mathbf{s}') \, d\mathbf{s}' \geq 0$ for all $\mathbf{s} \in \Omega$. This follows from an easy corollary of a maximum principle (Protter and Weinberger (1967)) guaranteeing that the Green's function G of the corresponding differential operator [see Definition 8.2.1] is positive for $\mathbf{s}, \mathbf{s}' \in \Omega$, $\mathbf{s} \neq \mathbf{s}'$. In other words, the inverse $\mathbf{T}^{-1}$ carries any non-negative function into a non-negative function. This preservation of positiveness together with some compactness properties (recall that $\mathbf{T}^{-1}$ belongs to some of the classes $\mathscr{C}_s$ defined in section 19.3.6) imply that the largest eigenvalue of $\mathbf{T}^{-1}$ is simple and the corresponding eigenfunction is positive in Ω. These facts have important consequences in some problems of mathematical physics, mechanics, engineering, economics etc.

In finite dimensional spaces the corresponding theory is known as the Perron–Frobenius theory of non-negative matrices [see I, § 7.11].

19.4.2.   Cones [see also § 15.2.4]

DEFINITION 19.4.1.   Let Y be a Banach space over $\mathbb{R}$. A non-empty set $K \subset Y$ is called a *cone* if

   (i) $\mathbf{x}, \mathbf{y} \in K$ implies that $\mathbf{x} + \mathbf{y} \in K$ (that is $K + K \subset K$);
   (ii) $\mathbf{x} \in K$, $\lambda \in \mathbb{R}$, $\lambda \geq 0$, implies that $\lambda \mathbf{x} \in K$ ($\lambda K \subset K$, $\lambda \geq 0$);
   (iii) $\mathbf{x} \in K \cap (-K)$ implies that $\mathbf{x} = \mathbf{0}$ ($K \cap (-K) = \{\mathbf{0}\}$).

We define a partial order in Y [see I, § 1.3.2] by putting $\mathbf{x} \leq \mathbf{y}$, or equivalently, $\mathbf{y} \geq \mathbf{x}$ if and only if $\mathbf{y} - \mathbf{x} \in K$. We should write $\mathbf{x} \leq^K \mathbf{y}$ but we omit the superscript.

DEFINITION 19.4.2.   A cone K is called *closed* if $\mathbf{x}_n \in K$, $\|\mathbf{x}_n - \mathbf{x}\| \to 0$ implies that $\mathbf{x} \in K$ ($\bar{K} = K$); K is called *normal* if there is a $\delta > 0$ such that $\|\mathbf{x} + \mathbf{y}\| \geq \delta \|\mathbf{x}\|$ for every $\mathbf{x}, \mathbf{y} \in K$; a cone K is called *generating* or *reproducing* if for every $\mathbf{y} \in Y$ there are $\mathbf{y}^+$ and $\mathbf{y}^-$ in K such that $\mathbf{y} = \mathbf{y}^+ - \mathbf{y}^-$, i.e. $K - K = Y$. We also say that the space Y is *generated* by the cone K.

DEFINITION 19.4.3.   A cone K is called a *lattice cone* if to every pair $\mathbf{x}, \mathbf{y} \in K$ there exist $\mathbf{z} = \inf \{\mathbf{x}, \mathbf{y}\}$ and $\mathbf{w} = \sup \{\mathbf{x}, \mathbf{y}\}$, that is, $\mathbf{z} \in Y$ is such that $\mathbf{z} \leq \mathbf{x}$, $\mathbf{z} \leq \mathbf{y}$ and $\mathbf{u} \leq \mathbf{x}$, $\mathbf{u} \leq \mathbf{y}$ implies that $\mathbf{u} \leq \mathbf{z}$; similarly for $\mathbf{w}$. If K is a lattice cone and the norm in Y is such that $\delta = 1$ in the definition of normality, the space Y is called a *Banach lattice* or a *Riesz space* (Schaefer (1971)).

EXAMPLE 19.4.1.   Let $X = \mathbb{R}^n$ and $\|\mathbf{x}\| = \max_k |\xi_k|$ for $\mathbf{x} = (\xi_1, \ldots, \xi_n)$. Let $K = \{\mathbf{x} \in \mathbb{R}^n : \xi_k \geq 0, k = 1, \ldots, n\}$.

This cone is called *standard* and is usually denoted by $\mathbb{R}^n_+$. It is evident that $\mathbb{R}^n_+$ is a lattice cone.

EXAMPLE 19.4.2.   Let $X = \mathbb{R}^3$ and $\|\mathbf{x}\|^2 = \xi_1^2 + \xi_2^2 + \xi_3^2$. The cone $\hat{K} = \{\mathbf{x} \in R^3 : \xi_3 \geq (\xi_1^2 + \xi_2^2)^{1/2}\}$, $\mathbf{x} = (\xi_1, \xi_2, \xi_3)$, fulfils all previously mentioned conditions but one: $\hat{K}$ is not a lattice cone. It is sometimes called the 'ice cream' cone.

EXAMPLE 19.4.3.   Let $X = L^2(\Omega)$ [see Example 19.2.3] with the inner product

$$(u, v) = \int_\Omega u(s)v(s)\, d\mathbf{s}, \qquad \Omega \subset \mathbb{R}^n.$$

Let

$$K = \{[u] \in L^2(\Omega) : u \geq 0 \quad \text{a.e. in } \Omega\},$$

where $[u]$ denotes the class of equivalent functions with respect to the Lebesgue measure on Ω; a.e. means almost everywhere [see Definition 4.9.3]. Then K is a closed normal reproducing lattice cone.

19.4.3. Positive Operators

DEFINITION 19.4.4.   A linear operator $\mathbf{T} \in \mathcal{B}(X)$, where X is a Banach space generated by a closed normal cone K, is called *positive*, or more precisely *K-positive*, if $\mathbf{x} \in K$ implies that $\mathbf{Tx} \in K$, i.e. $\mathbf{T}K \subset K$.

Note that the zero operator and the identity operator are K-positive operators with respect to any cone K.

EXAMPLE 19.4.4.   Let $X = \mathbb{R}^n$, $K = \mathbb{R}^n_+$, and $\mathbf{T}$ be represented by a matrix with real non-negative entries. Then $\mathbf{T}$ is $\mathbb{R}^n_+$-positive [see I, § 5.11].

EXAMPLE 19.4.5.   Let $X = \mathbb{R}^3$ and $K = \hat{K}$ be the ice cream cone. Let $\mathbf{T}$ be represented by the matrix

$$\begin{pmatrix} \cos\varphi & \sin\varphi & 0 \\ -\sin\varphi & \cos\varphi & 0 \\ 0 & 0 & 1 \end{pmatrix},$$

where φ is any real number. We see that $\mathbf{T}$ is K-positive although the matrix of $\mathbf{T}$ contains negative elements if φ is chosen appropriately, e.g. $\varphi = \pi + \varepsilon$ with $0 < \varepsilon < 1$.

EXAMPLE 19.4.6.   Suppose $t = t(\mathbf{s}, \mathbf{s}')$ such that $t \in L^2(\Omega \times \Omega)$ and $t(\mathbf{s}, \mathbf{s}') \geq 0$ a.e. in $\Omega \times \Omega$. Let

$$(\mathbf{Tu})(\mathbf{s}) = \int_\Omega t(\mathbf{s}, \mathbf{s}') \mathbf{u}(\mathbf{s}') \, d\mathbf{s}'.$$

Then $\mathbf{T} \in \mathcal{B}(L^2(\Omega))$ and is K-positive, where K is the cone of Example 19.4.3.

Note that the Green's function [see Definition 8.2.1] of the differential operator defined in Example 19.3.1 shares this property.

DEFINITION 19.4.5.   Suppose $K \subset Y$ is a closed normal and generating cone. Let

$$K' = \{\mathbf{y}' \in Y' : \mathbf{y}'(\mathbf{x}) \geq 0 \text{ for all } \mathbf{x} \in K\}.$$

Then K' is a closed normal and generating cone in the dual space Y'. The cone K' is called *dual* with respect to K.

DEFINITION 19.4.6.   A positive operator $\mathbf{T} \in \mathcal{B}(Y)$ is called *indecomposable* (K-indecomposable) if for every pair $\mathbf{x} \in K$, $\mathbf{x} \neq \mathbf{0}$, $\mathbf{x}' \in K'$, $\mathbf{x}' \neq \mathbf{0}$, there is a positive integer $p = p(\mathbf{x}, \mathbf{x}')$ such that $\mathbf{x}'(\mathbf{T}^p \mathbf{x}) > 0$. A positive operator $\mathbf{T}$ is called *primitive* (K-*primitive*) if for every $\mathbf{x} \in K$, $\mathbf{x} \neq 0$, there is a positive integer $p = p(\mathbf{x})$ such that $\mathbf{x}'(\mathbf{T}^k \mathbf{x}) > 0$ for every $\mathbf{x}' \in K'$, $\mathbf{x}' \neq \mathbf{0}$ and $k \geq p$.

EXAMPLE 19.4.7.   Let $\mathbf{T}$ be the operator of Example 19.3.1. This operator is not only positive but also primitive with respect to the cone of Example 19.4.3. We recall that this operator belongs to a class $\mathscr{C}_s$ [see Definition 19.3.14] for some s dependent on the dimension n. For such operators we have a complete analogue of the finite dimensional Perron–Frobenius theory [see I, § 7.11].

We notice that if a complex structure of the vector space is needed, we consider the corresponding complexification X in place of Y.

THEOREM 19.4.1.   *Suppose* $\mathbf{T} \in \mathscr{C}_s$ *for some* $s \geq 1$ *and* $\mathbf{T}$ *positive. Then*

  (i) *the spectral radius* $r(\mathbf{T}) \in \sigma(\mathbf{T})$;
 (ii) *there exist* $\mathbf{x}_0 \in K$, $\mathbf{x}_0 \neq \mathbf{0}$ *and* $\mathbf{x}_0' \in K'$, $\mathbf{x}_0' \neq \mathbf{0}$ *such that* $\mathbf{T}\mathbf{x}_0 = r(\mathbf{T})\mathbf{x}_0$ *and* $\mathbf{T}'\mathbf{x}_0' = r(\mathbf{T})\mathbf{x}_0'$.
      If $\mathbf{T}$ *is indecomposable, then*
(iii) $\mathbf{T}\mathbf{y} = r\mathbf{y}$, $\mathbf{y} \in K$, *implies that* $\mathbf{y} = c\mathbf{x}_0$ *for some* $c \in \mathbb{R}$, $c \geq 0$;
 (iv) *the eigenvectors* $\mathbf{x}_0$ *and* $\mathbf{x}_0'$ *are such that* $\mathbf{x}'(\mathbf{x}_0) > 0$ *for all* $\mathbf{x}' \in K'$, $\mathbf{x}' \neq \mathbf{0}$, *and* $\mathbf{x}_0'(\mathbf{x}) > 0$ *for all* $\mathbf{x} \in K$, $\mathbf{x} \neq \mathbf{0}$;
  (v) *the linear hull of the union* $\bigcup_{k=1}^\infty \mathscr{N}((r(\mathbf{T})\mathbf{I} - \mathbf{T})^k)$ *is one dimensional.*
      If $\mathbf{T}$ *is primitive then*
 (vi) $\lambda \in \sigma(\mathbf{T})$, $|\lambda| = r(\mathbf{T})$, *implies that* $\lambda = r(\mathbf{T})$;

If K is such that Y is a Riesz space and $\mathbf{T}$ is indecomposable then
(vii) there is a positive integer $d \geq 1$ such that

$$\sigma(\mathbf{T}) \cap \{\lambda: |\lambda| = r(\mathbf{T})\} = \{r(\mathbf{T}) \exp(2\pi ki/d): k = 1, \ldots, d\}.$$

Property (vii) does not hold if K is a general cone: let $\mathbf{T}$ be as in Example 19.4.5. We see that $\sigma(\mathbf{T}) = \{1, e^{i\varphi}, e^{-i\varphi}\}$ and it is enough to choose φ/π irrational.

THEOREM 19.4.2.   *Let $\mathbf{T} \in \mathscr{C}_s$ for some $s \geq 1$ and $\mathbf{T}$ be indecomposable. Then*

$$r(\mathbf{T}) = \max\{r_{\mathbf{x}}(\mathbf{T}): \mathbf{x} \in K^d\} = \min\{r^{\mathbf{x}}(\mathbf{T}): \mathbf{x} \in K^d\}$$

where

$$r_{\mathbf{x}}(\mathbf{T}) = \sup\{\nu \in \mathbb{R}: \mathbf{Tx} - \nu \mathbf{x} \in K\},$$
$$r^{\mathbf{x}}(\mathbf{T}) = \inf\{\tau \in \mathbb{R}: \tau \mathbf{x} - \mathbf{Tx} \in K\}$$

and

$$K^d = \{\mathbf{x} \in K: \mathbf{x}'(\mathbf{x}) > 0 \quad \text{for all } \mathbf{x}' \in K', \mathbf{x}' \neq \mathbf{0}\}.$$

Consequently,

$$r_{\mathbf{T}_{\mathbf{x}}^k}(\mathbf{T}) \leq \ldots \leq r_{\mathbf{x}}(\mathbf{T}) \leq r(\mathbf{T}) \leq r^{\mathbf{x}}(\mathbf{T}) \leq \ldots \leq r^{\mathbf{T}^k \mathbf{x}}(\mathbf{T}).$$

EXAMPLE 19.4.8.   Let $Y = \mathbb{R}^n$, $K = \mathbb{R}^n_+$ and $\mathbf{T}$ be represented by a matrix with real non-negative entries t_{jk}. Let $\mathbf{x} = (\xi_1, \ldots, \xi_n)'$ and let $\xi_j > 0$, $j = 1, \ldots, n$. Then

$$r_{\mathbf{x}}(\mathbf{T}) = \min_j \left[\sum_{k=1}^n (t_{jk}\xi_k)/\xi_j \right]$$
$$r^{\mathbf{x}}(\mathbf{T}) = \max_j \left[\sum_{k=1}^n (t_{jk}\xi_k)/\xi_j \right].$$

The results of Theorem 19.4.2 can be used for locating the spectrum in a manner shown for matrices in Volume I, section 7.11. The following example shows how to apply Theorem 19.4.2 in an infinite dimensional case; no symmetry assumptions are needed.

EXAMPLE 19.4.9.   Let $Y = C([0, 1])$ and $K = \{x \in C([0, 1]): x(s) \geq 0$ for all $s \in [0, 1]\}$. Let $t = t(s, s')$ be a real-valued function continuous and positive on $[0, 1] \times [0, 1]$. Let

$$(Tx)(s) = \int_0^1 t(s, s')x(s')\, ds'.$$

Then T is a compact positive operator. It can be shown that $x(s) > 0$ for all $s \in [0, 1]$

$$r_x(T) = \inf \left\{ \frac{(Tx)(s)}{x(s)}; s \in \Delta \right\}$$

and

$$r^x(T) = \sup \left\{ \frac{(Tx)(s)}{x(s)}; s \in \Delta \right\}$$

where Δ is a dense subset of $[0, 1]$.

19.5.   LINEAR AND NON-LINEAR EQUATIONS

A problem which is often met in practice can be described by equations of the type

$$G(x) = 0, \tag{19.5.1}$$

where G is a given operator mapping a given space X into another space Y. The problem is to find $x \in X$ for which (19.5.1) holds in some sense. Of particular interest are equations of the form

$$F(x) = b, \tag{19.5.2}$$

where F is a given operator and b a given vector. To solve practical problems it is desirable to know some characteristic properties of the set of vectors b for which at least one solution exists ('admissible right-hand side vectors'); also some knowledge of the set of solutions is useful. In other words, one may ask whether F is injective and surjective [see I, § 5.11] on the spaces considered.

19.5.1.   Linear Operator Equations

EXAMPLE 19.5.1.   Let $t = t(s, s')$ be a complex or real-valued function continuous on the square $[a, b] \times [c, d]$. Define the operator T by putting

$$Tx = \int_c^d t(s, s') x(s') \, ds'. \tag{19.5.3}$$

Let $b \in C([a, b])$ and λ be a complex or real scalar. Then

$$G(x) = F(x) - b = x - \lambda Tx - b = 0, \tag{19.5.4}$$

or

$$x(s) - \lambda \int_c^d t(s, s') x(s') \, ds' = b(s), \qquad s \in [a, b], \tag{19.5.5}$$

is a *Fredholm integral equation of the second kind*. Its solubility is described in Theorem 19.3.17 [see also III, § 10.2].

EXAMPLE 19.5.2.   Let T be as defined by (19.5.3). Then
$$G(x) = Tx - b = 0, \tag{19.5.6}$$
or
$$\int_c^d t(s, s')x(s')\, ds' = b(s'), \qquad s \in [a, b]. \tag{19.5.7}$$

This is a *Fredholm integral equation of the first kind* [see III, § 10.4].

Note that one extreme situation appears if (19.5.7) has infinitely many solutions and a different one if (19.5.7) has no solution at all. Moreover, small perturbations of b and t may have a very drastic influence on the properties of the solution. This type of problem belongs to a class of *ill-conditioned* (not well-posed) problems. Computational procedures for solving such problems require special care [see Lavrentiev (1967)].

Ill-conditioned problems also occur in finite-dimensional spaces:

EXAMPLE 19.5.3.   Let $\mathbf{T} = (t_{jk})$ be an $m \times n$ matrix with scalar entries t_{jk}. Then the equation (19.5.2) becomes
$$\mathbf{Tx} = \mathbf{b}, \tag{19.5.8}$$

and a classical solution exists if and only if the ranks of $\mathbf{T}$ and $(\mathbf{T}, \mathbf{b})$ are equal [see I, § 5.10]. However, a reasonably small perturbation either of $\mathbf{T}$ or of $\mathbf{b}$ may destroy the solubility of (19.5.8). Occasionally in practice one meets just such perturbed equations and so one needs some other suitable form of 'solution'. This is achieved by generalizing the concept of a solution.

Suppose $\mathbf{b} = (b_1, \ldots, b_m)'$ and for $\mathbf{x} \in \mathbb{R}^n$, say $\mathbf{x} = (\xi_1, \ldots, \xi_n)'$, define
$$\Phi(\mathbf{x}) = \sum_{j=1}^m \left| \sum_{k=1}^n t_{jk}\xi_k - b_j \right|^2. \tag{19.5.9}$$

A vector $\bar{\mathbf{x}} \in \mathbb{R}^n$ for which $\Phi(\bar{\mathbf{x}}) = \min \Phi(\mathbf{x})$ is called a *least squares solution*.

DEFINITION 19.5.1.   Let $\mathbf{T} \in \mathscr{B}(X, Y)$, $\mathbf{b} \in Y$ and the range $\mathscr{R}(\mathbf{T})$ be closed in Y, where X and Y are Hilbert spaces over $\mathbb{F}$. A vector $\bar{\mathbf{x}} \in X$ is called a *generalized solution* to
$$\mathbf{Tx} = \mathbf{b} \tag{19.5.10}$$
if
$$\|\mathbf{T\bar{x}} - \mathbf{b}\| = \inf \{\|\mathbf{Tx} - \mathbf{b}\| : \mathbf{x} \in X\}.$$

A generalized solution $\mathbf{x}^*$ of (19.5.10) is called a *normal solution* if
$$\|\mathbf{x}^*\| = \inf \{\|\bar{\mathbf{x}}\| : \bar{\mathbf{x}} \in \mathcal{M}\}, \tag{19.5.11}$$

where $\mathcal{M}$ is the set of all generalized solutions of (19.5.10).

Note that the set $\mathcal{M}$ is closed and convex [see § 15.2.2].

DEFINITION 19.5.2. (Some alternatives and generalizations are in Nashed (1971).) Let $\mathbf{T} \in \mathcal{B}(X, Y)$ and the range $\mathcal{R}(\mathbf{T})$ be closed in Y, where X and Y are Hilbert spaces over $\mathbb{F}$. An operator $\mathbf{Z} \in \mathcal{B}(Y, X)$ is called a *pseudo-inverse* of $\mathbf{T}$ if $\mathbf{Z}$ obeys the following four rules

   (i) $\mathbf{TZT} = \mathbf{T}$,
   (ii) $\mathbf{ZTZ} = \mathbf{Z}$,
   (iii) $(\mathbf{ZT})^* = \mathbf{ZT}$,
   (iv) $(\mathbf{TZ})^* = \mathbf{TZ}$.

Note that the adjoints in (iii) and (iv), although denoted by the same symbol belong, in general, to different spaces.

THEOREM 19.5.1. *Let* $\mathbf{T}$ *be as in Definition* 19.5.2. *Then the pseudo-inverse of* $\mathbf{T}$ *exists and is uniquely determined; it is denoted by* $\mathbf{T}^+$. *In addition, the vector* $\mathbf{x}^* = \mathbf{T}^+\mathbf{b}$ *is the unique normal solution to* (19.5.10) *in* X.

Obviously, if $X = Y$ and $\mathbf{T}$ has an inverse $\mathbf{T}^{-1} \in \mathcal{B}(X)$, then $\mathbf{Z} = \mathbf{T}^{-1}$, i.e. in the regular case the pseudo-inverse coincides with the ordinary inverse.

We see that Theorem 19.5.1 offers a sufficient tool for solving problems of the type determined, for example by Fredholm integral equations of the first kind. A more general theory is elaborated in Nashed (1971).

19.5.2. Variational Methods

There is a close relation between solutions to the equation

$$\mathbf{Tx} = \mathbf{b}, \qquad \mathbf{b} \in X, \tag{19.5.12}$$

where $\mathbf{T}$ is a positive definite operator, and the extremal vectors of the quadratic function F, where

$$F(\mathbf{x}) = (\mathbf{Tx}, \mathbf{x}) - (\mathbf{x}, \mathbf{b}) - (\mathbf{b}, \mathbf{x}). \tag{19.5.13}$$

THEOREM 19.5.2. *Let* $\mathbf{T}$ *be a positive definite operator on a domain* $\mathcal{D}(\mathbf{T}) \subset X$ *into* X. *Then any solution* $\mathbf{u}$ *of* (19.5.12) *corresponds to a minimal value of* F *defined in* (19.5.13). *Conversely, each vector* $\mathbf{u}$ *for which* F *attains its minimal value is a solution of* (19.5.12).

It should be noted that the results of Theorem 19.5.2 strongly depend upon the fact that any positive definite operator can be extended to a self-adjoint operator [see Theorem 19.3.15].

The results of Theorem 19.5.2 have important applications both to the theory and to the numerical treatment of elliptic boundary value problems [see § 8.5 and III, § 9.5].

19.5.3. Elements of Non-linear Analysis

Derivatives of scalar functions play an important role in calculus. It is natural that this concept is also needed in the theory of abstract functions. It seems that most of the results of classical analysis can easily be transferred to a more abstract level. Actually non-linear functional analysis arose from the solution of some problems in the calculus of variations (Chapter 12).

In this section we present some concepts and results of non-linear analysis as a sample of tools used in the non-linear theory and practice. More details can be found in Vainberg (1972).

Suppose $\mathbf{f}$ is a (possibly non-linear) map of a Banach space X into a Banach space Y.

DEFINITION 19.5.3.   Let $\mathbf{x}, \mathbf{h} \in X$. If there exists an element $\mathbf{Vf}(\mathbf{x}; \mathbf{h}) \in Y$ such that

$$\lim_{t \to 0} \left\| \frac{1}{t}(\mathbf{f}(\mathbf{x} + t\mathbf{h}) - \mathbf{f}(\mathbf{x}) - \mathbf{Vf}(\mathbf{x}; \mathbf{h}) \right\| = 0$$

then for fixed $\mathbf{x} \in X$ the map $(\mathbf{x}, \mathbf{h}) \to \mathbf{Vf}(\mathbf{x}; \mathbf{h})$ is called the *Gateaux differential* (or *G-differential*) of $\mathbf{f}$ at $\mathbf{x}$ in the direction $\mathbf{h}$.

The Gateaux differential $\mathbf{Vf}(\mathbf{x}; \mathbf{h})$ is a *homogeneous map* of X into Y, that is: $\mathbf{Vf}(\mathbf{x}; \alpha\mathbf{h}) = \alpha \mathbf{Vf}(\mathbf{x}; \mathbf{h})$, but it need not be linear [see I, § 5.11].

DEFINITION 19.5.4.   If the *G*-differential $\mathbf{Vf}(\mathbf{x}; \mathbf{h})$ is a linear bounded operator, then it is called a (first) *Gateaux derivative*, or *G-derivative* of $\mathbf{f}$ at $\mathbf{x}$ and is denoted by $D_G\mathbf{f}(\mathbf{x})$.

DEFINITION 19.5.5.   Let $\mathbf{f}$ be as before and $\mathbf{x}, \mathbf{h} \in X$. If there is an element $\mathbf{df}(\mathbf{x}; \mathbf{h}) \in Y$ such that

$$\lim_{\|\mathbf{h}\| \to 0} \frac{1}{\|\mathbf{h}\|} \|\mathbf{f}(\mathbf{x} + \mathbf{h}) - \mathbf{f}(\mathbf{x}) - \mathbf{df}(\mathbf{x}; \mathbf{h})\| = 0,$$

then it is called the *Fréchet*, or *F-differential* of $\mathbf{f}$ at $\mathbf{x}$. If $\mathbf{df}(\mathbf{x}; \mathbf{h}) \in \mathcal{B}(X, Y)$, we call it the (first) *Fréchet derivative*, or *F-derivative* of $\mathbf{f}$ at $\mathbf{x}$ and denote it by $D_F\mathbf{f}(\mathbf{x})$.

Obviously, if $\mathbf{df}(\mathbf{x}; \mathbf{h})$ exists then so does $\mathbf{Vf}(\mathbf{x}; \mathbf{h})$, and $\mathbf{Vf}(\mathbf{x}; \mathbf{h}) = \mathbf{df}(\mathbf{x}; \mathbf{h})$. The converse is not always true.

THEOREM 19.5.3.   *If $D_G\mathbf{f}(\mathbf{x})$ exists in some neighbourhood $U(\mathbf{x}_0)$ of $\mathbf{x}_0 \in X$ and is continuous there as a function of $\mathbf{x}$, then $D_F\mathbf{f}(\mathbf{x}_0)$ exists and $D_F\mathbf{f}(\mathbf{x}_0) = D_G\mathbf{f}(\mathbf{x}_0)$.*

Higher derivatives of abstract functions can be defined in a manner very similar to that in calculus, i.e. the second derivative is a derivative of the first derivative etc. [See § 5.5.]

EXAMPLE 19.5.4. Suppose $X = \mathbb{R}^n$, $Y = \mathbb{R}^m$ and $\mathbf{f}$ is a mapping on $\mathcal{D}(\mathbf{f}) \subset X$ into Y which is differentiable at $\mathbf{s} = (s_1, \ldots, s_n)$. Then

$$
\mathbf{Vf(s; h)} =
\begin{bmatrix}
\dfrac{\partial f_1}{\partial s_1}(s_1, \ldots, s_n), & \ldots, & \dfrac{\partial f_m}{\partial s_1}(s_1, \ldots, s_n) \\
& \cdots & \\
\dfrac{\partial f_1}{\partial s_n}(s_1, \ldots, s_n), & \ldots, & \dfrac{\partial f_m}{\partial s_n}(s_1, \ldots, s_n)
\end{bmatrix}
\begin{bmatrix}
h_1 \\ \vdots \\ h_n
\end{bmatrix}
$$

where $\mathbf{h} = (h_1, \ldots, h_n)'$ and $\mathbf{f(s)} = (f_1(\mathbf{s}), \ldots, f_m(\mathbf{s}))$. We have obtained a classical formula. In particular, by choosing $\mathbf{h} = \mathbf{e}_1 = (1, 0, \ldots, 0)'$ [see I, Example 5.4.2] we find that

$$
\mathbf{Vf(s; e_1)} = \left(\frac{\partial f_1}{\partial s_1}(s_1, \ldots, s_n), \ldots, \frac{\partial f_m}{\partial s_1}(s_1, \ldots, s_n) \right).
$$

EXAMPLE 19.5.5. Let F be a real-valued function which maps $J = [a, b] \times (-\infty, +\infty) \times (-\infty, +\infty)$ into $\mathbb{F}$. Let F have partial derivatives continuous on J. A typical functional of the calculus of variations [see § 12.3]

$$
I(y) = \int_a^b F(s, y(s), y'(s)) \, ds,
$$

where $y'(s) = (d/ds)y(s)$, considered as a functional on $C^1([a, b])$, has a G-derivative

$$
D_G I(y)h = \int_a^b \left\{ \frac{\partial}{\partial y} F(s, z, w) \Big|_{\substack{z = y(s) \\ w = y'(s)}} h(s) + \frac{\partial}{\partial w} F(s, z, w) \Big|_{\substack{z = y(s) \\ w = y'(s)}} h'(s) \right\} ds.
$$

DEFINITION 19.5.6. Let X and Y be Banach spaces and $\mathbf{T}$ a, possibly non-linear, operator mapping X into Y. We call $\mathbf{T}$ *bounded* if $\mathbf{T}(\mathcal{B})$ is bounded whenever $\mathcal{B}$ is bounded, i.e. there is a constant $\mathcal{H}$, independent of $\mathbf{x} \in \mathcal{B}$, such that $\|\mathbf{x}\| \leq \mathcal{H}$ implies $\|\mathbf{Tx}\| \leq \mathcal{H}$; here $\mathbf{T}(\mathcal{B}) = \{\mathbf{y} \in Y : \mathbf{y} = \mathbf{Tx}, \mathbf{x} \in \mathcal{B}\}$.

DEFINITION 19.5.7. A possibly non-linear operator $\mathbf{T}$ mapping a Banach space X into its dual X' is called *monotone* if the relation

$$
\langle \mathbf{x} - \mathbf{y}, \mathbf{Tx} - \mathbf{Ty} \rangle \geq 0 \tag{19.5.14}
$$

holds for arbitrary $\mathbf{x}, \mathbf{y} \in X$; here $\langle \mathbf{x}, \mathbf{x}' \rangle$ denotes the value $\mathbf{x}'(\mathbf{x})$.

If in (19.5.14) the strict inequality takes place for $\mathbf{x} \neq \mathbf{y}$, then $\mathbf{T}$ is called *strictly monotone*.

DEFINITION 19.5.8. An operator $\mathbf{T}$ mapping a Banach space X into its dual X' is called *coercive* if

$$
\frac{\langle \mathbf{u}, \mathbf{Tu} \rangle}{\|\mathbf{u}\|} \to \infty \quad \text{as } \|\mathbf{u}\| \to \infty.
$$

DEFINITION 19.5.9. An operator **T** mapping a Banach space X into its dual X' is called *potential* if there exists a functional **F** mapping X into $\mathbb{F}$ such that $\mathbf{T(u)} = \mathbf{F'(u)}$, where **F'** denotes the Gateaux derivative of **F** at $\mathbf{u} \in X$.

THEOREM 19.5.4. *Let X be a reflexive Banach space and* **T** *a map of X into the dual X'. Let* **T** *be potential, bounded, monotone and coercive. Then the range of* **T** *is the whole of X', i.e.* $\mathbf{T}(X) = X'$. *Thus, the equation*

$$\mathbf{Tu} = \mathbf{b} \qquad (19.5.15)$$

possesses, for arbitrary $\mathbf{b} \in Y$, *at least one solution* $\mathbf{u} \in X$.
If, in addition, **T** *is strictly monotone, then there is exactly one solution to* (19.5.15) *for a given* $\mathbf{b} \in Y$.

An application of Theorem 19.5.4 is given in the following example:

EXAMPLE 19.5.6. Suppose $f \in L(0, 1)$ and $g \in C^1(\mathbb{R})$ is non-decreasing in $\mathbb{R}$. Then the problem

$$\begin{cases} -u''(s) + g(u(s)) = b(s), & s \in [0, 1] \\ u(0) = u(1) = 0 \end{cases} \qquad (19.5.16)$$

has at least one solution $u_0 \in W_0^{1,2}(0, 1) = H_0^1(0, 1)$. If g is increasing in $(-\infty, +\infty)$ then there is exactly one solution u_0 of (19.5.16), i.e.

$$\int_0^1 [u'v' + g(u)v]\, ds = \int_0^1 vb\, ds \quad \text{for every } v \in W_0^{1,2}(0, 1).$$

The operator T is defined by

$$\langle v, Tu \rangle = \int_0^1 v'(s)u'(s)\, ds + \int_0^1 g(u(s))v(s)\, ds,$$

and it satisfies all the hypotheses of Theorem 19.5.4.

I. M.

REFERENCES

Achieser N. I. and Glasman I. M. (1954). *Theorie der Linearen Operatoren im Hilbert–Raum*, Akademie Verlag. (German; second Russian edition 1966).

Dunford, N. and Schwartz, J. T. (1958). *Linear Operators; Part 1, General Theory*, Interscience.

Dunford, N. and Schwartz, J. T. (1963). *Linear Operators; Part 2, Spectral Theory, Selfadjoint operators in Hilbert space*, Interscience.

Dunford, N. and Schwartz, J. T. (1971). *Linear Operators; Part 3, Spectral Operators*, Interscience.

Gelfand, I. M. and Shilov, G. E. (1964). *Generalized Functions*, Academic Press (Russian original edition 1958).

Halmos, R. (1964). *A Hilbert Space Book*, Van Nostrand.

Hille, E. and Phillips, R. S. (1957). *Functional Analysis and Semigroups*, Interscience.

Lavrentiev, M. M. (1967). *Some Improperly Posed Problems of Mathematical Physics*, Springer.

Ljusternik, L. A. and Sobolev, S. L. (1955). *Elements of functional analysis*, Ungar, N.Y. (English, original Russian edition 1951).

Najmark, M. A. (1960). *Normed Rings*, Nordhoff Ltd. (Original Russian edition 1956).

Nashed, M. Z. (1971). Generalized Inverses, Normal Solvability and Iteration for Singular Operator Equations, In *Nonlinear Functional Analysis and Applications*, Academic Press.

Protter, M. H. and Weinberger, H. F. (1967). *Maximum Principles in Differential Equations*, Prentice Hall.

Riesz, F. and Nagy, B. Sz. (1955). *Functional Analysis*, Frederic Unger Publ.

Rudin, W. (1973). *Functional Analysis*, McGraw-Hill.

Rudin, W. (1966). *Real and Complex Analysis*, McGraw-Hill.

Schaefer, H. H. (1974). *Positive Operators in Banach Lattices*, Springer.

Taylor, A. E. (1958). *Introduction to Functional Analysis*, J. Wiley.

Vainberg, M. M. (1972). *Variational Method and Method of Monotone Operators in Nonlinear Operator Theory* (Russian), Nauka.

CHAPTER 20

Fourier Series

20.1. PERIODIC FUNCTIONS

Nature abounds with phenomena that repeat themselves with remarkable regularity: the motion of the moon round the Earth is an obvious example, the period of revolution being a fixed interval of time (about 28 days). Other events show a regular pattern in space, like the shape of a vibrating string (Figure 20.1.1), which is repeated when a particular distance, the wavelength, has been traversed.

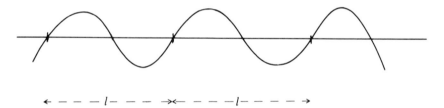

Figure 20.1.1

The mathematical description of these observations calls for a special class of functions, which is of great importance in many branches of mathematics and in the physical sciences. For the sake of simplicity we shall confine ourselves to functions of a single real variable, which may denote time, space or any other measurable quantity.

DEFINITION 20.1.1.   The function $f(x)$ of a real variable x is said to be *periodic* if there exists a non-zero number p, independent of x, such that the equation

$$f(x+p) = f(x) \tag{20.1.1}$$

holds for all values of x.

We may express (20.1.1) by saying that $f(x)$ is periodic with period p. But this statement needs to be made more precise. For on replacing x by $x+p$

we obtain that

$$f(x+2p)=f(x+p)=f(x)$$

and by repeating this substitution we find that

$$f(x+kp)=f(x),$$

where k is an arbitrary positive integer.

Similarly on replacing x by $x-p$ in (20.1.1) we deduce that

$$f(x)=f(x-p)$$

and, more generally,

$$f(x)=f(x-kp).$$

Hence if p is a period of $f(x)$, so is any number of the form

$$kp \qquad (k=\pm1, \pm2, \ldots).$$

Accordingly, we call p the *minimal period*, or simply *the* period of $f(x)$, if p is the least positive number for which (20.1.1) holds.

It is clear from (20.1.1) that the function f is completely determined by its values in the interval

$$0 \le x < p, \qquad\qquad\qquad\qquad (20.1.2)$$

the values outside this interval being generated by translations through $\pm p, \pm 2p, \ldots$. Indeed, the graph of an arbitrary function of period p may be constructed by drawing any graph in the interval $0 \le x < p$ and then moving it rigidly an indefinite number of times to the left and to the right through a distance p.

The defining property (20.1.1) requires that

$$f(0)=f(p)=f(2p)=\ldots=f(-p)=f(-2p)=\ldots.$$

The common value of $f(kp)$ ($k=0, \pm1, \pm2, \ldots$) is marked by a cross in Figure 20.1.2.

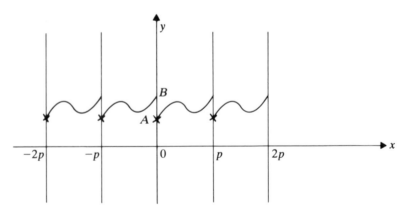

Figure 20.1.2

We see that the resulting graph is in general discontinuous at the points kp; for the points A and B need not coincide. Thus discontinuous functions arise naturally in the study of periodic functions. In the next section we shall discuss the type of discontinuity we are prepared to accept in the subsequent development.

20.2.   PIECEWISE DIFFERENTIABLE FUNCTIONS

Let $f(x)$ be a function which is defined in an interval $[a, b]$. At an interior point c, that is when $a < c < b$, the function is said to be continuous if

$$\lim_{x \to c} f(x) = f(c) \qquad (20.2.1)$$

[see Definition 2.3.1]. The definition of the limit process stipulates that the value of the limit does not depend on the manner in which x approaches c; in particular we may let x tend to c from the left or from the right only, and in either case the value of the limit will be equal to $f(c)$.

When f is not continuous at c, so that (20.2.1) ceases to hold, it may still happen that the 'one-sided' limits at c exist separately. More precisely, we say that f has a *left-hand limit* at c if $f(x)$ tends to a finite limit as x tends to c subject to the restriction $x < c$. Denoting this limit by $f(c - 0)$ we write

$$\lim_{x \to c - 0} f(x) = f(c - 0). \qquad (20.2.2)$$

Similarly, the *right-hand limit* of f at c, if it exists, is obtained by letting x tend to c subject to the restriction $x > c$, and we write

$$\lim_{x \to c + 0} f(x) = f(c + 0). \qquad (20.2.3)$$

It should be observed that the three numbers

$$f(c - 0), \quad f(c + 0), \quad f(c)$$

are, in general, distinct. This situation is indicated in Figure 20.2.1, where

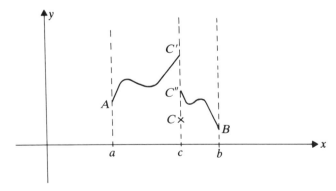

Figure 20.2.1

the points C', C'' and C have co-ordinates

$$(c, f(c-0)), \quad (c, f(c+0)), \quad (c, f(c))$$

respectively. In fact, f is continuous at c if and only if

$$f(c-0) = f(c+0) = f(c). \qquad (20.2.4)$$

If (20.2.4) fails to hold, then f is discontinuous at c. It will be necessary to restrict ourselves to a rather mild form of discontinuity by assuming that the limits (20.2.2) and (20.2.3) always exist whether or not they are equal.

As regards differentiability we postulate that f should have a (finite) derivative at every interior point, with at most a finite number of exceptions. When c is such an exceptional point, we assume that f possesses both a *right-hand derivative* and a *left-hand derivative*; by this we mean that as h tends to zero through positive values, then the limits denoted by

$$\lim_{h \to 0+0} \frac{f(c+h) - f(c+0)}{h} = f'_+(c) \qquad (20.2.5)$$

and

$$\lim_{h \to 0+0} \frac{f(c-h) - f(c-0)}{-h} = f'_-(c) \qquad (20.2.6)$$

exist and are finite. In Figure 20.2.1 the gradients at C' and C'' are equal to $f'_-(c)$ and $f'_+(c)$ respectively. The function f is differentiable at c, in the ordinary sense [see § 2.9 or § 3.1.1] if it is continuous at c, that is if (20.2.4) holds and if, in addition

$$f'_+(c) = f'_-(c). \qquad (20.2.7)$$

Since the function need not be defined outside the interval $[a, b]$, we may have to content ourselves with one-sided limits at the end-points. Thus we can assume only the existence of

$$f(a+0), \quad f'_+(a), \quad f(b-0), \quad f'_-(b). \qquad (20.2.8)$$

We summarize the hypotheses we are making about the functions that are discussed in this chapter:

DEFINITION 20.2.1.   The function f is said to be *piecewise differentiable* in $[a, b]$ if f is differentiable at all interior points with at most a finite number of exceptions. If c is such an exceptional point, then the one-sided limits

$$f(c+0), \quad f(c-0), \quad f'_+(c), \quad f'_-(c)$$

exist, as do the one-sided limits (20.2.8) which refer to the end-points.

Suppose there are altogether m points, say, $c_1, c_2, \ldots, c_m$, satisfying

$$a < c_1 < c_2 < \ldots < c_m < b,$$

at which f is either discontinuous, or is continuous but fails to have a (unique) derivative, that is has a 'kink'. Then f is differentiable throughout each of the open intervals $(a, c_1), (c_1, c_2), \ldots, (c_m, b)$, which explains the term 'piecewise' differentiable.

Let us now return to the discussion of periodic functions with a prescribed period p. We have already observed that such a function is completely determined by its values in the interval $0 \le x < p$ or, more generally, by any interval of length p, for example $-\frac{1}{2}p < x \le \frac{1}{2}p$. Notice that we include one but not both end-points, because the function is bound to take equal values at both end points. Any such 'half-open' interval of length p will be called a *fundamental interval* for the function.

The following restriction will apply to all functions treated in this chapter: *if $f(x)$ is periodic, then it is assumed to be piecewise differentiable in a fundamental interval.*

EXAMPLE 20.2.1. The function $f(x)$ has period 2 and in the fundamental interval $0 \le x < 2$ is given by the formula

$$f(x) = 1 - |x - 1|,$$

that is

$$f(x) = \begin{cases} x & \text{when } 0 \le x \le 1 \\ 2 - x & \text{when } 1 \le x < 2. \end{cases}$$

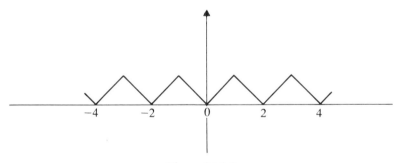

Figure 20.2.2

This function, which for obvious reasons is called the '*saw function*', is continuous for all values of x, but it is not differentiable at $x = 0, \pm 1, \pm 2, \ldots$. For example, when $x = 1$, we have that $f'_-(1) = 1$, $f'_+(1) = -1$.

EXAMPLE 20.2.2. The function $f(x)$ has period unity and in the fundamental interval $0 \le x < 1$ is given by the formulae

$$f(x) = \begin{cases} 1 & \text{when } 0 \le x < \frac{1}{2} \\ 0 & \text{when } x = \frac{1}{2} \\ -1 & \text{when } \frac{1}{2} < x < 1. \end{cases}$$

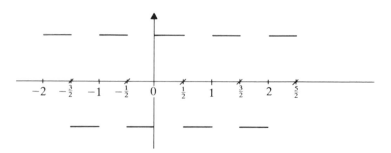

Figure 20.2.3

Note that $f'_+ (\tfrac{1}{2}) = f'_- (\tfrac{1}{2}) = 0$; nevertheless f is not differentiable at $x = \tfrac{1}{2}$ because it fails to be continuous at that point.

20.3.   TRIGONOMETRIC POLYNOMIALS AND SERIES

There is a simple device for constructing new periodic functions from given ones provided that they share the same period (which need not be the minimal period). Thus suppose that each of the functions

$$g_0(x), g_1(x), g_2(x), \ldots, g_n(x) \tag{20.3.1}$$

has period p. Then any linear combination of the form

$$f(x) = c_0 g_0(x) + c_1 g_1(x) + \ldots + c_n g_n(x) \tag{20.3.2}$$

has period p, where $c_0, c_1, \ldots, c_n$ are arbitrary constants. This is known as the *principle of superposition*. Since a constant function is periodic whatever the value of p, it is often convenient to put $g_0(x) = 1$.

The most familiar examples of periodic functions are the trigonometric functions

$$\left. \begin{array}{l} 1, \cos x, \cos 2x, \cos 3x, \ldots \\ \sin x, \sin 2x, \sin 3x, \ldots \end{array} \right\}. \tag{20.3.3}$$

Each of these functions has period 2π, but only $\cos x$ and $\sin x$ have 2π as minimal period [see § 2.12]. We regard (20.3.3) as the building material with which a wide class of periodic functions may be constructed by virtue of the principle of superposition. Of course, any function so obtained has period 2π. Later on [see § 20.5.4] we shall modify the formulae so as to cover the case of an arbitrary period.

A linear combination, $P(x)$, of finitely many members of (20.3.3) is called a *trigonometric polynomial*, thus

$$P(x) = \tfrac{1}{2}\alpha_0 + \sum_{r=1}^{n} (\alpha_r \cos rx + \beta_r \sin rx). \tag{20.3.4}$$

If at least one of the coefficients α_n, β_n is non-zero, we say that $P(x)$ is of degree n. We notice that the constant term is written as $\frac{1}{2}\alpha_0$, rather than α_0; this is a convention which has formal advantages, as we shall see presently.

It is tempting to generalize (20.3.4) by allowing infinitely many terms to appear on the right-hand side. This leads to a *trigonometric series*

$$F(x) = \frac{1}{2}\alpha_0 + \sum_{r=1}^{\infty} (\alpha_r \cos rx + \beta_r \sin rx), \qquad (20.3.5)$$

with the important proviso that the series converges for all values of x under consideration [see § 1.7]. Conditions have been established which are known to ensure the convergence of the series. One of the simplest sets of conditions is expressed by the inequalities

$$|\alpha_r| \leq \frac{M}{r^\lambda}, \qquad |\beta_r| \leq \frac{M}{r^\lambda}, \qquad (20.3.6)$$

$(r = 0, 1, 2, \ldots)$, where M and λ are constants such that $M > 0$ and $\lambda > 1$.

Clearly, if $F(x)$ is the sum of a convergent trigonometric series, then $F(x)$ is a periodic function with period 2π. We are, however, much more interested in the converse problem: given a periodic function $f(x)$, can it be 'represented' as a trigonometric series? If the answer is in the affirmative, we say that $f(x)$ can be expanded as a *Fourier series*. The coefficients of the series are related to $f(x)$ in a surprisingly simple manner, as we shall show in the next section.

20.4. ORTHOGONAL FUNCTIONS

We denote by $\mathscr{F}$ the collection of real functions which are bounded and piecewise differentiable in the interval $(0, 2\pi)$. It is readily verified that $\mathscr{F}$ forms a vector space [see I, § 5.2] over the real field $\mathbb{R}$: indeed, if $f_1(x)$ and $f_2(x)$ belong to $\mathscr{F}$, so does

$$\gamma_1 f_1(x) + \gamma_2 f_2(x),$$

where γ_1 and γ_2 are arbitrary real numbers, and it is not difficult to check that all the axioms of a vector space are satisfied [see I, § 5.1].

We recall [see I, § 5.4] that V is a vector space of finite dimension n if there exists a basis

$$\mathbf{e}_1, \mathbf{e}_2, \ldots, \mathbf{e}_n \qquad (20.4.1)$$

such that every vector $\mathbf{f}$ of V can be uniquely expressed as

$$\mathbf{f} = \alpha_1 \mathbf{e}_1 + \alpha_2 \mathbf{e}_2 + \ldots + \alpha_n \mathbf{e}_n. \qquad (20.4.2)$$

Suppose now that, furthermore, V is equipped with an inner product structure. This situation is discussed in Volume I, section 10.1, but it is convenient to recapitulate the main facts, bearing in mind that, at least for the time being, we are concerned with real vector spaces. If $\mathbf{f}$ and $\mathbf{g}$ are arbitrary vectors of V we denote their inner product by $(\mathbf{f}, \mathbf{g})$. For a given vector space the inner product may be defined in different ways; but whatever choice is made, the following general laws must hold for a real vector space:

$$(\mathbf{f}, \mathbf{g}) = (\mathbf{g}, \mathbf{f}) \tag{20.4.3}$$

$$(\alpha_1\mathbf{f}_1 + \alpha_2\mathbf{f}_2, \mathbf{g}) = \alpha_1(\mathbf{f}_1, \mathbf{g}) + \alpha_2(\mathbf{f}_2, \mathbf{g}) \tag{20.4.4}$$

$$\left.\begin{array}{l} (\mathbf{f}, \mathbf{f}) \geq 0 \\ (\mathbf{f}, \mathbf{f}) = 0 \quad \text{if and only if } \mathbf{f} = 0 \end{array}\right\}. \tag{20.4.5}$$

The *distributive law* (20.4.4) is easily extended to the case in which $\mathbf{f}$ or $\mathbf{g}$ or both consist of an arbitrary, but finite, number of terms. Thus if

$$\mathbf{f} = \alpha_1\mathbf{f}_1 + \alpha_2\mathbf{f}_2 + \ldots + \alpha_m\mathbf{f}_m, \qquad \mathbf{g} = \beta_1\mathbf{g}_1 + \beta_2\mathbf{g}_2 + \ldots + \beta_n\mathbf{g}_n,$$

then

$$(\mathbf{f}, \mathbf{g}) = \sum_{i=1}^{m} \sum_{j=1}^{n} \alpha_i\beta_j(\mathbf{f}_i, \mathbf{g}_j). \tag{20.4.6}$$

The vectors $\mathbf{f}$ and $\mathbf{g}$ are said to be *orthogonal* if

$$(\mathbf{f}, \mathbf{g}) = 0. \tag{20.4.7}$$

We say that $\mathbf{f}$ is *normalized* if

$$(\mathbf{f}, \mathbf{f}) = 1, \tag{20.4.8}$$

that is, $\mathbf{f}$ is a *unit vector*. A particularly favourable situation arises when the basis vectors (20.4.1) are orthogonal in pairs and each of them is normalized. In that case we say that (20.4.1) is an *orthonormal system*, that is

$$(\mathbf{e}_i, \mathbf{e}_j) = \delta_{ij} \qquad (i, j = 1, 2, \ldots n), \tag{20.4.9}$$

where $\delta_{ii} = 1$ $(i = 1, 2, \ldots, n)$ and $\delta_{ij} = 0$ when $i \neq j$ [see I, (6.2.7)]. We are now able to derive a simple formula for the coefficients $\alpha_1, \alpha_2, \ldots, \alpha_n$ in (20.4.2): take the inner product of $\mathbf{e}_1$ and either side of (20.4.2); by virtue of the distributive law we obtain that

$$(\mathbf{f}, \mathbf{e}_1) = \alpha_1(\mathbf{e}_1, \mathbf{e}_1) + \alpha_2(\mathbf{e}_2, \mathbf{e}_1) + \ldots + \alpha_n(\mathbf{e}_n, \mathbf{e}_1).$$

By (20.4.9) we have that

$$(\mathbf{e}_1, \mathbf{e}_1) = 1, \qquad (\mathbf{e}_2, \mathbf{e}_1) = \ldots = (\mathbf{e}_n, \mathbf{e}_1) = 0.$$

Hence $(\mathbf{f}, \mathbf{e}_1) = \alpha_1$. Generally, if we multiply (20.4.2) by $\mathbf{e}_m$ where m is any integer satisfying $1 \leq m \leq n$ we find that

$$\alpha_m = (\mathbf{f}, \mathbf{e}_m) \qquad (m = 1, 2, \ldots, n). \tag{20.4.10}$$

Our principal aim is to extend these ideas to the vector space $\mathscr{F}$ defined at the beginning of this section. Many of the foregoing arguments still hold, but the main difference stems from the fact that the space $\mathscr{F}$ is no longer finite-dimensional. It will now be necessary to seek an infinite basis consisting of functions

$$e_0(x), e_1(x), e_2(x), \ldots. \tag{20.4.11}$$

Next, we have to choose a convenient form of the inner product. It turns out that all conditions will be met if we put

$$(f, g) = \int_0^{2\pi} f(x)g(x)\, dx. \tag{20.4.12}$$

Thus $f(x)$ and $g(x)$ are orthogonal if

$$\int_0^{2\pi} f(x)g(x)\, dx = 0, \tag{20.4.13}$$

and $f(x)$ is normalized if

$$\int_0^{2\pi} \{f(x)\}^2\, dx = 1. \tag{20.4.14}$$

It is a remarkable fact that the members of the set (20.3.3) are mutually orthogonal in this sense; for we have the following integral formulae [see Example 4.3.12]

$$\int_0^{2\pi} \cos rx \cos sx\, dx = \begin{cases} 0 & \text{if } r \neq s \\ \pi & \text{if } r = s > 0 \\ 2\pi & \text{if } r = s = 0 \end{cases} \tag{20.4.15}$$

$$\int_0^{2\pi} \sin rx \sin sx\, dx = \begin{cases} 0 & \text{if } r \neq s \\ \pi & \text{if } r = s > 0 \end{cases} \tag{20.4.16}$$

$$\int_0^{2\pi} \cos rx \sin sx\, dx = 0 \quad \text{(all } r, s). \tag{20.4.17}$$

In the inner product notation these results are expressed as follows:

$$(\cos rx, \cos sx) = \begin{cases} \pi\delta_{rs} & (r = 0, 1, 2, \ldots; s = 1, 2, \ldots) \\ 2\pi & (r = s = 0) \end{cases} \tag{20.4.18}$$

$$(\sin rx, \sin sx) = \pi\delta_{rs} \quad (r = 1, 2, \ldots, s = 1, 2, \ldots) \tag{20.4.19}$$

$$(\cos rx, \sin sx) = 0. \tag{20.4.20}$$

Note that the above functions, though mutually orthogonal, are not normalized.

The ultimate aim of the theory is the assertion (with some proviso) that the functions

$$1, \cos x, \sin x, \cos 2x, \sin 2x, \ldots, \cos rx, \sin rx, \ldots. \tag{20.4.21}$$

form a basis of the vector space $\mathscr{F}$; in other words we should like to be assured that an arbitrary function $f(x)$ of $\mathscr{F}$ can be expressed uniquely as a trigonometric series

$$f(x) = \tfrac{1}{2}\alpha_0 + \sum_{r=1}^{\infty} (\alpha_r \cos rx + \beta_r \sin rx). \tag{20.4.22}$$

Actually, as we shall see presently, this equation may fail to hold at a finite number of points, where f is discontinuous.

In the meantime, let us assume that (20.4.22) is true and that the inner product of $f(x)$ with $\cos nx$ and $\sin nx$ can be computed by applying the distributive law (20.4.4) to (20.4.22) although infinitely many terms are now involved; thus

$$(f(x), \cos nx) = \tfrac{1}{2}\alpha_0(\cos 0x, \cos nx) + \sum_{r=1}^{\infty} \alpha_r(\cos rx, \cos nx)$$

$$+ \sum_{r=1}^{\infty} \beta_r(\sin rx, \cos nx).$$

Suppose first that $n > 0$. By virtue of (20.4.18) only one of the inner products on the right-hand side is non-zero, namely when $r = n$ in the first sum. Thus

$$(f(x), \cos nx) = \alpha_n \pi \qquad (n > 0). \tag{20.4.23}$$

When $n = 0$, it is the first term that remains non-zero and we find that

$$(f(x), \cos 0x) = \tfrac{1}{2}\alpha_0 2\pi = \alpha_0 \pi.$$

We observe that the formula (20.4.23) remains valid when $n = 0$; this is due to the device of attaching the factor $\tfrac{1}{2}$ to α_0. Similarly, it is found that

$$(f(x), \sin nx) = \beta_n \pi \qquad (n \geq 1). \tag{20.4.24}$$

For reference it is convenient to recast these results by substituting the definition (20.4.12) of the inner product. Thus if $f(x)$ possesses an expansion (20.4.22) with the additional property mentioned, then the coefficients of the expansion are given by

$$\alpha_n = \frac{1}{\pi} \int_0^{2\pi} f(x) \cos nx \, dx \qquad (n \geq 0) \tag{20.4.25}$$

and

$$\beta_n = \frac{1}{\pi} \int_0^{2\pi} f(x) \sin nx \, dx \qquad (n \geq 1). \tag{20.4.26}$$

So far, the existence of the expansion is still a matter of speculation for us. In 1807 the French engineer J. Fourier asserted that an 'arbitrary' periodic function can be expressed as a trigonometric series. Fourier was led to this astonishing conjecture by his research into the theory of heat rather than by mathematical reasoning. Eminent mathematicians at that time questioned the validity of Fourier's arguments and a prolonged controversy ensued. It became necessary to define more precisely what was meant by an arbitrary function, and certain conditions had to be imposed. The first rigorous treatment was given by G. L. Dirichlet in 1829.

20.5. FOURIER SERIES

20.5.1. An Expansion Theorem

Throughout this section, except for subsections 20.5.4 and 20.5.5 it will be assumed that $f(x)$ has period 2π. Moreover, we suppose that the integral

$$\int_0^{2\pi} |f(x)|\, dx$$

exists [see § 4.1]. Then the integrals on the right-hand sides of (20.4.25) and (20.4.26) also exist. We define the *Fourier coefficients* of $f(x)$ by

$$a_n = \frac{1}{\pi} \int_0^{2\pi} f(x) \cos nx\, dx \qquad (n = 0, 1, 2, \ldots) \qquad (20.5.1)$$

and

$$b_n = \frac{1}{\pi} \int_0^{2\pi} f(x) \sin nx\, dx \qquad (n = 1, 2, \ldots) \qquad (20.5.2)$$

without prejudice as to whether $f(x)$ possesses an expansion (20.4.22) with these or any other kind of coefficients—hence the change in notation. It is, however, convenient to set up the *formal Fourier series* associated with $f(x)$ and to write

$$f(x) \sim \tfrac{1}{2}a_0 + \sum_{n=1}^{\infty} (a_n \cos nx + b_n \sin nx), \qquad (20.5.3)$$

where a_n and b_n are given by (20.5.1) and (20.5.2). The following questions arise [see § 1.7]:

(i) for what values of x does the series converge, and
(ii) if the series converges at $x = x_0$, is the sum equal to $f(x_0)$?

When x_0 is a point of discontinuity so that

$$f(x_0+0) \neq f(x_0-0), \qquad (20.5.4)$$

the answer to (ii) cannot possibly be in the affirmative; for a convergent series can, of course, have only one value as its sum, and this may in general differ both from $f(x_0+0)$ and $f(x_0-0)$.

We shall now quote, without proof, a result which answers the above questions in a manner sufficient for most applications.

THEOREM 20.5.1 (*Expansion theorem*).   *Let $f(x)$ be a function of period 2π which is piecewise differentiable in the interval $[0, 2\pi)$ and which has*

Fourier coefficients a_n ($n = 0, 1, 2, \ldots$) and b_n ($n = 1, 2, \ldots$) defined in (20.5.1) and (20.5.2). Then, if x_0 is a point of continuity,

$$f(x_0) = \tfrac{1}{2}a_0 + \sum_{n=1}^{\infty} (a_n \cos nx_0 + b_n \sin nx_0) \tag{20.5.5}$$

while at a point of discontinuity

$$\tfrac{1}{2}(f(x_0+0) + f(x_0-0)) = \tfrac{1}{2}a_0 + \sum_{n=1}^{\infty} (a_n \cos nx_0 + b_n \sin nx_0). \tag{20.5.6}$$

Remarks. 1. It is only for emphasis that we distinguished between points of continuity and discontinuity. In fact, (20.5.5) is a special case of (20.5.6), when $f(x_0+0) = f(x_0-0) = f(x_0)$ [see (20.2.4)]. It is noteworthy that in the event of discontinuity the Fourier series neatly settles down at the average

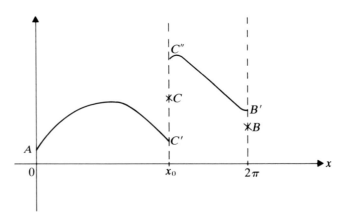

Figure 20.5.1

of the conflicting values (20.5.4). Thus Figure 20.5.1 shows the graph of a function $f(x)$ for which

$$f(0+0) = A, \qquad f(x_0-0) = C', \qquad f(x_0+0) = C'', \qquad f(2\pi-0) = B'.$$

At x_0 the Fourier series converges to $\tfrac{1}{2}(C' + C'')$ ($= C$).
   2. By virtue of the periodicity we have that

$$f(0+0) = f(2\pi+0), \qquad f(0-0) = f(2\pi-0).$$

Hence when $x = 0$ or $x = 2\pi$, the Fourier series converges to

$$\tfrac{1}{2}\{f(0+0) + f(2\pi-0)\},$$

indicated by the letter B in Figure 20.5.1.
   3. Other expansion theorems can be proved under less stringent assumptions about $f(x)$. But no convenient conditions are known which are both

necessary and sufficient for a function to be represented by its Fourier series [see Whittaker, E. T. and Watson, G. N. (1927), Chapter IX].

20.5.2. Examples of Fourier Series

If $f(x)$ satisfies the conditions of Theorem 20.5.1, then the problem of obtaining its Fourier series reduces to the evaluation of the integrals (20.5.1) and (20.5.2).

EXAMPLE 20.5.1. Obtain the Fourier expansion of the function $f(x)$ of period 2π, defined in the fundamental interval $[0, 2)$ by

$$f(x) = x \quad \text{when } 0 \le x < 2\pi, \tag{20.5.7}$$

see Figure 20.5.2.

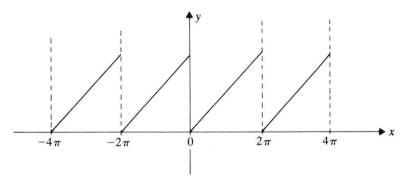

Figure 20.5.2

When $n > 0$, we use integration by parts [see § 4.3] to obtain that

$$a_n = \frac{1}{\pi} \int_0^{2\pi} f(x) \cos nx \, dx = \frac{1}{\pi} \int_0^{2\pi} x \cos nx \, dx$$

$$= \frac{1}{\pi} \left[\frac{x}{n} \sin nx \right]_0^{2\pi} - \frac{1}{\pi} \int_0^{2\pi} \sin n\pi \, dx = 0.$$

Next,

$$a_0 = \frac{1}{\pi} \int_0^{2\pi} x \, dx = \frac{1}{\pi} \left[\tfrac{1}{2} x^2 \right]_0^{2\pi} = 2\pi.$$

Again,

$$b_n = \frac{1}{\pi} \int_0^{2\pi} x \sin nx \, dx = \frac{1}{\pi} \left[\frac{-x}{n} \cos nx \right]_0^{2\pi} + \frac{1}{\pi n} \int_0^{2\pi} \cos nx \, dx$$

$$= \frac{2}{n}.$$

Since the only points of discontinuity occur when $x = 0, \pm 2\pi, \pm 4\pi, \ldots$ the expansion theorem asserts that the formula

$$x = \pi - 2\left(\frac{\sin x}{1} + \frac{\sin 2x}{2} + \frac{\sin 3x}{3} + \ldots\right) \qquad (20.5.8)$$

is valid when $0 < x < 2\pi$; it is obviously false when $x = 0$, but in that case the left-hand side must be replaced by

$$\tfrac{1}{2}\{f(0+0) + f(0-0)\} = \tfrac{1}{2}\{0 + 2\pi\} = \pi,$$

which agrees with the right-hand side.

An interesting result is obtained if we put $x = \pi/2$. Since

$$\sin \frac{\pi}{2} = 1, \qquad \sin \frac{2\pi}{2} = 0, \qquad \sin \frac{3\pi}{2} = -1, \ldots$$

we find that

$$\frac{\pi}{4} = 1 - \tfrac{1}{3} + \tfrac{1}{5} - \ldots. \qquad (20.5.9)$$

This formula is known as *Gregory's* or *Leibniz's* series. It is however only of theoretical value, as the series on the right of (20.5.9) converges very slowly.

EXAMPLE 20.5.2.   Find the Fourier expansion of the function $f(x)$ defined by

$$f(x) = \begin{cases} 1 & \text{when } 0 \le x \le \pi/2 \\ 0 & \text{when } \pi/2 < x < 2\pi \end{cases}$$

[see Figure 20.5.3].

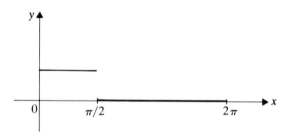

Figure 20.5.3

Thus

$$a_0 = \frac{1}{\pi} \int_0^{2\pi} f(x)\, dx = \frac{1}{\pi} \frac{\pi}{2} = \frac{1}{2}; \quad \text{and if } n > 0$$

$$a_n = \frac{1}{\pi} \int_0^{2\pi} f(x) \cos nx\, dx = \frac{1}{\pi} \int_0^{\pi/2} \cos nx\, dx = \frac{1}{\pi n} \sin \frac{n\pi}{2},$$

$$b_n = \frac{1}{\pi} \int_0^{2\pi} f(x) \sin nx\, dx = \frac{1}{\pi} \int_0^{\pi/2} \sin nx\, dx = \frac{1}{\pi n}\left(1 - \cos \frac{n\pi}{2}\right).$$

Hence

$$f(x) \sim \frac{1}{4} + \frac{1}{\pi}\left(\frac{\cos x}{1} - \frac{\cos 3x}{3} + \frac{\cos 5x}{5} - \ldots\right)$$

$$+ \frac{1}{\pi}\left(\frac{\sin x}{1} + \frac{2\sin 2x}{2} + \frac{\sin 3x}{3} + \frac{\sin 5x}{5} + \frac{2\sin 6x}{6} + \ldots\right). \tag{20.5.10}$$

20.5.3.   Sine and Cosine Series

If $g(x)$ has period 2π and if c is any real number, then

$$\int_0^{2\pi} g(x)\,dx = \int_c^{c+2\pi} g(x)\,dx, \tag{20.5.11}$$

because the areas represented by the two integrals are clearly equal on account of the periodicity of $g(x)$. Hence without loss of generality we may take $[c, c+2\pi)$ as the fundamental interval. In particular, when $c = -\pi$,

$$\int_0^{2\pi} g(x)\,dx = \int_{-\pi}^{\pi} g(x)\,dx.$$

The Fourier coefficients of f may therefore be expressed as follows;

$$\left.\begin{array}{ll} a_n = \dfrac{1}{\pi}\displaystyle\int_{-\pi}^{\pi} f(x)\cos nx\,dx & (n = 0, 1, 2, \ldots) \\[2mm] b_n = \dfrac{1}{\pi}\displaystyle\int_{-\pi}^{\pi} f(x)\sin nx\,dx & (n = 1, 2, \ldots). \end{array}\right\} \tag{20.5.12}$$

These formulae become much simpler when the function f is either even or odd:

PROPOSITION 20.5.2

(i) *If f is even, that is, $f(-x) = f(x)$, and of period 2π, then*

$$f(x) \sim \tfrac{1}{2}a_0 + \sum_{n=1}^{\infty} a_n \cos nx \tag{20.5.13}$$

where

$$a_n = \frac{2}{\pi}\int_0^{\pi} f(x)\cos nx\,dx \qquad (n = 0, 1, \ldots). \tag{20.5.14}$$

(ii) *If f is odd, that is, $f(-x) = -f(x)$, and of period 2π, then*

$$f(x) \sim \sum_{n=0}^{\infty} b_n \sin nx, \tag{20.5.15}$$

where

$$b_n = \frac{2}{\pi}\int_0^{\pi} f(x)\sin n\pi\,dx. \tag{20.5.16}$$

The proof of the proposition follows at once from the observation that, if F is an even function, then

$$\int_{-c}^{c} F(x)\, dx = 2 \int_{0}^{c} F(x)\, dx,$$

while for an odd function $G(x)$ we have that

$$\int_{-c}^{c} G(x)\, dx = 0.$$

We say that the Fourier expansion of an even function is a pure *cosine series* while the expansion of an odd function is a pure *sine series*.

EXAMPLE 20.5.3.   Obtain the Fourier expansion of the function $f(x)$ of period 2π, defined in the fundamental interval $[-\pi, \pi)$ by

$$f(x) = x \quad \text{when } -\pi \leq x < \pi,$$

see Figure 20.5.4. Notice that this example differs from Example 20.5.1 merely in the choice of the fundamental interval, throughout which the function is initially defined and then periodically extended.

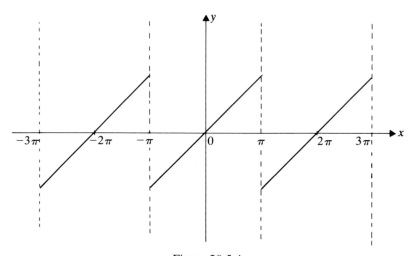

Figure 20.5.4

This is an example of an odd function; so $a_n = 0$ $(n = 0, 1, 2, \ldots)$ and by (20.5.16)

$$b_n = \frac{2}{\pi} \int_{0}^{\pi} x \sin nx\, dx = \frac{2}{\pi} \left[\frac{-x}{n} \cos nx \right]_{0}^{\pi} + \frac{2}{n\pi} \int_{0}^{\pi} \cos nx\, dx$$

$$= -\frac{2 \cos n\pi}{n} + \frac{2}{n^2 \pi} [\sin nx]_{0}^{\pi} = \frac{2}{n} (-1)^{n-1},$$

because

$$\cos n\pi = (-1)^n. \tag{20.5.17}$$

Thus the expansion

$$x = 2\left(\frac{\sin x}{1} - \frac{\sin 2x}{2} + \frac{\sin 3x}{3} - \frac{\sin 4x}{4} + \ldots\right) \tag{20.5.18}$$

is valid for $-\pi < x < \pi$. At the end points of the interval $[-\pi, \pi)$, the function is discontinuous, and the Fourier series has the sum

$$\tfrac{1}{2}\{f(-\pi+0)+f(\pi-0)\} = \tfrac{1}{2}(-\pi+\pi) = 0,$$

which agrees with the right-hand side when $x = \pm\pi$.

EXAMPLE 20.5.4.   Find the Fourier expansion of the function $f(x)$ of period 2π, defined in the fundamental interval $[-\pi, \pi)$ by

$$f(x) = x^2,$$

see Figure 20.5.5.

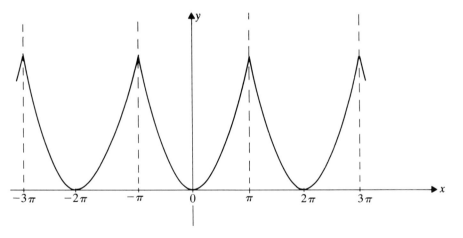

Figure 20.5.5

This is an even function. Hence the expansion will involve cosine-terms only. First,

$$a_0 = \frac{2}{\pi}\int_0^\pi x^2\,dx = \tfrac{2}{3}\pi^2. \tag{20.5.19}$$

Next, when $n > 0$ we use two successive integration by parts [see § 4.3]

$$a_n = \frac{2}{\pi} \int_0^\pi x^2 \cos nx \, dx = \left[\frac{2}{\pi} \frac{x^2}{n} \sin nx \right]_0^\pi - \frac{2}{n\pi} \int_0^\pi 2x \sin nx \, dx$$

$$= -\frac{4}{n\pi} \left[\frac{-x \cos nx}{n} \right]_0^\pi - \frac{4}{n^2\pi} \int_0^{2\pi} \cos nx \, dx$$

$$= \frac{4}{n^2} (-1)^n.$$

Since the function $f(x)$ is continuous for all values of x, including the end-points of the fundamental interval, the expansion

$$x^2 = \frac{\pi^2}{3} - 4\left(\frac{\cos x}{1^2} - \frac{\cos 2x}{2^2} + \frac{\cos 3x}{3^2} - \ldots \right) \tag{20.5.20}$$

is valid for $-\pi \leq x \leq \pi$. As a matter of fact, on putting $x = \pi$ in (20.5.20) we obtain that

$$\pi^2 = \frac{\pi^2}{3} - 4\left(\frac{1}{1^2} + \frac{1}{2^2} + \frac{1}{3^2} + \ldots \right),$$

that is

$$\frac{1}{1^2} + \frac{1}{2^2} + \frac{1}{3^2} + \frac{1}{4^2} + \ldots = \frac{\pi^2}{6}, \tag{20.5.21}$$

a result that can also be established by other methods [Example 9.11.1]. Again, when $x = 0$, we find that [see (9.11.2)]

$$\frac{1}{1^2} - \frac{1}{2^2} + \frac{1}{3^2} - \frac{1}{4^2} + \ldots = \frac{\pi^2}{12}. \tag{20.5.22}$$

We conclude this section with a slightly more elaborate example, which has interesting theoretical implications.

EXAMPLE 20.5.5.   Let p be a real number such that

$$0 < p < 1,$$

and define $f(x)$ by

$$f(x) = \cos px, \quad \text{when } -\pi \leq x < \pi$$

together with the equation $f(x) = f(x + 2\pi)$, which renders $f(x)$ periodic, see Figure 20.5.6.

The function is even and therefore has a pure cosine expansion. We find that

$$a_0 = \frac{2}{\pi} \int_0^\pi \cos px \, dx = \frac{2}{\pi} \frac{\sin p\pi}{p},$$

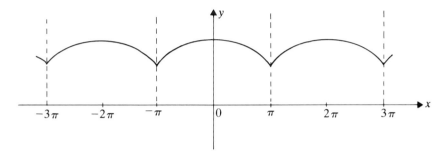

Figure 20.5.6

and when $n \geq 1$,

$$a_n = \frac{2}{\pi} \int_0^\pi \cos px \cos nx \, dx$$

$$= \frac{1}{\pi} \int_0^\pi \{\cos (p+n)x + \cos (p-n)x\} \, dx,$$

[see Table 4.2.1 and V (1.2.19)]. Hence

$$a_n = \frac{1}{\pi} \left\{ \frac{\sin (p+n)\pi}{p+n} + \frac{\sin (p-n)\pi}{p-n} \right\},$$

$$a_n = \frac{1}{\pi} (-1)^{n-1} \frac{2p \sin p\pi}{n^2 - p^2},$$

because $\sin (p \pm n)\pi = (-1)^n \sin p\pi$. Since the function $f(x)$ is continuous at the end-points of the interval $[-\pi, \pi)$, we deduce that the expansion

$$\cos px = \frac{\sin p\pi}{p\pi} + \frac{2p \sin p\pi}{\pi} \sum_{n=1}^\infty (-1)^{n-1} \frac{\cos nx}{n^2 - p^2} \qquad (20.5.23)$$

is valid for $-\pi \leq x \leq \pi$. In particular, on putting $x = \pi$ and dividing throughout by $\sin p\pi$ we obtain the interesting formula

$$\cot p\pi = \frac{1}{\pi p} - \frac{2p}{\pi} \sum_{n=1}^\infty \frac{1}{n^2 - p^2}, \qquad (20.5.24)$$

where we have used (20.5.18). Again, when $x = 0$, we derive the result that

$$\pi \operatorname{cosec} p\pi \left(= \frac{\pi}{\sin p\pi} \right) = \frac{1}{p} + \sum_{n=1}^\infty (-1)^{n-1} \frac{2p}{n^2 - p^2}. \qquad (20.5.25)$$

The equations (20.5.24) and (20.5.25) may be described as the *partial fraction expansions* of $\cot p\pi$ and $\pi \operatorname{cosec} p\pi$ [see I, § 14.10].

We stress again that the expansion formulae obtained in the above examples are valid only in the fundamental intervals. When x lies outside the fundamental interval we have to determine an integer k such that $x + 2\pi k$ lies

inside the fundamental interval. On the left-hand side of the expansion formula
we have to replace x by $x + 2\pi k$, no change being needed on the right-hand
side, since the Fourier series is periodic with period 2π. For example, if
$\pi \le x < 3\pi$, equation (20.5.20) becomes

$$(x - 2\pi)^2 = \frac{\pi^2}{3} - 4\left(\frac{\cos x}{1^2} - \frac{\cos 2x}{2^2} + \frac{\cos 3x}{3^2} - \cdots\right).$$

20.5.4.   Half-range Series

If $f(x)$ is neither even or odd, then its Fourier series in $(-\pi, \pi]$ contains
both cosine and sine terms. But if we are content with an expansion that is
valid only in the half-range $[0, \pi]$, then we can represent $f(x)$ as a pure cosine
or a pure sine series. To this end we define two new functions

$$F(x) = \begin{cases} f(x) & (0 \le x \le \pi) \\ f(-x) & (-\pi < x < 0) \end{cases}$$

and

$$G(x) = \begin{cases} f(x) & (0 \le x \le \pi) \\ -f(-x) & (-\pi < x < 0). \end{cases}$$

Clearly $F(x)$ is even and $G(x)$ is odd in the interval $(-\pi, \pi]$. Hence by
Proposition 20.5.2 $F(x)$ has a pure cosine expansion and $G(x)$ has a pure
sine expansion in $(-\pi, \pi]$. In the half-range $[0, \pi]$ these expansions represent
the original function $f(x)$, except at points of discontinuity. Moreover,

$$\int_{-\pi}^{\pi} F(x) \cos nx \, dx = 2 \int_0^{\pi} f(x) \cos nx \, dx$$

and

$$\int_{-\pi}^{\pi} G(x) \sin nx \, dx = 2 \int_0^{\pi} f(x) \sin nx \, dx.$$

Thus we have the following

PROPOSITION 20.5.3 *(Half-range expansion). Suppose that $f(x)$ is piecewise differentiable in $[0, \pi]$ and put*

$$a_n = \frac{2}{\pi} \int_0^{\pi} f(x) \cos nx \, dx \quad (n \ge 0); \qquad b_n = \frac{2}{\pi} \int_0^{\pi} f(x) \sin nx \, dx \quad (n \ge 1).$$

Then the expansions

$$f(x) = \tfrac{1}{2}a_0 + \sum_{n=1}^{\infty} a_n \cos nx \tag{20.5.26}$$

and

$$f(x) = \sum_{n=1}^{\infty} b_n \sin nx \tag{20.5.27}$$

are valid in $[0, \pi]$, *except at points of discontinuity; if f is discontinuous at* x_0, *then the left-hand sides of* (20.5.26) *and* (20.5.27) *have to be replaced by* $\frac{1}{2}(f(x_0+0)+f(x_0-0))$.

EXAMPLE 20.5.6.   Find the sine-series for the constant function $f(x) = 1$, valid for $0 < x < \pi$. The function is extended so as to be odd and of period 2π, see Figure 20.5.7.

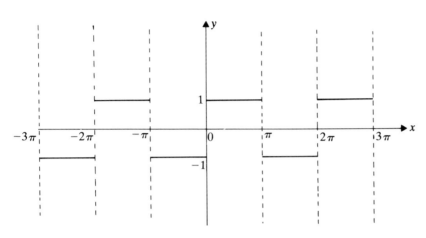

Figure 20.5.7

We have that

$$b_n = \frac{2}{\pi} \int_0^{2\pi} \sin nx \, dx = \frac{2}{\pi} \left[-\frac{1}{n} \cos nx \right]_0^{\pi}$$

$$= \frac{2}{\pi n} [-(-1)^n + 1]$$

whence

$$b_n = \begin{cases} 4/n\pi & \text{when } n \text{ is odd} \\ 0 & \text{when } n \text{ is even.} \end{cases}$$

Thus, on multiplying throughout by $\pi/4$, we obtain the result

$$\frac{\pi}{4} = \frac{\sin x}{1} + \frac{\sin 3x}{3} + \frac{\sin 5x}{5} + \ldots, \qquad (20.5.28)$$

valid for $0 < x < \pi$.

EXAMPLE 20.5.7.   Find the cosine-series for the function $f(x) = x$, valid for $0 < x < \pi$.

This time the function is extended to be even, and in fact is identical with $|x|$ in the interval $(-\pi, \pi)$. When continued periodically, $f(x)$ is seen to be continuous for all values of x (Figure 20.5.8).

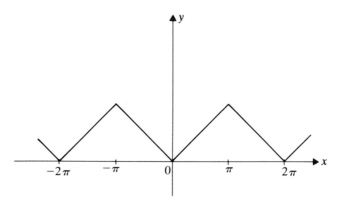

Figure 20.5.8

Now

$$a_0 = \frac{2}{\pi} \int_0^\pi x \, dx = \frac{\pi}{2},$$

and, when $n \geq 1$,

$$a_n = \frac{2}{n} \int_0^\pi x \cos nx \, dx = \begin{cases} \dfrac{-4}{n^2 \pi}, & \text{when } n \text{ is odd} \\[2ex] 0, & \text{when } n \text{ is even.} \end{cases}$$

[cf. Example 4.3.1]. Hence we have the expansion

$$|x| = \tfrac{1}{2}\pi - \frac{4}{\pi}\left(\frac{\cos x}{1^2} + \frac{\cos 3x}{3^2} + \frac{\cos 5x}{5^2} + \ldots\right), \qquad (20.5.29)$$

valid for $-\pi \leq x \leq \pi$. When $x = 0$, we obtain that

$$\frac{1}{1^2} + \frac{1}{3^2} + \frac{1}{5^2} + \ldots = \frac{\pi^2}{8}, \qquad (20.5.30)$$

a result which could have been deduced from (20.5.21) and (20.5.22).

20.5.5. Change of Variable

We shall now modify the theory so as to include functions $f(x)$ of period p, where p is an arbitrary positive number. Moreover, we shall assume that the function is initially given in the fundamental interval $[a, b)$, where $p = b - a$, and is then extended periodically. This situation is dealt with by the simple device of changing the variable. Thus we put

$$x = \frac{p}{2\pi} t + a,$$

thereby ensuring that the interval $a \le x < b$ is mapped onto the interval $0 \le t < 2\pi$, and we define the function $F(t)$ by

$$F(t) = f(x) = f\left(\frac{p}{2\pi} t + a\right).$$

Evidently, $F(t)$ is periodic with period 2π, and if $f(x)$ is piecewise differentiable with respect to x, then $F(t)$ is piecewise differentiable with respect to t, and conversely. Thus in accordance with Theorem 20.5.1 we have that

$$\tfrac{1}{2}(F(t+0) + F(t-0)) = \tfrac{1}{2}a_0 + \sum_{n=1}^{\infty} (a_n \cos nt + b_n \sin nt), \quad (20.5.31)$$

where

$$a_n = \frac{1}{\pi} \int_0^{2\pi} F(t) \cos nt \, dt \quad (n = 0, 1, \ldots) \quad (20.5.32)$$

and

$$b_n = \frac{1}{\pi} \int_0^{2\pi} F(t) \sin nt \, dt \quad (n = 1, 2, \ldots). \quad (20.5.33)$$

The expansion (20.5.31) and the integrals (20.5.32), (20.5.33) can be expressed in terms of x by reversing the substitution (20.5.29), namely

$$t = \frac{2\pi}{p}(x - a). \quad (20.5.34)$$

We summarize the result for the simpler situation, in which $a = 0$:

THEOREM 20.5.4 (*Expansion Theorem*).   *Let $f(x)$ be a function of period p. Suppose that $f(x)$ is piecewise differentiable in every finite interval. Then*

$$\tfrac{1}{2}\{f(x+0) + f(x-0)\} = \tfrac{1}{2}a_0 + \sum_{n=1}^{\infty} \left(a_n \cos\left(\frac{2\pi n}{p} x\right) + b_n \sin\left(\frac{2\pi n}{p} x\right)\right),$$

$$(20.5.35)$$

where $0 \le x < p$ and

$$a_n = \frac{2}{p} \int_0^p f(x) \cos\left(\frac{2\pi n}{p} x\right) dx \quad (n = 0, 1, \ldots), \quad (20.5.36)$$

$$b_n = \frac{2}{p} \int_0^p f(x) \sin\left(\frac{2\pi n}{p} x\right) dx \quad (n = 1, 2, \ldots). \quad (20.5.37)$$

EXAMPLE 20.5.8.   Find the Fourier expansion of the function $f(x)$ satisfying

$$f(x+p)=f(x), \quad \text{for all } x,$$

and

$$f(x)=x(p-x), \quad \text{when } 0\le x\le p,$$

[see Figure 20.5.9].

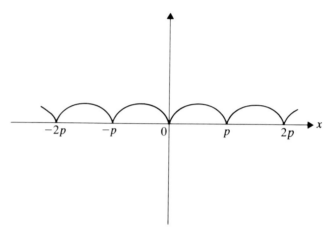

Figure 20.5.9

We have to evaluate the Fourier coefficients given in (20.5.36) and (20.5.37), thus when $n=0$, we have that

$$a_0=\frac{2}{p}\int_0^p x(p-x)\,dx=\tfrac{1}{3}p^2,$$

and when $n>0$

$$a_n=\frac{2}{p}\int_0^p x(p-x)\cos\left(\frac{2\pi nx}{p}\right)dx=-\left(\frac{p}{\pi n}\right)^2$$

as can be seen by two successive applications of integration by parts [see § 4.3]. Also,

$$b_n=\frac{2}{p}\int_0^p x(p-x)\sin\left(\frac{2\pi nx}{p}\right)dx=0;$$

this can be shown most easily by making the substitution $y=p-x$ [see § 4.3], which readily leads to the conclusion that $b_n=-b_n$. Hence

$$x(p-x)=\tfrac{1}{6}p^2-\left(\frac{p}{\pi}\right)^2\sum_{n=1}^{\infty}\frac{1}{n^2}\cos\left(\frac{2\pi nx}{p}\right), \qquad (20.5.38)$$

provided that $0 \le x \le p$, the end-points of the interval being included because the function remains continuous there.

20.5.6. Fourier Series in Complex Form

In this section we extend the theory to functions $f(x)$, where x, as hitherto, is a real variable but the values of $f(x)$ may be complex numbers. Let

$$f(x) = f_1(x) + if_2(x),$$

where $f_1(x)$ and $f_2(x)$ are the real and imaginary parts of $f(x)$ [see I, § 2.7.1]. If $f_1(x)$ and $f_2(x)$ separately satisfy the conditions of the expansion theorem 20.5.1, so does $f(x)$. More precisely, let a'_n and b'_n be the Fourier coefficients of $f_1(x)$ and let a''_n and b''_n be the Fourier coefficients of $f_2(x)$. Then

$$\tfrac{1}{2}\{f_1(x+0) + f_1(x-0)\} = \tfrac{1}{2}a'_0 + \sum_{n=1}^{\infty}(a'_n \cos nx + b'_n \sin nx)$$

and

$$\tfrac{1}{2}\{f_2(x+0) + f_2(x-0)\} = \tfrac{1}{2}a''_0 + \sum_{n=1}^{\infty}(a''_n \cos nx + b''_n \sin nx).$$

On multiplying the second equation by i and then adding it to the first equation we obtain that

$$\tfrac{1}{2}\{f(x+0) + f(x-0)\} = \tfrac{1}{2}a_0 + \sum_{n=1}^{\infty}(a_n \cos nx + b_n \sin nx), \quad (20.5.39)$$

where

$$a_n = \tfrac{1}{2}(a'_n + ia''_n), \qquad b_n = \tfrac{1}{2}(b'_n + ib''_n).$$

The result can be expressed more simply if we use the exponential function instead of sine and cosine. Generally, [see (9.5.10) and (9.5.11)]

$$\cos\theta = \tfrac{1}{2}\{\exp(i\theta) + \exp(-i\theta)\},$$

$$\sin\theta = \frac{1}{2i}\{\exp(i\theta) - \exp(-i\theta)\}.$$

Hence on substituting for $\cos(2\pi inx/p)$ and $\sin(2\pi inx/p)$ we can write (20.5.39) concisely by employing a series that extends from $-\infty$ to ∞, thus

$$\tfrac{1}{2}\{f(x+0) + f(x-0)\} = \sum_{n=-\infty}^{\infty} c_n \exp(2\pi inx/p), \quad (20.5.40)$$

where

$$c_0 = \tfrac{1}{2}a_0, \qquad c_n = \tfrac{1}{2}(a_n - ib_n), \qquad c_{-n} = \tfrac{1}{2}(a_n + ib_n) \qquad (n = 1, 2, \ldots).$$

$$(20.5.41)$$

Using (20.5.36) and (20.5.37) we can express the coefficients c_n directly in terms of the function $f(x)$. For when $n \geq 0$ we have that

$$c_n = \tfrac{1}{2}(a_n - ib_n) = \frac{1}{p} \int_0^p f(x) \exp\left(\frac{-2\pi inx}{p}\right) dx \qquad (20.5.42)$$

and

$$c_{-n} = \tfrac{1}{2}(a_n + ib_n) = \frac{1}{p} \int_0^p f(x) \exp\left(\frac{2\pi inx}{p}\right) dx. \qquad (20.5.43)$$

So the formula

$$c_m = \frac{1}{p} \int_0^p f(x) \exp\left(\frac{-2\pi imx}{p}\right) dx$$

holds in all cases ($m = 0, \pm 1, \pm 2, \ldots, \pm n, \ldots$). We summarize these results in the following

THEOREM 20.5.5   (*Expansion theorem—complex version*).
Let $f(x) = f_1(x) + if_2(x)$ be a complex-valued function of a real variable x and suppose that $f_1(x)$ and $f_2(x)$ are piecewise differentiable in the interval $[0, p)$. Then the expansion

$$\tfrac{1}{2}\{f(x+0) + f(x-0)\} = \sum_{n=-\infty}^{\infty} c_n \exp(2\pi inx/p) \qquad (20.5.44)$$

holds in $[0, p)$, where

$$c_n = \frac{1}{p} \int_0^p f(x) \exp\left(\frac{-2\pi inx}{p}\right) dx. \qquad (20.5.45)$$

Remark. If $f(x)$ is in fact real-valued, so that $f_2(x) = 0$, then (20.5.45) implies that

$$c_{-n} = \bar{c}_n, \qquad (20.5.46)$$

that is c_n and c_{-n} are complex conjugate numbers [see I, § 2.7].

EXAMPLE 20.5.9.   The function $f(x)$ satisfies $f(x + 2\pi) = f(x)$ for all x and $f(x) = e^x$ when $-\pi \leq x < \pi$ (see Figure 20.5.10). Find the Fourier expansion of $f(x)$, valid when $-\pi < x < \pi$.
   Using (20.5.45) with $p = 2\pi$ and the range of integration as $(-\pi, \pi)$ [see (20.5.11)] we find that

$$c_m = \frac{1}{2\pi} \int_{-\pi}^{\pi} e^x e^{-imx} \, dx = \frac{1}{2\pi} \int_{-\pi}^{\pi} e^{(1-im)x} \, dx$$

$$= \frac{1}{2\pi} \left[\frac{1}{1-im} e^{(1-im)x} \right]_{-\pi}^{\pi} = \frac{1}{2\pi} \frac{1+im}{1+m^2} (-1)^m (e^{\pi} - e^{-\pi}),$$

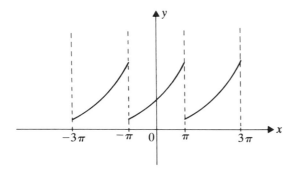

Figure 20.5.10

because $e^{-im\pi} = e^{im\pi} = (-1)^m$ [see (9.5.8)]. Hence if $-\pi < x < \pi$, we have the expansion

$$e^x = \frac{1}{2\pi}(e^\pi - e^{-\pi}) \sum_{m=-\infty}^{\infty} (-1)^m \frac{1+im}{1+m^2} e^{imx},$$

or equivalently by collecting pairs of terms with indices m, $-m$,

$$e^x = \frac{1}{2\pi}(e^\pi - e^{-\pi})\left\{1 + 2 \sum_{m=1}^{\infty} (-1)^m \frac{\cos mx - m \sin mx}{1+m^2}\right\}.$$

$$(20.5.47)$$

Interesting results are obtained when $x = 0$ or $x = \pi$; in the latter case the left-hand side of (20.5.47) has to be replaced by $\frac{1}{2}(e^\pi + e^{-\pi})$ on account of the discontinuities at the end-points of the interval $(-\pi, \pi)$. Thus

$$\sum_{m=1}^{\infty} (-1)^{m-1} \frac{1}{1+m^2} = \frac{1}{2}\left\{1 - \frac{2\pi}{e^\pi - e^{-\pi}}\right\}, \qquad (20.5.48)$$

$$\sum_{m=1}^{\infty} \frac{1}{1+m^2} = \frac{1}{2}\left\{\pi\left(\frac{e^\pi + e^{-\pi}}{e^\pi - e^{-\pi}}\right) - 1\right\}. \qquad (20.5.49)$$

20.6.   GENERAL PROPERTIES OF FOURIER SERIES

The mathematical theory of Fourier series involves investigations of great subtlety and depth. Some of the most fruitful ideas in modern analysis originated from problems posed in connection with Fourier series. It is not our intention to give an account of material that is chiefly of interest to the pure mathematician. In particular we shall confine ourselves to piecewise differentiable periodic functions, defined in section 20.2; much of the more advanced theory is concerned with an extension to a wider class of functions, which is beyond the scope of this book.

20.6.1. The Magnitude of Fourier Coefficients

Even when it is known that a function $f(x)$ can be represented by a convergent Fourier series, it would be useful to have some information about the speed of convergence: for example, in the notation of § 2.3 can we say whether $a_n = O(1/n)$ or $b_n = O(1/n^2)$, and so on? We begin with a celebrated result which can be proved under rather mild conditions.

THEOREM 20.6.1 *(Riemann–Lebesgue lemma).* *If* $f(x)$ *is absolutely integrable in* $(0, 2\pi)$ *that is if*

$$\int_0^{2\pi} |f(x)|\, dx$$

is finite, then

$$\lim_{n \to \infty} \int_0^{2\pi} f(x) \cos nx\, dx = 0,$$

and

$$\lim_{n \to \infty} \int_0^{2\pi} f(x) \sin nx\, dx = 0.$$

This theorem ensures that a_n and b_n tend to zero as n tends to infinity, but it does not tell us anything about the speed of convergence. The next proposition contains useful results in this respect; for the sake of simplicity we state the conditions in a more stringent form than is necessary and we confine ourselves to periodic functions whose sole discontinuities, if any, occur at the end-points of the fundamental interval.

PROPOSITION 20.6.2. *Let* $f(x)$ *be a function with period* 2π, *and suppose that* $f(x), f'(x), f''(x)$ *and* $f'''(x)$ *are continuous at all interior points of the interval* $-\pi < x < \pi$. *Then*

(i) *if* $f(\pi) \neq f(-\pi)$, *we have that*

$$\lim_{n \to \infty} na_n = 0, \qquad \lim_{n \to \infty} |nb_n| = \frac{1}{\pi} |f(\pi) - f(-\pi)|;$$

(ii) *if* $f(\pi) = f(-\pi)$, *we have that*

$$\lim_{n \to \infty} |n^2 a_n| = \frac{1}{\pi} |f'(\pi) - f'(-\pi)|, \qquad \lim_{n \to \infty} n^2 b_n = 0.$$

We omit the proof, but we shall review two typical examples of the preceding section in the light of this proposition.

EXAMPLE 20.6.1

(1) The function [see Example 20.5.6]

$$f(x) = \cos px \qquad (0 < p < 1)$$

satisfies all the conditions of the proposition; moreover, $f(\pi) - f(-\pi) = 0$, but $|f'(\pi) - f'(-\pi)| = 2p \sin p \neq 0$. Hence, by part (ii), $a_n = O(1/n^2)$.

(2) When

$$f(x) = x$$

[see Example 20.5.4], we have that $|f(\pi) - f(-\pi)| \neq 0$ and by inspection of (20.5.18) we verify that $b_n = O(1/n)$, as stated in part (i).

Thus a discontinuity of the function is more damaging to convergence than a discontinuity of its derivative.

20.6.2. Integration of Fourier Series

It is known [see § 1.12] that it is not, in general, admissible to integrate a convergent series

$$S(x) = \sum_{n=0}^{\infty} u_n(x) \qquad (20.6.1)$$

'term-by-term', that is (20.6.1) does not imply that

$$\int_a^b S(x) \, dx = \sum_{n=0}^{\infty} \int_a^b u_n(x) \, dx.$$

It is, however, a remarkable fact that, to put it briefly, a Fourier series may always be integrated term by term.

PROPOSITION 20.6.3.   *Let $f(t)$ be a piecewise differentiable function of period 2π having the Fourier expansion*

$$\tfrac{1}{2}\{f(t+0) + f(t-0)\} = \tfrac{1}{2}a_0 + \sum_{n=1}^{\infty} (a_n \cos nt + b_n \sin nt).$$

Then, for an arbitrary value of x,

$$\int_0^x f(t) \, dt - \tfrac{1}{2}a_0 x = \sum_{n=1}^{\infty} \frac{1}{n} \{a_n \sin nx + b_n (1 - \cos nx)\}. \qquad (20.6.2)$$

Remark.   We have transposed the term $\tfrac{1}{2}a_0 x$ in order to render each side of (20.6.2) periodic; indeed the left-hand side remains unchanged when x is replaced by $x + 2\pi$, because

$$\int_x^{x+2\pi} f(t) \, dt = \int_0^{2\pi} f(t) \, dt = \pi a_0.$$

As an illustration, replace x by t in (20.5.29) where $0 \le t \le \pi$, thus

$$t = \tfrac{1}{2}\pi - \frac{4}{\pi}\left(\frac{\cos t}{1^2} + \frac{\cos 3t}{3^2} + \frac{\cos 5t}{5^2} + \ldots\right).$$

Integrating from 0 to x, where x is an arbitrary number satisfying $0 \le x \le \pi$, we obtain that

$$\tfrac{1}{2}(x^2 - \pi x) = -\frac{4}{\pi}\left(\frac{\sin x}{1^3} + \frac{\sin 3x}{3^3} + \frac{\sin 5x}{5^3} + \ldots\right),$$

or

$$\sum_{n=1}^{\infty} \frac{\sin(2n-1)x}{(2n-1)^3} = \frac{\pi}{8}\, x(\pi - x), \tag{20.6.3}$$

valid for $0 \le x \le \pi$; both end-points are included because the equation obviously holds when $x = 0$ and $x = \pi$.

The function on the left of (20.6.3) is clearly odd; so the expression on the right has to be replaced by

$$-\frac{\pi}{8}\, x(\pi + x),$$

when $-\pi \le x \le 0$. The result may be contrasted with (20.5.38) when $p = \pi$. Apart from a constant factor the same function as in (20.6.3) is represented in the interval $[0, \pi)$. But when the validity is extended to the whole interval $(-\pi, \pi)$, different formulae emerge because (20.6.3) is odd while (20.5.38) is even.

Putting $x = \pi/2$ in (20.6.3) we obtain the remarkable result that

$$\frac{1}{1^3} - \frac{1}{3^3} + \frac{1}{5^3} - \ldots = \frac{\pi^3}{32}. \tag{20.6.4}$$

In contrast to Proposition 20.6.3 it is not always permissible to *differentiate* a Fourier series term by term. The conditions under which such an operation is legitimate, are cumbersome and will not be given here.

20.6.3. Parseval's Formula and Bessel's Inequality

In this section we continue the study of periodic complex-valued functions for which the expansion theorem 20.5.5 holds. To simplify the formulae we assume that the functions are of period 2π $(p = 2\pi)$ and that the fundamental interval is $(-\pi, \pi]$. Suppose that $f(x)$ has Fourier coefficients

$$c_n = \frac{1}{2\pi} \int_{-\pi}^{\pi} f(x) \exp(-inx)\, dx \qquad (n = 0, \pm 1, \pm 2, \ldots). \tag{20.6.5}$$

We denote the Fourier series of $f(x)$ by

$$F(x) = \sum_{n=-\infty}^{\infty} c_n \exp(inx).$$

The expansion theorem [see (20.5.44)] states that

$$\tfrac{1}{2}\{f(x+0)+f(x-0)\} = F(x) \qquad (-\pi < x \leq \pi).$$

In particular, if f is continuous at x_0, then $f(x_0) = F(x_0)$. Now Definition 20.2.1 implies that f has at most a finite number of discontinuities in the fundamental interval. Hence the equation

$$f(x) = F(x) \tag{20.6.6}$$

holds throughout $(-\pi, \pi]$ with at most a finite number of exceptions.

Similarly, let $g(x)$ be a function with Fourier coefficients

$$c'_m = \frac{1}{2\pi} \int_{-\pi}^{\pi} g(x) \exp(-imx)\, dx \qquad (m = 0, \pm 1, \pm 2, \ldots) \tag{20.6.7}$$

and put

$$G(x) = \sum_{m=-\infty}^{\infty} c'_m \exp(imx).$$

Then the equation

$$g(x) = G(x) \tag{20.6.8}$$

holds throughout with at most a finite number of exceptions.

When dealing with complex-valued functions we find it convenient to define the *inner product* of f and g by

$$(f, g) = \int_{-\pi}^{\pi} f(x)\overline{g(x)}\, dx, \tag{20.6.9}$$

where the bar denotes the conjugate complex number [see I, § 2.7.1]. For real-valued functions (20.6.9) reduces to (20.4.12).

It is our aim to express (f, g) in terms of the Fourier coefficients of f and g. As a consequence of (20.6.6) and (20.6.8), the equation

$$f(x)\overline{g(x)} = F(x)\overline{G(x)} \tag{20.6.10}$$

holds throughout $(-\pi, \pi]$ with at most a finite number of exceptions; despite these exceptions the two sides of (20.6.10) have the same integral [see Theorem 4.1.3], that is

$$\int_{-\pi}^{\pi} f(x)\overline{g(x)}\, dx = \int_{-\pi}^{\pi} F(x)\overline{G(x)}\, dx, \tag{20.6.11}$$

and we proceed to evaluate the integral on the right. We have that

$$\overline{G(x)} = \sum_{m=-\infty}^{\infty} c'_m \exp(-imx),$$

and the integrand can be written as a double series [see § 1.11], namely

$$F(x)G(x) = \sum_{n=-\infty}^{\infty} \sum_{m=-\infty}^{\infty} c_n c'_m \exp\{i(n-m)x\}. \tag{20.6.12}$$

It can be shown that the conditions we have imposed justify term by term integration of the last equation. Hence the problem is reduced to evaluating the integral

$$\int_{-\pi}^{\pi} \exp\{i(n-m)x\}\, dx,$$

where n and m are arbitrary integers.

Now, if k is any integer, we have that

$$\exp(k\pi i) = \exp(-k\pi i) = (-1)^k$$

[see (9.5.8)]. Hence if k is non-zero we find that

$$\int_{-\pi}^{\pi} \exp(ikx)\, dx = \left[\frac{1}{ik}\exp(ikx)\right]_{-\pi}^{\pi} = 0,$$

while the integral is obviously equal to zero when $k = 0$. Hence

$$\int_{-\pi}^{\pi} \exp\{i(n-m)x\}\, dx = \begin{cases} 2\pi & \text{if } n = m \\ 0 & \text{if } n \neq m. \end{cases}$$

When (20.6.12) is integrated, only those terms for which $n = m$ yield a non-zero contribution and the double sum is reduced to a single sum. Using (20.6.11) we obtain that

$$\frac{1}{2\pi} \int_{-\pi}^{\pi} f(x)\overline{g(x)}\, dx = \sum_{n=-\infty}^{\infty} c_n \bar{c}'_n. \tag{20.6.13}$$

When $f = g$, we can write [see I, § 2.7.2]

$$f(x)\overline{f(x)} = |f(x)|^2, \qquad c_n \bar{c}_n = |c_n|^2.$$

The formula (20.6.13) holds also for real-valued functions, but in that case it is more appropriate to express the result in terms of the Fourier coefficients a_n and b_n defined in (20.5.1) and (20.5.2). When f is real-valued, by (20.5.46) $c_{-n} = c_n$, and it is convenient to pair off terms that correspond to n and $-n$ $(n = 1, 2, \ldots)$. Thus

$$\sum_{n=-\infty}^{\infty} c_n \bar{c}'_n = c_0 c'_0 + \sum_{n=1}^{\infty} (c_n \bar{c}'_n + \bar{c}_n c'_n).$$

Using (20.5.41) we find after a short calculation that

$$c_0 c_0' = \tfrac{1}{4} a_0 a_0', \qquad c_n \bar{c}_n' + \overline{c_n} c_n' = \tfrac{1}{2}(a_n a_n' + b_n b_n').$$

We summarize the result in the following theorem:

THEOREM 20.6.4   (*Parseval's formula*)

(1) *Complex case: Let $f(x)$ and $g(x)$ be complex-valued functions of period 2π, whose Fourier series are*

$$\sum_{n=-\infty}^{\infty} c_n \exp(inx) \quad and \quad \sum_{m=-\infty}^{\infty} c_m' \exp(imx)$$

respectively, where c_n and c_m' are defined in (20.6.5) and (20.6.7). Then

$$\frac{1}{2\pi} \int_{-\pi}^{\pi} f(x) g(x)\, dx = \sum_{n=-\infty}^{\infty} c_n c_n'.$$

In particular

$$\frac{1}{2\pi} \int_{-\pi}^{\pi} |f(x)|^2\, dx = \sum_{n=-\infty}^{\infty} |c_n|^2. \qquad (20.6.14)$$

(2) *Real case: Let $f(x)$ and $g(x)$ be real-valued functions of period 2π, whose Fourier series are*

$$\tfrac{1}{2} a_0 + \sum_{n=1}^{\infty} (a_n \cos nx + b_n \sin nx) \quad and \quad \tfrac{1}{2} a_0' + \sum_{m=1}^{\infty} (a_m' \cos x + b_m' \sin mx)$$

respectively, where a_n and b_n are given in (20.5.12), and a_m' and b_m' are similarly defined in terms of g. Then

$$\frac{1}{\pi} \int_{-\pi}^{\pi} f(x) g(x)\, dx = \tfrac{1}{2} a_0 a_0' + \sum_{n=1}^{\infty} (a_n a_n' + b_n b_n').$$

In particular,

$$\frac{1}{\pi} \int_{-\pi}^{\pi} \{f(x)\}^2\, dx = \tfrac{1}{2} a_0^2 + \sum_{n=1}^{\infty} (a_n^2 + b_n^2). \qquad (20.6.15)$$

EXAMPLE 20.6.2.   In the expansion (20.5.20) we have that

$$f(x) = x^2, \qquad a_0 = 2\pi^2/3, \qquad a_n = -4/n^2 \qquad (n = 1, 2, \ldots).$$

Applying Parseval's formula (20.6.15) we obtain that

$$\frac{1}{\pi} \int_{-\pi}^{\pi} x^4\, dx = 2\pi^4/5 = \tfrac{1}{2}(2\pi^2/3)^2 + 16 \sum_{n=1}^{\infty} n^{-4},$$

whence

$$\sum_{n=1}^{\infty} n^{-4} = \pi^4/90.$$

Since all terms of the infinite series (20.6.14) and (20.6.15) are non-negative, the values of the series are diminished if some of the terms are dropped. We note the result explicitly only in the real case; for the more general case see Theorem 11.7.1.

COROLLARY 20.6.5   (*Bessel's inequality*).   *Let $f(x)$ be a real-valued function of period 2π having Fourier coefficients a_n $(n \geq 0)$ and b_n $(n \geq 1)$. Then for every positive integer N we have that*

$$\tfrac{1}{2}a_0 + \sum_{n=1}^{N} (a_n^2 + b_n^2) \leq \frac{1}{\pi} \int_{-\pi}^{\pi} \{f(x)\}^2 \, dx.$$

20.6.4.   Approximation by Trigonometric Polynomials

Even the most powerful computer can calculate only a finite number of terms. Hence in most practical situations the infinite Fourier series will have to be replaced by a Fourier polynomial [see (20.3.4)] $P(x)$ of degree n, say, where n is a suitable integer. Of course, if $f(x)$ requires for its representation an infinite Fourier series, we cannot expect that $f(x)$ is equal to a finite number of terms; in other words, $P(x)$ will only be 'approximately' equal to $f(x)$; or that the difference between $P(x)$ and $f(x)$ is 'small'.

Therefore, before we can proceed we must decide what we mean by the difference between two functions $f(x)$ and $g(x)$, both defined over a finite interval, which we take to be $[0, 2\pi)$. Also, we shall confine ourselves to real-valued functions.

In the present context it is convenient to adopt the *metric* [see § 11.1]

$$d_2(f, g) = \left[\int_{0}^{2\pi} \{f(x) - g(x)\}^2 \, dx \right]^{1/2} \tag{20.6.16}$$

to express the 'distance' or 'difference' between the functions f and g.

We now ask the following question: given a positive integer n, what trigonometric polynomial

$$S_n(x) = \tfrac{1}{2}\alpha_0 + \sum_{r=1}^{n} (\alpha_r \cos rx + \beta_r \sin rx) \tag{20.6.17}$$

furnishes 'the best' approximation to f in the above metric? More precisely, we seek a specific trigonometric polynomial $S_n(x)$ such that

$$d_2(f, S_n) \leq d_2(f, Q_n), \tag{20.6.18}$$

where $Q_n(x)$ is any trigonometric polynomial of degree not exceeding n. The answer is contained in a remarkable theorem which we quote here without proof:

THEOREM 20.6.6.   *Let $f(x)$ be a function whose square is integrable over*

[0, 2π] *and suppose that the Fourier coefficients*

$$a_r = \frac{1}{\pi} \int_0^{2\pi} f(x) \cos rx \, dx \qquad (r = 0, 1, 2, \ldots, n)$$

$$b_r = \frac{1}{\pi} \int_0^{2\pi} f(x) \sin rx \, dx \qquad (r = 1, 2, \ldots, n)$$

exist, that is f(x) possesses Fourier coefficients at least up to the nth order. Then the nth partial sum of the formal Fourier series for f(x), namely

$$S_n(x) = \tfrac{1}{2}a_0 + \sum_{r=1}^{n} (a_r \cos rx + b_r \sin rx) \qquad (20.6.19)$$

provides the best approximation to f in the sense that

$$d_2(f, S_n) \le d_2(f, Q_n),$$

where Q_n is an arbitrary trigonometric polynomial of degree not exceeding n.

It can be shown that

$$d_2(f, S_n)^2 = \pi \sum_{r=n+1}^{\infty} (a_r^2 + b_r^2). \qquad (20.6.20)$$

By (20.6.15) the right-hand side of (20.6.20) is the tail-end of a convergent series and hence can be made smaller than any pre-assigned number provided that n is large enough; in other words

$$d_2(f, S_n) \to 0, \quad \text{as } n \to \infty. \qquad (20.6.21)$$

If, instead of d_2, we use the metric

$$d_\infty(f, g) = \text{l.u.b.} |f(x) - g(x)| \qquad (x \in [0, 2\pi))$$

[see § 1.12], then different conclusions will be reached about the overall approximation of $f(x)$ by $S_n(x)$. It can be shown that the sequence $S_n(x)$ ($n = 1, 2, \ldots$) does not tend uniformly to $f(x)$ in $[0, 2\pi)$ if f fails to be continuous, that is [see Proposition 1.12.1]

$$d_\infty(f, S_n) \not\to 0, \quad \text{as } n \to \infty,$$

in contrast to (20.6.21). This result is sometimes referred to as *Gibb's phenomenon*. The reader will find details in Lanczos (1966), p. 51 and Sneddon (1961), p. 40.

20.7. MULTIPLE FOURIER SERIES

The theory of Fourier series can be extended to functions of several variables. However, as one would expect, the conditions for expansion theorems are more complicated than for functions of a single variable, and difficulties of another kind may arise when the domain of the function is of an awkward shape.

We shall confine ourselves to the simplest case and state an expansion theorem for a function $f(x, y)$ of two real variables, defined in a rectangle

$$R: \begin{cases} 0 \leq x \leq h \\ 0 \leq y \leq k. \end{cases} \qquad (20.7.1)$$

It is convenient to use the complex formulation of the theory [see § 20.5.6]. First we must introduce the Fourier coefficients, which now form a double sequence.

DEFINITION 20.7.1.   Suppose that

$$\iint_R |f(x, y)| \, dx \, dy$$

exists. Then the *multiple Fourier coefficients* of f are defined by

$$a_{rs} = \frac{1}{hk} \iint_R f(u, v) \exp\left\{-2\pi i\left(\frac{ru}{h} + \frac{sv}{k}\right)\right\} du \, dv \qquad (20.7.2)$$

$(r = 0, \pm 1, \pm 2, \pm 3, \ldots; s = 0, \pm 1, \pm 2, \pm 3, \ldots)$. We mention without proof the following expansion theorem

THEOREM 20.7.1.   *Suppose that the function $f(x, y)$ is continuous and has continuous partial derivatives* [see § 5.3]

$$\frac{\partial f}{\partial x}, \quad \frac{\partial f}{\partial y}, \quad \frac{\partial^2 f}{\partial x \, \partial y}$$

throughout the rectangle R defined in (20.7.1). Then if (x_0, y_0) is an interior point of R we have that

$$f(x_0, y_0) = \sum_{r=-\infty}^{\infty} \sum_{s=-\infty}^{\infty} a_{rs} \exp\left\{2\pi i\left(\frac{rx_0}{h} + \frac{sy_0}{k}\right)\right\}. \qquad (20.7.3)$$

We observe that if

$$f(x, y) = F(x)G(y),$$

then

$$a_{rs} = A_r B_s$$

where A_r and B_s are the complex Fourier coefficients of F and G respectively.

W. L.

REFERENCES

Henchey, F. A. (1980). *Introduction to Applicable Mathematics*, Wiley Eastern.

Lanczos, C. (1966). *Discourse on Fourier Series*, Oliver & Boyd.

Sneddon, I. (1961). *Fourier Series*, Routledge & Kegan Paul.

CHAPTER 21

Inequalities

21.1. MANIPULATIVE RULES

In this chapter we are concerned mainly with real numbers. Elsewhere in this work [see I, § 2.6.2] it was pointed out that the set of real numbers is totally ordered [see I, § 1.3.2], that is if a and b are any real numbers, then exactly one of the following relations holds:

$$a < b \quad \text{or} \quad a = b \quad \text{or} \quad a > b.$$

We can depict the real numbers as points on a straight line, thus

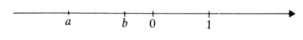

Figure 21.1.1

Then $a < b$ or, equivalently, $b > a$ means that a lies to the left of b or, equivalently, that b lies to the right of a, attention being paid to the signs of a and b.

EXAMPLE 21.1.1
  (i) $-\frac{1}{2} < 5$,
  (ii) $0 < \pi$,
  (iii) $-\pi < -2\sqrt{2}$.
The relation $a < b$ expresses a *strict* inequality between a and b. If a is less than or equal to b we write

$$a \le b \quad \text{or} \quad b \ge a, \tag{21.1.1}$$

and we say that (21.1.1) denotes a *weak* inequality between a and b. In particular, we write $a > 0$ when a is *positive*, and $a \ge 0$ when a is *non-negative*.
  The three relations

$$a < b, \qquad b \le c, \qquad c < d$$

837

can be condensed into a single chain of inequalities, namely,

$$a < b \leq c < d.$$

But the information that $a < b$ and $b > c$ must not be compressed into $a < b > c$, which is inadmissible, because the relationship between a and c is uncertain.

Manipulation with inequalities obeys certain general rules, which are however much more limited than the analogous rules for equations [see I, § 2.4.2]: if

$$a < b \qquad\qquad\qquad (21.1.2)$$

and

$$c < d, \qquad\qquad\qquad (21.1.3)$$

it follows that

$$a + c < b + d;$$

thus inequalities may be added, provided that they go in the same direction. But subtraction of inequalities is not allowed. As regards multiplication we have the following

PROPOSITION 21.1.1.   *Let $a < b$. Then*

$$ka < kb, \quad if \ k > 0 \qquad\qquad (21.1.4)$$

and

$$ka > kb, \quad if \ k < 0. \qquad\qquad (21.1.5)$$

EXAMPLE 21.1.2.   If we multiply the inequality $-1 < 3$ by 4 we obtain that $-4 < 12$, while multiplication by -4 yields $4 > -12$. The inequality $a < b$ does not imply that $a^2 < b^2$; for example $-3 < 2$, but $(-3)^2 > 2^2$. However, if

$$0 < a < b, \qquad\qquad\qquad (21.1.6)$$

we may infer that

$$a^2 < b^2; \qquad\qquad\qquad (21.1.7)$$

indeed by Proposition 21.1.1 we deduce from (21.1.6) that $a^2 < ab$ and $ab < b^2$, which proves (21.1.7). More generally, if (21.1.6) holds, then

$$a^n < b^n \quad (n = 1, 2, \ldots). \qquad\qquad (21.1.8)$$

Again, on dividing (21.1.6) throughout by ab, which is positive, we obtain that

$$0 < \frac{1}{b} < \frac{1}{a}.$$

This inequality may be raised to the nth power, thus (21.1.6) implies that, in contrast to (21.1.8),

$$a^{-n} > b^{-n} \qquad (n = 1, 2, \ldots). \qquad (21.1.9)$$

When a and b are positive, we have that

$$b - a = (\sqrt{b} - \sqrt{a})(\sqrt{b} + \sqrt{a}),$$

where, as always, the symbol $\sqrt{}$ denotes the positive square root. Hence (21.1.6) implies that

$$\sqrt{b} - \sqrt{a} > 0,$$

that is

$$0 < \sqrt{a} < \sqrt{b}. \qquad (21.1.10)$$

The formulae (21.1.8), (21.1.9) and (21.1.10) are instances of the statement

$$\text{`}a < b \quad \text{implies} \quad f(a) < f(b)\text{'}. \qquad (21.1.11)$$

Of course (21.1.11) does not hold for all functions; but it is certainly true when f is increasing [see § 2.7] and this property is ensured if f has a non-negative derivative in the interval (a, b) [see Corollary 3.3.4].

EXAMPLE 21.1.3

(i) $\exp a < \exp b$, if $a < b$, because $(\exp x)' = \exp x$, which is positive for all values of x [see § 2.11].

(ii) $\dfrac{a-1}{a+1} < \dfrac{b-1}{b+1}$ if $-1 < a < b$ or $a < b < -1$;

for

$$\left(\frac{x-1}{x+1}\right)' = \frac{2}{(x+1)^2},$$

and this expression is positive in every interval that does not include the point -1.

The same argument can be used to establish the results (21.1.8), (21.1.9) and (21.1.10) by choosing for f the functions x^n or x^{-n} or $x^{1/2}$ and observing that their derivatives are positive or negative when x is positive or negative respectively [see (3.1.1) and Example 3.1.2].

Care is needed in handling inequalities between quotients; the simplest case is explained in the following

PROPOSITION 21.1.2. *Let*

$$0 < u \leq U \quad \text{and} \quad v \geq V > 0.$$

Then

$$\frac{u}{v} \leq \frac{U}{V},$$

that is the value of a quotient is increased if either the numerator is increased or the denominator is decreased; similarly, the value of a quotient is decreased if either the numerator is decreased or the denominator is increased provided that all numbers involved are positive.

The technique described in this proposition is useful for discussing the order of magnitude of rational functions [I, § 14.9]. We illustrate the method by considering a typical example. Let

$$F(x) = \frac{3x^2 + x - 1}{4x^3 - 5x^2 + 1}. \tag{21.1.12}$$

We wish to examine the magnitude of $F(x)$ when x becomes large. First, we must warn against some abuses of mathematical language that are common in this context: it is sometimes argued that, since x is large, the terms x and -1 in the numerator can be 'neglected' in comparison with $3x^2$ and that likewise the terms $-5x^2$ and $+1$ can be 'dropped' from the denominator. Thus $F(x)$ is 'approximately' equal to $3x^2/4x^3 = 3/4x$. But the procedure of neglecting terms must be rejected as too vague or even illogical. Nor is it permissible to write

$$\lim_{x \to \infty} F(x) = \frac{3}{4} \frac{1}{x}. \tag{21.1.13}$$

This is wrong because in a limit relation, as x tends to infinity, the value of the limit must not involve x [see § 2.3].

It is, however, correct to state that, as $x \to \infty$,

$$F(x) = O\left(\frac{1}{x}\right). \tag{21.1.14}$$

[See Definition 2.3.2]; more precisely, this means that there exist constants, k, K and X (>0) such that

$$\frac{k}{x} < F(x) < \frac{K}{x} \quad \text{if } x \geq X. \tag{21.1.15}$$

Now in some circumstances it is desirable to find appropriate values for k, K and X, even if they are not the 'best possible'.

EXAMPLE 21.1.3.   Let $F(x)$ be the function defined in (21.1.12). We shall obtain upper and lower bounds for the numerator and for the denominator, valid for 'large' x; as a matter of fact it suffices for our purpose to assume that $x \geq 2$, whence $1 \leq x/2$, $x \leq x^2/2$, and so on. We then obtain the following inequalities:

$$3x^2 + x - 1 > 3x^2 - 1 > 3x^2 - (x/2)^2 = 11x^2/4,$$
$$3x^2 + x - 1 < 3x^2 + x < 3x^2 + (x^2/2) = 7x^2/2.$$

Similarly,

$$4x^3 - 5x^2 + 1 > 4x^3 - 5x^2 > 4x^3 - 5x^3/2 = 3x^3/2$$
$$4x^3 - 5x^2 + 1 < 4x^3 + 1 < 4x^3 + (x^3/8) = 33x^3/8.$$

Hence, by Proposition 21.1.2,

$$\frac{11x^2/4}{33x^3/8} < F(x) < \frac{7x^2/2}{3x^3/2},$$

that is

$$\frac{2}{3}\frac{1}{x} < F(x) < \frac{7}{3}\frac{1}{x}.$$

21.2. THE FUNDAMENTAL INEQUALITIES

21.2.1. The Inequality of Bernoulli

PROPOSITION 21.2.1. *If $h > -1$ and if n is any positive integer, then*

$$(1+h)^n \geq 1 + nh. \tag{21.2.1}$$

This result was discussed in I, Example 2.6.7.

21.2.2. Chebyshev's Inequality

Let m and n be arbitrary positive integers. If a and b are distinct real numbers, the differences

$$a^m - b^m \quad \text{and} \quad a^n - b^n$$

have the same sign, which is positive when $a > b$ and negative when $a < b$. Hence in all cases

$$(a^m - b^m)(a^n - b^n) \geq 0, \tag{21.2.2}$$

with equality if and only if $a = b$. A simple transcription of (21.2.2) yields the following

PROPOSITION 21.2.2. *If a and b are real numbers and m and n are positive integers, we have that*

$$\frac{a^{m+n} + b^{m+n}}{2} \geq \frac{a^m + b^m}{2} \cdot \frac{a^n + b^n}{2},$$

with equality if and only if $a = b$.

21.2.3.   The Arithmetic and Geometric Means

For any set of numbers

$$a_1, a_2, \ldots, a_n$$

we define the *arithmetic mean* by

$$A = \frac{a_1 + a_2 + \ldots + a_n}{n}, \tag{21.2.3}$$

and if the numbers are real and non-negative we can form their *geometric mean*, namely,

$$G = (a_1 a_2 \ldots a_n)^{1/n}. \tag{21.2.4}$$

The two types of mean are related by a famous inequality.

THEOREM 21.2.3.   *If $a_1, a_2, \ldots, a_n$ are real non-negative numbers, then*

$$\frac{a_1 + a_2 + \ldots + a_n}{n} \geq (a_1 a_2 \ldots a_n)^{1/n},$$

with equality if and only if $a_1 = a_2 = \ldots = a_n$.

Many proofs of this result have been published [see Beckenbach, E. F. and Bellman, R. (1961), pp. 3ff, where twelve different proofs are presented]. We shall indicate a proof in Example 21.3.1.

21.2.4.   The Cauchy–Schwarz Inequality

Let $a_1, a_2, \ldots, a_n$ and $b_1, b_2, \ldots, b_n$ be two sets of n real numbers. We may regard them as the components of two vectors $\mathbf{a}$ and $\mathbf{b}$ in the space $\mathbb{R}^n$ [see I, § 9.1]. Thus we shall write

$$\mathbf{a} = (a_1, a_2, \ldots, a_n), \qquad \mathbf{b} = (b_1, b_2, \ldots, b_n).$$

The *inner product* of $\mathbf{a}$ and $\mathbf{b}$ is given by

$$(\mathbf{a}, \mathbf{b}) = a_1 b_1 + a_2 b_2 + \ldots + a_n b_n \tag{21.2.5}$$

[see I (9.1.1)] and the *Euclidean length (norm)* of $\mathbf{a}$ is defined as

$$\|\mathbf{a}\| = \sqrt{(\mathbf{a}, \mathbf{a})} = (a_1^2 + a_2^2 + \ldots + a_n^2)^{1/2}, \tag{21.2.6}$$

where the square root is non-negative [see I (9.1.2)]. We have that

$$\|\mathbf{a}\| = 0 \quad \text{if and only if } \mathbf{a} = \mathbf{0}. \tag{21.2.7}$$

The result we wish to establish is given in the following

THEOREM 21.2.4   (*The Cauchy–Schwarz inequality: finite case*).   *Let*

$a_1, a_2, \ldots, a_n$ and $b_1, b_2, \ldots, b_n$ *be two sets of real numbers. Then*

$$(a_1b_1 + a_2b_2 + \ldots + a_nb_n)^2 \le (a_1^2 + a_2^2 + \ldots + a_n^2)(b_1^2 + b_2^2 + \ldots + b_n^2);$$

$$(21.2.8)$$

or more briefly

$$(\mathbf{a}, \mathbf{b})^2 \le (\mathbf{a}, \mathbf{a})(\mathbf{b}, \mathbf{b}). \qquad (21.2.9)$$

The equality sign in (21.2.8) and (21.2.9) holds if and only if there exists a real number k such that $b_i = ka_i$ $(i = 1, 2, \ldots, n)$, that is if the vectors are proportional.

Proof. The assertion (21.2.8) is trivial if $\mathbf{a}$ or $\mathbf{b}$ is the zero vector because in that case both sides of the formula are zero. Henceforth we shall assume that

$$\mathbf{a} \ne \mathbf{0},$$

and we shall use the abbreviations

$$\left. \begin{aligned} A &= a_1^2 + a_2^2 + \ldots + a_n^2 \quad (\ne 0), \\ B &= b_1^2 + b_2^2 + \ldots + b_n^2, \\ C &= a_1b_1 + a_2b_2 + \ldots + a_nb_n. \end{aligned} \right\} \qquad (21.2.10)$$

Consider the square of the length of the vector $t\mathbf{a} + \mathbf{b}$, where t is an arbitrary real number. By (21.2.6) we have that

$$\|t\mathbf{a} + \mathbf{b}\| \ge 0,$$

that is

$$\sum_{i=1}^{n} (ta_i + b_i)^2 \ge 0, \qquad (21.2.11)$$

which can be written as

$$At^2 + 2Ct + B \ge 0. \qquad (21.2.12)$$

This inequality holds for all values of t; in particular when $t = -C/A$ we find that

$$\frac{AC^2}{A^2} - \frac{2C^2}{A} + B \ge 0,$$

which, after a short calculation, reduces to

$$AB \ge C^2.$$

This proves (21.2.9). From (21.2.11) it is clear that the equality sign holds if and only if $b_i = -ta_i$ $(i = 1, 2, \ldots, n)$, as claimed.

EXAMPLE 21.2.1. If $p_1, p_2, \ldots, p_n$ are positive numbers, then

$$\left(\sum_{i=1}^{n} p_i \right)\left(\sum_{i=1}^{n} 1/p_i \right) \ge n^2.$$

This follows from (21.2.8) by putting $a_i = \sqrt{p_i}$ and $b_i = \sqrt{1/p_i}$ $(i = 1, 2, \ldots, n)$.

The notion of length defined in (21.2.6) satisfies the *triangle inequality*

$$\|\mathbf{a} + \mathbf{b}\| \le \|\mathbf{a}\| + \|\mathbf{b}\| \tag{21.2.10}$$

[see I, § 2.7.2], which is illustrated in Figure 21.2.1: in the triangle *OPQ* the sides *OP*, *PQ* and *OQ* are represented by the vectors **a**, **b** and **a**+**b** respectively; the length of *OQ* does not exceed the sum of the lengths of *OP* and *PQ*. However, the inequality (21.2.10) can (perhaps should!) be established by purely algebraic arguments. This was done in Volume I, Proposition 10.1.3 in the more general context of complex vectors. The proof is based on the Cauchy–Schwarz inequality.

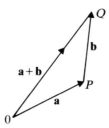

Figure 21.2.1

The Cauchy–Schwarz inequality can be generalized in several ways: when the components of the vectors are complex numbers, the inner product is defined as

$$(\mathbf{a}, \mathbf{b}) = a_1\bar{b}_1 + a_2\bar{b}_2 + \ldots + a_n\bar{b}_n,$$

where the bar denotes the complex conjugate [see I, § 2.7.1]. The Cauchy–Schwarz inequality then takes the form

$$|(\mathbf{a}, \mathbf{b})| \le \|\mathbf{a}\| \|\mathbf{b}\| \tag{21.2.11}$$

[see I, Proposition 10.1.2]. Similar extensions can be made in relation to other inequalities. But from now on we shall confine ourselves to real numbers.

Under suitable conditions we can generalize Theorem 21.2.4 to vectors with 'infinitely many components', that is to infinite series [see § 1.7]; thus

PROPOSITION 21.2.5   (*The Cauchy–Schwarz inequality: infinite case*). Let $a_1, a_2, \ldots$ and $b_1, b_2, \ldots$ be two infinite series of real numbers such that

$$A = \sum_{i=1}^{\infty} a_i^2 \quad and \quad B = \sum_{i=1}^{\infty} b_i^2$$

converge. Then the series

$$C = \sum_{i=1}^{\infty} a_i b_i$$

converges and

$$C^2 \leq AB,$$

with equality if and only if $b_i = ka_i$ $(i = 1, 2, \ldots)$ for some k.

Another version of the Cauchy–Schwarz inequality refers to spaces whose elements are functions; for simplicity we shall assume that we are concerned only with real-valued functions which are continuous in a fixed interval $[a, b]$ [see § 2.1]. The inner product is then defined as

$$(f, g) = \int_a^b f(x)g(x)\, dx \qquad (21.2.12)$$

[see (20.4.12)] and we have the following result.

PROPOSITION 21.2.6 (*The Cauchy–Schwarz inequality for function spaces*). *Let $f(x)$ and $g(x)$ be continuous in $[a, b]$. Then*

$$(f, g)^2 \leq (f, f)(g, g),$$

where the inner product (f, g) is defined in (21.2.12); the equality sign holds if and only if there exists a real number k such that $g(x) = kf(x)$ for $a \leq x \leq b$.

21.2.5. Hölder's Inequality

The Cauchy–Schwarz inequality can be generalized as follows:

THEOREM 21.2.7 (*Hölder's inequality*). *Let $a_1, a_2, \ldots, a_n$ and $b_1, b_2, \ldots, b_n$ be two sets of non-negative numbers and let p and q be positive numbers satisfying*

$$\frac{1}{p} + \frac{1}{q} = 1. \qquad (21.2.13)$$

Then

$$\sum_{i=1}^n a_i b_i \leq \left(\sum_{i=1}^n a_i^p \right)^{1/p} \left(\sum_{i=1}^n b_i^q \right)^{1/q},$$

with equality if and only if there exists a number k such that $b_i^q = ka_i^p$ $(i = 1, 2, \ldots, n)$.

The Cauchy–Schwarz inequality corresponds to the case in which $p = q = 2$.

Again, Hölder's inequality can be extended to infinite series (under suitable conditions) and to integrals. In its simplest form the result regarding the latter is given below:

PROPOSITION 21.2.8 (*Hölder's inequality*). *Suppose that $f(x)$ and $g(x)$ are continuous and non-negative in the interval $a \leq x \leq b$ and suppose that p* *and q satisfy* (21.2.13). *Then*

$$\int_a^b f(x)g(x)\,dx \le \left(\int_a^b \{f(x)\}^p\,dx\right)^{1/p}\left(\int_a^b \{g(x)\}^q\,dx\right)^{1/q},$$

with equality if and only if there exists a number k such that $\{g(x)\}^q = k\{f(x)\}^p$ *for* $a \le x \le b$.

21.2.6. Minkowski's Inequality

The Euclidean length defined in (21.2.6) is not the only possible way to assign a length to a vector in suitable circumstances. Thus if

$$\mathbf{a} = (a_1, a_2, \ldots, a_n)$$

is a vector with non-negative components and if p is a fixed real number satisfying $p > 1$, we may introduce the notion of length by the formula

$$\|\mathbf{a}\|_p = (a_1^p + a_2^p + \ldots + a_n^p)^{1/p}. \tag{21.2.14}$$

In order to substantiate this claim we have to show that the triangle inequality holds.

This is the content of the next theorem, which we quote without proof:

THEOREM 21.2.9 (*Minkowski's inequality*). *Let* $a_1, a_2, \ldots, a_n$ ·*and* $b_1, b_2, \ldots, b_n$ *be two sets of non-negative numbers and let* $p > 1$. *Then*

$$\left[\sum_{i=1}^n (a_i + b_i)^p\right]^{1/p} \le \left[\sum_{i=1}^n a_i^p\right]^{1/p} + \left[\sum_{i=1}^n b_i^p\right]^{1/p},$$

with equality if and only if there exists a real number k such that $b_i = ka_i$ $(i = 1, 2, \ldots, n)$.

21.3.   JENSEN'S INEQUALITY

The result of this section belongs to the theory of convex sets and functions, which is extensively discussed in Volume V, Chapter 4. For the present purpose only the simplest properties of convex functions are required.

DEFINITION 21.3.1.   A function $f(x)$, continuous on I: $a \le x \le b$, is said to be *convex* on I, if for every pair of points x_1 and x_2 in I, we have that

$$f\left(\frac{x_1 + x_2}{2}\right) \le \frac{f(x_1) + f(x_2)}{2}. \tag{21.3.1}$$

The geometric meaning of the definition is illustrated in Figure 21.3.1: without loss of generality we assume that $x_1 < x_2$. We consider the portion of the curve $y = f(x)$ between the points $P_1 = (x_1, f(x_1))$ and $P_2 = (x_2, f(x_2))$. We denote the

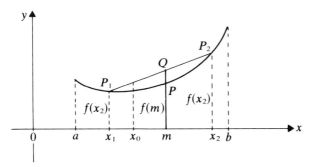

Figure 21.3.1

mid-point of x_1 and x_2 by

$$m = \tfrac{1}{2}(x_1 + x_2). \tag{21.3.2}$$

The line $x = m$ intersects the curve at

$$P = (m, f(m))$$

and it meets the chord P_1P_2 at

$$Q = (m, \tfrac{1}{2}\{f(x_1) + f(x_2)\}).$$

Thus the inequality (21.3.1) states that, for all values of x_1 and x_2, the point Q lies above or on the point P: the region above the curve is a *convex set* in $\mathbb{R}^2$ [see V, § 4.2.1].

Our definition of convexity does not presuppose the existence of a derivative; but for twice differentiable functions [see § 3.4] the following criterion is easy to apply:

PROPOSITION 21.3.1.   *If*

$$f''(x) \geq 0 \qquad (a \leq x \leq b), \tag{21.3.3}$$

then $f(x)$ is convex in $[a, b]$.

Proof.   Using the notation of Definition 21.3.1 we put $h = (x_2 - x_1)/2$. Then $m + h = x_2$ and $m - h = x_1$. By Taylor's formula [see § 3.6]

$$f(m + h) = f(m) + hf'(m) + \tfrac{1}{2}h^2 f''(m + \theta h),$$

where θ is a suitable number satisfying $0 < \theta < 1$. By virtue of (21.3.3)

$$f(m + h) \geq f(m) + hf'(m).$$

Similarly

$$f(m - h) \geq f(m) - hf'(m).$$

On adding these inequalities we obtain that

$$f(m+h)+f(m-h) \geq 2f(m),$$

which is equivalent to (21.3.1).

Any number that lies strictly between x_1 and x_2 can be written in the form

$$x_0 = px_1 + qx_2,$$

where

$$p>0, \qquad q>0, \qquad p+q=1.$$

Inspection of the Figure 21.3.1 suggests that the line $x = x_0$ will intersect the curve before it meets the chord. Indeed one can establish a more general result which involves n arbitrary points [see V, Theorem 4.3.8]; we omit the proof.

THEOREM 21.3.2   (*Jensen's inequality*).   *Suppose that $f(x)$ is a convex function defined for $a \leq x \leq b$. Let $x_1, x_2, \ldots, x_n$ be arbitrary points in $[a, b]$ and let $p_1, p_2, \ldots, p_n$ be non-negative numbers such that*

$$p_1 + p_2 + \ldots + p_n = 1.$$

Then

$$f(p_1 x_1 + p_2 x_2 + \ldots + p_n x_n) \leq \sum_{i=1}^{n} p_i f(x_i). \tag{21.3.4}$$

EXAMPLE 21.3.1.   Jensen's inequality affords a rapid proof of the result on the arithmetic and the geometric means [see Theorem 21.2.3]. We observe that the function

$$f(x) = -\log x \qquad (x > 0)$$

is convex because $f''(x) = 1/x^2$ [see Example 3.2.3 and (3.1.1)]. Let

$$a_1, a_2, \ldots, a_n$$

be a set of n positive numbers. If in Jensen's inequality we put

$$p_1 = p_2 = \ldots = p_n = \frac{1}{n},$$

we obtain that

$$-\log\left\{\frac{1}{n}(a_1 + a_2 + \ldots + a_n)\right\} \leq -\frac{1}{n}\log(a_1 a_2 \ldots a_n);$$

this is evidently equivalent to the assertion of Theorem 21.2.3.

21.4.   FUNCTIONAL INEQUALITIES

In this section we shall list a few inequalities of the form

$$f(x) \le g(x) \qquad (x \in I),$$

where I is a suitable interval.

21.4.1.   Trigonometric Functions

We recall the familiar inequalities [see § 2.12]

$$\left. \begin{array}{ll} -1 \le \sin x \le 1 & (x \in \mathbb{R}) \\ -1 \le \cos x \le 1 & (x \in \mathbb{R}) \end{array} \right\}, \tag{21.4.1}$$

where $\mathbb{R}$ denotes the set of all real values.

A rather less obvious result is given by

$$\sin x \le x \le \tan x \qquad \left(0 \le x < \frac{\pi}{2}\right), \tag{21.4.2}$$

with the understanding that x is measured in radians.

The comparison between $\sin x$ and x is made more precise in *Jordan's inequality*

$$\frac{2x}{\pi} \le \sin x \le x \qquad (0 \le x \le \tfrac{1}{2}\pi). \tag{21.4.3}$$

21.4.2.   Exponential and Logarithmic Functions

From the expansion (2.11.1) it is evident that

$$\exp x > x^k / k! \qquad (x \ge 0), \tag{21.4.4}$$

where k is an arbitrary positive integer; we say that $\exp x$ grows faster than any power of x. On the other hand, the function $\log x$ increases slowly. In Figure 21.4.1 we have sketched the graphs of $y = \log x$ and of $y = x - 1$. It

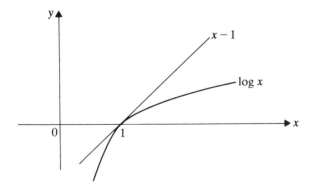

Figure 21.4.1

appears that the logarithmic curve touches the straight line at $x = 1$ but elsewhere lies below it. Thus

$$\log x \le x - 1 \qquad (x > 0). \tag{21.4.5}$$

This, and similar statements, can easily be proved by noting that a function increases if its derivative is positive and that it decreases if its derivative is negative [see Corollary 3.3.4]. Let

$$F(x) = x - 1 - \log x \qquad (x > 0).$$

We have that $F(1) = 0$ and

$$F'(x) = 1 - \frac{1}{x}.$$

Hence $F' > 0$, when $x > 1$, and F increases so that $F(x) > F(1)$; but $F'(x) < 0$ when $x < 1$ and F decreases so that, again, $F(x) > F(1)$. Thus in both cases $F(x) > 0$, which proves (21.4.5).

The inequality (21.4.5) may be translated into a statement about the exponential functions by means of the substitution $x = \exp u$. This yields

$$\exp u \ge 1 + u \qquad (u \in \mathbb{R}). \tag{21.4.6}$$

Of course the assertion is obvious when u is positive because the right-hand side consists merely of the first two terms of the expansion for $\exp u$ [see (2.11.1)].

21.5. HADAMARD'S INEQUALITY

In this section we shall quote a celebrated result which furnishes an upper bound for the determinant of an arbitrary real matrix [see I, § 6.9]. Let

$$\mathbf{A} = \begin{pmatrix} a_{11} & a_{12} & & a_{1n} \\ a_{21} & a_{22} & \cdots & a_{2n} \\ \cdots & \cdots & & \cdots \\ a_{n1} & a_{n2} & \cdots & a_{nn} \end{pmatrix} \tag{21.5.1}$$

be a real n by n matrix. Then we have

THEOREM 21.5.1   (*Hadamard's inequality*)

$$(\det \mathbf{A})^2 \le \left(\sum_{j=1}^{n} a_{1j}^2 \right) \left(\sum_{j=1}^{n} a_{2j}^2 \right) \cdots \left(\sum_{j=1}^{n} a_{nj}^2 \right), \tag{21.5.2}$$

with equality only either if one of the factors on the right of (21.5.2) *is zero or if*

$$a_{i1}a_{k1} + a_{i2}a_{k2} + \ldots + a_{in}a_{kn} = 0,$$

whenever $i \ne k$.

The theorem has an interesting geometric significance: the rows of **A** may be regarded as a set of vectors

$$\mathbf{a}_1, \mathbf{a}_2, \ldots, \mathbf{a}_n$$

in n-dimensional space; we suppose that their lengths are non-zero, but not necessarily equal. The vectors define a parallelepiped whose volume is given by $|\det \mathbf{A}|$. It follows from (21.5.2) that the volume attains its greatest value when the vectors are mutually orthogonal [I, Definition 10.2.1]. The case in which $n = 3$ is illustrated in Figure 21.5.1.

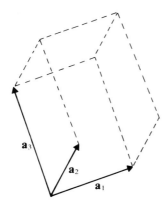

Figure 21.5.1

21.6. EIGENVALUES

There are numerous inequalities which involve the eigenvalues of a matrix. Some of these results are discussed in sections 7.9 and 7.11 of Volume I.

W.L.

BIBLIOGRAPHY

Beckenbach, E. F. and Bellman, R. (1961). *Inequalities*, Springer.
Hardy, G. H., Littlewood, J. E. and Polya, G. (1952). *Inequalities, Cambridge.*
Mitrinovic, D. S. (1970). *Analytic Inequalities,* Springer.

Index